THE AMERICAN COCKROACH

THE AMERICAN COCKROACH

edited by

William J. Bell
Professor of Entomology and Physiology & Cell Biology
University of Kansas

and

K. G. Adiyodi
Professor of Reproductive Physiology
and Director, Vatsyayana Centre of Invertebrate Reproduction
Calicut University

LONDON NEW YORK
CHAPMAN AND HALL

First published 1982 by
Chapman and Hall Ltd
11 New Fetter Lane, London EC4P 4EE
Published in the USA by
Chapman and Hall
in association with Methuen, Inc.
733 Third Avenue, New York NY 10017

Typeset in Great Britain by Scarborough Typesetting Services
and printed at The University Press, Cambridge

ISBN 0 412 16140 0

British Library Cataloguing in Publication Data

The American cockroach.

1. Cockroaches
1. Bell, William J. II. Adiyodi, K. G.
595.7'22 QL505.6

ISBN 0-412-16140-0

Contents

Contributors

K. G. Adiyodi, Department of Zoology and Vatsyayana Centre of Invertebrate Reproduction, Calicut University, PO Kerala 673635, India.

William J. Bell, Department of Entomology and Department of Physiology and Cell Biology, University of Kansas, Lawrence, Kansas 66045, USA.

D. E. Bignell,* University of Exeter, Department of Biological Sciences, Devon EX4 4PS, UK.

Fred Delcomyn, Department of Entomology and Program in Neural and Behavioral Biology, University of Illinois, Urbana, Illinois 61801, USA.

Roger G. H. Downer, Department of Biology, University of Waterloo, Waterloo, Ontario N2L 3GL, Canada.

P. Michael Fox, Department of Biological Sciences, State University of New York, Brockport, New York 14420, USA.

J. G. Kunkel, Department of Zoology, University of Massachusetts, Amherst, Massachusetts 01003, USA.

P. L. Miller, Department of Zoology, South Parks Road, Oxford OX1 3PS, UK.

Richard R. Mills, Department of Biology, Virginia Commonwealth University, Richmond, Virginia 23284, USA.

Donald E. Mullins, Virginia Polytechnic Institute and State University, Department of Entomology, Blacksburg, Virginia 24061, USA.

Rudolph Pipa, Department of Entomology and Parasitology, University of California, Berkeley, California 94720, USA.

Robert R. Provine, Department of Psychology, University of Maryland, Baltimore County, Catonsville, Maryland 21228, USA.

Louis M. Roth, US Army Natick Research and Development Command, Natick, Massachusetts 01760, USA.

Günter Seelinger, Institute of Zoology, University of Regensburg, D-8400 Regensburg, Federal Republic of Germany.

Barbara Stay, Department of Zoology, University of Iowa, Iowa City, Iowa 52242, USA.

* Present address, Department of Zoology, Westfield College, Hampstead, London NW3 7ST, UK.

Donald J. Sutherland, Department of Entomology and Economic Zoology, Rutgers University, New Brunswick, New Jersey 08903, USA.

Stephen S. Tobe, Department of Zoology, University of Toronto, Toronto, Ontario M5S 1A1, Canada.

Thomas R. Tobin, Department of Entomology, University of Kansas, Lawrence, Kansas 66045, USA.

Preface

This volume deals mainly with the biology of the American cockroach, *Periplaneta americana* (L.). Contributors were urged to emphasize recent findings, including unpublished data when possible, a goal that would not have been feasible if it were not for the two previously published books on the basic biology of cockroaches, *The Biology of the Cockroach* (1968) by D. M. Guthrie and A. R. Tindall and *The Cockroach*, Volume 1 (1968) by P. B. Cornwell. Those topics not included in *The American Cockroach*, such as external morphology, are well covered in the two preceding books. In addition, these books provided a broad background upon which contributors to *The American Cockroach* have been able to build with recent trends, new and established concepts and integration.

Although this book deals primarily with the American cockroach, many chapters offer a comparative approach in sections where the more recent and exciting research has been accomplished on other species. Most contributors place the cockroach in perspective with regard to its appropriateness or inappropriateness for various types of biological investigations. Many questions are realistically left unanswered when no acceptable or obvious solution is apparent; an invitation to new researchers to consider the cockroach as an experimental subject.

W. J. Bell, K. G. Adiyodi

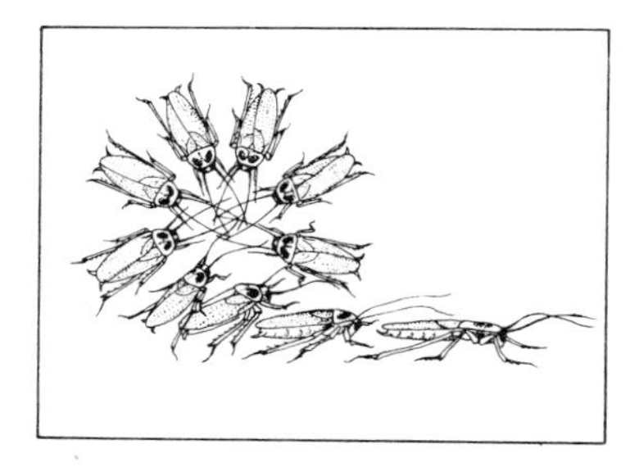

1

Introduction

Louis M. Roth

1.1 Distribution

Forty-seven species are included in the genus *Periplaneta* (Princis, 1966, 1971), none of which is endemic to America. The four species found in the United States, *P. americana* (Linnaeus), *P. australasiae* (Fabricius), *P. brunnea* (Burmeister) and *P. fuliginosa* (Serville) have all been introduced from foreign lands. *Periplaneta americana* originated in tropical Africa where it lives both inside and outside human dwellings. According to Rehn (1945), this species travelled to South America, the West Indies, and the southern United States on slave ships sailing from the west coast of Africa. This theory was tarnished, however, when Kevan, personal communication (1979) identified an oötheca found on the Spanish vessel *San Antonio*, sunk off Bermuda in 1625 (Peterson, 1977), as *P. americana*. Kevan believes that the species '. . . was certainly in the Americas before the slave trade reached large proportions, and . . . it was on vessels that were not slavers by the beginning of the 17th century.' The specific name *americana* is misleading because of the insect's African origin, and the generic name, which means to wander around, would have been more appropriate for this species because not all of the *Periplaneta* species are world travellers.

The American cockroach, also commonly called the ship cockroach, kakerlac, and Bombay canary, is virtually cosmopolitan in distribution nowadays, having been spread by commerce throughout the world. *P. australasiae* and *P. brunnea* are circumtropical, and *P. fuliginosa* is also found in Japan, China and Taiwan. Most of the remaining species of *Periplaneta* are more restricted in distribution, and probably not closely associated with man, which limits their spread.

1.2 Life cycle

Many authors have described the oviposition behaviour of the American cockroach. Outdoors, the female shows a preference for moist, concealed oviposition sites (Fleet and Frankie, 1974). Indoors, or in the laboratory, oviposition behaviour is influenced by the type of substrate available, but most often the oötheca is eventually concealed. In sand or loose soil the cockroach digs a hole in which she deposits an egg case, and covers it with substrate material (McKittrick, 1964). Ovipositing on wood or cardboard, she chews out a depression, depositing the oötheca, and covers it with substrate particles held in place by her saliva (Haber, 1920; Nigam, 1933; Rau, 1940). In stores, however, the oötheca may be attached to food wrappers or cans, and not necessarily covered (see Fig. 79 *in* Cornwell, 1976).

The deposited oötheca contains water sufficient for the eggs to develop without receiving additional water from the substrate. The principal function of the oötheca is to prevent the eggs or embryos from desiccating. It is covered by a waterproofing film that retards water loss, and eggs are able to develop even in a relative humidity as low as 15% (Roth and Willis, 1955a, 1955b).

The number of eggs per oötheca has been reported to vary from 6 to 18 (Lefroy, 1909; Fischer, 1928; Klein, 1933; Nigam, 1933; Gould and Deay, 1938, 1940; Rau, 1940; Gier, 1947a; Griffiths and Tauber, 1942a; Brunet, 1951; Pope, 1953). Usually, each ovary contains 8 ovarioles (Bonhag, 1959) so that the 'normal' number of eggs in an oötheca is about 16. Reports of *P. americana* oöthecae containing more than 18 eggs, or having more than that number of nymphs emerge from a single oötheca (e.g., Rosenfeld, 1910; Haber, 1919; Adair, 1923) probably were based on misidentified specimens. In a recent paper on the post-embryonic development and morphology of the antennal sense organs of *P. americana*, Schafer and Sanchez (1973) stated that their cockroaches came from 6 oöthecae, from which an average of 24 siblings per egg case emerged, but the number 24 was 'clearly a mistake', and the maximum number of eggs in an oötheca was 18 (Schafer, personal communication, 1979). Even if all of the mature ovarian oöcytes are placed in an oötheca, some may not develop or hatch. The number hatching has been reported to be 14 (Gould and Deay, 1938), 11 (Willis *et al.*, 1958), and 13 (Roth and Willis, 1956).

About one week after mating, the female begins to produce oöthecae and, at the peak of her reproductive period, she may form about 2 oöthecae per week. Then both fertility and fecundity decrease with age (Roth and Willis, 1956). During a female's lifetime, the total number of oöthecae produced has been reported to range from 10 to 84 (Takahashi, 1924; Klein, 1933; Nigam, 1933; Gould and Deay, 1938; Rau, 1940; Gould, 1941; Gier, 1947a; Griffiths and Tauber, 1942a; Pope, 1953). Certainly, some of this variability was due to varying experimental conditions, and very low values do not represent insects kept under ideal conditions. Regimens such as peptone or dextrose diets reduce the frequency of oötheca production as well as the number of eggs per oötheca (Gier, 1947b).

The number of times American cockroach nymphs moult varies from 6 to 14

(Fischer, 1928; Klein, 1933; Nigam, 1933; Gould and Deay, 1938, 1940; Griffiths and Tauber, 1942b; Gier, 1947a; Biellman, 1960; Willis *et al.*, 1958). Experimental amputation of antennae and/or cerci increases the number of times nymphs moult to 13–18 (females) and 15–17 (males) (Zabinski, 1936). This should occur in individuals that accidentally lose parts of or entire appendages, but the phenomenon is not confirmed by the data of Willis *et al.* (1958). The extent of injury or loss may influence the number of moults.

Scientists who investigated development time and adult longevity have found that these characteristics are also highly variable. Many studies were conducted at room temperatures which varied with the season in different geographic areas, and under rearing conditions differing in size of containers and in the numbers of individuals kept together. Variation is expected under these conditions, but even under relatively 'controlled' conditions of temperature (e.g. Griffiths and Tauber, 1942a), considerable variation occurs. In most of the studies, the number of individuals used was relatively small (4–30). The length of the nymphal period varies from 134 to 1031 days. Average values range from 160–666 days (Takahashi, 1924; Klein, 1933; Nigam, 1933; Zabinski, 1936; Rau, 1940; Gould and Deay, 1940; Griffiths and Tauber, 1942b; Pope, 1953; Willis *et al.*, 1958). The mean nymphal period for males is 171 ± 2 days (n = 225) and for females 160 ± 2 days (n = 243) (Willis *et al.*, 1958). Lifespans of the adults range from 69 to 1693 days with averages of 125–1212 days (Klein, 1933; Rau, 1940; Gould and Deay, 1940; Griffiths and Tauber, 1942a; Pope, 1953). The mean lifespan of mated females is 311 ± 13 days (n = 50) and 397 ± 13 days (n = 112) for virgin females (Roth and Willis, 1956).

1.3 Ecology

Outdoors, *P. americana* has been observed at night, in large numbers, on *Tribulus* blossoms on Johnston Island (ca 700 miles west-southwest of Honolulu) (Bryan, 1926). In the United States during the summer, alleyways and yards may be overrun by these cockroaches, and they have been found in large numbers in decaying maple trees (Gould and Deay, 1938, 1940), feeding on sap exuded from trees in Washington, D.C. (Anonymous, 1967), and in palm trees along the Gulf Coast of Texas (Zimmerman *in* Gould and Deay, 1940). Coconut palms near buildings provide harbourages for them (Fig. 11 *in* Cornwell, 1976).

Indoors, the American cockroach frequents restaurants, grocery stores, bakeries, meat-packing plants and other places where food is prepared or stored. The insects have been taken on aircraft and ships, and are sometimes associated with banana shipments. In greenhouses, *P. americana* can be troublesome because they damage orchids, *Cinchona*, and other plants. The cockroach is found in caves, mines, privies, latrines, cesspools, sewers, sewage treatment plants and dumps (see Roth and Willis, 1957, 1960, for original citations). Their presence in many of these habitats is of epidemiological significance. Deleporte (1976) has studied the ecology of a population of *P. americana* in caves in Trinidad, West Indies.

P. americana is an omnivorous and opportunistic feeder. In India, it eats paper, boots, hair, bread and fruit, and damages clothes, book bindings and cloth covers, and paste or sizing in books (Nigam, 1933). In Taipei, Takahashi (1924) reported its feeding on bananas and other fruit, fish, peanuts, bean cake, old rice, putrid sake, oil paper and other paper material, starchy paste, the soft part of the inner side of animal hides, crepe de Chine and other cloth, and insect corpses. By depositing excreta wherever it walks it spoils food, which may acquire a cockroach odour (Nigam, 1933).

Although this species can withstand a degree of starvation, it can survive even longer when only water is available, as shown below (Willis and Lewis, 1957):

	Longevity (days; 27°C, 36–40% r.h.)	
	Males	*Females*
No food or water	28	42
Dry food, no water	27	40
Water, no food	44	90

The longer survival rate for females probably is related to their greater fat reserves.

During the day the American cockroach, which responds negatively to light, rests in harbourages close to water pipes, sinks, baths, and toilets, for example, where the microclimate is suitable for survival. The insects become active at night when the temperature drops and the water-holding capacity of the air increases. In selecting a habitat, *P. americana* is strongly influenced by temperature, humidity, and air movement. As temperature increases, relative humidity drops and its drying power (saturation deficit) increases. Air movement may displace moist air with drier air, and this results in loss of water from the insect (Cornwell, 1968).

Necheles (1927) suggested that the relative humidity of the air, influenced by temperature, and its effect on the rate of water loss by the insects is the principal influence on cockroach activity. Gunn and Cosway (1938), on the other hand, questioned the importance of relative humidity in influencing activity rhythm. They espoused that danger of desiccation within the normal range of temperature and humidity (Gunn, 1935) is probably not great.

The temperature 'preferred' by *P. americana* lies between 24° and 33°C. The rate of weight loss, expressed as a percentage of the original weight is a measure of the rate of evaporation of water. The weight loss (%) in dry air at different centigrade temperatures is as follows: 3.1 (20°); 5.0 (25°); 8.6 (30°); 13.9 (33°); 21.3 (36°) (Gunn, 1935). Although some water is lost through the spiracles, at about 36°C and above, the insects lose water rapidly because of a change of state in the film of fatty material that covers the cuticle (Ramsay, 1935). Kept for 24 h in moist air (90% r.h.), *P. americana* can survive temperatures of 36–38°C, but most individuals die at 39°C; in dry air the number of deaths is somewhat higher at 37° and 38°C (Gunn and Notley, 1936). Klein (1933) found that the temperature preference of American cockroach nymphs was 24–26°C, and for adults, 28–30°C.

Usually, American cockroaches move around at temperatures between 15.5–31.7°C (nymphs) and 17.6–31.1°C (adults). Nymphs of various sizes become inactive at 3.6–7.1°C, and adults at 5.2°C. The insects become highly active above 34°C and succumb to heat paralysis and death at temperatures above 42°C (Klein, 1933). Gould (1941) collected American cockroaches on a bare wall of a meat packing plant where they had been 'numbed' by the 10°C temperature.

The antennae of *P. americana* bear sense organs sensitive to rapid temperature drops in the range of 25°C (Loftus, 1966, 1968, 1969). In adult cockroaches, there are, on the average, 33 of these temperature receptors on the ventral side of alternating segments in the distal third of each antenna. Although none of the receptors is found on the antennae of first and second instar nymphs, receptors occur in later instars (Schafer and Sanchez, 1973). In addition, sense organs on the arolium and euplantulae of the tarsus are sensitive to low temperatures and, at temperatures below 13°C, they can detect a drop of 1°C (Kerkut and Taylor, 1957). Sense organs on the antennae of the American cockroach respond to both moist and dry environments. They apparently transduce humidity changes to electric potential through mechanical changes in the sensillae induced by relative humidity (Yokohari *et al.*, 1975; Yokohari and Tateda, 1976; Yokohari, 1978).

Table 1.1 Viruses associated with cockroaches.*

Poliomyelitis viruses
Four unspecified strains†
Lansing strain
Brunhilde Type, Minnesota and Mahoney strains
Polio virus Type 1 (Klowden and Greenberg, 1974)
Columbia SK
Coxsackie A-12 (Lawson and Johnson, 1970, 1971a, 1971b)
Coxsackie Type 4, subgroup A
Coxsackie Type B5 (Klowden and Greenberg, 1974)
Echo Type 6 (Klowden and Greenberg, 1974)
Mouse Encephalomyelitis GD VII strain
Hepatitis B (Zebe *et al.*, 1972)
Newcastle disease (California strain No. NC 194–5–6–7)‡

* Only references to papers published after 1960 are listed; for earlier citations see Roth and Willis (1960).

† The four unspecified strains of poliomyelitis were found naturally, whereas all others were experimental infections.

‡ Not recovered from cockroach faeces after feeding experiment.

Necheles (1927) believed that light of different wavelengths does not exert direct influence on the activity of *P. americana*, which responds to light slowly. If they are hungry, the cockroaches emerge from hiding places during the day, even in sunlight, to search for food. Their escape response when a light is suddenly turned on in a dark room is primarily due, not to the light *per se*, but to drafts or air currents. On the cerci are sense organs responsive to vibration, sound, or air movement (Pumphrey and Rawdon-Smith, 1936) which mediate the escape response (Roeder, 1959) noted by Necheles (1927).

Table 1.2 Bacteria associated with *P. americana*.*

Bacteria of known, suspected, or questionable pathogenicity to man

Natural infections

Bacillus subtilis Cohn emend Prazmowski
Clostridium perfringens (Veillon and Zuber) Holland
Escherichia coli (Migula) Castellani and Chalmers† (Ryan and Nicholas, 1972)
Mycobacterium lacticola Lehmann and Neumann?
M. leprae (Armauer-Hansen) Lehmann and Neumann†
M. phlei Lehmann and Neumann
M. piscium Bergey *et al.*
M. sp.
Nocardia sp.?
Paracolobactrum coliforme Borman *et al.*
P. sp.
Proteus mirabilis Hauser
P. morganii (Winslow *et al.*) Rauss
P. rettgeri (Hadley *et al.*) Rustigian and Stuart
P. vulgaris Hauser
Pseudomonas aeruginosa (Schroeter) Migula† (Sauerländer and Köhler, 1961; Sauerländer and Ehrhardt, 1961; Zuberi *et al.*, 1969)
P. fluorescens Migula†
Salmonella anatum (Rettger and Scoville) Bergey *et al.*† (Rueger and Olson, 1969)
S. morbificans (Migula) Haupt
S. schottmuelleri (Winslow *et al.*) Bergey *et al.*†
S. sp. Types: Bareilley; Bredeny; Kentucky; Meleagris; Newport; Oranienburg†; Panama; Rubislaw; Tennessee (Rueger and Olson, 1969)
Serratia marcescens Bizio (Sauerländer and Ehrhardt, 1961; Sauerländer and Köhler, 1961; Steinhaus and Marsh, 1962)
Shigella alkalescens (Andrewes) Weldin
Straphylococcus aureus Rosenbach† (Rueger and Olson, 1969)
Streptococcus faecalis Andrewes and Horder
Veillonella parvula (Veillon and Zuber) Prévot

Experimental infections

Brucella abortus (Schmidt and Weis) Meyer and Shaw
Mycobacterium tuberculosis (Schroeter) Lehmann and Neumann
Pasteurella pestis (Lehmann and Neumann) Holland
Salmonella typhosa (Zopf) White (Rueger and Olson, 1969)
S. sp. Type Montevideo
S. sp. Type Typhimurium (Klowden and Greenberg, 1976, 1977a, 1977b)
Shigella dysenteriae (Shiga) Castellani and Chalmers
S. flexneri Castellani and Chalmers (Rueger and Olson, 1969)
S. paradysenteriae (Collins) Weldin
Vibrio comma (Schroeter) Winslow *et al.*

Bacteria non-pathogenic to man

Natural infections

Achromobacter hyalinum (Jordan) Bergey *et al.*
Aerobacter aerogenes (Kruse) Beijerinck
A. cloacae (Jordan) Bergey *et al.*
A. sp.
Alcaligenes faecalis Castellani and Chalmers
Bacillus cereus Frankland and Frankland†
B. megaterium De Bary
Bacterium alkaligenes Nyberg

Table 1.2—*cont.*

Natural infections—*cont.*

Clostridium sp.
Coliform bacteria (Frishman and Alcamo, 1977)
Colon bacilli
Diptheroid I and II†
Eberthella oedematiens Assis
Escherichia freundii (Braak) Yale
E. intermedium (Werkman and Gillen) Vaughn and Levine
Micrococcus sp.
Mycobacterium friedmannii Holland
Paracolobactrum aerogenoides Borman, Stuart, and Wheeler
Paracolon bacilli
Proteus sp.
Sarcina sp.†
Spirillum periplaneticum Kunstler and Gineste
Spirochaetes
Staphylococcus bacteria (Frishman and Alcamo, 1977)
Streptococcus non-hemolyticus II Holman
Streptomyces leidynematis Hoffman§
Tetragenous sp.†

Experimental infections‡

Bacillus thuringiensis Berliner
Corynebacterium sp. (Ryan and Nicholas, 1972)
Micrococcus nigrofaciens Northrup

* Only references to papers published after 1960 are listed; for earlier citations see Roth and Willis (1957, 1960).
† Also used in experimental infections.
‡ In addition to the above species marked with footnote†.
§ The bacterium is anchored to and grows on the nematodes *Hammerschmidtiella diesingi* and *Leidynema appendiculata* found in the cockroach's gut. The micro-organism probably obtains its nourishment from the intestinal contents of the cockroach.

1.4 Biotic associations and medical importance

A large number of plants and animals are associated with cockroaches (Roth and Willis, 1960). Viruses (Table 1.1), bacteria (Table 1.2), fungi (Table 1.3), protozoa (Table 1.4), helminths (Tables 1.5, 1.6) and insects (Table 1.7) are commensals or parasites on *P. americana* or have been introduced experimentally into these insects. The organisms which cause diseases of human beings and other animals are of particular interest. Because American cockroaches can feed on human faeces and migrate from sewers to homes where they contaminate human food, they are potential mechanical vectors of disease (see original citations in Roth and Willis, 1957).

That pathogens survive after experimental introduction into cockroaches does not prove that these insects are natural vectors; nor is the discovery of pathogens naturally on or in cockroaches necessarily acceptable evidence that they are vectors. Some writers believe that cockroaches are not a hazard to health because disease-causing organisms supposedly do not multiply inside them. But to anyone who gets dysentery from a cockroach, it is of little consolation that the pathogen was simply transmitted mechanically without first having increased in numbers inside the insect's body.

Table 1.3 Fungi associated with *Periplaneta americana.* *

Aspergillus flavus Link†	*Histoplasma capsulatum* Darling††
A. niger van Tieghem‡ (Zuberi *et al.*, 1969)	*Metarrhizium anisopliae* (Metschnikoff) Sorokin‡‡
A. sp.‡	*Mucor guilliermondi* Nadson and Filippov§§
A. usutus (Bainer) Thom and Church (Zuberi *et al.*, 1969)	*M.* sp.†
Beauveria bassiana (Balsamo) Vuillemin§	*Penicillium* sp.‡
Cephalosporium sp.‡	*Rhizopus* sp.‡
Herpomyces chaetophilus Thaxter¶	*Syncephalastrum* sp.‡
H. periplanetae Thaxter**	*Torula acidophila* Owen and Mobley

* Only references to papers published after 1960 are cited; for earlier citations see Roth and Willis (1957, 1960). *Beauveria bassiana* and *Histoplasma capsulatum* were experimental infections, whereas all other fungi occurred naturally.

† From oöthecae. To prevent fungal growth on egg cases, Griffiths and Tauber (1942a) autoclaved their rearing containers and dipped the oöthecae in 70% alcohol for 10 s. Gier (1947a) sterilized oöthecae by dipping them in 95% alcohol, then placed them in mercuric chloride alcohol solution for 15 min, and rinsed in 70% alcohol.

‡ From faeces.

§ Experimentally infected insects became paralysed and died.

¶ On spines of legs, antennae, and cerci.

** On spines, tegmina, integument, and antennae.

†† Not recovered in faeces after feeding experiment.

‡‡ From female genitalia.

§§ From intestine.

Among the pathogenic nematodes for which *P. americana* is an intermediate host (Table 1.6), *Gongylonema neoplasticum* is of particular interest. Fibiger and Ditlevsen (1914) reported that the nematode produced 'cancer' in the forestomach and tongue of wild rats, its primary hosts. In 1926, Fibiger was awarded the Nobel Prize for this study. But Hitchcock and Bell (1952), who failed to repeat Fibiger's results, showed that a vitamin A deficiency in the rats' diet exacerbated the harmful effects of the nematode; in the absence of concurrent nutritional deficiency, the parasite acted as a chronic irritant which produced only minimal effects on the forestomach epithelium. Although Fibiger's work stimulated research on the association of *G. neoplasticum* with cockroaches, the award was a mistake, one of few made by the Nobel Committee (Wade, 1978).

The close association of domiciliary cockroaches with man exposes him to cockroach allergens. Considerable evidence indicates that cockroaches may provoke symptoms of hay fever, asthma, and dermatitis in susceptible individuals (Roth and Alsop, 1978). *P. americana* is one of several species of cockroaches which can excite allergenic reactions. Cockroaches produce 4 classes of allergens (Bernton and Brown, 1964; 1969). (1) Contact allergens may develop when cockroaches crawl over a person, or when a person handles the insects. (2) Inhalant allergens may arise by inhaling substances emananting from cockroaches or their faeces. (3) Injectant allergens may result when cockroaches bite man (Roth and Willis, 1957). Rageau and Cohic (1956) and Sonan (1924) made specific references to *P. americana* biting

*Table 1.4 Protozoa associated with P. americana.**

Natural infections

Balantidium blattarum Ghosh
B. ovatum Ghosh
B. sp.
Bodo sp.
Coelosporidium periplanetae (Lutz and Splendore)
Diplocystis schneideri Kunstler
D. sp.
Dobellina sp.
Endamoeba blattae (Bütschli) (Hoyte, 1961a)
Endolimax blattae Lucas (Warhurst, 1963, 1967)
E. sp.
Entamoeba coli (Grassi)†
E. histolytica Schaudinn†
E. thomsoni Lucas
E. sp.†
Eutrichomastix sp.
Gregarina blattarum von Siebold (Desportes, 1966)
G. legeri Pinto
G. neo-brasiliensis Al. Cunha
Hexamita periplanetae (Bělăr)
Iodamoeba sp.
Isotricha caullery Weill
Lophomonas blattarum Stein (Hoyte, 1961a)
L. striata Bütschli (Hoyte, 1961a)
Monocercomonoides orthopterorum (Parisi) (Hoyte, 1961a)
Nyctotherus ovalis Leidy (Hoyte, 1961a, 1961b, 1961c)
Plistophora periplanetae (Lutz and Splendore)
Polymastix melolonthae (Warhurst, 1966)
P. periplanetae Qadri and Rao (Qadri and Rao, 1963)
Protomagalhaesia serpentula Magãlhaes
Rhizomastix periplanetae Rao (Rao, 1963)
Spirillum periplaneticum Kunstler and Gineste
Tetratrichomastix blattidarum Young

Experimental infections‡

§*Chilomastix mesnili* (Wenyon)
Entamoeba pitheci Prowazek
§*Giardia intestinalis* (Lambl)
Nosema sp. (Fisher and Sanborn, 1964)
Paramecium sp.
Tetrahymena pyriformis (Seaman and Tosney, 1967; Seaman and Robert, 1968; Seaman and Clement, 1970; Seaman *et al.*, 1972
§*Trichomonas hominis* (Davaine)

* Only papers published after 1960 are listed; references to the earlier literature are given in Roth and Willis (1957, 1960). Species preceded by § are known or suspected pathogens of vertebrates.
† Also used successfully in experimental infections.
‡ In addition to the above species marked with footnote†.

Table 1.5 Helminths served by *P. americana* as a primary host.*

Binema mirzaia (Basir) Basir
Gordius sp.
Hammerschmidtiella diesingi (Hammerschmidt) Chitwood (Jarry and Jarry, 1963; Leong and Paran, 1966; Kloss, 1966; Hominick and Davey, 1972)
Leidynema appendiculata (Leidy) Chitwood (Jarry and Jarry, 1963; Leong and Paran, 1966; Kloss, 1966; Pawlik, 1966; Feldman, 1972; Hominick and Davy, 1972)
Neoaplectana sp.
Protrellus aurifluus (Chitwood) Chitwood (Leong and Paran, 1966)
P. künckeli (Galeb) Travassos
Schwenckiella icemi (Schwenck) Basir (Jarry and Jarry, 1963; Leong and Paran, 1966)
Severianoia severianoi (Schwenck) Travassos (Leong and Paran, 1966)
Thelastoma singaporensis (Leong and Paran, 1966)
T. pachyjuli (Parona) Travassos
T. sp. (Jarry and Jarry, 1963)

* Only references to papers published after 1960 are listed; for earlier citations see Roth and Willis (1960). With the exception of *Neoaplectana* sp. which was an experimental infection, all other helminths occur naturally in the American cockroach.

Table 1.6 Pathogenic helminths for which *P. americana* may or may not serve as an intermediate host.*

Natural infections

Ancylostoma duodenale (Dubini) Creplin† [Old World hookworm: man]
Ascaris lumbricoides (Linnaeus)†‡ [giant intestinal roundworm: man]
Gongylonema neoplasticum (Fibiger and Ditlevsen) Ransom and Hall† [wild rats and laboratory rodents]
Gongylonema sp.
Hymenolepis sp.
Mastophorus muris (Gmelin)† [rats] (Campos and Vargas, 1977; Quentin, 1970)
Moniliformis dubius Meyer† [rodents, occasionally man] (Mercer and Nicholas, 1967)
Moniliformis moniliformis (Bremser) Travassos† [rodents; occasionally man] (Acholonu and Finn, 1974)
Necator americanus (Stiles)† [hookworm; man]
Protospirura muricola Gedoelst† [rats] (Campos and Vargas, 1977; Quentin, 1969)
Spirura gastrophila (Müller) Seurat [hedgehogs, foxes, lizards, chameleons, dogs, cats, mongooses]
Trichuris trichiura (Linnaeus) Stiles† [human whipworm: man and many other animals]

Successful experimental infections§

Ancylostoma caninum (Ercolini) Hall [hookworm: dogs, cats]
A. ceylanicum (Looss) Leiper [hookworm: dogs, rarely man]
Ascaris sp.
Echinococcus granulosus (Batsch) Rudolphi [tapeworm: dogs and man]
Hookworm ova
Oxyspirura mansoni (Cobbold) Ransom¶ [chicken eyeworm: chickens and other birds]
Rictularia coloradensis Hall [mice]
Schistosoma haematobium (Bilharz) Weinland [vesicle blood fluke: man]
Taenia saginata Goeze [beef tapeworm: cattle and related animals, man]

Unsuccessful experimental infections

Hymenolepis nana (von Siebold) Blanchard [tapeworm: man]
Physaloptera hispida Schell [cotton rat]
Prosthenorcis elegans (Diesing) Travassos [monkeys, lemur]
P. spirula (Olfers) Travassos [monkeys, lemurs, chimpanzees]
Protospirura bonnei Ortlepp [rats]
Spirocerca sanguinolenta (Rudolphi) Seurat [dogs]

---

* Common names and definitive hosts are given in square brackets. Only papers published after 1960 are listed; earlier citations are given in Roth and Willis (1957, 1960).
† Also used successfully in experimental infections.
‡ The eggs may have been of *Ascaris suum* Goeze, indicating the ingestion of pig faeces rather than human faeces.
§ In addition to the above species that are followed by footnote†.
¶ The normal intermediate host is the cockroach, *Pycnoscelus surinamensis* (Linnaeus).

Table 1.7 Insect parasites of *P. americana*.*

I. Egg parasites (Hymenoptera)

Anastatus floridanus Roth and Willis†
A. tenuipes Bolívar y Pieltain
Evania appendigaster (Linnaeus) (Piper *et al.*, 1978; Deleporte, 1976)
Melittobia chalybii Ashmead‡
Prosevania punctata (Brullé)
Szepligetella sericea (Cameron)
Tetrastichus hagenowii (Ratzeburg) (Fleet and Frankie, 1975; Piper *et al.*, 1978)
T. periplanetae Crawford
T. sp.

II. Nymph and/or adult parasites

Hymenoptera

Ampulex amoena Stål
A. compressa (Fabricius)

Diptera

Calodexia (?) *ventris* Curran

Coleoptera

Ripidius pectinicornis Thunberg§
Trirhogma caerulea Westwood

* Only references published after 1960 are listed; for earlier citations see Roth and Willis (1960). Unless otherwise indicated in footnotes, all of the parasites may occur naturally.
† Experimental parasitization; could not be maintained beyond one generation. The natural host is the cockroach, *Eurycotis floridana* (Walker).
‡ Normally a parasite of Coleoptera and Hymenoptera but will attack almost any insect to which it is exposed, and can be a serious pest of insect cultures of practically any insect order. Only one record is known of this species parasitizing the eggs of *Periplaneta*, and the wasp probably was brought into the laboratory in the nests of a mud dauber.
§ Based on one cockroach nymph and this may have been misidentified. The beetle is known to parasitize only adults and nymphs of *Blattella germanica*.

human beings. (4) Ingestant allergens may be caused by ingesting food partially eaten by cockroaches.

In addition to the insect parasites of *P. americana* shown in Table 1.7, *Pimeliaphilus cunliffei* is the only mite known to be a natural parasite of this cockroach. In the laboratory, entire colonies of cockroaches can be destroyed by the mite (Piquett and Fales, 1952). Its life cycle requires 28–32 days. The adult, which lives 2–3 weeks, produces 2 or 3 batches of eggs, from 1–20 in number, but usually 12. The just-hatched larva feeds on cockroaches (Cunliffe, 1952). This mite has been accused of biting people in addition to cockroaches (Baker *et al.*, 1956). Three other mites, *Caloglyphus* sp. (Piquett and Fales, 1952), *Rhizoglyphus tarsalus* Baker (Rau, 1940), and *Tyrophagus noxius* A.Z. (Roth and Willis, 1960), not normally parasitic on cockroaches, may attack *P. americana* and reduce the vigour of laboratory colonies. Mites have been controlled by spraying a 5% solution of *p*-chlorobenzene sulphonate on the outsides of cockroach cages, and by applying a 5% dust of the same compound inside the cages and on the insects (Fisk, 1951). Flowers of sulphur and general sanitary procedures have also been used to control mites in laboratory cultures of the American cockroach (Piquett and Fales, 1952; Qadri, 1938).

Among insect parasites of the American cockroach (Table 1.7), the most important destroyer of eggs in the oötheca is *Tetrastichus hagenowii*. This wasp parasitizes at least six species of Blattidae. Because of a wide geographic distribution

Table 1.8 Predators reported to feed on *P. americana*.*

Insects

Hemiptera

Spiniger domesticus Pinto (Reduviidae)

Coleoptera

Morion simplex (Carabidae) (Deleporte, 1976)

Amphibia

Bufo marinus (Linnaeus) (Deleporte, 1976)
Hyla cinerea (Schneider)

Reptiles

Ameiva exsul Cope
Anolis cristatellus Dumeril and Bibron

Birds

Chicken
Myna
Sparrow

Mammals

Rat
Mongoose
Opossum (Deleporte, 1976)
Cat
Man

* See Roth and Willis (1960) for literature references prior to 1960. Many of the records referred to as cockroaches or *Periplaneta* spp. may well have been *P. americana*. The American cockroach has been eaten by a number of animals in captivity including scorpions, house centipedes, mantids, toads, frogs, snakes, birds, unidentified batrachians and lizards in a zoo (Roth and Willis, 1960), and monkeys (Cornwell, 1976).

and a rather high rate of parasitization, it is undoubtedly an important biological control agent wherever it is found (Cameron, 1955). The number of wasps that emerge from a single oötheca varies and, in multiple attacks, more than a hundred individuals may develop per cockroach egg case. *Evania appendigaster* is probably the second most important parasite of American cockroach eggs. The wasp deposits a single egg in an oötheca, and its larva usually destroys all of the cockroach eggs. *Evania appendigaster* from Hawaii was introduced and established on Canton Island, and also on Samoa (Dumbleton, 1957). The wasp's presence in houses is a good indication of cockroaches too being present. Much larger and more conspicuous than the relatively minute *Tetrastichus*, *Evania* is readily noticed and may be killed by householders unaware of its role in controlling cockroaches (Edmunds, 1953).

Less well known in North America are species of *Ampulex* which parasitize the nymphs and adults of cockroaches. *Ampulex compressa*, introduced to the island of Oahu from New Caledonia (Williams, 1942), has become established there and on other islands of the Hawaiian Group (Pemberton, 1945). It does not appear to be very effective in reducing *P. americana* populations, however, either indoors or out. Probably it would not succeed in homes because the public has a greater fear of large conspicuous wasps than of the more secretive cockroaches (Hagen, 1979).

Whereas many creatures prey on cockroaches, unfortunately, most of the prey species have not been identified. The number of animals listed in Table 1.8 which feed on *P. americana* is undoubtedly smaller than the actual number of predators in nature.

American cockroaches themselves may be cannibalistic and predacious. In laboratory cultures, they have been known to eat members of their own species including oöthecae (Adair, 1923; Sonan, 1924; Klein, 1933; Gould and Deay, 1938; Griffiths and Tauber, 1942a). If only a small portion of the oötheca keel is eaten or damaged, cockroach eggs will desiccate and die (Roth and Willis, 1955a). Cannibalism occurs more frequently among young nymphs than among older nymphs and adults. The behaviour is stimulated by the presence of injured nymphs. Eating crippled nymphs is supposedly beneficial as a sanitary measure in laboratory colonies (Wharton *et al.*, 1967).

As a predator, *P. americana* has been reported to eat termites (Annandale, 1910), eggs of the hemipteran *Cantao ocellatus* Thunberg (Takahashi, 1924), young bedbugs (Gulati, 1930), parasitic mites *Pimeliaphilus cunliffei* (Cunliffe, 1952), egg clusters of first instar larvae of the moths *Prodenia litula* and *Attacus atlas* (Sonan, 1924), and reptile eggs in the aquarium at Frankfurt am Main (Lederer, 1952).

1.5 Economic and biological importance

It is impossible to assess the monetary value of foodstuffs and other articles damaged or destroyed by the American cockroach. The characteristic cockroach odour imparted to food, which remains even after cooking or processing makes the food unfit for human consumption. Loss of business and goodwill can result when customers find cockroaches or their parts in food sold by stores and restaurants. *P. americana* is one of the three most common species of cockroach found in buildings (Cornwell, 1976).

The presence of the American cockroach in dwellings can inspire entomophobia, distress, and embarrassment. In the United States, on the average, a person may tolerate a maximum of five cockroaches per week seen inside his home (Piper and Frankie, 1978). However, this number would probably vary with the economic status of a family and, in tropical cities and towns, people often consider cockroaches an inevitable part of living (Cornwell, 1976), and this may well be true among residents of large American cities. The ability of *P. americana* to acquire, harbour, and mechanically transmit several human pathogens, in addition to causing allergenic reactions, makes this cockroach a potential health hazard.

On the other hand, where it occurs out-of-doors, it plays an important role in the food cycle of animals that prey upon it. In addition, *P. americana* is of value to: pest control operators who attempt to control it; biology teachers who use it in the classroom; insecticide chemists who study the mode of action of insecticides and test the efficacy of new insecticides on the species; and last, as the following chapters show, to research physiologists who, because of its size and ease of rearing, employ the American cockroach to elucidate their knowledge of physiological systems and processes in a group of animals of great significance to man.

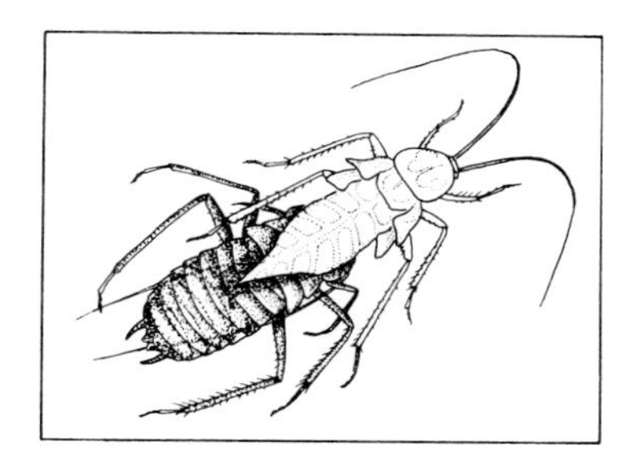

2

Integument

Richard R. Mills

2.1 Introduction

The integument of insects has been extensively reviewed during the past two decades [e.g. Neville (1975) and Andersen (1979); for earlier work see Richards (1951)]. Morphological aspects of integument have been reviewed both at the histochemical level (Wigglesworth, 1972) and ultrastructural level (Smith, 1968; Neville, 1970). Biochemistry of insect cuticle is the subject of a recent article by Andersen (1979). Although numerous references are made to *Periplaneta americana* in these reports, not all aspects of cuticular chemistry and deposition of cuticle have been studied using this species as the experimental animal. In the following account, the results from studies of other insect species are noted and, in some cases, the information is extrapolated to *P. americana*.

The cockroach, like other insects, possesses an integument which acts as a barrier to external absorption, and limits water loss to prevent desiccation. Some degree of protection is afforded via the sclerotized exocuticle, and the dark brown colouration allows it to blend into its surroundings.

2.2 Structure of integument

The structure of the integument is divided into three general areas. First, the epidermal cells, with the associated basement membrane, secrete and maintain the cuticle. Numerous specialized cells occur within the epidermal layer and perform specific functions. The second area is the procuticle itself, which is composed of the endocuticle and the sclerotized exocuticle. The outer area is the epicuticle which is composed of thin distinct layers. The structure of cockroach integument is shown diagrammatically in Fig. 2.1.

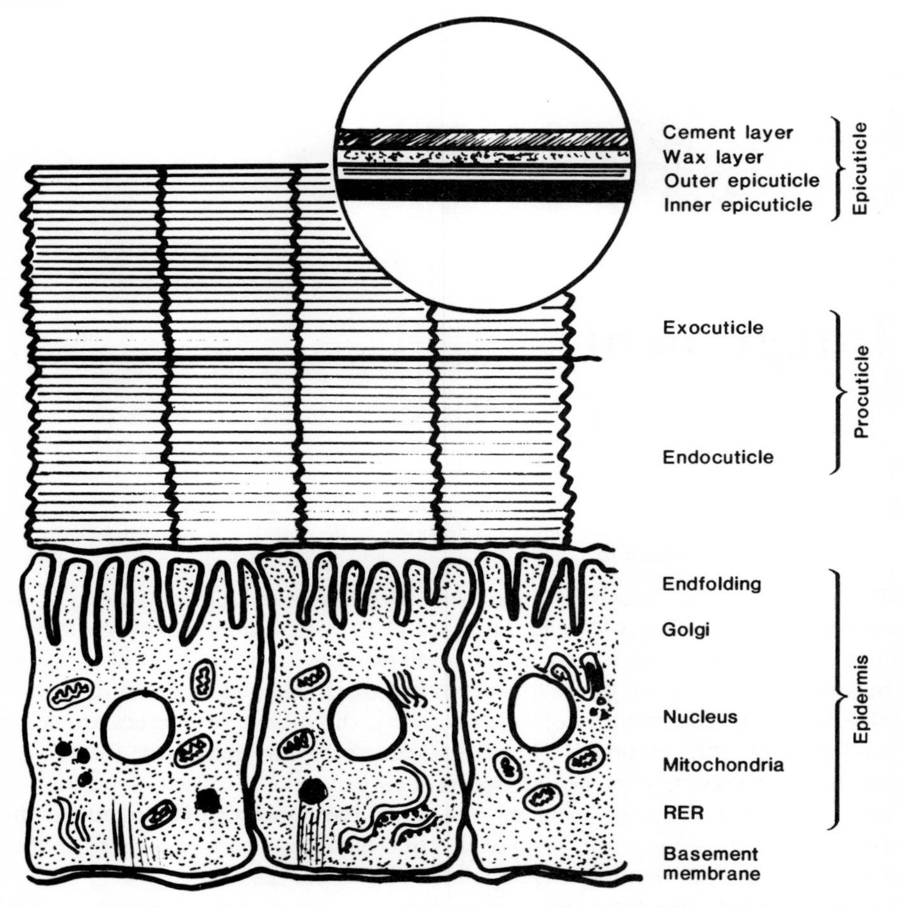

Fig. 2.1 Diagramatic representation of the cockroach integument. The integument is composed of three major areas, which consist of the epidermal cell layer, the procuticle and the thin, four-layered epicuticle. The epidermis secretes and maintains the cuticle while the epicuticle acts as a base of deposition and later as a waterproofing mechanism. The bulk of the cuticle is composed of the procuticle which is subdivided into the pre-ecdysial sclerotized exocuticle and the post-ecdysial endocuticle.

2.2.1 Macromolecular structure

The epicuticle is made up of a number of distinct layers. The inner epicuticle is approximately 0.5–2.0 μm thick and is composed of lipids, proteins and phenols. This layer is probably the polyphenol layer of Wigglesworth (1947) and corresponds to the 'dense layer' or protein epicuticle of Brück and Stocken (1972). The outer epicuticle is usually 150–250 Å and appears to be trilaminar (Locke, 1974; Krolak and Mills, 1981a). Wigglesworth (1947) referred to this layer as cuticulin.

Muscle fibres attach to the outer epicuticle and it is the first part of the new cuticle to be secreted by the epidermal cells. It plays an important role in the presumptive architecture of the newly forming cuticle and is instrumental in the determination of surface patterns.

The inner epicuticle stains for phenoloxidase in *P. americana* and a positive argentaffin reaction apparently denotes the presence of diphenols (Mills, Androuny and Fox, unpublished). Laminae appear to be present in the inner epicuticle (Krolak and Mills, 1981a) and are quite prominent in the spermathecal duct (Gupta and Smith, 1969).

The wax layer is difficult to visualize in *P. americana* electron micrographs of integument, but various staining procedures show a thin line of lipid-positive material on the external surface. This probably includes the cement layer which has been reported to be a shellac-like substance (Beament, 1955). The function of the cement layer is to protect the wax layer which in turn prevents desiccation of the insect.

The wax layer (or perhaps both the wax and cement layers) is known to prevent water evaporation from the cockroach. Ramsey (1935) showed that, at 30°C, the rate of evaporation of water from *P. americana* suddenly increases, a phenomenon that is attributed to molecular rearrangement of the cuticular lipids. Further evidence proved that the surface epicuticular lipids were responsible since appropriate solvents or an abrasive dust removed the waterproofing properties (Wigglesworth, 1945).

The deposition of wax on the external surface may occur via the pore canals, although other insects secrete large quantities of wax and have no pore canals. It has been speculated that the wax is soluble in volatile solvents and is deposited as the volatile components evaporate (Beament, 1955). However, Gilby (1962) found no short-chain volatile compounds present which could fulfill this function. The pore canals are believed to be formed by the fusion of wax filaments which could act like a wick of a candle to transport wax to the surface. The pore canals do not open onto the surface in *P. americana* while the dermal glands do (Scheie *et al.*, 1968). Morphological evidence suggests that the pore canals secrete the wax layer while the dermal glands deposit the cement layer.

The bulk of the cuticle is made up of procuticle (terminology of Neville, 1975) which is subdivided into exocuticle and endocuticle. The exocuticle is composed primarily of chitin microfibrils embedded in a protein matrix. The proteins are thought to be sclerotized, a process that involves a variety of physical transformations. *N*-acetyldopamine (NADA) is essential for hardening and normal darkening, and enters the epidermis as the free 4-*o*-*β*-D-glucoside or bound to carrier proteins (Koeppe and Mills, 1972). Dehydration of the cuticle occurs during the immediate post-ecdysial periods (Krolak and Mills, 1981b), and various lipids are present which may function in reduction or oxidation reactions. The tanning process is discussed in Section 2.3.

The endocuticle is made of the same basic chemical components but the proteins are not sclerotized. It is deposited after ecdysis while the exocuticle is elaborated prior to ecdysis. The endocuticle is therefore not subjected to the post-ecdysial diphenolic interactions nor to the general body dehydration. It is laid down in daily growth layers by the epidermis which is controlled by a clock mechanism (Neville, 1965).

Other types of cuticle that are somewhat more elastic and not as hard include resilin, arthrodial membrane and transitional cuticle. Resilin is a pliable rubber-like cuticle which is a true elastomer protein. It contains specific di- and trityrosine cross-links and has no pore canals. It functions primarily to assist muscles and probably as a phasic mechanoreceptor. The major use is as elastic ligaments in insect flight systems and it performs this function in the American cockroach (Andersen and Weis-Fogh, 1964). It also occurs in the spermathecal duct wall (Gupta and Smith, 1969) and probably in many other areas of *P. americana*. The arthrodial membrane (previously known as the intersegmental membranes) is a soft, colourless, and flexible cuticle located between sclerites. It can be seen at the base of sensory hairs, gills, wings, etc., and forms the basis of the spectacular stretching capacity of queen termites. Transitional cuticle is intermediate between sclerotized and rubber-like cuticle. A precise definition has never been published, but both the flexibility and hardness are transitional. The compound eye and ocellus are examples of this type of cuticle.

2.2.2 Chemical composition of the cuticle

The chemical components that make up the cuticle are primarily chitin, proteins and lipids. Chitin is a *N*-acetylglucosamine polymer with a β-1,4-linkage (Fig. 2.2). The acetyl group may be missing in a few residues so that the chain as a whole may be positively charged and the missing group may be instrumental in the orientation of chitin chains (Neville, 1967). Other carbohydrates such as glucose, galactose, mannose, xylose and arabinose can be extracted from the cuticle (Lipke *et al.*, 1965a) although the significance of their presence is unclear. The basic scheme of chitin synthesis is well known and there is fairly conclusive evidence that

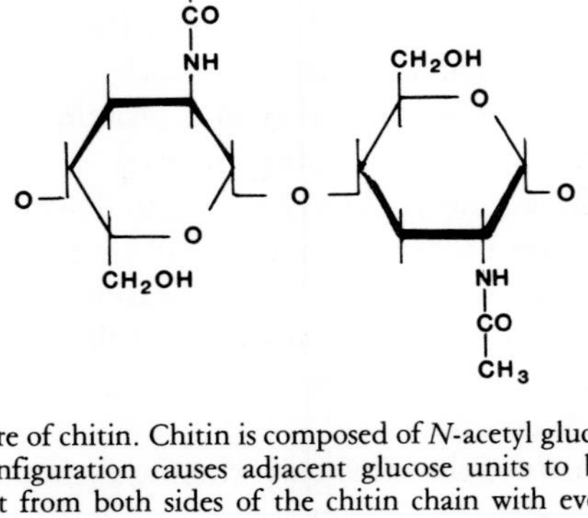

Fig. 2.2 The basic structure of chitin. Chitin is composed of *N*-acetyl glucosamine moieties linked by β-1,4 bonds. The beta configuration causes adjacent glucose units to be inverted. Therefore the *N*-acetyl group extends out from both sides of the chitin chain with every other glucosamine unit oriented in the same direction.

glucose is converted to chitin by the cockroach in a similar manner (Lipke *et al.*, 1965b).

The lipid components of the cockroach cuticle have been extensively investigated. The major unsaturated hydrocarbon is *cis*, *cis* 6,9 heptacosadiene (Baker *et al.*, 1963) while both 15-hentetracontene and 15-tritetracontene are also present in appreciable quantities (Conrad and Jackson, 1971). Apparently, no long-chain alcohols occur although 7–11% of the wax from *P. americana* consists of C_{13}–C_{18} *n*-alkanoic acids and 8% of the total lipids are C_{13}–C_{25} aliphatic aldehydes (Gilby and Cox, 1963). A branched-chain hydrocarbon, 3-methypentacosane (Baker *et al.*, 1963) is also present and may serve to elevate the melting point of the hydrocarbon mixture. This melting point appears to be only about 20°C in *P. americana* (Tartivita and Jackson, 1970). Sterols are also found in the cuticle (Gilby and Cox, 1963) but the major sterol, cholesterol, is unesterfied (Vroman *et al.*, 1964).

The synthesis of various hydrocarbons can actually take place in the integument itself when incubated *in vitro* with labelled palmitate and acetate (Nelson, 1969). The actual synthesis in teneral adults probably takes place by the condensation of two fatty acids followed by decarboxylation and reduction (Conrad and Jackson, 1971). On the other hand, these same authors find that during the synthesis of the branched-chain 3-methylpentacosane, propionate is incorporated during the elongation process in adult cockroaches.

Structural proteins make up the major bulk of the cuticle although no specific protein has yet been completely purified and sequenced. The cuticular proteins of *P. americana* have been subjected to acrylamide gel electrophoresis and approximately 14 different fractions have been resolved (Fox and Mills, 1969).

2.2.3 Supermolecular structure of the cuticle

The cuticle, as noted above, is made up of proteins, chitin and an assortment of lipids. The precise structure of the cuticular matrix has not yet been elucidated but recent studies using modern techniques have resulted in the construction of various models. Neville's (1975) detailed review delineates what is presently known about arthropod cuticles.

Specific information on the supermolecular structure of cockroach cuticle is lacking but, since all arthropod cuticle studied appears to have a number of common characteristics, current knowledge will be extrapolated to *P. americana*.

Insect cuticle is made up of a series of microfibrils (Rudall, 1965, 1967) with a diameter of approximately 28–31 Å (Neville, 1975). The microfibril is composed of chitin embedded in a protein matrix (Rudall, 1967; Neville, 1970; Weis-Fogh, 1970).

The unit cell of extracted α-chitin is 18.84 Å × 4.76 Å or 89.678 $Å^2$ (Carlstrom, 1957). This cell contains two anti-parallel chains, so that one chain would have a cross-sectional area of one half the unit cell or approximately 44.84 $Å^2$. If one assumes the microfibril is square (roughly 28 Å × 28 Å with an axial height of 5.14 Å) then the volume is approximately 403 $Å^3$. Three sheets, each sheet being

9.42 Å wide, (3 × 9.42 Å = 28.26 Å) with each sheet having 6 chains of chitin (4.76 Å per chain × 6 = 28.56 Å) would fit the approximate dimensions of a microfibril. Using the mean volume of 230 $Å^3$ for a unit cell, it is possible that as many as 18 chains of chitin could pass through a crystallite at any one level (Neville, 1975).

The chitin crystallites have been shown to have a nearly perfect hexagonal packing in pre-ecdysial rhinoceros beetle cuticle (Neville, 1975). From an analysis of synthetic systems such as plastics, one can deduce that the hexagonal chitin crystallites can arise via crystallization. Given that the microfibrils of insect cuticle are pure crystalline chitin, one can assume that only the peripheral chains would be available for bonding to the matrix (Neville, 1975).

The microfibrillar orientation is in one direction creating a layer or lamellae with the microfibrils parallel to each other. The thickness varies according to the insect species. A single lamella is usually laid down at a slight angle to the one before it, creating a helicoidal arrangement of the layers which rotates counter-clockwise. The physical character of this arrangement causes a parabolic pattern when the cuticle is sectioned at an oblique angle to the surface. The original discovery of cuticle deposition was made by Bouligand (1965) and extensively reviewed by Neville (1975). Electron micrographs of both nymphal and adult cuticle of *P. americana* show similar architecture including the characteristic parabolic patterns (Krolak and Mills, 1981a). The pattern alternates between a helicoidal layer and a unidirectional layer in cockroaches and this alternate deposition is controlled by a circadian clock (Neville, 1965, 1967).

In summary, it appears that the cuticle of *P. americana* is similar to other insects and various models used to explain the supermolecular architecture can be extended to the American cockroach.

2.3 Integument during ecdysis

At each ecdysis, a new cuticle is elaborated by the epidermal cells. A complex series of events precedes the actual ecdysis and specific morphological changes occur in an ordered sequence. After ecdysis, the pliable cuticle becomes sclerotized through a series of timed events. The integumental changes have been divided into three stages as suggested by other reviewers. The pre-apolysis stage is the period between the initiation of ecdysis and apolysis which is the separation of the old and new cuticle (Jenkin and Hinton, 1966). The post-apolysis stage is the two to three day period between apolysis and ecdysis. The post-ecdysial stage is denoted as the immediate 0–24 h period after the shedding of the old cuticle. The term inter-ecdysis is used as the time between post-ecdysis and the following pre-apolysis.

The ecdysial cycle in the American cockroach has been studied by many earlier investigators and by the present author. If sixth to last instar nymphs (400 mg or larger) are kept at 28 °C with greater than 40% r.h. and a 12 : 12 LD photoperiod, the time from one ecdysis to another is 28 ± 2.3 days. Events at ecdysis or during the immediate post-ecdysial period can be analysed with confidence. As the inter-ecdysial period progresses, considerable variation occurs and the initiation of ecdysis is difficult to ascertain precisely. At apolysis (or within six hours after apolysis), the eyes become an opaque grey (described by Flint and Patton, 1959) and so the post-apolysis period can be easily determined. The following account of integumental changes are based on this 28-day cycle.

On approximately day 24, the epidermal cells tend to change configuration, and some cytoplasmic inclusions occur. The nuclei tend to enlarge or become more oblong in shape. Presumably, two different peaks of β-ecdysone occur sometime during days 23–25. On day 25, the ecdysial membrane is laid down between the epidermis and endocuticle (Krolak and Mills, 1981a). Within several hours, spaces can be seen between the ecdysial membrane and the epidermis which denotes the initial stages of apolysis. A short time later, the epidermal cells begin to divide, and the exuvial space becomes correspondingly larger. It appears that many microtubules are formed during mitosis and these remain after cell division terminates. The outer epicuticle is elaborated almost immediately from the apex of the apical microvilli. The dense layer (inner epicuticle) follows. The precise timing of exocuticle deposition, old endocuticle hydrolysis and increases in the exuvial space have been impossible to determine. Electron micrographs taken from a variety of insects (0–18 h post-apolysis) show almost complete exocuticle formation with little or no loss of old endocuticle while others show almost complete hydrolysis of the old cuticle. In all cases, however, at 30 h after apolysis, the exocuticle is formed, the exuvial space widens and most of the untanned endocuticle is no longer present. A number of protein granules remain in the epidermal cell cytoplasm, but a very reduced apical exocytosis and basal pinocytosis is observed. Ecdysis occurs between 42 and 60 h after apolysis. The average of 23 animals was 51.3 h (s.d. = 4.1). The correlation of integumental structure with the ecdysial cycle is summarized by Krolak and Mills (1981a). The post-apolysis or pharate adult period compares favourably with that proposed by Hopkins and Wirtz (1976).

As these structural changes take place, various biochemical transformations can also be observed. The total haemolymph protein begins to build up on approximately day 25. Just prior to this time (early day 25), the haemolymph volume as well as total body water increases (Krolak and Mills, 1981b). The haemolymph protein concentration (mg μl^{-1}) and the cells per unit volume remain relatively constant during pre-apolysis and post-apolysis, but the vast increase in haemolymph volume causes additional cells and protein to accumulate. This maintains the normal concentration, yet allows the build-up of numerous cuticle precursor molecules. An antidiuretic hormone (Goldbard *et al.*, 1970) is responsible for the post-apolysis water retention (Mills and Whitehead, 1970).

Polyacrylamide gel electrophoresis of the haemolymph and cuticle proteins shows a number of fractions with similar mobility (Fox and Mills, 1969). Immunodiffusion techniques revealed that some haemolymph proteins are antigenically similar to cuticular proteins, suggesting that haemolymph proteins can be incorporated into the cuticular matrix relatively unchanged (Fox *et al.*, 1972). The presence of basal pinocytosis and apical exocytosis in the epidermal cells during pharate cuticle deposition also tends to support this contention (Krolak and Mills, 1981a).

Haemolymph proteins bind NADA-β glucoside during post-apolysis and post-ecdysis. The protein-bound phenol conjugate is not incorporated until after ecdysis suggesting bursicon mediation (Koeppe and Mills, 1972). Further work has substantiated these results and suggests that bursicon stimulation of epidermal transport may be cyclic AMP-dependent (Foldesi and Mills, in preparation).

The diphenolic tanning compounds do not appear to be present in any quantity prior to ecdysis. Tyrosine-metabolizing enzymes, or inactive proenzymes, are primarily synthesized during the 51-h post-apolysis period. Tyrosine hydroxylase is active in haemolymph cells, fat body, and epidermis if the tissue is homogenized and allowed to stand for a short time. On the other hand, intact haemolymph cells will not metabolize tyrosine in the absence of bursicon (Mills and Whitehead, 1970). It was initially thought that bursicon-directed tyrosine uptake was solely responsible for this phenomenon, but recent studies suggest that a cAMP-dependent protein kinase phosphorylates the tyrosine hydroxylase proenzyme (Shafer and Mills, 1981). Thus, it appears that bursicon-induced, cAMP-dependent protein kinases act, not only to potentiate tyrosine transport, but also to activate the initial step in tyrosine metabolism (see Section 12.4.2).

The dopa-decarboxylase activity increases in both the haemolymph and epidermis during post-apolysis (Hopkins and Wirtz, 1976). Most of the activity is present by ecdysis but the actual peak of activity is at 6 h post-ecdysis. They found tyrosine decarboxylase to be present in low concentrations, but post-apolysis analysis failed to show an increase in enzyme titre (Hopkins and Wirtz, 1976). These results paralleled those found for *in vitro* experiments using labelled tyrosine as the substrate (Hopkins *et al.*,1971a). Similar results have been obtained by Whitehead and Mills (unpublished) except that considerably more tyrosine decarboxylase activity could be demonstrated. Partial purification of crude haemocyte preparations of P-60 polyacrylamide gel resulted in the loss or dilution of necessary co-factors. The addition of pyridoxyl phosphate to this preparation resulted in a highly active conversion of labelled tyrosine to tyramine. The conversion of dopa to dopamine was approximately equal. These data are summarized by Whitehead (1969) and Mills and Whitehead (1970).

The *N*-acetyl transferase activity follows the basic pattern of dopa decarboxylase. Peak titres occur during post-ecdysis although at least 80% of the activity is synthesized *de novo* between apolysis and ecdysis. Enzyme activity can be found in the fat body and epidermis, but the subsequent conversion to the 4-*o*-β-D-glucoside does not occur. Contrary to previous postulations, the *N*-acetyl transferase does not occur in the haemolymph cells. The serum contains this transferase as well as glucose transferase. When ^{14}C-dopamine (plus acetyl-CoA) is incubated with the serum, both NADA and its glucoside were found (Siegrist *et al.*, 1981).

The overall pathway involving the metabolism of tyrosine to *N*-acetyl dopamine NADA has been known to occur for some time (Fig 2.3). Mills *et al.* (1967) observed the presence of essential enzymes including those that could metabolize tyrosine to tyramine to *N*-acetyl tyramine to NADA. Murdock *et al.* (1970b) performed the conversion of tyrosine to NADA *in vivo* showing that the pathway

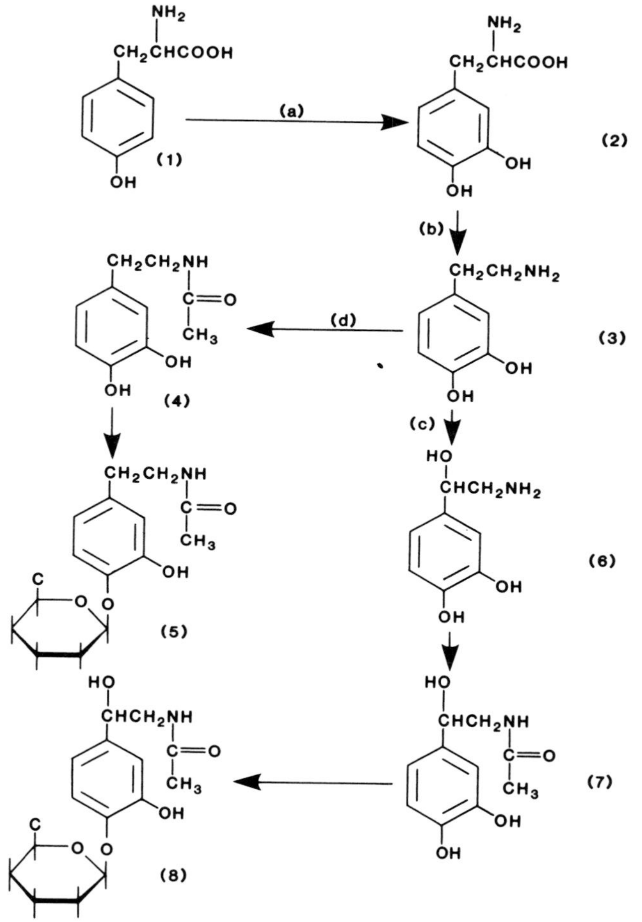

Fig. 2.3 Metabolism of tyrosine to *N*-acetyldopamine and *N*-acetylnoradrenalin. Tyrosine (1) is converted to dopa (2) via tyrosine hydroxylase (a). Dopa decarboxylase (b) removes the carboxy group from dopa to produce dopamine (3). Dopamine metabolism is a key metabolite since *N*-acetyl transferase (d) will catalyse the formation of *N*-acetyl dopamine (4) while dopamine-β-hydroxylase (c) will metabolize dopamine to noradrenalin (6). *N*-acetyl transferase (d) will also convert noradrenalin to *N*-acetyl noradrenalin (7). Both of the end metabolites are generally found as their beta-glucosides (5,8). *N*-acetyl dopamine can also be detected as a 3-*o*-phosphate or 3-*o*-sulphate derivative. Dopamine can be acted on by monoamine oxidase to form dihydroxyphenyl acetic acid and dopamine can be found as the 3-*o*-sulphate. Both dopa and dopamine are readily converted to melanin. Tyrosine and dopa can be transaminated to their corresponding phenyl pyruvic acid moiety with further metabolism to other acids.

actually existed. Koeppe and Mills (1975) found that both dopamine and noradrenalin are *N*-acetylated and converted to their corresponding 4-*o*-β-D-glucosides.

Other conjugates of NADA can occur in the cockroach. Both a 3-*o*-phosphate and a 3-*o*-sulphate are synthesized after ecdysis (Bodnaryk *et al.*, 1974). It was proposed that these two moieties are the major NADA esters involved in the transportation to the cuticle. More recent studies suggest that most of the NADA occurs as the glucoside although there are appreciable quantities of the phosphate ester present (Krolak *et al.*, 1981).

Tanning precursor molecules such as phenylalanine or tyrosine dipeptides do not occur in the American cockroach. β-alanyl-tyrosine, glutamyl-phenylalanine, and tyrosine phosphate could not be found in cockroach blood (Whitehead and Mills, unpublished). The concentration of tyrosine builds up rapidly just before ecdysis and declines as the new cuticle is tanned (Wirtz and Hopkins, 1974).

Melanin synthesis and concurrent polymerization is a well-known phenomenon in insects. Production of melanoproteins and their incorporation into the cuticle is well documented. In some insects, melanin may also be instrumental in the sclerotization process. In *P. americana*, ^{14}C-indole (formed from dopa) is not incorporated into the cuticle during the first few hours after ecdysis (Mills and Fox, 1972) but some label is deposited after several hours. These data suggest that melanin is not a sclerotization agent but may function in secondary darkening.

Tyrosine can be metabolized to various diphenols with acidic side-chains (Mills and Lake, 1971). Tyrosine amino-transferase (TAT) converts tyrosine to *p*-hydroxyphenyl pyruvic acid which in turn is decarboxylated to *p*-hydroxyphenyl acetic acid or hydrogenated via a NADH + H^+-dependent lactic dehydrogenase to *p*-hydroxyphenyl lactic acid. The latter can be further metabolized to *p*-hydroxyphenyl propionate (phpp) which appears to be the end-product. These reactions are commonly found during the inter-ecdysial period, but may not function at ecdysis. Hopkins *et al.* (1971a) found that TAT was present but the equilibrium was towards the conversion of phpp to tyrosine. Recently, Hume and Mills (unpublished) have found that TAT activity declines at apolysis which suggests a specific control mechanism. The acids apparently do not play an important role in the sclerotization process and may be of little consequence to the integument.

In summary, the available evidence suggests that tyrosine is not metabolized appreciably prior to ecdysis. It is not until after apolysis (when the new cuticle is elaborated), that the enzymes become available and sufficient tyrosine builds up. Bursicon initiates the metabolism of tyrosine by activating adenyl cyclase receptor sites on cell surfaces. Cyclic AMP, via specific protein kinases, increases tyrosine transport into the cell and activates tyrosine hydroxylase. Tyrosine is rapidly converted to NADA conjugates which occur in, or are transported to, the epidermal cells. Free NADA is incorporated into the cuticle where it is required for normal tanning.

2.3.1 Post-ecdysial changes in the cuticle

The complicated series of reactions involved in the hardening and darkening of the cockroach cuticle are not completely understood. The complex incorporation of proteins, lipids and carbohydrates followed by dehydration creates a number of possible biochemical combinations and several models have been proposed for different insects based on what is known of the physical and chemical interactions. A summary is presented here, using available evidence from cockroach studies and information extrapolated from investigations of other insects.

The enzymolysis of sclerotized cuticle from *P. americana* results in the hydrolysis of only 3–7% of the available linkages despite the use of a variety of enzymes (Lipke and Geoghegan, 1971). These same authors (after a rigorous cleansing of the cuticle) extracted with 6N HCl in 25% ethyl alcohol for 12 h under nitrogen. An insoluble and soluble fraction were obtained which, in turn, were subjected to various proteases and carbohydrases. The very low cleavage suggested that the β-linked chitin, the α-linked neutral sugars and the proteins were cross-linked in such a way that the enzymes could not reach the hydrolytic site. From these data, Lipke and Geoghegan (1971) suggest that dehydration and the subsequent hydrogen bond formation is the basis for resistance to enzymolysis.

The exchange of water with ethyl alcohol in *Rhodnius* cuticle reduces the amount of material extractable with urea or with water (Hillerton, 1978). Other solvents such as 1, 2-diaminoethane remove the brown pigment (possibly the diquinones) from blowfly puparia and from the oötheca of *P. americana* (Hillerton and Vincent, 1979). These types of study tend to delineate the nature of molecular cross-linking in cuticles (Hackman and Goldberg, 1977). In general, the popular theory of diquinone tanning of insect cuticle via ortho-substituted amines (originally postulated by Pryor, 1940), is no longer universally accepted. As suggested by Lipke and Geoghegan (1971), dehydration produces hydrogen bonding or perhaps a more closely packed configuration which resists enzyme attack. The loss of water may also cause rearrangement of protein structure as envisaged by Hillerton and Vincent (1979). The overall effect of dehydration (which accompanies sclerotization) would be to cause closer spatial arrangement between proteins which, in turn, by virtue of their closer proximity, could create an array of hydrogen bonding and possibly other forces. This general theory was put forth by Fraenkel and Rudall (1940, 1947).

The alternate hypothesis is that of diquinone cross-links between cuticular proteins (Pryor, 1940). NADA could be oxidized to its corresponding diquinone via phenoloxidase and undergo substitution on the ortho position of the ring. The electron-withdrawing characteristics of the oxo group would produce electrophilic character at the ortho position allowing a nucleophilic or electron-donating primary amine (such as lysine) to bind. The replacement of a hydrogen on the ring and the loss of a hydrogen from the nitrogen returns the diquinone to the diphenol. The phenoloxidase subsequently repeats the process with the second protein binding to the other ortho position. This type of reaction occurs in the quinone tanning of leather and has been universally accepted for years as explaining sclerotization in insects.

The enzymes for the synthesis of NADA are present in *P. americana* (Mills *et al.*, 1967) and NADA has been synthesized *in vivo* from tyrosine (Murdoch *et al.*, 1970a) and from dopamine (Koeppe and Mills, 1975). Phenoloxidase is present in the cuticle (Mills *et al.*, 1968) and is known to oxidize NADA. Furthermore, NADA is essential for normal sclerotization in the cockroach cuticle (Mills, unpublished). The evidence that NADA is the sclerotization agent unfortunately does not provide specific data as to its role in the cuticle.

Other reactions involving NADA may also occur. Andersen has found various compounds with the β carbon altered (Andersen, 1970, 1971, 1972, 1974; Andersen and Barrett, 1971). He proposes that various proteins may link to the β position in lightly sclerotized cuticle while the conventional ortho substitution occurs in more heavily sclerotized sections (Andersen, 1974). These experiments also provide some proof of ortho substitution since ortho-tritiated tyrosine gives up label when reacted with his system (Andersen, 1974).

At this time, the evidence is not conclusive for any model system. The NADA is essential and must perform some specific function during sclerotization. In addition, it is well known that cuticle dehydration occurs during the same process. Cuticle water loss occurs during the second hour post-ecdysis in the cockroach (Krolak and Mills, 1981b) which is the same time that NADA (Koeppe and Mills, 1975) and NADA-bound protein (Koeppe and Mills, 1972) enter the cuticle. The almost completely white cuticle begins to tan at this point (Mills, 1965), and so all the processes seem to occur simultaneously in *P. americana*.

One could speculate that NADA binds to various cuticular proteins, linking them together in such a way that, during probable self-assembly, the hydrophobic amino acids line up opposite each other creating an apolar matrix. Leucine and other apolar amino acids occur in the cuticle of *P. americana* in relatively high numbers (Lipke and Geoghegan, 1971). Numerous lipids, as noted in Section 2.2.2, could also impregnate these apolar regions. The chitin crystallites lying within the protein matrix would bind with the polar protein surface containing reactive groups such as histidine or aspartate. The general hexagonal arrangement of chitin crystallites and their spacing from each other produces only a specific amount of area between them. If an apolar region exists at equidistant points between the crystallites, the stereochemical configuration required for enzymolysis via proteases or other enzymes may not be present.

2.3.2 Bursicon-mediated control of cuticle sclerotization

Bursicon is released by the American cockroach immediately after ecdysis and becomes detectable in the haemolymph within fifteen minutes. Peak titres of the hormone occur at approximately 90 min and trace amounts last until seven or eight hours post-ecdysis (Mills, 1965). The hormone is released in the vicinity of the terminal abdominal ganglion (Mills *et al.*, 1965) and more recent studies show that small neurohaemal organs located posteriorly to the last four ganglia actually sequester and release the peptide (Mills, unpublished).

During the immediate post-ecdysial period (the initial 0–5 h), bursicon is released into the haemolymph, and subsequently mediates sclerotization of the cuticle (Mills *et al.*, 1965). The precise mode of action of bursicon has not been completely elucidated, but it appears that the hormone increases membrane permeability to essential components such as tyrosine (Mills and Whitehead, 1970). The pattern of hormone titre in the haemolymph mimics that of diuretic hormone (DH) during the immediate post-ecdysial period and it has been speculated that bursicon and DH may be the same. The uptake of tyrosine by the haemocytes is enhanced by the addition of bursicon, and comparable haemolymph cell samples from post-ecdysial animals take up tyrosine at a much greater rate than similar haemolymph cell preparations from inter-ecdysial cockroaches. These data are summarized by Mills and Whitehead (1970). A more refined study, using haemocyte stabilization agents, confirmed the bursicon-stimulated uptake of tyrosine by haemolymph cells from a lepidopteran larva (Post, 1972).

The mechanism by which bursicon enhances the transport of tyrosine appears to be mediated by cAMP. The injection of cAMP into the isolated thorax of *P. americana* (see Mills *et al.*, 1965 for the assay) results in visible tanning, and the incorporation of ^{14}C-tyrosine into the cuticle (Vandenberg and Mills, 1974). Similar results have been obtained with other insects (von Knorre *et al.*, 1972; Seligman and Doy, 1972). Thus it appears that cAMP mimics bursicon action and one could speculate that cAMP is a second messenger released when bursicon stimulates adenylate cyclase (Vandenberg and Mills, 1974, 1975).

The results of initial experiments did not support this contention since haemolymph cells incubated with active bursicon preparations failed to produce measurable cAMP levels (Vandenberg and Mills, unpublished data). However, similar experiments conducted at depressed temperatures do result in high levels of cAMP (Compton and Mills, 1981). At room temperature, phosphodiesterase activity cleaves cAMP and prevents its accumulation. These studies show that bursicon does activate adenylate cyclase which in turn converts ATP into cAMP. Further experimentation revealed that cAMP also enhanced the uptake of both ^{14}C-tyrosine and ^{14}C-leucine by the haemocytes (Compton and Mills, 1981).

Because specific cAMP-dependent protein kinases are known to exist in various vertebrate tissues, the same type of system has been sought in cockroach haemocytes. Molecular sieve chromatography of soluble haemolymph cell proteins, previously incubated with ^{3}H-cAMP, results in the binding of label to the larger proteins, and inhibition of phosphodiesterase leads to a reduced level of labelled protein. These preliminary data suggest the presence of both phosphodiesterase and protein kinase (Compton and Mills, 1981).

Further studies using polyacrylamide gel electrophoresis of the soluble haemolymph cell fractions show three protein fractions binding cAMP. Two of these fractions are capable of transferring the γ ^{32}P of ATP to protein (Shafer and Mills, 1981).

In summary, it appears that bursicon stimulates cockroach haemocytes by activating adenylate cyclase. The resultant cAMP allows the phosphorylation of specific proteins via a cAMP-dependent protein kinase. The phosphorylated proteins could be either incorporated into the plasma membrane or stimulate

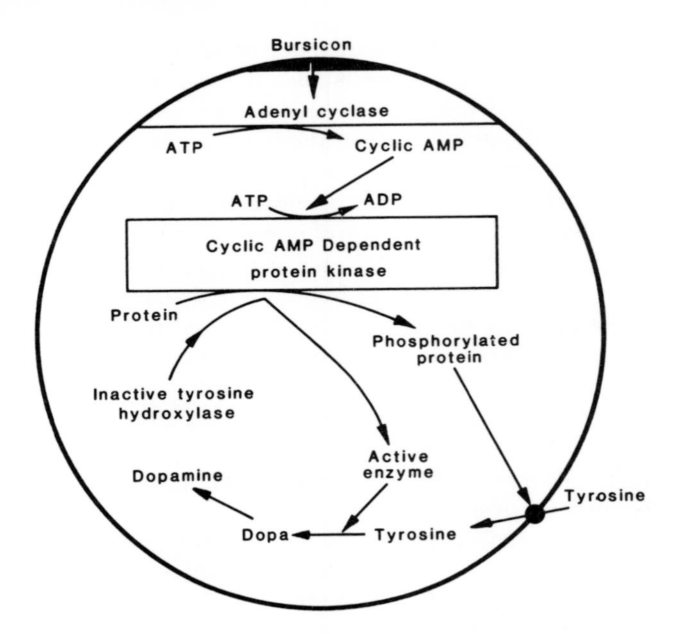

Fig. 2.4 Probable mechanism of bursicon action. The peptide hormone, bursicon, is thought to add to a specific receptor site which results in the activation of adenyl cyclase. The cyclase converts ATP into cyclic AMP (cAMP). The cAMP is essential for cAMP dependent protein kinases which in turn phosphorylate specific proteins. These phosphorylated proteins theoretically increase amino acid transport (including tyrosine) into the cell. Tyrosine hydroxylase also appears to be activated via phosphorylation. Again cAMP dependent protein kinase is instrumental. In summary, bursicon not only acts via cAMP to increase tyrosine transport into the cell but also activates tyrosine hydroxylase to convert tyrosine to dopa. It appears that bursicon can stimulate haemocytes and epidermal cells during the immediate post-ecdysial period. Preliminary evidence suggests fat body cells may also be stimulated and it would not be surprising to find that many other cell types are susceptible to bursicon action.

a receptor site internally. These phenomena would serve to enhance the transport of amino acids across the membrane and into the haemolymph cell. In addition, the cAMP-dependent protein kinase appears to phosphorylate tyrosine hydroxylase, converting the enzyme into its active form (Shafer and Mills, 1981) as in mammalian systems.

Bursicon also appears to act upon the epidermal cells. Diphenolic compounds such as NADA are bound to serum proteins during the immediate post-ecdysial period and are incorporated into the cuticle at peak bursicon titres (Koeppe and Mills, 1972). Radioactive protein-bound phenols are transported across the epidermis in the presence of bursicon and cAMP (Foldesi and Mills, 1981).

Preliminary experiments using neck and waist-ligated freshly ecdysed cockroaches show that injection of bursicon or cAMP into the isolated thorax results in thoracic sclerotization and increased incorporation of ^{14}C-tyrosine (Vandenberg and Mills, 1974). Dopamine also has a stimulatory effect but NADA does not enhance the uptake of ^{14}C-tyrosine into the cuticle (Mills, unpublished). One could speculate that dopamine is stimulating adenylate cyclase receptor sites on the epidermal cell membranes. Some very preliminary data duplicating the repressed temperature experiments of Compton and Mills (1981) show that cockroach epidermal cells do have bursicon-sensitive adenylate cyclase sites on the epidermis (Mills, unpublished). This is not surprising since it has been shown that these sites occur in epidermal systems of *Tenebrio* (Delachambre *et al.*, 1979). A probable mechanism summarizing the mode of action of bursicon is shown in Fig. 2.4.

2.3.3 Accumulative control mechanisms associated with ecdysis

In *P. americana*, ecdysis begins several days prior to the actual shedding of the old cuticle. As described in Chapter 12, specific neurosecretory cells in the brain release prothoraciotropic hormone which in turn induces the prothoracic glands to synthesize ecdysone. Ecdysone is subsequently converted to ecdysterone (20-hydroxyecdysone), the apparent actual moulting hormone (MH). Juvenile hormone (JH) is secreted by the paired corpora allata and modifies the action of MH to control the maturation of the insect (see Chapters 12 and 16).

The integument, like other organs is specifically affected by the titre of ecdysone and JH in the haemolymph. Moulting hormone (20-hydroxyecdysone) initiates cuticle synthesis and acts to control its deposition by the epidermal cells. Ecdysone is present in *Tenebrio* through the beginning of procuticle formation (Delbecque *et al.*, 1978). Specific evidence for sequential determinations of ecdysone titre during the ecdysial cycle in *P. americana* is not available but it appears that two major peaks of ecdysone occur in other insects (Delbecque *et al.*, 1978; Hwang-Hsu *et al.*, 1979; Wielgus *et al.*, 1979). In *Manduca sexta*, the synthesis of DNA occurs in two sequential steps elicited by the two peaks of ecdysone. Mitosis occurs some time later followed by apolysis and cuticle deposition (Wielgus *et al.*, 1979).

In *P. americana*, a series of electron micrographs (based on a 28-day ecdysial cycle of large nymphs held at 28 °C) show epidermal cell division on or around day 24 or 25. Apolysis and cuticle deposition occur during days 26 or 27; a day or two later, ecdysis occurs (Mills, unpublished). Ecdysone peaks would presumably occur prior to day 25.

The cellular programming or reprogramming of the epidermal cells occurs during the first period of ecdysone release. JH must be present during this intial peak in order to modify epidermal cuticle deposition. These data are extensively reviewed by Riddiford (1980), and one could expect a similar sequence to occur in *P. americana*.

2.4 Stabilization of the oötheca

The oötheca, a carpet bag-shaped object approximately 7 mm in length, is elaborated by the asymmetrical pair of colleterial glands. This hardened bag, formed by structural protein and the diphenolic tanning agent, acts as a vehicle for the deposition of the cockroach eggs. The initial studies involving oöthecal synthesis and deposition are summarized by the elegant early work of Pryor (1940).

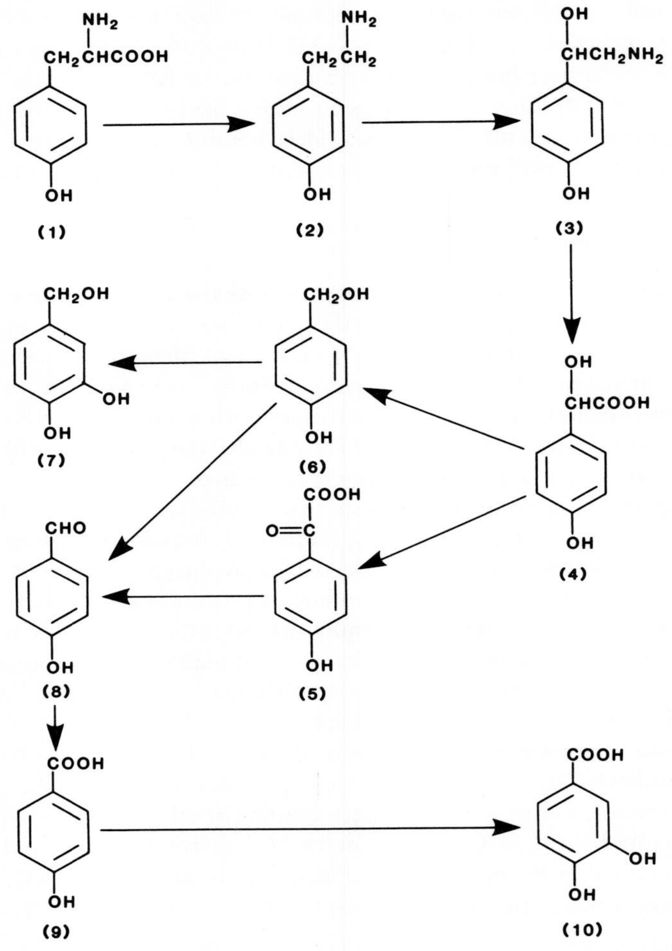

Fig. 2.5 Metabolism of tyrosine to protocatechuic acid and dihydroxybenzyl alcohol. The oöthecal tanning agents, 3,4-dihydroxybenzoic acid and 3,4-dihydroxybenzyl alcohol are derived from tyrosine. Tyrosine (1) is decarboxylated to tyramine (2) which in turn undergoes beta-hydroxylation to octopamine (3). Monoamine oxidase converts octopamine to an aldehyde which is subsequently oxidized to *p*-hydroxymandelic acid (4). The mandelic acids (both *p*-hydroxy and 3,4-dihydroxy) can be decarboxylated to their corresponding benzylalcohol derivatives (6,7). Mandelic acid can be metabolized via a NAD dependent dehydrogenase to the phenylglyoxylic acid (5). Further decarboxylation leads to the benzaldehyde (8) which is oxidized to the benzoic acid (9,10). Each of the compounds can occur as the *p*-hydroxy or the 3,4-dihydroxy moiety and be metabolized to protocatechuic acid (3,4-dihydroxybenzoic acid). The final form of both the benzyl alcohol or benzoic acid diphenol is the 4-*o*-β-D-glucoside.

The primary sclerotization agent, protocatechuic acid or 3,4-dihydroxybenzoic acid (Pryor *et al.*, 1946) is contained in the larger left gland as a 4-*o*-β-D-glucoside (Kent and Brunet, 1959). The left gland also synthesizes the protein and phenoloxidase (Brunet, 1952). The right gland liberates a β-glucosidase which, upon mixing with the secretions of the left gland, hydrolyses the 4-*o*-β-D-glucoside into glucose and protocatechuic acid (Brunet and Kent, 1955).

Four regions of cytologically different cells in the left colleterial gland have been described (Brunet, 1951, 1952). The more numerous 'type 4' cells which lie in the innermost ends of the tubules, secrete the structural proteins, while the taller (15–50 μm) 'type 3' cells appear to elaborate the phenoloxidase system. Type 2 cells, which are located in the posterior branches of the tubules, may also secrete protein and/or the phenoloxidase. The ultrastructure is described in the report of Mercer and Brunet (1959).

The synthesis of the protocatechuic acid utilizes tyrosine or glucose as the precursor (Brunet, 1963). Lake and Mills (1975) found that cockroach haemolymph was capable of converting tyrosine, through the β-hydroxylation of tyramine, to the protocatechuic acid. The method of collecting haemolymph in these experiments, however, involved the squeezing of the animals and probably resulted in some fat body tissue being used. Parts of the pathway could not be shown conclusively in this study, but later work revealed that both mandelic acid dehydrogenase and benzoyl formate decarboxylase occur in the cockroach (Taylor and Mills, 1976). Additional studies have involved the injection of ^{14}C-carbinol noradrenaline into female cockroaches. After an *in vivo* incubation of several hours, ^{14}C-protocatechuic acid β glucoside can be extracted from the left colleterial gland and other radioactive metabolites are found in the haemolymph (Lake *et al.*, 1975). The pathway as shown in Fig. 2.5 is probably the main pathway for tyrosine conversion to the 3-hydroxy, 4-*o*-β-D-glucosidobenzoic acid.

A second oöthecal hardening agent appears to be dihydroxybenzyl alcohol. Pau and Acheson (1968) identified this compound in the cockroach *Blaberus discoidalis* occurring naturally as the 4-*o*-β-D-glucoside. Stay and Roth (1962) had previously found an unidentified glucoside in *P. americana* which corresponds to the glucoside of dihydroxybenzyl alcohol. Positive identification of the 3,4-dihydroxybenzyl alcohol has now been attained (Lake and Mills, 1975; Lake *et al.*, 1975).

The oöthecal proteins have been isolated by Pau *et al.* (1971). Three water-soluble proteins with molecular weights of 13 000, 13 500 and 39 000 have been separated by SDS-gel electrophoresis. Three water-insoluble proteins of 15 500, 21 000 and 39 000 daltons were characterized in a similar manner. The phenol oxidase may be found in both of the fractions and could become insoluble after activation. Characterization of the phenol oxidase system has been accomplished (Whitehead *et al.*, 1960, 1965a, 1965b).

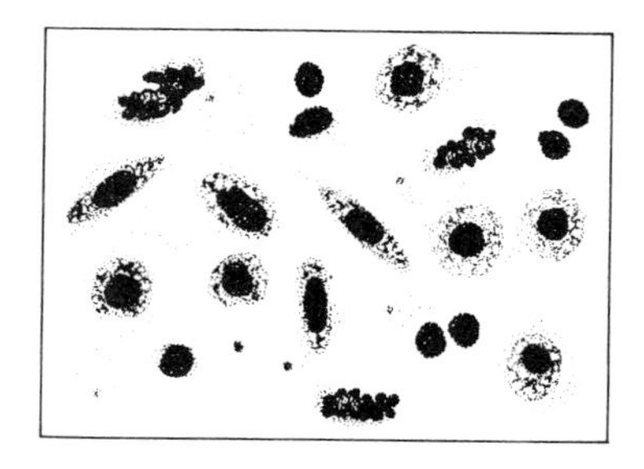

3

Circulatory system

P. Michael Fox

3.1 Introduction

The primary function of the circulatory system in cockroaches, as in other insects, is to transport nutrients from the site of absorption to the cells of the body, metabolic waste products to organs of excretion, and hormones from secretory to target organs. Respiratory needs are met by the tracheal system as discussed in Chapter 5, therefore gaseous transport is not an important function of the circulatory system.

The present review has been restricted to investigations which deal with *Periplaneta americana* directly. This article is not intended as an exhaustive survey of all work concerning the circulatory system, but rather as an introduction to areas that have been reviewed fairly extensively (heart, circulation of haemolymph, haemocytes – Crossley, 1975; Miller, 1973, 1974, 1975a; Jones, 1964, 1977; Gupta, 1979), and a more detailed treatment of the major problems of interest in the last decade in relation to the constituents of the haemolymph. For further information on topics relating to the circulatory system refer to Section 10.4 (rhythmic fluctuations in haemolymph constituents and circulation), Section 16.3.4a (role of haemocytes in wound healing), Chapter 6 (osmoregulation and excretion) and Section 2.3 (involvement of haemocytes and haemolymph constituents in ecdysial events).

3.2 Heart and circulation

Haemolymph circulation of *P. americana*, as in other insects, depends on an 'open system' in which the haemolymph moves freely within the body cavity or haemocoel. The dorsal vessel, a straight tube extending from the head through the abdomen, is usually referred to as the heart in the abdomen and the aorta in the thoracic and cephalic parts of the body (Jones, 1964).

Haemolymph enters the 12-chambered heart through 12 pairs of lateral

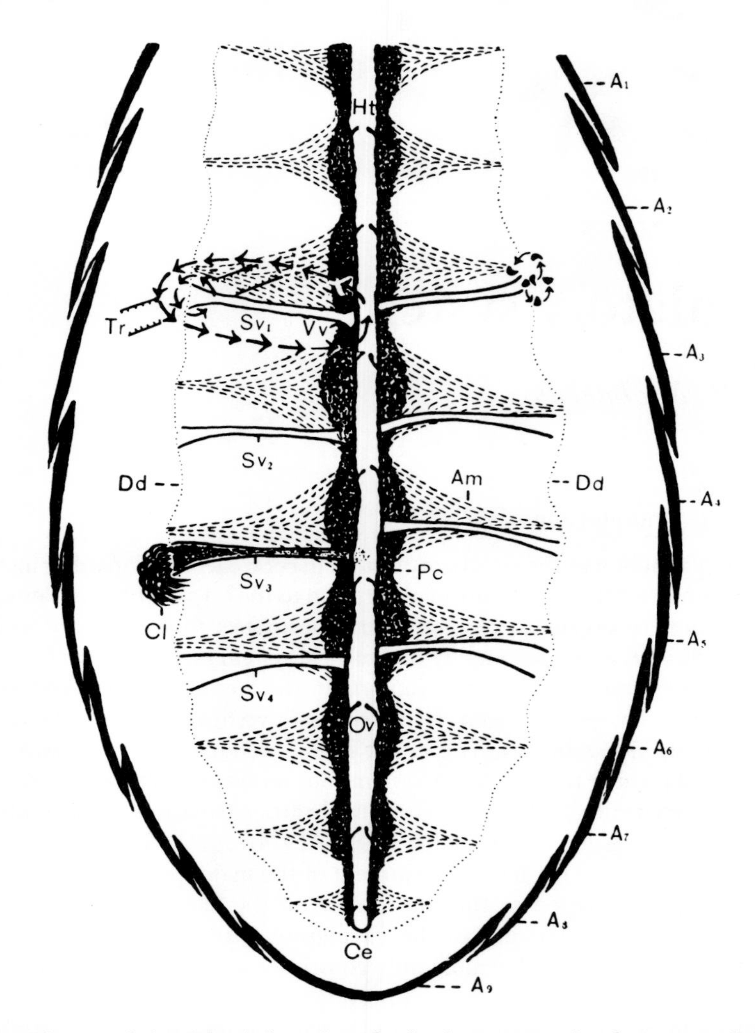

Fig. 3.1 Diagram of the abdominal portion of a live-heart preparation from a nymphal female cockroach, × 8. The preparation (whole mount) is shown ventral (dorsal diaphragm) side up. The outlines of the heart (Ht) and four pairs of segmental vessels (Sv_1 to Sv_4) were drawn with the aid of a camera lucida while the heart was beating. The heart is represented as relaxed. Particles of india ink suspended in saline solution, dropped upon the distal ends of vessels, did not go into vessels, although much of it reached the pericardial cells (Pc) surrounding the heart. Ink dropped upon the caudal end (Ce) of the heart is shown to have passed into the heart and into a segmental vessel (Sv_3) from the distal end of which it is being ejected as a black cloud (Cl). Arrows at left represent movements of ink particles and fat globules floating in the saline solution, arrows at right represent movements of two ink particles also in the saline solution which covered the entire heart preparation. (From McIndoo, 1939.)

openings, the ostia, and passes forward towards the head through the aorta and laterally through the segmental vessels (McIndoo, 1939) (Fig. 3.1). The segmental vessels, one pair each in the mesothorax and metathorax and three pairs in the abdomen, are orientated at right angles to the dorsal vessel between the dorsal diaphragm and cuticle. In *Blaberus craniifer*, these vessels extend to the coxal muscles (Nutting, 1951), whereas in *P. americana* the vessels are minute (3 mm long) (McIndoo, 1939).

The dorsal vessel is attached along the midline to the epidermis of the dorsal body wall by double strands of connective tissue. A pericardial septum is composed of twelve pairs of wing-shaped alary muscles and a membranous dorsal diaphragm (see Fig. 3.1) (Nutting, 1951).

P. americana has pulsatile sacs at the bases of the antennae that open into blood vessels passing up the shaft of each antenna. Other pulsatile organs are located at the base of each wing to relay the haemolymph from the tegmina and hindwings to the heart (Yeager and Hendrickson, 1934). Haemolymph passes outwards by way of costal, subcostal and proximal parts of radial, medial and cubital veins and returns by distal parts of radial, cubital and vannal veins.

Pulsations of the heart are generated primarily by contractions of the muscular wall, possibly assisted by the alary muscles. The entire abdominal section of the dorsal vessel contracts as a single functional unit, and not as a peristatic wave from posterior to anterior (Miller, 1967). Characteristics of heart contraction, passing haemolymph anteriorly, was demonstrated early in *P. americana* (Yeager and Hager, 1934) to include three phases (Fig. 3.2): contraction (systole), relaxation or dilation (diastole) and a rest period (diastasis). Discharge of haemolymph into the head leads to flow back into the body cavity.

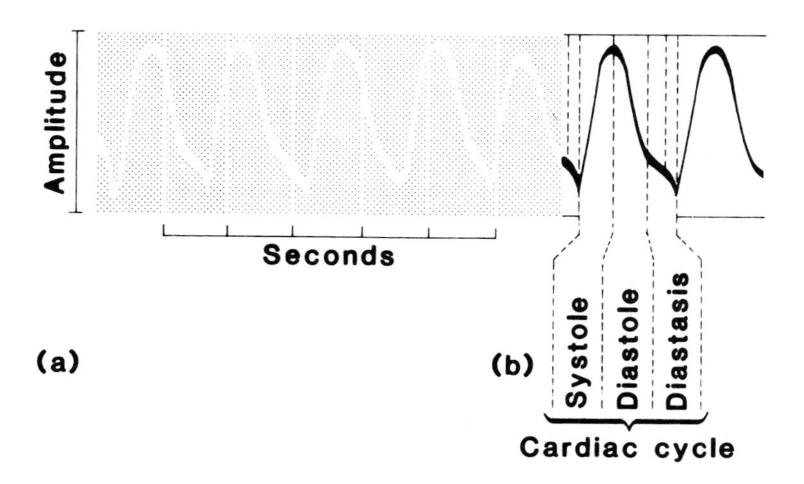

Fig. 3.2 A mechanocardiogram from the third abdominal segment of *P. americana*, (a) showing frequency of beats and amplitude, (b) the component parts of the cardiac cycle showing the 'presystolic notch' at the end of diastasis. (From Cornwell, 1968.)

The cockroach heart was first considered to be a 'neurogenic' organ (Krijgsman, 1952; Senff, 1966). Miller (1967) showed, however, that the heart of *P. americana* continues to beat rhythmically after removal of the cardiac nerves. Smith (1969) later demonstrated that the heart continues to beat after treatment with tetrodotoxin, a drug known to paralyse the nerves associated with the heart. A part of the pacemaker of the cockroach heart is apparently localized within each chamber, as Yeager (1939) found that a single isolated chamber can pulsate rhythmically and with normal amplitude.

Cardiac nerves do not specifically control the heart beat with regard to cycles of systole and diastole (Miller, 1968a, 1969, 1973; Smith, 1969), although the rate is influenced by the cardiac nerves. Lateral nerves are probably responsible for co-ordinating the contractions of heart chambers such that the chambers contract in a synchronous manner (Miller, 1979a; Miller and Usherwood, 1971).

The cockroach heart, especially that of *P. americana*, has been subject for bioassays of pharmacological agents and hormones (Jones, 1974). Most often a semi-isolated heart preparation is employed after removing the head, thorax and internal organs. The exposed heart attached to the dorsal cuticle is bathed in saline that may contain test compounds. Contractions of the semi-isolated heart preparation are completely inhibited in Na-free saline (Miller and Metcalf, 1968b), but will continue for up to one hour in isotonic sodium chloride. The heart is very sensitive to the K^+/Ca^{2+} ratio (Ludwig *et al.*, 1957; Miller and Metcalf, 1968b; Miller, 1969).

Acetylcholine accelerates the heart rate of *P. americana*, probably by acting on the lateral nerves (Richter and Stürzebecher, 1969), since it has no effect on the denervated heart (Miller and Metcalf, 1968a). Dopamine (Miller, 1968b) and 5-hydroxytryptamine (Twarog and Roeder, 1957; Davey, 1964; Brown, 1956) increases the rate of beating of the semi-isolated heart. γ-Amino butyric acid (GABA) has no effect.

Secretions of the corpus cardiacum (CC) are known to stimulate the heart of *P. americana* (Brown, 1965; see also Chapter 12). One pair of CC in saline can increase the rate of heart beat by 50% (Cameron, 1953a,b). The effect begins within 5 min (Davey, 1961a,b), lasts up to 9 h (Kater, 1968a,b), and requires continuous action of the secretions to maintain the effect (Kater, 1968b). Several neurohormones, water-soluble peptides, which have been isolated from both the CC and corpora allata (CA), have an effect on the heart. Neurohormone C (from the CA) and D (from the CC) probably act antagonistically (Gersch and Unger, 1957; Unger, 1957) (see also Chapter 12).

3.3 Haemolymph volume

Haemolymph volume varies in nymphs and adults relative to time before or after ecdysis (Mills and Whitehead, 1970), period of dehydration (Edney, 1968) and stage of females in the reproductive cycle (Bell, 1969b; Verrett and Mills, 1973, 1976). The haemolymph volume in *P. americana* has been measured using a variety of techniques. The results of these measurements, regardless of method, are in general agreement (see Table 3.1).

Table 3.1 Haemolymph volume in *P. americana*.

Life stage and state of hydration	*Volume as % of body weight*	*Volume in μl*	*Measurement technique*	*Source*
Nymphs				
hydrated	17.5	62.35	Inulin dilution	Edney (1968)
dehydrated	13.5	56.70	Inulin dilution	Edney (1968)
Adult males				
hydrated	18.0	134 ± 6.3	Dissection Absorption on paper	Wall (1970)
dehydrated (8 days)	9.6	53 ± 5.3	Absorption on paper	Wall (1970)
Adult females	—	ca. 34	Cardiac puncture	Verrett and Mills (1976)
Adult males				
'normal'	—	200	^{14}C-carboxyinulin dilution	Heit *et al.* (1973)
Adults, sex unspecified				
hydrated	18.9	157.9	Puncture of dorsal aorta	Hyatt and Marshall (1977)
dehydrated (8 days)	9.3	55.8	Paper absorption	

The data of Wall (1970) and of Hyatt and Marshall (1977) seem to be in close agreement with respect to their values of 18.0 ± 1.1 S.E. and 18.9 ± 0.9 S.E. respectively for the percentage contribution of haemolymph to total body weight. Nymphs have a similar percentage value according to Edney (1968), but due to their smaller size contain less haemolymph.

The haemolymph volume seems to undergo a rather striking reduction under conditions of water deprivation with decreases of about 50% commonly reported for an 8-day dehydration period (Wall, 1970; Hyatt and Marshall, 1977). The haemolymph, then, represents a major reservoir of water which can be drawn upon by other tissues in periods of dehydration stress. This sharp reduction in haemolymph volume during dehydration does not lead to a large increase in haemolymph osmolality. A number of investigators of osmoregulation in *P. americana* have shown that the osmotic concentration of the haemolymph is controlled so that the rise in osmolality is much lower than would be expected without solute removal from the haemolymph (Edney, 1968; Wall, 1970; Hyatt and Marshall, 1977) (see Table 3.2). Wall (1970) has calculated that the degree of reduction in volume encountered with dehydration would be expected to lead to an

Table 3.2 Haemolymph osmotic pressure in *P. americana*.

Life stage and state of hydration	*Osmotic pressure (mosmol l^{-1})*	*Source*
Nymphs		
hydrated	410 ± 0.32 S.E.	Edney (1968)
dehydrated	467 ± 5.44 S.E.	
Adult males		
hydrated	ca. 400 (255–420 range)	Wall (1970)
dehydrated (8 days)	Little increase (375–425 range)	
Adult males		
hydrated	380 ± 4.1 S.E.x2	Wall and Oschman (1970)
Adult males		
hydrated	413 ± 7.8 S.E.	Hyatt and Marshall (1977)
dehydrated (8 days)	459 ± 7.9 S.E.	
Adult males		
(drinking 0 mosmol medium)	395 ± 21 S.E.	Heit *et al.* (1973)
(drinking 475 mosmol medium)	464 ± 32 S.E.	
(drinking 700 mosmol medium)	497 ± 23 S.E.	

osmolality of 960 mosmol in the haemolymph. Her maximum osmolality of 447 mosmol shows that the haemolymph is strongly regulated with respect to osmolality.

The regulation of osmolalty was further demonstrated by experiments in which drinking media of increasing osmolalities were given to cockroaches (Heit *et al.*, 1973). Good regulation of haemolymph osmolality was encountered all across the range from 0–700 mosmol (drinking water medium) although mortality in the 700 mosmol group was 50% over the 10-day experimental period. These investigators did find that Na^+ in haemolymph increased markedly when NaCl was added to increase osmolality of drinking water. K^+ concentration was unchanged.

Cockroaches given drinking medium higher in osmolality tend to drink more of these solutions – compare drinking 'isotonic' medium (400 mosmol), imbibition rate, 28.1 μl day^{-1}, with drinking 700 mosmol medium, imbibition rate, 95.5 μl day^{-1} (Heit *et al.*, 1973). Haemolymph volume seemed to show an increase which followed the increase in imbibition rate. The mechanisms involved in water balance are discussed in detail in Chapter 6 on osmoregulation.

3.4 Haemocytes

Much of the recent work on the haemocytes of cockroaches has concerned the nearly century-old problem of the origin and classification of haemocyte types. Arnold (1972) has provided an historical perspective on this problem in his comparative study of 16 species of cockroaches representing each family in the sub-order.

The essence of the classification problem seems to be whether the haemocyte types are actually all one type of cell which passes through a number of stages during its existence (according to the hypothesis of Cuenot (1896, 1899)) or whether each identifiable type originates from a stem or germinal cell and, thereafter, remains distinct. The literature on this subject retains influence from Cuenot's hypothesis as recently as the work of Schlumberger (1952), Day (1952) and Gupta and Sutherland (1966, 1967).

Gupta and Sutherland (1966) chose the plasmatocyte (amoebocyte) as the type of cell which becomes modified into other types (granular cells, spherule cells, adipohaemocytes, cystocytes, vermiform cells, podocytes, oenocytoid-like cells or lamellocyte-like cells) as they are needed. The later work by these authors on *P. americana* simplified the classification to include only plasmatocytes, granular cells, cystocytes and spherule cells (Gupta and Sutherland, 1967).

Other investigators have taken the second viewpoint as described above and have identified germinal cells, the prohaemocytes, and various categories of cells which are thought to be derived from them – the plasmatocytes and cystocytes (Jones, 1957), or the plasmatocytes, cystocytes and granular cells (Wheeler, 1963).

Arnold (1972) upon re-investigating the haemocytes of *P. americana* found in the haemolymph: prohaemocytes (present but rare), plasmatocytes, granular haemocytes (without prominent cytoplasmic granules). No spherule cells were encountered in this study. The granular cells were by far the predominant cell type at all stages examined. Price and Ratcliffe (1974) disagreed with the conclusions of Arnold (1972). Based upon their work with insects from 15 different orders, they erected a classification which included prohaemocytes, plasmatocytes, granular cells, spherule cells, cystocytes and oenocytoids. They found in *P. americana* all types except for the oenocytoids and possibly the spherule cells.

Another perspective on this issue has come from the examination of cockroach haemocytes from several species including *P. americana* with the electron microscope (Moran, 1971; Scharrer, 1972). Scharrer (1972) has remarked on the functional versatility of insect haemocytes and the 'many transitional features' found in the ultrastructural appearances of the cockroach haemocytes. She seems to view the classification of these cells into immutable categories as being artificial and based upon the false impression of structural uniformity that the light microscope has allowed. She believes that the ultrastructural variation among the haemocytes encountered in her work indicates that 'the capacity of haemocytes to undergo ready transformations in preparation for or in response to changing physiological requirements' has gone unrecognized (Scharrer, 1972). This opinion was also voiced by Price and Ratcliffe (1974).

No definitive ultrastructural study on *P. americana* haemocytes has been found in the literature. The major issue of the derivation and classification of these cells must await further research for its resolution. The application and refinement of tissue culture methods to such problems may prove helpful. Ratcliffe and Rowley (1975) have studied the phagocytosis of latex, chick erythrocytes and bacteria in a short-term culture arrangement. These authors were able to show that only the plasmatocytes phagocytized the bacteria.

The study of the physiological functions of cockroach haemocytes has not been an active area of research in recent years. Only a few reports have been found which relate the haemocytes to the basic physiology of the organism and these have been discussed in the appropriate sections above and in relevant chapters in this volume (e.g. Integument, Chapter 2). This is not to say, however, that little is known of the functions of blood cells in insects in general. A volume *Insect Hemocytes* edited by Gupta (1979) contains reviews on development, differentiation and the functions of haemocytes.

3.5 Constituents of haemolymph

3.5.1 Ionic composition

The investigation of the ionic composition of insect haemolymph has had a long history. The information accumulated indicates that there exists a variation in ionic content from one order to another. It is well-known, for example, that the Na^+/K^+ ratios of insect haemolymphs cover the entire range from high Na^+ to high K^+ types (Pichon, 1970).

Unexpectedly, we find that the haemolymph ionic content is highly variable even within a single species of insect. Analysis of haemolymph Na^+/K^+ in *P. americana* (using diets of tap water–milk powder) gave values for this ratio ranging from 1 to 14 (Pichon, 1963; Pichon and Boistel, 1963). The Na^+/K^+ ratio can be experimentally manipulated by changing diets, allowing the cockroaches to drink KCl solutions, and by various water deprivation regimens (Tobias, 1948; Pichon, 1963). This extreme variability renders the interpretation of measurements of the ionic content open to question. The conditions of diet and dietary ionic content as well as the state of hydration of the experimental animals therefore have to be carefully specified. Only under such strictly controlled conditions can meaningful statements be made in relation to ionic composition of the haemolymph in *P. americana*.

A complicating factor relative to this discussion arises from the results of sampling the ionic content of the haemolymph from different regions of the body of these cockroaches (Pichon, 1970). The localized variation one encounters within the body of a single insect is unexplained, but might be the result of the ionic requirements of the predominant cell types in certain regions of the body, localized sluggishness of the circulation within the sinuses of the haemocoel, or differential permeability of cuticular areas to water that could lead to concentrating of haemolymph as it moves through a particular body region such as the antenna or leg.

The Na^+, K^+ and Ca^{2+} ions of the haemolymph have been shown to vary independently of each other even though trends seem to be similar from one body part to another (see Pichon, 1970; Fig. 3) with the exception of the Na^+/K^+ ratio of the antennal haemolymph. The antennae, being so highly innervated, might produce a high Na^+/K^+ ratio as the result of intense active transport by the nerves in a small compartment of the haemolymph.

There seems to be good agreement among various authors concerning the haemolymph ionic content of *P. americana* (see Table 3.3).

Table 3.3 Ionic content of *P. americana* haemolymph

Stage	*Na^+*	*K^+*	*Ca^{2+}*	*Mg^{2+}*	*Cl^-*	*PO_4^{3-}*	*Source*
	(mM kg^{-1} haemolymph)						
Nymphal	114.8	61.8	4.3	—	—	—	Pichon (1970)
Adult males	132	8.7	3.7	4.7	107	1.04	Weidler and Sieck (1977)
Adult males	185	16.0	—	—	—	—	Hyatt and Marshall (1977)
Adult males	179	—	—	—	—	—	Treherne *et al.* (1975)
Adult males	161	—	—	5.6	—	—	Van Asperen and Van Esch (1956)
Adult females	—	—	—	—	112	—	Wendt and Weidler (1973)

Much of the effort in analysing the ionic content of cockroach haemolymph has been expended towards answering questions related to the influence of haemolymph ions on the functional state of the nerves (Chapter 8) and muscles (Chapter 11) and in studies of osmoregulation (Chapter 6).

3.5.2 Haemolymph sugars: trahalose and glucose

The recent research on haemolymph sugars of *P. americana* has centred around two major issues: (1) stress and exercise influences on haemolymph levels of these sugars, and (2) the enzymatic and/or hormonal control of haemolymph sugar concentrations.

The basal level of haemolymph glucose was reported by Matthews *et al.* (1976) as 0.025–0.029 $\mu g\ ml^{-1}$. Matthews and Downer (1974) have found trehalose levels for resting, isolated *P. americana* to be 56.5 ± 2.25 S.E. mM kg^{-1}.

A number of stresses applied to cockroaches have been found to influence the haemolymph sugar levels. Matthews and Downer (1973), for example, found that the simple act of handling the animals to remove haemolymph in an experimental setting affected both the fat body glycogenolysis and the release of trehalose into the haemolymph with the maximum hypertrehalosemic response appearing 10–15 min after stress application.

A similar response of glucose levels in the haemolymph in cockroaches stressed for 5 min by shaking in a glass jar was reported by Wilson and Rounds (1972). These authors speculated that the oligosaccharides of the haemolymph might act as a carbohydrate storage reserve as they seem not to penetrate easily the nerves or the muscle tissues. Conversion of trehalose to glucose could lead to faster penetration and utilization in these tissues.

The stress of anaesthesia seems to be capable of inducing haemolymph hyperglycemic response when the anaesthetics used were nitrogen, chilling or diethyl ether. Anaesthesia with carbon dioxide, in contrast, did not produce hyperglycemic response (Matthews and Downer, 1973).

The fat body glycogen storage seems to be a likely source of the carbohydrate precursors for haemolymph trehalose (Matthews and Downer, 1974). Steele (1963) has also proposed that the glycogen of fat body was the source of the haemolymph trehalose which appeared upon injection of cockroaches with corpus cardiacum (CC) extracts. Matthews and Downer (1974) were able to show that the incubation of fat body in Ringer's solution led to a decline of glycogen in amounts equivalent to the amounts of trehalose which appeared in the medium (36 $\mu g\ min^{-1}$ trehalose production of fat body; 343–624 $\mu g\ h^{-1}$ released from fat body during one hour incubation in Ringer's solution) (see also Chapter 7).

The origin of this hypertrehalosemic response seems to be somewhat obscure at this time. Steele (1963) demonstrated the influence of aqueous extracts of the CC in causing a profound hypertrehalosemia in male *P. americana* and concomitantly, a significant decrease in fat body glycogen content occurred. Highnam (1961) described the emptying of the CC in response to stress. Taken together, these two reports might be considered to implicate the neurosecretion(s) released by the CC in the hypertrehalosemic response.

There are some differences in the time course of the haemolymph hypertrehalosemic response brought on by the injection of CC extracts (maximum 150% increase after 5 h according to Steele, 1961) and the hyperglucosemic response due to stress (which requires only minutes according to Wilson and Rounds, 1972). This probably implies different initiators for the two responses.

The conversion of trehalose to glucose and the control of this event is of importance in the discussion of the levels of these sugars in the haemolymph. Matthews *et al.* (1976) have found two electrophoretically distinguishable α-glucosidases in the serum component of *P. americana* haemolymph. One of these enzymes seems to be a trehalase. The trehalase activity was found in haemocytes as well. The electrophoretic mobility and the pH sensitivity of both enzymes as well as the EDTA inhibition of the two were similar. These authors have calculated the haemocyte contribution to this activity to be around two-thirds of the total (21 nM glucose produced h^{-1} per 10^6 cells as opposed to 32 nM glucose $h^{-1}\ \mu l^{-1}$ haemolymph).

With respect to the control of trehalase activity and, through this, the control of haemolymph glucose levels, Matthews *et al.* (1976) suggested that the Ca^{2+} activation of trehalase (by the removal of citrate inhibition?) might be of importance physiologically. These authors also pointed to the strong pH effect on the haemocyte trehalase in the range pH 6–7.5 reflecting that pH changes of this magnitude occur in the haemolymph of cockroaches as they go from resting to exercising modes of activity. The shift to more acidic haemolymph during exercise could be expected to strongly influence trehalase activity and trehalose hydrolysis as has been discussed previously.

Van Handel (1978) was not able to confirm the dilution activation of haemolymph trehalase which Matthews *et al.* (1976) encountered earlier. The absence of similar dilution effect would counter the hypothesis that haemolymph trehalase is normally under the influence of an inhibitor. Comparison of the maximal activity of the trehalases under study shows that they were equivalently active (ca. 5 mg glucose produced ml^{-1} haemolymph h^{-1}) although different enzyme incubation temperatures were used in the two laboratories.

The co-existence of trehalose and trehalase in the cockroach haemolymph and the large difference seen upon comparison of *in vitro* trehalose hydrolysis rate (4–8 mg ml^{-1} h^{-1}) with the *in vivo* turnover (1.3 mg ml^{-1} h^{-1}) (Van Handel, 1978) presents a dilemma. Utilization of trehalose seems to occur at a faster rate than the entry of trehalose into the haemolymph. As this is an obvious impossibility without a constant decrease in the trehalose level, it implies that some experimental procedure was altering the physiological situation. Either an unphysiological activation of the trehalase occurs with *in vitro* assays or the trehalose turnover data must be in error.

Even though the weight of evidence seems to be to the contrary, it must be taken into consideration that Katagiri (1977) reported that trehalase in *P. americana* was measureable only in the haemocytes with no activity occurring in the serum. The resolution of the dilemma may be related to the degree to which trehalase is confined to the haemocytes. Perhaps under physiological conditions more of the trehalase activity than has been previously thought occurs in the haemocytes, and *in vivo* control of trehalose hydrolysis could then be due to changes in haemocyte permeability to trehalose. The study of haemocytes and their trehalase content as well as their membrane permeability control would seem to be important in dealing with this problem. One might expect great methodological problems, however, due to the notorious difficulty in handling cockroach haemocytes without altering their functional states.

Glucose levels in the haemolymph seem to be controlled by conversion of glucose either to haemolymph trehalose or to fat body glycogen, or both. Spring *et al.* (1977) have used injections of ^{14}C-glucose into the haemolymph to study the fate of this sugar. These investigators have found evidence for an initial burst of trehalose synthesis in haemolymph lasting for about 20 min; this was followed by an increased rate of glycogen synthesis in fat body between 20 and 30 min after injection. The increase in glycogen synthesis coincided with the levelling off of trehalose synthesis. This has been taken as evidence for a 'metabolic switch' changing the direction of glucose movement from trehalose towards glycogen.

The possibility of hormonal substances acting as glycogenic agents or inhibitors of glycogenesis was suggested from experiments involving ligation at the neck. In such animals, Spring *et al.* (1977) have shown that trehalose synthesis did not level off after 10 min and that glycogen synthesis was almost absent for the initial 20 min after glucose injection. The agent(s) which mediate these effects do not seem to have been characterized at this time and clarification of this control mechanism awaits further research.

3.5.3 Haemolymph lipids

Nelson *et al.* (1967) have noted the lack of information on the fatty acid composition of the neutral lipid fractions in tissues of *P. americana*. These authors studied haemolymph and fat body lipids in an effort to supply such information and were able to show that, in both haemolymph and fat body, the predominant fatty acids were C_{16}, $C_{18:1}$ and $C_{18:2}$ in all three types of glycerides. This pattern was fairly stable between two and three months and occurs similarly in both males and females. Nelson *et al.* (1967) made some speculative statements about the reasons for the differences they measured between animals of varying ages and sexes although the lack of statistical analysis of their data make convincing statements difficult.

The triglycerides of haemolymph and fat body seemed to be by far the dominant lipid class at the 8th instar while, in the young adults, there was a shift towards the dominance of the diglycerides in both sexes (Nelson *et al.*, 1967). Using young adults (2–3 weeks old), Cook and Eddington (1967) also found that normal haemolymph contained far more diglyceride than triglyceride lipid (0.22 ± 0.63 S.E. mg ml^{-1} triglyceride; 1.9 ± 0.167 S.E. mg ml^{-1} diglyceride; 0.23 ± 0.017 S.E. mg ml^{-1} monoglyceride).

The exchange of lipid materials between the haemolymph and the fat body has been studied with conflicting results. Chino and Gilbert (1965) have contended that diglyceride was the major lipid class released upon incubation of cockroach fat body with haemolymph. Cook and Eddington (1967), in contrast, have evidence that fat body glycerides are predominantly triglycerides, and that these were released into the haemolymph in proportion to the 'concentration' of haemolymph bathing the fat body. In their experiments, when haemolymph was diluted more than 20%, triglyceride release dropped markedly below the diglyceride release. There also seemed to be a difference in the time course of release in that initial releases of diglyceride (10–20 min incubation) from fat body into 50% haemolymph were greater than for triglyceride. Triglyceride release became dominant only after 20 min of incubation.

This problem was also considered by Downer and Steele (1972), who used 50% haemolymph in their incubation medium and followed lipid release from fat body into haemolymph for 0–60 min. These investigators found, consistently with Gilbert *et al.* (1965), that diglycerides were the major lipids released into the medium.

It is possible, then, from these comments to speculate that the source of the controversy was experimental variation as a result of the use of different incubation times and haemolymph concentrations. The question of possible synthesis of triglycerides after release into the haemolymph has been brought up, but, as yet, has not been answered. Another important factor encountered by Cook and Eddington (1967) was that loss of diglyceride from fat body into buffer was greater than that of triglyceride, so a diffusion control was necessary to correct for this and to measure the true relative glyceride loss stimulated by the presence of haemolymph.

Regardless of whether, in fact, triglycerides or diglycerides are the major lipid class released into the haemolymph of adult *P. americana*, triglycerides seem to be the dominant class stored in the fat body. This possible necessity for conversion of tri- to diglyceride forms has stimulated the search for enzymes in the haemolymph which have the capability of cleaving and/or synthesizing glycerides. Downer and Steele (1973) have found what they considered to be 'true' lipase activity in the haemolymph of *P. americana*. The lipase was characterized by dual activity maxima at pH 4.9 and 7 with the low pH peak activity being the highest (0.5 μM free fatty acids produced 0.1 ml^{-1} haemolymph h^{-1} at 30°C). This haemolymph lipase showed some enhancement of activity in the presence of 5% bovine serum albumen possibly due to its action as a free fatty acid acceptor. It was inhibited by fluoride ion (65% at 1×10^{-3} M). Downer and Steele (1973) believe that the difficulty which some workers (i.e. Gilbert *et al.*, 1965) have encountered in demonstrating 'true' lipase activity is the result of the sensitivity of this type of enzyme to the nature of the lipid–water interface in the assay system. In studies such as their own and that of Wlodawer and Baranska (1965), where such activity was measured, Ediol was used as the lipid substrate.

The source of the haemolymph lipase is unknown at present. Downer and Steele (1973) speculated that the fat body or the gut cells might be the source. These authors pointed to the similarity in pH response of the haemolymph enzyme and the gut lipase (as reported by Gilbert *et al.*, 1965).

Certain haemolymph proteins have been reported to act as carriers for the lipids released from the fat body. Chino and Gilbert (1964, 1965), for example, claim that diglycerides (and possibly unesterified fatty acids) are bound to haemolymph proteins. The subject of lipid binding and transport by the haemolymph proteins is considered below in Section 3.5.4.e. The information available at present on the regulation of lipid levels in the haemolymph seems very sparse. The hormonal influences on lipid metabolism and movements within the cockroach will require much more study before a clear understanding of this subject can emerge.

Vroman *et al.* (1965) have shown that allatectomy resulted in a general large increase in the total lipid and especially in the triglyceride fraction, which more than doubled in response to the operation. More recently, the CC of *P. americana* has been found to contain a substance which produced a decrease in the haemolymph levels of both tri- and diglycerides (Downer and Steele, 1972). These investigators also demonstrated concomitant increases of these glycerides in the fat body. Such effects were produced by dilute extracts with the equivalent of 0.02 pairs of glands leading to a 54% decrease in haemolymph triglycerides. The corpus allatum (CA) also showed this potential although this gland gave only about one half the effect of similar amounts of CC materials. As the CA in *P. americana* are intimately associated with the CC, one cannot rule out cross-contamination of these extracts.

Hoffman and Downer (1974) showed that the administration of CC extracts retarded $^{14}CO_2$ formation from ^{14}C-acetate injected into the haemocoel. This result is consistent with the shift towards lipid storage in the fat body and with lipid movement out of the haemolymph as described by Downer and Steele (1972).

3.5.4 Haemolymph proteins

(a) *General information*

The thorough investigation of the protein components of the haemolymph in *P. americana* began with the work of Siakotos (1960a,b) who analysed the plasma proteins and tested them for phospholipid, neutral lipid, carbohydrate and sterol content. Unfortunately, the technique availible at the time (paper electrophoresis) has poor resolving power and reveals, at the most, only general classes or groups of proteins with similar relative mobilities. As a result, he was only able to detect five protein fractions in the plasma. These fractions possessed differing amounts of protein and different relative amounts of the conjugates mentioned above. Siakotos (1960b) extended these investigations to show some changes in the protein contents of these fractions during the phases of moulting and upon clotting of the haemolymph. The fact that the electrophoretic technique used by Siakotos (1960a,b) resolved only five groups of proteins and no individual ones renders the information on conjugated material ambiguous. It would seem of value to repeat this work with the more powerful electrophoretic techniques currently in use – polyacrylamide gel electrophoresis (PAGE) and isoelectric focusing in polyacrylamide gels.

A problem which, in my opinion, hampers work with the haemolymph proteins is the lack of a standard reference method for identifying haemolymph proteins or even a thorough study of the plasma proteins to which other studies might refer. Researchers use electrophoretic separations designed for the particular needs of the study at hand with varying conditions of buffer, gel concentrations, currents and time courses. The result is that it is usually difficult or impossible to relate and compare work done in different laboratories on the haemolymph of this animal.

The most recent study which analysed the haemolymph proteins of *P. americana* along with proteinaceous extracts from many other parts of the body was that of Singer and Norris (1973). Using PAGE, these authors were able to separate 23 protein bands from the haemolymph of male cockroaches. Many proteins seemed to be present in the haemolymph and several other tissues as well. Only one band in the haemolymph pattern (R_f 33) had no counterpart in any other tissue.

This does not imply, of course, that equivalency of R_f or R_m means identity of the proteins migrating to a certain position in the gels. Other studies such as immunological ones are required to establish true identity of these proteins. Singer and Norris (1973), in fact, pointed out a number of variables which must be taken into consideration in evaluating electrophoretic results – protein concentration and its influence on aggregation, effects of desalting procedures and alteration of protein molecules during extraction.

Only a few haemolymph proteins of *P. americana* have been studied in a manner which would allow their identities to be established from tissue to tissue or from animal to animal. A good example of such identification was reported by Kunkel and Lawler (1974) in their work with a larval specific serum protein in the Dictyoptera. These experiments involved the preparation of rabbit antiserum to a larval specific protein from *Blattella germanica* and the demonstration of cross-reactivity throughout the order. The list of cross-reactive species included *P. americana*. Unfortunately, no analytical electrophoresis was used, so one does not know which protein from the normal haemolymph complement was being studied.

Other immunological identifications have been made with respect to the vitellogenic proteins of *P. americana* (Bell, 1970; Clore *et al.*, 1978) and the haemolymph proteins which are incorporated into the cuticle (Fox *et al.*, 1972). Both types of protein are discussed in sections to follow and in Chapters 2 and 13.

The electrophoretic protein patterns of *P. americana* haemolymph seem to be relatively constant in adults from a few days after the imaginal moult on into advanced age. Even though there was much individual variation especially with respect to the minor protein constituents, no new bands were found to appear with age in adult males (Fox, unpublished).

One can perform experimental manipulations on *P. americana* to bring about alterations in haemolymph proteins. Adiyodi and Nayar (1968), for example, showed that prolonged starvation of female cockroaches resulted in an hypoproteinemia and the disappearance of some protein bands along with a diminution of stainable material in others. Their fractions 3 and 4 were most affected, the bands they termed the 'common insect protein' and the 'sex-specific protein', respectively. Conjugated materials associated with the haemolymph proteins also changed with starvation. The amounts of Sudan Black B (SBB)-stainable material decreased, but never entirely disappeared, and there was a marked decrease in PAS-stainable substances indicating the reduction in protein-bound carbohydrates. The loss of specific proteins from the haemolymph has been confirmed as the result of food deprivation of male cockroaches and using a different electrophoretic system. The major serum lipoprotein was less affected than the other major haemolymph proteins by this treatment (Fox, unpublished).

Adiyodi and Nayar (1968) demonstrated stress-induced changes in electrophoretic protein patterns and in the conjugated lipid and carbohydrate material. Hypothermia (48 h at 15 °C) resulted in the increase of PAS-stainable material in their fraction 4 while stress by 24 h immobilization seemed to cause a depletion of this fraction.

Experimental surgery such as the ablation of endocrine centres which influence protein metabolism (i.e. CA and CC) or protein utilization is known to affect haemolymph protein levels in *P. americana* (Mills *et al.*, 1966; Bell, 1969a). Adiyodi and Nayar (1967) have performed cardiaca-allatectomies on *P. americana* and observed that amount of total haemolymph protein found in their electrophoretic fractions 3 and 4 changed from 66.21% in 4-day post-operative females to as much as 80% in 7–18-day post-operatives. There was a strong decrease in the amount of protein material which migrated fastest towards the anode. Lipid material of operated females tended to become restricted to fraction 3 after 30 days, and carbohydrates became more widely associated during this same period with the major amount being associated with their fraction 4.

Prabhu and Nayar (1972b) treated male *P. americana* with the JH analogue, farnesyl methyl ether, and detected a striking increase in the serum proteins within one day after treatment. They found that their fractions 4, 6 and 7 accounted for the largest increase. These authors concluded that the hormone analogue had caused stimulation of synthesis of proteins in the male which are normally present in much lower concentrations than in the female. The changes were reversible, however, and within 3 days the electrophoretic protein pattern was reverting to the more nearly normal pattern for males.

Treatment of *P. americana* females with sterilizing chemicals produced changes in the haemolymph proteins which were qualitatively different from those resulting from surgical sterilization (Prabhu and Nayar, 1972a). Metepa-sterilized females showed increased concentrations of many haemolymph proteins including the less mobile ones which correspond to the soluble oöcyte proteins. The surgical sterilization led to increases only in the less mobile proteins. As these authors pointed out, chemosterilization with an agent like metepa results in a 'deep-seated derangement of protein metabolism' and in addition, drastic damage to various body tissues. For these reasons, alterations in haemolymph protein patterns should not be unexpected after such treatment.

(b) *Vitellogenic proteins*

Considerable effort has been expended in recent years in studying the origin of the yolk proteins of insect eggs and the relationship between the ovarian proteins and those of the haemolymph. *P. americana* has been well-studied in this respect and, as this subject is covered in detail in other sections of this book, the reader is referred to sections 7.6.4 and 13.2.2.

(c) *Haemolymph proteins in cuticle formation*

Fox and Mills (1969) extended previous work done with a variety of insects to demonstrate in *P. americana* the presence of a large number of proteins with a similar electrophoretic mobility in the cuticle and the haemolymph. These authors found that haemolymph proteins declined, and soluble cuticle proteins were highest during the moult. This indicates that the haemolymph proteins could serve as a source of protein to be incorporated into the cuticular material. This suggestion has since gained some support through the demonstration of antigenic similarity between cuticle proteins and at least two cockroach haemolymph proteins as measured by immunodiffusion in Ouchterlony plates (Fox *et al.*, 1972).

Further evidence for incorporation of haemolymph proteins came through the work of Koeppe and Mills (1972) who showed that 2-^{14}C-dopamine injected into the haemocoel of the cockroach led to the binding of diphenols to haemolymph proteins. The peak of bound label in haemolymph proteins occurred at the stage of apolysis, and thereafter declined as the peak appearance of label in the cuticle was reached at ecdysis. Label in both types of proteins declined during the post-ecdysial period.

Geiger *et al.* (1977) reported labelling of haemolymph serum and haemocyte proteins upon injection of newly ecdysed immature cockroaches with UL-^{14}C-leucine. Serum proteins, thus labelled, were purified using gel filtration and injected into other animals that had recently moulted. Within one hour the serum haemocyte and cuticular proteins were showing label. Soluble haemocyte proteins used experimentally in this fashion gave similar results.

On the basis of these results, the incorporation of haemolymph serum proteins and perhaps haemocyte proteins as well into the cuticular matrix seems to be well supported. The question which still remains is the degree of selectivity of this process. Are the haemolymph proteins in general withdrawn for use in the cuticle or are only certain specific proteins used in this way? The electrophoretic similarity of many serum proteins to proteins in the cuticle would lead one to suspect that there are more proteins involved than the two confirmed by Fox *et al.* (1972) on the basis of antigenic similarity. This would seem to be an important area for further research.

(d) *Insecticide binding to haemolymph proteins*

The possible importance of the haemolymph proteins to problems of insecticide effects and in relation to the movement of insecticides within the insect body has recently been recognized. Gerolt (1969, 1972) de-emphasized the role of the haemolymph in insecticide movement and asserted that contact insecticides move through the cuticle and then the trachael system to the level of the individual cells. This view of insecticide transport has, however, been challenged by other researchers.

Olsen (1973), for example, who tested Gerolt's (1969, 1972) hypothesis, found that the haemolymph of *P. americana* will absorb the insecticide dieldrin to a maximum of 57 μM ml^{-1} at 22°C. This finding indicates that dieldrin is 142–400 times more soluble in haemolymph than in water (depending upon values used for water solubility). Olsen (1973) observed that it was the haemolymph proteins which were responsible for this high dieldrin solubility. The proteins which bind dieldrin consisted of a low molecular weight fraction (ca. 18 900 daltons) and two fractions of high molecular weight (ca. 160 000 daltons). The haemocytes contained 37% of the dieldrin found in the haemolymph but the state of the insecticide (whether bound or unbound) was not determined. This report indicates that translocation of dieldrin by haemolymph is a possibility, but does not rule out other modes of insecticide transport, which might not involve the haemolymph.

The binding of the insecticide DDT to *P. americana* proteins was demonstrated by Winter *et al.* (1975) using PAGE and isolectric focusing on polyacrylamide gels for haemolymph protein separation. When haemolymph was electrophoresed on 3.75% gels, DDT was bound to a single protein peak which also stained for lipid with SBB. Further experiments showed that haemolymph could be fractionated on discontinuous gels of 4%, 5% and 7% segments so that two lipoprotein groups (of differing R_m) became differentiated – one group of three SBB-staining lipoproteins retained near the origin of the 4% segment and a single band lying between the 4 and 5% segments. Both areas contained ^{14}C-DDT. Isoelectric focusing in 5% gels revealed that the bulk of the haemolymph protein migrated as four protein peaks with pI's between 5.9 and 6.8. This area also contained most of the ^{14}C-DDT.

Skalsky and Guthrie (1975) investigated the binding of labelled DDT, dieldrin and parathion by the haemolymph. The DDT and dieldrin, consistent with the previously mentioned studies were bound, while parathion was not. Some separation of the radioactive protein containing bound DDT and dieldrin (labelled) as eluted from Sephadex G-200 column was accomplished by application of this peak to a DEAE-Sephadex column. Using step-wise elution with 0.1–1.0 M NaCl, it was (sometimes) possible to separate a predominantly DDT-binding peak eluting with 0.4 M NaCl from a predominantly dieldrin-binding peak which eluted with 0.45 M NaCl. Analytical electrophoresis (PAGE) revealed that the 0.40 M peak possessed two Oil Red O-staining lipoproteins and that the 0.45 M peak contained only a single lipoprotein. The latter protein was evidently quite large and gel filtration studies yielded a preliminary molecular weight value of 520 000.

No work seems to exist which demonstrates a specificity of binding of any insecticide to haemolymph proteins in *P. americana* except for the 'preferred' binding of lipid-soluble insecticides to plasma lipoproteins. Boyer (1975) has, in contrast, shown that the binding of tetrachlorvinphos (Gardona, SD 8447) was not specific. At least seven haemolymph proteins contributed to the binding of 65% of this insecticide recovered from PAGE separations of haemolymph proteins. This study is of interest from a toxicological viewpoint because 80% of the LD_{50} dose of this insecticide can bind to the haemolymph of an individual cockroach, perhaps leading to errors in the determination of true lethality. Even though extensive non-specific binding was obvious in this study, most of the insecticide in PAGE gels seemed to be in an area also containing the R_m 0.09 protein. It is possible that this protein should be studied more closely to look for a high affinity binding of tetra-chlorvinphos.

Taken as a series, these studies demonstrate that binding of insecticidal substances to haemolymph proteins of *P. americana* is not unusual. There seem to be two major types of binding phenomena, the hydrophobic attraction of lipid-soluble substances for the serum lipoproteins and a non-specific binding of the more hydrophilic compounds to many of the haemolymph proteins. Even though non-specific, some of the binding proteins evidently allow for high capacity storage of chemicals of both types in the haemolymph. The importance of such binding to the toxic process seems to be undetermined at present. As Olsen (1973) has shown, the haemolymph proteins undoubtedly increase the total amount of dieldrin which can pass across the cuticle but whether the dieldrin bound to lipoprotein is available to participate in toxicity is not known.

(e) *Lipid binding and transport by haemolymph proteins*

The question of lipid binding by haemolymph proteins in *P. americana* is of importance in three major areas of research: (1) lipid metabolism and lipid movement within the insect, (2) insecticide effects and transport, and (3) the metabolism and transport of lipoidal hormones. In my opinion, these three lines of research intersect, at least in the case of *P. americana*, in the discussion of binding and transport by the haemolymph proteins.

In the literature concerning lipids and lipid movement in insects, one encounters repeated references to the work of Chino and Gilbert (1964, 1965) used to support the assertion that 'in insects' diglycerides (and perhaps other lipids as well) are bound to haemolymph lipoproteins. Although these reports and the later ones by Chino *et al.* (1965, 1969) have provided strong evidence for binding and transport of diglycerides in *Hyalophora cecropia* and *Philosamia cynthia*, in this series of articles, these authors seem only to have suggested that this type of binding might be a general phenomenon in other insects including *P. americana*. They have referred back to Siakotos's (1960a,b) work as an indication that neutral lipids are associated with haemolymph proteins of *P. americana* (see Chino and Gilbert, 1964).

With the exception of the work of Reisser-Bollade (1976) who found label associated with lipoproteins having no or low electrophoretic mobility after feeding labelled tripalmitin to *P. americana*, I have not been able to locate any published report in which the binding of specific lipid materials was actually tested using proteins from cockroach haemolymph. It would seem, then, that statements made up to now concerning lipid binding in *P. americana* haemolymph have been cross-specific extrapolations or conjecture and were not based upon actual experimentation with this insect.

The work of several investigators (discussed in detail in Section 3.5.4.d) has shown that plasma lipoproteins of *P. americana* bind lipid-soluble insecticidal compounds (i.e. DDT and dieldrin) and can transport them in the haemolymph (Winter *et al.*, 1975; Olsen, 1973; Skalsky and Guthrie, 1975). Skalsky and Guthrie (1975) have used PAGE to show the presence of two or three Oil Red O-staining lipoproteins in fractions eluted from DEAE-Sephadex fractionation of cockroach haemolymph. These lipoproteins bound DDT and dieldrin.

It has been speculated, although not proven, that the lipoidal hormones of insects require 'carrier' proteins to transport them in the haemolymph. This is especially true of the JH which is commonly thought to have a very low water solubility. Kramer *et al.* (1974), for example, have reported solubility of 5×10^{-5} M in tris-HCl buffer for C_{18}-synthetic cecropia hormone.

Results from several laboratories have shown JH binding to haemolymph proteins from a variety of species – saturniid moths such as *H. cecropia*, *H. gloveri* and *Antheraea polyphemus* (Whitmore and Gilbert, 1972), *Tenebrio molitor* (Trautmann, 1972), *Locusta migratoria* (Emmerich and Hartmann, 1973) and *Manduca sexta* (Kramer *et al.*, 1974). In all of these studies, some degree of specificity of binding was indicated by the fact that all lipoproteins present in the haemolymph did not bind JH or did not bind the hormone as avidly.

The haemolymph proteins of *P. americana* have been studied recently in my laboratory in an attempt to detect lipid and JH binding proteins. This unpublished material is presented here to demonstrate a relationship between various lines of research on lipid binding in this insect.

PAGE analysis of the haemolymph proteins of both male and female adults and late nymphs has revealed the presence of a single lipoprotein band, the most slowly migrating major haemolymph protein, which can be visualized with Oil Red O-staining (but not with Lipid Crimson or SBB). This band has always appeared broad and somewhat dispersed but, in my hands, has never subdivided into separate bands (compare to the results of Skalsky and Guthrie (1975)). This lipoprotein migrates only very slowly in polyacrylamide gels of 7% concentration, but moves into 6% gels much more easily. Because this slowly migrating protein band seems to be the sole Oil Red O-staining material in these gels, and in view of the similarity in technique between the two methods used, I consider this band to contain the same lipoprotein(s) that Skalsky and Guthrie (1975) detected and found capable of DDT and dieldrin binding. Incubation of haemolymph with ^{3}H-JH *in vitro* followed by PAGE separation of proteins demonstrated that JH was bound only to this lipoprotein band in signficant amounts (Fig. 3.3). This statement is true of adult males and females. The major haemolymph protein of late nymphs seemed to carry lesser amounts of JH under these conditions.

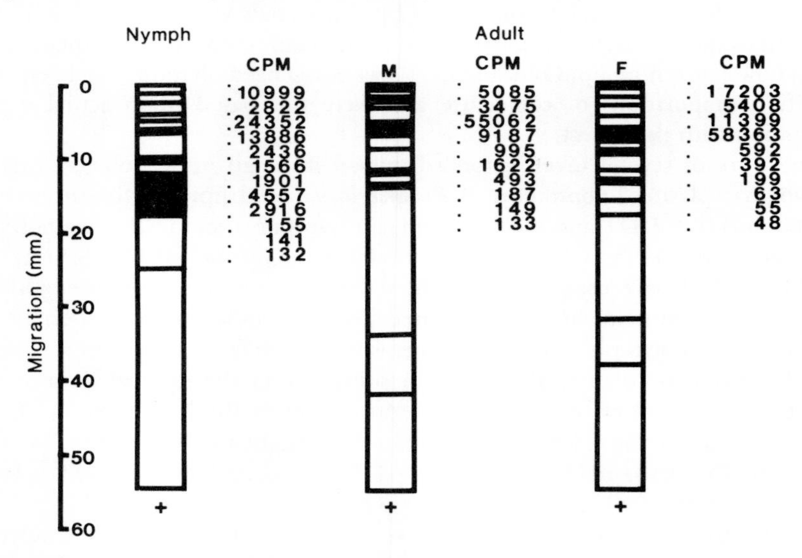

Fig. 3.3 Juvenile hormone binding to haemolymph proteins of late nymphs, adult males and females of *P. americana*. ^{3}H-C-18-cecropia juvenile hormone (0.5 μC) added to 30 μl haemolymph and incubated at RT for 60 min. PAGE separation was by modification of Davis (1964) method using 7% gels, 120 min run, 3mA/gel, 340 V, 5 μl samples. Methyl green stain for proteins was used on some gels. Other gels were sectioned at 2 mm intervals, digested in H_2O_2, and counted in Aquasol. (M = male, F = female.)

Partly as a test for specificity of binding, haemolymph has been incubated with ^{14}C-palmitic acid and subjected to the same kind of electrophoretic analysis. The palmitic acid detected in the gels was bound predominately to the same lipoprotein band which binds JH (Fig. 3.4).

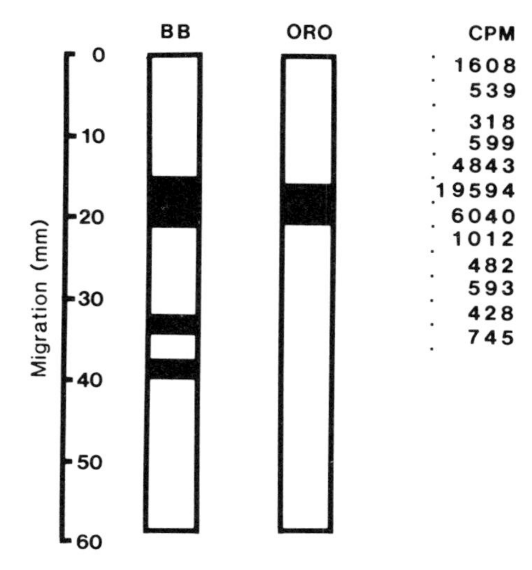

Fig. 3.4 Palmitic acid binding to haemolymph proteins of adult male *P. americana*. 1-^{14}C-palmitic acid (0.5 μC) added to 20 μl adult male haemolymph and incubated at RT for 15 min. PAGE separation was by modification of Davis (1964) method using 6% gels, 120 min run, 3 mA/gel, 260 V, 10 μl samples. Buffalo Black stain for proteins was used on some gels. Other gels were sectioned at 3 mm intervals, digested in H_2O_2, and counted in Aquasol. (BB = Buffalo Black, ORO = Oil Red O.)

On this basis, it seems that the Oil Red O-staining lipoprotein is responsible for the binding of lipid-soluble insecticides, JH and, at least, one fatty acid.

Two hypotheses could be formulated at this point from the evidence at hand. First, if the lipoprotein band represents a single lipoprotein one might consider the 'binding' of all three classes of substances to be non-specific and probably a generalized hydrophobic interaction of some type. Or second, if the lipoprotein band actually contains a number of polyeptides of similar size and charge, one might be justified in searching for separate binding proteins for each substance and a more specific type of binding. The nature of this lipoprotein material must be clarified before the binding of these substances can be understood.

Attempts at purification have encountered difficulty with trace contamination from other major haemolymph proteins when preparative electrophoresis was used. The purest preparations obtained from small-scale PAGE separations and elution from gel slices still showed JH-binding ability. Large-scale preparation has been attempted using ultracentrifugation in sucrose gradients. In this way, it has been possible to obtain fractions containing large amounts of the lipoprotein material which were only slightly contaminated with other proteins (Fig. 3.5 and Table 3.4).

Molecular weight studies with SDS-PAGE, according to Weber *et al.* (1972), using the purest preparations of the lipoprotein material indicate that the band either contains several proteins or else has an extremely complex subunit structure. When 10% gels were used one protein band was found with a molecular weight of ca. 80 000. SDS-PAGE using 6% gels on the same material revealed 10 protein bands with the heaviest stain in those at 24 000, 29 000, 33 000 and 41 000.

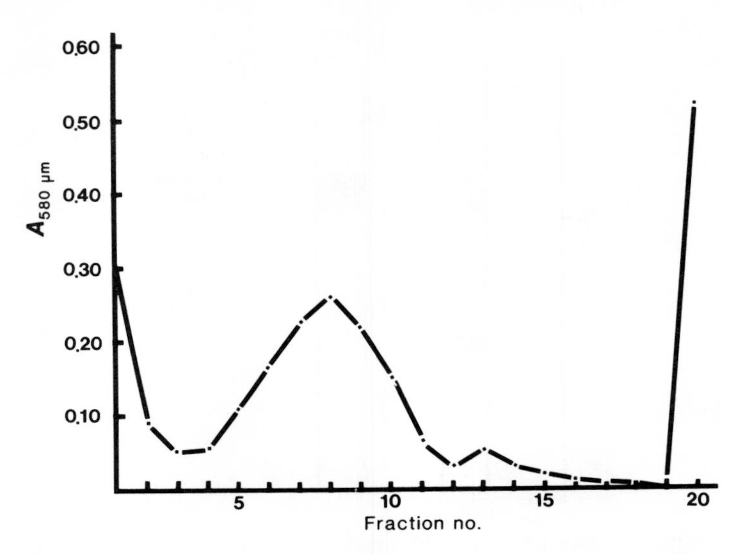

Fig. 3.5 Protein determinations on sucrose density gradient fractions – adult male *P. americana* haemolymph. Gradients used were 5–20% sucrose in 0.2 M NaCl. Haemolymph sample was 80μl of 75% male haemolymph centrifuged at 45 960 revs/min in SW 50.1 rotor for 22 h. Twenty fractions of ten drops were collected. Absorbance at 580 μm was used for protein estimation.

Weaker bands ranging up to 110 000 and 135 000 daltons were also present. It is of interest that Skalsky and Guthrie (1975) found a molecular weight of 520 000 for one of their lipoproteins using gel filtration. A protein of the size reported by Skalsky and Guthrie (1975) would certainly not enter the matrix of a 10% gel and possibly not even that of a 6% gel. If the lipoprotein in its native state has such a high mol.wt., the polypeptides found in my gels must represent subunits of sizes small enough to enter gels of 10% to 6% concentrations.

Table 3.4 PAGE analysis of sucrose gradient fractions

			Fraction number				
1 ...	6	7	8	9	10 ...	13 ...	20
—	*70 (vl)	70 (vl)	—	—	—	—	—
—	†178 (h)	199 (h)	188 (vh)	172 (vh)	156 (h)	—	—
266 (h)	277 (vl)	254 (vl)	269 (vl)	265 (vl)	270 (vvl)	258 (vl)	258 (h)
327 (h)	330 (vl)	305 (vl)	336 (vl)	318 (vl)	319 (vvl)	317 (vl)	317 (h)
457 (vl)	—	—	—	—	—	—	437 (vl)
—	741 (vl)	729 (vl)	730 (vl)	—	—	—	763 (vl)

* Relative mobilities (dye front = 1000). Band density codes (vh = very heavy, h = heavy, vl = very light, vvl = trace).

† Lipoprotein band.

Gradient fractions described in Fig. 3.3 electrophoresed using modification of Davis (1964) method for PAGE using 6% gels, 60 min run, 3 mA/gel, 230 V, 20 μl samples. Buffalo Black used for protein staining.

Further work presently underway in my laboratory is expected to yield lipoprotein preparations of greater purity and in quantities which will allow it to be studied more thoroughly. When this goal has been reached the questions of specificity of binding and of the possible identity of the proteins involved in binding various lipoidal substances can be answered.

The reports that haemolymph can exert a protective influence on JH in the presence of degradative enzymes (Sams *et al.*, 1978) and the earlier reports that haemolymph is necessary for the transport of glycerides out of the fat body, considered in light of the ability of the haemolymph lipoprotein(s) to bind such substances, make it of interest to test the purified lipoproteins for ability to produce these effects.

4

Nutrition and digestion

D. E. Bignell

4.1 Introduction

The nutrition and digestion of cockroaches have been previously reviewed by Guthrie and Tindall (1968) and by Cornwell (1968). Excellent general accounts of these topics, drawing on the entire insect literature, have been offered by Dadd (1970a,b) and by House (1974a,b), and are strongly recommended for background reading. In the account that follows some basic information is drawn from other species where none is available for *Periplaneta americana*, and is acknowledged as such. Statements which do not specifically refer to other cockroaches are drawn from work on the American cockroach.

4.2 Morphology of the alimentary canal and salivary glands

4.2.1 General organization

The alimentary canal is a coiled tube, approximately twice the length of the body connecting the mouth (cibarium) and the anal sphincter (Fig. 4.1). The foregut is the largest single structure, occupying almost one half of the total gut volume (Bignell, 1977a). It consists of an anterior pharynx extending to the posterior dorsal head region, a longitudinally folded oesophagus extending to the posterior prothorax and a large crop which, when filled, may occupy a substantial part of the anterior abdomen. The foregut is terminated posteriorly by a conical gizzard which tapers to a narrow tube and projects 2–3 mm into the midgut lumen. This tube is termed, misleadingly, the oesophageal invagination. In addition to a number of

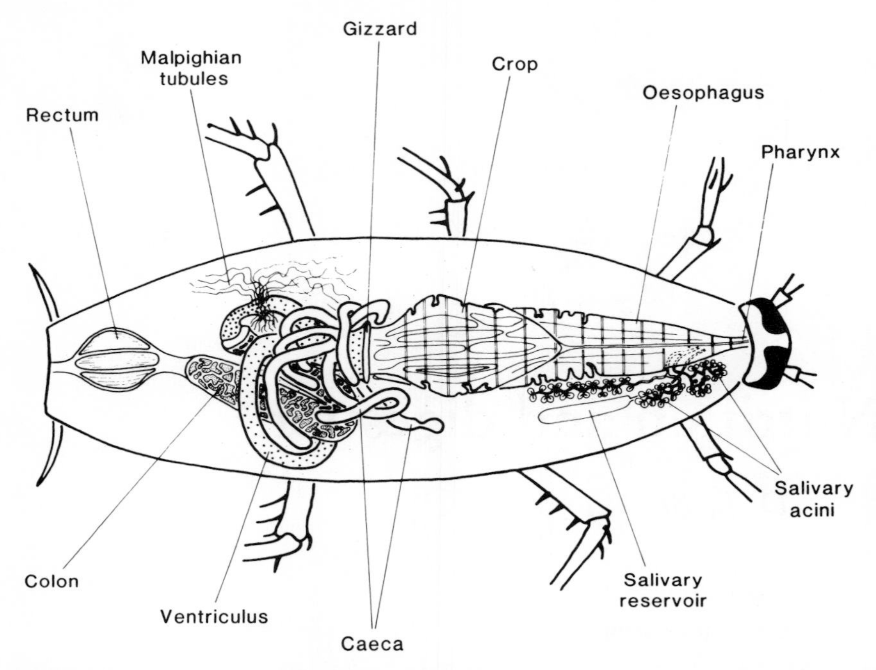

Fig. 4.1 *In situ* morphology of the alimentary canal and salivary glands. For clarity, the full ramifications of the Malpighian tubules are not shown. Also, only a single salivary gland and reservoir are shown, displaced ventro-laterally to show the acini in four groups. Normally, the paired glands cling loosely to the dorso-lateral walls of the oesophagus and crop. The total length of the gut in an adult cockroach averages 6.7 cm. (From Sanford, 1918.)

external dilator muscles, the pharynx shows a sphincter-like thickening of intrinsic muscle at the junction with the oesophagus which can close the lumen (Davey and Treherne, 1963a).

The salivary glands are paired and lie dorsolaterally to the oesophagus and the crop. In each gland, the secretory acini are grouped into four lobes arranged on either side of a salivary reservoir. Ducts from the acini and reservoirs pass forwards and join together ventrally to the pharynx to form a common salivary duct opening at the base of the hypopharynx (Sutherland and Chillseyzn, 1968).

The midgut consists of eight anterior caeca and the ventriculus (posterior midgut) which connects with the hindgut. Each caecum is about one-third the diameter of the ventriculus, but the length and volume vary with feeding (Guthrie and Tindall, 1968; O'Riordan, 1968). The ventriculus is coiled underneath (i.e. ventral to) the colon, while the majority of the caeca are grouped dorsally, covering much of the ventriculus and colon, and bound together by fat body and numerous Malpighian tubules.

The hindgut consists of a muscular ileum, a large sac-like colon separated into anterior and posterior regions by an *in situ* fold, a short connection (the posterior colon) and finally the rectum.

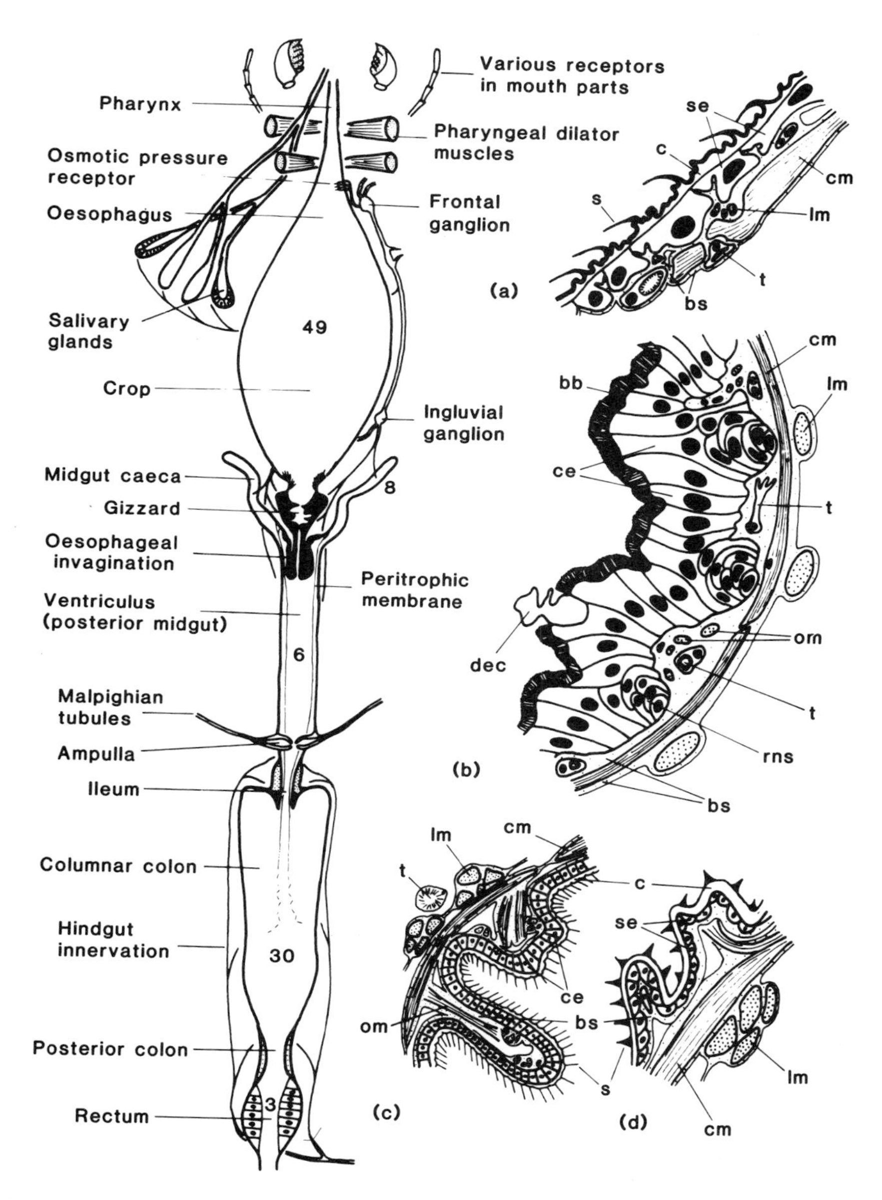

Fig. 4.2 Generalized organization of the alimentary canal, with numbers giving the mean volumes of the crop, midgut caeca, ventriculus, colon and rectum respectively as percentages of the volume of the whole tract. (From Bignell, 1977a.) The structure and mode of action of the gizzard (sometimes termed proventriculus) are discussed by Sanford (1918), Davey and Treherne (1963a) and Lee (1968). The entire structure is conical, about 3.5 mm long excluding the oespophageal invagination which projects as a narrow tube 2–3 mm into the midgut. The anterior half of the gizzard is equipped with 6 large sclerotized teeth which fit closely together to occlude the lumen when the valve is closed by circular muscle contraction. The posterior gizzard bears setal-like spines grouped into 6 pairs of pulvilli.

These are said to permit the passage of only fine particles into the midgut (Cornwell, 1968). A second valve, also equipped with 6 sclerotized lobes, separates the ileum and colon. This may be termed the pyloric or enteric valve. The extreme posterior part of the colon is strongly muscularized and also appears to function as a valve, controlling the movements of the colonic contents into the rectum.

The insets show the light-microscope level structure of the gut wall. (a) Foregut; this consists of well-separated squamous epithelial cells 20–25 μm wide and up to 15 μm in height, overlaid by a prominent cuticle 6–8 μm wide. The surface of the cuticle is elaborated into spinous projections (oesophagus) or regular ridges (pharynx and crop). (b) Midgut; the basic pattern of a folded columnar epithelium is common to the caeca and the ventriculus (posterior midgut), but mature ventricular cells, approximately 5 μm wide by 100 μm tall, are twice the height of those in the caeca. Gross folding of the epithelium is less in the ventriculus. Regenerative cells are present in small clusters (nidi) regularly spaced on the basal side of the epithelium; these divide by mitosis and attenuate to replace older columnar cells which degenerate and are expelled into the lumen. Day and Powning (1949) estimated that a complete replacement of the midgut epithelial cells of *B. germanica* occurred every 40–120 h. (c) Columnar colon; this consists of an extensively folded epithelium of columnar cells averaging 15–25 μm in height and 6–12 μm wide, with an overlying cuticle 1–2 μm thick. Contrary to some previous reports, the cuticle bears numerous spines, projecting up to 20 μm into the lumen and serving as attachment sites for a semi-permanent hindgut flora (Bracke *et al.*, 1979). (d) Posterior colon; here the columnar epithelium gives way to smaller, flattened cells with a thicker cuticle up to 3 μm wide and bearing short sclerotized spines. The general structure of the ileum resembles that of the posterior colon.

All the gut epithelia stand on a basement sheath of connective tissue in which muscles, tracheolar cells and nerve axons are found. The maximum thicknesses of the sheath complex are: foregut, 25–30 μm; midgut caeca, 10 μm; ventriculus, 12–15 μm; columnar colon, 15–30 μm; posterior colon, 80 μm. The muscle elements are not always clearly separable into circular and longitudinal groups, but, in general, the circular muscles are outermost (i.e. closest to the haemolymph) in the foregut and midgut caeca, while the longitudinal muscles are outermost in the ventriculus and hindgut. Although the muscular envelope of the gut is not continuous (except in the rectum), the basement sheath isolates all basal extracellular space from the haemolymph.

Key: bb, brush border of microvilli; bs, basement sheath; c, cuticle; ce, columnar epithelium; cm, circular muscle; dec, degenerating epithelial cells; f, fibroblast; lm, longitudinal muscle; om, oblique muscle; rns, regenerative nidus; s, spine; se, squamous epithelium; t, tracheal cell.

4.2.2 Microscopy

Fig. 4.2 shows the detailed organization and light microscopy of the gut, summarizing the work of Sanford (1918), Gresson (1934), Day and Powning (1949), Lee (1968), O'Riordan (1968) and Bignell (1980). Two issues raised by earlier studies now require discussion in the light of more recent information available from the electron microscope. These are that (1) despite the relative thickness of the foregut cuticle and the apparently unspecialized nature of its epithelium, lipid droplets increase in size and number within the cytoplasm when fats or sterols are ingested (Sanford, 1918; Eisner, 1955; Joshi and Agarwal, 1976), and (2) the columnar cells of the midgut epithelium are functionally differentiated into secretory and absorptive types (Gresson, 1934; Shay, 1946). A third possible issue centres on the contention of Gresson (1934) that digestive enzymes are secreted into the lumen within cytoplasmic extrusions or the remnants of whole cells. However, these phenomena have been satisfactorily explained as fixation artefacts or as instances of cellular degeneration (Day and Powning, 1949; Srivastava, 1962) and need not detain us further. A detailed account of the ultrastructure of the midgut is given by O'Riordan (1968) and aspects are discussed by Smith (1968), Berridge (1970a) and Oschman and Wall (1972).

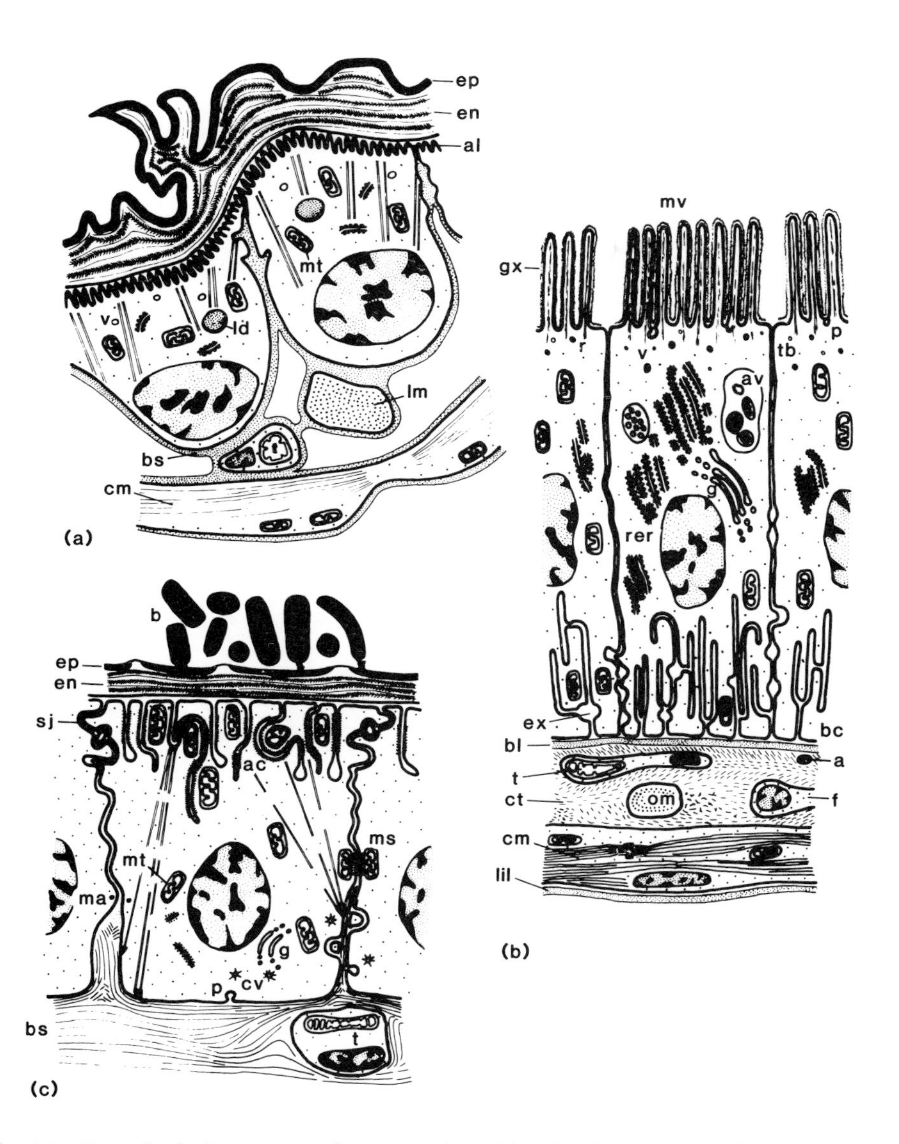

Fig. 4.3 Generalized ultrastructure of representative epithelial cells and associated tissues (not to scale). (a) Crop, (b) midgut and (c) columnar colon. Key: a, axon; ac, apical complex comprising infolding of the plasma membrane with internal particulate coating and associated mitochondria; al, apical lappets; av, autophagic vacuole; b, bacteria; bc, basal complex comprising extensive infolding of the basal plasma membrane and a number of associated mitochondria; bs, basement sheath; bl, basal lamina; cm, circular muscle; ct, connective tissue; cv, coated vesicle; en, endocuticle; ep, epicuticle; ex, extracellular compartment; f, fibroblast; g, Golgi complex; gx, glycocalyx; ld, lipid droplet; lil, limiting lamina; lm, longitudinal muscle; ma, macula adherens; ms, mitochondrial–scalariform junction; mt, mitochondrion; mv, microvilli; om, oblique muscle; p, pinocytotic profile; r, rootlets of microvilli; rer, rough endoplasmic reticulum; sj, septate junction; t, tracheolar cell; tb, terminal bar junctional complex; v, vesicle of secretion.

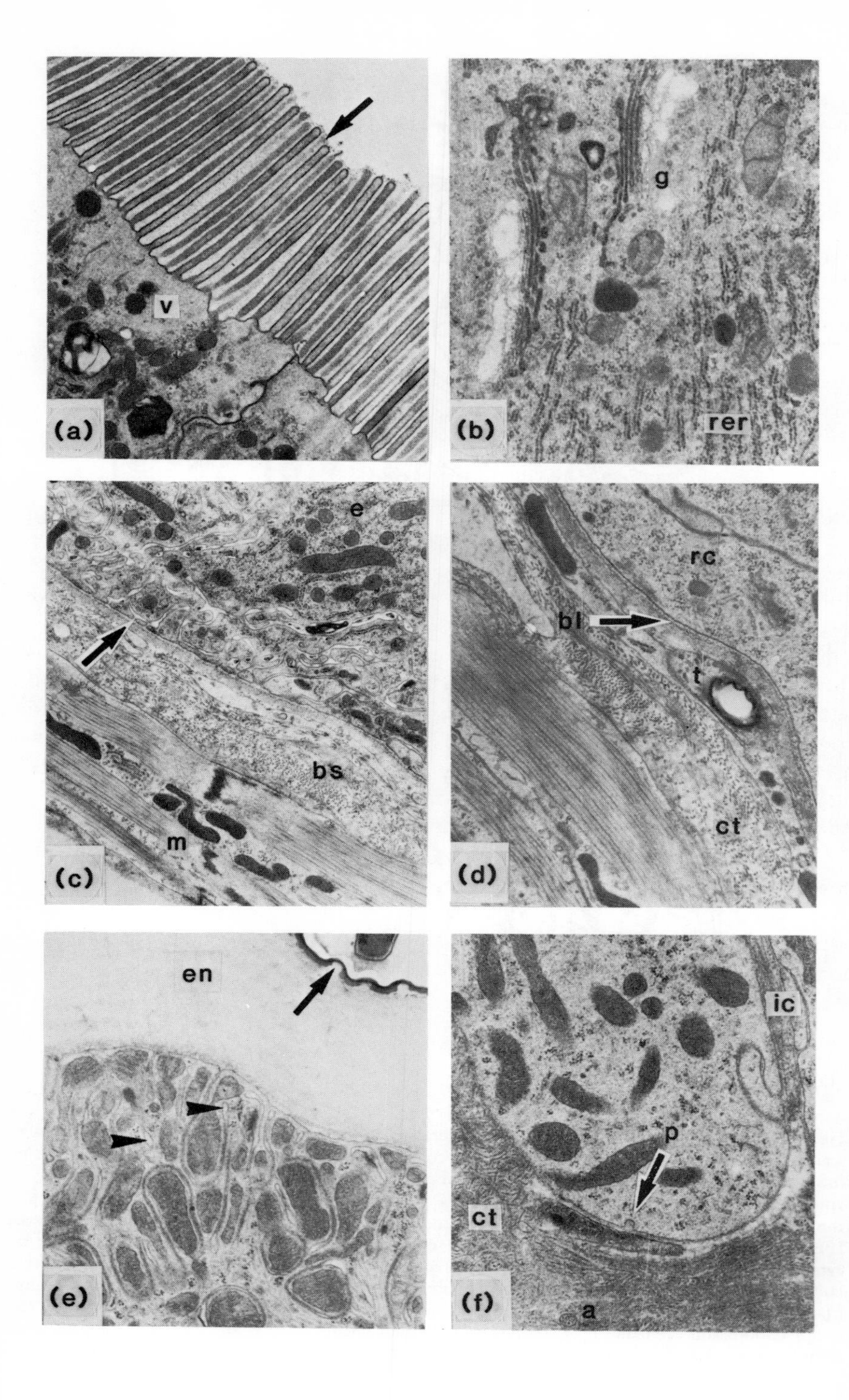
(a)
v
(b)
g
rer
(c)
e
bs
m
(d)
rc
bl
t
ct
(e)
en
(f)
ic
p
ct
a

Fig. 4.4 Ultrastructure of the midgut and colonic epithelia. (a) Apical portion of a caecal epithelial cell showing folding of the plasma membrane into numerous parallel microvilli, each of which is coated externally with an amorphous glycocalyx (arrow). Vesicles (v) containing secretory products are present in the cytoplasm, in addition to mitochondria. × 9500. (b) Cytoplasm of a ventricular cell showing stacks of rough endoplasmic reticulum (rer) and Golgi complexes (g) elaborating vesicles. × 17 100. (c) Base of a caecal cell showing extracellular compartments (arrow) formed within the epithelium by infolding of the plasma membrane; bs, basement sheath; e, epithelial cell; m, muscle cell. × 9025. (d) Nidal regenerative cell (rc) showing scatter of free ribosomes and absence of infolding of the basal plasma membrane; the specialized nature of the basement sheath is evident in this micrograph which shows the granular basal lamina (bl) and underlying fibrillar connective tissue (ct). t, tracheolar cell. × 9025. (e) Apical cytoplasm and overlying cuticle of the epithelium of the anterior columnar colon showing invaginations of the plasma membrane closely associated with mitochondria. In many cases, the infoldings are dilated at the tip to form extracellular compartments (arrowheads). Note also that the epicuticle is thinned periodically to form dome-like protrusions into the lumen (arrow). en, endocuticle. × 12 350. (f) Basal half of the same tissue showing intercellular channel (ic) and the absence of membrane elaboration. a, axon; ct, connective tissue; p, pinocytotic profile. × 12 350.

No published accounts on the ultrastructure of the foregut are known to the author but preliminary observations have shown little evidence of elaboration of the plasma membrane or the presence of extracellular compartments within the epithelium that would be expected in a permeable tissue (Fig. 4.3a). Bound ribosomes, Golgi complexes and associated vesicles were relatively rare, arguing against an active secretory function, but the presence of lipid droplets within the epithelial cytoplasm was confirmed. Microtubules were the most prominent organelles, generally orientated perpendicular to the plane of the epithelium and suggesting that the cells are principally designed for mechanical rather than absorptive purposes. The prominent, continuous epicuticle indicates a low permeability to aqueous fluids, but lipids must be assumed to penetrate from the foregut lumen, possibly assisted by transient pulses of hydrostatic pressure generated within the contents (Davey and Treherne, 1964).

A striking contrast is presented in the midgut epithelium (Fig. 4.3b) which shows a dense array of apical microvilli, of uniform diameter and spacing (Fig. 4.4a), together with elaborate infoldings of the basal plasma membrane which form a series of extracellular channels and compartments within the basal half of the cell (Fig. 4.4c). Berridge (1970a) noted that the number of openings from these basal compartments to the haemolymph was limited and that mitochondria were quite frequently associated with the invaginated membranes. This arrangement may enable the cells to control solute concentrations within the extracellular compartments by means of ion pumps located in the bounding plasma membranes, and thus permit additional solutes or fluids to enter the epithelium along concentration or osmotic gradients that are independent of the overall gradients between the lumen of the gut and the haemolymph. The lateral plasma membranes of adjacent midgut cells are closely apposed forming a mixture of gap junctions and septate junctions of a variety peculiar to the tissue (Oschman and Wall, 1972). An intercellular route for solute uptake seems unlikely (Smith, 1968; O'Riordan, 1968).

An abundance of parallel stacks or whorls of rough endoplasmic reticulum is a feature of all mature midgut epithelial cells. Golgi complexes are present near the nucleus (Fig. 4.4b) and also elaborate vesicles of electron-dense granular material, which accumulate in the apical cytoplasm close to the microvilli (Fig. 4.4a). O'Riordan (1968) described pinocytotic profiles which could occasionally be seen at the bases of intermicrovillar clefts, but no information is available on the direction of this transport or its importance in the digestive functions of the cells.

Regenerative cells within the nidi are undifferentiated, containing large numbers of unattached ribosomes and lacking both basal infoldings and microvilli (Fig. 4.4d). Somewhat surprisingly, no ultrastructural evidence of heterogeneity has been reported amongst mature epithelial cells, which all appear to conform to the description given above. This is in conflict with earlier light microscopical studies in which maturation was said to be accompanied by the appearance of secretory granules in the apical cytoplasm of some, but not all, of the cells. Gresson (1934) and Shay (1946) considered cells lacking stainable granules to be absorptive, while Day and Powning (1949) suggested that individual cells might go through cycles of alternating secretion and absorption. While all authors describing secretory granules in the light microscope reported that they were more abundant in the anterior ventriculus and caeca, O'Riordan (1968) could find no differences between the caecal and ventricular epithelia other than those of cell size, nidal frequency and tissue folding noted in Fig. 4.2. Bignell (1980 and unpublished) analysed the midgut epithelium by stereology and showed that the surface densities of apical plasma membranes and rough endoplasmic reticulum were significantly greater in caecal cells than in those of the ventriculus (Table 4.1). The balance of evidence from the electron microscope is therefore that both secretion and absorption may take place in any mature cell, but that these activities are more intensive in the caeca. Degenerating epithelial cells contain large autophagic vacuoles which are centres of acid phosphatase accumulation (Couch and Mills, 1968).

Folding of the apical plasma membrane is also a feature of the colonic epithelium (Figs. 4.3c and 4.4e), but, in this case, the folds are associated with mitochondria in configurations that resemble the apical complexes of rectal pad cells (Oschman and Wall, 1969). In the anterior colon, infoldings at the apex of the epithelial cells form, not only narrow extracellular channels, but also extracellular compartments up to almost 1 μm in width (Bignell, 1980). Stereological analysis showed that extracellular space within the anterior colon, as a proportion of the total epithelial volume, was not significantly different from that of the midgut or rectal pads (Table 4.1). Thus, the colon may have an absorptive function resembling that of the midgut or rectum, although this has yet to be identified exactly. Significantly, the colonic cuticle is relatively thin and appears to be permeable to aqueous fluids (Bignell, 1977b, 1980). No specialized folding of the basal plasma membrane occurs at the base of the colonic epithelium, contrasting with the situation in *Blattella germanica* (Ballan-Dufrançais, 1972).

Francois (1978) has drawn attention to the complexity of the basement sheath and connective tissues which together invest the intestinal epithelia (e.g. Fig. 4.3b). Fibroblasts are present in this matrix in addition to the muscle tissues,

Table 4.1 Analysis of some features of the midgut, colonic and rectal pad epithelia by quantitative electron microscopy (stereology). Means and standard deviation are shown of stereological parameters measuring the proportion of each epithelium occupied by extracellular compartments, the relative degrees of folding of cell surfaces and the prominence of mitochondria and rough endoplasmic reticulum: $n = 6$ (insects). Full details are given in Bignell, 1980. Means with the same superscript are not significantly different ($p > 0.05$).

Parameter	*Interpretation*	*Caeca*	*Posterior midgut*	*Anterior columnar colon*	*Posterior columnar colon*	*Rectal pads*
Volume ratio extracellular space/cytoplasm	Volume of extracellular compartments per unit volume of cytoplasm	0.041*† ± 0.028	0.049* ± 0.020	0.036* ± 0.012	0.015† ± 0.005	0.041* ± 0.010
Surface density of apical plasma membranes	Area of apical plasma membrane per unit volume of cytoplasm	2.650 ± 0.939	1.251* ± 0.275	1.295* ± 0.102	1.220* ± 0.459	1.852 ± 0.243
Surface density of basal and lateral plasma membranes	Area of basal and lateral plasma membranes per unit volume of cytoplasm	1.920* ± 0.399	1.949* ± 1.089	0.261† ± 0.048	0.262† ± 0.073	0.620 ± 0.058
Surface density of rough endoplasmic reticulum	Area of rough endoplasmic reticulum per unit volume of cytoplasm	3.108 ± 0.858	1.297* ± 0.552	0.497* ± 0.414	0.189* ± 0.096	0.053 ± 0.004
Volume density of mitochondria	Volume occupied by mitochondria per unit volume of cytoplasm	0.167* ± 0.032	0.145* ± 0.026	0.252† ± 0.024	0.281† ± 0.029	0.279† ± 0.024

tracheocytes and nerve axons which have been noted by previous authors. Histochemical tests have shown the presence of collagen and acid mucopolysaccharide in the basement sheath (O'Riordan, 1968; Francois, 1978). Acid mucopolysaccharide is also present in the glycocalyx coating the external face of the microvilli (O'Riordan, 1968; Fig. 4.4a).

The salivary glands were first investigated histologically by Day (1951), and subsequently studied in the electron microscope by Kessel and Beams (1963). These accounts agree that two quite distinct types of cell can be distinguished in the secretory acini: zymogen cells (sometimes called central or secretory cells), which accumulate and discharge large coalescent granules in a cyclical manner, and parietal cells (peripheral cells), which are characterized by infoldings of the basal membranes confronting the haemolymph and outfoldings of the apical membranes to form numerous microvilli bordering an intracellular duct. A third type of cell makes up the epithelium of extracellular ducts draining the acini, showing extensive folding of the apical and basal membranes and many associated mitochondria. From this work and parallel studies of other cockroaches, it has been concluded that the parietal cells and the extracellular duct cells are jointly responsible for producing the salivary fluid, including its ionic and mucous components, while the enzymes originate from the zymogen cells.

4.2.3 The peritrophic membrane

The peritrophic membrane is the most enigmatic structure in the insect gut, there being no completely satisfactory explanation of its existence and function. The primary source of the membrane is a ring of columnar cells at the base of each caecum and in the adjacent epithelium, connecting the opening of each caecum with the foregut. Several laminae arise individually from separate groups of cells and are pressed out from clefts formed between adjacent lobes of folded epithelium. Additional laminae are added from the anterior ventriculus so that the fully formed membrane contains about 15 rather well-separated sheets. These are enclosed between two amorphous layers to make a complete assembly about 20 μm thick (Lee, 1968).

In the electron microscope, whole or tangentially sectioned laminae can be seen to be composed of strands of parallel fibrils arranged in three directions to form a hexagonal lattice. Individual fibrils were said to have a diameter of about 100 Å (Mercer and Day, 1952), but O'Riordan (1968) gives the figure as 350 Å. Adjacent parallel strands lie about 0.15 μm apart, a spacing that corresponds approximately to that of the epithelial microvilli. It is therefore suggested that the laminae are formed initially by secretion into the intermicrovillar spaces, followed by crystallization or polymerization and lifting off in a non-fluid form as perforated sheets (Mercer and Day, 1952). The histochemistry of the peritrophic membrane indicates that it contains acid mucopolysaccharides in addition to chitin and protein (O'Riordan, 1968). There are fewer laminae in starved insects, but production of the membrane does not stop.

Lee (1968) stated that the lumina of the caeca and the ventriculus were continuous and described a delicate, poorly consolidated membrane secreted by the caecal epithelium which was connected to the main cylinder of the peritrophic membrane in the ventriculus. The membrane does not therefore function to keep ingested food particles out of the caeca – this is accomplished by the oesophageal invagination. Other possible functions are a barrier to microbial infection (Mercer and Day, 1952), as protection against abrasion (Day and Powning, 1949), or as a molecular sieve that facilitates the access of soluble digestion products to absorptive sites independently of rearward food passage (Waterhouse, 1957). Evidence of a function is, unfortunately, meagre.

Fragments of the peritropic membrane were recovered from saliva by Mercer and Day (1952). No explanation has been offered for this extraordinary observation, but it may be cited as evidence that portions of the midgut contents are refluxed into the foregut. Normally, the fully formed membrane moves posteriorly in the ventriculus and ileum as a sheath enclosing ingested food material, and disintegrates in the colon, possibly as the result of shredding by the sclerotized teeth of the valve separating the ileum and colon (Bignell, 1977a). It is possible that membrane constituents are then degraded by hindgut bacteria.

4.2.4 Innervation

The stomatogastric nervous system (SNS) has been described by Willey (1961). Additional details of the innervation of the posterior crop region were given by Guthrie and Tindall (1968), while the distribution of nerves to the salivary glands has been minutely documented by Whitehead (1971). Very little can be added to these accounts, which are summarized in Fig. 4.5, however, it is worth pointing out that both the cibarial and pharyngeal musculature are innervated from the frontal ganglion and labral nerves (Fig. 4.5b). The division of fibres in these tracts between motor efferents and sensory afferents is unclear, but Davey and Treherne (1963b) described bipolar sense cells in the wall of the pharynx supplied from the n5 branch of the frontal ganglion, while Guthrie and Tindall (1968) reported the presence of many stretch receptors in the inner muscularis of the crop. On this evidence it seems clear that stomatogastric nerves are responsible for gathering sensory information from the foregut, but whether this extends to the chemoreceptors present on the epipharyngeal and hypopharyngeal surfaces adjacent to the cibarial cavity (Davey, 1962c) is unclear. An elegant study of the corresponding receptors in *Blaberus craniifer*, by Moulins (1971), has shown that their innervation is drawn in part from the labral nerves, but not from the frontal ganglion.

In contrast to the SNS, the innervation of the hindgut arises directly from the central nervous system and is characterized by the absence of peripheral cell bodies amongst its motor elements (Brown and Nagai, 1969; Fig. 4.5c). However, sensory nerve cells are present in the protodaeal wall below the circular muscles. These evidently correspond to the nerve net described by Davey (1962b). Rather little is known about the morphology and function of the midgut innervation, although

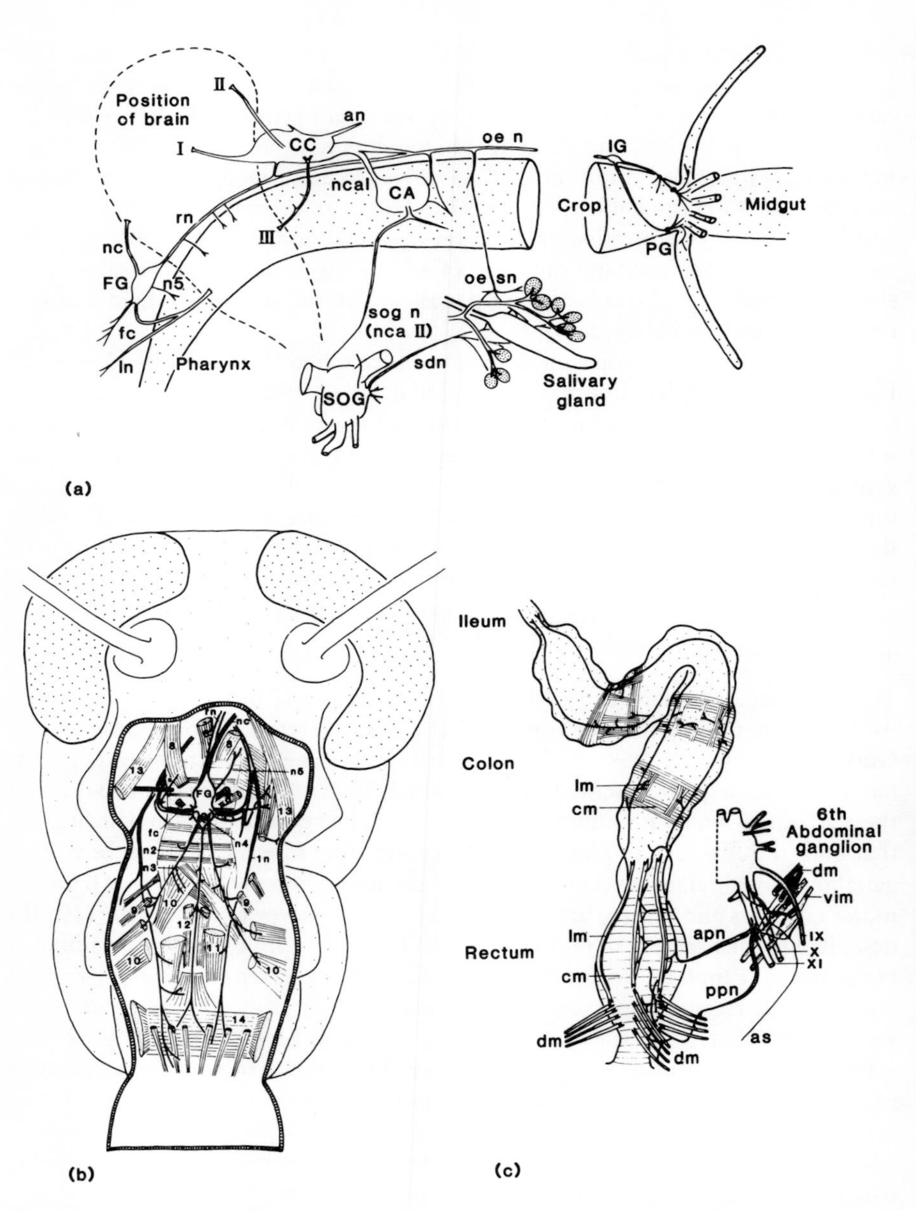

Fig. 4.5 Innervations of the gut wall. (a) Generalized arrangement of the SNS showing the major connections with higher centres and endocrine glands. The basic arrangement is a system of nerves, with four ganglia, extending over the foregut wall and connecting to the midgut musculature. An additional ganglionic area (hypocerebral ganglion) is represented by a slight swelling of the recurrent nerve underneath the CC and connects directly with them. Guthrie and Tindall (1968) have described a second ingluvial ganglion of about 20 nerve cell bodies on the branch connecting with the dorsal proventricular ganglion. For clarity the CA is shown displaced ventrally. (b) Innervation of the pharyngeal and cibarial musculature from the frontal ganglion and associated nerves. The numbers are those given to muscle

groups by Carbonell (1947, cited and reproduced by Guthrie and Tindall, 1968). (c) Proctodaeal nervous system, redrawn and slightly simplified from Brown and Nagai (1969). Dorsal view showing right side innervation of rectum only.

Key: CC, corpora cardiaca; CA, corpus allatum; FG, frontal ganglion; IG, ingluvial ganglion; PG, proventricular ganglia (two); SOG, sub-oesophageal ganglion; I–III, nervi corporis cardiaci. an, aortic nerve; apn, anterior proctodaeal nerve; as, branch to anus; cm, circular muscle; dm, dilator muscle; fc, frontal connective (paired); lm, longitudinal muscle; ln, labral nerve (paired); nc, nervus connectivus; nca, nervus corporis allati; oen, oesophageal nerve; oesn, oesophageal salivary nerve; ppn, posterior proctodaeal nerve; rn, recurrent nerve; sdn, salivary duct nerve; sog n, sub-oesophageal to corpus allatum connective; vim, ventral intersegmental muscle.

---

the majority of fibres apparently originate from the SNS (Day and Powning, 1949). Axons present in connective tissue at the base of the epithelium are said to be associated with the muscles only (Wright *et al.*, 1970), and may be contrasted, therefore, with nerve fibres in the salivary acini and extracellular ducts, which are directly applied to the surfaces of secretory cells (Whitehead, 1971).

It is clear that the SNS incorporates pathways that regulate some gut movements independently of higher centres (see Section 4.5.1), an observation consistent with the presence of peripheral nerve cells in the main ganglia, but the microtopography of the fibres is of interest for a number of other reasons. These are (1) that B-type neurosecretory neurons are included within the system (Whitehead, 1971; Aloe and Levi-Montcalcini, 1972a,b) raising the possibility that trophic relationships exist with visceral muscles (Miller, 1975a,b) or, conceivably, with secretory epithelia. (2) There is some evidence that stomatogastric nerves, particularly those

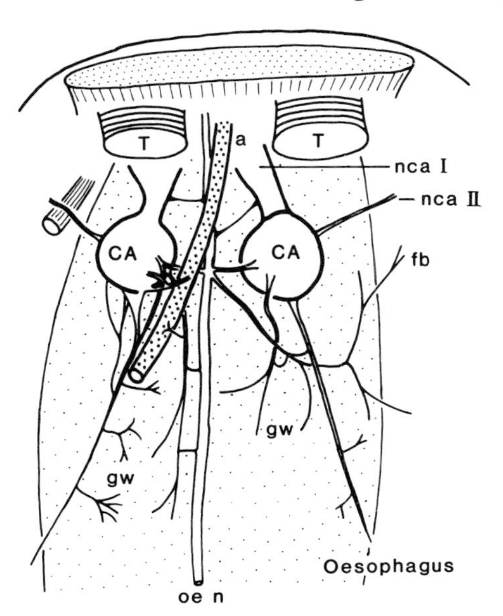

Fig. 4.6 Microtopography of connections between stomatogastric nerves and the corpora allata. Dorsal view. The arrangement of the finest tracts varies from animal to animal. Key: CA, corpus allatum; T, trunk trachea. a, aorta; fb, connections to fat body; gw, gut wall; nca, nervus corporis allata; oen, oesophageal nerve.

associated with the frontal ganglion, regulate the release of some endocrine factors from the corpora cardiaca (CC) by relaying afferent information from foregut receptors (Davey, 1962c; Penzlin, 1971; Verrett and Mills, 1976). (3) Stomatogastric nerves make complex and variable connections with the CC and corpora allata (Figs. 4.5 and 4.6), raising the possibility that experimental extirpation of these glands may necessarily disrupt sympathetic nervous pathways with subsequent uncontrolled effects on nutrient uptake. Similar reasoning has been used to explain the growth-arresting effects of frontal ganglionectomy and recurrent nerve severance (Roome, 1968; Prabhu and Hema, 1969). Finally, it can be noted that the SNS stains positively for aminergic neurons (Chanussot and Pentreath, 1973).

4.3 Nutrition

Despite the obvious nutritional versatility of the American cockroach, there have been relatively few critical studies of its requirements. This must be attributed to the longer generation interval, which makes *B. germanica* a more convenient subject of investigation, and consequently some extrapolations must be made from this species to *P. americana*.

Under ideal circumstances, the assessment of any specific diet or dietary component would require measurements of growth rate, moulting frequency, fecundity and mortality over several generations, but such complete data are rarely available for the American cockroach. More often, investigations have spanned shorter periods and have relied on determinations of fresh weight, dry weight and mortality only, although carcass analysis, diet consumption and fecundity have been chosen as additional criteria in recent work (e.g. Table 4.2).

Table 4.2 Consumption of some synthetic diets, and oöthecal production in adult female American cockroaches. Mean and standard deviation are shown for experimental groups of 25 insects maintained in individual petri-dish cultures for 18 weeks at 25 ± 2°C and ambient photoperiod. Except where stated, all diets contained yeast extract (8% by weight), Hawk–Oser salt mixture (3%) and cellulose (17%). Data are drawn from Bignell and Mullins, 1977, and Bignell, 1978.

Diet composition (% by weight)	*Consumption g/18 weeks/insect*	*Oötheca/18 weeks/insect*
Milled Purina dog food*	2.22 ± 0.64	15.4 ± 5.3
21% casein protein, 51% dextrin	2.11 ± 0.65	12.3 ± 2.1
38% casein protein, 33% dextrin	1.91 ± 0.40	10.2 ± 3.0
36% casein hydrolysate, 1% tryptophan, 33% dextrin	1.72 ± 0.38	7.1 ± 3.2
36% casein hydrolysate, 16% dextrin, 37% cellulose	1.28 ± 0.52	4.4 ± 2.3
36% casein hydrolysate, 53% cellulose (no dextrin)	0.65 ± 0.50	3.9 ± 1.6
72% casein protein (no dextrin)	1.32 ± 0.38	7.2 ± 2.9

* Crude protein 24%.

4.3.1 Natural and synthetic diets

A wide variety of natural foods and materials are said to support the growth of cockroaches. These include raw meats, cooked egg, dried milk, skimmed milk, whole wheat, flour, bread, confectionary, apple, potato, banana, cod-liver oil, corn oil, brewer's yeast, and, on a less savoury note, the faeces of humans or other animals. Cannibalism also occurs frequently. Laboratory stock cultures are commonly maintained on commercial dry dog foods which are cheap, parasite-free, indefinitely stable and can also be milled very readily into powder for quantitative work. Very high fecundity is given by a ground-up mixture of oats (54% by weight), unroasted unsalted peanuts (12%), dog food (31%) and dried baker's yeast (Pratt, 1972, cited by Hominick and Davey, 1972).

Synthetic diets have been devised to study quantitative aspects of nitrogen metabolism and carbohydrate nutrition (Haydak, 1953; Mullins and Cochran, 1975a,b; Bignell, 1976, 1978) and gut pathology (Bignell and Mullins, 1977). These contained casein protein (or hydrolysate), dextrins and Hawk–Oser inorganic salt mixture as defined nitrogen, carbohydrate and mineral sources respectively, and brewer's yeast extract as a source of lipids and vitamins. Cellulose powder was used as a supposedly inert filler. Diets based on this formulation are relatively well-defined, easily handled and stable but, nevertheless, do appear to be deficient in some respects, when compared to dog food of equivalent crude protein content (Table 4.2). A simple technique for handling synthetic diets in nutritional studies with cockroaches is given in Fig. 4.7.

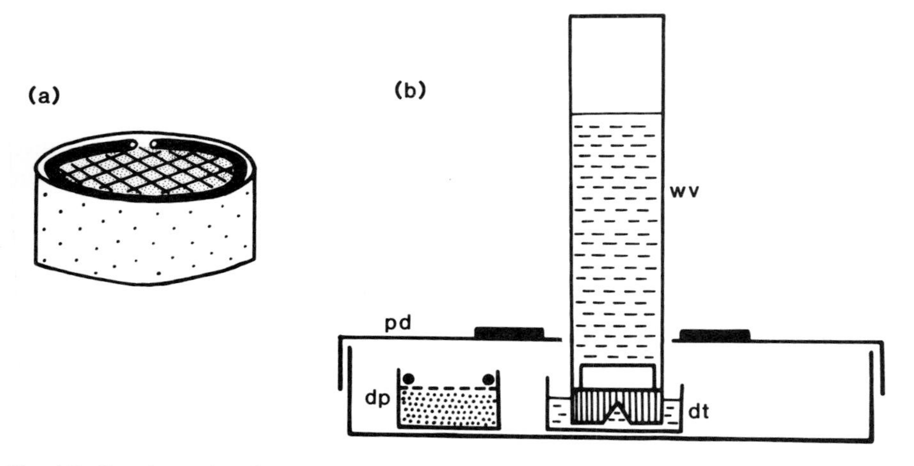

Fig. 4.7 Experimental techniques for quantitative nutrition. (a) Stainless steel diet planchet, 0.7 × 2.5 cm, devised by Mullins and Cochran (1975a) to hold test diets. A circular piece of ⅛ in wire mesh screen is placed on top of the diet and secured in place by a plastic café ring, 2.5 cm outside diameter, pressed in against the inner side of the planchet. This arrangement allows cockroaches to feed *ad libitum* but prevents them from dispersing the diet or depositing oötheca within it. (b) Culture chamber incorporating the diet planchet (dp). Key: dt, drinking trough formed from an inverted hard plastic cap; pd, plastic petri-dish, 10 × 1.5 cm with weighted lid; wv, glass or plastic water vial with notched plastic cap.

4.3.2 Specific requirements

Essential nutrients are those required for indefinite growth and reproduction, and are conventionally determined by deletion or substitution in completely defined synthetic diets. However, the interpretation of such experiments is not always straightforward since the effectiveness of a nutrient depends not only on its being ingested, but also on satisfactory absorption and metabolism. Hence, the growth performance of an insect may be strongly influenced by the presence or absence of substances which are not themselves essential nutrients according to the above definition, and also by the balance of such nutrients in the diet (House, 1974a). If the rate at which essential nutrients are converted into non-essential ones is limited, the addition of the latter to the diet may improve growth. Consequently, the distinction between essential and non-essential nutrients can be blurred.

A particular problem in nutritional work with cockroaches is posed by the presence of microbial symbionts in specialized cells (mycetocytes) within the fat body. There is also a substantial gut flora. It has been known for many years now, that cockroaches deprived of their intracellular symbionts by treatment with antibiotics or other substances, show poor growth and reproductive capacity (Richards and Brooks, 1958), but inability to sustain the microbes *in vitro* has restricted our knowledge of their exact function. Aposymbiotic *P. americana* were unable to synthesize methionine and cysteine from inorganic sulphate (Henry and Block, 1960), and also showed reduced levels of ascorbic, folic and pantothenic acids in the fat body (Ludwig and Gallagher, 1966). Brooks and Kringen (1972) have presented evidence that mycetocyte symbionts are responsible for providing polypeptide growth factors in *Blattella germanica*, while in *P. americana* they may assist in mobilizing nitrogen from deposits of crystalline urate (Brooks, 1970; Mullins and Cochran, 1975b).

Information gathered largely from *B. germanica* suggests that arginine, histidine, isoleucine, leucine, lysine and valine are essential amino acids in cockroaches. Methionine and cysteine are synthesized from inorganic sulphate by symbionts while phenylalanine, threonine and tyrosine can be furnished by intestinal micro-organisms if tryptophan is present. Proline, serine and alanine can apparently be synthesized by the insect, but in limited amounts and must be added to the diet for optimum growth. Arginine and glutamic acid promote growth but are not considered essential. Only the L forms of amino acids are utilizable by insects (Gordon, 1959).

It is characteristic of insects that mixtures of amino acids cannot substitute completely for the parent proteins in the diet, even when the total nitrogen content is the same in each case. This phenomenon is also shown by the American cockroach (Table 4.2). It has been suggested that the amino acid mixtures may inhibit ingestion, cause osmotic disruption within the gut or place strain on interconversion pathways (Dadd, 1970a). It is by no means certain that ingested proteins are broken down completely into separate amino acids before absorption.

Since insects lack the ability to synthesize carbohydrates from lipids, the former are regarded as essential for meeting normal energy requirements. Amongst hexose sugars, glucose, fructose, sucrose and maltose are completely adequate for this purpose; lactose, melibiose, sorbose and galactose are less effective and rhamnose and cellobiose are unsatisfactory (Gordon, 1959). There is little or no ability to metabolize pentoses. Although lipids may take part in oxidative metabolism, the degree to which they can substitute for dietary carbohydrate is unclear. Oily or fatty ingredients will alter the physical properties of diets and therefore affect the mechanisms of ingestion, digestion and absorption.

Long-chain saturated fatty acids may be readily synthesized from simple precursors, e.g. acetate or glucose, but Gordon (1959) was able to show a long-term requirement for the unsaturated linoleic and linolenic acids in *B. germanica*. The presence of lipids in the diet is also presumably necessary for the efficient absorption of the lipid-soluble A and E group vitamins, carotene, α-tocopherol and retinol.

The requirement of insects for sterols is well-known, stemming from an inability to synthesize cholesterol, the key precursor of many membrane components and the steroid hormones. The synthesis is readily accomplished by gut micro-organisms (Clayton, 1960) but it remains unclear whether this absolves the insect completely from a requirement to ingest sterols. Clayton found that asceptic *B. germanica* (i.e. containing mycetocyte symbionts, but no gut flora) could not grow when cholestanol (dihydrocholesterol) was provided as the sole sterol source, but responded to supplementation with 0.004% cholesterol, a level below that required for growth when cholesterol only was present. In cases like this, cholestanol, or substances with the same property, are said to exert a sparing action, i.e. to substitute for some, but not all, of the functions of cholesterol. Both *B. germanica* and *P. americana* are able to convert C_{28} and C_{29} plant sterols (e.g. β-sitosterol, stigmasterol) to cholesterol or 7-dehydrocholesterol by dealkylation (Robbins *et al.*, 1971). 7-dehydrocholesterol itself, an intermediate in the pathway of ecdysone synthesis, is said to be an inadequate sterol source when supplied alone in the diet (Gordon, 1959).

A variety of water-soluble vitamins are required as co-enzymes, including thiamin, riboflavin, niacin, nicotinic acid, pyridoxin, folic acid, biotin and pantothenic acid, but some of these at least are supplied by symbionts. Inositol and choline, whose functions include participation in phospholipid synthesis are also essential (Noland and Baumann, 1949; Forgash and Moore, 1960). Very little information is available on mineral nutrition but Mg, Fe, Zn, Cu and Mn are required by *B. germanica* (Gordon, 1959; Brooks, 1960). The important electrical and osmotic roles of Na^+, K^+ and Ca^+, and the key position of P in metabolism must obviously lead to their designation as essential nutrients, but there are few instances in which this has actually been demonstrated by standard nutritional techniques. It must be assumed that requirements for these elements can be met from impurities in other components of experimental diets (House, 1974a).

Mullins and Keil (1980) have shown that radio-actively labelled uric acid secreted by males of *B. germanica* around the spermatophore subsequently appeared in oöthecae for use as a nitrogen source during embryogenesis. Although direct assimilation by the genital tract of the female is a possibility, it was thought more likely that the spermatophore was eaten after voiding and that the labelled material was therefore acquired from the gut. Earlier work by Mullins and Cochran (1975a,b) had shown that excesses of nitrogen (as casein protein) in the diet of *P. americana* led to corresponding increases in uric acid content of the fat body. Conversely, nitrogen-deficient diets caused a depletion of these urate stores, the highest rates occurring in insects completely deprived of a protein source. The clear implication of these studies is that stored urates function as a nitrogen reserve which can be mobilized under conditions of dietary stress. The exact metabolic pathways followed during the process of mobilization are not yet known, but it must be assumed that the mycetocyte symbionts are mediators. The phenomenon may also explain why *P. americana* and other cockroaches are able to complete growth on diets containing a very wide range of protein content, relative to carbohydrate (Haydak, 1953; Gordon, 1959). However, the quantitative balance between protein and carbohydrate does affect the rate of growth (Haydak, 1953) and also oöthecal production (Table 4.2).

4.3.3 Long-term regulation of intake

B. germanica has been shown to maintain a constant intake of phagostimulatory sugars (e.g. glucose, sucrose) over a large interval of time (Gordon, 1968). Insects deprived of food for short periods consumed larger than normal amounts on the first and subsequent days of free access, although the compensation declined with time. In the same species, van Herrewege (1971) found that the cumulative intake of a balanced diet based on wheat-germ was the same over 14 days in insects deprived of food for three days during the experiment and in controls which were able to feed *ad libitum*. In *P. americana*, Bignell (1978) measured the consumption over 18 weeks of a series of diets, based on casein hydrolysate, in which the digestible carbohydrate (dextrin) was progressively replaced with cellulose. Except in the case of a dextrin-free diet, the insects responded by consuming greater volumes (but not greater weights) of diluted diets; however, the compensation was insufficient to offset the dilution completely and cumulative dextrin intake was therefore less on the diluted diets. Interestingly, a dextrin-free diet did not elicit the response, and dietary stress ensued as determined by the rate of oöthecal production. Moderate dilution, on the other hand, caused no change in the rate of oöthecal production. It is possible, therefore, that the compensation mechanism is peripheral, for example determining the rate and/or extent of crop-filling or crop-emptying. Central monitoring of internal metabolite levels seems unlikely.

The only insects in which the mechanism of hunger has been resolved to any extent are the fly, *Phormia regina*, and the locust, *Locusta migratoria* (reviewed by Barton-Browne, 1975). In these species, there is evidence that elevated haemolymph osmotic pressure retards crop-emptying (*P. regina*) and reduces the size and duration of meals (*L. migratoria*). Changes in haemolymph osmotic pressure are evidently integrated with distension of the foregut, assessed by stretch receptors, to terminate feeding. While information on the relationship between feeding and haemolymph osmotic pressure is lacking in cockroaches, it is known that the rate of crop-emptying increases with decreasing osmotic pressure of ingested solutions (Treherne, 1957) and that the proportion of a meal released to the midgut in a given time is a constant, independent of meal size (Davey and Treherne, 1963a). These relationships, together with the assumption that each meal approximately replaces material emptied from the crop in the preceding inter-meal period, could account for the consumption of greater volumes of diluted food (Bignell, 1978).

The choice of cellulose as a dietary diluent rests on the assumptions that it is neither a phagostimulant nor a feeding deterrent, and also that it is not degraded in the gut to any significant extent. While there is no evidence that cellulose or cellulosic fibres are selectively consumed by cockroaches to compensate for deficiencies in more readily digested carbohydrates (Bignell, 1976 and unpublished), some digestion of cellulose by gut micro-organisms does occur (Bignell, 1977b). Cellulose ingested with other substances as a component of synthetic diets has a number of physical effects within the gut which include increases in both the fluid content and volume of the hindgut (Bignell, unpublished). By analogy with vertebrates, this might lead to a reduction of the transit time and thus an indirect effect on consumption. The characteristic gnawing of wood, cardboard, paper and other fibrous materials by cockroaches is not accompanied by significant uptake and appears to be behavioural correlate of oöthecal deposition (Bignell, 1976).

4.4 Enzymology

4.4.1 pH and redox potentials

Intestinal pH is an important determinant of digestive enzyme action, but may also influence the osmotic activity of gut fluids and the process of absorption. Determinations of pH by Warhurst (1964) and Greenberg *et al.* (1970) are summarized in Fig. 4.8. Although the results of Warhurst (1964) are consistently higher, the two accounts agree that the pH of the midgut and hindgut contents is higher and also less variable than that of the crop contents. The acidity of the foregut has been attributed to microbial degradation of ingested sugars (Wigglesworth, 1965), but little evidence of this exists. It is widely assumed that the midgut and hindgut fluids of insects have a buffering capacity that maintains a constant pH, unaffected by the nature of ingested foods, but practically no information is available concerning this mechanism. However, it is known that the potassium-rich fluid secreted by the Malpighian tubules enters the midgut of *P. americana* as well as contributing significantly to the fluid phase of the hindgut contents (O'Riordan, 1968; Wall, 1970). Since this excretory fluid may contain amino acids, proteinaceous materials and other nitrogenous organic materials (Mullins and Cochran 1973a), an effective buffering capacity is to be expected.

Redox potential is chiefly of interest as a determinant of microbial activity, but may also be a factor in the destabilization of proteins containing disulphide bonds. Using indicators, Day and Powning (1949) and Warhurst (1964) showed that E_0' decreased *in vivo* from the crop to the colon; however, the technique can be criticized on the grounds that indicators must be concentrated to be visible, and the system is thereby overpoised. Fortunately, Warhurst (1964) also made an electrometric determination with a Pt–Ir electrode and gives the E_0' of the adult colon as −84 to −240 mV, suggesting a low oxygen tension that may approach anaerobic conditions. It is interesting to note that Ritter (1961) analysed the hindgut of the wood-feeding cockroach, *Cryptocercus punctulatus*, by a polarographic technique sensitive to 10^{-6}M, but failed to detect the presence of oxygen. Ritter attributed these putatively anaerobic conditions to the presence of a strong reducing agent which he identified as the tripeptide glutathione. Glutathione is present in the Malpighian tubules of *P. americana* and may therefore contribute to the low E_0' of the colon (Metcalf, 1943).

Low redox potentials are sometimes said to be associated with relatively poor tracheation in some parts of the insect gut (e.g. Dadd, 1970b), but nothing is known about the diffusion of oxygen from the epithelium to the lumen contents. Since the ultrastructure of the colonic epithelium suggests that it supports a strong oxidative metabolism (Section 4.2.2 above), the low oxygen tension in the hindgut contents must result from microbial activity or the secretion of reducing agents.

4.4.2 Digestive enzymes

The variety of digestive enzymes described in *P. americana* is shown in Table 4.3. In most cases, identification has depended on the activity of crude extracts against selected substrates, with few attempts at purification. Consequently, many questions concerning bond specificity, site of action and the major hydrolytic products of digestive enzymes remain unanswered. The most frequently investigated property of enzymes is pH optimum, from which some inferences may be drawn. For example, crude extracts of the midgut show dual pH optima for lipase at 5.2 and 7.2 (Gilbert *et al.*, 1965). While the latter suits the conditions in the midgut lumen, the former is appropriate for the more acidic environment of the crop. Since significant lipolytic activity can be shown in the foregut contents (Eisner, 1955; Bollade *et al.*, 1970), but the enzyme does not apparently occur in the salivary glands or the foregut wall, it must be derived from the midgut by regurgitation through the gizzard. Although optimum proteolysis requires mildly alkaline conditions, proteinase is also present in the crop of fed insects (Day and Powning, 1949).

Wharton *et al.* (1965b) investigated cellulase and concluded that only the salivary glands contained endogenous enzyme, all other cellulolytic activity being attributable to gut micro-organisms. The logic of this arrangement is hard to understand, however, and further difficulty is introduced by the observation that while cellulase activity is strong in the contents of the midgut (Wharton *et al.*, 1965b;

Table 4.3 Summary of selected digestive enzymes: occurrence, assay, authorship and some characteristics.

Enzyme	Site identified	Basis of assay	Citation	Comments
Amylase	Saliva	Starch cleavage	Wigglesworth, 1927	pH optimum 7.0
Amylase	Midgut	Starch cleavage	Day and Powning, 1949	pH optimum 5.7–7.0; more enzyme activity in caeca than ventriculus
Invertase	Midgut	Sucrose cleavage	Day and Powning, 1949	pH optimum 5.0–6.5; enzyme activity equal in caeca and anterior ventriculus
Maltase	Midgut	Glucose production from maltose	Wigglesworth, 1927	pH optimum 5.0–6.5
β-Glucosidase	Caeca	Glucose production from cellobiose	Newcomer, 1954	pH optimum 4.2–6.5
Cellulase	Salivary glands and midgut contents	Reducing sugar production from carboxymethylcellulose	Wharton *et al.*, 1965b	Probably of microbial origin (except salivary glands)
Chitinase	Saliva, foregut, midgut, hindgut	*N*-acetylglucosamine, diacetylchitobiose and triacetylchitobiose production from purified chitin	Waterhouse and McKellar, 1961	Not of microbial origin; pH optimum 5.4–6.0; most activity in gut contents with midgut > foregut > hindgut
Non-specific esterase*	Midgut	α-naphthol production from 1-naphthyl acetate; likewise 1-naphthyl propionate and 1-naphthyl butyrate	Hipps and Nelson, 1974	7 electrophoretically separable fractions may form a thermally dependent molecular aggregate; little lipase activity*
Lipase*	Crop contents	Oleic acid production from triolein	Eisner, 1955	Slow reaction, but proceeds faster in fed insects
Lipase*	Foregut regurgitate	Free fatty acid production from 1,3 dipalmityl 2-oleoyl glycerol	Bollade *et al.*, 1970	Slow reaction producing first 1,2 diglycerides, then 2-monoglycerides and finally glycerol
Lipase*	Caecal homogenate	Free fatty acid, diglyceride and monoglyceride production from tripalmitin	Cook *et al.*, 1969b	Activated by Co and Mn; may also attack short-chain esters
Lipase*	Whole midgut homogeneate	Free fatty acid production from triolein	Gilbert *et al.*, 1965	Dual pH optima at 5.0 and 7.2
Proteinase	Whole midgut or midgut contents	Degradation of sulphanilomide azocasein	Day and Powning, 1949	pH optimum 7.5–8.5
Proteinase	Whole caecal homogenate	Degradation of sulphanilomide azocasein plus reduction of gelatin viscosity	Powning *et al.*, 1951	pH optimum 7–9; resembles trypsin in lack of inhibition by CN or thioglycollate and lack of activation by enterokinase; unaffected by E_0' in range −460 to +460 mV
Acid phosphatase	Whole midgut homogenate and autophagic vacuoles in epithelium	Inorganic PO_4 production from β-glycerophosphate	Couch and Mills, 1968	pH optimum 5.2; enhanced by Mg
Neutral phosphatase	Whole caecal homogenate	α-naphthol production from α-naphthol phosphate	Cook *et al.*, 1969b	pH optimum 7.0; two electrophoretically separable bands
Alkaline phosphatase	Caecal, ventricular and colonic epithelia	Inorganic PO_4 production from Na β-glycerophosphate	Srivastava and Saxena, 1967	Active at pH 9.2
Lysozyme	Whole gut homogenate	Solubilization of *Micrococcus* cell walls	Powning and Irzykiewicz, 1967	pH optimum 3.5; probably a separate enzyme from chitinase

* Although the terms esterase and lipase are frequently used without distinction, lipase is used here to characterize enzymes capable of splitting long-chain glycerides (i.e. neutral fat) which form the bulk of dietary lipids. Esterases are therefore those enzymes which attack simpler soluble esters of low molecular weight, although this criterion is not always satisfactory (see Cook *et al.*, 1969b).

Cruden and Markovetz, 1979), greater numbers of bacteria occur in the hindgut (Bignell, 1977a). From work with diets containing ^{14}C-cellulose and ^{14}C-hemicellulose, Bignell (1977b) considers the colon the most likely site of attack on ingested plant polysaccharides, the reducing conditions leading to a fermentative degradation yielding organic acids (cf. Bracke and Markovetz, 1980). Some explanation of these conflicting results may be deduced from the technique of cellulase assay, which employs carboxymethylcellulose as a substrate. This determines only the C_x component of the cellulase complex, and does not necessarily show that native cellulose would be degraded to any significant extent.

Chitinase is distributed throughout the gut, but is not of microbial origin and cannot be wholly attributed to moulting fluid in the cuticle-secreting epithelia of the foregut and hindgut. It must therefore serve a true digestive function which may be related to the cockroach's habit of consuming cast skins or to cannibalism. It is not known how the gut cuticle or the peritrophic membrane escape digestion under these circumstances (Waterhouse and McKellar, 1961).

Very little information is available on the fate of enzymes secreted into the gut lumen. In a few cases, activities have been determined simultaneously in the midgut and hindgut, and these results show that hindgut activities are generally lower. The exception appears to be cellulase, which shows high activity in the faeces of adults (Wharton *et al.*, 1965b).

4.4.3 Control of secretion

The possibility of direct nervous control of digestive enzyme secretion was raised by Day and Powning (1949), who noted the innervation of both the midgut wall and the salivary glands from the SNS. It is now clear that axonal ramifications in the midgut are associated with the musculature only (Wright *et al.*, 1970), but in the salivary glands nerve terminals are applied directly to the external (i.e. basal) surfaces of both zymogen and parietal cells in each acinus. Electrical stimulation of the salivary duct nerve induced fluid secretion from the glands (Whitehead, 1971), but the exact mechanism by which this response occurs has not been studied in *P. americana*.

Fluid produced by the secretory acini and/or the acinar ducts fills the salivary reservoirs, unless direct access to the cibarial cavity is made available by raising of the hypopharynx (Sutherland and Chillseyzn, 1968). Thus, reservoir fluid contains the same enzymes as are found in the acini, but their activity depends on the volume of the reservoir contents. There is no evidence that the walls of the salivary reservoirs are permeable to any component of the fluid they contain, but the volume of stored fluid is drastically reduced when cockroaches are dehydrated. Emptying of the reservoirs is probably accomplished by haemolymph pressure or by the foregut expanding with swallowed food or air, but since salivation can occur without actual ingestion there may be an involvement of chemoreceptors associated with the mouth parts. Although the salivary glands have an intrinsic musculature, in *P. americana* this serves only to occlude the mouth of the reservoir (Sutherland and Chillseyzn, 1968).

The flow of fluid from the secretory acini may also be induced *in vitro* by the application of a bathing solution containing 10^{-9}M 5-hydroxytryptamine, suggesting that this substance may be a natural neurotransmitter in the salivary glands (Whitehead, 1971). However, more extensive research in *Nauphoeta cinerea* has implicated dopamine in this role, since the manner of the response to this amine corresponds more closely to that obtained when the salivary duct nerves are stimulated electrically (Bowser-Riley and House, 1976). The principal response to dopamine seems to occur in the parietal cells, where there is an increase in membrane K^+ conductance and active Na^+ transport, which together generate fluid secretion (House, 1973; Smith and House, 1977). The fluid itself contains Na^+, K^+ and Cl^-, with Na^+ the major cation, and also a mucous component derived from the extracellular duct cells (Bland and House, 1971; Smith and House, 1977). No information is yet available on the relationship between fluid secretion and enzyme synthesis, but both the parietal and the zymogen cells respond to electrical stimulation of the salivary duct nerve by transient increases in resting membrane potential (House, 1975).

Except for the preliminary work of Day and Powning (1949), *P. americana* has scarcely featured in any investigation of the control of enzyme synthesis in the insect midgut. These authors found that feeding increased the number of mitoses in the caecal epithelium and suggested that this could be explained by an endocrine mechanism. In other cockroaches, however, nearly all the available evidence favours a secretogogue, i.e. direct chemical stimulation of the epithelium by ingested food. For example, in *Leucophaea maderae*, Englemann (1969a) found that midgut proteinase activity was proportional to the quantity of certain proteins passing from the crop. Casein, fibrin, elastin and gluterin were effective in producing this proportional response even if the brain, CC or CA were removed. Other substances, such as keratin, gelatin, haemoglobin, casein hydrolysate and starch failed to influence proteinase activity. Endocrine mechanisms or mere mechanical stimulation of the epithelium thus appear to be ruled out in this species. Further evidence against endocrine involvement may be drawn from the observation that severence of the recurrent nerve in *L. maderae* (which inhibits crop-emptying, but does not directly depress feeding) reduced midgut protease activity to the level of starved controls (Engelmann, 1966). Rounds (1968) showed that ethanolic extracts of the whole midgut of *P. americana* caused a significant increase of midgut proteinase when injected into intact insects, provided that the extracts were made at the onset of darkness during an ambient photoperiod. While this may in theory indicate an endocrine influence on enzyme secretion, it evidently has no direct relationship with the ingestion of food.

A low enzyme activity is usually detectable in the midguts of starved cockroaches (e.g. Day and Powning, 1949; Rao and Fisk, 1965). It is therefore assumed that synthesis in the epithelium and discharge to the gut lumen are interdependent (Dadd, 1970b). The possibility that the tissue continues to accumulate an inactive precursor seems to be eliminated by the observation that when starved midguts were homogenized and then mixed with rat food, starch-casein paste or the crop contents of fed insects, there was no increase in protease activity (Engelmann, 1969a). Engelmann also found that ^{14}C-leucine, injected into the haemolymph, was not incorporated into intestinal protease in starved insects. The secretogogue mechanism may therefore extend to enzyme synthesis as well as secretion.

4.5 Physiology of digestion

4.5.1 Gut movements

Transit times were determined by Snipes and Tauber (1937) who used a banana diet containing coloured dyes. The mean transit time in both adults and nymphs of *P. americana* was 20.6 h. Although a portion of each meal is retained in the crop for up to 100 h after feeding, the rate of passage generally slows with progress through the gut. Consequently, the greater part of the transit period is spent in the hindgut. Movement of food within the colon is complex: materials entering from the ileum are carried to the midregion of the organ within the remnants of the peritrophic membrane before dispersing anteriorly and posteriorly to fill it (Bignell, 1977a). When fed insects were fasted, a cyclical change of particulate and fluid contents, and of total volume was observed in the colon (Hominick and Davey, 1975). If the fasting was continued beyond 7 days the particulate component of the contents became permanently reduced relative to fluid.

Contractions of the gut musculature, although frequently rhythmical, are difficult to analyse by conventional mechanographic techniques since they involve simultaneous movements in different directions by separate regions. However, Yeager (1931) managed to distinguish three types of foregut movement in *P. fulginosa*: peristaltic waves originating in the oesophagus as constrictions extending completely around the foregut; antiperistaltic waves originating in the anterior crop and unified contractions involving the whole posterior third of the foregut. Cook *et al.* (1969a) described very similar patterns in *Blaberus giganteus*, but also noted the existence of pacemaker regions in which contraction continued after *in vitro* preparations of the gut were cooled with ice. The hindgut movements of *L. maderae* were analysed cinematographically by Cook and Reinecke (1973). Four types of motility were recognized: (1) segmentation, i.e. the formation of annular constrictions followed by relaxation without progression; (2) compression or shortening of the hindgut by contraction of the longitudinal muscles; (3) peristalsis and (4) antiperistalsis. Each showed some rhythmicity but peristalsis and antiperistalsis were relatively infrequent movements. No permanent pacemaker regions were identified, but excitability increased progressively from the ileum to the rectum. Movements of the cockroach midgut appear to consist only of slow weak peristalsis, but the passage of materials through the lumen is probably assisted by pressure applied from contractions of the crop when the gizzard is open (O'Riordan, 1968). Constriction of the caeca can be seen in freshly dissected specimens, but agitation of the caecal contents may also be assisted by the writhing of the Malpighian tubules. The oesophageal invagination ensures that materials entering the midgut are first discharged into the anterior ventriculus; after this, the caeca acquire fluids which presumably contain the products of digestion. The posterior ventriculus is filled last (O'Riordan, 1968; Bignell, unpublished observations).

Recent investigation of the physiology of the gut muscles has been dominated by a controversy concerning the identity of the neurotransmitter. Two candidates have been proposed: proctolin, a pentapeptide first isolated from *P. americana* (Brown, 1975) and glutamate, which closely mimics the effects of the natural transmitter in *L. maderae* and *P. americana* (Holman and Cook, 1970; Cook and Holman, 1979a,b). Although discussion of the issue would be beyond the scope of this chapter, it has emerged from a number of studies that foregut and hindgut muscles are both myogenic, but control and co-ordination of contraction are superimposed by the stomatogastric and proctodaeal nervous systems respectively (Cook *et al.*, 1969a; Nagai and Brown, 1969; Holman and Cook, 1970) (see Section 12.4.1). The only exception to this rule is the gizzard, where interruption of the nerve supply leads to complete paralysis of the musculature (Davey and Treherne, 1963b). In this case, the sympathetic innervation forms part of a pathway which is designed to regulate the rate of crop-emptying in response to the properties of ingested materials, notably osmotic pressure and meal size. The mechanism controls the extent and frequency of opening of the gizzard valve and influences the generation of transient pressure pulses within the crop such that as meal size increases, the rate of crop-emptying also rises and the proportion of a meal passed from the crop in a given time therefore remains constant. In addition to this an increase in the viscosity of the crop contents increases, and an increase in osmotic pressure decreases the rate of crop-emptying (Davey and Treherne, 1963a, 1964). Clearly, the mechanism is, in part, homeostatic, ensuring a uniform short-term rate of uptake from the midgut, but it might also serve to reduce general metabolism by the insect when food is scarce. The sensory end of the system appears to be a group of cells in the wall of the pharynx innervated by branch n5 of the frontal ganglion. Cutting these nerves greatly slowed crop-emptying (Davey and Treherne, 1963b).

A hormonal influence over visceral muscle contraction may be inferred from the observations of Davey (1964) that extracts of the CC stimulate movements of the isolated foregut and hindgut. The brain of *L. maderae* synthesizes, stores and releases a myotropic peptide which is chromatographically distinguishable from proctolin, but has very similar effects on the hindgut, producing a prolonged excitation (Holman and Cook, 1979). Thus, there is some reason to doubt that proctolin is a neurotransmitter. Holman and Cook consider both peptides to be hormones, whose site of action is the muscle cell membrane in general, rather than the neuromuscular junction in particular. The difference between them seems to be that proctolin is released locally from the terminal regions of some proctodaeal nerve fibres.

4.5.2 Absorption of inorganic ions and water

An analysis of ionic concentrations in caecal and ventricular fluids by O'Riordan (1968) showed that Na^+ and K^+ normally maintain steep, but opposed, gradients across the midgut wall (the gut fluid being K^+-rich). In the epithelium, the cellular fluids differ from the haemolymph in having a high K^+ and low Na^+ content. When insects were made to ingest experimental solutions of NaCl and KCl the first response was an increase in the volume of both the caeca and the ventriculus, but this declined after reaching a peak some 45 min from the moment of consumption. Accompanying changes in ionic concentrations and the osmotic pressure of gut fluids indicated that a net efflux of Na^+ and water was the principal absorptive phenomenon, and this was confirmed in a tracer experiment with ^{22}Na. The caeca appeared to be the main sites of efflux and little uptake of the isotope could be demonstrated from the crop, ventriculus or colon. Ingested dye solutions were concentrated in the caeca, indicating that this was also the site of absorption of water. Although the concentration of Na^+ is normally low in the gut lumen, it rose rapidly when the haemolymph side of the midgut wall was irrigated with a Ringer's solution containing 2,4-dinitrophenol or ouabain. Consequently, O'Riordan (1968) proposed the existence of a metabolically driven Na^+ pump in the gut wall, which was directly responsible for the transport of Na^+ against an overall concentration gradient and which would also account for the accompanying passive movements of water.

A further investigation of the pump was made by double perfusion of isolated midguts (O'Riordan, 1968, 1969). In a simple Ringer's solution containing Cl^-, a stable transepithelial potential of 12 ± 1 mV (lumen side negative to the haemolymph side) rapidly developed and was sustained for several hours, but substitution of Mg^{2+} or choline for Na^+ on the lumen side of the preparation led to a decline on the potential. A similar result was obtained when K^+ was increased on the haemolymph side, but little effect resulted from changes in K^+ concentration in the luminal compartment. The model favoured by O'Riordan (1968) thus envisages a linked Na^+/K^+ pump located on or near the haemolymph (basal) surface of the midgut epithelium which removes Na^+ from the cytoplasm to the haemolymph in exchange for K^+, the latter ion being thereby maintained at a high intracellular concentration (Fig. 4.8). The differential effects of K^+ on each side of the tissue are explained by proposing that the apical membrane (confronting the lumen) has a low permeability to this ion. Passive pathways are open to both Na^+ and Cl^- across the apical and the basal membranes, and to K^+ at the basal side. The transepithelial potential therefore appears to result from the sum of a lumen-side Na^+ diffusion potential and a haemolymph-side K^+ diffusion potential. Significantly, the concentration of K^+ in the midgut lumen falls if the posterior ventriculus is ligatured off *in vivo*, or if the Malpighian tubules are irrigated with N_2-saturated (i.e. anoxic) Ringer's solution. Much of the K^+ normally present in the midgut lumen may therefore be derived from the primary excretory fluid, secreted by the Malpighian tubules, and carried anteriorly from the midgut/hindgut junction in the ectoperitrophic space (see Berridge, 1970a).

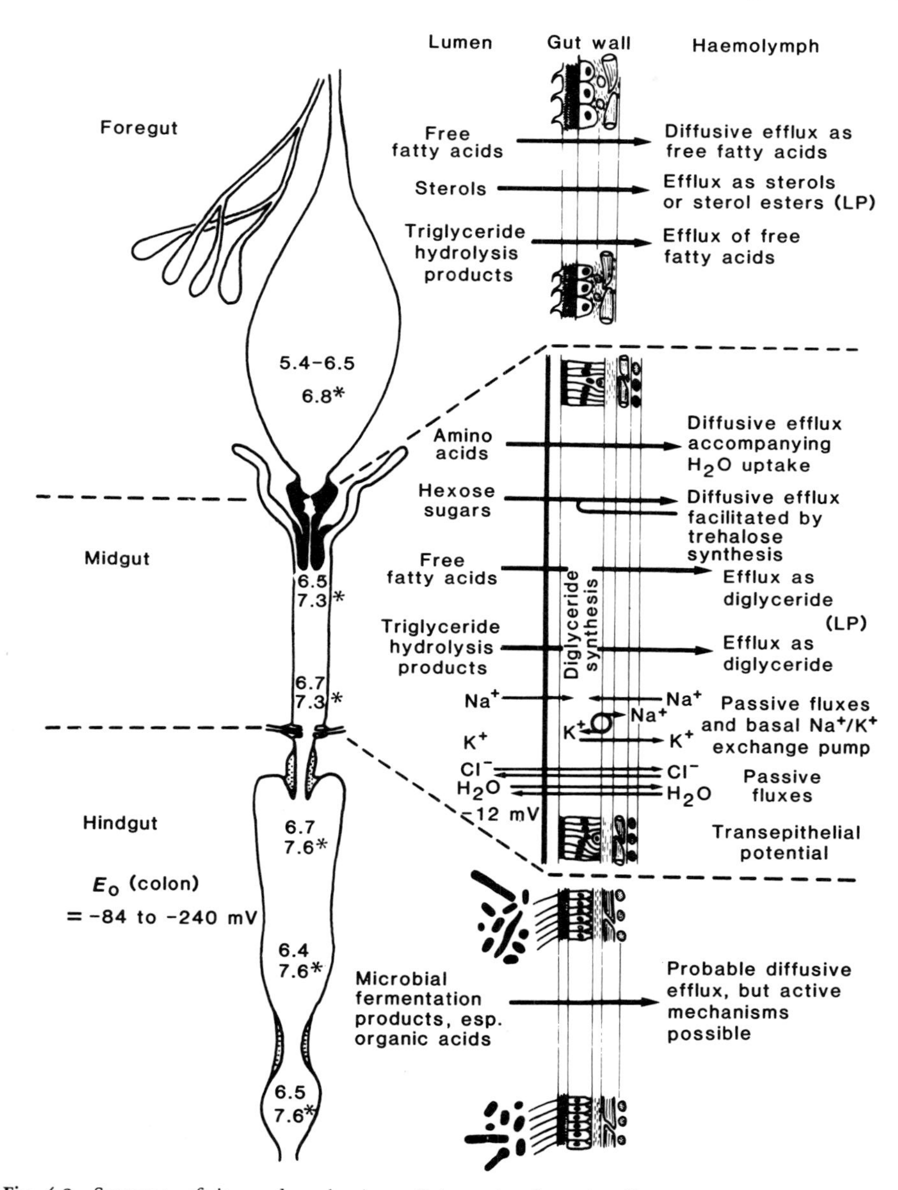

Fig. 4.8 Summary of sites and mechanisms of absorption from the alimentary canal, excluding the rectum. Figures within the gut refer to pH and are averaged from the potentiometric data of Greenberg *et al.* (1970) and Warhurst (1964). Warhurst's data are indicated by an asterisk. LP, associated lipoprotein carrier.

O'Riordan (1968) obtained further evidence for her model by showing that ^{22}Na efflux *in vitro* exceeded influx, but was sensitive to both 2,4-dinitrophenol and ouabain, when these were applied to the haemolymph side of the midgut.

Essentially similar conclusions were reached by Sauer *et al*. (1969), although from less extensive experimentation. These authors proposed a major role for the midgut in controlling haemolymph osmolality, but there is little evidence that it has a regulatory capacity comparable with that of the rectum.

4.5.3 Absorption of organic materials

The demonstration by O'Riordan (1968) and others of net Na^+ efflux from the caeca, with an accompanying passive flux of water appears to provide the basis of an absorption mechanism for organic solutes, which would move with the solvent to the basal side of the epithelium or into the extracellular compartments within it. There is little doubt that the midgut is the site of absorption for sugars. For example, by comparing the ratio of ^{14}C-glucose to non-absorbed dye during the passage of an experimental fluid meal, Treherne (1957) was able to show that most of the uptake occurred in the anterior midgut and caeca. However, parallel experiments in the locust, *Schistocerca gregaria*, showed that glucose uptake was unaffected by the application of KCN and iodoacetate to the haemolymph side of the midgut *in vitro*, and further, that fructose and mannose were absorbed less rapidly than glucose under comparable conditions (Treherne, 1958a,b). These results suggested that the uptake of sugars was more closely related to their rates of metabolism once absorbed than to any active efflux from the midgut lumen. Consequently, Treherne (1968a,b) concluded that monosaccharides were absorbed by passive diffusion down a concentration gradient that was facilitated by their conversion to trehalose in the haemolymph and fat body. This thesis enables us to explain a number of other observations, for instance that ingested fructose is converted to glucose prior to absorption in *P. americana* (Pillai and Saxena, 1959), a transformation which would accelerate uptake. It also accounts for the presence of trehalase in the midgut wall (Friedman, 1978) which presumably ensures the re-cycling of any sugar lost from the haemolymph down the trehalose concentration gradient into the gut lumen. Evidence against the existence of a Na^+-non-electrolyte glucose carrier is provided by the observation of O'Riordan (1968) that the transepithelial potential of the midgut was unaffected by phlorizin, a substance that inhibits the formation of this carrier complex in the vertebrate intestine.

Little is known about amino acid absorption in *P. americana*, but the work of Treherne (1959) in *S. gregaria* has shown that the caeca are once again the primary sites of uptake. The mechanism also depends on the formation of a concentration gradient, but, in this case, by the prior efflux of water from the gut lumen. Thus, the absorption of amino acids follows rather than accompanies the active fluxes of inorganic ions and the attendant movement of water.

Lipid absorption in *P. americana* has received more attention than the absorption of sugars and amino acids and is in some ways now better understood. Early work showing that ingested lipids were readily accumulated by the cytoplasm of the foregut epithelium, and therefore implicating this organ in absorption (e.g. Sanford, 1918; Eisner, 1955) was challenged by Treherne (1958c) who showed that ^{14}C-tripalmitin or its digestion products were not absorbed to any significant extent from the crop, but instead disappeared rapidly from the caeca and anterior ventriculus. Since this observation was made, a number of investigators have examined lipid absorption in the midgut with a view to discovering whether the mechanism shows any parallels with the vertebrates. This approach has been fruitful. Although insects apparently lack emulsifiers the midgut epithelium secretes a lipase which resembles mammalian pancreatic lipase and is able to hydrolyse triglycerides in the lumen into a mixture of monoglycerides and free fatty acids which are then available for absorption (Bollade *et al.*, 1970; Hoffman and Downer, 1979a,b). Once incorporated into midgut tissue, the products of triglyceride hydrolysis are then resynthesized into diglyceride which enters the plasma component of the haemolymph, in all likelihood attached to a lipoprotein carrier (Hoffman and Downer, 1979a,b; Chino and Downer, 1979). Synthesis into diglyceride and efflux from the midgut in this form seems also to be the fate of unesterified fatty acids in the diet (Bollade and Boucrot, 1973; Chino and Downer, 1979). It is not yet known whether phospholipids act as transcellular carriers within the midgut epithelium, as in the mammalian intestine, but it may be significant that label from ingested triglycerides and free fatty acids was rapidly incorporated into the phospholipid fraction of the midgut wall, although not into the equivalent fractions of the haemolymph (Bollade and Boucrot, 1973).

The idea that the crop may be an absorptive site for lipids has been rehabilitated by Hoffman and Downer (1976) who showed that label from ingested 1-^{14}C palmitic acid and glyceryl tri(1-^{14}C) palmitic acid was able to cross the wall of the crop *in vitro*. The efflux consisted chiefly of free fatty acids and was enhanced by the addition of midgut extracts containing lipase to the crop contents. Some diglyceride synthesis occurred in the crop, but the importance of this in lipid absorption by the crop *in vivo* remains unclear. Parallel incubations of the midgut showed that absorption of the same labelled materials was more rapid at this site than in the foregut. The crop wall is also permeable to short-chain acids (e.g. acetate) which appear to be more rapidly absorbed than the long-chain variety (Hoffman and Downer, 1976).

Joshi and Agarwal (1976) reported that a substantial proportion of ingested cholesterol was incorporated into the epithelia of the oesophagus, crop and caeca, and that it subsequently entered the haemolymph from some or all of these locations. Some of the incorporated sterol was present as cholesterol ester, an observation consistent with the report of Casida *et al.* (1957) that cholesterol esterase occurred in the epithelia of the crop and midgut, and was active in both the esterification of cholesterol and the subsequent hydrolysis of its esters. Esterifiction of cholesterol in *Eurycotis floridana* is said to facilitate absorption (Clayton *et al.*, 1964).

Preliminary work on the colon has shown that it is permeable to small organic acids (e.g. acetate, butyrate, propionate) of the kind generated by microbial fermentation (Bignell, 1977b; Bracke and Markovetz, 1980). Although the process is probably diffusive, the structure of the epithelium is consistent with an active transport process (Bignell, 1980).

4.6 Microbiology and pathology of the gut

Since comprehensive accounts of the biotic associations of cockroaches have been given by Roth (Chapter 1), only those recent investigations which bear on the organization and function of the gut will be discussed here. Hominick and Davey (1973, 1975) studied the spatial distribution of oxyuroid nematodes in the anterior colon and found that the mouths of two species, *Leidynema appendiculata* and *Hammerschmidtiella diesingi* were radially segregated such that *L. appendiculata*, which ingests relatively large particles, occupied the central portion of the lumen while *H. diesingi*, a fluid feeder ingesting mainly bacteria, was restricted to a position near the gut wall. This offers indirect support for the suggestion of Bignell (1977a) that the periphery of the anterior colon is a relatively stable area in which large, semi-permanent populations of bacteria may become established. The morphology of bacteria attached to the colon wall and to the cuticular spines arising from it has been described very recently by Bracke *et al.* (1978, 1979). Organisms isolated from this site included facultative and obligate anaerobes representing the genera *Bacterioides*, *Clostridium*, *Eubacterium*, *Fusobacterium*, *Peptococcus* and *Peptostreptococcus* in addition to the methane-producing bacillus *Methanospirillum*. Bacteria isolated both aerobically and anaerobically from the midgut and hindgut were capable of carboxymethylcellulose degradation *in vitro* (Cruden and Markovetz, 1979), but no organisms were found adhering to the walls of the foregut or midgut (Bracke *et al.*, 1979).

The growth and viability of anaerobic bacteria in the cockroach hindgut must increase concern about the hazards to health posed by this insect. Early work showing the survival of pathogens in the alimentary canal (see Bignell, 1977a) has been confirmed by Burgess and Chetwyn (1979, cited by Burgess, 1979), who found that cockroaches of several species captured in hospital kitchens harboured pathological serotypes of *Escherichia coli*, *Klebsiella pneumoniae*, *Serratia marcescans* and *Pseudomonas aeruginosa*, in addition to many other potentially dangerous strains. Clearly, we cannot regard the cockroach as merely an unpleasant nuisance.

Abnormal microbial metabolism is probably the cause of melanotic and other non-neoplasmic lesions which occur in the gut wall (Bignell and Mullins, 1977). Such lesions occur most frequently in the colon, but may be induced in other regions of the gut by treatments such as severence of the recurrent nerve or anal blockage which restrict or abolish the passage of food (reviewed by Harshbarger and Taylor, 1968).

Acknowledgements

I thank Dr Anne O'Riordan for making her doctoral thesis available to me during the preparation of this chapter. I am also grateful to Dr Donald E. Mullins and Professor John E. Steele, who together introduced me to the sophistication of the American cockroach.

5 Respiration

P. L. Miller

5.1 Introduction

Periplaneta americana is a flat, broad insect specialized for hiding in narrow crevices during the day and for being highly active at night. It survives well in dry places, runs extremely rapidly, but flies only rarely in the wild. It can also jump, and it can swim on the water surface at considerable speed. Some of these features of its life are reflected in specializations of the tracheal system which is able to meet the high demands of active locomotion, restrict water loss to a minimum when it is quiescent, and perform a number of non-respiratory functions. Tracheal morphology is based on a repeating segmental plan linked by longitudinal trunks which unite the whole system. The basic plan is homologous throughout the Dictyoptera and is most highly developed in the Blaberidae (Baudet and Sellier, 1975). For an insect so popular in biological research, there have been surprisingly few studies of its respiratory system. No detailed account of the tracheal system is known, and its spiracular and ventilatory mechanisms have received less attention than those of many other species. I shall therefore include some mention of studies on other cockroach species, particularly of Blaberidae, to supplement what is known of *P. americana*, since the basic anatomy and physiology are probably similar throughout the order.

5.2 Rates of gaseous exchange

In many organisms, oxygen uptake is proportional to $m_b^{0.75-0.8}$ (m_b = mass), and this is thought to hold true for *P. americana* (Gunn, 1935). Resting rates of oxygen consumption in this species have been measured at 30°C as 0.38 ml g^{-1} h^{-1}, or about twice this value per gram of metathorax (Gunn and Cosway, 1942; Kubǐsta, 1966). However, basal metabolic rates are difficult to measure accurately in insects (Keister and Buck, 1974), and they vary with such factors as time of day, stage of development, proximity to moulting and previous activity. Rates are sometimes given in terms of the surface area, estimated to be 8.2 cm^{-2} g^{-1} in *P. americana*, which gives a value of 1.6 mg 0_2 cm^{-2} day^{-1}, but the same problems are encountered (Gunn, 1935). At 30°C, Richards (1963c) found that females consumed 30% more oxygen than males, but this probably varies considerably with the reproductive state. He measured a Q_{10} for oxygen uptake of 2.25 (15–25°C), but marked temperature adaptation may occur (Dehnel and Segal, 1956; cf. Sanchez, 1975). Kestler (1971) measured rates of CO_2 production in an inactive male *P. americana* weighing 677 mg which increased from 36μl h^{-1} at 10°C to 443 μl h^{-1} at 35°C. The mean Q_{10} over the range for this insect was 2.74.

Oxygen uptake has not been measured during the most characteristic form of locomotion of this species, fast running, but a few measurements have been made on flying cockroaches. If measurements are made on tethered insects in a wind tunnel, the rates may be subnormal because the cockroach may fail to generate positive lift. Moreover, the rate may vary considerably with wingbeat frequency, and this in turn alters both with the ambient and the thoracic temperature, there being no evidence for thermoregulation in flight in this species (Farnworth, 1972; Kammer and Heinrich, 1978). Increases in oxygen uptake by about 100 times have been measured in 15 min flights during which extensive glycogen breakdown occurred (Downer and Matthews, 1976a,b), and values of up to 93 ml (g flight muscle)$^{-1}$ h^{-1} have been recorded in *Blatta orientalis* (Hofmanová *et al.*, 1966; Poláček and Kubǐsta, 1960).

A considerable problem in measuring metabolic rates has recently been underlined by the work of Downer and his colleagues (Matthews *et al.*, 1976; Spring *et al.*, 1977; Downer, 1979a). Cockroaches which have been handled for a few seconds show stress-induced metabolic changes which may persist for over an hour. For example, there are marked increases in haemolymph levels of trehalose and glucose. Dosing insects with CO_2 may have similar effects, the CO_2 both causing hyperactivity and itself elevating glucose levels through action on haemocyte trehalases (see Chapter 3). Kestler (1971) found that CO_2 production took more than an hour to return to a resting rate after very intensive struggling (see Section 5.6), and his results given here and in Section 5.6 are from insects undisturbed for several hours.

5.3 Morphology of the tracheal system

5.3.1 The tracheal plan

Miall and Denny (1886) have given a useful account of the tracheal system of *B. orientalis*, and Haber (1926) has described that of *Blattella germanica*.

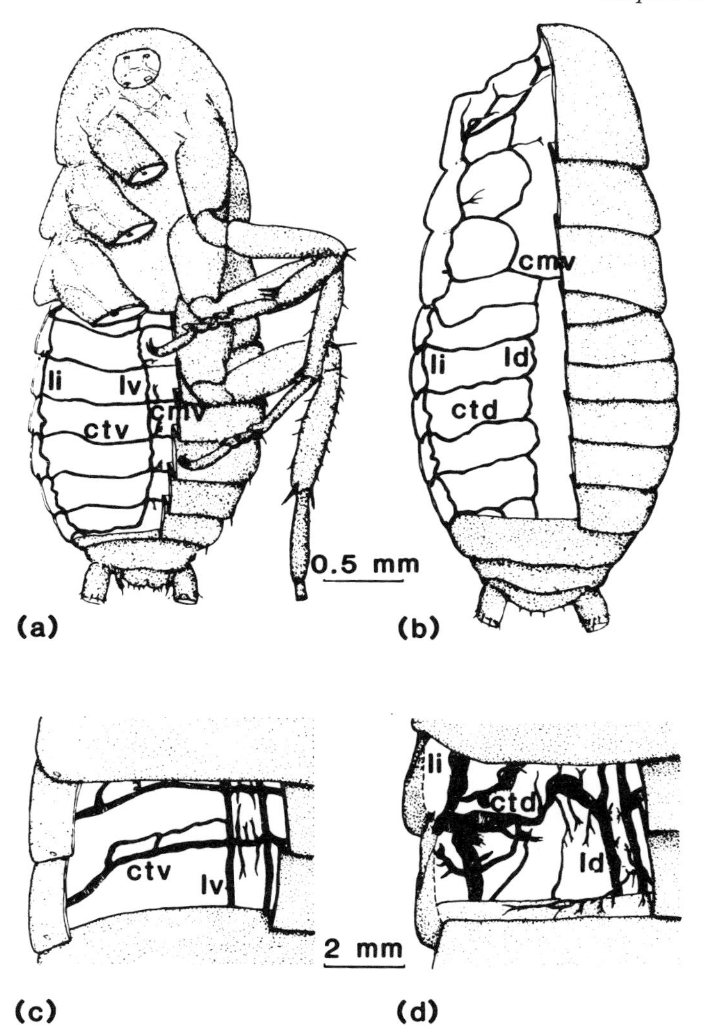

Fig. 5.1 Tracheal morphology of *P. americana*. (a) Ventral view of a partly dissected first-instar nymph. (b) Dorsal view of a first-instar nymph. (c) Ventral view of parts of two abdominal segments in an adult. (d) dorsal view of the same as in (c). cmv, ventral commissure; ctd, dorsal transverse connective; ctv, ventral transverse connective; ld, ll and lv, dorsal, lateral and ventral longitudinal trunks. (From Baudet and Sellier, 1975.)

Only recently, in the studies of Baudet and Sellier (1975), has tracheal morphology in *P. americana* received much attention. They showed that the blattid tracheal system was the most primitive cockroach type, blatellids had an intermediate form, and blaberids a more advanced and specialized one. They distinguished tracheae which formed trunks, commissures and connectives joining other tracheae together, from those which led to tracheoles and supplied tissues. As with most insects, interconnections in *P. americana* permit the whole system to be supplied from any single open spiracle.

On either side, a lateral longitudinal trunk connects with each spiracle through a short side branch (Fig. 5.1). A ventral trunk runs on each side of the CNS and the two ventral trunks are joined by segmental interconnections. A dorsal trunk runs likewise on either side of the heart, but here the two trunks have few interconnections. Over all, the three longitudinal pairs of trunks are well interconnected and no isolated region of the system, comparable to those in the thorax of locusts, occurs in cockroaches. Bilateral interconnecting trunks are strongly developed ventrally and dorsally in blaberids, and in all cockroaches, the dorsal ones are dilated to provide a voluminous abdominal air reservoir (Fig. 5.2). Other tracheae in many regions, including the legs, are also dilated, and they may act mainly as ventilatory bellows, being deformed during expiration as can be seen through regions of transparent cuticle. Air-sacs, defined as swollen regions of the tracheal system without organized taenidia, are lacking in cockroaches (Faucheux and Sellier, 1971).

Baudet (1974a) and Baudet and Sellier (1975) have examined the tracheal morphology of nine species of cockroach belonging to three families. They find the complexity of the system to be related both to taxonomic position and to adult size. All the basic features of the adult can be found in the first instar, and during development more tracheal branches, many more tracheoles, and further regions of dilated tracheae are added. The adults of small blattids may show less complexity than the first instars of advanced families such as blaberids. Likewise, the systems in similarly sized adults of *B. orientalis* and *Diploptera punctata* differ considerably, in line with their different taxonomic positions. Tracheal development and the number of dilations are not related to the powers of flight, nor to the existence of ovoviviparity or oviparity, but only to taxonomic position and size.

Little is known of the mechanism and control of tracheal development in cockroaches, but it is probably dependent, as in other species, both on genetic instructions and on responses to local demands acting through ecdysterone, lactate levels, tissue factors or by active competition between epithelial cells for tracheoles (Locke, 1958; Pihan, 1971, 1972; Wigglesworth, 1977; Ryerse and Locke, 1978). Young cockroaches are known for their considerable powers of limb regeneration, and tracheal invasion of regenerates has been examined by O'Farrell and Stock (1958; see also O'Farrell, 1959), who showed that new tracheae are generated from small existing branches and not from the stump of the main trunk (see Chapter 16).

Estimates of the total volume of the tracheal system, made by measuring the rate of wash-out of an inert gas, argon, have been made for *Byrsotria fumigata* and give a value of 82.0 ± 7.0 $\mu l\ g^{-1}$ for a male weighing 1.82 g, and 127 $\mu l\ g^{-1}$ in a nymph of 0.9 g. In *Gromphadorhina brunneri*, a value of 83 $\mu l\ g^{-1}$ has been calculated, a surprisingly high value considering that it is an apterous species (Bridges and Scheid, personal communication). These values can be compared to the value of 48 $\mu l\ g^{-1}$ (range 39–59) in the pupa of *Hyalophora cecropia* calculated by the same method (Bridges, Kestler and Scheid, in preparation). Published data indicate that

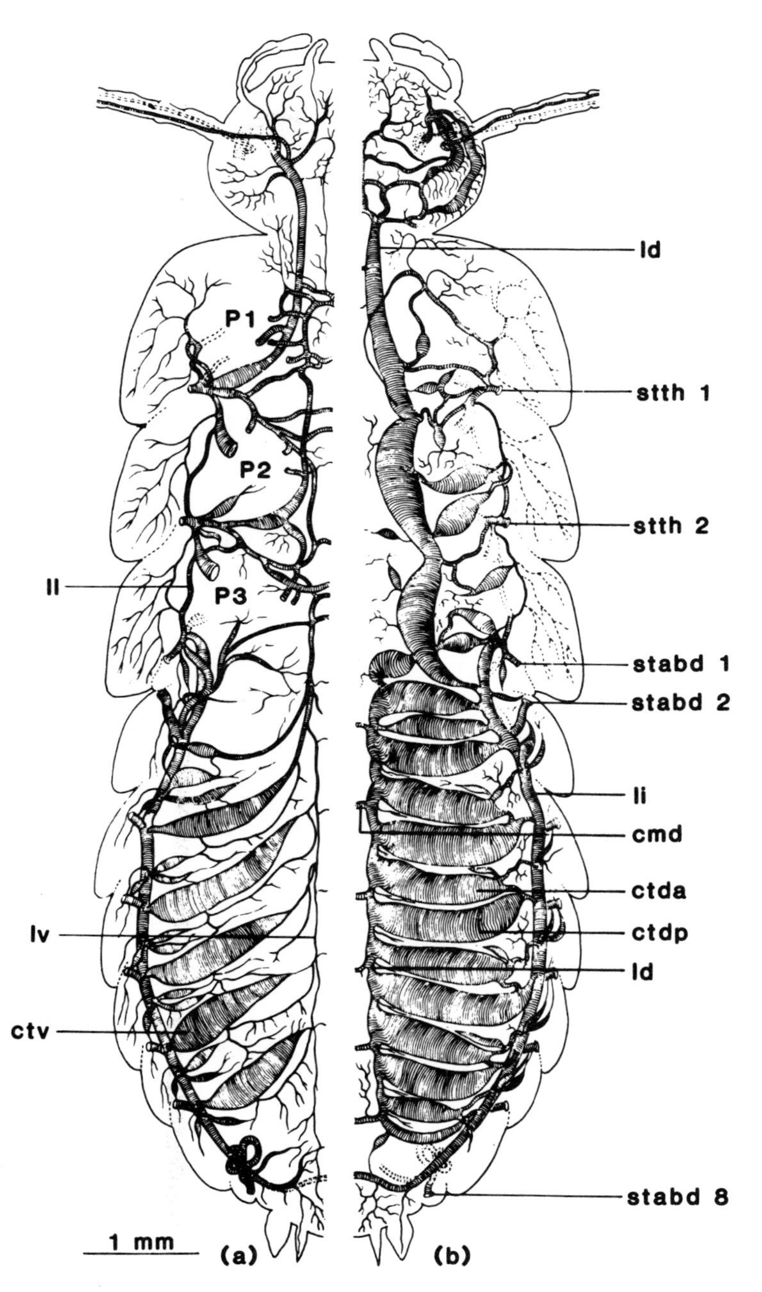

Fig. 5.2 Tracheal morphology of first-instar *Gromphadorhina portentosa*. (a) ventral view; (b) dorsal view. cmd, dorsal commissure; ctda, ctdp, anterior and posterior dorsal transverse connectives; ctv, ventral transverse connective; ld, ll and lv, dorsal, lateral and ventral longitudinal trunks; P1, P2 and P3, first, second and third legs; stabd., abdominal spiracle; stth, thoracic spiracle. (From Baudet and Sellier, 1975.)

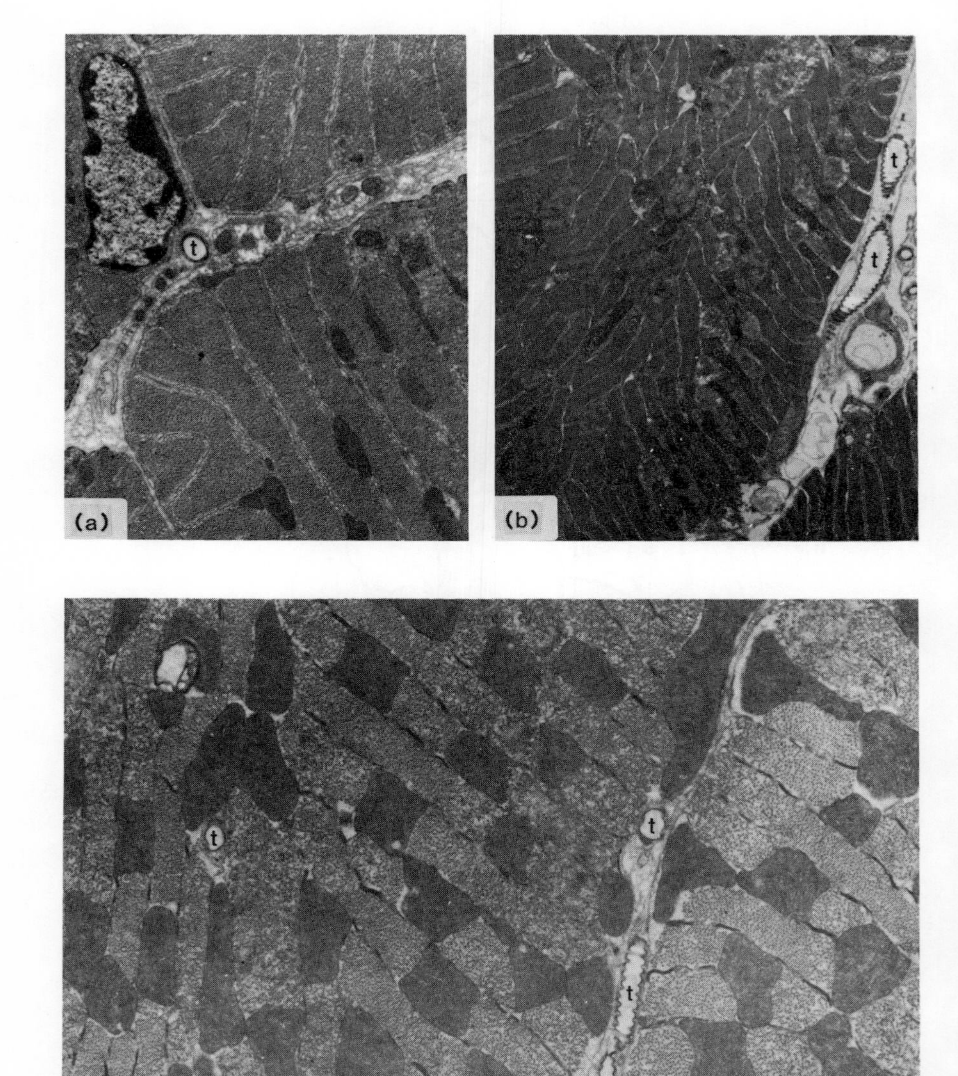
t
(a)
t
t
(b)
t
t
t
(c)

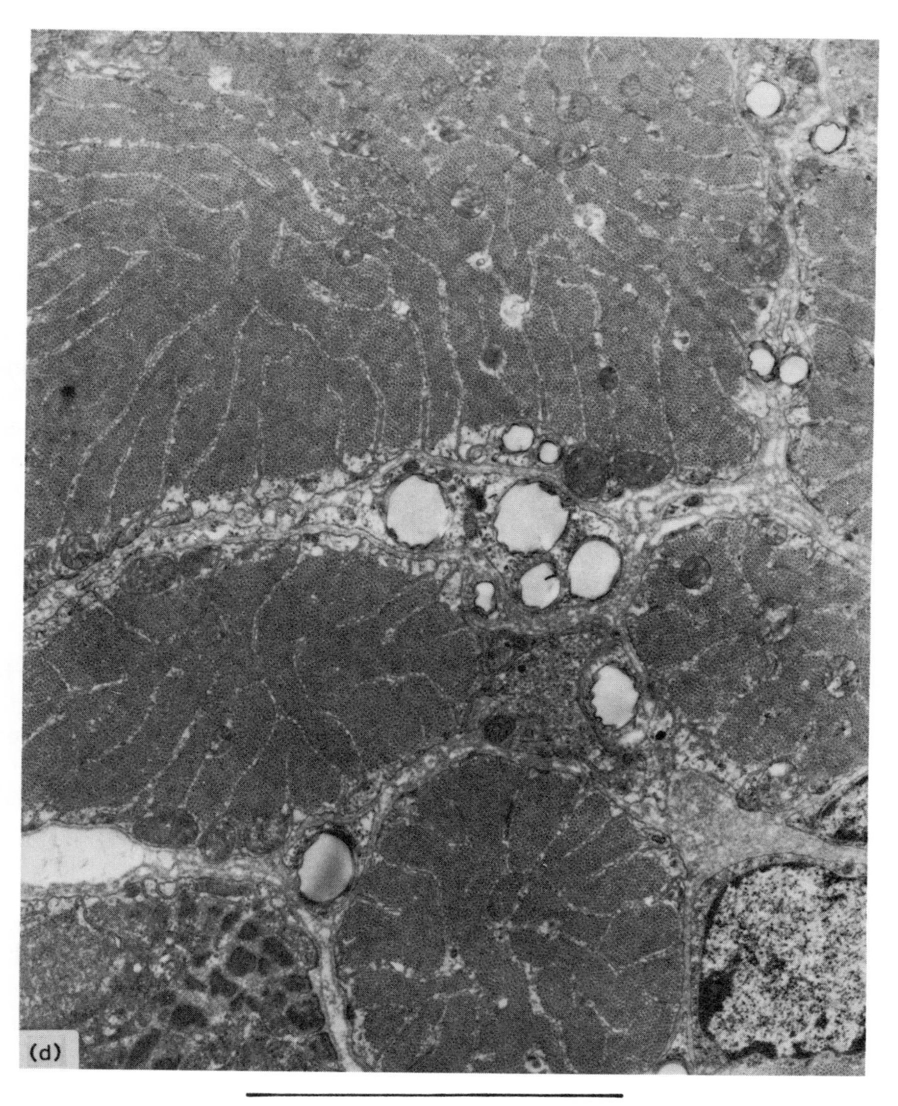

Fig. 5.3 Electron micrographs of muscles of *P. americana* to show features of the tracheal supply. (a) Dorso-ventral abdominal muscle. (b) Tibial muscle. (c) Thoracic tergo-sternal muscle. Only in the last example, a flight muscle, are tracheoles (t) found indenting the muscle fibres. Elsewhere they run between fibres. (d) Closer muscle of spiracle 1, showing parts of several fibres and the abundant tracheae and tracheoles which ramify between them. The rich tracheal supply of this muscle may be associated with its responsiveness to tracheal CO_2 levels. Scale: a and c, 5 μm; b, 8 μm; d, 5 μm.

the tracheal volume of 5th instar *Locusta migratoria* is 265 μl g^{-1} (Clarke, 1958). In general, the figures of air volume per unit body mass for flying insects are higher than those for the lung volumes of vertebrates, probably because of (a) the presence in vertebrates of an efficient blood circulation with a respiratory pigment, (b) the high oxygen demands of insect flight, and (c) non-respiratory benefits derived by insects from filling internal space with air (Section 5.3.3).

5.3.2 Fine structure

General reviews of the structure and distribution of tracheoles have been given by Whitten (1972) and Locke (1974). Tracheoles taper to about 0.4 μm in *P. americana* and end blindly either among tissues or by indenting the plasma membrane of cells with high metabolic demands such as in thoracic musculature (Richards and Korda, 1950). This brings oxygen close to the mitochondria. In detail, the structure of cockroach tracheoles does not differ from that of other species, the tracheoles being ensheathed in a thin tracheole end-cell and strengthened by helical or circular taenidia (Fig. 5.3). Thoracic spiracle-closer muscles have a rich but non-indenting tracheal supply, perhaps related to their responsiveness to tracheal gases (Fig. 5.3d: Section 5.4.1).

5.3.3 Non-respiratory functions of the tracheal system

The large dorsal abdominal air reservoir is well developed in most cockroach species. It may serve as ventilatory bellows, but also probably allows for volume changes and movements of internal organs without deforming the cuticle or compressing other regions. For example, food intake, fat deposition and the rapid development of the oötheca in females can be accommodated by the partial collapse of the dorsal tracheae. Air-filling of internal space also allows insects to become large without necessarily incurring a corresponding weight penalty, an advantage not only to flying species. In starved cockroaches, and also at moulting, the gut may become inflated with air which further helps to maintain abdominal volume. Other non-respiratory functions include sound production and the expulsion of noxious chemicals through spiracles (see Section 5.5.6). The possible role of the tracheal system in temperature regulation has not been explored in cockroaches.

5.3.4 Spiracle structure

P. americana possesses ten pairs of functional spiracles. In some other cockroach species spiracles 3 and 10 may be reduced in size and possibly non-functional. All spiracles are closed by valves or lips operated either by an opener and a closer muscle, or by a closer muscle which acts against cuticular elasticity. Each of the large thoracic spiracles (1 and 2) has two projecting valves which are closed inwards (Fig. 5.4). Abdominal spiracles (3–10) are all recessed and possess valves or lips which

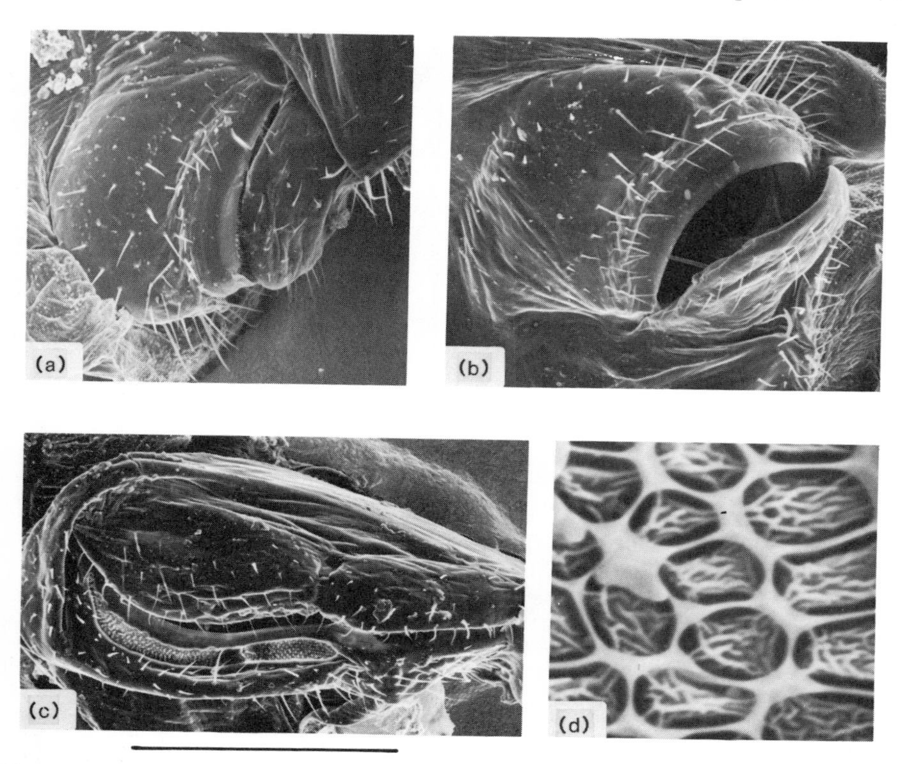

Fig. 5.4 Scanning electron micrographs of external views of thoracic spiracles of *P. americana*. (a) Spiracle 2 with the valves closed. (b) Spiracle 2 with the valves open. (c) Spiracle 1. (d) Part of the honeycomb structure inside the posterior valve of spiracle 1 (seen in c). Scale: a, b and c, 500 μm; d, 31 μm.

open inwards (Fig. 5.5). Every spiracle has a honeycomb-like structure lining the atrium either inside or outside the valves. It takes the form of a network of elevated ribs defining small areas more or less hexagonal in shape, within which the cuticle appears to be thrown into many minute folds (Fig. 5.5d). It is a hydrofuge structure and retains a silvery layer of trapped air when other parts are wetted. It may act to resist water entry into the tracheal system. Spiracle valves all open to reveal a narrow slit, and small variations in position probably have a large effect on gas flow past them. Closed valves can sometimes be seen to be sealed by a lipid-like material.

Spiracle 1 opens between the pro- and mesothorax on a peritreme which rests on membranous cuticle. It has two bi-lobed convex valves which project above the surface, open elastically and are both closed inwards by the action of a muscle which lies inside the anterior valve and is attached to the peritreme. The valves open into a tripartite atrium from which five major tracheal trunks are given off, three anteriorly and two posteriorly.

Spiracle 2 lies on a separate peritreme between the meso- and metathorax.

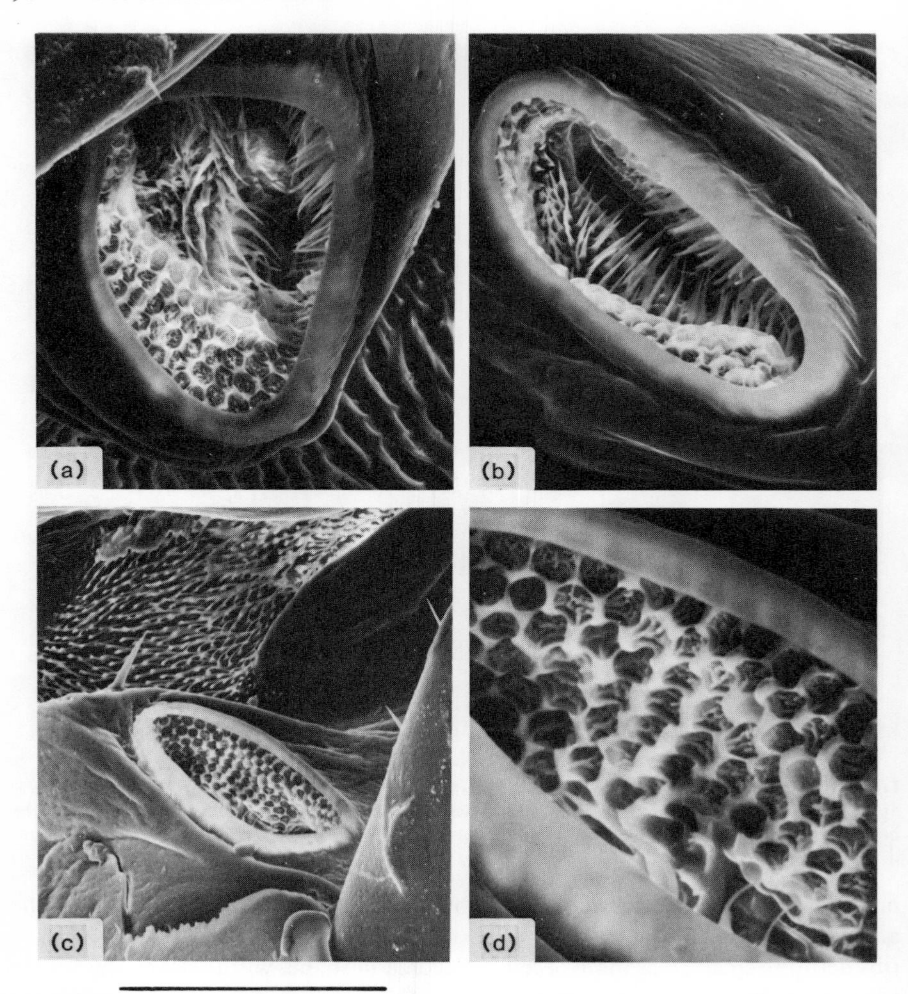

Fig. 5.5 Scanning electron micrographs of external views of abdominal spiracles of *P. americana*. (a) Spiracle 10: note the honeycomb structure and the bristles. (b) Spiracle 7: note the valves below the outer atrium. (c) Spiracle 4, an angled view to show the honeycomb structure on the atrial wall. (d) An enlarged view of the honeycomb structure. Scale: a and b, 125 μm; c, 250 μm; d, 62.5 μm.

The two projecting convex valves open elastically and are closed by a ventrally situated muscle which arises on the peritreme. They lead into a single atrium from which three major tracheae arise.

Spiracle 3 lies dorsally on the first abdominal segment and is considerably larger than the remaining abdominal spiracles. It is normally concealed under the wings in the adult, or under the wing buds in later nymphal instars. In principle, its closing mechanism is like that of other abdominal spiracles, but only a closer muscle is present and it opens elastically. Shankland (1965) described an opener muscle, but this seems to be incorrect. The closer muscle draws the two lips together mainly by pulling the outer one inwards.

Spiracles 4–10 lie in the anterior lateral part of their segments between the anterior and posterior parts of the paratergites. All open via a short atrium through a nozzle which faces posteriorly. In contrast, blaberid abdominal spiracles have an opening which faces more ventrally. The outer atrium is lined with honeycomb structure, and a thick felt of bristles (Fig. 5.4) partly covers the external parts of the lips. The anterior lip has a sclerotized bar which extends into a manubrium to which a short closer muscle and a longer and broader opener are attached (Fig. 5.6). The opener in some blaberid spiracles is divided into two morphological regions which may represent tonic and phasic parts of the muscle (Miller, 1973; Kaars, 1979; Nelson, 1979).

5.3.5 Spiracle innervation

Information on spiracle innervation in cockroaches is based on physiological and anatomical examination of several species, but in none has every spiracle been examined (Schmitt, 1954; Case, 1957; Shankland, 1965; Miller, 1969, 1973; Kaars, 1979; Nelson, 1979). Fig. 5.7 gives a tentative summary of what is known with details assembled from different species.

The closer muscles of all spiracles receive two unpaired axons which arise in the ganglion lying anterior to the spiracle, emerge in a posterior median nerve and then travel in right and left transverse nerves (Fig. 5.8). By recording and by the introduction of cobalt chloride, each axon has been shown to divide so that a segmental pair of spiracle closers receives identical excitation. Impulses in the axons produce EPSPs which may vary from 1–10 mV in amplitude. Paired inhibitory axons also supply the closer muscles of spiracles 1 in *Blaberus discoidalis*, and of spiracles 4 and 5 in *Gromaphadorhina portentosa*, and this may be a general feature of cockroach spiracle innervation. In *G. portentosa*, each arises from a contralateral soma in the ganglion posterior to that supplying excitatory axons to the same muscle. Their unilateral firing may allow right and left closers of a segmental pair to have some independence of action.

The opener muscles of spiracles 4–10 are each supplied by one paired 'fast' axon which in *G. portentosa* arises close to the inhibitory neurone of the same spiracle (Nelson, 1979). There may also be a second excitatory axon which produces much smaller responses in the muscle. The abdominal median nerves are associated with neurosecretory cells whose neurohaemal organs lie at the base of the transverse nerves (Case, 1957; Smalley, 1970). In *P. americana* it has been suggested that neurosecretions have some significance for spiracle activity (Bhatia and Tonapi, 1968). Innervated bristles lie in the vicinity of thoracic spiracles and their stimulation causes reflex valve closure. Sensory axons enter the segmental ganglion in a lateral nerve and excitation is relayed anteriorly to the next ganglion where it excites the closer motor neurons.

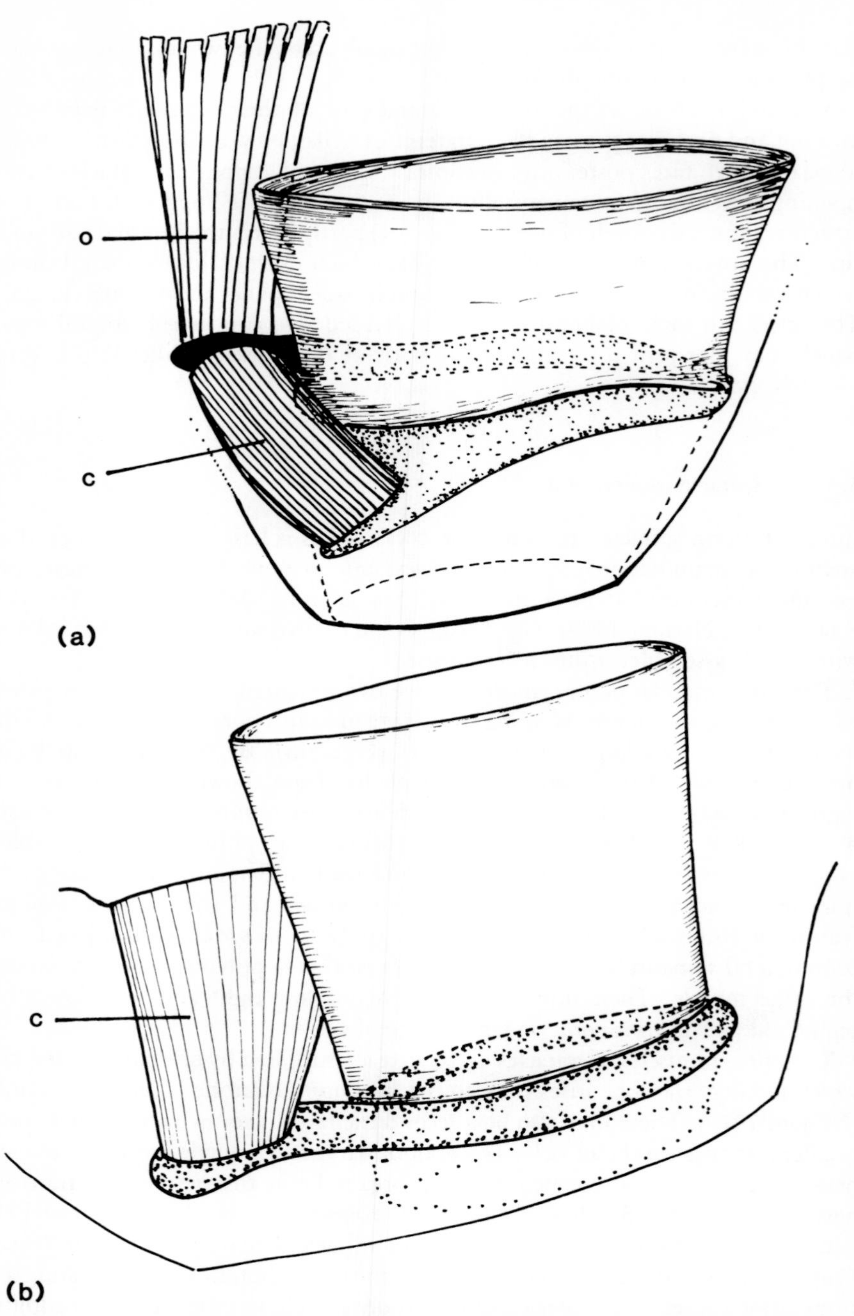

Fig. 5.6 Diagrams of abdominal spiracles of *P. americana* to show the opener (o) and closer (c) muscles which act on the manubrium to control valve position. (a) Spiracle 7. (b) Spiracle 3 which lacks an opener muscle.

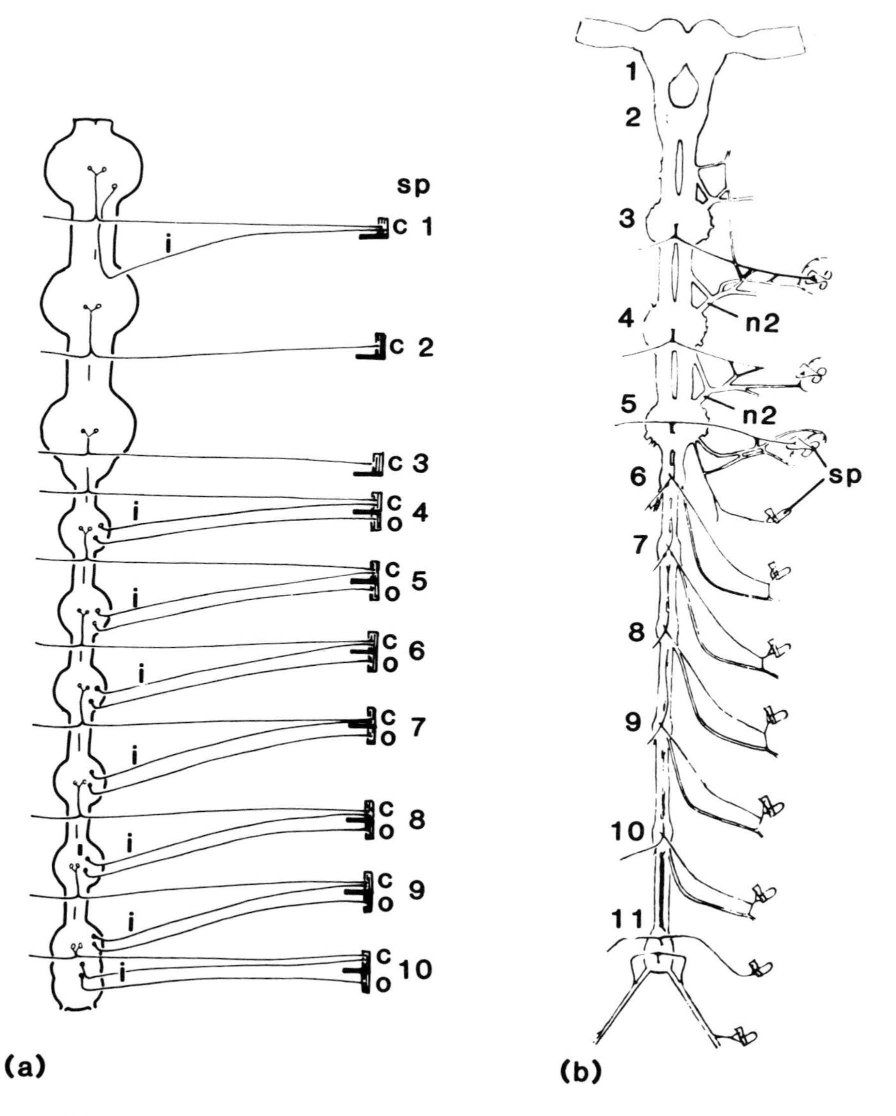

Fig. 5.7 The postulated innervation of the spiracles. (a) A summary of the supposed excitatory and inhibitory axons (i) supplying the opener (o) and closer (c) muscles of spiracles (sp) 1–10. A pair of excitatory axons leaves the median nerve and supplies the closer muscle in each segment. A single, paired excitatory axon is shown supplying the opener. The existence of inhibitory axons to spiracles in *P. americana* is unconfirmed, but spiracle 1 of *Blaberus discoidalis* receives an inhibitory axon arising in the prothoracic ganglion (Miller, 1969), while the closers of spiracles 4 and 5 of *Gromphadorhina* are known to receive paired inhibitory axons (Nelson, 1979). It is suggested in the diagram that all abdominal spiracles receive such an inhibitory supply. (b) A summary of the nerves supplying spiracles and associated regions in *P. americana*. n. 2, nerve 2; 1–11, ganglia of the CNS; sp, spiracle. (From Guthrie and Tindall, 1968.)

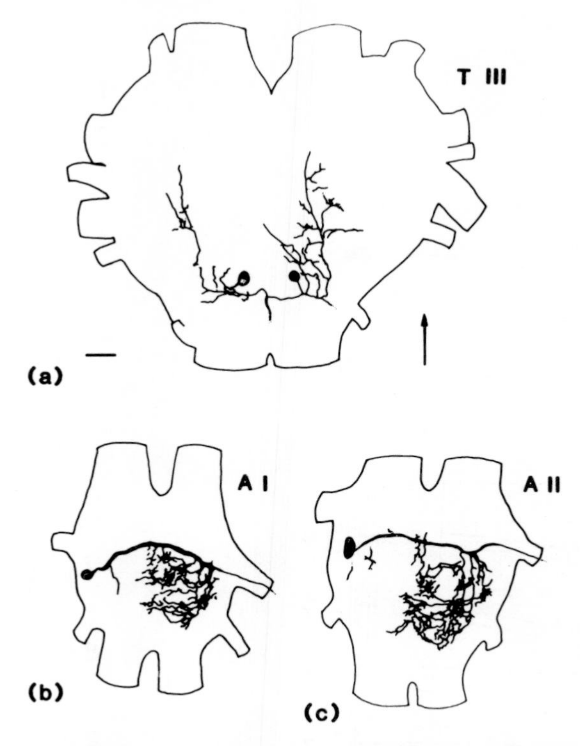

Fig. 5.8 Camera lucida drawings of motor neurons stained with cobalt sulphide to spiracles in *Gromphadorhina portentosa.* (a) Metathoracic ganglion, showing the two excitatory neurons which emerge in the median nerve and supply the closer muscle of the left and right spiracle 4. (b) A paired motor neuron in the first abdominal ganglion which supplies spiracle 4. (c) A similar motor neuron in the second abdominal ganglion which supplies spiracle 5. These neurons which have contralateral somata represent either the excitor to the opener, or the inhibitor to the closer muscle. Scale: 100 μm. (From Nelson, 1979.)

In summary, closer muscles receive excitation in two unpaired axons which emerge in the median nerve, and they may also receive inhibition from a paired neuron. Opener muscles receive excitation in one or two paired axons. The control of thoracic spiracles is co-ordinated by one ganglion (the inhibitory and excitatory neurons of *B. craniifer* spiracle 1 both arise in the prothoracic ganglion), but that of abdominal spiracles is shared between two adjacent ganglia.

5.4 Spiracle activity

Most mechanical or optical methods of registering spiracle valve movements, which include the use of small mirrors, transducers and reflected laser beams and photocells, give readings which are hard to relate precisely to valve position. Moreover, since cockroach thoracic spiracles ride on soft cuticle they are often moved by pressure changes without opening. A better method may therefore be to observe valves directly through a microscope, or to use a video-camera and monitor movements subsequently on a screen (Kestler, personal communication).

Spiracle control can be divided into three types according to whether responses arise independently and locally, at the segmental level, or are integrated throughout the insect.

5.4.1 Independent spiracle action

One-muscle spiracles in many species are known to be able to respond independently to CO_2 in spite of maintained tonic motor impulses from the CNS (Hazelhoff, 1926; Kitchel and Hoskins, 1935; Miller, 1962; Burkett and Schneiderman, 1974). Hoyle (1960, 1961) investigated the mechanism in spiracle 2 of the locust where CO_2 seems to act both at the neuro-muscular junction and also on the muscle membrane causing the muscle to relax and the valve to open elastically. Spiracles 1–3 in *P. americana* may also act in this way (cf. Schreuder and de Wilde, 1952).

Some spiracles of this type are able to remain active after denervation. In *P. americana*, Case (1956, 1957) noted that spiracle 2 closed 24 hours after section of the transverse nerve and it remained closed indefinitely. Treatment with more than 15% CO_2 was needed to cause spiracle opening and subsequent reclosure occurred after a long interval. He described visible contractions and relaxations of groups of muscle fibres (fasciculation) which started 4–6 days after denervation and may have been caused by neural degeneration. Unlike some other denervated muscles, the closer remained electrically excitable throughout. Nerve regeneration and the return of normal function took 16–32 days in 36% of his preparations, but in others it was not complete after 3 months. In the locust, Hoyle (1960, 1961) suggested that maintained closure was due to a potassium contracture, but van der Kloot (1963), working on denervated spiracles of *Hyalophora cecropia* pupae, believed that regenerative muscle potentials were responsible for the maintained activity.

5.4.2 Segmental control

Much spiracle activity may be controlled locally from the segmental and possibly an adjacent ganglion (Hazelhoff, 1926). In dragonflies (Miller, 1964), and in saturniid pupae (Burkett and Schneiderman, 1974), ganglia may respond directly to lowered oxygen levels and cause the spiracle valve to make fluttering movements. Fluttering is characteristic of spiracle activity at certain phases of respiratory cycles in pupae and in quiescent cockroaches (Section 5.6), and it may perhaps be brought about by a central response to hypoxia interacting with spiracle-mediated responses to CO_2. No separate tracheal receptors responding to O_2 or CO_2 levels are known, and all responses seem to occur at the ganglion or spiracle levels.

5.4.3 Intersegmental control

During ventilation most spiracles display distinct movements accurately synchronized with certain phases of the abdominal pump. These may result in an at least partially directed flow of air through the tracheal system (see Section 5.5.5). Coupled activity in spiracle 2 is brought about by high-frequency bursts of motor impulses in both excitatory axons to the closer, which coincide with the expiratory stroke (Case, 1957). A burst may be followed by a silent interval, perhaps due to post-excitatory depression, with consequent valve opening, before the tonic 'resting' pattern of firing is resumed.

Other forms of intersegmentally controlled spiracle activity include the sudden opening of thoracic spiracles in response to strong stimuli, and various co-ordinated patterns of movements which accompany moulting.

5.5 Ventilation mechanisms

5.5.1 The organization of pumping movements

No detailed examination has been made of the participation of abdominal muscles in the ventilatory cycle of *P. americana*. Two types of ventilatory movement can be distinguished. The first consists of dorso-ventral expiratory strokes brought about by contractions of pairs of tergo-sternal muscles (29) in each segment, followed by inspiratory elastic recoil due mainly to the resilience of the bowed sterna. The second type occurs in the longitudinal plane synchronized with the first in hyperventilating insects. Longitudinal expiration is brought about by up to 5 dorsal and 2 ventral longitudinal muscles in each segment, while inspiration may be partly elastic, and partly by the synchronized contractions of the dorsal and ventral lateral external muscles (18 and 26) in each segment. These extend one segment on the next and so protract the abdomen (Shankland, 1965). Activity in the longitudinal muscles probably also accompanies dorso-ventral pumping and helps to resist abdominal extension. In *B. craniifer*, dorso-ventral and dorsal longitudinal muscles are both rhythmically active during normal expiration. The expiratory stroke is well synchronized in all segments during both slow and fast ventilation.

5.5.2 Ventilation in quiescent insects

Hazelhoff (1926) believed that resting *P. americana* did not ventilate and depended only on gaseous diffusion for oxygen. However, Paulpandian (1959) reported the occurrence of intermittent bouts of pumping in quiescent cockroaches at 29°C. Bouts lasted 3–4 min, comprised 20–30 strokes and occurred every 10 min. More recently, Kestler (1971) has shown that intermittent ventilation is characteristic of undisturbed *P. americana* and other large insects at a variety of temperatures (see Section 5.6). Each pumping cycle consists of a rapid expiratory stroke, a maintained compression or plateau and then inspiratory recoil. The pumping frequency was about 3.8 min^{-1} at 20°C, and this rose to 5.7 min^{-1} at 30°C. Intermittent ventilation leads to the release of CO_2 in bursts whose periodicity reflects the metabolic rate.

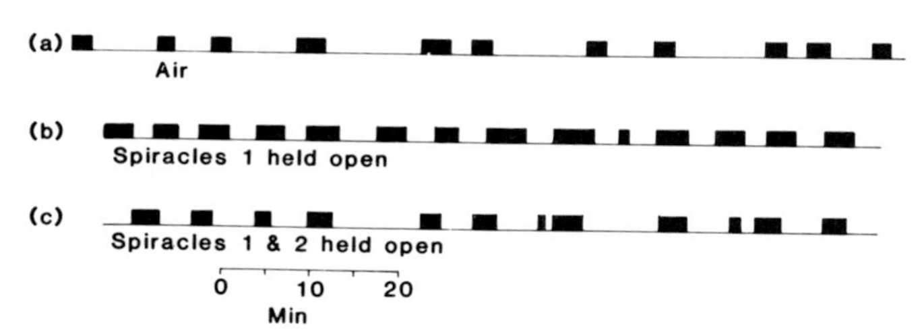

Fig. 5.9 (a) Records based on electromyograms from an abdominal dorso-ventral muscle of an unrestrained and quiescent last-instar *Blaberus craniifer*, showing periods of abdominal ventilation (shaded). (b) The same with spiracles 1 fixed open. (c) The same with spiracles 1 and 2 fixed open. In b and c, intermittent ventilation persists in spite of the opened thoracic spiracles. These records are not from the same insect, and changes in the duration or frequency of ventilatory bouts may not be significant.

Myers and Retzlaff (1963) measured intermittent ventilation in the Cuban blaberid, *Byrsotria fumigata*, and found that about 5 slow ventilation cycles were regularly followed by a 7-min non-ventilating interval. Similarly, in *Blaberus giganteus* and *B. craniifer* (Miller, 1966 and unpublished), intermittent ventilation occurs in all instars when undisturbed. The use of long leads to record electromyograms from dorso-ventral muscles has enabled ventilatory patterns to be recorded from unrestrained insects which are free to bury themselves (Fig. 5.9a). The proportion of time spent pumping in *B. craniifer* characteristically fell from about 35% in first instars to 18% in last nymphal instars at 20–22°C. It may rise again in the adult (Hogg and Rhodes, personal communication). Each bout of ventilation consisted of 5–30 strokes and the intervals separating bouts of ventilation were variable. Expiratory bursts tended to be either short (ca. 200 ms) or long (2–4 s) with few of intervening duration. Bouts of pumping sometimes started with short bursts and then switched to long ones towards the end, or they consisted only of long ones. Perfusion of buried cockroaches with about 5% oxygen in nitrogen caused most bursts to be short, but they were still grouped into bouts and the last few bursts were usually long (Fig. 5.10). Sometimes groups of short and long bursts alternated, the short ones occurring in pairs at twice the frequency of the long ones, apparently due to the splitting of each long burst to give two short ones. Perfusion with CO_2 mixtures, or with lower concentrations of oxygen, usually produced continual pumping with only short bursts appearing.

5.5.3 Ventilation in active and in stressed cockroaches

Pumping cycles stimulated by CO_2 or by activity are short and lack the tonic compression phase (expiratory plateau). An expiratory stroke is immediately followed by inspiration, and there is then an inspiratory pause before the next cycle.

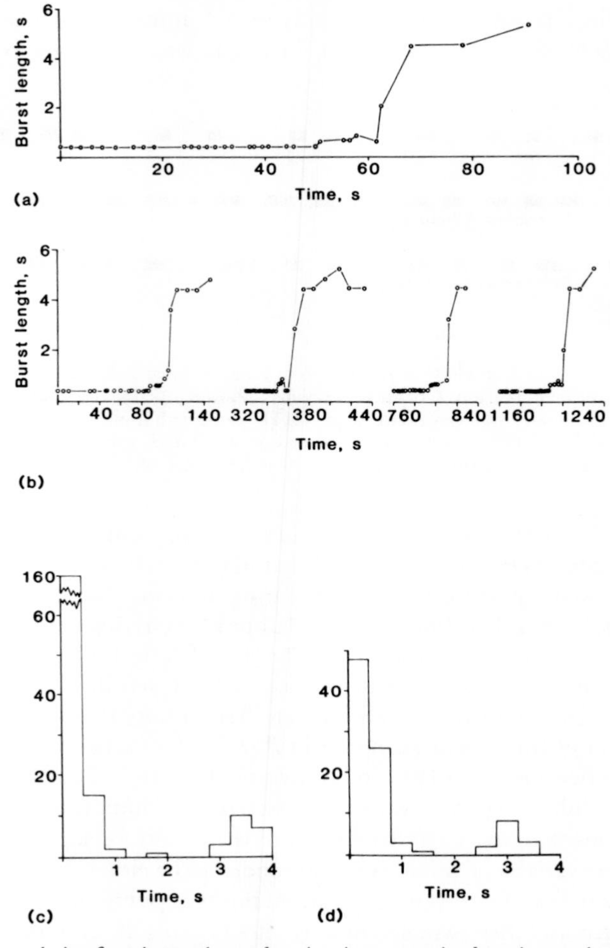

Fig. 5.10 The analysis of expiratory-burst duration in unrestrained, quiescent last-instar *Blaberus craniifer*, based on electromyographs from a dorso-ventral abdominal muscle. The cockroach was perfused with 5% oxygen in nitrogen and most bursts were short, but towards the end of each bout there was an abrupt transition to long bursts. (a) A single ventilatory bout. Each point represents one burst. (b) Four consecutive ventilatory bouts. (c and d) Histograms of the occurrence of bursts of different length from two insects.

According to Hazelhoff (1926), 5% CO_2 excites ventilation at 5 min^{-1}; in 10% the rate doubles; in 15% it reaches 90–120 min^{-1}, and in 20–30% it achieves a maximum of 150–180 min^{-1}, all at 28°C. In higher concentrations, the rate is reduced to 60–80 min^{-1} (Schreuder and de Wilde, 1952).

Records of neural activity show a delay of about 50 ms between the motor burst and the start of expiration. Expiration lasts 50–70 ms while inspiration takes 140–150 ms (Farley *et al.*, 1967), and, even at the highest frequencies, inspiratory pauses between cycles are nearly twice as long as the pumping cycle itself.

Rapid ventilation is also stimulated by handling or other forms of 'stress'. In *B. craniifer*, stress-ventilation can reach frequences of 4–5 Hz, and it may be more prolonged after minor tissue damage: it tends to occur at a relatively fixed frequency and then stops suddenly. It can be evoked by electrical stimulation of lateral nerves or a connective (Fig. 5.11). In a preparation consisting of the isolated metathoracic and abdominal ganglia (T3–A6), single shocks given to a lateral nerve evoked single bursts, but more prolonged stimulation initiated bursting which persisted at a high frequency for several minutes (Miller, unpublished). In *B. craniifer*, and also in *B. discoidalis* (Kaars, 1979), stress-ventilation is associated with altered spiracle coupling and it may be significant for defensive behaviour (Section 5.5.6), as well as for respiration. Thus stress-ventilation seems normally to be triggered by strong mechanoreceptor input and is less dependent on chemical stimulation.

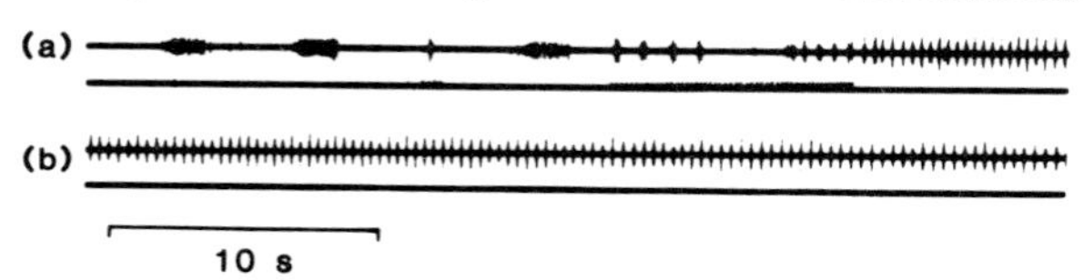

Fig. 5.11 (a) and (b) represent a continuous record of ventilatory activity, recorded from a dorso-ventral abdominal muscle of a last-instar nymph of *Blaberus craniifer* (upper records). Electrical stimulation was applied to a lateral nerve of the first abdominal ganglion (lower records). Four 1-ms shocks produced a short burst, interpolated among long bursts and apparently re-setting the rhythm. An 8-s burst of shocks (at 5.5 Hz) initiated a long period of short bursts which occurred at 3 Hz.

5.5.4 Auxiliary ventilation and autoventilation

Auxiliary forms of pumping are well developed in locusts where they include additional abdominal movements with coupled head and prothoracic activity. In *P. americana*, hyperventilation normally includes longitudinal abdominal strokes supplementing those in the dorso-ventral plane. High concentrations of CO_2 or severe activity may induce additional coupled movements in a variety of thoracic muscles and in those moving the head, but this is of uncertain ventilatory significance.

Autoventilation, or ventilation resulting indirectly from locomotor or other movements, is important in many insects including *P. americana* during flight. Fraenkel (1932) reported that abdominal ventilation did not occur during flight in this species and appeared afterwards only briefly. The abdomen stays in the expiratory position, but is elongated more than normal (Kestler, personal communication). During flight dorso-ventral excursions of the meso- and metanota with the wingbeats probably autoventilate thoracic tracheae through opened spiracles (cf. Weis-Fogh, 1967). Running and some other activities may also be to some extent autoventilatory.

5.5.5 Spiracle activity coupled to ventilation

In some Orthoptera, spiracles 1–4 open during inspiration and 5–10 do so with expiration, thereby directing air posteriorly through longitudinal trunks (Lee, 1924;

McArthur, 1929; Fraenkel, 1932). In the Hawaiian cockroach, *Nyctobora noctivaga*, Kitchel and Hoskins (1935) used a divided-chamber technique to measure the posteriorly directed airstream produced by coupled spiracle activity. The flow rate was greatest in young instars, reaching 9.6 ml g^{-1} h^{-1} in 0.4–0.8 g insects, but falling to 7 ml g^{-1} h^{-1} in 1.6–2.0 g insects. Flow could be maintained against a back pressure of 20 cm water implying efficient action by the spiracles, and it could increase by six times in 15% CO_2. Unlike CO_2, hypoxic conditions (induced by perfusion with nitrogen) caused a reversal of the flow in that species. *B. giganteus* may similarly reverse its airflow in nitrogen by altering the phase of coupling of some abdominal spiracles (Miller, 1973). Airflow measurements have been made only on *N. noctivaga*, but interpretations of supposed flow based on observations of spiracles have been made in a number of cockroach species.

With fast, CO_2-induced ventilation in *P. americana*, strong expiratory closing can be observed in spiracle 2 and expiratory opening in 4–10, while spiracles 1 and sometimes 3 remain open. During slow ventilation, however, spiracles 1 and 3 may become coupled, closing with expiration and opening with inspiration. Spiracle 2 usually opens immediately after expiration during the maintained compression (Kestler, 1971). Thus, airflow may be partly tidal and partly directed posteriorly, as occurs in ventilating mantids (Miller, 1973).

Fig. 5.12 Electromyograms from the right (R) and left (L) spiracles 10 of *Blaberus giganteus* showing the unilateral coupling of the left opener muscle to ventilation. Towards the end of the expiratory stroke, a high-frequency burst (large potentials) occurs in the left opener, while in the right there are only one or two potentials. Note also the cessation of small potentials (from the closer) in the right spiracle during the left opener burst. Closer potentials come from the median nerve and cannot differ on the left and right, but opener potentials arise in paired nerves and differ markedly on the two sides. s, seconds.

Thoracic spiracles are strongly coupled to ventilation in blaberids, and a posteriorly directed airflow may be a normal feature of their ventilation. In several species, unilateral abdominal spiracle coupling has been observed (Miller, 1973; Kaars, 1979). On the 'dominant' side, certain spiracles open towards the end of expiration emitting a blast of air, but their contralateral segmental partners remain closed. Dominance may change from side to side spontaneously, or may be induced by unilateral mechanical stimulation. Dominant spiracles exhibit high-frequency motor bursts in their opener muscles, while the contralateral spiracles show only abortive bursts or remain silent (Fig. 5.12). This type of coupling may ensure that air currents are directed across the abdomen in transverse tracheae with a significant reduction of the dead space (Kaars, 1979). Unilateral activity has not been observed in thoracic spiracles.

In *B. discoidalis*, spiracles 6, 7, 8 and 10 all function unilaterally as expiratory

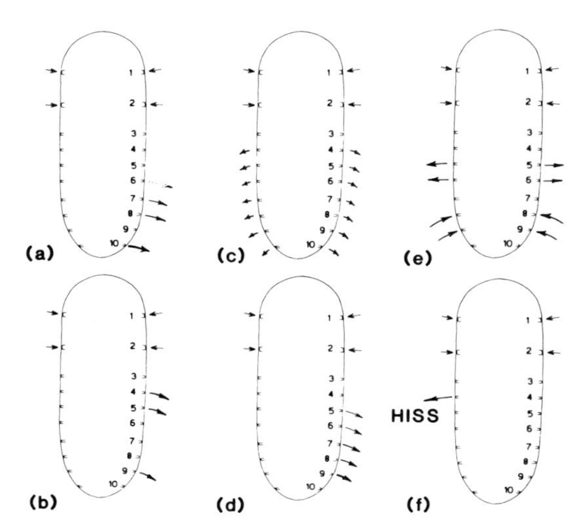

Fig. 5.13 A summary of the directions of airflow thought to be produced by spiracle activity coupled to ventilation in *Blaberus discoidalis* and *Gromphadorhina portentosa*. In all cases, spiracles 1 and 2 act bilaterally as inspiratory spiracles. (a) *B. discoidalis*, quiescent ventilation: spiracles 10, 8, 7 and 6 act unilaterally as expiratory spiracles, being recruited in that order. (b) *B. discoidalis*, stress-ventilation: spiracles 4, 5 and 9 open unilaterally with expiration. (c) *B. discoidalis*, CO_2-induced hyperventilation: spiracles 4–10 open bilaterally with expiration. (d) *G. portentosa*, quiescent ventilation: spiracles 9, 8, 7, 6 and 5 open unilaterally with expiration. (e) *G. portentosa*, reversed coupling: spiracles 8 and 9 open bilaterally with inspiration, and 5 and 6 bilaterally with expiration. (f) *G. portentosa*, stress-ventilation: spiracle 4 opens uni- or bilaterally with expiration (hissing). (Information mainly from Kaars, 1979.)

spiracles, but they have different thresholds and are recruited in the reverse order, spiracle 10 functioning alone in inactive insects (Fig. 5.13). During CO_2-induced hyperventilation, all spiracles 4–10 may be bilaterally coupled, but at other times spiracles 4, 5 and 9 are separately recruited in connection with defensive behaviour (Kaars, 1979; Section 5.5.6).

In *G. portentosa*, spiracles 5–9 normally function unilaterally as expiratory spiracles, again exhibiting differential thresholds in the reverse order. Spiracles 3 and 10 are reduced in size, while spiracle 4 is enlarged and is used in sound production (Dumortier, 1965; Section 5.5.6). Reversed coupling of spiracles 8 and 9 in which they open with inspiration has been reported. At such times, air may enter spiracles 1, 2, 8 and 9, and leave from 5 and 6 (Fig. 5.13e).

Thus, in blaberids, complex and varied patterns of spiracle coupling occur, and airflow may at different times be posteriorly or anteriorly directed, or tidal.

5.5.6 Non-respiratory functions of spiracle coupling

Some spiracles serve other functions in addition to acting as sites for gaseous exchange. During stress-induced ventilation (Section 5.5.3) in *B. discoidalis*, spiracles 6, 7, 8 and 10 remain closed while 4, 5 and 9 open unilaterally with expiration, the dominant spiracles being on the side mechanically stimulated (Fig. 5.13b).

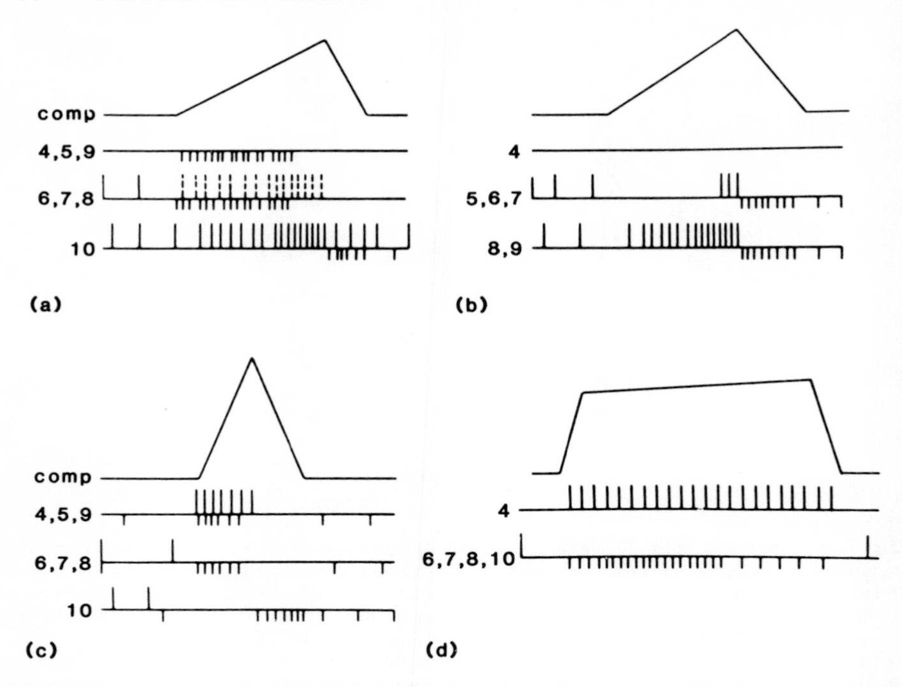

Fig. 5.14 A summary of the activity patterns in motor neurons of spiracles 4–10 of *Blaberus discoidalis* and *Gromphadorhina portentosa*. The time course of abdominal compression is shown on the top (comp), and the activity of motor neurons below. Upward strokes indicate opener spikes; downward strokes, closer spikes. Dashed opener spikes indicate variable activity. (a) Normal ventilation in *B. discoidalis*. (b) Normal ventilation in *G. portentosa*. (c) Stress ventilation in *B. discoidalis*. (d) Stress ventilation (hissing) in *G. portentosa*. (From Kaars, 1979.)

Spiracles 4, 5 and 9 are associated with long looped tracheal branches joining them to the longitudinal trunk. Glands may liberate noxious chemicals into the tracheal lumina and these may then be expelled by vigorous expiratory srokes through the opened spiracles (Kaars, 1979). Similarly, spiracle 4 in *D. punctata*, which is joined to the trunk by a 7 mm tracheal loop, opens to allow quinones to be expelled through it by ventilation (Roth and Stay, 1958; Roth and Eisner, 1962), and these have a defensive role.

Several cockroach species stridulate, for example, *Archiblatta hoevenii* does so by rubbing the wings against the abdomen (Dumortier, 1963) and *L. maderae* by rubbing the pronotum against the forewing (see Guthrie and Tindall, 1968, p. 249). *Gromphadorhina* spp. hiss by expelling air through the enlarged spiracles 4. Defensive hissing occurs in nymphs and adults of both sexes in this genus. Adult males also hiss in several ways during courtship, copulation and territorial behaviour (Dumortier, 1965). Hisses are broad-band sounds occurring at 2–20 kHz with a peak at 9 kHz (Nelson, 1979). They are produced by forcing air through the narrow origin of the 1 cm-long tracheal branch which joins the spiracle to the longitudinal trunk (Fig. 5.15; Baudet, 1974b). This sets up a turbulence and the

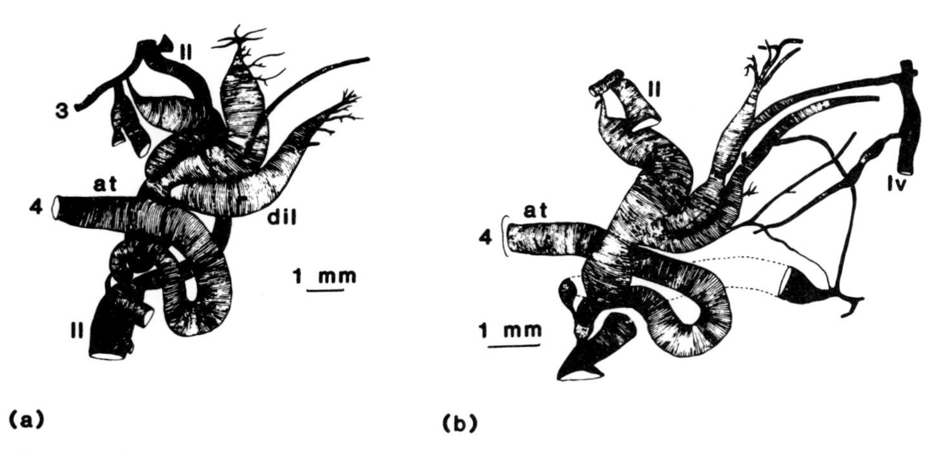

Fig. 5.15 Tracheae supplying spiracle 4 in *Gromphadorhina portentosa*. (a) Ventral view in an adult. (b) Dorsal view in a late nymph. at, branch to spiracle 4; dil, dilated tracheae; ll, lateral longitudinal tracheal trunk; lv, ventral longitudinal tracheal trunk; 3,4, spiracles 3 and 4. Note the narrow origin and the expansion of the looped branch to spiracle 4, important features in the production of sound. (From Baudet, 1974b).

sound is amplifed in the conical trachea which expands from 50–850 μm in diameter. Hissing is produced by a prolonged expiratory stroke during which all other spiracles are closed. Either one or both spiracles 4 may open. During normal ventilation, spiracles 4 receive a weak copy of the motor patterns to other spiracles and remain closed (Nelson, 1979) (Figs 5.13f and 5.14).

The similarity of the long branch tracheae and of the pattern of activation suggests that the *Gromphadorhina* spp. spiracle 4 may have evolved from a glandular spiracle acting in defensive behaviour as occurs in *D. punctata*. Moreover, odour release may accompany hissing in *Gromphadorhina*.

5.5.7 Origin and control of the ventilatory rhythm

The organization of insect ventilatory rhythms has recently been discussed by Kammer (1976). As in many species, the isolated CNS of *P. americana* can sustain a pattern of motor activity which can be identified as ventilatory by its form and by its responsiveness to CO_2 (Farley *et al.*, 1967). It may occur partly because the tracheal system in such preparations is damaged and the ganglia become hypoxic. Farley, Case and Roeder measured bursts at up to 180 min^{-1} in whole nerve cords (Br–A6) or in T3–A6 preparations, and sometimes they found that there were two simultaneous rhythms, one at 100–180, and the other at 8–12 min^{-1}. Similar observations have been made on *B. fumigata* (Myers and Retzlaff, 1963) and on *B. craniifer* (Miller, unpublished). In *P. americana* the activity was recorded as bursts of spikes either in lateral nerves or in the connectives: interneurons and motor neurons may both contribute to the latter as occurs in locusts (Lewis *et al.*, 1973). Similar activity was recorded occurring either spontaneously or in response to CO_2 from isolated metathoracic (T3) or second abdominal (A2) ganglia, and A3 was also rhythmically active at lower frequencies in a few preparations. Analysis of spikes in connectives between T3 and A2 indicated that, in two-thirds of the preparations, most were propagated posteriorly, but, in the remaining third, most travelled anteriorly. In one preparation, the direction of propagation showed spontaneous reversals. Farley *et al.* (1967) concluded that T3 normally acted as the dominant pacemaker, but infrequently A2 took on that role. In *B. craniifer*, Case (1961) assigned major roles to A1 and either adjacent ganglion (cf. Smalley, 1963). Thus, one of several ganglia in cockroaches may act at different times as the dominant pacemaker and occasionally two can do so simultaneously, when two rhythms appear.

To examine how activity is distributed to other ganglia so as to produce a nearly synchronous motor output in each segment, Farley *et al.* (1967) made simultaneous recordings from the T3–A1 and A4–A5 connectives. They recorded spike pairs a few ms apart and deduced that they were in interneurons which propagated impulses along the abdominal cord at 3 m s^{-1}, and thereby co-ordinated the ventilatory rhythm in every segment. Such interneurons could possibly be activated in more than one segment thus allowing different oscillators to control the rhythm.

Isolated cords not only generate ventilatory activity, but they can also organize it into bouts which resemble the patterns of intermittent ventilation seen in quiescent cockroaches. Myers and Retzlaff (1963) found that a Br–A6 preparation from *B. fumigata* showed intermittent patterns of bursting whereas, after removal of the brain and sub-oesophageal ganglion, bursting became continual. Removal of A6 then allowed the intermittent pattern to reappear. In *B. craniifer*, isolated ganglion chains can show similar patterns of intermittent bursting. In a T3–A6 preparation the pattern resembles that of ventilation in intact insects, but is usually more uniform, and the bout of ventilation occupies a greater part of the whole cycle (Fig. 5.16).

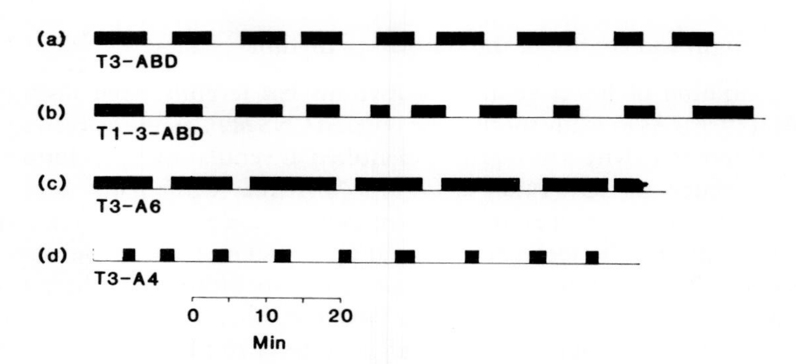

Fig. 5.16 Traces based on records from lateral abdominal nerves in various preparations of last-instar nymphs of *Blaberus craniifer*, showing in each the continued appearance of bouts of presumed ventilatory activity (shaded). (a) Metathoracic ganglion and intact abdomen. (b) All thoracic ganglia with lateral nerves cut, but abdomen intact. (c) Isolated metathoracic and all abdominal ganglia. (d) Isolated metathoracic and first four abdominal ganglia.

Thus isolated cords of blaberids apparently have the capacity to generate both the fast cycles of ventilation and the slow cycles of periodic ventilation, and the same has been shown to be true for *P. americana* (Hustert and Kestler, personal communication). This is discussed further in Section 5.6.

5.5.8 Sensory co-ordination of ventilation

The proprioceptive control of walking in cockroaches is quite well understood (Pearson, 1972) and it may be analogous in some respects to that of ventilation. Inspiration is primarily due to elastic recoil of the abdomen, but it may be assisted in the longitudinal axis by external lateral muscles. It is comparable to the swing phase or recovery stroke in walking and as such it is likely to be monitored by position detectors. Expiration, brought about by dorso-ventral and by longitudinal muscles, is comparable to the power-stroke of walking and it may therefore excite load receptors which can modify the power output. The power required for expiration and the duration of inspiration may both vary with the resilience of the cuticle, with the gut contents (food or air), and with the growth of internal organs such as the oötheca in the female. Stimulation of the load receptors may amplify the expiratory motor bursts, possibly by monosynaptic action on motor neurons, and it may also slow the ventilatory oscillator.

Farley and Case (1968) recorded activity in a lateral abdominal nerve which came from two proprioceptors in *P. americana*. One fired during expiration, and the other was active throughout expiration and inspiration. The latter showed phasic-tonic properties, responding both to velocity and position, and is probably one of the stretch receptors described by Finlayson and Lowenstein (1958) in this species. By mechanically imposing momentary inward or outward movements of the abdominal wall, they were able to evoke brief motor responses after which the ventilatory rhythm was reset. When the forced movements were repeated at near the natural ventilatory frequency, the rhythm became phase-locked to them. Essentially, the same result was obtained by applying electrical shocks to a lateral nerve of an isolated Br–A6 preparation, and the spontaneous rhythm could be either slowed or accelerated by appropriately timed shocks. These experiments indicate that the ventilatory oscillator can respond to phasic inputs and this may constitute part of its normal control mechanism. Such input can adjust the endogenous rhythm to the resonant frequency of the abdomen which may vary at different times as already described. It is interesting to note that both stress-ventilation in response to electrical stimulation (Section 5.5.3) and CO_2-induced hyperventilation can occur at higher frequencies in dissected insects or in isolated cords than in intact insects. For example, CO_2 can produce bursting at up to 250 min^{-1} in a Br–A6 preparation from *P. americana* (Farley *et al.*, 1967). This may be partly accounted for by the absence of normal proprioceptive input which, in responding to load, normally acts to slow the endogenous rhythm.

5.6 Diffusive-convective gas exchange in quiescent insects

5.6.1 Saturniid pupae

Schneiderman and his colleagues (Schneiderman, 1960; Levy and Schneiderman, 1966; Burkett and Schneiderman, 1974; cf. Miller, 1974a) established that CO_2-release from saturniid pupae occurs in a series of bursts, separated by many minutes or several hours during which oxygen uptake and aerobic respiration continue steadily. CO_2 is banked in solution largely as bicarbonate ions until the next massive release, and the activity is controlled by the spiracle valves. After a period when they are constricted (C), they start to open slightly and reclose with fluttering movements (F); then they open widely (O) permitting the release of a burst of CO_2 in what is termed here the CFO cycle. During constriction, pressure in the tracheal system falls sharply as oxygen is consumed. Each flutter then permits a small inflow of air and pressure returns stepwise towards atmospheric. The low tracheal oxygen level (about 3.5%) means that oxygen entry will be both by diffusion as well as by suction of air. This has been termed passive suction ventilation (Miller, 1974a) or flow diffusion (Buck, 1958). Since there is probably no flow-diffusion steady-state, a better term may be diffusive-convective gas exchange (Kestler, personal communication). The outward diffusion of CO_2 and water vapour are probably barred by the inflow of air, and the mechanism is suggested to be an adaptation to minimize water loss (Buck, 1962).

Fluttering is probably stimulated by low oxygen levels acting at the ganglion, whereas mounting CO_2 probably acts at the spiracle where the wide opening which releases the burst of CO_2 is triggered. Thus, the whole CFO cycle is driven by exogenous means (Burkett and Schneiderman, 1974).

5.6.2 *Periplaneta americana*

CO_2 has been found to be released in discrete bursts in many insects (Punt, 1956). Using an infrared gas analyser, Wilkins (1960) measured bursts of CO_2-release in *P. americana* every 14 min at 21°C, and every 21 min at 11°C. Edwards (1970) made comparable measurements using a thermal conductivity method. Periodic ventilation had already been noted in this species (Paulpandian, 1959), and Kestler (1971) was able to show that in quiescent insects each burst of CO_2-release coincided with a bout of abdominal pumping (Fig. 5.17). In fact, there were many CO_2 volleys in a bout, each corresponding to a pumping stroke. He measured rates of CO_2-release at different temperatures for many hours and correlated them with a detailed examination of thoracic spiracle activity. More recently, Kestler (1978a,b; personal communication) has extended the analysis by measuring weight changes on an extremely sensitive balance, and by recording intratracheal and haemolymph pressures during the cycle.

As in pupae, the cycle in an adult cockroach starts with a constriction phase when there is no exchange through the spiracles. As oxygen is consumed, the tracheal

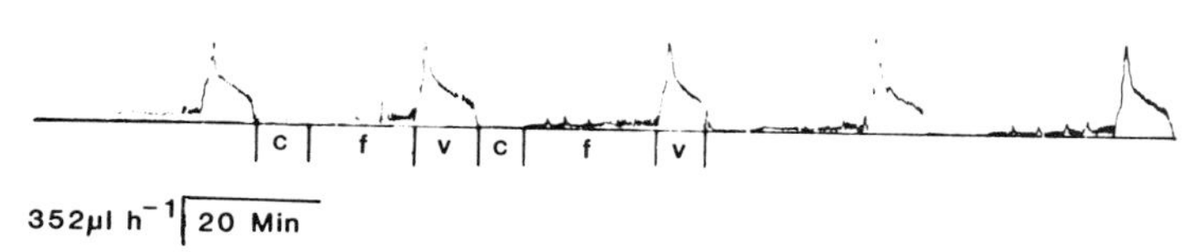

Fig. 5.17 Record of the rate of release of CO_2 from a quiescent *P. americana* made with an infrared gas analyser. CO_2 is released in a series of bursts during which abdominal ventilation (v) occurs. When the bout has finished thoracic spiracles are constricted (c) for a period during which no gas exchange occurs. Subsequently, spiracle valves start to flutter (f) and there is a very low rate of CO_2 release before the next bout of ventilation commences. Vertical scale indicates the rate of CO_2-release per hour. (Redrawn from Kestler, 1971.)

pressure falls rapidly. Cuticular elasticity and muscular activity probably help to resist abdominal deformation. As the spiracles start to flutter at about 1 min^{-1}, pressure returns stepwise towards atmospheric. Each flutter allows air to be sucked in and there is a simultaneous small loss of CO_2, which in one insect at 20°C amounted momentarily to an escape rate of 40–50 $\mu l\ h^{-1}$. Towards the end of the flutter period the escape rate rises slightly and during the subsequent ventilation bout it may peak at up to 700 $\mu l\ h^{-1}$. During a 5–7 min ventilation bout, CO_2 is lost at about 20–40 times the rate of loss which occurs during the 15–20 min CF period (Fig. 5.17).

Kestler has found that the constriction-flutter-ventilation (CFV) cycle occurs at rest in all the larger species of insect he has examined. Similar cycles (possibly of the CFO type) may be present in many small species, some of which have been shown to release CO_2 cyclically (Punt, 1956; Daynes *in* Edwards and Patton, 1967), while in others the periodic collapse and re-inflation of tracheae has been observed (Herford, 1938). Thus diffusive-convective gas exchange may be a dominant feature of the respiration of most insects at rest.

In active *P. americana* the CFV cycle is abolished. In one example, described by Kestler, the cycle began to reappear about 20 minutes after a period of intense activity, but it was still abnormal after 87 min. In quiescent and undisturbed insects, diurnal bouts of grooming or searching activity reset the cycle to the end of the ventilatory period, but the CFV cycle is undisturbed by less intense activities.

Many parameters of the cycle are affected by the metabolic rate. For example starved insects have a CFV cycle 1.43 times as long as well-fed ones. Kestler examined the cycle in detail at different temperatures and found its duration to be much shortened at high temperatures (Fig. 5.18), an effect mainly due to a curtailment of the pause (the CF period) between bouts of ventilation. In one insect, the CF period fell from 57 min at 9°C to 1 min at 35°C. The duration of the ventilatory bout was much less reduced at high temperatures; the pumping frequency rose, but the number of strokes per bout changed less consistently.

Measurements of weight changes at 100% r.h. allowed Kestler to calculate the changes due to metabolism. They confirmed that oxygen uptake continues during the CF period while there is very little CO_2 loss. He subtracted the values from total weight changes in dry air and calculated the water losses. Even the water loss from a single pumping stroke could be measured. During the CF period water is lost only

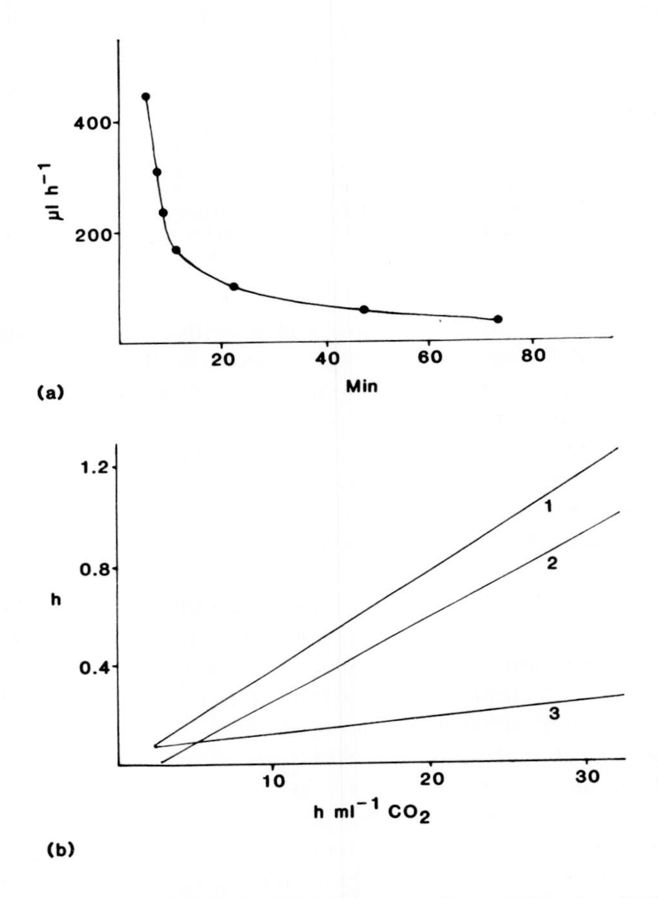

Fig. 5.18 (a) Plot of the metabolic rate (CO_2 released per cockroach h^{-1}) against the duration of the constriction-flutter-ventilation (CFV) cycle. Results are from one adult male *P. americana* at various temperatures. (b) Plots of the duration of the CFV cycle (1), the non-ventilating or CF period (2), and of the ventilation period (3), against the time taken to release 1 ml of CO_2 (h ml^{-1}). The CF phase shows a marked change in duration with temperature, while the ventilation period shows a smaller but significant change. (Redrawn from Kestler, 1971.)

from the cuticle. The rate of water loss rises greatly at the start of ventilation, but there is then an exponential decline during the ventilatory bout.

It can be concluded that in *P. americana* and probably in many other species, the CFV cycle allows diffusive-convective gas exchange to occur during the flutter period and that this has adaptive significance in minimizing water loss from the tracheal system. There is no dependence of cycling upon the relative humidity of the surrounding air, as has been suggested. *In vivo*, the cycle appears to be controlled by tracheal pO_2 and pCO_2, as is the CFO cycle in pupae.

5.6.3 The endogenous patterning of intermittent ventilation

The occurrence of periodic ventilation has already been noted in Section 5.5.2, and of similar patterns of motor activity in isolated cords in Section 5.5.7. It may be asked if this activity is the same as that which underlies the CFV cycle, or if it is an artifact of cord isolation. Intermittent ventilation can readily be recorded in intact *B. craniifer* (Miller, unpublished). Moreover, when both spiracles 1 and 2 are intubated so that they cannot close, periodic ventilation persists although the cycle may be altered (Fig. 5.9b,c). A similar experiment abolishes the CFO cycle in pupae. In cockroaches, a build-up of CO_2 may still occur under these conditions, and a better experiment might be to perfuse the tracheal system with air, but this has not been attempted so far. Endogenous patterns of activity in isolated cords resemble the CFV cycle in the intact insect. Their regularity is like that seen in undisturbed insects and their short 'interventilatory' periods are like the brief CF periods in intact insects with high metabolic rates. The changes which occur are perhaps due to hormones, to damage of the tracheal system, and to the deprivation of sensory input.

If the endogenous activity recorded does underlie periodic ventilation in intact insects, what function does it play when exogenous mechanisms alone would seem to be adequate? By comparison with pupae, it is probable that CO_2 levels in cockroaches rise to about 6.5%, at which value a bout of ventilation is triggered, but this reduces the level more effectively than in pupae to values of about 1% (Kestler, 1971). This implies that the level of CO_2 required to trigger ventilation is higher than that needed to sustain it. One possible explanation is that sensory feedback resulting from the pumping strokes, together with a level of CO_2 above 1%, are adequate to keep it going. Another is that the CO_2 sensors, presumably in the CNS, are in some way shielded from the rising pCO_2. Alternatively, the feedback loop might have a long latency which could cause the system to oscillate. However, there is no evidence for these, nor do they explain the presence of an endogenous oscillator. It may be proposed therefore that 6.5% CO_2 levels trigger an oscillation of a slow pacemaker whose intrinsic properties account for the duration of the subsequent ventilatory bout. Kestler (1971) has shown that the mean metabolic rate is accurately related to the duration of the ventilatory bout. This means that if the pacemaker times the ventilatory bout, it must itself be responsive to the metabolic rate, or to some other factor which also acts on metabolism. The function of pacemaker participation in the generation of the CFV cycle may be to prevent ventilation from becoming stabilized at a slow regular frequency of pumping in which each stroke temporarily lowered the CO_2 level to just below threshold.

The significance of the oscillator may therefore be to ensure that ventilation occurs in a series of bouts. The benefit from employing this pattern is that it allows the system to have a high triggering threshold and permits long periods to occur between bouts, when diffusive-convective gas exchange can occur, with a consequent saving of water. Thus, the postulated slow oscillator may promote this form of oxygen uptake. The fact that there is apparently no endogenous control of the CFO cycle in pupae may be connected with the peripheral location of the CO_2 sensor, at the spiracle, in that system. During diffusive-convective gas exchange the spiracle may experience a lower pCO_2 than prevails elsewhere in the tracheal system.

5.6.4 One-muscle spiracles and diffusive-convective gas exchange

In diffusive-convective gas exchange the spiracles play a key role. According to Burkett and Schneiderman (1974), low oxygen levels act on the spiracles of pupae via the segmental ganglion and CO_2 acts directly on the closer muscle. Similar conclusions were reached about spiracle control in adult dragonflies (Miller, 1964). It may be that both gases interact with the system to account for the fluttering which underlies diffusive-convective gas exchange. Only one-muscle spiracles with an elastic opening system are known to be capable of responding to local CO_2 levels (Hazelhoff, 1926; Beckel, 1958; Hoyle, 1961; Miller, 1974a). A survey of 31 families representing 17 orders (information based mainly on Maki, 1938) indicates that all insects possess at least one pair of spiracles of this type. Of the families surveyed, 88% possess spiracles 1 with elastic opening; all have elastic opening in spiracles 2; 42% in spiracles 3; 52% in spiracles 4; and 55% in spiracles 5–10. Thus elastic opening is universal in at least one thoracic spiracle and it occurs in the abdomen in about half the families. If the flutter is controlled by the interaction of ganglionic and peripheral responses, the evolution of elastic openers can be seen as being essential for diffusive-convective gas exchange. The widespread occurrence of such spiracles suggests that this type of gas flow is also widespread in insects. However, detailed measurements of neural activity during fluttering are needed, and until they are available this proposal must remain speculative.

Acknowledgement

I am most grateful to Dr Paul Kestler both for permission to use his unpublished results, and for much valuable discussion and many useful comments.

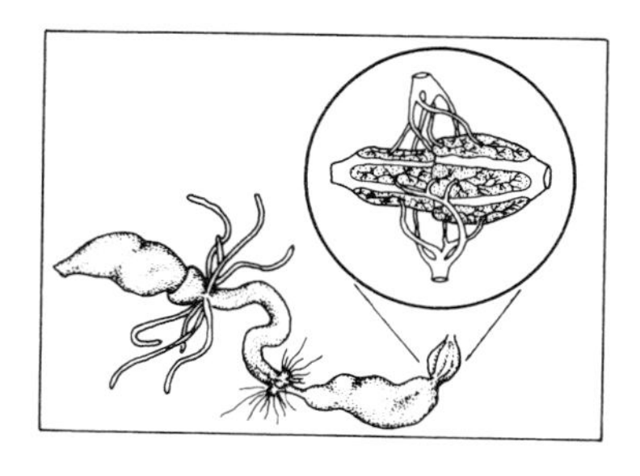

6

Osmoregulation and excretion

Donald E. Mullins

6.1 Introduction

6.1.1 General comments and reference to previous treatments of osmoregulation and excretion in cockroaches and other insects

Cockroaches can be viewed as generalists, capable of thriving under conditions where food resources are limited; the terrestrial habitat in which they live may be subject to great variation. A considerable amount of general information on these insects attesting to this view has already been well documented (Cornwell, 1968). Detailed studies of certain species have provided information regarding the physiological parameters under which they may successfully function. These investigations have led to discoveries which have provided workers with new facts that must now be developed into a working hypothesis describing the processes that are involved in maintaining homeostasis. For example, recent work has indicated that water conservation appears to involve changes in substrate metabolism, tissue exchanges of solute materials, ion exchanges with stored urates and excretory processes, all of which are quite complex. These activities appear to be well orchestrated by mechanisms which are as yet unclear.

In this chapter, topics relating to water balance in the American cockroach will be considered in as much detail as possible. Information on water, ions, excretory products and their regulation or production will be presented in a manner focusing on the water relationships which are apparently involved.

Previous treatments of cockroach osmoregulation and excretion are found in texts by Cornwell (1968) and Guthrie and Tindall (1968). Since their publication, much new information has become available on these topics in the American cockroach.

There have been numerous recent reviews published on topics relating to the maintenance of homeostasis in insects (Maddrell, 1971; Bursell, 1974a,b; Stobbart and Shaw, 1974; Cochran, 1975; Edney, 1977). Additional relevant information in other reviews will be referred to in appropriate sections of this chapter.

6.1.2 Basic concepts in osmoregulation and excretion

Water is an essential constituent of all living organisms and it is in the aqueous phase in which most of the metabolic reactions necessary for life occur. Insects have made the transition from aquatic habitats to land more completely than many other groups (Prosser, 1973). One of the major problems encountered by organisms establishing themselves on land has been to elaborate mechanisms enabling them to maintain water balance. Water balance in an organism is related to the total water content and the actual water activity (viewed thermodynamically as the effective concentration). The total effective water concentration of all solutes present can be expressed in terms of the total number of moles of solute per litre of solvent designated as osmoles (Prosser, 1973). The osmolal concentration is determined by the colligative properties of the solution which depend upon or vary as a function of the number of solute molecules in solution and not upon the nature of the molecules (Hammel, 1976). Osmotic pressure expressed in terms of osmolality is used to measure water and solute concentrations which occur in biological systems (Prosser, 1973). The water balance established by an organism then reflects its ability to maintain homeostasis by striking a delicate balance between uptake and loss of solutes and water (Berridge, 1970b).

Terrestrial organisms which have a large surface to volume ratio would be quite susceptible to desiccation were it not for their ability to minimize water losses. Potential sources of water loss may include simple diffusion from the respiratory and general body surfaces, secretion of various materials (digestive fluids, reproductive products, etc.) defaecation and excretion (Arlian and Veselica, 1979). Reduction of water loss may include mechanisms involving such things as behavioural responses (humidity preference), control of transpiration losses, control of excretory processes, and possession of water stores from which water can be withdrawn under conditions of desiccation (Berridge, 1970b).

Another important aspect of homeostasis in an organism is the elimination of materials produced in excessive quantities or toxic materials taken up as a consequence of life processes. These materials must be eliminated before they cause problems. The concentration at which such materials become obtrusive may vary widely depending, of course, on their chemical and physical nature (Cochran, 1975). Maddrell (1971) and Cochran (1975) have discussed definitions of two terms, secretion and excretion, which relate to the elimination of materials from the metabolic pools of an organism. Secretion is defined as the removal of substances from metabolic pools in order to achieve some useful purpose outside a particular pool. On the other hand, excretion is defined as the process by which substances which may interfere with ordered metabolism, because of their presence (at excessive levels) or because of their toxic properties (at relatively or exceedingly low concentrations), are removed from a particular pool. These excreted materials may be voided from the organism or simply isolated within it by confinement to a particular compartment or special physico-chemical state (Maddrell, 1971).

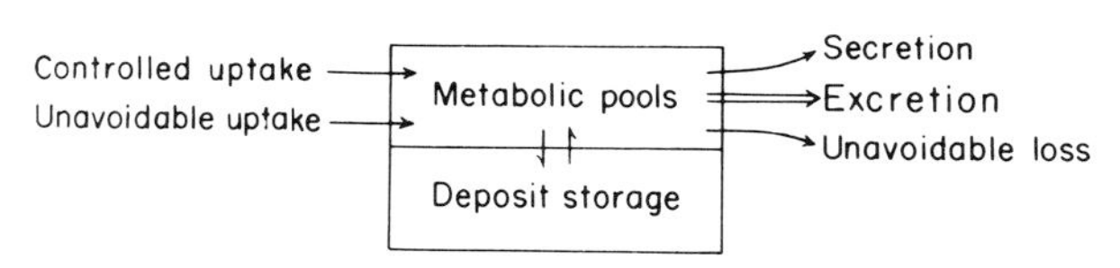

Fig. 6.1 Flows of materials into and out of the metabolic and storage pools of *P. americana* (L.) (Adapted after Maddrell, 1971.)

Figure 6.1 illustrates the general flow of materials into and out of cockroaches. The processes involved in maintaining metabolic and osmotic homeostasis include controlled and unavoidable uptake of materials into various metabolic pools and subsequent removal of materials from these pools via secretory, excretory and unavoidable loss routes. These relationships with the metabolic pools are similar to those described for other insects (Maddrell, 1971). However, certain metabolic pools in cockroaches are closely associated with processes involving deposit storage of urates and perhaps other materials. It now appears that these stored materials are in dynamic equilibrium with various metabolic pools and in addition may provide the cockroach with an alternate or additional means for osmoregulation under certain conditions. The relationships between metabolic pools, deposit storage and metabolic and osmotic homeostasis will be considered in more detail later.

6.2 Water relations

6.2.1 Water content

Water content in insects can vary considerably within and between a species, and in fact, may differ with the age or developmental stage of a particular species. Factors such as ambient temperature, and relative humidity, diet (including its water content), availability of drinking water and general physiological condition of the insect, may influence total body water content (Arlian and Veselica, 1979). Edney (1977) has pointed out that water content alone does not reveal much information on either the immediate water status or the water balance mechanisms of an arthropod. However, its biological significance may be clarified when total water content is examined in relation to components involved in water balance (i.e. storage sites and water reserves) and in relation to the limits of water depletion which can be tolerated.

Values on water content in insects may range from about 45–93.5% wet weight (Arlian and Veselica, 1979). Studies on *P. americana* have indicated that their water content ranges from about 55–74%. The reported values differ to some extent, but the experimental conditions also differed and obviously influenced the results. Some specific information which has been reported on the water content in *P. americana* will now be considered.

(a) *Water content of oöthecae*

Roth and Willis (1958) found the water content of oöthecae and eggs at the time of oviposition to average 60–64%. The intact oötheca is capable of retaining sufficient water to allow the eggs to develop in dry environments. However, if the oötheca is damaged, allowing water loss, the eggs will develop only under conditions where the relative humidity is high (Roth and Willis, 1955a). Munson and Yeager (1949) reported the water content of male and female nymphs to be 64.9% (undefined dietary, temperature and relative humidity conditions). Edney (1968) reported that nymphs maintained in a hydrated condition (high r.h., 25 °C, with access to food and water) had a water content of 68.5%, whereas dehydrated nymphs deprived of water (dry air, 25 °C, with access to food) for 4 days contained 55.4% water.

(b) *Water content of adults*

The water content of adult male *P. americana* under various conditions of hydration has been shown to vary (Tucker, 1977a). She found normal 1–3-week-old males (uncontrolled r.h., 27 °C, with access to food) contained 68.6% water, and dehydrated males (similar age, r.h., and temperature; no food and water for 14 days) contained 62.1% water. Dehydrated males maintained under dehydration conditions for 6 days before being allowed access to drinking water for 1 day contained 74.3% water. Studies designed to examine the influence of age and starvation on haemolymph volume and water content of males have revealed that water content changes very little with age, but haemolymph volume decreases markedly during the first 4 days after nymphal/adult ecdysis (Wharton *et al.*, 1965a). These studies also indicated that starvation results in a significant increase in water content (70.5% water after 5 days starvation and 74.1% water after 12 days starvation), whereas haemolymph volume decreased (from 34.9% to 32.5%). This information suggests that food intake influences the balance of assimilated solutes and metabolic products between body tissues and haemolymph.

Water balance in adult female *P. americana* is influenced significantly by the vitellogenic cycle (Verrett and Mills, 1973). As pointed out above, at the time of oviposition, sufficient water is present in oötheca for complete embryonic development. Water is translocated into oöcytes during the latter portion of the 6-day vitellogenic cycle, requiring relatively large volumes of water (Verrett and Mills, 1973). Water content in females has been found to decrease from about 72.5% (2nd day of the vitellogenic cycle) to 67.0% (6th day of the vitellogenic cycle) (Verrett and Mills, 1975b). These workers presented evidence that water consumed the first two days of the cycle contributes the most water to the oötheca, and various tissues are involved in storage and subsequent water incorporation into oöcytes during the cycle (Verrett and Mills, 1975a,b).

6.2.2 Water loss

(a) *Integument*

The water content of cockroaches is influenced by water loss which may occur primarily by transpiration, excretion and defaecation. Water loss by transpiration occurs along two major routes; through the cuticle of the general body surface and through the tracheal system. Transpiration occurring through the arthropod integument has been an active area of investigation for the past 40 years since the discovery that integumental permeability could be affected significantly by temperature (Ramsay, 1935). Ramsay (1935) found that living and dead *P. americana* displayed an increase in transpiration rate at about 30°C. Waterproofing has subsequently been shown to depend upon the presence of lipids in the epicuticle which, as the cuticular temperature increases, undergo changes producing marked increases in water permeability (Bursell, 1974b; Edney, 1977). Beament (1958b) has proposed that these lipids form a monolayer film over the integument, which becomes disoriented at a specific transition temperature (i.e. the temperature at which water permeability markedly increases). However, recent work on the cuticular hydrocarbons of several Orthoptera (including *P. americana*) has provided evidence in support of an alternate hypothesis suggesting that lipids may be present in the cuticulin layer and participate in waterproofing of the cuticle (Lockey, 1976; see also Section 2.2.1).

Two other properties associated with insect integument have been proposed in reducing water loss. These include the presence of asymmetrical cuticular permeability and epidermal cells functioning as water barriers (Edney, 1977). Their involvement in reducing transpirational water loss across cockroach integument is not clear, but some evidence suggests there is an active process involved. Winston and Beament (1969) have found that water activity in *P. americana* cuticle is substantially lower than in haemolymph. They suggest that an active, energy-requiring mechanism is involved in maintaining an osmotic pressure difference between cuticle and haemolymph (based on measurements made from cuticles excised from living cockroaches). Results from studies on transpiration rates in living and dead *P. americana* held at constant vapour pressures, but increasing temperatures also suggest an active reduction of cuticular permeability (Coenen-Stass and Kloft, 1976b, 1977).

(b) *Transpiration*

Little information is available on water loss in *P. americana* due to transpiration from the insects respiratory surfaces (tracheal system) (Cornwell, 1968; Guthrie and Tindall, 1968; Tucker, 1977d). Bursell (1974b) has pointed out that investigations on water loss from the insect tracheal system are difficult to perform, primarily due to variable levels of activity and its relationship to humidity. Control of the spiracles is viewed as a means for reducing water loss, and in some insects (tsetse flies and locusts), the degree of spiracular opening is influenced by atmospheric humidity (Edney, 1977). It is quite likely that the discontinuous CO_2-release cycle observed in *P. americana* (Wilkins, 1960) represents a mechanism for water conservation, but evidence supporting this view is lacking (Edney, 1977). Indeed, a detailed examination of transpirational losses and spiracular control, coupled with studies on *P. americana* activity upon placement in environments of variable temperature and humidity, would be quite useful.

(c) *Excretion*

The processes involved in excretion and defaecation may provide for significant water loss in *P. americana* depending on factors such as environmental conditions and diet. Tucker (1977a) has observed that when males are well-hydrated the faecal material produced may be greater than 80% water, but when drinking water is disallowed the water content of the faeces is rapidly reduced. Wall (1970) has found that in addition to a reduced faecal water content, the number and weight of faecal pellets decrease as the cockroaches dehydrate. She found that after 7 days of dehydration, faecal pellets remained in the rectum from 2 to 4 days and became highly concentrated (up to 1600 mosmol kg^{-1} H_2O). Under these conditions, excretion accounted for only a small part of the daily water loss (Wall, 1970). Some information on the effects of diet on excretion is available. Mullins (1974) has shown that as the dietary nitrogen increases, the drinking water requirement significantly increases. For example, male *P. americana* fed on dextrin (0% nitrogen) and 76% casein protein (12% nitrogen) had *ad libitum* water drinking requirements of 59 ± 3 and 179 ± 8 μl/male/day, respectively. This is presumably due to elevation of faecal ammonium cation levels in response to high nitrogen diets, and as a consequence the faeces produced are quite high in water content (Mullins, 1974). Heit *et al.* (1973) have shown that *ad libitum* water consumption in *P. americana* increases in response to increasing concentrations of NaCl in their drinking medium. It appears that this species can and does regulate the amount of water contained in its faeces. Under normal conditions (free access to drinking water), faeces may contain a substantial amount of water, a large portion of which can be removed if access to water becomes limited. On the other hand, when there is a need for removal of excess solutes (potentially toxic cations) increased uptake of water may be required to facilitate the process.

(d) *Other routes*

Other routes of water loss from *P. americana* are possible. These may include regurgitation when they become excited (Wall, 1970), and loss by secretions (which may contain water) of integumentary glands (Guthrie and Tindall, 1968; Brossut *et al.*, 1975). In addition, water losses may occur during the reproductive processes, which may be significant in the production of viable oöthecae (Verrett and Mills, 1973; see also Section 6.2.1).

6.2.3 Tolerance to stress

Various responses to stress may be observed when an organism encounters changes in or approaches its physiological limitations to, environmental conditions, diet, activity and exposure to foreign compounds. In many situations, stress may induce changes in water balance as discussed below.

(a) *Food and water*

P. americana can survive for quite lengthy periods of time when food and water sources are limited. Willis and Lewis (1957) examined survival rates of this species under conditions of partial or complete starvation at 27 °C and 36–40% r.h. They found that the mean length of survival in adult females was 40 days (dry food–no water), 90 days (water–no food) and 42 days (no food–no water) compared to 190 days for controls (food and water). Survival rate of adult males was not as high: 27 days (dry food–no water), 44 days (water–no food), and 28 days (no food–no water) compared to 97 days for controls. From these data, it appears that deprivation of water is most critical (reducing longevity by about 78% for females and 71% for males), while deprivation of food only is much less critical (reducing longevity by about 43% and 55%, respectively). The presence of food had no effect (water not available) on longevity. This was most likely due to their inability to feed when dehydrated (Wall, 1970; Tucker, 1977a).

The amount of protein in the diet has been shown to influence longevity (Gier, 1947b; Haydak, 1953; Mullins and Cochran, 1975a). Haydak (1953) observed lowest mortality and most rapid nymphal development rate to occur in *P. americana* maintained on diets containing 49–79% protein. He also found that as dietary nitrogen fed to adults increased above 49% there was increased mortality. Increased mortality due to enforced feeding on high protein diets correlates with significant build-up of uric acid/urates and a decrease in non-nitrogenous materials (i.e. lipids and carbohydrates) in fat body (Gier, 1947b; Haydak, 1953; Mullins and Cochran, 1975a). The effectors involved in this type of dietary stress are not clear. The storage of urates and production of excretory products by *P. americana* on high protein diets will be considered in Section 6.4.1.

(b) *Tolerance to water loss*

Water loss tolerances of insects and mites may vary considerably, ranging from 17–89% of the species normal water content (Arlian and Veselica, 1979). Coenen-Stass and Kloft (1976a) have found adult *Periplaneta* held at 26 °C at various relative humidities (33–98% r.h.) lose weight and die when they reach a weight loss ranging from 25–35% of their normal weight. Physiological responses to dehydration stress are of considerable interest because numerous studies have shown that cockroaches are able to maintain a relatively constant haemolymph osmolality while undergoing severe dehydration (Edney, 1968; Wall, 1970; Laird and Winston, 1975; Hyatt and Marshall, 1977). With loss of haemolymph volume (water) there is a corresponding removal of solutes to some site(s) within the cockroach. These solutes are apparently returned to haemolymph upon rehydration (Wall, 1970). In addition to dehydration-hydration experiments, such things as injection of distilled water or salt solutions into *P. americana* haemocoels (Van Asperen and Van Esch, 1956), and drinking of hyperosmotic saline solutions by

Periplaneta (Tobias, 1948; Heit *et al.*, 1973) indicate that this species can withstand osmotic stress quite well by mechanisms which are not yet well understood.

(c) *Temperature and evaporative cooling*

Temperature tolerance levels of an organism fall within a fairly well-defined temperature range (Bursell, 1974a). Since water loss increases with increasing temperatures and declining relative humidities, temperature tolerance may be closely linked to water loss tolerance. Considerations in determination of the upper limits of temperature tolerance must include the duration of exposure. Short-term exposures (about 1 h) indicate that upper lethal limits for most insects range between 40–50°C (Bursell, 1974a). The thermal death points of *P. americana* under short-term and long-term exposure and different relative humidities have been determined by Gunn and Notley (1936). They found that long-term (24 h) thermal death points both at 0% and 90% r.h. were similar ranging from 37–39°C. The thermal death points were different in short-term exposures (1 h), being 45°C at 0% r.h. and 42°C at 95% r.h. The increased heat tolerance observed in short-term exposure at lower relative humidities is thought to be due to evaporative cooling, a mechanism which may assist various arthropods in surviving brief exposures to elevated temperatures (Edney, 1977).

Recent work indicates that evaporative cooling may also be of significance to cockroaches under conditions less stressful than temporary protection from elevated temperatures. Simultaneous measurements of *P. americana* body temperatures and transpiration rates over a range of air temperatures in dry air (r.h. <15%) showed no differences between body and air temperatures up to 32°C; however, at temperature ranges from 32–39°C, body temperatures were lower than those of air (at 39°C by 0.8°C) (Coenen-Stass and Kloft, 1977). By contrast, *P. americana* in moist air (r.h. >95%) had body temperatures higher than air temperatures (at 26°C by 0.1°C; at 39°C by 0.7°C). Coenen-Stass and Kloft (1977) concluded that in dry air heat loss due to cooling by transpiration exceeds that heat produced by metabolism, whereas in humid air, transpirational cooling may not compensate for body heat gains due to metabolism. Farnworth (1972) has examined the relationships of ambient temperature and humidity on internal temperatures in *P. americana* at rest and during flight. He found that internal temperatures at 50% and 95% r.h. were routinely lower than ambient temperatures before flight. During flight, maximum increases observed in thoracic temperatures were less at 50% than at 95% r.h., and these differences were attributed to evaporative cooling. It appears that, in *P. americana*, heat may be lost due to evaporative cooling, when exposed to high temperatures, or when they become physically active (i.e. flight), allowing for a limited degree of thermoregulation. Unless there is access to water sources, allowing the maintenance of a stable water balance, water losses resulting from evaporative cooling can reach the maximum tolerance levels.

(d) *Behavioural responses*

Behavioural responses to unfavourable temperature and humidity conditions may reduce exposure of *P. americana* to these stressful situations (Cornwell, 1968). When *Periplaneta* have had access to food and water, their preferred temperature range falls between 24–33 °C (Cornwell, 1968). Recent work (Coenen-Stass, 1976) on combined temperature and humidity preferences in *P. americana* and *Blaberus trapezoideus* has indicated an acclimatization effect. He found that after spending one month at 20 °C, *P. americana* preferred temperatures ranging from 26–29 °C and after subsequent exposure to 35 °C, its preference was about 29 °C. Hydrated and fed *P. americana* preferred a low relative humidity (18%). However, upon desiccation (15–18% of their body weights), they prefer a temperature 1–3 °C lower than before desiccation and, in addition, desiccated individuals show a preference for high humidities (Coenen-Stass, 1976).

(e) *Fatigue and chemical exposure*

There is a considerable amount of information available indicating that excessive stress on *P. americana* can lead to the internal release of pharmacologically active substances. Severe forms of stress, whether produced by chemical stimulation (Hawkins and Sternburg, 1964; Shamburgh, 1969; Flattum *et al.*, 1973), mechanical stimulation (Cook and Holt, 1974), or mild electrical stimulation (Beament, 1958a), results in release of neuro-active substance(s) into the haemolymph, which causes paralysis and may lead to death. Sternburg (1963) suggested that, whatever the source, physiologically active compounds liberated during stress must have a normal function, but when excessive amounts are released due to intensive stress, hyperactivity, paralysis and death may occur.

There is some evidence that water balance may be disturbed by various forms of stress which contributes significantly to symptoms and consequences of severe stress (Maddrell and Casida, 1971; Cook and Holt, 1974). The chemical identity of the active factor(s) is not known. However, various studies have shown that they are distinct from all commonly known neuropharmacologically active agents (Cook, 1967). Flattum *et al.*, (1973) found that an 'autoneurotoxin' obtained from stressed *P. americana* accelerated the contraction frequency of *Periplaneta* and *Schistocerca gregaria* Malpighian tubule preparations *in vitro*. These observations suggest a function of this active agent may involve the excretory system. Stressed *P. americana* may undergo a considerable weight loss during the intoxication process (Cook and Holt, 1974) which could be due, in part, to activities involving the excretory apparatus. In support of this view, Maddrel and Casida (1971) have shown that insecticide-poisoned *Rhodnius prolixus* release a diuretic factor which is secreted during the paralytic stage of poisoning.

New information on such things as the chemical identity, site of release, normal titres, biological fate, and the precise role of the 'autoneurotoxin(s)' in relation to stress is needed. This is an area in which further investigation should be encouraged.

6.3 Components of water and ion metabolism

6.3.1 Sources

(a) *Uptake of liquids*
Edney (1977) has described three distinct ways in which arthropods obtain water. These are: (1) uptake of liquid water, which includes water contained in food and water imbibed by drinking, (2) absorption of water from a vapour phase and (3) production of metabolic water. Uptake of liquid water in *P. americana* is indicated by their omnivorous nature. They have been reported to feed on materials including maimed individuals, their own oötheca, fruits, vegetables and flower petals (Cornwell, 1968; see also Section 1.3). All of these materials contain variable amounts of water and, therefore, little information is available on the amount of liquid water obtained from consumed foods. Uptake of liquid water by drinking can be viewed as a means by which *P. americana* can maintain water balance when other means of acquisition are inadequate. Wall (1970) has observed their large crop enables them to drink enough water in 15–30 min to last for 2–3 weeks, and dehydrated individuals may imbibe 100–200 μl at one time. Starved *P. americana* (27 °C, 30% r.h.) allowed access to tap water drank 21 μl $male^{-1}$ day^{-1} (Heit *et al.*, 1973). Fed individuals (28 °C, 64% r.h.), maintained on dog food, consumed 104 μl $male^{-1}$ day^{-1}, but on diets containing various levels of nitrogen and salts the amounts consumed ranged from 59–179 μl $male^{-1}$ day^{-1} (Mullins, 1974).

(b) *Uptake of water vapour*
The possibility that *P. americana* is capable of absorbing water vapour has been examined. In a comparative study, Edney (1966) found that dehydrated *Arenivaga* spp. nymphs and adult females were capable of absorbing water vapour above 82.5% r.h. (10–30 °C), whereas adult *Arenivaga* males, and *P. americana* and *Blatta orientalis* nymphs could not. Recently, Coenen-Stass and Kloft (1976a) reported that, even at 95% r.h., male *P. americana* lose weight. In experiments using tritiated water, a small amount of water uptake occurred (which increased disproportionately with rising humidity), but this probably represented passive transport because narcotized individuals showed the same rate of absorption as controls. It appears from these reports that, although water vapour exchange may occur, there is no significant evidence in *P. americana* to indicate an active water vapour absorption process.

(c) *Metabolic water*
The production of metabolic water in organisms has been known to occur for a long time. However, assessment of its importance in the maintenance of water balance is difficult (Bursell, 1974b). Production of metabolic water depends on the particular substrates which are oxidized. For example, the water produced from complete

oxidation of each of the three major groups of substrates is: (1) carbohydrates 0.56 g H_2O g^{-1}, (2) fats 1.07 g H_2O g^{-1} and (3) proteins 0.40 g H_2O g^{-1} (Edney, 1977). An animal which is fasting would be capable of actually gaining weight if it were not for unavoidable water losses (i.e. transpiration, defaecation, etc.). Certain insects living in quite dry habitats, where there is no drinking water and the food is low in water content, depend to a large extent on metabolic water as a resource (Bursell, 1974b; Edney, 1977).

Information on production and significance of metabolic water in *P. americana* is meagre. Observations on longevity of starved, dehydrating cockroaches (Willis and Lewis, 1957) indicate that they have sufficient food reserves which they can use as an energy source. Tucker (1977c) pointed out that catabolism of these reserves might be important to *P. americana* deprived of water and food. Under dehydration conditions, she found that last-instar nymphs (which have a greater lipid content than adult males) showed significantly slower rates of weight loss than adults. It appears that catabolism of these lipid stores may contribute to maintaining the water balance under certain conditions (Tucker, 1977c). The precise roles of substrate metabolism in production of metabolic water and its relationship to dehydration and starvation conditions have not been determined. Investigations designed to clarify these relationships would be quite useful indeed. Studies employing two different radio-labelled species (i.e. radio-labelled organic substrates ^{14}C-carbohydrates, ^{14}C-lipids and $^{3}H_2O$) and dual-label radio-isotope methods might provide some of this information.

6.3.2 Pools and structures involved in osmoregulation and excretion

Before discussing pools and structures which are involved in osmoregulation and excretion, it should be pointed out that the relationships of stored uric acid/urates in various tissues include a dimension that most other insects do not have. The possibility that urates play a role in osmotic homeostasis will be considered later. One should keep in mind that under variable water balance conditions, if solutes can be sequestered/released in/from tissues in accordance with certain physiological demands, the net osmotic impact of dehydration/rehydration could be reduced.

Bursell (1974b) defined water reserves of an insect as the quantity of water which it can lose before death supervenes. In *P. americana*, various tissues and systems contain water pools which, under desiccation conditions, may function as water reserves. Some of the more important tissues and systems which may serve as such a water reserve will now be considered.

(a) *Salivary glands*

The cockroach salivary gland has been identified as a water reservoir (Tucker, 1977d). Sutherland and Chillseyzn (1968) pointed out that although the primary function of the salivary reservoir is to store salivary secretions used in moistening and digesting food, it contains sufficient water (up to 100 μl) that is available for use under conditions of dehydration. The osmolality of salivary fluid is much lower than haemolymph (Wall, 1970) and it is thought that these secretions may be absorbed and used to maintain water balance in other tissues when *P. americana* dehydrates (Tucker, 1977d).

(b) *Alimentary canal as a water reserve*

Portions of the alimentary canal may serve as a water reserve. The crop represents the largest volume among the components of the alimentary canal (Chapter 4) and may therefore be of major importance in this regard. It was stated earlier that the crop can be filled with 100–200 μl water when dehydrated animals are allowed to drink (Wall, 1970). Verrett and Mills (1975b) found that the gut releases large volumes of water during the 6th day of the vitellogenic cycle. The water present in the normally hydrated cockroach alimentary canal is absorbed and used early as a water resource, since after one day of dehydration, the faecal water content is greatly reduced (Wall, 1970; Tucker, 1977a). In addition to the gut, other tissues including the fat body, integument and adhering tissue also may serve as a water reserve (Verrett and Mills, 1975b).

(c) *Haemolymph*

One of the most obvious water reserves in *P. americana* is haemolymph itself as discussed in Chapter 3. Early work on cockroaches indicated that the haemolymph volume was indeed quite variable, and, therefore, implicated as having a major role in water balance (Cornwell, 1968; Guthrie and Tindall, 1968). Under various conditions of dehydration, haemolymph may be significantly reduced (Edney, 1968) and, in fact, when the animal is severely dehydrated, it is quite difficult to obtain samples for analysis (Tucker, 1977a). Wall (1970) found after 9 days of dehydration, the haemolymph volume of males was reduced by 81 μl (resulting in a total volume of 53 μl). When they were allowed to rehydrate the haemolymph volume increased by 93 μl (resulting in a total volume of 146 μl). Although haemolymph volume may fluctuate, its osmolality remains relatively constant and is therefore subject to strong osmoregulation (Edney, 1968; Wall, 1970).

(d) *The alimentary canal–osmoregulation and excretion*

The alimentary canal and its associated structures (salivary glands and Malpighian tubules) is most directly involved in osmoregulation and excretion (see Fig. 6.2). However, the fat body also plays a significant role because it stores/releases uric acid (urates). Due to lack of space, details of the structure of salivary glands, crop, midgut caeca, ventriculus and hindgut will not be included here. The structure and function of cockroach salivary glands has been reviewed recently (House, 1977). In addition, information regarding the crop, midgut caeca, ventriculus and hindgut can be found in Chapter 4. The major excretory apparatus in the cockroach, forming and elaborating the final excretory products that are eliminated in the faeces, involves the Malpighian tubules, hindgut (ileum and colon) and the rectum.

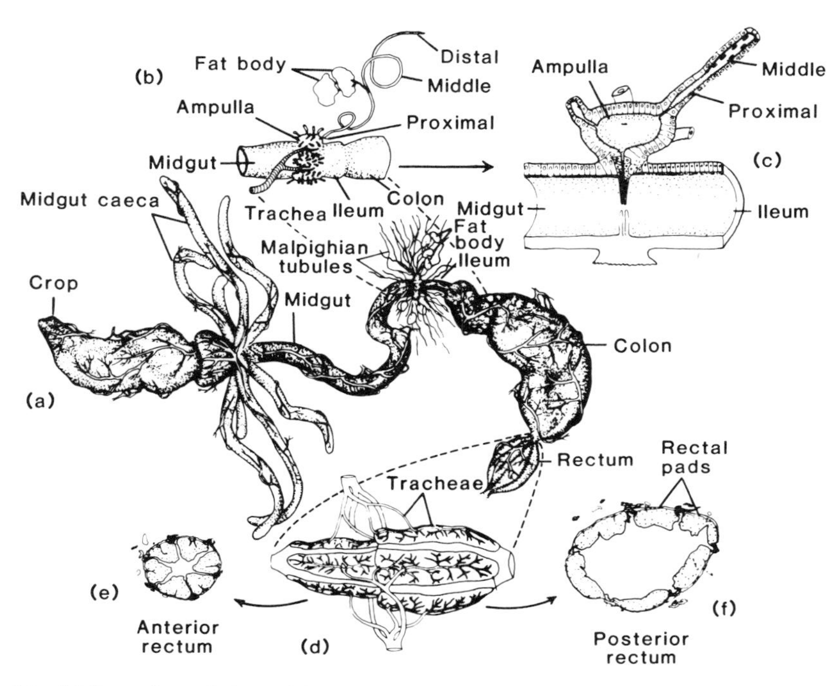

Fig. 6.2 (a) General morphology of the alimentary canal of *P. americana*. (b) The arrangement of the structures associated with the Malpighian tubules. The full length of only one of the Malpighian tubules is shown. The middle region is the largest part. The short thin distal region is highly contractile and the short proximal region inserts into the ampulla. Each of the six ampullae drains about 24–32 tubules. The ampullae are contractile enlargements on the intestinal surface at the junction between midgut and ileum. Tracheae branch over the midgut surface and send processes into the ampullae. Fine tracheoles also attach the tubules to other organs such as fat body. (From Wall *et al.*, 1975.) (c) Summary illustration of an ampulla and associated structures. Muscles are not included. Proximal region of tubule is composed of cells similar to ampulla. Tubule lumen narrows at region where it drains into ampulla. Cavity of ampulla drains via a narrow slit into gut. Apical surfaces of tubules, ampullae and midgut are folded to form microvilli. Ileum is lined with cuticle. (From Wall *et al.*, 1975.) (d–e) The general organization of the rectum, illustrating (d) the rectum as seen *in vivo*; (e) a cross-section of the anterior region; and (f) a cross-section of the posterior region. Two main tracheal trunks (from the marginal abdominal tracheae) send numerous branches over and into the rectal pads. There is usually a radial fold between the anterior and posterior parts of the rectum, with the anterior portion slightly contracted so that the pads are flat. (From Oschman and Wall, 1969.)

(e) *Malpighian tubules*

The Malpighian tubules produce what is essentially an ultrafiltrate of the haemolymph which passes into the hindgut. This fluid is mixed with food residues which have been processed in the midgut. The hindgut contains microflora which have an opportunity to further process these combined materials and perhaps remove certain substances (metabolites produced by the microflora) and water. The role of the hindgut is such that it reabsorbs those useful constituents needed by the insect, rejecting the others and, in this way, homeostasis is maintained.

Recent studies on Malpighian tubules of nymphal and adult *P. americana* have provided detailed information on their structure and function (Wall *et al.*, 1975, Schmidt, 1979a,b). Portions of their work and that of others will be summarized here.

The general arrangement of *P. americana* Malpighian tubules and associated structures is found in Fig. 6.2b. The numbers of tubules range from 144–192 per insect (Wall *et al.*, 1975). Each tubule is from 2–3 cm long, 40–60 μm thick, and collectively they extend throughout the abdomen, but are held in close contact with fat body tissue and tracheae (Wall *et al.*, 1975). They are highly contractile due to two parallel, closely apposed muscles (5–10 μm in diameter) which are wrapped helically around each tubule (Crowder and Shankland, 1972a). Four distinct regions have been identified in these tubules (distal, middle, lower middle and proximal) (Wall *et al.*, 1975; Schmidt, 1979a). The distal region is short and thinner than the other regions. It is highly contractile, exhibiting rapid bending movements and contains large numbers of mitochondria. Originally, the middle region (the longest portion of the tubules) was considered to be one region, but, recently, the more transparent portion of the middle region was found to be ultra-structurally distinct from the other adjoining regions (Schmidt, 1979a). The middle regions are composed of two cell types: primary cells which contain mineral concretions, and stellate cells which have smaller nuclei, fewer organelles, simpler brush borders and numerous multivesicular bodies. Symbiont bacteria are found in the lumen of the lower middle region, but their function in the tubules is not clear. The proximal region is short, releasing tubular fluid into one of six ampullae. Each ampulla (Fig. 6.2c) is composed of a layer of epithelial cells which surrounds a cavity opening into the gut lumen (at the junction of midgut and ileum) via a narrow slit (Wall *et al.*, 1975). When fluids accumulate in the ampullae, they contract vigorously, forcing fluid into the gut. This arrangement may have two consequences: (1) as the fluid is forced through the labyrinthine drainage canal, back-flow from the gut into the ampulla may be prevented, and (2) the fluid may be retained sufficiently long enough to allow for modification of the tubule secretions (Wall *et al.*, 1975).

The exact mechanisms of ultrafiltrate (primary urine) production are not clear. However, available evidence suggests that water movement across insect Malpighian tubule epithelium involved ion movements generated by a cation pump and concomitant flow of anions which follow positively charged solutes. The tubule lumen typically has a transepithelial potential about 30 mV positive to experimental bathing solutions (haemolymph-side) indicating that movement of cations (K^+) into the lumen is thermodynamically uphill (Maddrell, 1977). In most cases, the rate of fluid secretion is dependent upon the K^+ concentration of the bathing solution. When K^+ is not included in the bathing solution, the secretory rate is usually around 10% of maximum (Maddrell, 1977). Wall (1970) observed that the K^+/Na^+ ratio of fluid secretions collected (*in vivo*) from the ampullae was higher when the rate of fluid secretion was higher. In addition, the K^+ concentration was slightly higher than that of Na^+. It is important to note that the haemolymph K^+/Na^+ (1/11) ratio is low (Wall, 1970), indicating that K^+ is actively secreted into the tubule lumen. The osmolality of the secreted fluid was slightly higher (32 mosmol) than haemolymph when fluid was collected from ampullae (Wall, 1970) and 38 mosmol when collected from tubule lumens above ampullae (Wall *et al.*, 1975). Maddrell (1977) has recently summarized evidence that the water flux is a secondary consequence of ion movements, and that secreted fluids are usually slightly hyperosmotic to experimental bathing solutions which vary over a wide range of osmotic concentrations. Furthermore, fluid flow rates into the Malpighian tubules show a fairly close inverse relationship to the osmolality of the bathing solution. This indicates that the rate of solute movement is nearly constant, but water movements change so that secreted fluid is slightly hyperosmotic. Several hypotheses have been proposed to explain how this water flux is produced, but none of them are completely satisfactory. Maddrell (1977) has concluded that Malpighian tubule ion transport is so well coupled to water movements that their secretions are essentially isosmotic. This view lends support to the hypothesis that the main function of Malpighian tubules is to provide a fluid flux into which solutes from the haemolymph can diffuse or be transported (Maddrell, 1977).

Removal of organic materials appears to involve two types of processes which are active secretion and passive flow. It has been known for quite some time that Malpighian tubules are capable of concentrating acidic dyes from dilute solutions (Maddrell, 1971; 1977). Although recent interest in the processes of organic anion and cation excretion has produced some useful information, no clear patterns have been established. It appears that active processes resulting in excretion of these materials undoubtedly occur, but the various systems may not be universal in insects (Maddrell and Gardiner, 1976). Passive permeability of Malpighian tubules is viewed as a means by which toxic materials may be easily removed from haemolymph (Maddrell, 1977). However, useful materials (i.e. metabolites of low molecular weights) may also be removed from haemolymph, which must then be reabsorbed (requiring expenditure of energy). Both active and passive excretory processes in Malpighian tubules are in need of extensive investigation, particularly because of their importance in understanding how insects deal with toxic compounds.

The fluid secretion produced by Malpighian tubules passes into the ileum and on into the colon (Fig 6.2a). Wall *et al.*, (1975) have provided a brief description of the ileum, and Bignell (Chapter 4) has discussed the structure and function of the colon. Information on the physiological role of the insect hindgut is limited (Stobbart and Shaw, 1974). In cockroaches, its role has been described as being concerned primarily with the absorption of water and diffusion of certain ions (Cornwell, 1968), but few details regarding these activities in the ileum and colon are available. However, the observation that the colon represents 30% of the alimentary canal volume (Chapter 4) and harbours a variety of micro-organisms (Hominick and Davey, 1975) suggests that it may play an important role.

(f) *Rectum*

The insect rectum has been subject to study by many workers because it is a good system for examining fluid transport. In addition, the success of many terrestrial insects living in dry environments depends on the ability of this structure to conserve water. Studies on *P. americana* have improved our understanding on the structure and function of the insect rectum. Two detailed reports on rectal morphology have been published (Oschman and Wall, 1969; Noirot and Noirot-Timothée, 1976) that are basically in general agreement, with one exception involving identification of one cell type, as will be discussed below.

The rectum is that portion of the hindgut located between the colon and anus (Fig. 6.2a). It is about 4 cm long with an anterior portion about 1 mm in diameter and a posterior region ranging from about 1.5–2.0 mm in diameter (Fig. 6.2d). A distinguishing characteristic is the presence of an extensive tracheal supply consisting of 2 lateral abdominal tracheal trunks. From these trunks, 6 main tracheae emerge each forming a separate branching system extending over and into one of the 6 rectal pads. These pads are lined with lamellate cuticle bearing epicuticular depressions (lumen side). The rectal pads were originally described by Oschman and Wall (1969) as having radially arranged cushion-shaped thickenings of epithelium composed of a single layer of tall columnar cells. However, recently Noirot and Noirot-Timothée (1976) have reported that the rectal papillae (pads) are composed of two layers of cells. The first layer is one of principal cells situated beneath the cuticle. They are columnar and correspond to the rectal pad cells described by Oschman and Wall, 1969 (Fig. 6.3). The second layer consists of the much flatter or compressed basal cells that form a continuous layer at the base of the principal cells. Each of the pads is surrounded by a sclerotized frame supported by long, narrow junctional cells which ensures association of the pads with the general rectal epithelium. The epithelium is surrounded by a layer composed of circular and longitudinal muscles and connective tissue (musculo-connective sheath), but is separated from the rectal pad surface by a subepithelial sinus. The musculo-connective sheath is innervated (neurosecretory fibres) and tracheated. In addition, other tracheal trunks penetrate into the epithelium, insinuated between the basal and principal cells. Fluid flowing through the sinus enters the haemolymph through openings in the musculo-connective sheath at points where large tracheae penetrate. It is thought that these openings can be sealed by muscular contractions which appress the muscle around the openings against the pad surface (Oschman and Wall, 1969).

The principal cells are about 100 μm in height and 10–15 μm in width, have a lobed nucleus and two important types of differentiation (Noirot and Noirot-Timothée, 1976). These are: the apical complex, comprising numerous folds of the plasma membrane associated with numerous elongate mitochondria, and the lateral complex, lying beneath the septate junctions and extending to the basal region. Adjacent principal cells have very convoluted interdigitations whose membranes are linked by scalariform junctions and associated with mitochondria (Fig. 6.3). The apposition of these mitochondria to the plasma membrane is very precise (separated by a regular space of 105 Å) forming a characteristic structure identified as the mitochondria–scalariform junction complex (MS).

The basal cells form a continuous single layer at the base of the principal cells

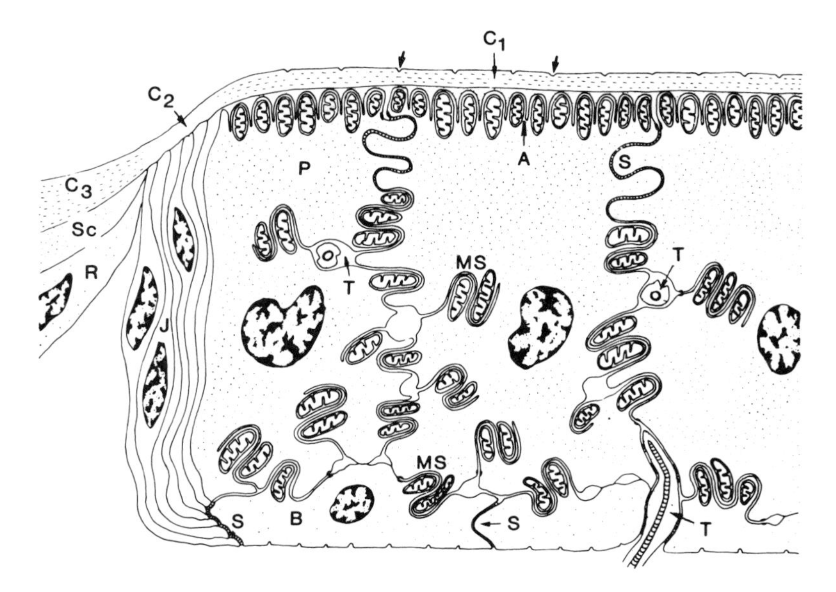

Fig. 6.3 General organization of the rectum of *P. americana*. The rectal papillae are composed of two layers of cells; principal cells (P), and basal cells (B), covered with lamellate cuticle (C_1) bearing epicuticular depressions (arrows). Each papilla is limited along its edge by very narrow junctional cells (J), supporting the sclerotized cuticle of the frame (C_2). Between the papillae, the rectal epithelium (R) supports a lamellate cuticle (C_3) including a thick subcuticle (Sc). On the principal cells, the apical complex (A) is provided by folds of the plasma membrane, associated with elongated mitochondria; on the lateral and basal faces, mitochondria–scalariform junction complexes (MS) are very well developed. MS complexes also occur between principal and basal cells. Tracheae and tracheoles (T) are insinuated between the basal cells, and ramify between the principal cells. The intercellular junctions are quite varied: note the presence of septate junctions (S), both at the apical and basal poles of the papilla. (From Noirot and Noirot-Timothée, 1976.)

and are quite flattened (less than 1 μm thick in some places). Their basal face is regular, but their apical face (which is in contact with the principal cells) is quite convoluted (Noirot and Noirot-Timothée, 1976).

More recent studies on the organization and isolating functions of the rectal pads using freeze-fracture and lanthanum impregnation techniques, have revealed that sheath cells (formerly termed junctional cells), septate junctions, and tight junctions (found at the basal intercellular clefts) are structured in a manner which form permeability barriers (Lane, 1978; Noirot-Timothée *et al.*, 1978; Noirot *et al.*, 1979). While some details are not clear, it is obvious that these structures are designed to function in facilitating water flow from the rectal lumen to the haemolymph.

As one can see, the structural relationships of the *P. americana* rectum have been studied in sufficient detail allowing for development of a working model on how this structure functions in osmoregulation. Furthermore, a series of studies by Wall and Oschman (1970) have attempted to obtain information to test this model. Micro-measurements of the osmotic concentrations (rectal lumen, subepithelial

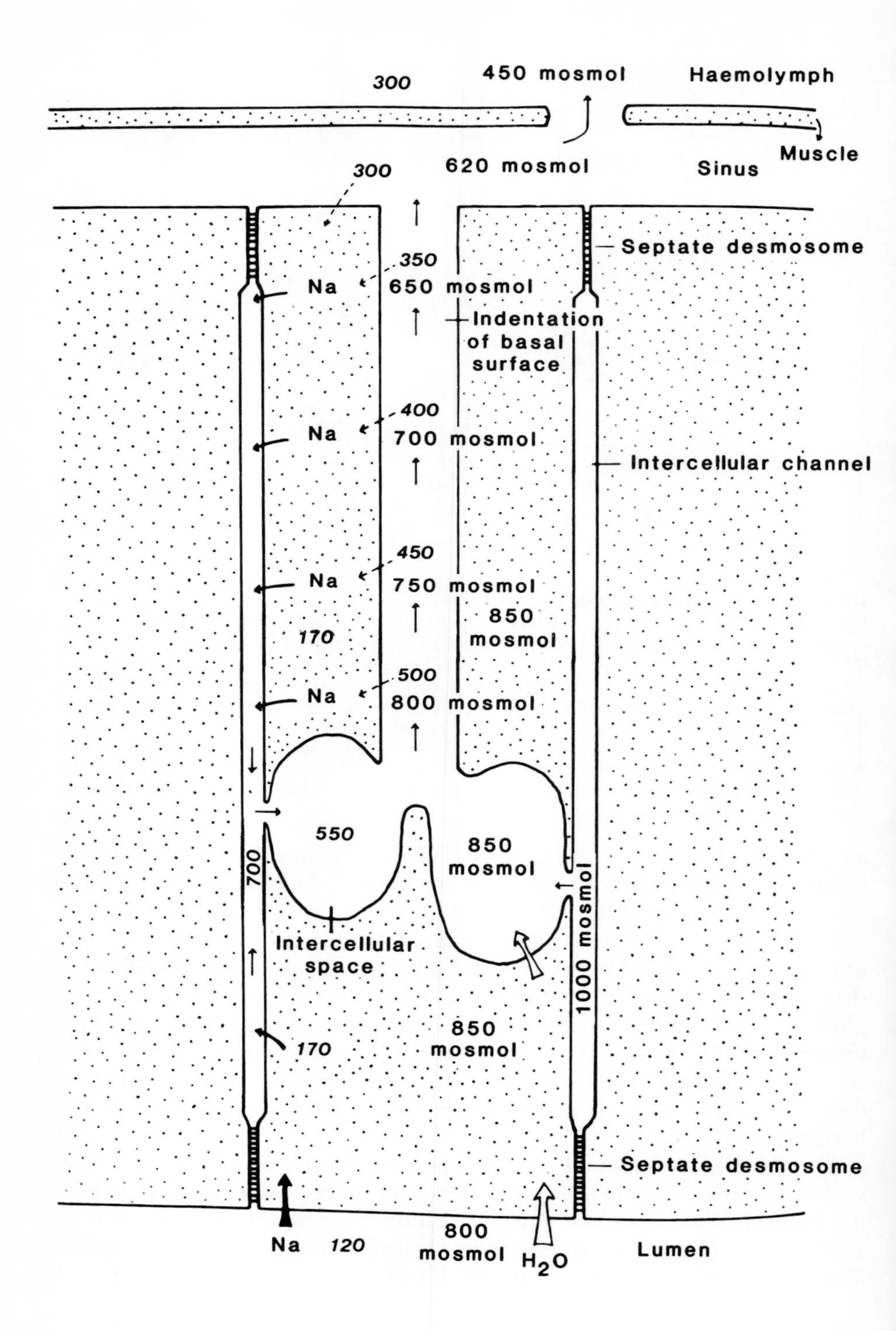

300
450 mosmol
Haemolymph
Muscle
300
620 mosmol
Sinus
Septate desmosome
350
Na
650 mosmol
Indentation of basal surface
400
Na
700 mosmol
Intercellular channel
450
Na
750 mosmol
850 mosmol
170
500
Na
800 mosmol
550
850 mosmol
700
1000 mosmol
Intercellular space
170
850 mosmol
Septate desmosome
Na
120
800 mosmol
H_2O
Lumen

Fig. 6.4 Diagrammatic representation of rectal pads, with the scheme proposed for solute recycling. To illustrate how a solute such as Na^+ could be recycled, concentration of Na^+ (plus anion, mosmol) that might be expected in various regions is given on left (italics), and expected total osmolality (mosmol) is shown on right. Bold-faced values (lumen, sinus and haemolymph are average determinations made on dehydrated animals. Other values indicated are highly speculative, and are intended only as examples. To generate a flow of water into cell, it is presumed that the cell is more concentrated than lumen. This may be accomplished by solute pumps located along the membrane-facing lumen. It is suggested that solutes are pumped into narrow intercellular channels so that the latter become quite concentrated. This fluid flows into larger intercellular spaces and becomes diluted because of the osmotic influx of water from the cell. Larger intercellular spaces may become diluted to the same total osmolality as the cell, but the concentration of transported solute would remain quite high relative to that in the cell. Solute could then diffuse down the concentration gradient back into the cell. This might occur (dashed arrows) as fluid flows along indentations of the basal surface and perhaps also from the subepithelial sinus. Presumably, basal plasma membrane is relatively impermeable to H_2O. Solutes may then enter narrow intercellular channels to be returned to the apical region, completing the cycle. (From Wall and Oschman, 1970.)

---

sinus and haemolymph) in hydrated and dehydrated *P. americana* have revealed that water is removed from the rectum under certain physiological conditions (Wall and Oschman, 1970). Under dehydration conditions (antidiuresis), the average *in vivo* osmotic gradient across the posterior portion of rectal pads was 972 mosmol (posterior lumen), 620 mosmol (posterior sinus) and 436 mosmol (haemolymph). During antidiuresis, the rectal lumen contents may reach 1600 mosmol (Wall, 1971). Clearly, the sinus fluid becomes hypoosmotic to the lumen. In hydrated insects, the average osmotic gradient across this structure was 275 mosmol (posterior lumen) 391 mosmol (posterior sinus) and 379 mosmol (haemolymph). Under these conditions, the rectal lumen is hypoosmotic to the haemolymph, but the sinus is hyperosmotic to the lumen. Measurement of Na^+, K^+ and amino acids (determined by a ninhydrin colour reaction) indicated that net water and solute movement across rectal pads are not stoichometric (Wall and Oschman, 1970).

A general means for achieving transepithelial water transport is to develop an osmotic gradient by pumping solutes into compartments within the epithelium. It would appear that, in maintaining a constant water flux, a continual supply of solutes would be required, which would presumably result in dilution of solutes by water flow. However, Wall and Oschman (1970) have found that the rectum is capable of removing water from the lumen even in the absence of transportable solutes in the lumen. Two sources of solutes needed for water transport have been suggested by these workers. These are: recruitment of solutes from haemolymph, and, solute recycling in the rectum. Due to the overall structure it is thought that the musculo-connective sheath forms a barrier which does not allow for adequate recruitment of solutes to occur. On the other hand, Wall and Oschman (1970) present a model and some evidence on how solute recycling within the rectal pad could occur. Fig. 6.4 summarizes their proposed scheme for solute recycling, which could provide a continuous water flux across this tissue. In presenting this scheme, various concentrations of Na^+ (plus anions) mosmol kg^{-1} H_2O which might be expected in the different regions are designated (in italics) along with the expected total osmolality. The values given for lumen, sinus and haemolymph represent averages of determinations on fluids obtained from dehydrated *P. americana*. The other values are speculative, since they were not based on actual measurements. Thus, these values and the use of Na^+ (plus anion) are intended only to provide an example of how an anion could be recycled.

Wall (1977) summarized some of her more recent work on this problem. She observed that amino acids may be actively absorbed by rectal pads and may be recycled along with other solutes (sugars, glycerol, glycoproteins, etc.) within the rectal pad. One point regarding the possibility that amino acids are involved in the recycling process should be made. Amino acids have been determined on the basis of a ninhydrin reaction (Wall and Oschman, 1970) which also gives a positive reaction to ammonia. Ammonia (probably as ammonium ions) has been found to be a major excretory product along with some other amino nitrogen-containing compounds (Mullins and Cochran, 1973a). In addition, injections of Na^+ into *P. americana* result in a reduction in the amount of ammonia produced when compared to that produced by equimolar ammonia and K^+ injections (Mullins, 1974). The role of ammonium ions in the excretory process is not clear, but studies designed to further examine this process should include determinations of both total amino nitrogen and ammonia.

Published work has not clearly established the mechanisms involved in the rectal fluid transport. However, uptake of water by the rectum of *P. americana* can be reduced or abolished by dinitrophenol (Sauer *et al.*, 1970; Wall, 1967). Tolman and Steele (1976) have identified ouabain-sensitive, (Na^+–K^+)-activated ATPase in the rectal tissue of *P. americana*. In *Leucophaea maderae*, lumen K^+, but not Na^+ or Ca^{2+} stimulates tritiated water absorption from the rectal lumen and lumen Na^+ increases water influx into the lumen (Hopkins and Srivastava, 1972). In addition, CO_2 anaesthesia reduces water efflux and influx, and ouabain reduces water efflux from the rectal lumen. All of these observations do not clarify the mechanisms which may be involved, but they do indicate that water and ion movements are energy-dependent.

It can be surmised that sodium and other solutes are probably transported into rectal cells via the apical complex. Depending upon physiological conditions, water (diuresis) or solutes (antidiuresis) may be absorbed. The structures of these cells appear to be designed such that high concentrations of solutes in one part of the apparatus combined with membranes, which may have variable permeabilities to water may produce a net water flux. Solutes may be absorbed or recycled in the structure depending upon the requirements of the insect.

As mentioned previously, the rectum is thought to absorb various materials (other than ions and water) from the lumen. Wall and Oschman (1970) have shown that amino nitrogen is absorbed, but otherwise little else is known about this process in insects (Maddrell, 1971; Stobbart and Shaw, 1974). Phillips and Dockrill (1968) have shown that the cuticular lining of the locust is designed in a manner such that only small molecules and ions are permeable. Presumably, this restricts exposure of large molecules to the rectal tissues. Whether this situation exists in *P. americana* has not been reported. Hopkins *et al.*, (1971b) suggested that, in *L. maderae*, the rapid turnover and cycling of water through the rectal wall and lumen, in contrast to the relatively long time required to dehydrate a faecal pellet, may be a means by which small metabolically useful solutes, filtered from the haemolymph or occurring as digestion residues, are absorbed. Studies designed to explore the processes of absorption of various materials in the rectum would undoubtedly provide some useful information, and perhaps some surprises.

6.4 Excretory products

It was pointed out in the introduction that certain materials produced in insects might themselves be toxic, or if allowed to build up in the insect, could interfere with normal metabolic or physiological functions. Homeostasis is maintained by removal or storage of these products. Some of these materials may be readily removed. For example, CO_2 is lost by gaseous exchange (diffusion) through the cuticle or by the tracheal system (Stobbart and Shaw, 1974). The precise mechanisms of CO_2 transport are not clear, but they may well involve carbonic anhydrase since this enzyme has been found in several insect tissues (Johnston and Jungreis, 1979) including *P. americana* (Anderson and March, 1956). Some of the aspects of water loss due to transpiration or diffusion have been presented in Section 6.2.2. Elimination of nitrogen is viewed as a biochemical and physiological necessity since more protein is usually consumed than needed to meet an insect's metabolic requirements. Ammonia is the primary degradation product of nitrogen-containing compounds. Since it is quite water-soluble and toxic, its disposal exerts an influence on an organism's water balance. As a consequence, terrestrial insects usually incorporate excess amino nitrogenous materials into a less soluble, non-toxic form prior to excretion. Insects are generally considered to be uricotelic, since many species may excrete 80% or more of this nitrogen as uric acid (Stobbart and Shaw, 1974). In addition, a wide variety of nitrogen-containing compounds may be excreted (Cochran, 1975). The nitrogen excretion patterns of cockroaches have been found to be quite different from the scheme established from studies on other terrestrial insects. These differences are discussed below.

6.4.1 Nitrogenous excretory products

Studies on nitrogenous excretory products and nitrogen balance of *P. americana* maintained on diets containing various levels of dietary nitrogen have revealed two important aspects regarding nitrogen metabolism in this species. The American cockroach may be characterized as being externally ammonotelic and internally uricotelic. Both the rate of ammonia excretion and urate retention are dependent on the dietary nitrogen level at which the insect is maintained (Mullins, 1974; Mullins and Cochran, 1974).

(a) *Internal storage*

Internal storage of uric acid or urates occurs widely in cockroaches (Cornwell, 1968; Guthrie and Tindall, 1968), where it is found primarily in specialized fat body cells, termed urate cells (Cornwell, 1968; Cochran *et al.*, 1979; see also Section 7.3.2). These urate reserves are in a dynamic state and fluctuate in response to nitrogen balance. When maintained on positive nitrogen diets ranging from 24%–91% casein protein for 16 weeks, females synthesize and store from 0.18–1.12 mg uric acid cockroach^{-1} day^{-1} and males synthesize and store 0.06–0.73 mg uric acid cockroach^{-1} day^{-1} (Mullins and Cochran, 1975a). In addition, individuals maintained on high protein diets for extended periods of time appear to preferentially synthesize and store uric acid apparently at the expense of other metabolically useful compounds (Gier, 1947b; Haydak, 1953; Mullins and Cochran, 1975a).

If *P. americana* adults whose urate reserves have been allowed to build up (maintenance on a high protein diet), are placed on diets low in nitrogen, the stored urates are mobilized. The rate of urate mobilization is related to the amount of carbohydrate and nitrogen present in the low nitrogen diets. For example, females lost body urates at the following rates when placed on these diets: cellulose + 5% protein (low carbohydrate, low nitrogen) −0.04 mg urate cockroach^{-1} day^{-1}; dextrin + 5% protein (high carbohydrate, low nitrogen) −0.32 mg urate cockroach^{-1} day^{-1}; and dextrin (high carbohydrate, no nitrogen) −0.45 mg urate cockroach^{-1} day^{-1} (Mullins and Cochran, 1975b). Urate nitrogen appears to be incorporated into oöthecae when females are on a negative nitrogen balance. When nitrogen consumption plus urate nitrogen mobilization was measured in females maintained on the 3 diets over a 17-week interval, 23% (cellulose + 5% protein) 52% (dextrin + 5% protein) and 89% (dextrin) of the dietary and mobilized nitrogen appeared in oöthecae produced. The rates of urate mobilization for males was not as large as those observed in females. They were: 0.04 mg urate cockroach^{-1} day^{-1} (cellulose + 5% protein diet), 0.15 mg urate cockroach^{-1} day^{-1} (dextrin + 5% protein diet) and 0.20 mg urate cockroach^{-1} day^{-1} (dextrin diet). The differences in rates of urate mobilization between sexes appears to be related with their nitrogen demands, females having a higher demand to satisfy their reproductive role (Mullins and Cochran, 1975b). The metabolic processes involved in mobilization are not clear, but fat body mycetocyte bacteriods may play a significant role (*see* Section 7.4).

Urate storage in the body appears to involve formation of a uric acid/urate complex with protein (peptides) or its association with monovalent cations (H^+, K^+, Na^+ and NH_4^+). The precise storage form of the complex is not yet known, although some information is available. Hopkins and Lofgren (1968) found that *L. maderae* fat body urates appeared to be associated with proteins or peptides. Urate build-up in *P. americana* is accompanied by increased K^+ and Na^+ retention. Females on a 50% protein diet (16 weeks) excrete only 40% of the K^+ and 39% of the Na^+ consumed in their diets (Mullins and Cochran, 1974). On the other hand, urate mobilization results in a concomitant decrease in K^+ and Na^+. Females on a dextrin diet (17 weeks) excrete 157% of the K^+ and 105% of the Na^+ they consumed in their diets (Mullins and Cochran, 1974). Characterization of isolated and partially purified urate spherules obtained from *P. americana* have been found to contain 42 μg non-urate N(proteins, NH_4, etc.) 49 μg mg^{-1} K^+ and 9 μg mg^{-1} Na^+ (Mullins, 1979a). Infrared spectrophotometric studies on these spherules have revealed that, although they display a characteristic i.r. spectrum, it does not correspond with certain urate standards (urate salts of K^+, Na^+, NH_4^+). It is possible that the crystalline spherules represent a mixture of uric acid and the various salts (K^+, Na^+ and perhaps NH_4^+), which may be variable. Support for this hypothesis includes the following observations: (1) when males are maintained on high K^+ diets (variable nitrogen levels), they show a tendency to store greater amounts of K^+ and urates than those fed on similar diets containing less K^+ (Mullins, 1979b), (2) males fed on low K^+ diets showed a tendency for increased Na^+ retention in the fat body (Mullins, 1979b) and (3) dehydration has been found to increase the K^+/urate ratio and the total fat body urates (Tucker, 1977b). From this discussion, it can be seen that nitrogen and ion metabolism appear to be inter-related. Urate storage and mobilization are influenced by the diet and presumably by the metabolic and physiological state of this insect.

(b) *External excretion*

The major nitrogenous excretory materials which are eliminated from *P. americana* have been studied (Mullins and Cochran, 1973a,b). A survey of 36 potential nitrogenous materials which might occur in faeces from insects maintained on three diets revealed that few of the classical terrestrial insect excretory products (including uric acid) could be detected. Instead, ammonia, three tryptophan metabolites, unidentified (5–9, depending on the diet) ninhydrin-positive (amino group-containing), water-soluble materials and unidentified water-insoluble materials constitute most of the nitrogen excreted. Ammonia was the largest single material which was excreted (Mullins and Cochran, 1973a).

Ammonia excretion is of particular interest because the classical view holds that terrestrial organisms generally do not eliminate nitrogenous wastes in this form. It has been found that male *P. americana* may eliminate from 14–91% of their externally excreted nitrogen as ammonia (Mullins and Cochran, 1972). Generally diets containing high levels of nitrogen, produce higher levels of ammonia release. Ammonia appears to be eliminated exclusively via the faeces (presumably as NH_4^+), since little is released across the respiratory surfaces (Mullins, 1974). Ammonium ions were found to be the major cations present in faeces obtained from males maintained on 12 diets containing various levels and sources of nitrogen. Therefore, this insect can be described as being ammonotelic (based on external nitrogen excretion), and this fact seems to influence its water balance (required uptake). Indeed, increased *ad libitum* water consumption appears to be closely correlated with nitrogen contained in the insect's diet and nitrogen excreted (Mullins, 1974).

Although the mode of ammonia release is associated with the gut, its precise origin is unknown. There is some evidence that microbes residing in the gut may be involved. However, it is quite probable that ammonia can be released by gut tissues since various ammonia-generating enzymes have been found in insect gut tissues. Ammonia excretion is most likely the result of several activities including gut microflora, nitrogen status and ion balance of *P. americana*. Some possible functions of gut ammonia include: (1) it may serve as a *bona fide* excretory product, (2) it may serve to buffer the gut contents for various enzymatic and microbial activities, (3) it may be the result of de-amination of the D-amino acids which might be produced by mycetocyte bacteriods and the gut microflora, (4) it may be released into the gut so as to provide gut microflora with a nitrogen source with which they might produce nitrogen-containing materials, that if absorbed could prove useful to the host, and (5) under certain conditions, ammonia excretion may be useful in water and ion conservation. These possible functions have been discussed in more detail elsewhere (Mullins, 1974).

Three tryptophan metabolites have been found in faeces of *P. americana* and are excreted in increasing concentrations as dietary nitrogen is increased (Mullins and Cochran, 1973b). These are kynurenic (1–2 $\mu g\ mg^{-1}$ faeces), xanthurenic (1–3 $\mu g\ mg^{-1}$ faeces) and 8-hydroxyquinaldic (2–4 $\mu g\ mg^{-1}$ faeces) acids. It has been found that cockroaches fed antibiotics produce little 8-hydroxyquinaldic acid. Presumably, antibiotics inhibit dehydroxylation of xanthurenic acid to 8-hydroxyquinaldic acid by gut microflora. In addition, faeces from two eye colour forms of *P. americana* (wild-type and lavender-eye mutants) contain all three metabolites, but pearl-eye mutants excrete none of them (Mullins and Cochran, 1973b). Cochran (1976) has suggested that pearl-eye mutants do not have an active tryptophan oxygenase, preventing production and excretion of these three metabolites.

6.4.2 Miscellaneous materials

The elimination of miscellaneous compounds which an insect may absorb through its integument or that it may ingest is a complex topic. The metabolic excretory products or unmetabolized materials found in faeces (or eliminated via gaseous exchange) are dependent on the metabolic state of the insect and the specific nature of the compounds involved. Information on types of materials excreted by insects has usually resulted from studies directed towards learning about the fate of specific compounds. It is quite likely that excreted materials contain large numbers of waste products present in amounts which are below the limits of detectability (Cochran, 1975).

With these points in mind, specific information dealing with the fate of individual compounds will not be discussed. However, a few general comments should be made. Cochran (1975) has discussed the general biochemical and excretory mechanisms which insects use to deal with foreign compounds (xenobiotics). Dyes and insecticides have been used to examine detoxication and excretion mechanisms in an attempt to gain knowledge of elimination processes involved. The products of many detoxication reactions are usually rendered more water-soluble which is often due to their conjugation with simple sugars, amino acids, peptides, glutathione, sulphate or phosphate (Cochran, 1975). Aside from the fact that certain materials occur in the faeces, little is known about the excretory processes involved. However, water solubility (of an unmetabolized compound) or increased solubility (due to metabolic transformation) of toxic or unneeded compounds may increase flow from the haemolymph into the Malpighian tubules by passive diffusion. If these materials are not reabsorbed in the hindgut, they will be eliminated in the faeces.

6.5 Regulation and kinetics of osmoregulation

It is the purpose of this section to focus on the general whole body responses, hormonal influences, kinetics, energetics and inter-relationships of the components which provide *P. americana* with its ability to osmoregulate. These activities are undoubtedly complex, particularly since nitrogen and ions may be stored internally.

6.5.1 Regulation of ionic composition

Information on the ability of *P. americana* to osmoregulate has been obtained primarily from studies involving dehydration, dehydration and rehydration, injections of saline solutions and ingestion of salts in food or water. Edney (1968) observed that dehydration caused a significant loss in haemolymph water, and an elevation in haemolymph osmotic pressure (410 mosmol l^{-1} for controls to 467 mosmol l^{-1} for dehydrated insects). Based on total water loss, the expected rise should have been about 607 mosmol l^{-1} (Edney, 1968). Evidence for strong osmoregulation of *P. americana* haemolymph under dehydration and rehydration conditions was further supported by Wall (1970). She found that extreme dehydration and subsequent rehydration resulted in less than expected changes in haemolymph osmolality with excess solutes apparently being sequestered and later released by some tissue(s). The observed changes could not be explained by the excretory functions as they were understood. Figure 6.5 illustrates these findings. Mullins and Cochran (1974) suggested that stored urates might be involved in ion exchanges (ion sink hypothesis) useful to maintaining homeostasis. Support for this hypothesis has been obtained by subsequent studies (Hyatt and Marshall, 1977; Tucker, 1977a,b,c). These workers have found that during dehydration and rehydration the fat body (and stored urates) was apparently the major tissue involved in haemolymph ion exchanges. It should be noted, however, that the results from these studies were variable. It appears that other tissues and mechanisms may also be involved in maintaining haemolymph osmotic homeostasis (Tucker, 1977d).

Injections of various saline solutions into *P. americana* haemocoels has indicated that rapid increases in ionic content are accommodated with little effect. For example, Munson and Yeager (1949) found that injection of high levels of NaCl

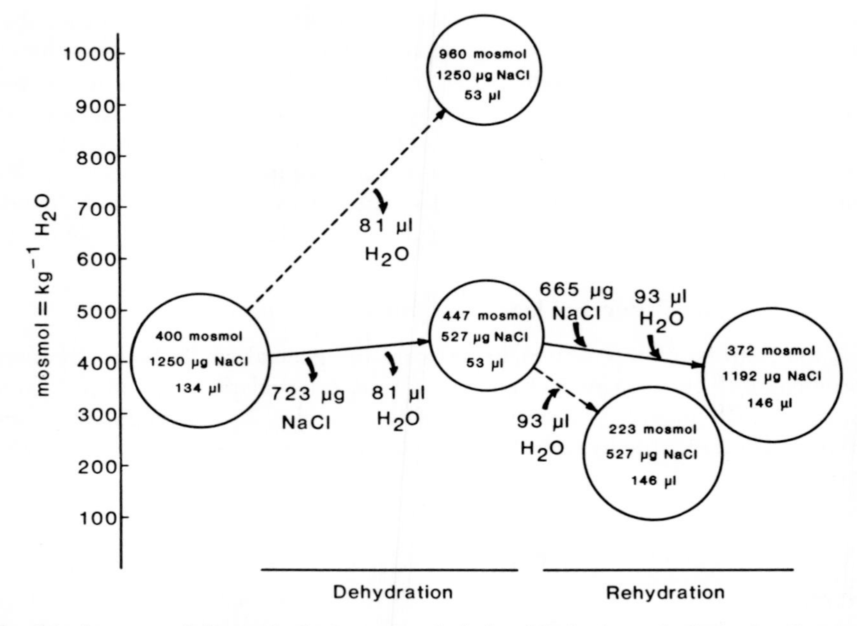

Fig. 6.5 Summary of changes in the haemolymph during dehydration and rehydration. In normal animals (left circle), the haemolymph has an osmolality of 400 mosmol. This consists of 1250 µg of NaCl, and the total volume is about 134 µl. After 9 days of dehydration, the haemolymph volume decreases to an average of 53 µl, a loss of 81 µl of water. If no solutes were removed from the haemolymph, the haemolymph osmolality should increase to 960 mosmol (upper middle circle). Since haemolymph osmolality did not increase to this extent about 723 µg of NaCl were removed from the haemolymph. When a dehydrated animal is given water but no food, haemolymph volume increases to the normal level within 24 h. If no solutes were added to the haemolymph during rehydration, the haemolymph osmolality should decrease to 223 mosmol (lower right circle). Since haemolymph osmolality in rehydrated animals was actually 372 mosmol (upper right circle), about 665 µg NaCl were added. The source of these solutes is not known. (From Wall, 1970.)

appeared to result in release of tissue water into the haemocoel and some of the Cl^- passed from the haemolymph into the tissues. Van Asperen and Van Esch (1956) conducted a series of experiments including injections of distilled water, and a series of saline solutions followed by analysis of samples obtained at different post-injection times. They found that normal haemolymph ion concentrations were restored rather rapidly.

When *P. americana* is presented with water or foods containing various levels of salts it demonstrates a capacity to regulate its haemolymph composition to some extent. Tobias (1948) found that large oral doses of KCl solutions (8 times the osmotic pressure of the haemolymph) did not depress haemolymph Na^+/K^+ below unity. When *P. americana* were maintained (7 days) on a series of NaCl solutions ranging from 0–700 mosmol the haemolymph Na^+ concentration varied from 115 mmol l^{-1} (0 mosmol saline) to 310 mmol l^{-1} (700 mosmol saline) (Heit *et al.*, 1973). The osmolality of the haemolymph increased in individuals maintained on salines hyperosmotic to their haemolymph, but displayed some regulation (i.e. the observed osmolality was less than that expected, based on the saline imbibed). Pichon (1970) observed that, when *P. americana* are fed different diets containing various concentrations of salts, the haemolymph concentrations were regulated quite effectively. Starvation, however, resulted in a decrease in haemolymph ion concentration which showed an increase when starvation was associated with dehydration (Pichon, 1970). From this discussion, it is evident that regulation of ionic content does occur in *P. americana*. The discussion of how this regulation might be achieved will follow.

6.5.2 Hormonal influences and kinetics of ion transport

Integration and control of excretion in insects involves ion, water and other metabolic processes which are mediated by hormones. The state of knowledge in this area has expanded considerably in recent years, using certain insects as models (Gee, 1977) (see also Chapter 12). In this regard, *P. americana* and other cockroach species have been utilized to some extent, but not all aspects have been examined either equally or exhaustively. Information on general hormonal influences affecting water and ion balance has been obtained primarily from studies using dehydrated/rehydrated or saline-injected insects coupled with exposure to tissue extracts and various pharmacological agents. In addition, removal of glands and severance of certain nerves have served to provide evidence of some hormone sources and sites of hormonal action. Studies designed to determine the general influences have included the whole insect, followed by examination of *in vivo* preparations of salivary glands (secretion), Malpighian tubules and the rectum.

(a) *Whole insect*

The influences of hormonal factors in whole body *P. americana* have included some observations which implicate the brain and associated glands as playing a major role in controlling water balance. For example, Treherne and Willmer (1975) have presented evidence that integumentary water loss is influenced by haemolymph-borne factor(s) which originate from the brain and CC. Penzlin and Stolzner (1971) have found that removal of the frontal ganglion or severance of the frontal connectives results in water loss. Keeley (1975) reported that allatectomized and cardiacectomized *Blaberus discoidalis* suffered significant water loss over the controls. These findings suggest that water balance is affected by certain factor(s) which are present in the brain or associated glands, but other metabolic mechanisms could also be involved.

(b) *Malpighian tubules*

Investigations presently indicate that insect hormones controlling excretion by the Malpighian tubules and rectum are produced by neurosecretory cells (Gee, 1977). It is generally thought that neurohormones are synthesized in the cell bodies and are transported as neurosecretory granules down the neurosecretory axons to specific sites of storage and points of release. Diuretic factors or hormones (DH) are materials which affect the removal of water from the insect via Malpighian tubule secretion and by reduced rectal reabsorption. Contrarily, antidiuretic factors or hormones (ADH) are materials which decrease tubular secretion and increase rectal absorption. Early work on cockroaches indicated that regulation of water content in *Blaberus giganteus* was influenced by specific brain neurosecretory cells (Wall and Ralph, 1962). In *P. americana*, the brain was found to contain ADH (Nayar, 1962), and Wall and Ralph (1964) proposed that it was released from the corpora allata (CA).

There is evidence that certain insects store excretory hormonal factors in their respective neurohormonal areas before release. Furthermore, the presence of these materials at their release sites is of value to the insect because it allows for their rapid release in response to appropriate stimuli without requiring rapid hormone synthesis and axonal transport (Gee, 1977). It appears that, in some insects, both ADH and DH are released in response to stimuli associated with feeding and the osmotic pressure in the haemolymph (Gee, 1977). Studies conducted by Penzlin and Stolzner (1971) and Wall and Ralph (1964) indicated that ADH in *P. americana* was released in response to hyperosmotic haemolymph (dehydration, salt injections). This presupposed the presence of haemolymph osmoreceptors (Gee, 1977).

The action of hormonal factors on Malpighian tubules in *P. americana* has been examined by several workers. Antidiuretic factors which decrease Malpighian secretory rates (dye excretion) *in vitro* were found in brain extracts from dehydrated animals and CA extracts from normal animals, but not in CA extracts from dehydrated animals (Wall and Ralph, 1964). These workers suggest that the 'antidiuretic principle' may be produced in the brain, transported to the CA (for storage) and released when needed in water conservation. Furthermore, its absence in CA of dehydrated or salt-loaded *P. americana* suggests that it is utilized as rapidly as it is produced in insects undergoing hyperosmotic stress. In addition to the ADH found in the brain and CA, extracts from the three thoracic and the six abdominal ganglia have also been found to produce an antidiuretic response in *in vitro* Malpighian tubule preparations (Wall, 1965). Mills (1967) confirmed the presence of ADH in the tissues examined by Wall (1965) and Wall and Ralph (1964), but also found that the terminal abdominal ganglion contained a DH that promoted Malpighian tubule secretion. Mills (1967) also reported some evidence that ADH and DH are antagonistic, since combinations of these extracts result in decreased Malpighian tubule diuresis.

The chemical nature of these factors is unknown, although Goldbard *et al.* (1970) and Mills and Whitehead (1970) partially purified the DH and suggested that it is a peptide of molecular weight about 30 000. Maddrell *et al.* (1969) found that, in *Rhodnius prolixus* and *Carausius morosus*, 5-hydroxytryptamine (5-HT) is an active stimulator of Malpighian tubule fluid secretion. The DH hormones and 5-HT behave in a similar manner under some conditions, including separation in Sephadex G-25, but other evidence indicates they are not the same. The conditions which lead to the destruction of DH in *R. prolixus* do not effect 5-HT; thus, they are not the same (Maddrell *et al.*, 1969). 5-Hydroxytryptamine has been found in various tissues of *P. americana* (Colhoun, 1963) and has been shown to stimulate Malpighian tubule muscles (Flattum *et al.*, 1973; Crowder and Shankland, 1974). However, Flattum *et al.*, (1973) have also found that 'autoneurotoxins' (see Section 6.2.3) collected under different stress conditions caused increased Malpighian tubule coiling which was not solely due to 5-HT.

One can see that a considerable amount of information is needed regarding mechanisms which control hormonal releases, the precise nature of the compounds involved and the mode of action on Malpighian tubules. For example, it is possible that a second messenger (i.e. cAMP) plays a role in this system as indicated in *R. prolixus* (Aston, 1975).

(c) *Rectum*

Hormonal factors have been found to influence rectal uptake of water (measured by uptake of water-soluble dyes) in *P. americana* (Wall, 1965). Wall (1967) found that water absorption from recta of dehydrated animals compared with those of hydrated animals was influenced by tissue extracts. *In vitro* experiments using extracts of the corpora allata, metathoracic ganglia and the terminal abdominal ganglia from hydrated animals produced an increase in the water absorption rate, but similar extracts from dehydrated animals resulted in less activity. However, brain extracts from dehydrated animals were found to produce an increase in water absorption. Goldbard *et al.*, (1970) have partially purified an ADH which is thought to be a polypeptide with an estimated molecular weight of about 8000. Wall (1967) postulated that a single ADH could exert an antidiuretic effect on both Malpighian tubules and rectum, restricting water movement into tubules and increasing water absorption in the rectum. Further studies are needed in order to clarify the hormonal control and kinetics of Malpighian tubule and rectal functions.

(d) *Midgut*

The insect midgut has been studied extensively with regard to transport mechanisms involved in water and solute transport (see also Section 4.5.2 and Fig. 4.8). In fact, recently, Blankemeyer and Harvey (1977) and Zerhan (1977) have discussed work on the midgut that has included an examination of kinetics and modelling of the processes involved. Some work on *in vitro* preparations of cockroach midgut has been reported. Movements of both K^+ and Na^+ across the *P. americana* midgut epithelium appears to be affected by the concentration and/or the nature of the lumen contents (Sauer and Mills, 1969b). The net directional flow of K^+ was found to be from the haemolymph side to the lumen and was abolished in the presence of dinitrophenol. Both haemolymph- and lumen-side dinitrophenol-sensitive mechanisms seem to be involved in Na^+ transport, but the direction and magnitude of these movements appear to be dependent upon the nature of the lumen contents. In addition, non-dinitrophenol mechanisms may also be important (Sauer and Mills, 1969b). Similar studies on Ca^{2+} and Mg^{2+} have been done (Sauer and Mills, 1969a). Again, movements of both ions are affected by the concentration of the midgut contents. The influx movements of both Ca^{2+} and Mg^{2+} increase lumen osmolality, but the presence of dinitrophenol tends to increase influx of Ca^{2+}, while Mg^{2+} influx is decreased (Sauer and Mills, 1969a).

O'Riordan (1969) has examined the electrical potential differences across *P. americana* midgut perfusions *in vitro*. He found that these preparations maintained a potential difference of about 12 mV (lumen negative to haemolymph side) in chloride Ringer composed primarily of a lumen-side Na^+ diffusion potential and a haemolymph-side K^+ diffusion potential. The potential difference was enhanced in sulphate Ringer, but reduced when perfused with metabolic inhibitors (nitrogen, dinitrophenol and ouabain). Inhibition of the potential difference by ouabain was effective only when applied to the haemolymph side. O'Riordan (1969) suggested that a Na^+-K^+ ion pump is indirectly responsible for the observed potential differences and directly involved in Na^+ transport into the haemolymph. Recently, Sacchi and Giordana (1979), using midguts obtained from *L. maderae*, found that they exhibit spontaneous transepithelial potential differences ranging between −30 mV (lumen negative to the haemolymph side) and +35 mV (lumen positive to the haemolymph side). These potential differences were strongly affected by low Na^+ or K^+ ion concentrations in the perfusion media. Here, low Na^+ or low K^+ caused a drop in the potential difference towards lumen negative values.

In vitro preparations of the *P. americana* midgut have indicated that hyperosmotic lumen concentrations fail to promote movement of water from the haemolymph side to the lumen (Sauer *et al.*, 1969). However, under these conditions, the osmolality decreases of the midgut lumen are indicative of solute absorption. Subsequent work by this group showed that solute molecules move out of the hyperosmotic lumen without concurrent water movement to the haemolymph side, and a dinitrophenol-sensitive process is correlated with the exclusion of water molecules (Mills *et al.*, 1970).

The midgut appears to be influenced by hormones. Sauer and Mills (1971) have isolated fractions from terminal abdominal ganglion which may cause increases as well as decreases in midgut lumen volume of *P. americana*. There appear to be at least three substances capable of producing these responses. Two may be proteins, one with a mol wt of about 10 000 that effects water removal from the lumen; the other with a mol wt above 30 000 that effects an increase in lumen volume. The third factor which these workers have isolated appears to be a small molecule that potentiates water movements into the midgut lumen. In addition to this study, ultrastructural examination of the midgut has indicated the presence of numerous axonal ramifications which contain neurosecretory material (Wright *et al.*, 1970). These axons are located within a mass of connective tissue interdispersed with muscle bundles and it has been postulated that they may release materials which act upon localized areas in the control of midgut water movement (Wright *et al.*, 1970).

(e) *Visceral musculature*
In addition to hormonal factors described in previous discussions on the Malpighian tubules, rectum and midgut, one other aspect of hormonal control should be mentioned. This includes the general class of agents which affect the activities of visceral muscles. Some of the known agents are classified as neurotransmitters or neurohormones (Cook and Holman, 1979b). Some of these compounds may be involved in controlling movements of the alimentary canal and Malpighian tubules, facilitating the excretory process by moving materials through the digestive tract and compacting dehydrated faecal pellets. In *P. americana*, one such material (proctolin) described as either a neurotransmitter or neurohormone has been isolated, and found to be a pentapeptide with an amino acid sequence of Arg-Try-Leu-Pro-Thr (Starratt and Brown, 1975). Proctolin is associated with the innervation of the viscera and elicits a myotropic response from hindgut muscles at levels as low as 10^{-9}M. Recently, Holman and Cook (1979) have presented evidence for proctolin and a second myotropic peptide neurohormone in *L. maderae*. This is an active area of research and further study should provide additional information on the excretory processes, as well as neural transmission.

(f) *Urate storage and mobilization*
In previous sections it has been emphasized that storage or mobilization of uric acid as urates does exert an impact on ion and nitrogen balance in cockroaches. Since many metabolic processes in insects are considered to be subject to hormonal regulation, one could assume that uric acid metabolism must be regulated in such a manner (see also Section 7.6.5). Steele (1976) has provided an excellent review of insect hormones and metabolism in which the general aspects on known controls of nitrogen metabolism may be found. He has reviewed the findings of several workers on the regulation of urate metabolism in *P. americana*. It appears that the CA is somehow involved in controlling the rate of urate synthesis and storage in fat body, but the manner in which it exerts its influence is not clear. For example, studies have shown that allatectomy of young *P. americana*, but not adults, results in a reduction of uric acid (Steele, 1976). On the other hand, the function of CC appears to differ from that of CA. The removal of CC results in loss of fat body urates, but this trend can be reversed by implantation of CC (Steele, 1976). From these reports, it would appear that much could be learned regarding regulation of nitrogen and ion metabolism in this species and it would undoubtedly provide much needed and useful information.

6.5.3 Energetics of excretion

The energetics of water balance and excretion is a topic which is difficult to address. Throughout this chapter, reference has been made to processes involved in providing this insect with the advantages it uses to deal successfully with its environment. From an ecological standpoint, one might ask the question – how much do all of these processes cost? The relationship of the kinetics and energetics of biochemical and physiological processes occurring in whole insects is indeed complex. Although there are many obvious gaps in our understanding of the specifics relating to ion transport and metabolic transformations, perhaps focus should be brought to bear on the energy relationships involved in some of the activities that are known to occur. Since cockroaches have been used extensively as physiological and biochemical models, some general comments concerned with water balance and excretion are appropriate. Woodland *et al*., (1968) studied the gross bioenergetics of *Blattella germanica*, but this study did not examine excretory metabolism in any detail. It is generally thought that terrestrial arthropods excrete uric acid because it is advantageous to void a substance which is non-toxic and insoluble. While it is often suggested that this is done at considerable expense to the insect (Edney, 1977), it is possible that this hypothesis may not be entirely true. In 1954, Pilgrim discussed the waste of carbon and energy in nitrogen excretion and pointed out that a careful review of the energetics of formed excretory products (NH_4^+, urea and uric acid) indicated the energy input (losses) is (are) not very great. Cochran (1975) pointed out that xanthine dehydrogenase, which completes the final two steps of uric acid formation, is a dehydrogenase; if this enzyme is coupled with an appropriate link to the electron transport system a total of six moles of ATP could be recovered for every mole of uric acid formed. Since *P. americana* is externally ammonotelic, and the *ad libitum* drinking requirements increase in response to elevated dietary nitrogen and nitrogen excretion, the metabolic costs of this process could prove useful in the determination of the energy budget. Similarly, the fact that uric acid is stored internally in association with various cations that might otherwise be excreted should also prove useful.

6.6 Conclusions

Recent studies have revealed that the American cockroach does not fit into the general osmoregulatory and excretory schemes established for many insects. Under appropriate conditions, they are quite capable of synthesizing and storing uric acid or urate salt forms internally. These stored materials may be mobilized or changed in response to dietary or osmotic stress. Therefore, it now appears that urates play a central role in nitrogen metabolism and osmoregulation.

Although some work has been done on ion and water transport in the Malpighian tubules and the alimentary canal, several areas need more extensive examination. These include studies designed to evaluate the importance of organic ions (carbonic, ammonia, amines, etc.), hormonal regulation and related metabolic activities associated with the excretory and osmoregulatory processes. In addition, information is needed on the biochemistry and physiology of the hindgut plus microflora in modifying the primary urine and formation of those materials which are actually excreted.

Acknowledgements

Appreciation is expressed to Dr D. G. Cochran for his interest and the many helpful discussions leading up to and including the preparation of this chapter. Recognition and thanks are extended to both Drs Cochran and J. L. Eaton for critical review of the final manuscript. Thanks are also due to Ms K. J. Mullins who assisted in the preparation of the figures and Miss B. J. Waller who typed the manuscript.

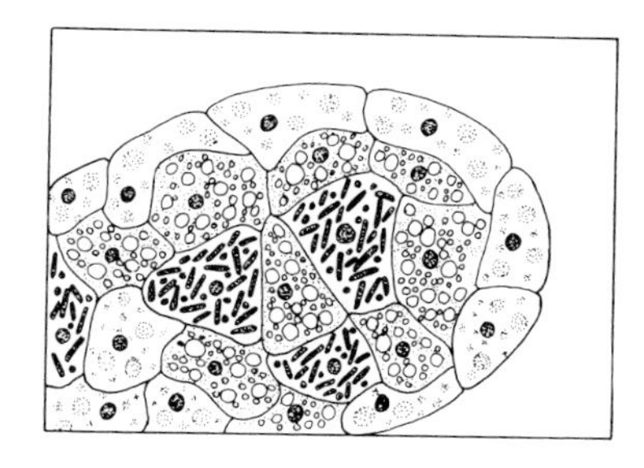

7

Fat body and metabolism

Roger G. H. Downer

7.1 Introduction

Fat body is the major site of intermediary metabolism in insects and also serves as a storage organ for nutrient reserves. These functions indicate that the organ is essential for homeostatic maintenance and occupies a central position in a number of physiological processes. The ubiquitous involvement of fat body in the physiology of insects ensures that a full account of the organ would encroach upon the domain of other contributors to this volume; therefore, the present account is restricted to those topics that appear to be germane to an understanding of the basic structure and biochemistry of fat body. Where possible, the chapter is limited also to consideration of the American cockroach, *Periplaneta americana*, although, in certain sections, the lack of information on this species necessitates reference to related species.

General accounts of fat body may be found in Kilby (1963), Cornwell (1968), Guthrie and Tindall (1968), Price (1973), Richards and Davies (1977) and the reader is referred to these accounts for early references and information on other insect species.

7.2 Morphology

Fat body comprises a diffuse, irregular mass of cells that are suspended in and occupy a major portion of the abdominal haemocoel. The organ extends into the thoracic cavity, thoracic appendages, head capsule and surrounds many nerve axons. The cells are of mesodermal origin being formed by differentiation of the inner walls of the coelomic cavity; however, the prodigious growth of the structure during post-embryonic development obscures the metameric origin. The organ is well-tracheated and is enclosed within an outer sheath of connective tissue which, in the abdomen, surrounds compact aggregations or lobes of cells. Abdominal fat body may be divided into two distinct regions, the inner or *visceral* layer which invests and lies between other abdominal organs, and the outer or *parietal* layer located beneath the body wall. Outside the abdomen, the organ is less conspicuous, being composed of sheets or strands of cells that are interspersed among more obvious tissues and organs.

The form and extent of fat body varies considerably between individuals and, although there appear to be morphological differences that are attributable to sex and stage of development, the extreme variability makes it difficult to ascribe particular differences to specific factors. Thus, the generalized account that is presented in this paragraph should be interpreted with caution and accommodation made for the vagaries of individual insects. The visceral layer appears in early post-embryonic development as a small tissue consisting of only a few hundred cells – the three cell types that will be described in Section 7.3 are represented at this early stage. The organ grows rapidly during nymphal development and, in well-fed nymphs, soon becomes abundant. During early nymphal development, there is no obvious sexual dimorphism but, at the time of the final moult and during the early stages of adulthood, the visceral fat body of males decreases and assumes a chalky white appearance due to deposition of urate/uric acid. By contrast, female visceral fat body remains abundant and does not show marked differences between late nymphal and adult stages.

The cause(s) of the individual variation is (are) not known although some gross morphological changes can be effected experimentally. For example, the nutritional state of the animal influences the appearance of fat body (Haydak, 1953; Mullins and Cochran, 1975a,b; Cochran, 1977) and 'shining white translucent lobules' may be induced in abdominal and thoracic fat body by exposure to cold (15°C) (Singh and Das, 1978). These observations suggest that the gross morphology of fat body reflects the physiological condition and history of the individual, and such factors together with innate individual differences contribute to the observed variability.

7.3 Histology and ultrastructure

Three cell types, trophocytes, urate cells and mycetocytes, have been recognized in cockroach fat body, although the distribution of these cells varies among different anatomical locations. In *Blattella germanica*, mycetocytes are confined to the visceral fat body of abdominal segments 2–6 (Brooks, 1970) and they have not been detected in the parietal fat body of either the abdomen or thorax of *P. americana* whereas trophocytes and urate cells are evident in tissue from these regions (D. G. Cochran, personal communication). Studies on the histology and ultrastructure of cockroach fat body have concentrated on visceral fat body, and the discussion that follows is based upon observations of this tissue.

Visceral fat body is characterized by a lobulated appearance with individual lobes enclosed by a three-layered 'membrane' of connective tissue. In *Blaberus discoidalis*, the 'membrane', which may be up to 3 μm in thickness, comprises an outer layer of closely packed granules and fibrils, a more loosely packed middle layer of banded fibrils that lie perpendicular to those of the outer layer, and an inner layer that is structurally similar to the outer layer (Walker, 1965). The lobe 'membrane' does not invade the interstitial spaces of the lobe cells. These cells are arranged typically with an outer layer of trophocytes surrounding an inner core of cells (D. G. Cochran, personal communication). The typical arrangement of cells within a lobe is illustrated in Fig. 7.1. Urate cells are closely associated with and tend to surround mycetocytes (Cochran *et al.*, 1979) as shown in Fig. 7.2. Cochran *et al.* (1979) identified the principal characteristics of the three cell types and these are summarized below.

7.3.1 Trophocytes

These are the most abundant cells in cockroach fat body and are equivalent to the common fat body cells of most insect species. They function as important sites for storage of fats, carbohydrates and lipids and their structure reflects this storage role. The cells are characterized by the presence of numerous lipid vacuoles (Fig. 7.3). Carbohydrate storage is indicated by the granular nature of the cytoplasm, whereas dense granules characteristic of protein deposits (Gharagozlou, 1972) suggest a protein-storage function (Fig. 7.3). Trophocytes also contain an abundant rough endoplasmic reticulum and numerous mitochondria, both of which serve to emphasize the central role played by these cells in the synthesis, storage and mobilization of metabolic substrates.

7.3.2 Urate cells

These irregularly shaped cells with distinct nuclei serve as sites for deposition and mobilization of uric acid/urates. They are characterized by the presence of crystalloid spherules of urate (Fig. 7.2) which exhibit pronounced birefringence under plane-polarized light. The urate spherules contain a central darkly staining region (Fig. 7.4) that may serve as a nuclear seed around which urate synthesis and/or deposition can occur (Cochran *et al.*, 1979). The characteristic nature of urate deposition in the cockroach is demonstrated by comparison of infrared spectra from urate spherules with those from regular urate crystals and it seems probable that the central 'urate structural unit' contributes to this distinctive mechanism (Mullins, 1979a).

7.3.3 Mycetocytes

These cells are characterized by the presence of intracellular micro-organisms contained within individual membrane-bound vacuoles. The nature and function of

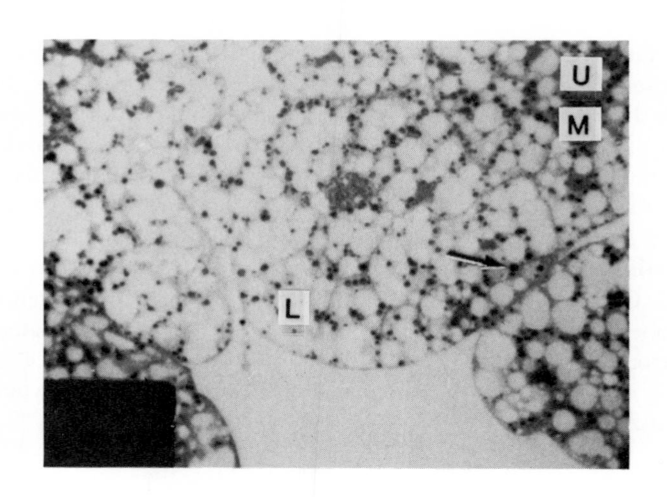
U
M
L

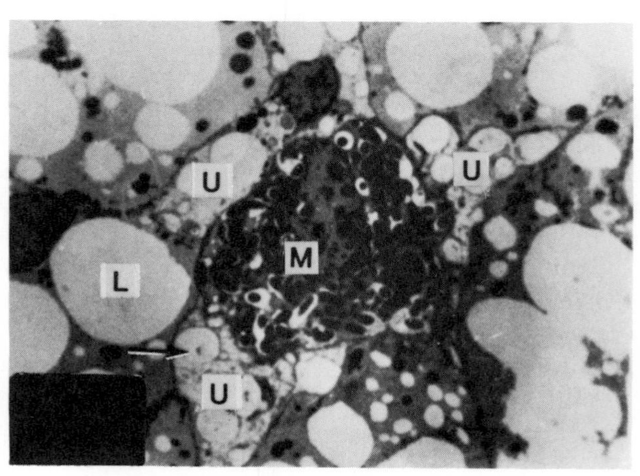
U
U
M
L
U

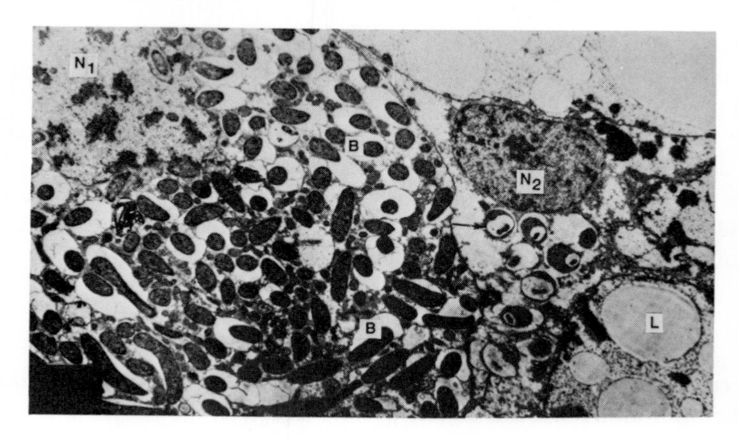
N1
B
N2
B
L

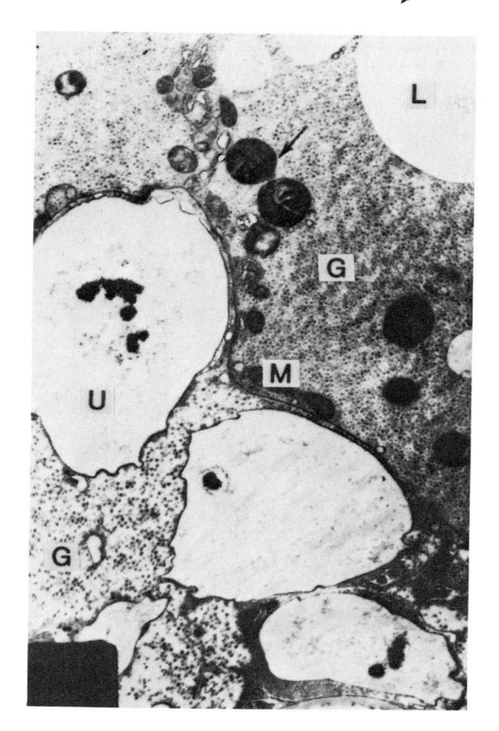

Fig. 7.1 Section of fat body lobes of adult male *P. americana* maintained on diet of dog food. Lipid vacuoles (L); small dark spheres (arrow); mycetocyte (M); urates (U). × 200. (From Cochran *et al.*, 1979.)
Fig. 7.2 Section of fat body of adult male *P. americana* maintained on diet of dog food. U, urate cell; M, mycetocyte; L, lipid vacuole of trophocyte; urate spherule with darkly staining interior (arrow). × 850. (From Cochran *et al.*, 1979.)
Fig. 7.3 Section of fat body of adult male *P. americana* maintained on diet of dextrin. Urate structural units with small, dark-centred spherules (arrows); B, bacteroids; L, lipid vacuole of trophocyte; N_1, nucleus of mycetocyte; N_2, nucleus of urate cell. TEM × 1940. (From Cochran *et al.*, 1979.)
Fig. 7.4 Section of fat body of adult male *P. americana* maintained on a diet of 42% casein protein. U, urate spherule with darkly staining interior; G, granular cytoplasm; M, trophocyte boundary mitochondrion; L, lipid vacuole; protein granule (arrow) TEM × 5100. (From Cochran *et al.*, 1979.)

---

the micro-organisms will be discussed in a subsequent section. The nucleus of the mycetocyte is larger than that of the urate cell (Fig. 7.3) and, in *Nauphoeta cinerea*, the cytoplasm contains glycogen, but little protein and no lipid (Wuest, 1978).

7.4 Bacteroids

7.4.1 General description

Bacteria-like bodies were first reported in insect fat body by Blochmann (1887) and are often named Blochmann bodies (Lanham, 1968) although the more commonly employed trivial term, bacteroid, will be used in the present account. The early literature on bacteroids has been reviewed (Lanham, 1968; Guthrie and Tindall, 1968; Brooks, 1970) and these excellent accounts document the considerable confusion that has arisen concerning the nature and classification of fat-body symbionts. The bacteroids are prokaryotes and resemble rickettsiae or rickettsia-like bacteria (Milburn, 1966). The failure of investigators to establish a pure culture of the bacteroids has resulted in many misidentifications (Gier, 1947a), and Brooks (1970) discounts all early claims of bacteroid culture. She directs attention to the work of Hollande and Favre (1931) as providing the first cytological characterization of the micro-organisms as they exist in the insect and proposes adoption of the classification presented by these workers, namely *Blattabacterium cuenoti*, Hollande and Favre, 1931.

The bacteroids are non-motile, non-spore-forming, faintly gram-positive or gram-variable, straight or slightly curved rods that often appear to be dividing by transverse fission; they range from 0.8–1.0 μm in diameter and 1.5–8.0 μm in length with size and shape varying according to host species, tissue site within the host and the developmental stage of the host (Gier, 1947a; Milburn, 1966; Brooks, 1970). Bacteroids in ovaries tend to be shorter than those in the mycetocytes, suggesting that they may be modified to a 'transmission form' before they infect developing oocytes by the process described by Gier (1936) (see Section 13.2.4). Bacteroids are non-pathogenic to other animal species; indeed natural hosts that have been rendered aposymbiotic cannot be infected by oral administration of bacteroids (Brooks and Richards, 1956).

7.4.2 Functional significance

Ascription of specific physiological and biochemical functions to fat-body bacteroids may be based on observations of bacteroids in culture or on studies that compare xenic insects with aposymbiotic animals. Cockroach bacteroids are difficult to culture and considerable caution must be exercised to prevent contaminating organisms being misidentified and characterized as the natural bacteroids of fat-body (Brooks, 1970). Several reports on the biochemistry of bacteroidal cultures (Pierre, 1964; Donellan and Kilby, 1967; Dubowsky and Pierre, 1967; Tarver and Pierre, 1967) have been challenged on the basis of contamination (Brooks and Richards, 1966; Brooks, 1970) and more recent accounts may be subject to similar criticism. The discounting of such reports results in a sparseness of data that describe direct measurements of the physiological and biochemical features of bacteroids. A few observations on micro-organisms that were isolated from the host tissue by differential centrifugation indicate the presence of the transamination enzymes, L-aspartate: 2-oxoglutarate amino-transaminase and L-alanine: 2-oxoglutarate aminotransaminase, and several steps in the tricarboxylic acid cycle have also been demonstrated (Brooks, 1970).

Bacteroids are sensitive to a number of experimental manipulations and population changes may be induced by heat treatment (Glaser, 1946; Brooks and

Richards, 1956), or exposure of the host insect to sub-lethal doses of the organochlorine insecticide, lindane (Harshbarger and Forgash, 1964). However, of particular interest to the present account is the observation that prolonged feeding or injection of antibiotics into host females reduces the population of bacteroids and results in production of a filial generation of aposymbiotic cockroaches (Glaser, 1946; Brooks and Richards, 1955; Brooks, 1970). The metabolism of such aposymbiotic cockroaches may be compared with that of normal insects and, from these studies, the contributions of bacteroids to fat body function may be inferred.

Aposymbiotic cockroaches are unable to synthesize the amino acids tyrosine, phenylalanine, isoleucine, valine, arginine and probably threonine from carbohydrate precursor, and require a dietary source of these amino acids (Henry and Block, 1962). Tyrosine is an essential precursor in the tanning process and reduced amounts of this amino acid in aposymbiotic cockroaches may explain the lighter colouration that has been reported in such insects (Brooks and Richards, 1955). Tyrosine is also the precursor of several putative neurotransmitters in the cockroach, and lack of these biogenic amines probably contributes to the high mortality of newly emerged aposymbiotic nymphs that are not maintained on an enriched diet. In addition to their possible role in the synthesis of some aromatic amino acids, bacteroids may also be involved in aromatic-ring degradation (Murdock *et al.*, 1970a). Studies with aposymbiotic cockroaches have also led to the conclusion that bacteroids may be involved in the synthesis of certain vitamins. Thus, the concentrations of ascorbic, folic and pantothenic acids are lower in fat bodies of aposymbiotic insects than in those of normal cockroaches (Ludwig and Gallagher, 1966). It is important to recognize in all studies of aposymbiotic insects, that the observed manifestation of the aposymbiotic condition does not imply that bacteroids are solely responsible for the normal state. For example, bacteroids may contribute an essential co-factor to a biosynthetic process, but may not represent the site of synthesis. The resolution of such uncertainties will be possible only when bacteroidal cultures are available.

Considerable circumstantial evidence implicates bacteroids in the recycling of stored urates. Cockroaches that are maintained on a high-protein diet deposit large amounts of urate/uric acid in the urate cells (Haydak, 1953; Mullins and Cochran, 1975a) and these stores may be mobilized and serve as a nitrogen reserve when the animal is placed on a low-protein diet (Mullins and Cochran, 1975b). Elimination of bacteroids results in an accumulation of urate stores (Brooks and Richards, 1956; Harshbarger and Forgash, 1964; Pierre, 1964; Malke and Schwartz, 1966a) and, if bacteroid-containing mycetocytes are transplanted into aposymbiotic individuals, regions that become infected with bacteroids lose their dense urate deposits in contrast to aposymbiotic regions (Brooks and Richards, 1956). These observations, together with the close morphological association of urate cells and mycetocytes (Section 7.3) support the suggestion that bacteroids are involved in uric acid metabolism (Malke and Schwartz, 1966a) and facilitate the maintenance of a desirable nitrogen balance (Cochran, 1975). The relationship of bacteroid activity to uric acid deposition may explain also the change in Na^+/K^+ ratio that has been demonstrated in the haemolymph of aposymbiotic cockroaches (Wharton and Lola, 1970). A correlation between K^+ and urate concentrations has been reported in fat bodies of nymphal and adult cockroaches (Mullins and Cochran, 1974a,b; Tucker, 1977b); therefore, the increased urate deposition that occurs in aposymbiotic insects may be expected to influence the Na^+/K^+ ratio of haemolymph.

7.4.3 Culture of bacteroids

It is apparent from the foregoing account that unequivocal demonstration of many of the inferred bacteroidal functions requires study of isolated bacteroids. Among several factors that may be invoked to explain the difficulties that have been experienced in culturing the micro-organisms are the high degree of specificity (non-pathogenicity) and possible requirement for intercellular associations. In addition, it has been suggested that the bacteroids suffer lysozymal damage during isolation procedures (Malke and Schwartz, 1966b; Daniel and Brooks, 1972), although the development of a new technique for separating mycetocytes from the other cell types (Kurtti and Brooks, 1976) offers new promise for eventual colonization of the bacteroids.

7.5 Composition of fat body

7.5.1 Organic constituents

A major function of fat body is the accumulation, storage and mobilization of carbohydrate, lipid, protein and uric acid. Absolute amounts of these compounds are influenced by basic principles of supply and demand; thus such parameters as diet, age, reproductive state and degree of excitation may be expected to affect the composition of the tissue. Each organic constituent responds differently to a particular set of conditions; therefore, the ratio of one compound to another is variable. This variation prevents adoption of a single organic compound as a standard against which the concentration of other substances may be expressed. The problem is exacerbated by the incidence of ploidy in insect fat body which precludes the use of DNA as an acceptable standard for comparative studies. Most determinations of fat-body constituents are expressed in terms of wet (fresh) weight or dry weight, but it is important to recognize the problems inherent in this form of expression, particularly in the case of long-term comparative investigations (see Gilbert, 1967a). An additional problem associated with estimation of fat-body content arises from the anatomical arrangement of fat body and its suspension in haemolymph. The sponge-like nature of the tissue results in freshly dissected fat body retaining at least 16% haemolymph (Matthews and Downer, 1974) and great care must be taken to remove this source of organic and inorganic constituents.

Table 7.1 reports typical amounts of the major organic constituents of fat body from adult male cockroaches. The values are typical of those obtained for the colony of cockroaches maintained on a diet of dog chow and water in this laboratory, but the capricious nature of cockroach fat body should not be underestimated and different values may be expected under different rearing conditions. Dramatic reductions in the concentrations of all reserves occur during starvation, although the extent of mobilization varies according to the reserve, with lipid providing 66% of the metabolic energy used by starving cockroaches, whereas glycogen and protein account for 22% and 11.8% respectively (Cervenkova, 1960).

Table 7.1 Major organic constituents of fat body of adult male *P. americana*

Constituent	*Concentration* μg/mg fresh weight
Glycogen	11.44
Trehalose	5.47
Triacylglycerol	107.58
Diacylglycerol	4.16
Protein	29.40
Uric acid	40.51*

* From data of Cochran *et al.*, (1979) assuming arbitrary fresh weight of 100 mg.

Glycogen contents ranging from 4.02–21.86 μg/mg fat body were reported in fat bodies of insects taken from different colonies (Steele, 1963) and a value of 5.4 mg glycogen has been reported in a single abdominal fat body (Hanaoka and Takahashi, 1976). The glycogen content can be enhanced by supplementing the diet with glucose and/or sucrose (R. G. H. Downer, unpublished observations) and developmental changes in glycogen content occur, particularly at the time of moulting when carbohydrate reserves are required for chitin synthesis (Lipke *et al.*, 1965a,b). Haemolymph trehalose serves as a supplementary source of metabolic energy for thoracic musculature and, during prolonged locomotory activity, this reserve is replenished by mobilization of fat-body glycogen (Downer and Matthews, 1976b; Downer and Parker, 1979). The rate at which carbohydrate is provided to the haemolymph, and ultimately to other tissues, is facilitated by the relatively high concentration of trehalose within fat body and the rapidity with which the glycogenolytic/trehalogenic pathways are stimulated (Matthews and Downer, 1974). The cyclic non-reducing sugars *chiro*-, *myo*- and *scyllo*- inositol and the enzymes required for interconversion of the various forms have been detected in fat-body homogenates (Hipps *et al.*, 1972, 1973), and mannose, glucosamine and galactosamine have also been reported in association with newly synthesized or stored glycoproteins (Lipke *et al.*, 1965a).

Triacylglycerol is the dominant lipid component of fat body although considerable variation is apparent among insects from different colonies (Cook and Eddington, 1967; Downer and Steele, 1969, 1972). Nymphal fat body tends to contain more triacylglycerol than that of adults (Nelson *et al.*, 1967; Tucker, 1977c) and it has been suggested that the greater resistance to desiccation demonstrated by nymphs over adults may result from the increased lipid stores (Tucker, 1977c); the rationale behind this proposal is that a unit weight of lipid generates more metabolic water than an equivalent amount of carbohydrate or protein (Downer and Matthews, 1976a). Gilbert (1967b) described changes in the lipid content of fat body of *Leucophaea maderae* during the female reproductive cycle; it is probable that similar changes occur in *P. americana*. The fatty-acid composition of fat-body acylglycerols shows a preponderance of oleic acid, which comprises about 40% of the total fatty acids; palmitic acid (28%), linoleic acid (18%) and stearic acid (8%) are also present in appreciable amounts (Nelson *et al.*, 1967). There have been no reports of longer chain polyunsaturated fatty acids in cockroach tissues, although further studies in this regard are required in light of recent findings that demonstrate the presence of prostaglandin-positive material in fat body (J. W. D. Gole and R. G. H. Downer, unpublished observations). Other lipid components that occur in fat body include diacylglycerols, monoacylglycerols, free fatty acids, phospholipids, sterols and sterol esters (Casida *et al.*, 1957; Agarwal and Casida, 1960; Ishii *et al.*, 1963; Nelson *et al.*, 1967; Downer and Steele, 1969, 1972; Downer and Chino, 1979).

Protein storage granules have been identified in male and female fat body (Section 7.3), but most studies of fat body protein have concentrated on changes related to the female reproductive cycle. Pronounced fluctuations in protein content occur in fat body of the viviparous cockroach, *Diploptera punctata* during the reproductive cycle (Stay and Clark, 1971), but this may be a feature of viviparity as less extreme changes were observed in *P. americana* (Subramoniam, 1973). Nevertheless, haemolymph protein (Mills *et al.*, 1966) and vitellogenin (Bell, 1969a,b) concentrations vary during the reproductive cycle of *P. americana* and, as fat body is the site of vitellogenin synthesis (Pan *et al.*, 1969), the fat body may be expected to reflect these fluctuations. Female *P. americana* feed during the first part of the reproductive cycle but fast during the period of oöthecal construction (Mills *et al.*, 1966; Bell, 1969a). Such behavioural changes in feeding pattern may also be expected to cause fluctuations in the amount of stored protein and, indeed, other reserves. The role of fat body as a primary site of protein synthesis indicates that the tissue contains a variety of free amino acids but there are few reports of the amino acid composition of fat body. The total tyrosine content of a single fat body rises from a pre-apolysis level of 31 μg to 397 μg at ecdysis and falls to 79 μg within 24 h of ecdysis; by contrast the phenylalanine concentration does not change markedly during the moult, remaining at levels of 0.18–0.30 μg/mg tissue (Wirtz and Hopkins, 1974).

Fat body occupies a central role in the synthesis of purines (McEnroe and Forgash, 1958) and substantial amounts of uric acid may accumulate in the tissue under particular dietary regimens (Sections 7.3 and 7.4). Sexual dimorphism occurs with females demonstrating a greater rate of accumulation than males when both are fed on high-protein diets (Mullins and Cochran, 1975a). The deposition of uric acid/urate is affected by dehydration with greater accumulation occurring in dehydrated insects (Bodenstein, 1953d; Anderson and Patton, 1955; Tucker, 1977b), probably due to the decreased amount of water that is available for the excretion of excess nitrogen in the form of ammonia (Mullins and Cochran, 1973a).

7.5.2 Inorganic constituents

Adult male *P. americana* maintained on dog chow and water *ad libitum* have an intracellular water content of 27.7 ± 2.63% fresh weight (mean ± s.d.; n = 10) (Matthews and Downer, 1974). This value declines to 22.1 ± 0.6% fresh weight (mean ± s.e. mean; n = 4) following 8 days of dehydration (Hyatt and Marshall, 1977). The water content of fat body also varies during the vitellogenic cycle, with the fluctuations attributable to the cyclic nature of water ingestion during the first three days of the cycle and the transfer of water to the developing oötheca in the final stage of the cycle (Verrett and Mills, 1975b). Intracellular concentrations of potassium, sodium, magnesium and calcium have been reported for fat bodies containing varying amounts of uric acid (Mullins and Cochran, 1974a) and are summarized in Table 7.2. The data indicate a correlation between the concentration of potassium ion and that of uric acid/urate, an observation that has been confirmed in males and females at nymphal and adult stages of development (Tucker, 1977b). Tucker (1977a,b,c,d) also studied the influence of diet, age and state of hydration on the concentration of Na^+ and K^+ in fat body. She confirmed earlier reports (Edney, 1968; Wall, 1970) that some ions are taken up by tissues during dehydration and identified fat body as the major tissue involved. A similar conclusion was derived independently by Hyatt and Marshall (1977). The results of these studies and those of Steele (1969) on fat-body tissues pre-incubated in Na^+- or K^+-free medium, are consistent with the suggestion that a uric acid/urate-associated ion sink operates in fat body to sequester or release certain ions and thereby maintain a constant ionic environment (Mullins and Cochran, 1974a) (see Section 6.5.1). In addition to the uptake of cations, fat body also takes up ^{36}Cl by a process that functions independently of the state of hydration (Tucker, 1977d). Carbonic anhydrase (Anderson and March, 1956) has also been detected in fat body, but no attempts have been made to define a physiological role for the enzyme or to correlate the enzyme activity with levels of HCO_3^- or Cl^-.

Table 7.2 Relationship between concentrations of uric acid and major cations in fat body of adult male *P. americana*

Concentration (μM/mg dry weight) *				
Uric acid	K^+	Na^+	Mg^{2+}	Ca^{2+}
0.27	0.30	0.25	0.04	0.04
1.40	0.47	0.20	0.03	0.03
4.58	0.98	0.20	0.02	0.02

* From the data of Mullins and Cochran (1974).

7.6 Intermediary metabolism

7.6.1 Synthesis of glycogen and trehalose

The biosynthetic pathways for glycogen and trehalose have been described in a number of reviews (Kilby, 1963; Chefurka, 1965; Wyatt, 1967; Bailey, 1975; Chippendale, 1978) and are summarized in Fig. 7.5. Several of the enzymes have been identified in *P. americana* including UDPG-pyrophosphorylase, UDPG-glycogen transglucosylase (Dutton, 1962; Vardanis, 1963) and trehalose-6-phosphate phosphatase (Friedman and Hsueh, 1979).

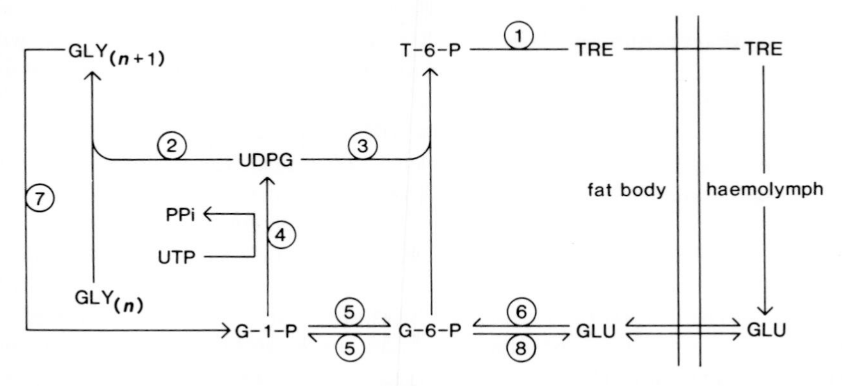

Fig. 7.5 Metabolic pathways of trehalose and glycogen synthesis. Key: GLY, glycogen; UPDG, uridine diphosphate glucose; G-1-P, glucose-1-phosphate; G-6-P, glucose-6-phosphate; GLU, glucose; T-6-P, trehalose-6-phosphate; TRE, trehalose. (1) trehalose-6-phosphate phosphatase; (2) UDPG-glycogen transglucosylase; (3) UDPG-trehalose phosphate transglucosylase; (4) UDP-glucose pyrophosphorylase; (5) phosphoglucomutase; (6) hexokinase; (7) glycogen phosphorylase; (8) glucose-6-phosphatase.

Incorporation of radio-labelled glucose into glycogen and/or trehalose has been demonstrated in *P. americana* (Treherne, 1960; Chefurka *et al.*, 1970; Spring *et al.*, 1977), but the regulatory mechanisms that favour the production of one compound over another have not been determined. Trehalose occurs in fat body as a readily available source of metabolic fuel that is produced and released during periods of metabolic demand, whereas glycogen represents a more permanent storage form of carbohydrate. These functional differences suggest that the two synthetic pathways may compete with each other, the competition for UDPG representing a probable site of antagonism. In the silkmoth *Hyalophora cecropia*, the K_m for UDPG in the trehalose-6-phosphate synthetase reaction is about five times lower than in the glycogen synthetase reaction; therefore, trehalogenesis assumes precedence over glycogenesis when concentrations of UDPG are limited (Murphy and Wyatt, 1965). However, increased trehalose concentrations decrease the affinity of trehalose-6-phosphate synthetase for glucose-6-phosphate and make UDPG available for glycogen synthesis (Murphy and Wyatt, 1965). It is tempting to suggest that a similar mechanism operates in the cockroach, and experiments in which insects were injected with glucose and with glucose plus trehalose indicate that trehalogenesis is suppressed to some extent by the presence of excess trehalose (J. H. Spring and R. G. H. Downer, unpublished observations). However, the high levels of trehalose that occur in the haemolymph following injection of aqueous extracts of corpora cardiaca (Steele, 1961) indicate that regulation involves more than simple feedback inhibition.

7.6.2 Gluconeogenesis

Conversion of certain amino acids and glycerol into carbohydrate by reversal of glycolysis has been described in several insect species (Bailey, 1975). In order for the glycolytic pathway to be reversed, additional enzymes are required to catalyse reactions in the glycolytic sequence that are essentially irreversible. These are pyruvate carboxylase and phosphoenolpyruvate carboxykinase which catalyse the conversion of pyruvate to phosphoenolpyruvate, and fructose-1,6-diphosphatase which mediates the conversion of fructose-1,6-diphosphate to fructose-6-phosphate. These enzymes have been detected in fat body of *P. americana* (Storey and Bailey, 1978b), and it seems likely that gluconeogenesis is a feature of fat body metabolism in this species. Gluconeogenesis may represent a route of carbohydrate synthesis for supply of the nervous system during periods of prolonged starvation, but no definitive studies have been conducted to test this hypothesis.

The enzymes malate synthetase and isocitrate lyase are absent from cockroach fat body, thus negating, for this species, the possibility that lipid may be converted to carbohydrate by the glyoxylate pathway (Storey and Bailey, 1978b).

7.6.3 Synthesis of lipid

(a) *Fatty acids*

Fat body is a major site of fatty acid synthesis in insects, and it is reasonable to assume that this tissue contains the complement of enzymes necessary for effecting the biosynthetic sequence described for a number of species (Gilbert, 1967a; Bailey, 1975; Downer, 1978a). In this pathway, fatty acids are formed by the initial condensation of acetyl-CoA with malonyl-CoA and subsequent successive condensations of acetyl-CoA with the newly formed intermediate compounds. The first step in the pathway is the activation of acetate by formation of a thiol ester with the nucleotide co-enzyme A, a reaction catalysed by acetyl CoA-synthetase. The activated acetate then undergoes carboxylation to malonyl-CoA in the presence of acetyl-CoA carboxylase. Both enzymes have been detected in the cytosol of cockroach fat body together with ATP-citrate lyase, NADP-dependent isocitrate dehydrogenase, glucose-6-phosphate dehydrogenase and malic enzyme, all of which may provide the NADPH required for lipogenesis (Storey and Bailey, 1978b). Other enzymes of fatty-acid synthesis are believed to be associated with an acyl carrier protein complex, but no such complex has yet been isolated in the cockroach.

Radio-labelled acetate, administered either by injection or feeding, is rapidly incorporated into the fatty-acid component of cockroach fat body (Louloudes *et al.*, 1961; Bade, 1964; Vroman *et al.*, 1965); thus any compound capable of yielding acetate or acetyl-CoA may undergo conversion to fatty acid. Incorporation of label from injected U-^{14}C-glucose into fat-body lipid has been demonstrated (Lipke *et al.*, 1965b) with most of the label detected in the fatty-acid fraction (Kallapur *et al.*, 1980). Storey and Bailey (1978a,b) studied the intracellular distribution of enzymes that are likely to function in the conversion of carbohydrate to lipid and concluded that the metabolic pathway for this conversion is probably the same as that proposed for the locust and mammalian lipogenic tissues (Bailey, 1975). Extrapolation from other systems indicates that the inner mitochondrial membrane may be impermeable to acetyl-CoA, and it is suggested that transfer of acetyl-CoA to the cytosol for fatty-acid synthesis may involve citrate as proposed for the locust (Bailey, 1975; Storey and Bailey, 1978b).

(b) *Acylglycerols*

Incubation of cockroach fat body in medium containing 1-^{14}C-palmitate results in the incorporation of label into the acylglycerol fraction (Cook and Eddington, 1967) and a similar incorporation occurs *in vivo* (Bhakthan and Gilbert, 1970; Bollade and Boucrot, 1973). Acylglycerol synthesis may proceed by acylation of glycerol-3-phosphate or by the transfer of a fatty-acid moiety to monoacylglycerol (Downer, 1978a). Monoacylglycerol acyltransferase has been demonstrated in fat-body homogenates of adult male cockroaches (Hoffman and Downer, 1979a) and evidence for the occurrence of the glycerol-3-phosphate pathway has also been obtained (Kallapur *et al.*, 1980). Little information is available concerning the relative contribution of each pathway to acylglycerol synthesis or the physiological conditions under which each pathway operates.

Similarly, few data are available to define the neuroendocrine mechanisms that are known to influence acylglycerol accumulation. Allatectomy causes an accumulation of acylglycerol in fat body of *P. americana* (Bodenstein, 1953d) and, in *L. maderae*, the incorporation of radio-labelled fatty acids into the acylglycerol fraction is suppressed by corpora allata (Gilbert, 1967b). Although not discounting a direct effect of corpus allatum hormone on acylglycerol synthesis, Steele (1976) has suggested that the depressed synthesis of protein that results from allatectomy may contribute to the accumulation of fat-body lipid. Thus, amino acids and carbohydrates become available to serve as lipogenic precursors and no apoprotein is available to facilitate lipid transport from the tissue. Aqueous extracts of CC cause a slight, but statistically significant, elevation of fat body acylglycerol levels *in vivo* with maximal increases evident at about 2 h after injecting the extract into cockroach haemocoele (Downer and Steele, 1972; Downer, 1972; Goldsworthy *et al.*, 1972). CC contains several metabolically active factors including the hypertrehalosemic hormone which promotes trehalogenesis and increased oxidation of fatty acid (Steele, 1976; Section 7.6.14). It seems likely that the injection of CC extracts results initially in a hypertrehalosemic state but, as the resting condition is restored, a lipogenic flux may be favoured. Indeed, the difficulty in demonstrating a pronounced hypolipemic effect with corpora cardiaca extracts, and the lipolytic effect that is observed with fat body *in vitro*, may be due to the hypolipemic effect being masked to some extent by the preceding hypertrehalosemic condition and the associated lipolysis.

(c) *Phospholipids*
Fat body is the major site of phospholipid synthesis and the biosynthetic pathways have been discussed in several reviews (Fast, 1964, 1970; Gilbert, 1967a; Downer, 1978a). It is probable that the biosynthetic pathways in cockroaches are the same as those reported in other insects but, apart from a single study on choline kinase in gut-free homogenates (Kumar and Hodgson, 1970), there is little information available on phospholipid synthesis in cockroaches.

(d) *Sterols*
Cockroaches lack the capacity to synthesize sterols and require a dietary source to satisfy their sterol needs (Clayton, 1964; Robbins *et al.*, 1971; Svoboda *et al.*, 1975). However, dietary sterols can be modified and the conversion of ergosterol to 22-dehydrocholesterol (Clark and Bloch, 1959) and β-sitosterol to cholesterol (Robbins *et al.*, 1962) has been demonstrated in *B. germanica*. The latter transformation has been shown also in *P. americana* with demesterol identified as an intermediate (Svoboda and Robbins, 1971). Dietary cholesterol tends to accumulate in fat body (Ishii *et al.*, 1963) and this tissue is the probable site of the interconversions. Cholesterol may also undergo esterification although most remains in the non-esterified form (Vandenheuvel *et al.*, 1962; Downer and Chino, 1979).

(e) *Ketogenesis*
The ketone bodies, D-3-hydroxybutyrate and acetoacetate occur in haemolymph of adult male cockroaches with the concentration of the latter compound increasing under conditions of starvation (Shah and Bailey, 1976). Two enzymes of ketogenesis, hydroxy-methylglutaryl-CoA-synthase and hydroxy-methylglutaryl-CoA-lyase have been described in cytosolic and mitochondrial fractions of fat body (Shah and Bailey, 1976), thus demonstrating a ketogenic capacity for this tissue. The production of ketone bodies offers a further analogy between cockroach fat body and vertebrate liver and indicates a potentially important mechanism for the sparing of glucose during periods of starvation.

7.6.4 Protein synthesis

The fat body of female cockroaches synthesizes and secretes into haemolymph a specific lipophosphoprotein, vitellogenin, which is taken up by the oöcyte and incorporated into yolk protein. Synthesis of vitellogenin following its induction by JH, provides a convenient model for investigation of the mechanism and regulation of protein synthesis and has been studied in *P. americana* (Mills *et al.*, 1966; Pan *et al.*, 1969; Subramoniam, 1973; Sams *et al.*, 1981) and *L. maderae* (Dejmal and Brookes, 1972; Engelmann, 1974b; Koeppe and Ofengand, 1976). Vitellogenin synthesis in *L. maderae* is located within a particular group of membrane-associated polysomes, comprising 35–40 ribosomes, which occur in fat bodies of vitellogenic females but not nymphs or males (Engelmann, 1974a; 1977a,b). However, identification of female fat body as the exclusive site of vitellogenin synthesis may require revision in light of the recent report that fat body from male *Diploptera punctata* can synthesize vitellogenin in response to treatment with high doses of JH (Mundall *et al.*, 1979). The primary translation product of vitellogenin synthesis in *L. maderae* is believed to be about 250 000 daltons, with assembly of the final molecule occurring within rough endoplasmic reticulum in a manner similar to that reported for other protein-secreting systems (Engelmann, 1977b). The total SDS-phenol-extracted RNA increases in the presence of JH and this extract stimulates vitellogenin synthesis (Engelmann, 1972). Studies with fat body of *P. americana* also indicate that the primary translation product for vitellogenin is a large polypeptide with a molecular weight of approximately 250 000 (Sams *et al.*, 1981). However, the vitellogenin secreted by fat body in organ culture, and that isolated from haemocytes and oöcytes, comprises three polypeptide subunits with approximate mol wts of 123 000, 118 000 and 57 000; these subunits are organized to form a natural vitellogenin with a mol wt of about 600 000. Few studies are available to describe the synthesis of non-vitellogenic proteins, although rapid incorporation of ^{14}C-leucine into the major diacylglycerol-carrying lipoprotein of adult male *P. americana* (Chino *et al.*, 1981) can be demonstrated *in vivo* (R. G. H. Downer and C. Kapron, unpublished observations).

The general picture that emerges from these studies is that protein synthesis in the cockroach proceeds according to the pathway identified in other insect species (Price, 1973; Wyatt, 1975; Chen, 1978).

7.6.5 Uric acid metabolism

There are two possible routes for uric acid formation in insects: the uricotelic pathway that involves *de novo* synthesis from protein nitrogen, and the nucleicotelic pathway which uses nucleic acids, purines or pyrymidines, as precursors (Bursell, 1967; Cochran, 1975). The uricotelic pathway involves a reaction squence in which glutamine, aspartate, glutamate, glycine, formate, carbon dioxide and ribose-5-phosphate combine in the presence of ATP to yield inosine. This is then converted to hypoxanthine, xanthine and finally to uric acid. The sequence is well-established in birds and appears to function also in *P. americana* (McEnroe and Forgash, 1957, 1958). Allusion has been made previously to the close correlation between fat body uric acid levels and the dietary intake of protein (Sections 7.3 and 7.4), and these observations suggest that the uricotelic pathway is of primary importance in the cockroach (Cochran, 1975). However, the nucleicotelic pathway should not be neglected and several of the enzymes required to effect this metabolic sequence have been reported. Enzymes of pyrymidine metabolism have been detected in cockroach abdomen (Bruno and Cochran, 1965) and the purine deaminases, adenase and guanase, have been described in fat body of *P. americana* (Anderson and Patton, 1955) and *L. maderae* (Lisa and Ludwig, 1959). The final step in both uric acid-synthesizing pathways is conversion of xanthine to uric acid, and the enzyme xanthine dehydrogenase, which catalyses this reaction, has been demonstrated in cockroach fat body (Anderson and Patton, 1955).

Early suggestions that uric acid serves as a permanent storage product of excretion have been discounted and there is little doubt that the metabolic substrate may be mobilized under conditions of negative nitrogen balance (Mullins and Cochran, 1974; 1975a,b; Cochran, 1975). The pathway of uric-acid breakdown has not been fully elucidated, although uricase, which converts uric acid to allantoin, has been reported in fat body of *P. americana* (Cordero and Ludwig, 1963) and *L. maderae* (Lisa and Ludwig, 1959). The liberation of $^{14}CO_2$ from 2- and 4-^{14}C-uric acid was also demonstrated (McEnroe, 1966). Several workers have implicated fat-body symbionts in the catabolism of uric acid (Malke and Schwartz, 1966a,b; Donnellan and Kilby, 1967) but, although much of the evidence is highly compelling, unequivocal acceptance of such a mechanism must await its demonstration in bacteriodal cultures (Section 7.4).

7.6.6 Amino acid metabolism

The amino group of an amino acid may be cleaved by oxidative deamination to yield ammonia and the corresponding keto acid, which can then act as a receptor for the transfer of amino groups from other amino acids. D-Amino acid oxidase activity has been reported in cockroach fat body (Auclair, 1959; Blaschko *et al.*, 1961), and glutamate dehydrogenase and transaminase activities are also present (McAllan and Chefurka, 1961a,b). McEnroe and Forgash (1958) demonstrated the synthesis of serine from glycine and formate by addition of the formyl carbon into the β-position of serine.

The tryptophan–ommochrome pathway has been extensively studied in insects (Linzen, 1974) and several intermediates have been detected in the faeces of *P. americana* including kynurenic acid, xanthurenic acid and 8-hydroxyquinalic acid (Mullins and Cochran, 1973b). Involvement of fat body in this pathway is indicated by the high activity of kynuredine formamidase reported in this tissue (Cochran, 1976).

At this point, it is convenient also to indicate detection of isoxanthopterin deaminase in fat body (Gyure, 1975), thus implicating this tissue in pteridine metabolism.

7.6.7 Glycogenolysis

The α-glycosidic bonds of glycogen are cleaved to glucose-1-phosphate in the presence of glycogen phosphorylase (Fig. 7.5) (Chefurka, 1965; Chippendale, 1978). In mammalian liver, glycogen phosphorylase is an allosteric enzyme which undergoes interconversion between active (a) and inactive (b) forms according to the relative activities of the enzymes phosphorylase kinase and phosphorylase phosphatase. The equation for mammalian glycogen phosphorylase is shown below:

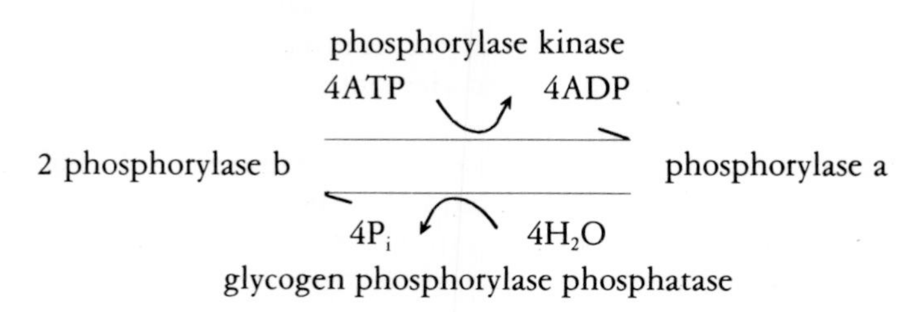

Glycogen phosphorylase in cockroach fat body also appears to exist in active and inactive forms (Steele, 1963; Wiens and Gilbert, 1967; Hanaoka and Takahashi, 1978). The enzyme is present primarily in the soluble fraction of fat-body homogenates with little activity being precipitated in association with glycogen particles (Storey and Bailey, 1978a), a situation which contrasts with that reported for the glycogen phosphorylase of locust fat body (Applebaum and Schlesinger, 1973). Enzyme activity is enhanced by the presence of AMP (Steele, 1976; Storey and Bailey, 1978a) and extrapolation from mammalian systems suggests that at least part of this enhancement may result from AMP binding to phosphorylase a, rendering the enzyme resistant to the action of glycogen phosphorylase phosphatase. The enzyme also appears to be sensitive to changes in the intracellular ionic balance (Steele, 1969), but until additional data are available to more fully characterize the several enzymes that contribute to phosphorylase activity, further discussion of this interesting observation is unwarranted.

7.6.8 Glycolysis and the pentose cycle

Conversion of glucose to pyruvate may proceed by glycolysis and/or the pentose cycle, the essential features of which have been described (Kilby, 1963; Chefurka, 1965). Complete oxidation of glucose results in the six constituent carbon atoms being liberated as carbon dioxide; however, the sequence in which they are released depends upon their location within the glucose molecule. Thus, following glycolysis, C_3 and C_4 become the carboxyl carbon of pyruvate and will be liberated during the first turn of the carboxylic acid cycle, whereas C_1 and C_6 form the methyl carbon of pyruvate and are liberated in the third turn of the cycle. If glucose enters the pentose cycle, C_1 will be released at an early stage during oxidative decarboxylation of 6-phosphogluconate to the five-carbon compound, ribulose-5-phosphate. Chefurka *et al.* (1970) monitored the evolution of $^{14}CO_2$ from specifically labelled glucose molecules and determined the relative contributions of the glycolytic and pentose pathways to glucose degradation. The studies revealed sexual dimorphism in the pattern of glucose breakdown with the pentose cycle accounting for about 21% of the $^{14}CO_2$ evolved in males and about 3% in females. Differences due to nutritional state, substrate pool size and sex were also observed (Ela *et al.*, 1970). In interpreting these findings, it is important to recognize that the pentose cycle represents more than an alternative pathway of glucose breakdown. It provides a source of reduced NADP for lipogenesis and other biosynthetic pathways, in addition to offering a pathway of pentose metabolism. Thus, the relative contribution of either pathway to the degradation of glucose may be expected to reflect the activities of these related processes. A key enzyme of the pentose cycle, glucose-6-phosphate dehydrogenase, has been detected in fat-body cytosol together with other enzymes of the glycolytic pathway (Storey and Bailey, 1978a,b).

7.6.9 Tricarboxylic acid cycle

There is little doubt that the tricarboxylic acid cycle occurs as the terminal event of biological oxidation in cockroach fat body. Details of the cycle and evidence in support of its occurrence in fat body have been presented in previous reports (Kilby, 1963; Chefurka, 1965) and need not be recounted at this time.

7.6.10 Respiratory chain and oxidative phosphorylation

Extensive studies on isolated mitochondria from fat body of *B. discoidalis* have provided much useful information on the development and properties of the organelles (Keeley, 1970, 1971, 1972, 1977). Equivalent data are not available for *P. americana*, although some enzymes of the respiratory chain have been identified (Sacktor and Bodenstein, 1952; Sacktor and Thomas, 1955).

7.6.11 Lipolysis

Lipolytic activity has been demonstrated in fat-body homogenates of *P. americana* (Gilbert *et al.*, 1965; Downer and Steele, 1973; Hoffman and Downer, 1977, 1979a). Incubation of tissue homogenates in the presence of triacylglycerol results in preferential accumulation of diacylglycerol (Fig. 7.6) (Hoffman and Downer, 1979a). Stereospecific analysis of the diacylglycerol products indicates accumulation of *sn*-1, 2-diacylglycerol as the major product of lipolysis and suggests the presence of a stereospecific triacylglycerol lipase. Diacylglycerol is the major form in which neutral lipid is released from fat body under a variety of physiological conditions (Downer and Steele, 1972; Chino and Downer, 1979; Chino *et al.*, 1981)), and it is tempting to speculate that the triacylglycerol lipase serves to liberate specific diacylglycerol for release to haemolymph. However, evidence exists to suggest that the regulation of lipid mobilization in cockroaches is more complex than has previously been suggested. Fatty acids and triacylglycerols may be released under particular conditions (Cook and Eddington, 1967; Chang and Friedman, 1971) and

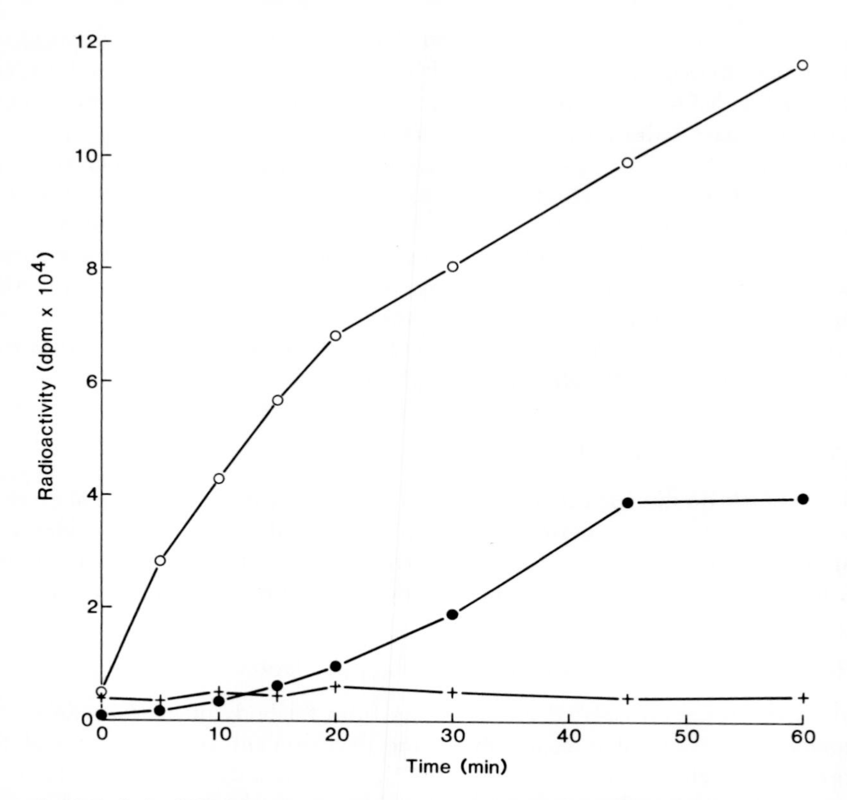

Fig. 7.6 Hydrolysis of [^{3}H]-glycerol triolein by fat body homogenate of *P. americana* [Each sample contained 300 μl tissue homogenate, 600 μl Tris-HCl buffer (0.1 mM, pH 7.0) containing 5% BSA and 100 μl triacylglycerol emulsion containing 3.18×10^6 dpm ^{3}H-glycerol triolein]. (From data of Hoffman and Downer, 1979b.) Key: o———o, diacylglycerol; +———+, monoacylglycerol; ●———●, glycerol.

a metabolic switch to lipid oxidation appears to accompany the induction of trehalogenic flux (Section 7.6.14). Acid and alkaline lipases with different specificities have been reported in locust fat body (Tietz and Wientraub, 1978) and it is also reasonable to propose that more than one species of lipase occurs in cockroach fat body. Clearly, elucidation of the mechanism(s) by which lipolytic activity is regulated requires purification and detailed characterization of the enzyme(s) involved.

7.6.12 Cyclic nucleotide metabolism

The elevation of haemolymph trehalose by aqueous extracts of CC is simulated by adenosine-3,5-monophosphate (cyclic AMP) (Steele, 1964) and the gland extract also elevates intracellular concentrations of the cyclic nucleotide and stimulates adenylate cyclase (Hanaoka and Takahashi, 1977; Gäde, 1977). The cyclic nucleotide serves in the well-established role of a 'second messenger' and activates a protein kinase, which in turn initiates a cascade of reactions that culminate in the formation of active glycogen phosphorylase (Hanaoka and Takahashi, 1977). Two cyclic AMP-dependent protein kinases have been isolated and partially characterized from fat bodies of male and female cockroaches, and a third protein kinase of unknown dependency was also detected in female fat body (Hanaoka and Takahashi, 1977, 1978). A cGMP-dependent protein kinase was reported in abdominal fat body of female *B. discoidalis* (Kuo *et al.*, 1971), and it is possible that the third kinase of female *P. americana* may be equivalent to the GMP-dependent enzyme described for *B. discoidalis*.

Adenylate cyclase and cAMP are also responsible for mediating the octopamine-induced hypertrehalosemic response, and preliminary attempts to characterize the octopamine-sensitive adenylate cyclase receptor of fat body indicate that it is pharmacologically similar to the α-receptor of vertebrates (Gole and Downer, 1979; Downer, 1980). Cyclic nucleotides mediate a variety of physiological responses throughout the animal kingdom and there is no reason to believe that they will be any less ubiquitous in insects. Characterization of the activating enzymes and the phosphodiesterases responsible for cyclic nucleotide breakdown are needed in order to gain a fuller understanding of short-term metabolic control processes in the cockroach.

7.6.13 Other metabolic activities

Fat body has often been likened to mammalian liver and adipose tissue and it is not surprising to find that the tissue contains a varied complement of enzymes that enable it to effect many metabolic reactions in addition to those described above. Of particular importance in contributing to this enzymic diversity are the mixed-function oxidases, which are capable of catalysing such reactions as hydroxylation, epoxidation, alkyl-group oxidation and *N*-demethylation. Mixed-function oxidases have been recognized in fat body of several cockroaches (Benke *et al.*, 1972; Turnquist and Brindley, 1975) and are of unquestioned importance in enabling cockroaches to develop resistance to particular insecticides. Also of interest from the viewpoint of insect control is the mixed-function oxidase that catalyses the desulphuration of the organophosphate, parathion to the potent acetylcholinesterase inhibitor, paraoxon (Kok and Walop, 1954; Nakatsugawa and Dahm, 1965; Brindley and Dahm, 1970). Enzymes that activate other organophosphates are also present (Nakatsugawa and Dahm, 1962). The mixed-function oxidases of cockroach fat body appear to be similar to those described for other insects, having a requirement for NADH and molecular oxygen, and precipitating with the microsomal fraction during differential centrifugation.

The ability of the synthetic steroid 22,25-dideoxyecdysone to induce cuticle deposition in cultured leg regenerates is enhanced greatly in the presence of fat body, suggesting that this tissue can convert the steroid to a biologically active form. This possibility has been tested in *L. maderae* and analysis of the metabolic products indicates preferential hydroxylation of the 25-C position (Thompson *et al.*, 1978). Other C-positions may also be hydroxylated but no enzymes appear to be present for hydroxylation of the 22-position.

Metabolism of JH by fat body is indicated by studies using cultured tissue (Sams *et al.*, 1978). The fat body appears to secrete enzymes that hydrolyse the hormone in the surrounding medium. Addition of cockroach haemolymph, which is presumed to contain JH-binding protein, protects the hormone but this protection does not persist when the hormone enters the tissue.

7.6.14 Physiological factors influencing fat body metabolism

The neuroendocrine system of the cockroach contains a number of active factors that elicit a diversity of biochemical and physiological responses in fat body. These compounds and their actions have been discussed in recent reviews (Gilbert, 1976; Steele, 1976; Riddiford and Truman, 1978) and these accounts, together with other contributions to the present volume (Chapter 12) make further treatment unnecessary. However, it is appropriate in this account to consider the integrated response(s) of fat body metabolism to particular physiological events and to recognize that the aforementioned neuroendocrine factors may be involved in triggering or mediating such responses.

The metabolic state of an animal, and a tissue, varies according to the energy balance that prevails at any particular moment. Thus, during conditions that require expenditure of stored reserves, for example, exercise, starvation and/or oögenesis, fat body may be expected to mobilize these reserves; by contrast, a newly fed animal is likely to be in a state of positive energy balance and may be expected to accumulate reserves in the fat body. These considerations emphasize that fat body must be recognized as a dynamic tissue that responds to prevailing physiological conditions, and the metabolic state of the tissue may be expected to vary accordingly (Downer, 1981).

The appearance of excess glucose in haemolymph may result in the synthesis of trehalose and/or glycogen by fat body (Section 7.6.1). The relative amount of each product is influenced by the state of excitation of the experimental animal, with the excited animal favouring the production of trehalose, whereas under resting conditions the extra glucose is stored as glycogen (Spring *et al.*, 1977). The switch from a resting glycogenic state to a trehalogenic state appears to be mediated by a neural factor and the biogenic amine, octopamine, has been proposed as a possible effector of the metabolic switch (Downer, 1978b, 1979a,b, 1980). The excitation-induced response may be expected to influence other metabolic pathways. The respiratory quotient (RQ) of resting fat body of *L. maderae* is close to 1.0, indicating the predominance of carbohydrate oxidation for resting metabolism, but the addition of hypertrehalosemic factor(s) lowers the RQ, decreases the oxidation of glucose and enhances palmitate oxidation (Wiens and Gilbert, 1967). The results suggest that in the trehalogenic state, lipid oxidation provides energy for metabolic maintenance and synthesis of trehalose. Such a metabolic switch would also serve to decrease the glycolytic demand for glucose-6-phosphate and free this substrate for trehalogenesis. The technical problems associated with handling highly excitable animals like cockroaches result in most studies on short-term metabolic effects being conducted on insects in the excited (trehalogenic) state. It is apparent from the above discussion that fresh methodologies may be required in order to determine the metabolic fluxes of resting animals.

Studies on vertebrates have recognized important metabolic differences between animals in absorptive and post-absorptive states. In the former condition, the metabolic pathways tend to favour anabolic fluxes with the major fuel for basal metabolism being newly absorbed glucose; in the post-absorptive state the accumulated reserves tend to be catabolized and carbohydrates are conserved for support of nerve metabolism. Such metabolic distinctions are likely to occur also in the cockroach, and preliminary studies from this laboratory indicate differences in glycogenic and lipogenic capacities according to the nutritional state of the animal. However, no detailed studies have been reported on this subject.

7.7 Conclusions

The American cockroach is a popular laboratory insect for investigations of insect physiology and biochemistry, yet the present account demonstrates that many fundamental gaps exist in our knowledge of the major organ of intermediary metabolism. Our appreciation of the basic anatomy of the tissue suffers from a lack of information on innervation, the homogeneity/heterogeneity of fat body from different anatomical sites and the nature and direction of haemolymph flow through the different regions under different physiological conditions. The contribution of intracellular symbionts to metabolic and physiological function remains to be determined and can only be resolved when pure bacteroidal cultures are maintained. Although many pathways of intermediary metabolism have been established in fat body, few attempts have been made to consider metabolism as an integrated, holistic process in which the stimulation or inhibition of one pathway is likely to impinge upon several other metabolic sequences. Inadequate attention has been given to the physiological condition of experimental animals, and it is probable that this has contributed greatly to the considerable variation reported for substrate and enzyme levels in the cockroach. Very few studies have examined the role and nature of fat body in immature insects when the metabolic demands are likely to be different from those of an adult.

On the positive side, cockroach fat body offers a challenging field for microscopists, physiologists and biochemists. The development of improved methods for organ and tissue culture together with the availability of increasingly specific pharmacological probes offer the prospect of much progress in the next decade. It is hoped that this brief account will stimulate interest in the subject.

Acknowledgements

I am grateful to Professor D. G. Cochran for valuable counsel and for providing me with many unpublished observations on the morphology of cockroach fat body. Professor Cochran and his co-worker, Dr Mullins, also provided plates from which Figs 7.1–7.4 were prepared and I acknowledge their generous assistance in this regard. The manuscript was read and criticized by Drs S. M. Smith and J. E. Steele, to whom my appreciation is extended. Studies from my laboratory were conducted in association with Drs H. Chino, J. W. D. Gole, A. G. D. Hoffman, V. L. Kallapur, J. R. Matthews, J. H. Spring, Misses L. Feniuk, V. Blok and C. Kapron, and were supported by grants from the Natural Research Council of Canada.

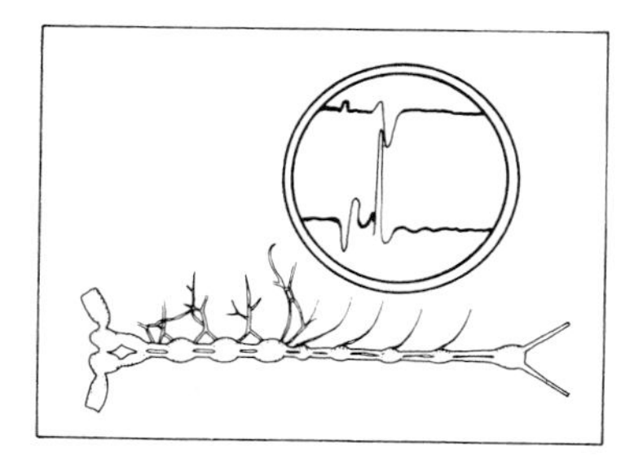

8

Nervous system

Rudolph Pipa and Fred Delcomyn

8.1 Introduction

The central nervous system (CNS) is exceedingly complex, both morphologically and physiologically. This is correlated with the capacity to fulfill three separate functions: receive sensory input from the insect's internal and external environment, integrate this input, and effect adaptive responses, either motor or glandular. As the insect's very survival depends on appropriate reactions to various internal or external environmental cues, the influence of the nervous system is pervasive, and few homeostatic processes can be comprehended apart from the neural mechanisms which regulate them. It is therefore inevitable that aspects of the cockroach nervous system, such as its structure and function, are treated in many places throughout this book in addition to this Chapter. Furthermore, the literature on the nervous system of *Periplaneta americana* has expanded rapidly. These considerations permit us to outline only certain of the advances that have been made since the subject was reviewed by Guthrie and Tindall (1968).

8.2 Structural organization of central and peripheral pathways

Nervous systems must be characterized anatomically if their physiology is to be understood. It is important to chart the distribution of the peripheral nerves, for they contain afferent and efferent pathways that can be monitored conveniently. But this, alone, is insufficient. To know how neurons interact it is essential that their central connections be defined. The goal is to locate equivalent members of a neuron population in different individuals. These 'identified' neurons occupy similar sites, and have branching patterns that are unique and relatively stable.

Once these homologous neurons have been mapped, their functional interrelationships can be examined, and the ways their properties are affected by genetic and environmental manipulations can be explored.

8.2.1 General anatomical features

In *P. americana*, as in other insects, the nervous system is bilaterally symmetrical and segmented. The peripheral nerves arise from a series of linearly arranged ganglia. In the CNS the ganglia are united by paired interganglionic connectives.

Each ganglion consists of a cluster of neurons, their supporting cells (neuroglia), and tracheal cells. These are encapsulated by an acellular connective tissue sheath (neural lamella). In a cross-section through a ganglion, two regions are apparent: a cortex containing the neuron somata (cell bodies), and a central neuropile composed of neuron processes. The larger neuron branches contribute to the longitudinal and transverse fibre tracts of the neuropile, while the finest arborizations end at synapses. Certain neuropiles seem to be more highly structured than others; the fibre pathways and dense concentrations of synapses form consistently recognizable areas. These zones, particularly evident in the brain, are called glomeruli.

In *P. americana* nymphs and adults, there are three separate thoracic ganglia and six distinct abdominal ganglia. The first abdominal neuromere is obscured when it fuses with the metathoracic ganglion during embryogenesis; the first discrete abdominal ganglion evident in nymphs and adults is therefore equivalent to the second abdominal ganglion of the embryo. Despite this transposition, the original distribution of the nerves from both the actual and apparent first abdominal ganglion is preserved, and the nerves can be traced to their corresponding abdominal segments (Shankland, 1965). The sixth (last) discrete abdominal ganglion lies in the seventh abdominal segment. This composite ganglion innervates abdominal segments seven to eleven and, correspondingly, it is derived from the embryonic neuromeres of those segments.

The relationships of the cerebral ganglia are not so clear. During arthropod evolution the process of cephalization has resulted in a loss of segmental boundaries, both in the skeleto-muscular system and in the nervous system. Consequently, the number and derivations of nerve centres that form the brain are undecided despite embryological studies. There is more agreement regarding the origin of the suboesophageal ganglion; it seems to have formed by fusion of centres that supplied the mandibular, maxillary and labial segments. Fusion of these three neuromeres is also evident during embryogenesis.

Located upon the pharynx, oesophagus and crop, and ending near the junction of the fore- and midgut, is the stomodaeal nervous system (SNS). Its name denotes its origin from neuro-ectoderm associated with the stomodaeum, the embryonic foregut. Other descriptors that have been applied to the SNS include 'visceral', 'stomatogastric', or 'sympathetic' (see Willey, 1961, for an historical review of this nomenclature).

The SNS of *P. americana* consists of five ganglia interconnected by four major nerves. The frontal ganglion is a primary component; the three nerves that issue from it join the SNS to the brain anteriorly. Posteriorly, the SNS and retrocerebral neuro-endocrine complex (RNC) are linked by a pair of cardiostomatogastric nerves. The systems are so close that Willey (1961) considers the CC (the intrinsic cells) to be an additional stomodaeal ganglion. The branches that leave the various ganglia and connecting nerves of the SNS are traceable to muscles of the clypeus, labrum, hypopharynx, foregut, and to the salivary glands. Afferent fibres from foregut sensory receptors enter the SNS. This nervous system is involved with regulation of ingestion and, possibly, with neuroendocrine reflexes (see Chapters 4 and 12).

8.2.2 The brain (supraoesophageal ganglion)

(a) *Principal nerves*

The major neural components of the brain of *P. americana* are sensory neurons, interneurons and neurosecretory cells. According to conventional wisdom, motoneurons are rare. This interpretation is based on the topographic anatomy of the cerebral nerves and may not be entirely correct. Certainly, there are nerves to the antennal muscles, and cerebral motoneurons that pass to other parts of the CNS before leaving probably remain to be discovered.

Nerves issue from three subdivisions of the brain: the protocerebrum, deutocerebrum and tritocerebrum (Figs. 8.1a,b). They can be traced to the compound eyes, ocelli, antennae and integument. Nerves or connectives from the brain also join the ventral nerve cord, SNS and RNC.

(b) *Protocerebrum: afferent pathways*

The compound eyes, ocelli and integumental sensilla are the most prominent sensory organs associated with the protocerebrum. The anatomy of these receptors is treated in Chapter 9, so we will be concerned, primarily, with certain features of their central projections.

A single compound eye of *P. americana* consists of approximately 2000 visual units (ommatidia) (see Section 9.2.1). Associated with each ommatidium are eight retinal cells (first-order neurons). Their axons (Fig. 8.2) pass through two basement lamellae (connective tissue sheaths) to terminate either in the first (lamina) or second (medulla) optic glomerulus (Butler, 1973b; Ribi, 1977). The third (innermost) optic glomerulus is the lobula. Two chiasmata composed of interconnecting neurons join the three optic glomeruli to each other. The outer chiasma connects the lamina with the medulla and the inner one, the medulla with the lobula. Several thick axon bundles join the lobula to the protocerebral lobes.

The retinal cell projections and interneurons of the lamina of *P. americana* are described by Ribi (1977), who notes that the neural elements are less precisely arranged than in certain Diptera and Hymenoptera. Individual optic cartridges with a constant number of first- and second-order neurons are absent. Arising from

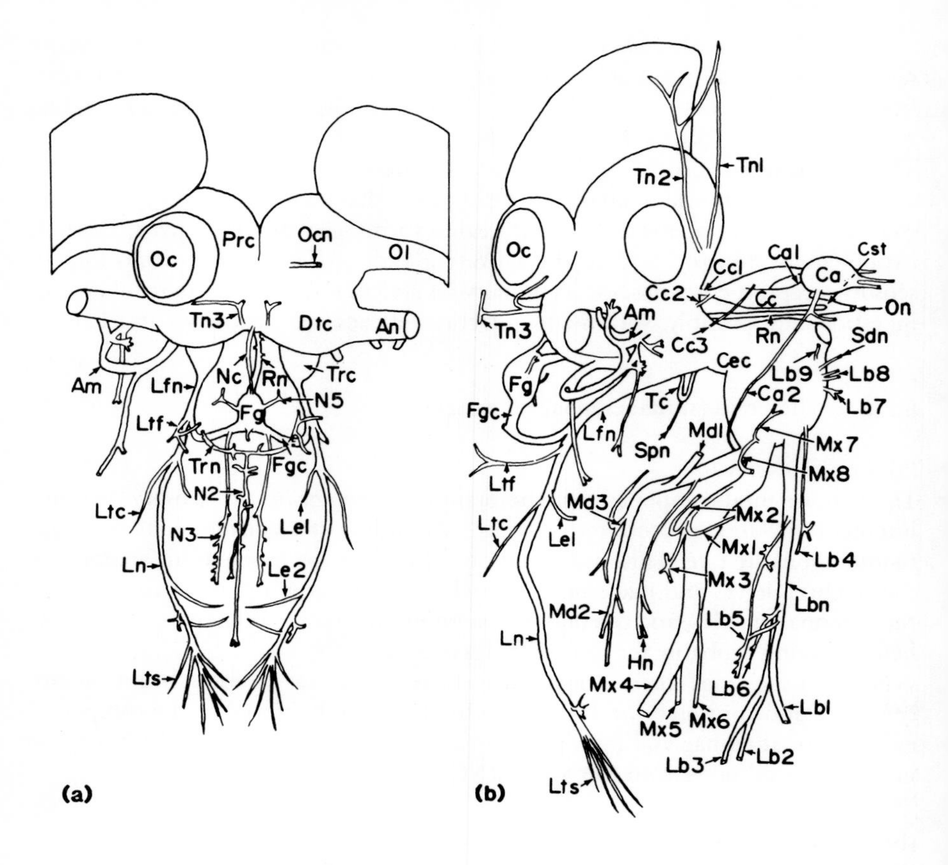

Fig. 8.1 (a) Anterior view of the brain and stomodaeal nervous system of *P. americana*. (After Arnold, 1960; Willey, 1961.) (b) Lateral view of the cephalic nervous system of *P. americana*. The left optic lobe has been removed. (After Arnold, 1960; Willey, 1961; Whitehead, 1971.)

ABBREVIATIONS	DESCRIPTIONS (Muscles named and numbered according to Snodgrass, 1944)
Am	Antennomotor nerves
An	Antennal nerve
Ca	Corpus allatum
Ca1	Nervus corporis allati 1
Ca2	Nervus corporis allati 2
Cc	Corpus cardiacum
Cc1	Nervus corporis cardiaci 1
Cc2	Nervus corporis cardiaci 2
Cc3	Nervus corporis cardiaci 3
Cec	Circumoesphageal connective
Cst	Cardiostomatogastric nerve

Dtc	Deutocerebrum
Fg	Frontal ganglion
Fgc	Frontal ganglion connective
Hn	Hypopharyngeal nerve to muscles of salivarium (17, 18) and sensilla at tip of hypopharynx
Lbn	Labial nerve
Lb1	Labial palp nerve
Lb2	Glossal nerve
Lb3	Paraglossal nerve
Lb4	To retractor muscle of prementum (45)
Lb5	To tentorial labial productor (43), tentorial hypopharyngeal productor (16), tentorial labial reductor (44), labial palp levator, and premental extensor muscles
Lb6	To labial muscle of the hypopharyngeal fulcrum (20), reductor of the hypopharynx (19), palpal depressor muscle, and muscles of glossa and paraglossa
Lb7	Tegumentary nerve of submentum
Lb8	Ventral cervical nerve; to prothoracic gland?
Lb9	To post-occiput and dorsal cervical muscles
Le1, 2	To epipharynx
Lfn	Labrofrontal nerve
Ln	Labral nerve
Ltc	Tegumentary nerve of clypeus
Ltf	Facial tegumentary nerve
Lts	To labral sensilla
Md1	To mandibular adductor muscle (28)
Md2	To hypopharyngeal muscle of mandible (29) and mandibular sensilla
Md3	To cranial abductor (27) and tentorial muscle (30) of mandible
Mx1	To adductor muscle of stipes (34b)
Mx2	To rotator muscle of cardo (31) and retractor muscle of maxilla (32)
Mx3	To reductor muscle of galea (42), palpal levator muscle, palpal depressor muscle, and palpal integument
Mx4	Maxillary palp nerve
Mx5	Lacinial nerve
Mx6	Galeal nerve
Mx7	To productor of lacinia (41), reductor of galea (42), one fascicle of maxillary protractor (33) and integument of cardo and stipes
Mx8	To maxillary protractors (33) and adductor of stipes (34a)
Nc	Nervus connectivus
N2	Median labral nerve to anterior muscles of labrum (3), compressor muscles of labrum (2) and clypeal dilator muscles
N3	Clypeal nerve to cibarial dilator muscle (5) and to oblique muscles of cibarium
N5	To precerebral dilator muscles of pharynx (6, 7)
Oc	Ocellus
Ocn	Ocellar nerve
Ol	Optic lobe
On	Oesophageal nerve
Prc	Protocerebrum
Rn	Recurrent nerve
Sdn	Salivary duct nerve
Spn	Subpharyngeal nerve to ventral cibarial muscles and some ventral pharyngeal muscles
Tc	Tritocerebral commissure
Tn1,2,3	Tegumentary nerves to cranium
Trc	Tritocerebrum
Trn	Transverse nerve to posterior frontal muscles of labrum (4) and frontal productor muscles of hypopharynx (13); possibly to productor muscles of hypopharyngeal oral arms (14)

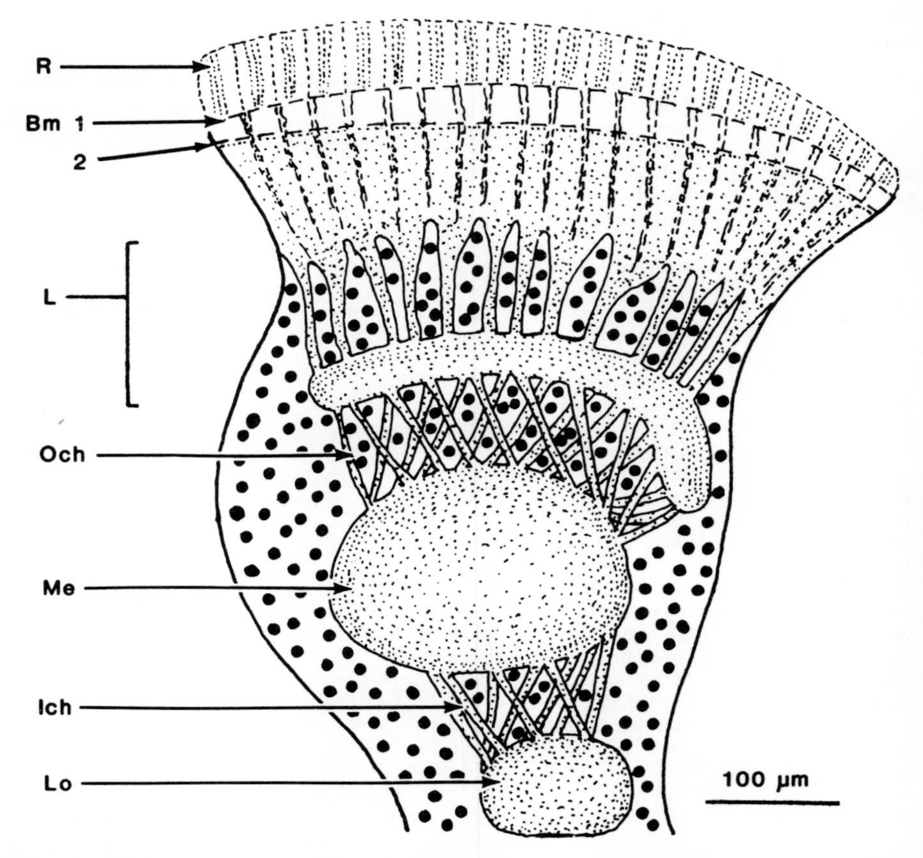

Fig. 8.2 Schematic horizontal section through the left optic lobe of *P. americana*. Bm 1, 2, basement lamellae; Ich, inner chiasma; L, lamina ganglionaris; Lo, lobula; Me, medulla; Och, outer chiasma, R, retina. (After Ribi, 1977.)

the retinal cells are three different kinds of short visual fibres that terminate in two separate layers in the lamina, and one long visual fibre type that ends in the medulla. Monopolar second-order neurons and horizontal fibres also arborize in the lamina, where they synapse with the visual fibres.

It has been proposed that the optic glomeruli on the right are associated with those on the left by direct neural pathways. This was claimed to be so in *P. americana* by Hanström (1928; his 'Sehcommissur') and in *L. maderae* by Roth and Sokolove (1975). Roth and Sokolove (1975) employed degenerative techniques, and also traced the retrograde transport of horseradish peroxidase injected into the left cerebral lobe. By using the cobalt diffusion and precipitation (CoS) method, Honegger and Schürmann (1975) detected, in the cricket, four tracts that connect the medulla on one side directly with the lobula and medulla on the opposite side. These pathways may contain interneurons involved with binocular vision, or they may serve to co-ordinate the output of neuronal oscillators that govern circadian locomotor activity (see Section 10.5).

Each of the two ocelli of *P. americana* is evident as a white patch located between the antero-medial border of the compound eye and the dorsal perimeter of the antennal socket (see Section 9.2.5). The axon bundles of the ocellar retinal cells pass through a reflecting tapetum and synapse immediately with dendritic branches from large interneurons (Cooter, 1975; Weber and Renner, 1976). In each ocellar nerve there are five of these large fibres (10–50 μm diameter) plus 40–80 smaller (ca. 1.5 μm diameter) ones. The larger fibres enter the brain near the midline in the pars intercerebralis (PI), to form a tract that passes postero-ventrally.

The two reports regarding the numbers and locations of ocellar interneurons demonstrable after CoS staining are in disagreement. Cooter (1975) found eight somata situated along the ipsilateral nerve tract in the PI, and one soma on the contralateral side. The larger fibres arborize extensively in the area of the ipsilateral posterior protocerebrum. Bernard (1976) could consistently stain 5–6 somata located at variable positions along the ocellar tract, all on the ipsilateral side. Different branches from each of the two largest neurons project to the vicinity of the ipsilateral optic lobe, towards the protocerebrum, and to the contralateral circumoesophageal connective where they descend to the ventral nerve cord. It is evident that more detailed studies are needed before meaningful comparisons can be made with the pattern in acridid grasshoppers (Goodman, 1974, 1976; Goodman and Williams, 1976; Patterson and Goodman, 1974).

The possibility that ocellar fibres enter the nervi corporis cardiaci 1 and 2 (NCC 1, NCC 2) in *Gryllus domesticus*, *Locusta migratoria* and *Schistocerca gregaria* is indicated by Brousse-Gaury (1971b), and Cooter (1975) describes in the ocellar tract of *P. americana* a fibre that may leave the brain via NCC 2. In whole-mount preparations stained with the CoS method and also with a neurosecretory stain, Bernard (1977) could not find connections between the ocellar interneurons and the neurosecretory cells of the PI.

The central pathways of the remaining sensory nerves of the protocerebrum, the integumentary nerves, have not been defined. In locusts, however, the axons project through the brain to the subesophageal ganglion (Aubele and Klemm, 1977; Tyrer *et al.*, 1979). In *S. gregaria* (Tyrer *et al.*, 1979), the integumentary nerves supply facial mechanoreceptor sensilla that are important in initiating and maintaining flight.

(c) *Protocerebrum: pars intercerebralis (PI) and pars lateralis (PL)*

These regions of the protocerebrum are best known for the neurosecretory cells they contain (see Section 12.2). Incorrectly, they are often discussed as though the neurons found there have no other function. As noted above, the somata of ocellar interneurons are located in the PI. Willey (1961) divided this region into five zones, several of which contain neurons that may not be neurosecretory. Williams (1975) made considerable progress in relating the topology of individual neurons to gross features of midbrain anatomy in *S. gregaria*. His study provides further evidence for a multiplicity of functions by the units in the PI.

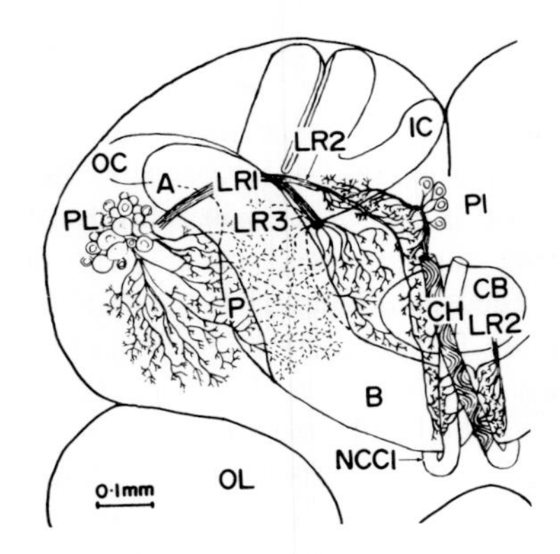

Fig. 8.3 Diagram showing relationships of pars intercerebralis (PI) and pars lateralis (PL) neurons to each other, and to cerebral glomeruli on right side; frontal view. A, α-lobe; B, β-lobe; CB, central body; CH, chiasma; IC, medial calyx; LR1, 2, 3, major PL nerve tracts. LR3 joins NCC2. NCC1, intracerebral part of nervus corporis cardiaci 1; OC, lateral calyx; OL, olfactory lobe; P, pedunculus. (From Pipa, 1978a.)

The locations and intracerebral projections of PI and PL neurons that enter the RNC of *P. americana* have been described (Fraser and Pipa, 1977; Pipa, 1978a). The CoS method showed that the CC on one side receives axons from 20–45 somata located in the ipsilateral PL and from another 240–260 in the contralateral PI (Fig. 8.3). In paraffin sections that were silver-intensified (Tyrer and Bell, 1974), there was no evidence that neurons in the PL cross into the opposite cerebral lobe (Pipa, 1978a). This agrees with the interpretation by Willey (1961).

Although many of these neurons are presumably neurosecretory, their identities have not been fully resolved. The somata of 80–107 putative neurosecretory neurons were counted in whole-mounts of each half of the PI stained with resorcin-fuchsin, but none was revealed in the PL (Pipa, 1978a). In paraffin sections stained with paraldehyde-fuchsin, however, such somata could be shown in the PL (Fraser and Pipa, 1977; Pipa 1978a; see Fig. 12.2).

All the neurons that enter NCC 1 and 2 may not have been demonstrated by the CoS technique, for many somata, particularly the smallest, were faintly stained in whole-mounts. Furthermore, by counting all the axons in electron micrograph reconstructions of transverse sections through these nerves, 550–650 units were revealed in NCC 1 and approximately 75 were seen in NCC 2 (Novak, 1980).

The distributions of CoS-stained PI and PL neurons and their relationships to the neuropile surrounding the corpus pedunculatum are shown in Fig. 8.3. Rami from the PL cluster are abundant below the somata, between the α-lobe and pedunculus of the mushroom body, and surrounding the central body. The axons in tract LR2 arise from the Pl somata and ramify extensively in the neuropile containing the 'dendritic arborizations' of PI neurosecretory cells (Adiyodi and Bern, 1968). Numerous branches are also found along the protocerebral chiasma. In these regions, the collaterals from the PL and PI neurons are close to one another, suggesting that they form synapses.

(d) *Corpora pedunculata (mushroom bodies)*

These conspicuous glomeruli were discovered by Dujardin (1850). Thinking that they were most highly developed in social ('intelligent') insects, he compared them to the convolutions of the mammalian cerebral cortex. Since that time, the corpora pedunculata have been the focus of numerous studies. This led to considerable speculation regarding their importance as centres for 'complex' behaviour. We now know that these structures are well developed in insects that are difficult to characterize as mentally acute (i.e. cockroaches), and a less anthropocentric view prevails.

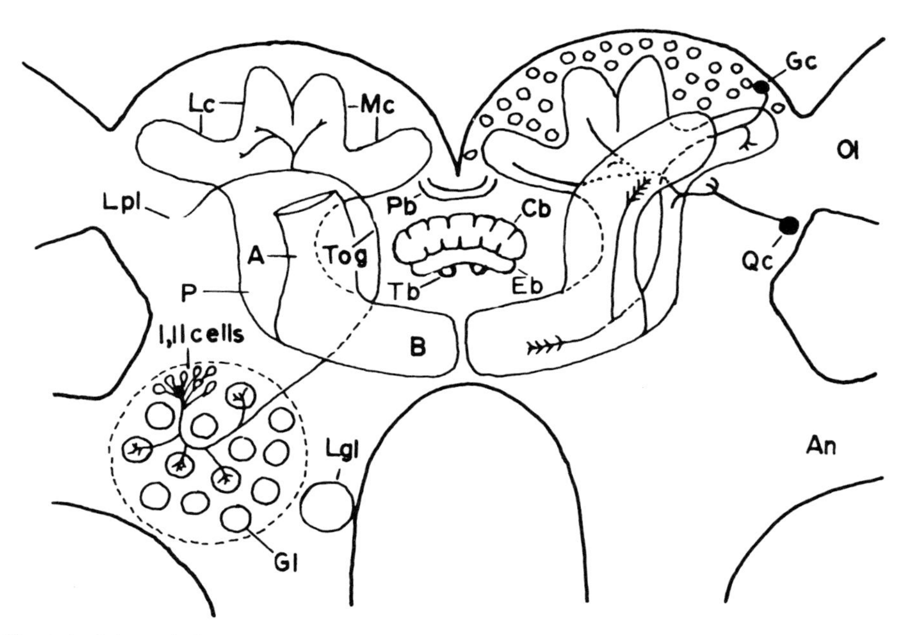

Fig. 8.4 Schematic frontal section through the brain, showing relationships among neurons of the deutocerebrum and corpora pedunculata. A, α-lobe; An, antennal nerve; B, β-lobe; Cb, central body. Cells I, II, deutocerebral neurons with processes to tractus olfactorio-globularis; Eb, ellipsoid body; Gc, globuli cell of corpus pedunculatum; Gl, glomerulus of deutocerebrum; Lc, lateral calyx; Lgl, lobus glomerulatus of tritocerebrum; Lpl, lateral protocerebral lobe; Mc, medial calyx; Ol, optic lobe; Pb, protocerebral bridge; Qc, quartet cell; Tb, posterior tubercle; Tog, tractus olfactorio-globularis. (After Weiss, 1974; Ernst *et al.*, 1977.)

There are two symmetrically arranged corpora pedunculata in the brain of *P. americana*. Each one consists of a pair of dorsal calyces that rest on a stalk (pedunculus) (Figs. 8.3, 8.4). The pedunculus bifurcates ventrally to produce the β-lobe that extends to the brain midline and the α-lobe that projects dorsolaterally, beneath the anterior surface of the protocerebrum and up to the level of the calyces. The two calyces on each side contain ca. 200 000 globuli cells (Neder, 1959). These neurons have spherical somata with very little cytoplasm, and they seem to be entirely intrinsic; certain of their branches ramify within the calyx, while others descend along the pedunculus where they diverge. One branch ascends into the α-lobe, and the other continues into the β-lobe (Fig. 8.4).

Interneurons with somata in the olfactory lobe (deutocerebrum) seem to constitute the major, but not exclusive, input to the corpora pedunculata (Hanström, 1928; Ernst *et al.*, 1977). In the olfactory glomeruli these interneurons synapse with sensory axons from the antennae, and then project via the tractus olfactorio-globularis to terminals in the calycal walls (Fig. 8.4). Weiss (1974) describes three other classes of extrinsic fibres that enter the outer synaptic layer of the calyces. Only one of these classes, the quartet cells (Fig. 8.4), were traced from their somata. Their four large cell bodies are located near the junction of the optic lobe and protocerebrum.

Connections between the optic lobes and calyces are reported by Hanström (1928), who hypothesizes that the primary function of the corpora pedunculata is to integrate optic and antennal information. This view is discounted by Weiss (1974), who, in agreement with Jawlowsky (1963), finds no evidence of optic input. Weiss (1974) suggests, instead, that these glomeruli are major second-order processing centres for antennal sensory signals.

Willey (1961) and Jawlowsky (1963) note the presence of tracts that may join the tritocerebral glomeruli to the corpora pedunculata of *P. americana*. Using the CoS method on locusts, Aubele and Klemm (1977) could follow single axons from the labral nerves into the inner calyx of the ipsilateral mushroom body.

The fine structure of the corpora pedunculata of *P. americana* has also been investigated (Mancini and Frontali, 1967; Frontali and Mancini, 1970). The axon processes of the globuli cells are uniformly thin, nearly parallel to each other, and often in direct contact. These fibres (type I) have numerous endings that contain clear vesicles and slightly larger semi-opaque granules. They seem to form axo-axonic synapses amongst themselves but, more often, they synapse on type 2 fibres. The type 2 fibres lack synaptic vesicles, and are probably the dendrites of extrinsic neurons.

(e) *Central body complex and protocerebral bridge*

The central body complex of *P. americana* resembles that found in saltatory Orthoptera, and it is illustrated in Fig. 8.4. It consists of a dorsal and a ventral glomerulus and a pair of posterior tubercles (Hanström, 1940; Arnold, 1960). The largest component, the central body, overlies the others. The central body is convex above and concave below, and it is divided by fibre tracts into a variable number (probably 8) of vertically parallel columns. The neuropile of the ventral glomerulus (ellipsoid body) is incompletely separated from the central body. It is more homogeneously dense, and a distinct palisade arrangement is not evident. The posterior tubercles are irregularly polygonal in transverse sections.

The protocerebral bridge (PB, Fig. 8.4) lies above the central body complex, with its dorsal surface in the PI. It is located near the tractus olfactorio-globularis and the ocellar tracts. Although Bretschneider (1913, 1914) describes an ocellar connection with PB, this is denied by Hanström (1940). Connections between PB and the following structures are also proposed by Bretschneider: medulla, contralateral antennomotor nerve, and contralateral circumoesophageal connective. In these cases, the evidence for neuron connections is only suggestive; it is based on the proximity of tracts passing between these centres, and individual axons were not traced.

A detailed account of the close association of the PB and central body complex of *S. gregaria* is presented by Williams (1975). In the pars intercerebralis 16 clusters of neurons, each consisting of four units, project into the PB. After producing arborizations in the PB, each quartet of fibres passes to the contralateral side of the central body via a chiasma. Within the central body the quartet traverses a second chiasma, and the fibres synapse again with units in the underlying ellipsoid body. A third chiasma is formed by the 16 quartets of fibres as they leave the ellipsoid body and pass to the lateral accessory lobes of the protocerebrum, anterior to the antenno-glomerular bundles. In effect, then, the three chiasmata relay the ordered zones of arborizations of the PB to the lateral accessory lobes on the opposite side. According to Williams (1975) this arrangement may permit activity within both sides of PB or both lateral accessory lobes to be communicated to other neurons with arborizations within the ellipsoid body.

(f) *Deutocerebrum*

Nerves arising from the two deutocerebral lobes supply the sensory receptors and muscles of the antennae. The neuropile of each centre is divided into two distinct regions: an anterior 'olfactory' or 'antennal' lobe containing about 125 (Ernst *et al.*, 1977) dense glomeruli, and a less highly organized posterior dorsal lobe with coarse fibres that can be traced into the antennomotor nerves (Arnold, 1960; Prigent, 1966). Prigent (1966) notes sexual dimorphism in the structure of the deutocerebrum of *P. americana*; a lenticular mass of neuropile anterolateral to the base of the antennal nerve occurs only in males. In *Blaberus craniifer*, the glomerular organization is remarkably constant, and individually identifiable glomeruli have been characterized (Chambille *et al.*, 1980).

The glomeruli of the 'olfactory' lobe (Fig. 8.4) consist of many neuron arborizations and synapses. These are contributed by antennal receptor cells, by deutocerebral neurons, and probably by neurons from other parts of the brain. Ernst *et al.* (1977) found that the axons from each antennal flagellum enter the glomeruli of the ipsilateral deutocerebrum, where they contact branches from several groups of interneurons. The somata of these interneurons are located in the deutocerebral cortex. The connections are both convergent and divergent; single deutocerebral neurons receive inputs from numerous antennal axons, and the antennal axons often branch to terminate at more than one deutocerebral neuron.

Although most of the processes from the deutocerebral neurons remain within the deutocerebrum, where they may serve as local interneurons, some of them enter the tractus olfactorio-globularis. This latter group of fibres apparently arborize within the calyces of the corpora pedunculata (see Section 8.2.2(d) and Fig. 8.4) before proceeding into a circumscribed region between the deutocerebrum and optic lobes (Ernst *et al.*, 1977).

(g) *Tritocerebrum*

Notwithstanding its small size, the tritocerebrum appears to be an important component. It connects the brain to the stomodaeal nervous system, to the CC (via the NCC 3), and to the ventral nerve cord (Fig. 8.1a,b). For the most part, however, the functional significance of this region can only be surmised, since the neuron connections are inadequately understood.

In *P. americana*, separate fibre tracts from the labral nerve and frontal ganglion connective unite in a glomerulus at the centre of the tritocerebrum (Willey, 1961). This neuropile apparently corresponds to the lobus glomerulatus (Lgl, Fig. 8.4) described by Hanström (1928). It seems to receive fibres from interneurons with somata in the deutocerebrum, and may not be separated entirely from the antennal glomeruli (Ernst *et al.*, 1977). Axons from sensilla on the maxillary palps also project into the lobus glomerulatus (Ernst *et al.*, 1977). As they pass towards the corpora pedunculata, fibres from this neuropile contribute to the tractus olfactorio-globularis (Section 8.2.2(d)).

Gundel and Penzlin (1978) find that after Co^{2+} is allowed to enter a single frontal ganglion connective (Fgc, Fig. 8.1a) towards the brain, only a few somata in the tritocerebrum stain. Although the majority of fibres seem to end in the anterior lobus glomerulatus, some pass through it and into the PI and PL of the protocerebrum, while others enter the suboesophageal ganglion. Fibres within the following nerves (Fig. 8.1a) can be followed to 'endings' (t) or to somata (s) in the tritocerebrum: (1) the labral nerve (Ln; t); (2) the median labral nerve (N2; s); (3) certain anterior and posterior branches from Trn (t); and (4) the nervus connectivus (Nc; t).

Each CC receives a nerve (NCC 3) from the tritocerebrum (Cc3, Fig. 8.1b), but the distributions of the component neurons are poorly understood. Willey (1961) counted about 10 fibres in NCC 3, and these could be traced into the chiasma that connects the two corpora cardiaca. Putative neurosecretory cells were not observed in the tritocerebrum by Willey, but Khan (1976) did find them. Attempts to demonstrate the NCC 3 neurons by using the CoS technique have not been successful (Fraser and Pipa, 1977). In locusts, however, that method has revealed three such cells (Mason, 1973; Aubele and Klemm, 1977).

8.2.3 Stomodaeal nervous system (SNS)

Various aspects of this so-called 'vegetative nervous system' are presented in Chapter 4, and in Sections 8.2.1 and 8.2.2(g). The cephalic components of the SNS are shown in Figs. 8.1a,b, and the extracephalic members are illustrated in Fig. 4.5.

The mapping of neuron pathways in the SNS has just begun. Axons from 50–60 somata in the frontal ganglion descend posteriorly, apparently beyond the hypocerebral ganglion (Gundel and Penzlin, 1978). Each soma seems to produce three rami: a median branch to the recurrent nerve, and one to each frontal ganglion connective. Some of the frontal connective fibres project to the tritocerebrum (Section 8.2.2(g)), while others bypass that region to enter the suboesophageal ganglion. Neurons with processes in nerve 2 (Fig. 8.1a) descend to the suboesophageal ganglion, as well.

Each of the two neurons that comprise the nervus connectivus (Nc, Fig. 8.1a) also displays the 'triaxonal' branching pattern (Gundel and Penzlin, 1978; Hertel *et al.*, 1978). From a soma located in the anterior protocerebrum an axon enters the frontal ganglion, where it produces numerous arborizations before dividing into three long branches. One of these branches enters the recurrent nerve, and, near the hypocerebral ganglion, it produces bifurcations that continue posteriorly in the esophageal nerve. Each of the other two branches enters a frontal ganglion connective and arborizes in the ipsilateral tritocerebral neuropile. Arborizations from the nervus connectivus synapse with neurosecretory fibres in the frontal ganglion. This suggests a role for neurosecretion in the electrical activity of the two nervus connectivus neurons (Ude and Agricola, 1979).

8.2.4 Retrocerebral neuro-endocrine complex (RNC)

The components of the RNC include the paired CC and CA, the nerves that join these glands to each other, and other nerves that connect them to the CNS and SNS. The diverse and important roles of the endocrines in this sytem are presented in Sections 12.3 and 12.5.

The pathways and affiliations of neurons in the cockroach RNC have been studied, but our understanding is far from complete. To a great extent this is because of the intricate and overlapping associations of the components. For example, unlike the condition in saltatory Orthoptera, in the cockroach, the CC do not consist of a discrete glandular zone that is readily distinguishable from adjacent neurohormone release sites (neurohaemal areas). In *P. americana*, putative neurohaemal areas occur along the NCC (Cc1 and 2, Fig. 8.1b; Fig. 12.3). They surround the CA, and extend into the nervi corporis allati 2 (NCA 2) (Ca2, Fig. 8.1b) and postallatal nerves (Dogra, 1968; Adiyodi, 1974; Pipa and Novak, 1979; Novak, 1981). The extent to which these neurohaemal areas are formed from extracerebral neurosecretory cells remains to be determined. A further difficulty is that the CC and CA are not connected to each other by a well-defined bundle of axons (Novak, 1981).

Axons contributing to the RNC have been traced to somata in the PI and PL (Section 8.2.2(c)). Other neurosecretory neurons to this region originate in the suboesophageal ganglion (Pipa and Novak, 1979). The somata of the latter occupy two separate locations adjacent to the sagittal plane of the ganglion and ipsilateral to NCA 2 (Figs. 8.5a,b). They apparently produce six of the axons that enter the CA via that nerve. From a transverse tract within the CA, various branches from these axons ramify amongst the gland cells, penetrate the cap-like union of the gland

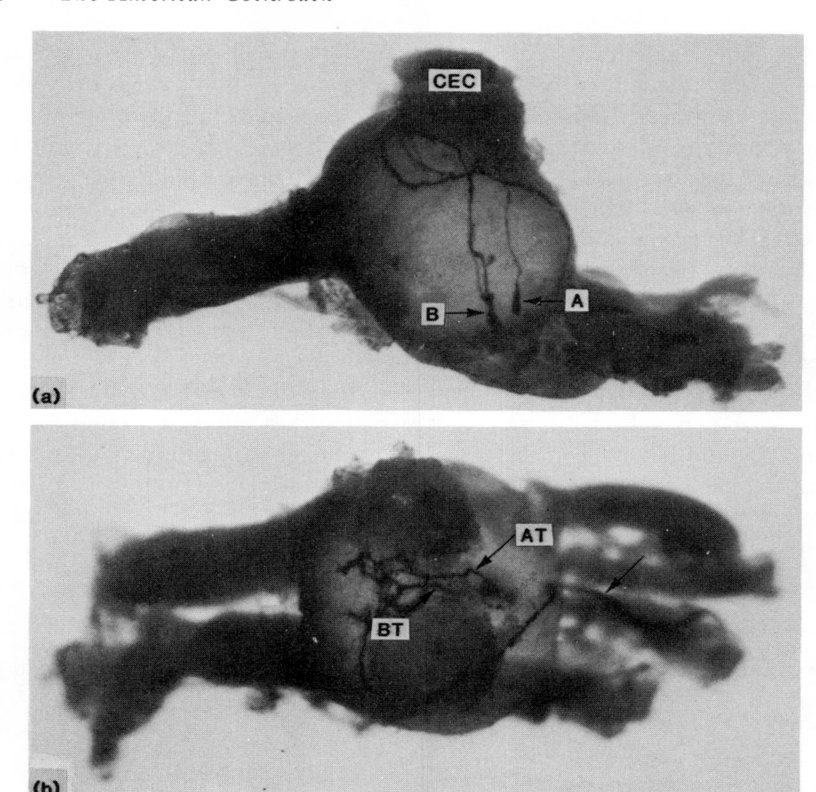

Fig. 8.5 (a) Lateral aspect of a suboesophageal ganglion. Co^{2+} was perfused retrogradely via severed right NCA2. Processes from two anterior (A) and five posterior (B) neurons issue dorso-posteriorly between the two circumoesophageal connectives (CEC), encompass the one on the right, unite, and proceed anteriorly towards NCA2. × 55. (b) Dorsal view of preparation shown in (a). Note pathways (AT, BT) of the two groups of neurons, locations of their separate synaptic fields between and behind the two CEC, and their projection into the right NCA2 (arrow), × 55. (From Pipa and Novak, 1979.)

with NCA 1, or enter the postallatal nerves. Other branches pass into the CA on the opposite side. Although the destinations of the neurosecretory axons that leave the CA via the postallatal nerves are largely undermined, Novak (1981) could trace some of them to synaptic contacts with foregut muscle fibres.

8.2.5 Ventral nerve cord (VNC)

(a) *Peripheral nerves*

Many of the principal nerves that stem from the ventral ganglia of *P. americana* have been described (Nijenhuis and Dresden, 1955; Pipa and Cook, 1959; Shankland, 1965), and certain aspects of these studies are summarized by Guthrie and Tindall (1968). Less well-known are the nerves from the last (sixth) discrete abdominal ganglion, particularly those to the genital and postgenital segments. Guthrie and Tindall (1968) present their observations of the pattern in males, but only a few of the secondary and tertiary branches are shown, and the innervation of many phallic muscles is not mentioned. An analysis of the phallic motoneurons of *P. americana* is given by Grossman and Parnas (1973). Aside from the work by Engelmann (1963) on *L. maderae*, the nerves supplying the female genitalia of cockroaches appear not to have been traced. Innervation of the hindgut musculature is presented by Brown and Nagai (1969).

Table 8.1 Descriptive summary of the thoracic nerves of *P. americana*. (Except for nerves 2C and 8, these are illustrated in Fig. 8.6.)

Nerve	*Destination and/or Function*
1	Interganglionic connective
2A	Spino-sternal intersegmental and ventral longitudinal muscles
2B	A branch from prothoracic 2B joins mesothoracic 2B to form a 'long nerve' that extends into the head, where it meets nerve PB from NCA 2 (Fig. 12.3) Meso- and metathoracic branches from 2B are sensory to spiracles, wings, and axillary region
2C	Dorsal longitudinal and oblique muscles
3A	Coxal adductor and promotor muscles
3B	Major sensory nerve of coxa
4	Main leg depressor muscle, certain coxal remotors, and muscles that probably function as coxal adductors and rotators: predominantly motor
5	Sensory to coxa and meron; the only sensory nerve distal to coxa Coxal branches of main leg depressor muscle; femoral reductor; tibial, tarsal and pretarsal flexors; tarsal extensor muscle
6A	Coxal remotor muscles
6B	Main leg levators, and muscles that probably function as coxal rotators or rotator-adductors
7	Sensory to basisternum and furcasternum
8 (Transverse nerve; its root is shown in Fig. 8.7(b))	Muscles that regulate spiracular aperture (see Chapter 5)

We cannot present here detailed accounts of the thoracic and abdominal nerves, so the references cited above should be consulted. However, we do illustrate those nerves that arise from the suboesophageal ganglion. Figure 8.1b is based largely on the study by Arnold (1960), who described more branches than did Guthrie and Tindall (1968). To Arnold's original figure we have added Ca2 (NCA 2; Willey, 1961) and the salivary duct nerve (Sdn; Whitehead, 1971). An overview of the thoracic innervation can be obtained from Fig. 8.6 and Table 8.1.

With regard to the abdominal division of the VNC, only five unfused ganglia are evident once embryogenesis is completed (see Section 8.2.1); the first (embryonic) abdominal ganglion is incorporated with the metathoracic ganglion, and embryonic

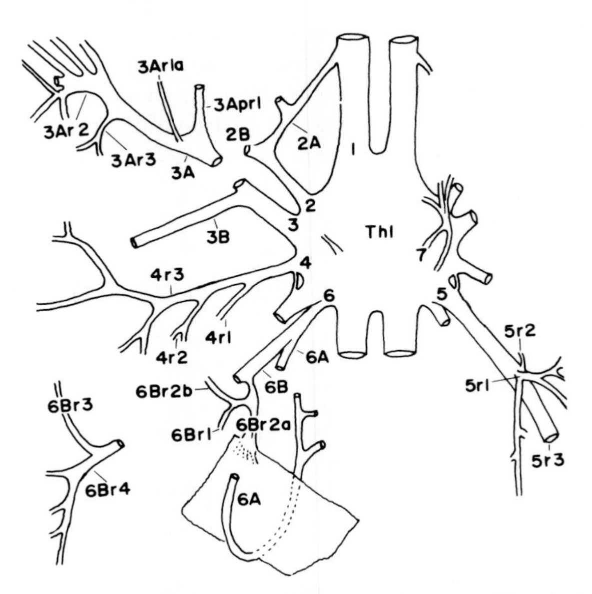

Fig. 8.6 Schematic ventral view of a prothoracic ganglion, showing some of the principal nerve branches. (After Pipa and Cook, 1959.) Refer to Tables 8.1 and 8.2 for a descriptive summary of the nerves.

ganglia seven to eleven are coalesced to form a terminal complex, the 'sixth ganglion' evident in larvae and adults. A pair of nerves issues from each of the five unfused ganglia. Their branches supply the integumental sensilla, the abdominal muscles and the heart (Shankland, 1965). A fine median nerve leaves the postero-dorsal surface of the fused metathoracic-first abdominal ganglia. It continues posteriorly between the two interganglionic connectives to join the anterior margin of the second abdominal ganglion. From the posterior rim of the second abdominal ganglion it passes into the anterior border of the third. This pattern is repeated until the 'sixth abdominal ganglion' (i.e. the terminal complex) is reached.

Most of the transverse nerves of the abdomen arise from these median nerves, occasionally very close to the ganglia. Each of the two transverse branches is dilated basally, where it joins a short, dorsally directed trunk. These two dilations and the short median nerve trunk define the location of the neurohaemal perisympathetic organ (Chapter 12). Peripherally, in abdominal segments two to seven, each transverse nerve is connected to a nerve that issues from the next ganglion (Shankland, 1965). Superficially, the assemblage resembles the 'link nerve' plexus found in the stick insect, *Carausius morosus* (Finlayson and Osborne, 1968; Fifield and Finlayson, 1978). It would be interesting to know whether in *P. americana*, as in *C. morosus*, multipolar neurosecretory cells occur there.

P. americana has 10 pairs of spiracles (two thoracic and eight abdominal), not 11 pairs as shown by Guthrie and Tindall (1968; their Fig. 8.27b). The innervation of the spiracular occlusor and dilator muscles by axons of the median and transverse nerves is described in Section 5.3.5.

(b) *Major ganglionic nerve tracts*

Pipa *et al.* (1959) identified seven pairs of longitudinal tracts and 10 transverse tracts (commissures) that cross the neuropile of each thoracic ganglion. These observations were confirmed, for the most part, and extended in a clearly illustrated study of the mesothoracic ganglion by Gregory (1974). The names and relative locations of the tracts are shown in Figs. 8.7a,b. Numerous other vertical and oblique tracts, some of which are useful landmarks, are evident in the ganglion core (Gregory, 1974).

Gregory (1974) finds that the axon bundles forming the roots of each mesothoracic nerve are usually segregated according to inferred function; the roots are either motor or sensory, and the former generally lie above the latter. Rarely, however, are the central projections of the motor roots strictly dorsal and the sensory roots ventral. Most of the thin (<5 μm diameter) sensory axons of the ventral nerve roots are associated with the ventral neuropile, but coarser sensory fibres project more dorsally, and thin motor ones extend ventrally. It is significant that Gregory (1974) was unable to establish a constant relationship between axon diameter and inferred function; there was much overlap in the sizes of motor and sensory axons.

(c) *Maps and projections of identified neurons*

With notable exceptions (i.e. the dorsal unpaired median neurons (Crossman *et al.*, 1971), certain neuron groups associated with the median nerves (Gregory, 1974; Ali and Pipa, 1978)), most of the somata in the thoracic and abdominal ganglia of *P. americana* occur in the ventral and ventro-lateral cortex. The somata are arranged in groups, and there have been attempts to identify those that send axons down particular peripheral nerves (Cohen and Jacklet, 1967; Young, 1969; Pearson and Fourtner, 1973; Gregory, 1974; Iles, 1976).

There is little agreement regarding the number of somata present in the thoracic ganglia. Iles (1976), using the CoS method, approximated a total of 248 motoneuron somata in the prothoracic ganglion. From whole-mounts, Gregory (personal communication related by Pearson, 1977) estimated that 1500 somata occur in the mesothoracic ganglion. About 300 of these belong to motoneurons (Gregory, 1974). Cohen and Jacklet (1967) counted 3422 somata from histological serial sections of the metathoracic ganglion, and proposed that 230 of these were motoneurons. They considered the motoneuron somata to be 20 μm in diameter or larger, but Gregory (1974) found that somata smaller than this belong to that category. Pearson (1977) thinks it reasonable to conclude that the total number of somata in a thoracic ganglion is about 2000. Up to 500 of these send axons from the ganglion, so he postulates that 1500 intraganglionic interneurons are present.

The somata of motoneurons that contribute axons to the same peripheral nerve are often clustered together. Usually, they occur on the side ipsilateral to the nerve they supply. The bilateral symmetry in the locations of paired somata is so regular that corresponding cells on either side of the mesothoracic and metathoracic ganglia have been identified (Cohen and Jacklet, 1967; Young, 1969),

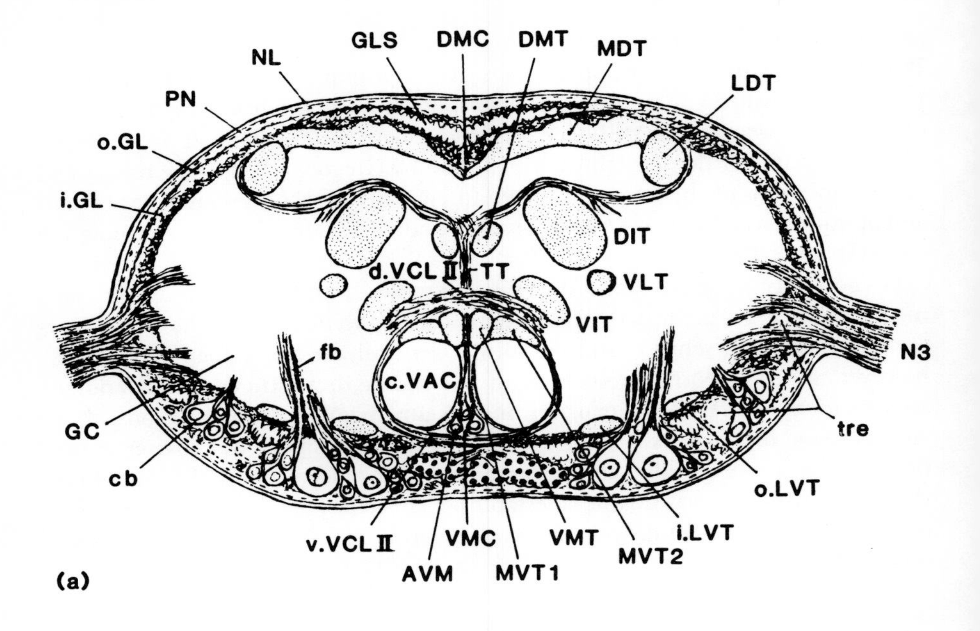

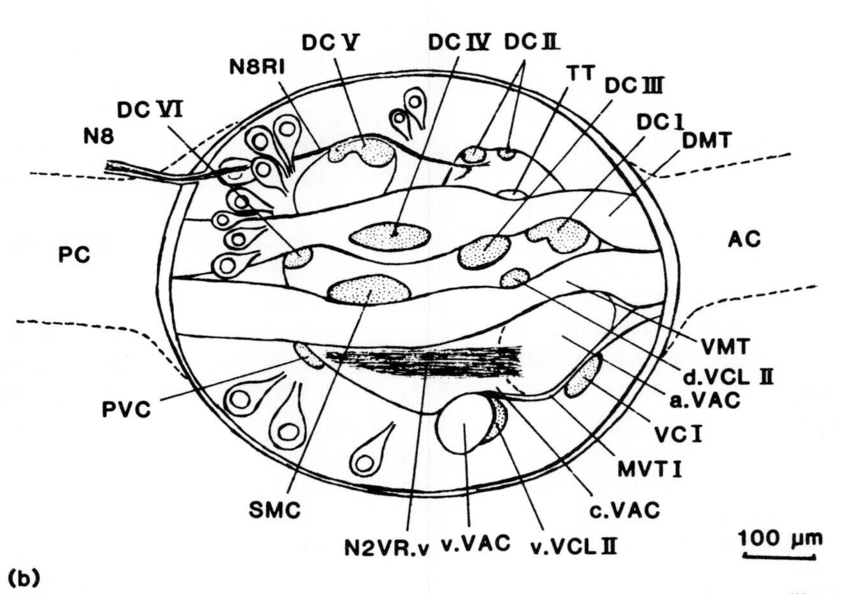

Fig. 8.7 (a) Schematic transverse section of a mesothoracic ganglion at level of nerve 3 illustrating general internal structure, and locations of longitudinal fibre tracts. Note ventral parts of ventral commissural loop II, v.VCLII, encircling cylinders of ventral association centre, c.VAC, and T-shaped fibre tract from antero-ventral medial cell body group, AVM. See (b) for magnification and explanation of abbreviations. (b) Schematic parasagittal section close to the midline of a mesothoracic ganglion. Positions of commissures, divisions of the ventral neuropile, VAC, and a root of nerve 8 are shown; viewed from midline.

AC, anterior interganglionic connective; a.VAC, anterior mass of ventral association centre; AVM, anterior ventral median cell body group; cb, neuron cell body; c.VAC, cylindrical region of ventral association centre; DC I to DC VI, dorsal commissures I to VI; DIT, dorsal intermediate tract; DMC, dorsal midline cleft; DMT, dorsal median tract; d.VCLII, dorsal part of ventral commissural loop II; fb, fibre bundle (of neuron cell body group); GC, ganglion core; GLS, glial lacunar system; i.GL, inner glial cell layer; i.LVT, inner lateral ventral tract; LDT, lateral dorsal tract; MDT, median dorsal tract; MVT1, median ventral tract 1, MVT2, median ventral tract 2; N2VR.v, ventral bundle of nerve 2, ventral root; N3, nerve 3; N8, nerve 8; N8R1, nerve 8, root 1; NL, neural lamella; o.GL, outer glial cell layer; o.LVT, outer lateral ventral tract; PC, posterior interganglionic connective; PN, perineurium; SMC, supra-median commissure; tre, tracheae; TT, T-shaped tract; VCI, ventral commissure I; VIT, ventral intermediate tract; VLT, ventral lateral tract; VMC, ventral midline cleft; VMT, ventral median tract; v.VAC, most ventral region of ventral association centre; v.VCLII, ventral part of ventral commissural loop II. (From Gregory, 1974.)

although homologous somata occupying different relative positions have also been observed (Pearson and Fourtner, 1973; Gregory, 1974). In addition, there is evidence that serially homologous neurons supply serially homologous muscles. Homologous somata may be recognizable topographically, but, unfortunately, a sterotaxic atlas with co-ordinates that would ensure identification of most of them has yet to appear.

Table 8.2 lists thoracic motoneurons whose locations and main branches have been mapped. The central arborizations of these neurons are widespread and complex,

Table 8.2 Motoneurons (MN) identified in the thoracic ganglia of *P. americana*. (After Fourtner and Pearson, 1977.)

MN	*Nerve containing axon (see Figs 8.6 and 8.8)*	*Function*	*References*
D_f (28*, 30†)	5r1	Fast extensor (depressor) of femur	Iles, 1972 Pitman *et al.*, 1972a Tweedle *et al.*, 1973 Young, 1972
D_s	5r1	Slow extensor (depressor) of femur	Pearson and Bradley, 1972 Pearson and Fourtner, 1975
I_1 and I_2	5r1–3; 4r2	Local common inhibitor	Iles, 1976 Pearson and Fourtner, 1973
I_3	Various branches of nerves 3–6	Widespread common inhibitor	Iles, 1976 Pearson and Bergman, 1969 Pearson and Iles, 1971 Pearson and Fourtner, 1973
4,5,6,7	6Br4	Slow flexor (levator) of femur	Pearson and Fourtner, 1975
FFFe	6Br4	Fast flexor (levator) of femur	Fourtner and Pearson, 1977
SETi	3B	Slow extensor of tibia	Fourtner and Pearson, 1977
SFTi	5	Slow flexor of tibia	Fourtner and Pearson, 1977
FPC (27*, 29†)	3Ar2, 3Ar3‡	Fast promotor of coxa	Young, 1972
FEFe (18*†)	4r3§	Fast extensor (depressor) of femur	Young, 1972
FT	5	Fast flexor of tarsus	Fourtner and Pearson, 1977
FPT	5	Fast flexor of pretarsus	Fourtner and Pearson, 1977

* Numbers from Cohen and Jacklet (1967).
† Numbers from Young (1969).
‡ Fourtner and Pearson (1977) indicated that the metathoracic muscles supplied by FPC (161 and 167) receive axons via 4r2. Pipa and Cook (1959) found that these muscles are innervated by 3Ar3 and 3Ar2, respectively.
§ In the metathorax, this muscle (177a) is innervated by 4r3 (Pipa and Cook, 1959), not 4r2 (Fourtner and Pearson, 1977).

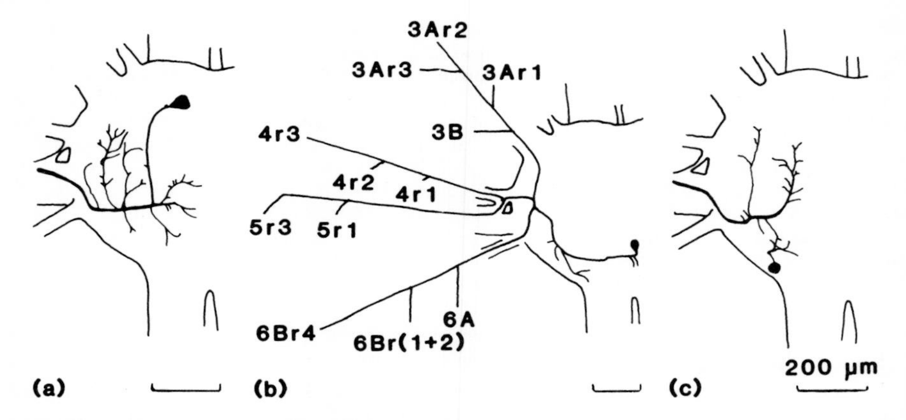

Fig. 8.8 (a) Dendritic morphology of an identified prothoracic D_f motoneuron. (b) Diagram of axon branches of a prothoracic widespread common inhibitor neuron (see Fig. 8.6 for relationships of the peripheral nerves containing the axon branches). (c) Dendritic morphology of identified prothoracic D_s motoneurone.
Scale bar in a, b and c represents 200 μm. (From Iles, 1976.)

and they have not been defined sufficiently to permit detailed comparisons. Three identified motoneurons and their largest branches are illustrated in Fig. 8.8. Physiological attributes of motoneurons are discussed in Section 8.7.

Pearson (1977) has surveyed the interneurons (INs) of certain Orthoptera. Some of the intraganglionic INs in *P. americana* are non-spiking, that is they fail to produce action potentials (Pearson and Fourtner, 1975). Nevertheless, when depolarized, some of these non-spiking INs will excite or inhibit the firing of motoneurons that supply the leg muscles (Section 8.7.4). Staining with CoS reveals that the soma of one of these INs is located in the postero-ventral ganglion cortex contralateral to the flexor motoneuron that it excites (Pearson and Fourtner, 1975). Some of the extensive terminal branches of this neuron lie close to the arborization of the flexor motoneurones (MNs 5 and 6, Table 8.2). This suggests that it may be monosynaptically connected to them.

Among the interganglionic INs, the ascending giant interneurons (GIs) of the VNC of *P. americana* have been studied most often, for they are accessible, their action potentials are easily identified, and they are associated with important sensory and motor pathways (see Section 8.7.3). It is not surprising that discrepant numbers of neurons have been assigned to this category, for the diameter of each axon varies greatly along the length of the VNC, and the measurements were obtained by using diverse histological methods. Usually, two well-separated bundles of giant fibres are seen in each abdominal connective: a dorsal group composed of 4–5 axons ranging from 20–35 μm in diameter, and a ventral group of four that measure 25–60 μm (Parnas and Dagan, 1971). Harris and Smyth (1971) assigned numbers to each of the four in the ventral group (GI 1–4) and Camhi (1976) extended this nomenclature to include the three largest axons in the dorsal group (GI 5–7; Fig. 8.9a).

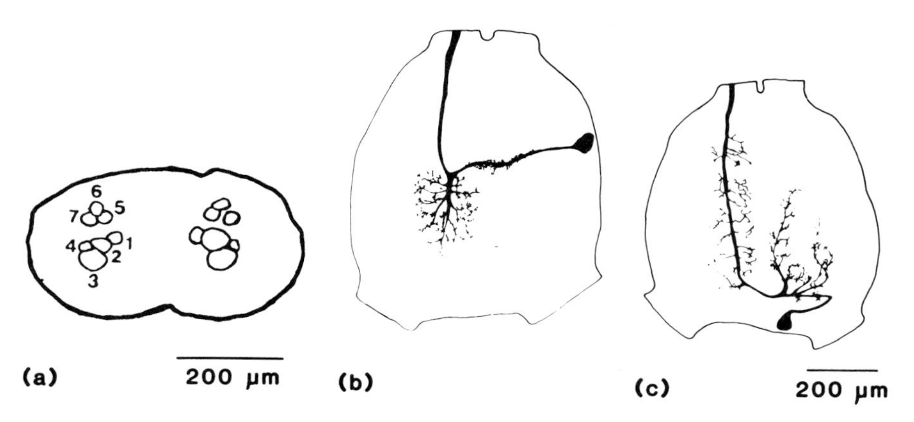

Fig. 8.9 (a) Diagrammatic cross-section of the fifth abdominal ganglion, showing positions and designations of the giant interneuron (GI) axons. Dorsal surface is at the top. (b) Camera lucida tracing of GI3 viewed dorsally. Magnification as in (c). Interganglionic connectives are at the top, and the two cercal nerves are at the bottom. (c) Camera lucida tracing of GI7. Orientation as in b. (From Daley *et al.*, 1981.)

Modern techniques, especially intracellular staining methods, have clarified the morphology of the GIs. Contrary to early reports (Roeder, 1948) it is now known that each axon arises from a single soma in the last abdominal ganglion and extends into the head (Hess, 1958b; Pipa *et al.*, 1959; Farley and Milburn, 1969; Spira *et al.*, 1969; Harris and Smyth, 1971; Milburn and Bentley, 1971; Pitman *et al.*, 1973; Füller and Vent, 1976; Harrow *et al.*, 1980; Daley *et al.*, 1981). Although apparently uninterrupted by septa, the giant axons do contribute lateral extensions to the ganglia of the VNC (Harris and Smyth, 1971; Delcomyn and Daley, unpublished). GIs 2–4 produce at least one branch in the centre of each abdominal ganglion, and GI 3 and 4 branch in the metathoracic ganglion, once posteriorly and again anteriorly (Harris and Smyth, 1971).

Cobalt preparations in which single GIs are stained indicate that the structure of each GI is unique, and that each of them can be identified on the basis of soma position and dendritic branching pattern (Harrow *et al.*, 1980; Daley *et al.*, 1981). The only known afferent pathways to these INs are axons from mechanoreceptor sensilla located on the two cerci (Section 9.3.1). The projections of these sensory axons are almost entirely ipsilateral, and closely overlap the dendritic fields of the seven GIs. Nonetheless, all of the GIs have some branches near the contralateral cercus (Daley *et al.*, 1981), and there is electrophysiological evidence for some contralateral excitatory input (Westin *et al.*, 1977). Furthermore, there is positive correlation between the effectiveness of ipsilateral or contralateral cercal stimulation and the anatomy of the GIs (Daley *et al.*, 1981). The ventrally located GIs (1–3) receive stronger excitation from the ipsilateral than from the contralateral cercus; correspondingly, their ipsilateral arborizations are greater (Fig. 8.9b). By contrast, GIs 4, 5 and 7 are strongly excited by input from the contralateral cercus, and have extensive contralateral branching (Fig. 8.9c).

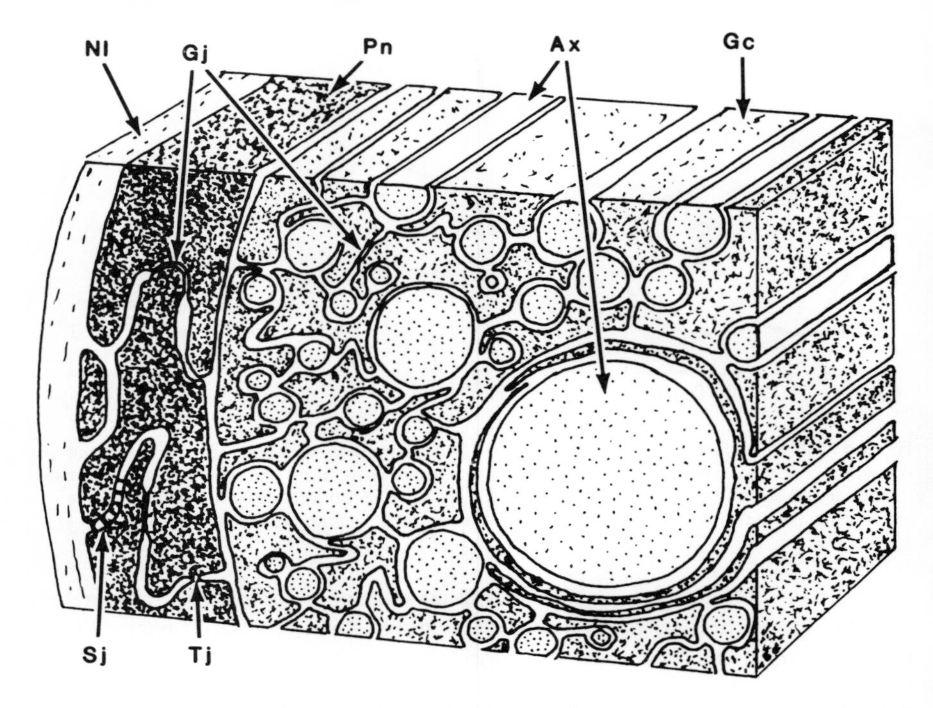

Fig. 8.10 Schematic view of peripheral part of an interganglionic connective. Ax, axons; Gc, glial cytoplasm; Gj, gap junctions; Nl, neural lamella; Pn, perineurial cell cytoplasm; Sj, septate junction; Tj, tight junction. (After Treherne and Schofield, 1979.)

8.3 The extraneuronal compartment

In addition to neurons, nervous systems include internal supporting cells (neuroglia) and a covering sheath (Figs. 8.7a and 8.10). The neuroglia are thought to serve a nutritive and supportive role in *P. americana*, as they do in other animals. The sheath, which has both cellular and acellular components, helps to protect the central neurons from variations in the external ionic environment (see Section 8.3.2) and from physical distension. Descriptions of the morphology follow.

8.3.1 Structural components

(a) *Neural lamella and perineurium*
The neural lamella is a tough but elastic acellular connective tissue sheath (Richards and Schneider, 1958). It surrounds the entire nervous system and becomes continuous with the basement lamellae of other organ systems peripherally. There is evidence that the cockroach neural lamella contains collagen (Rudall, 1955; Smith and Treherne, 1963; Harper *et al.*, 1967), together with chondroitin, dermatin and keratin sulphates (Ashhurst and Costin, 1971).

The cells immediately beneath the neural lamella form the perineurium.

They contain numerous perinuclear mitochondria (Hess, 1958a), and, in well-fed individuals, dense concentrations of glycogen (Wigglesworth, 1960; Ashhurst, 1961a). The perineurial cells are distinguishable from the underlying neuroglia by the less sinuous arrangement of their plasma membranes. Unlike the other cells of the nervous system, they accumulate vitally injected trypan blue (Scharrer, 1939; Pipa, 1961).

Certain of the junctional complexes that span the narrow intercellular channels (Fig. 8.10) are thought to constitute a barrier to rapid diffusion of water-soluble ions into the CNS (Treherne, 1974; Lane *et al.*, 1977). The penetration of electron-opaque tracer molecules through this barrier has been studied in *P. americana* (Lane and Treherne, 1972, 1973). In the interganglionic connectives, microperoxidase (mol.wt. 1900) and lanthanum pass through the neural lamella and are deposited in the periphery of the extracellular channels between adjacent perineurial cells. Microperoxidase incompletely penetrates the septate desmosomes, while lanthanum enters both the septate desmosomes and the gap junctions. The most effective barriers seem to be the tight junctions at the inner end of the perineurial cells; they restrict access to both substances. In marked contrast to these results, lanthanum readily penetrates to the intercellular spaces between axons and glia in the small peripheral nerves. This suggests that there may be different mechanisms for the exchange of water-soluble ions and molecules in the two locations.

(b) *Neuroglia and the glial lacunar system*

The neuroglia are classified into three types according to location, shape and internal structure (Pipa *et al.*, 1959; Wigglesworth, 1960; Pipa, 1961). The *first* type, those directly beneath the perineurium (o.GL, Fig. 8.7a), present a 'fibrous' appearance when viewed with the light microscope. This is due to their abundant cell membrane invaginations. Cytoplasmic processes from these neuroglia encapsulate the nerve cell bodies, sometimes producing deep indentations that constitute the neuron 'trophospongium'. Neuroglia of the *second* type form a compact layer immediately about the neuropile (i.GL, Fig. 8.7a). They lack extensively invaginated membranes, and their processes ramify within the cortex and extend into the neuropile. The cytoplasmic processes of the *third* type of neuroglia ensheath axons within the interganglionic connectives and peripheral nerves. These neuroglia are analogous to the Schwann cells of the vertebrate nervous system.

In the ganglion cortex, the extracellular spaces (GLS, Fig. 8.7a) can be extensive between the first two types of neuroglia. The enlarged sinuses form a 'glial lacunar system' (Wigglesworth, 1960) that contains an acid mucopolysaccharide, probably hyaluronic acid (Ashhurst, 1961b; Pipa, 1961; Ashhurst and Patel, 1963; Ashhurst and Costin, 1971). The electron microscopic evidence suggests that these large lacunae are continuous with the narrow extracellular spaces between the neuroglial sheaths and neurons (Smith and Treherne, 1963). It is probable that this system of channels extends into the interganglionic connectives and peripheral nerves, too, but so far we lack rigorous proof. It is likely that the acid mucopolysaccharide in these channels is a cation reservoir and/or molecular sieve (Treherne and Moreton, 1970; Ashhurst and Costin, 1971).

8.3.2 Ionic balance and the 'blood–brain' barrier

As described in Section 8.4, nervous conduction and synaptic transmission in the CNS of *P. americana*, as in other insects, requires a medium which is high in sodium, low in potassium, and with a relatively high Ca^{2+} to Mg^{2+} ratio. Measurements of the amounts of these ions in the haemolymph, however, have shown that there is less Na^+ and Ca^{2+} and more K^+ and Mg^{2+} than expected for normal nerve function. Other insects, notably phytophagous ones such as the stick insect. *C. morosus*, have even less Na^+ and more K^+ and Mg^{2+}. The obvious inference is that the neural sheath and its associated structures are somehow responsible for maintaining an appropriate ionic balance around the neurons. Nevertheless, in spite of a great deal of work on the problem, only recently have we been able to develop satisfactory hypotheses about the underlying mechanisms. Interested readers should consult the reviews by Treherne (1974) and Schofield (1979) for more detail than can be provided here.

The road to our recognition of the dynamic nature of ionic regulation in the nerve cord began with the finding that small monovalent cations such as K^+ and Na^+ were rapidly taken up by, or lost from, whole nerve cords. This was suggested by radioactive isotope studies (Treherne, 1961a,b), and led to attempts to explain ionic regulation in the CNS exclusively in terms of a passive Donnan equilibrium between the inside of the nerve cord and the haemolymph (Treherne, 1965). Yet it was also clear that individual axons in intact connectives of *P. americana* could withstand for considerable lengths of time a bathing solution high in K^+ (Treherne *et al.*, 1970) or low in Na^+ (Pichon and Treherne, 1970), without losing their ability to conduct action potentials. Several additional findings helped resolve these apparently conflicting results. First, Tucker and Pichon (1972) found a slow stage in the exchange of cations which follows the initial fast one from nerve cords (Treherne, 1961a). Second, a physical barrier was discovered. This is formed by the tight junctions between adjacent perineurial cells (see Section 8.3.1(a)). And, finally, it was found that the neuroglia constitute a cation reservoir (Schofield and Treherne, 1978).

Our current view of ionic regulation (Treherne and Schofield, 1979) is that the physical barrier formed by the perineurium and the reservoir formed by the neuroglia or extracellular mucopolysaccharide together act to maintain a precise ionic environment in the immediate vicinity of the neurons of the CNS while simultaneously allowing a dynamic, rapid exchange of ions with the haemolymph itself. Although this view is well supported by experimental data, it should be emphasized that there are many details which must yet be worked out. In addition, there is still a problem in understanding how peripheral nerves and muscles, which seem not to be significantly protected by sheaths, are able to function in the environment provided by insect haemolymph.

Apart from their function in regulating the ionic environment of neurons in the CNS, the physiology of the non-neural cellular elements of the nervous system has been little studied. It appears that one role of these elements is to act as a reservoir of glycogen and other nutritive materials for the neurons during periods of starvation (Treherne, 1960; Wigglesworth, 1960). They also may move nutrients into the CNS normally, since the insect nervous system is unusual in having no direct blood supply to its interior (Wigglesworth, 1960). This is an important topic that deserves further investigation. (cf. Smith and Treherne, 1963; Radojcic and Pentreath, 1979).

8.4 Electrophysiology

Just as the giant axons of the squid have been used extensively in studies of basic nerve function, the giant axons of *P. americana* have attracted considerable attention as model insect neurons, and nearly every study of nerve conduction and transmission in insects deals with these cells. The electrophysiology of insect neurons has been reviewed in depth on several occasions (e.g. Pichon, 1974).

8.4.1 Cable properties and resting potential

If a small electrical charge is applied at a point on the membrane of a neuron, the charge will spread around and along the length of the fibre to an extent, and at a rate, determined by the passive electrical properties of the cell. These passive or 'cable' properties, so called because the neuron acts like a leaky (poorly insulated) cable in sea water, are of interest to neurophysiologists, for they are the basis of several important functional characteristics of neurons.

Table 8.3 Mean measured or calculated electrical constants of *P. americana* giant axons

R_i (Ω cm)	R_m (Ω cm^2)	C_m (μF cm^{-2})	τ_m (ms)	λ (mm)	Reference
46	610			1.3	Boistel (1959)
130	800	6.3	4.2	0.86	Yamasaki and Narahashi (1959a)
	293	3.3	0.96	0.80	Pichon (1974)

The main cable properties of the neuron are reflected in two constants, τ_m, the time constant, and λ, the space (or length) constant. These constants can be calculated from the electrical characteristics of the neuron and its surrounding medium: R_m, the membrane resistance, R_i, the specific resistance of the axoplasm, C_m, the membrane capacitance, and R_o, the resistance of the external medium. (The last is, of course, independent of the neuron). The values of τ_m and λ are important to physiologists because they are measures of two important characteristics of neuronal membrane: how quickly the potential across the membrane will change when current flows across it, and how far an imposed potential will spread away from its source. The values of these constants and electrical properties for giant axons in *P. americana* are similar to those of neurons in other animals; values measured or calculated by various authors are listed in Table 8.3.

The resting membrane potential (RP), is due to the presence of a semi-permeable membrane and an unequal distribution of ions across it. In *P. americana* giant axons, the principal ion involved is K^+ and, to a much lesser extent, Na^+ (Yamasaki and Narahashi, 1959a). The RP of an intact giant axon *in situ* is about −70 mV, but this is uncertain since the nerve sheath not only regulates the ionic content of the fluid around the axon (see Section 8.3.2) but adds an electrical component of its own (Pichon and Boistel, 1967). Measurements of RP in desheathed preparations therefore differ from measurements in intact but isolated nerve cords (Pichon, 1974), and, of course, vary with the saline used. Since we do not know the normal concentrations of Na^+ and K^+ immediately around the axon, we cannot measure RP in a saline having these concentrations.

8.4.2 Active properties and action potential

In *P. americana*, the ion responsible for the rising phase of an action potential (AP) in a giant axon is Na^+. This statement is based on the observation that placing a desheathed giant axon in a Na^+-free medium abolishes the AP (Yamasaki and Narahashi, 1959b), while eliminating other ions has little or no effect on AP amplitude. The relative contribution of the movements of both Na^+ and K^+, the latter to the falling phase of the AP, has been studied using voltage clamp techniques, and is summarized by Pichon (1974). The results show that the curves of both the peak and delayed currents as functions of the clamped voltage have the same shapes as they do in other excitable tissues (Fig. 8.11, open symbols). The contribution of Na^+ to each of the two currents can be assessed by repeating the experiment in the presence of tetrodotoxin (TTX), which selectively blocks sodium channels. The result is complete elimination of the peak current (Fig. 8.11, closed symbols), which supports the original suggestion that Na^+ carries the inward current during the rising phase of the AP (Pichon, 1969a). Similar experiments carried out in the presence of tetraethylammonium (TEA), a selective blocker of potassium current, has verified that outward movements of K^+ are responsible for repolarizing the membrane after the depolarization caused by sodium influx (Pichon, 1969b).

8.4.3 Excitability

Although the axons of the GIs are eminently suitable for the study of general membrane properties, they are not necessarily representative of all neurons in the insect nervous system, since the membranes of dendrites often have electrical properties unlike those of the somata. It is generally thought that the somata of the insect CNS are not electrically excitable; that is, they cannot sustain APs (Hoyle, 1970). While this appears to be true for most neuron somata, there is a group which

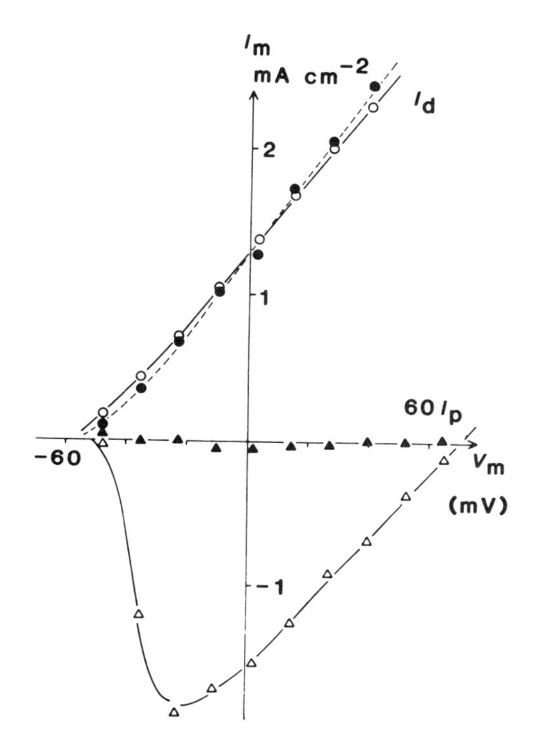

Fig. 8.11 A plot of maximum peak current (triangles, I_p) and maximum delayed current (circles, I_d) developed during passage of an action potential under voltage clamp in the absence (open symbols) and presence (filled symbols) of 10^{-6} g ml^{-1} TTX. The abscissa shows the clamp voltage. Leak current has been subtracted. Note that TTX, which blocks Na^+ channels, has no effect on the delayed current, but entirely eliminates the peak current, indicating that the latter, and therefore the rising phase of the AP, is due to Na^+ influx. (From Pichon, 1969a, reprinted by permission of the Academy of Science, Paris.)

does exhibit APs, and it is located in the mid-dorsal region of the thoracic ganglia of *P. americana* (Kerkut *et al.*, 1969; Crossman *et al.*, 1971). Electrically inexcitable somata can be induced to generate APs by treating them with colchicine or by axotomy (Pitman *et al.*, 1972b). The mechanism of this transformation is now known. The AP generated in this fashion appears to be based on conventional ionic movements, i.e. principally Na^+ and K^+ (Pitman, 1975a). The likelihood that Ca^{2+}; too, can become a major inward carrier of current in special circumstances (Pitman, 1975b), raises important questions about the properties of the soma membrane.

The relationship between membrane depolarization and firing rate varies from one motoneuron to the next (Fourtner and Pearson, 1977; Meyer and Walcott, 1979). Generally, smaller neurons tend to be more excitable than the larger ones. Since slow axons are usually smaller than fast axons, they tend to fire at lower voltages.

8.5 Synaptic transmission

Considering the importance that an understanding of synaptic transmission in insects has for studies of both the physiology of CNS circuits and the effects of toxins such as insecticides on the CNS, it is remarkable how little we know about the subject. The main reason for this dearth of information seems to be technical. The electrical inexcitability of most cell bodies (see above) and the electrical isolation of them and their relatively large axons from the synapses by narrow regions of high resistance makes it impossible to record synaptic potentials from them. Furthermore, the small size of identified pre- and post-synaptic elements close to the site of the synapse makes recording from these elements very difficult.

8.5.1 Electrical and chemical transmission

Except for one report of the presence of electrical transmission in *P. americana* (Harris and Garrison, 1976), all synaptic transmission in insects seems to be chemical. The physiology of chemical transmission has been studied primarily at the cercal afferent–giant interneuron synapse. Callec (1974) provides a thorough discussion of this work, including some of his own unpublished observations. Callec (1974) was able to record unitary EPSPs and IPSPs from neurites of single GIs. These varied from 0.5–2.3 mV in amplitude. Synaptic delay for EPSPs was about 0.68 ms, indicating a monosynaptic connection between the cercal afferents and the GI. IPSPs appeared about 1.33 ms after an AP in the afferent, indicating, a second synapse in the inhibitory pathway. The synapses appear to be conventional in requiring a relatively high ratio of Ca^{2+}/Mg^{2+} in the bathing medium for normal function.

Attempts to identify the ionic bases of EPSP and IPSP by measuring the reversal potential (i.e. that imposed membrane potential at which the post-synaptic potential reverses polarity) are complicated by the distance of the site of recording from the active synapse under study, and results of such investigations must be interpreted with caution. Pitman and Kerkut (1970) determined the reversal potential of EPSPs in cell bodies to be approximately −45 mV. However, Callec (1974) was unable to invert EPSPs in GIs, because the amplitude of EPSPs was not a linear function of imposed membrane potential except over a fairly narrow range. Extrapolating a line through this linear portion yielded an estimated reversal potential of −35 to −47 mV. This value is consistent with the idea that both Na^+ and K^+ contribute to the EPSP, although possibly K^+ movement has greater importance here than in other neurons, since the value of the EPSP is more negative than usual.

For IPSPs in cell bodies, Pitman and Kerkut (1970) obtained a reversal potential of about −79 mV. In the giant interneurons it was determined experimentally to be −80 mV, and seems likely to be produced by movement of Cl^- or K^+ ions alone or together (Callec, 1974).

8.5.2 Graded transmission

One of the most important recent developments in our understanding of insect nervous systems has been the discovery of INs which do not normally produce spikes. The main functional implications of this discovery are considered in Section 8.7, but some aspects of it are important in the context of a discussion of transmission as well. This is because these neurons release transmitter substances (and therefore influence the activity of post-synaptic neurons) in a graded fashion.

Non-spiking, local INs of insects were first described in the CNS of *P. americana* by Pearson and Fourtner (1975). These authors showed that small changes in the membrane potential of such neurons can produce significant changes in the spike activity of post-synaptic follower cells. Their conclusion that a non-spiking cell releases transmitter substance in a graded fashion as a function of the cell's membrane potential has been borne out by subsequent work on locusts (Burrows and Siegler, 1978; Burrows, 1979). For example, EPSPs as small as 2–3 mV, generated in a non-spiking neuron, can cause the release of sufficient transmitter substance to affect significantly the membrane potential of a post-synaptic follower cell. The effect might be either excitatory or inhibitory, depending on whether the transmitter depolarizes or hyperpolarizes the follower cell. But, in some cases, a single non-spiking neuron can have both excitatory and inhibitory effects on a specific follower cell. This is possible when the non-spiking cell releases transmitter substance continuously, thereby keeping its follower cells in a mild state of inhibition (hyperpolarization) or excitation (depolarization), depending on the specific effect of the transmitter substance. Excitation of such a non-spiking cell causes the release of more transmitter and augments its post-synaptic effect. But inhibition of the non-spiking neuron reduces the ongoing release of the transmitter, so that if the follower is normally partly inhibited, it will be released from this inhibition, i.e. be excited. The capability of a single local IN to excite or inhibit its follower cells, as its own input dictates, clearly adds considerable complexity to the integrative capabilities of the small networks to which these neurons belong.

8.5.3 Transmitter substances

Progress in identifying the transmitter substances used by central neurons in *P. americana* has been rather slow, not least because of the presence of an effective 'blood–brain' barrier around the CNS (see Section 8.3.2). Nevertheless, there is a substantial body of evidence that supports two substances as transmitters – acetylcholine (ACh) and γ-aminobutyric acid (GABA).

ACh appears to be the excitatory transmitter at synapses between cercal hair afferents and the GIs in the sixth abdominal ganglion (A6), as well as at many other synapses in the CNS. Early work showed that quite large concentrations ($>10^{-3}$ mol l^{-1}) of the substance are required to affect gross activity in A6 (Yamasaki and Narahashi, 1960). However, later studies revealed that much lower concentrations (10^{-7} mol l^{-1}) are effective if the ganglion is first soaked in ACh for several hours, and that pharmacological agents which interfere with cholinergic transmission in vertebrates have similar effects in A6 (Shankland *et al.*, 1971). The problem seems to be that ACh lacks free access to the interior of the nerve cord, at least partly because the sheath contains acetylcholine esterase (AChE), which degrades ACh (Smith and Treherne, 1965). Iontophoretic application of ACh to cell bodies of central neurons can evoke electrical responses at concentrations as low as 10^{-13} mol l^{-1} (Kerkut *et al.*, 1969), showing that these neurons are quite sensitive to the substance.

Not only do central neurons respond to low concentrations of ACh, but the responses resemble normal post-synaptic events. For example, if ACh is applied iontophoretically to the somata of central neurons (Kerkut *et al.*, 1969; Pitman and Kerkut, 1970) or the dendrites of GIs (Callec, 1974), electrical responses resembling EPSPs are evoked. In addition, the sizes of the ACh-evoked responses and of the EPSPs change similarly when the membrane potential of the responding cell is altered (Fig. 8.12; Callec, 1974). Furthermore, eserine (a blocker of AChE) potentiates the effects obtained by applying ACh (Yamasaki and Narahashi, 1960; Kerkut *et al.*, 1969; Callec, 1974).

These physiological results strongly suggest that ACh is an excitatory transmitter at some central synapses. But, in addition, a number of other criteria such as localization of the compound at synapses and the presence of appropriate precursors and enzymes for synthesis and de-activation, have also been met, as detailed by Callec (1974) and Leake and Walker (1980).

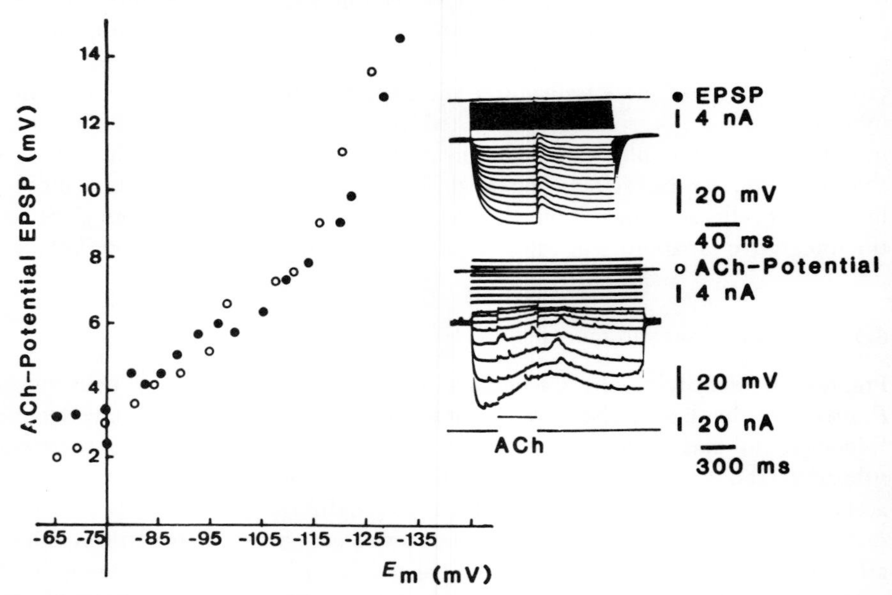

Fig. 8.12 Plot of the amplitude of the EPSP (filled circles) and the amplitude of the response to iontophoretically applied ACh (open circles) against the imposed membrane potential of a giant fibre in the 6th abdominal ganglion. Since only the shape and slope of the relationship is important, the amplitude of the ACh response was adjusted initially to equal the EPSP. Insets: the experimental records from which the data in the graph were derived. (From Callec, 1974.)

There is good evidence that GABA is the transmitter at some central *inhibitory* synapses. Kerkut *et al.* (1969) and Pitman and Kerkut (1970) have shown that iontophoretic application of GABA on to cell bodies of neurons in the third thoracic and sixth abdominal ganglia of *P. americana* will produce potential changes similar to IPSPs. The reversal potential of these responses, about −75 mV, is very close to that of the IPSP (−79 mV). In addition, drugs, such as picrotoxin, which block production of IPSPs, also block the effects of iontophoretically applied GABA (Pitman and Kerkut, 1970; Callec, 1974). Other evidence, reviewed by Callec (1974) and by Leake and Walker (1980), also supports the view that GABA is an inhibitory transmitter substance.

Do other compounds serve as transmitters in the CNS of insects in general, or *P. americana* in particular? Very likely, but it must be admitted that, at present, we have little more than suggestive evidence to support that likelihood. Octopamine has been found to occur in high concentration in cockroach brain (Evans, 1978a), along with a powerful mechanism for its uptake into neurons (Evans, 1978b), but since the action of this substance on muscles seems to be better documented, it is discussed in more detail with the proposed neuromuscular transmitters, glutamate and proctolin, in Section 11.5. The strongest remaining candidates are serotonin (5-hydroxytryptamine) and the catecholamines: adrenaline, noradrenaline and dopamine. The main lines of evidence supporting their candidacies are histochemical localization and pharmacological action. The evidence is thoroughly reviewed by Pitman (1971), Klemm (1976), and, most recently, Leake and Walker (1980). Since it is clear that a compound may be present in the CNS and even have pharmacological action there without necessarily being a transmitter substance, the available evidence must be interpreted cautiously. Judging from the astonishing recent increase in the number of compounds suspected of being transmitters in the vertebrate CNS, it would not be surprising if all of the substances mentioned above, and others as well, were neurotransmitters in *P. americana* and in other insects.

8.6 Toxicology

The literature on the effects and mode of action of toxic substances on the nervous system of insects is enormous, and much of it is concerned directly with *P. americana*. There are three main reasons for such intense interest in the subject. First, many natural or artificial toxins interfere with quite specific components of normal nerve function, such as the action of TTX in blocking the sodium channels in nerve membranes. These materials can therefore be used to study the basic properties of nerve membranes and synapses. We will not discuss this aspect of toxicology further than has already been considered in Section 8.5 (see general review by Narahashi, 1974). Second, many substances have been discovered which are lethal to insects at doses which are not toxic to vertebrates. Study of how these materials exert their lethal effects should allow us to design compounds that act with increased selectivity. And third, a number of animals produce venoms which have striking effects on insect nervous systems. Study of the active components of these venoms can provide excellent model systems for the development of new classes of chemical control agents.

Because only a small selection of papers on this topic can be mentioned here, readers interested in the subject should consult review articles or monographs, such as those by Narahashi (1971, 1976), Corbett (1974), O'Brien (1978) and Shankland *et al*. (1978).

8.6.1 Effects of insecticides on the nervous system

It may seem curious in view of the great interest in insect control by chemical means that we know relatively little about the exact molecular basis of the toxicity of many insecticides now in use. One reason for this is that much of the work with pesticides has been concerned more with the effects of dose than with molecular mechanisms of action. In addition, few well-trained biophysicists have attacked the problem. Nevertheless, while exact mechanisms remain to be worked out, at least the general sites of action of most insecticides are now known.

(a) *Effects on the nerve membrane*

Some insecticides act directly on the nerve membrane, and thus affect the conduction of APs. The best-known of these are the chlorinated hydrocarbon DDT and some of its analogues. The gross effect of this compound on nerves in *P. americana* (as well as in other arthropods) is hyperexcitability, characterized by repetitive bursts of activity (see reviews by Narahashi, 1971, 1976 for specific references). These bursts are caused, in part, by prolonged afterpotentials during which the membrane potential is closer to zero than normal. These afterpotentials therefore act as a continuing source of stimulation to the axon (Narahashi and Yamasaki, 1960). The prolonged afterpotential, in turn, is a manifestation of two phenomena, a decrease in the rate at which the sodium channels close during repolarization after an AP (thereby allowing a greater than normal inward flow of Na^+), and a partial suppression and slowing of potassium current (Narahashi and Haas, 1967; see Narahashi, 1979 for review). Precisely how DDT and its analogues exert these effects on Na^+ and K^+ movements is not known, although it is hypothesized that they partly wedge open channels through which the ions move (see Holan, 1969).

Allethrin and other synthetic analogues of the natural insecticide pyrethrum also act directly on the axonal membrane. Narahashi (1962, 1976) has shown that at low concentrations allethrin affects the rate of change of Na^+ and K^+ conductances in cockroach giant axons, so that an allethrin-poisoned neuron fires in bursts like one poisoned with DDT. At higher concentrations, allethrin, unlike DDT, will block conduction altogether.

A peculiar aspect of the response of an intact cockroach to poisoning by DDT or allethrin is that the compounds are more effective in killing the insect at low temperatures than at higher ones. With both compounds this seems to be because a given dose causes more pronounced effects on the nervous system at low temperatures (Gammon, 1978a,b). However, while DDT seems to affect all parts of the nervous system equally (e.g. Gammon, 1978b) at low temperatures, allethrin appears to cause hyperexcitability preferentially in the peripheral nervous system (Gammon, 1978a). Gammon (1979) postulates that this effect may somehow be at the root of the greater lethality of allethrin at low temperatures.

(b) *Effects on synapses*

Other insecticides have specific actions on synapses. This may be to interfere with any of three different steps in synaptic transmission: the release of transmitter from the pre-synaptic terminal, the binding of the transmitter with a post-synaptic receptor site, or the inactivation of the transmitter after it has functioned.

The insecticides which are thought to affect transmitter release are the chlorinated hydrocarbon, lindane, and the biocyclodiene compounds, aldrin and dieldrin. These compounds have an excitatory effect on transmission in ganglion A6 in *P. americana* (e.g. Wang *et al.*, 1971), and this effect is specific for cholinergic synapses (Shankland *et al.*, 1971; Shankland and Schroeder, 1973). Apparently, the primary mode of action of these toxicants is to cause an excess release of ACh from the synaptic terminals upon stimulation of the pre-synaptic neurons (see review by Shankland, 1979).

Only a few compounds have been shown to interfere with the post-synaptic binding of the natural transmitter. Nicotine, a long-known naturally occurring compound with insecticidal action, is one of these. Its effect on some vertebrate cholinergic receptors is well understood, and there is good evidence (reviewed by Corbett, 1974) that it acts similarly in insects. The synthetic compound, cartap, also appears to work by binding with ACh receptors, and represents an interesting case of an insecticide 'copied' from nature, since its structure is similar to that of the neurotoxic constituent of nereistoxin, a poison from a marine worm (see review by Sakai, 1969).

Finally, other insecticides act by binding acetylcholine esterase (AChE). The carbamate insecticide diazinon and the organophosphate insecticides, malaoxon and paraoxon fall into this category. The literature on the mode of action of the AChE inhibitor compounds is voluminous. (Corbett, 1974, devotes an entire chapter to the subject). This is mainly because we have the most complete understanding of the mechanisms involved, including the kinetics of the binding reactions (see Corbett, 1974, pp. 131–138). Advanced knowledge in this area has led to detailed studies of the relationship between the structure of a particular compound and its effectiveness in inhibiting AChE (e.g. Metcalf, 1971; Fukuto, 1979).

Before leaving the topic of pesticides, we should emphasize a point not often made: in spite of the volume of studies of the effects of these chemicals, we still know little about how a compound actually kills an insect. There are, for example, a host of problems raised by the biochemical transformations which many insecticides undergo in the insect. We also do not know how hyperexcitability can cause death, but there have been a number of suggestions that death may be due to the effects of neurohormones released by the massive neural activity (e.g. Granett and Leeling, 1971; Samaranayaka, 1974). Furthermore, there is evidence that death may be caused by a toxic substance produced endogenously by the insect in response to massive neural activity, and not to the direct effects of the activity itself (see below). In any event, it is clear that pesticide action still poses many problems which will take considerable effort to solve.

8.6.2 Autotoxicity

The phenomenon of autotoxicity was first described in *P. americana* by Beament (1958a), who showed that physical restraint for several days resulted in paralysis of the insect some hours or even days after it had been released. In some cases, the insect died, never recovering from the paralysis. Experiments showed that the paralysis was due to a haemolymph-borne factor; haemolymph from a restrained insect could induce paralysis when injected into one which had not been held down.

This phenomenon excited a flurry of interest, ably reviewed by Sternburg (1963). Two main points emerged from the work. First, it was shown that the phenomenon was stress-related; that is, any procedure which produced physical stress in the animal was effective in inducing paralysis, and often death. And, secondly, stress paralysis showed some symptoms in common with the effects of DDT poisoning (Heslop and Ray, 1959), for which it had been shown by Sternburg and Kearns (1952) that a haemolymph-borne toxic factor was produced. Hawkins and Sternburg (1964) demonstrated that application of other insecticides also produced a haemolymph-borne factor chemically similar to that generated in DDT-poisoned cockroaches, and that this factor would cause paralysis when injected into non-stressed or non-poisoned insects. Recent results suggest that the substance may act in part to block transmission at neuromuscular junctions (Cook and Holt, 1974), although it clearly has a strong effect on neural activity in the CNS as well (see Sternburg, 1963). Other recent work attempts to identify the source of the stress factor, and suggests that perhaps the suboesophageal ganglion plays a role in producing it (Rounds and Riffel, 1974).

It is unfortunate that relatively little attention has been paid to the phenomenon recently. The relationship between stress paralysis and the paralysis and death induced by pesticides such as DDT and allethrin, whose main effect is to produce a massive increase in neural activity, seems worth pursuing. This is especially true in view of Gammon's (1979) finding that administering a small (non-lethal) amount of the nerve-blocking agent, TTX, protects the insect from this barrage of input, and enables it to survive a dose of insecticide that otherwise would be lethal.

8.7 Functional organization

8.7.1 Reflexes and their central control

A reflex is a simple sterotyped behavioural response to a particular stimulus. In the past, reflexes were frequently studied, often with the supposition that, by investigating the properties and characteristics of reflex behaviour, important features of the neural organization underlying its expression could be revealed. However, as technical advances have allowed us to study such organization directly, the study of reflexes *per se* has become less important, and we will not discuss particular reflexes here. Readers interested in more detailed information should consult recent reviews (Finlayson, 1976; Wright, 1976) or original reports concerning *P. americana* (e.g. Wong and Pearson, 1976; Fraser, 1977; Reingold and Camhi, 1977; Krämer and Markl, 1978).

One of the important outcomes of recent work has been the recognition that the CNS has much greater influence than had previously been thought over what had been considered to be simple stereotypic behaviour. Some of this influence is manifest as variability of responses to repetitive stimuli. For example, Wilson (1965) and Delcomyn (1971c) studied the phase-locked activity of leg muscles induced by sinusoidal flexion and extension of a leg in *P. americana*. While the muscular response (and therefore the response of the controlling motoneurons) was often quite strong, it was also rather variable. Not only were there occasions during which it could not be elicited at all, but there were periods of stimulation during which the motor response would wax and wane, even though the stimulus would remain unchanged.

Delcomyn (1971c) suggested that this variability was due to the summation, at the motoneurons, of the sinusoidally changing sensory input with different levels of excitation from elsewhere in the CNS. Only if the excitation were sufficiently high would the sensory input be able to generate a motor output. Such a mechanism requires that we consider a motoneuron to be a complex integrative element whose output is a reflection of the instant by instant interplay of excitatory and inhibitory inputs acting on it, rather than as a simple relay of an excitatory sensory input to an effector. Intracellular recordings from motoneurons in locusts support the validity of this newer view (Hoyle and Burrows, 1973). Miller (1974b) and Pearson and Rowell (1977) discuss the broader aspects of such central control of excitability.

Another aspect of central control of reflexes is the ability of factors other than the triggering stimulus itself to change behavioural responses. For example, Camhi and Hinkle (1974) have shown in locusts that a yaw-correcting displacement of the abdomen in response to changes in wind direction on the head is a reflex that operates only during flight. When the insect is not flying, the same stimulus has no effect on the abdomen. And in the cockroach, *Gromphadorhina portentosa*, tactile stimuli on the dorsum produce either avoidance or a righting response, depending on whether the insect has tarsal contact (Camhi, 1977). While these particular phenomena have yet to be demonstrated in *P. americana*, there is every indication that, in this insect as well, reflexes are not as simple as previously thought, but are under considerable central control (see Chapter 14).

8.7.2 Intersegmental co-ordination

Unfortunately, our knowledge of the mechanisms by which movements in

Fig. 8.13 Records of neural activity in flexor motoneurons of a rear leg (top trace) and an interneuron from a connective joining the second and third thoracic ganglia (bottom trace). Note the nearly simultaneous termination of bursts in the two traces. (From Pearson and Iles, 1973.)

different segments of an insect are co-ordinated is quite limited. In *P. americana*, the only published work which bears on this problem (apart from work on escape, Section 8.7.3) is the study by Pearson and Iles (1973) on co-ordination of leg movements during walking. They demonstrated that in the connectives joining the meso- and metathoracic ganglia there is an axon whose activity is closely related to that of leg motoneurons in the metathoracic ganglion (Fig. 8.13). They inferred that this unit is responsible for ensuring that stepping of the middle leg does not occur before stepping of the leg behind it is completed. While many important details obviously still need to be worked out, their results demonstrate that central connections between neural centres for the control of motor output (which had been inferred from earlier studies) can be found.

8.7.3 The giant fibre system

The GIs described in Section 8.2.5(c) are among the most intensively investigated, identified neurons in any insect. The function of these neurons is still not clearly understood.

The generally accepted early idea was that the giant interneurons drive leg motoneurons during the initial stages of the insect's escape run (Roeder, 1948; see also Section 14.2.2). Only after 20 years was this view challenged. Dagan and Parnas (1970) suggested that, in fact, the giants did not drive escape movements, but might instead transmit a 'clear-all-stations' signal preparing the animal to run (see their review, Parnas and Dagan, 1971). Their rejection of the earlier idea was based principally on two results: their inability to elicit leg motor nerve activity if the GIs alone were stimulated electrically, and their ability to elicit such activity in the absence of GI stimulation. The conclusion that the largest giants, at least, did not drive motor activity was supported by subsequent work by Iles (1972), Harris (1977) and Schlue (1974), none of whom was able to elicit motor activity by stimulating only one or more giant axon. Nevertheless, an experiment designed to test the 'clear-all-stations' hypothesis failed to confirm it (Harris, 1977).

To confuse the picture further, Delcomyn (1976, 1977) reported that some of the GIs became active when the animal walked spontaneously (Fig. 8.14). This observation seemed completely at odds with any role of these neurons in driving escape. Furthermore, this excitation was centrally generated. It was not the result of reafference (i.e. self-stimulation due to movement through the air). Only some of the giants were affected; others were inhibited (Delcomyn and Daley, 1979; Daley and Delcomyn, 1980a,b).

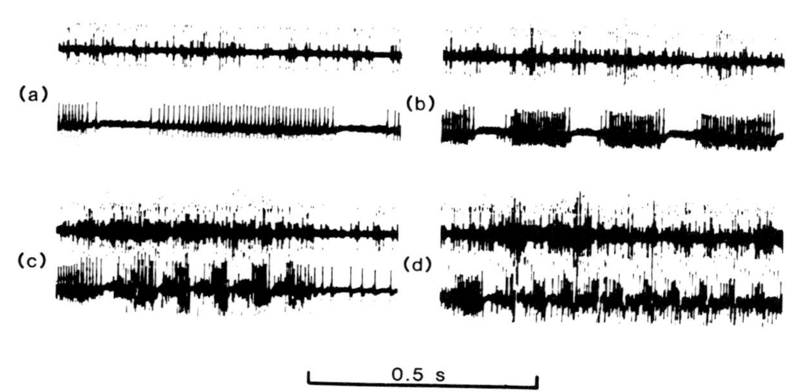

Fig. 8.14 Activity in the ventral nerve cord (VNC) of a free-walking cockroach at various speeds of walking. Upper traces: VNC activity. Lower traces: EMG of rear-leg extensor muscle. (a–d) Insect walking at 2, 4½, 12 and 15 steps per s, respectively. Note the increase in average firing frequency of all units in the VNC as speed of walking increases. (From Delcomyn and Daley, 1979.)

In contrast to this negative evidence, other work seemed to support at least some parts of Roeder's (1948) original concept. The most significant aspect of this work was the striking parallel between the directional sensitivity of some giant fibres to air disturbances (puffs) from different directions (Westin *et al.*, 1977) and the cockroach's behavioural response to similar disturbances: a turn away from the stimulus (Camhi and Tom, 1978; Camhi *et al.*, 1978 – see Section 14.2.2). But can the giant interneurons drive leg motoneurons? Camhi (1976) and Ritzmann and Camhi (1978) claimed that they can, if the frequency of stimulation is high enough, although the connection need not necessarily be monosynaptic. Furthermore, the latter authors showed that an individual GI tended to cause excitation of motoneurons of just those leg muscles whose contraction would tend to turn the insect away from the stimulus which would preferentially excite that giant.

Recent work by Fourtner and Drewes (1977) has shown, in addition, that some of the common inhibitor (CI) motoneurons are excited by air-puff stimulation. Since one function of the CIs seems to be to relax a muscle just before or after it contracts (to reduce residual tension which would oppose action of its antagonists – see Section 11.7), it would be useful for the cockroach to fire the CIs in preparation for escape. Although Fourtner and Drewes (1977) suggest that non-giant ascending fibres are responsible for the excitation, the conduction velocity they estimate for the excitatory pathway, about 3.7 m s^{-1}, is within the range calculated for the smaller, dorsal group of giant interneurons. (Calculations are based on an axon diameter of about 25 μm, using the formula of Pearson *et al.*, 1970.)

It is probably too early to resolve the apparent contradictions in all this work. It does seem likely that at least some of the giants are involved in an escape turn, and that the failure of Dagan and others to elicit motor responses by stimulating the GIs was due to the use of ineffective patterns of stimuli or to other technical problems. However, the motor output recorded by Ritzmann and Camhi (1978) seems too weak to be able to produce the vigorous thrust necessary for a rapid turn. The resolution of the puzzle may lie in investigations of the tracts of smaller axons that Dagan and Parnas (1970), Iles (1972) and Schlue (1974) showed could elicit leg motor activity when stimulated alone or with one or more giants. In light of the obvious physiological complexity of the giants and of the escape response, it may be fruitful to expand our search for the neural basis of escape beyond the group of large-diameter INs which have been classified as giant fibres. After all, the classical 'giants' are not the only INs which respond to puffs of air on the cerci (Westin *et al.*, 1977; Daley and Delcomyn, unpublished).

Finally, the reader should recognize that the GIs do not constitute a homogeneous group (Ritzmann and Camhi, 1978; Delcomyn and Daley, 1979). Not only may some of them play roles which are quite distinct from others, but some giants may play different roles in different circumstances. For example, Ritzmann *et al.* (1980) have now shown that, in the absence of tarsal contact, appropriate stimulation of any dorsal giant will evoke activity in flight motoneurons (and flight behaviour) rather than leg movements. Some of the giants also show vigorous activity during flight induced by loss of tarsal contact (Delcomyn and Daley, unpublished observations). We can only conclude that there is much more to be learned about the function of these GIs.

8.7.4 Neural basis of rhythmic behaviour

Much of the work in the new field of neuro-ethology (Hoyle, 1970), the study of the neural basis of behaviour, has been done on rhythmic behaviours such as swimming, chewing and flying. (The studies of escape behaviour in *P. americana*, which are described above, form an important exception.) An important generalization to emerge from the work is that *all* rhythmic behaviours can be generated by the CNS; sensory feedback is not necessary to provide timing cues (see review, Delcomyn, 1980). Miller (1974b) has reviewed much of the literature supporting this generalization in insects.

In *P. americana*, research has focused on walking. Pearson and Iles (1970, 1973) have provided the main evidence supporting the idea that the rhythmic stepping movements of individual legs are driven by appropriate motoneurons via a network of INs (termed a central pattern generator, or CPG) located in each half ganglion, and that these CPGs are interconnected by co-ordinating INs, as described in Section 8.7.2.

With this hypothesis well-established, it has become important to find a CPG and describe its functional properties. Pearson and Fourtner (1975) made a significant contribution to the solution of this problem by showing that, in *P. americana*, a system of small, intraganglionic, non-spiking neurons plays an important role in generating stepping movements of individual legs. They found several types of these INs, each capable of driving or suppressing activity in specific groups of leg flexor or extensor motoneurons.

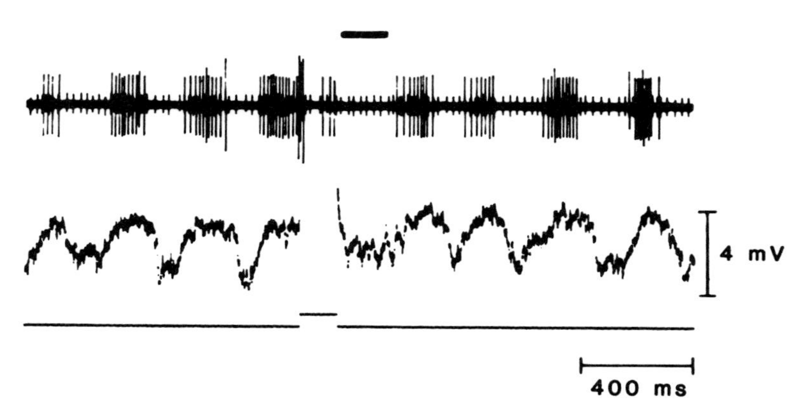

Fig. 8.15 Recording of burst activity in flexor motoneurons (top trace) and activity in a non-spiking interneuron (middle trace). Note that depolarization of the interneuron is accompanied by firing of the flexor motorneurons, and that when a brief pulse of current (bottom trace) is injected into the interneuron the on-going rhythm is reset. The black bar above the motoneuron trace shows the expected timing of the next burst. (From Pearson and Fourtner, 1975.)

Pearson and Fourtner (1975) invoked several lines of evidence to support their claim that these cells are part of the rhythm-generating network. First, the membrane potential of each non-spiking neuron fluctuates in phase with the alternating bursts of flexor and extensor activity which signal rhythmic motor output (e.g. Fig. 8.15); secondly, driving a non-spiking cell sinusoidally produces bursts in either the flexor or extensor motoneurons; and finally, injecting a brief current pulse into a spontaneously oscillating non-spiking cell (Fig. 8.15) during production of rhythmic motor output will reset the rhythm of the output. This strongly suggests that the non-spiking IN is part of the network which generates the motor pattern, not just another follower cell.

Subsequent work has shown that some INs do not act upon motoneurons directly, but rather at higher levels of an apparent hierarchy (Fourtner, 1976). Sustained depolarization of a single IN of this kind produces reciprocal bursting in flexor and extensor motoneurons (Fig. 8.16). Arguing by analogy from the work of Burrows (1979, 1980) on non-spiking INs in locusts, the unit Fourtner (1976) describes seems likely to be a high-level integrating element which can control the activity of several other INs, and, through them, a whole set of motoneurons. (Burrows, 1980, has reported finding such 'command' INs for control of leg movements in locusts.)

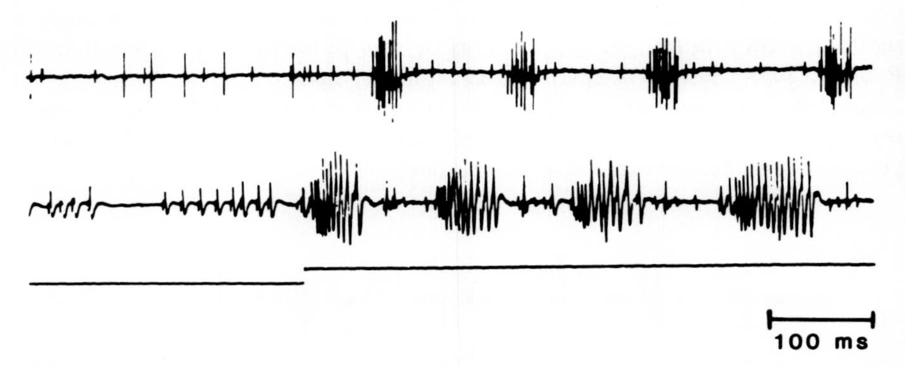

Fig. 8.16 Recording of burst activity in flexor motoneurons (top trace) and extensor muscles (middle trace) resulting from depolarization (shown in bottom trace) of a non-spiking interneuron. Bursting occurred only while the interneuron was depolarized. (From Fourtner, 1976.)

In view of the many tantalizing questions raised by this recent work, it is especially unfortunate that *P. americana* has proved to be a difficult preparation from a technical point of view. Aside from a brief mention by Pearson (1977) that he has found interganglionic INs which appear to be able to turn on the locomotor system, no further progress along these lines has been reported using this species. The field is wide open for an enterprising and skilful investigator.

8.8 Learning

The topic of learning, *per se*, is clearly beyond the scope of this Chapter. Yet cellular correlates of a simple conditioned avoidance task have been studied intensively in *P. americana*. Although we can do little more here than identify good reviews, and briefly mention a few of the most recent findings, this might be useful. The reader should also refer to Section 14.6 for an account of behavioural aspects of learning.

The basic phenomenon, and some of the early literature on the subject, have been outlined by Eisenstein (1972). In the experimental arrangement, two cockroaches are suspended so that the leg of one, the test animal, will make contact with a saline solution when the leg is extended. The animals are then connected to a stimulator in such a way that when the test animal extends its leg into the saline both animals receive a shock. In successful experiments the test animal learns to keep its leg flexed, while the control does not.

This preparation has attracted attention as a simple model system in which the cellular and biochemical bases of learning might be investigated. Miller (1974b) has reviewed this work up to the early 1970s. The conflicting results which have been reported leave the impression that experimental difficulties abound, and that valid generalizations are not yet possible. For example, Kerkut *et al*. (1972) and Oliver *et al*. (1971) have claimed that in *P. americana* the levels of many biologically important materials (e.g. choline esterase) change as a result of training.

However, neither Woodson *et al.* (1972) nor Willner and Mellanby (1974) were able to verify this claim. There are other areas of controversy as well, such as the possibility that the learned task may be stored as protein-encoded memory (Pak and Harris, 1975), or even the possibility that *P. americana* really cannot remember training for longer than 15 min, and, therefore, that it is an unsuitable subject for studies of memory (Willner, 1978).

Acknowledgements

We thank Drs Derek W. Gammon and J. E. Treherne for their suggestions regarding sections of the manuscript, and Mr Frank J. Novak, who reproduced most of the illustrations. Preparation of this chapter was supported, in part, by NIH grant NS 15632.

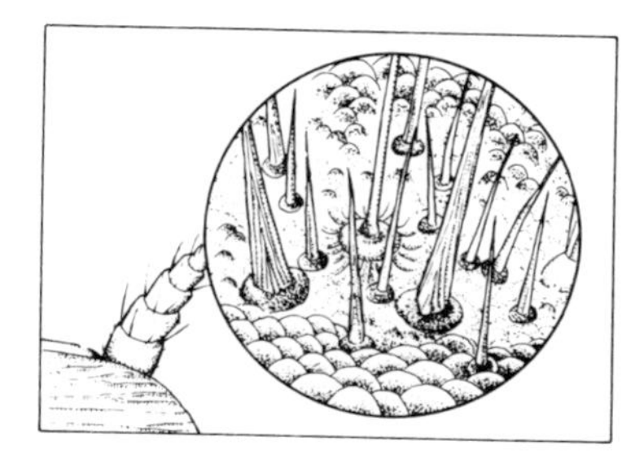

9

Sense organs

Günter Seelinger and Thomas R. Tobin

9.1 Introduction

The thrust during the last decade of investigations of the *Periplaneta americana* sensory system has been to answer two major questions; how is a particular sensory unit structured to perceive differences in stimulus modality or quality, and how are these properties encoded within the nervous system? Much of the recent work has required coupling physiological recordings of the sensory input with motor output or behavioural observations, in order to determine the functions of the sensory system. Each sensory system is composed of sensory units, as are ommatidia or sensilla, which contain one or several receptor cells. The specificity of a receptor cell is determined by the surrounding structures and by the stimulus-perceiving membrane of the receptor cell itself. Within each modality, receptors are further specialized to respond to only a certain range of stimulus qualities.

Each sensory modality has been primarily investigated on different parts of the insect by different groups of investigators. Mechanoreception has been studied on the legs and cerci; chemo-, hygro-, and thermoreception on the antennae. This approach resulted in a poor understanding of additional sensory regions and areas of overlap between modalities, such as the important mechanoreceptive input from the antennal base. However, this approach has helped to focus on the spatial organization of a particular modality and the central integration of large sensory areas.

This chapter reviews each of the major sensory systems which have been investigated. The structure, functional specialization and spatial organization of the sensilla in each region are covered, and, when possible, the physiological responses and sensory encoding are discussed with reference to their behavioural significance.

Additional references on other sensory organs may be found in Chapter 4 (Digestion), 5 (Respiration) and 14 (Behaviour).

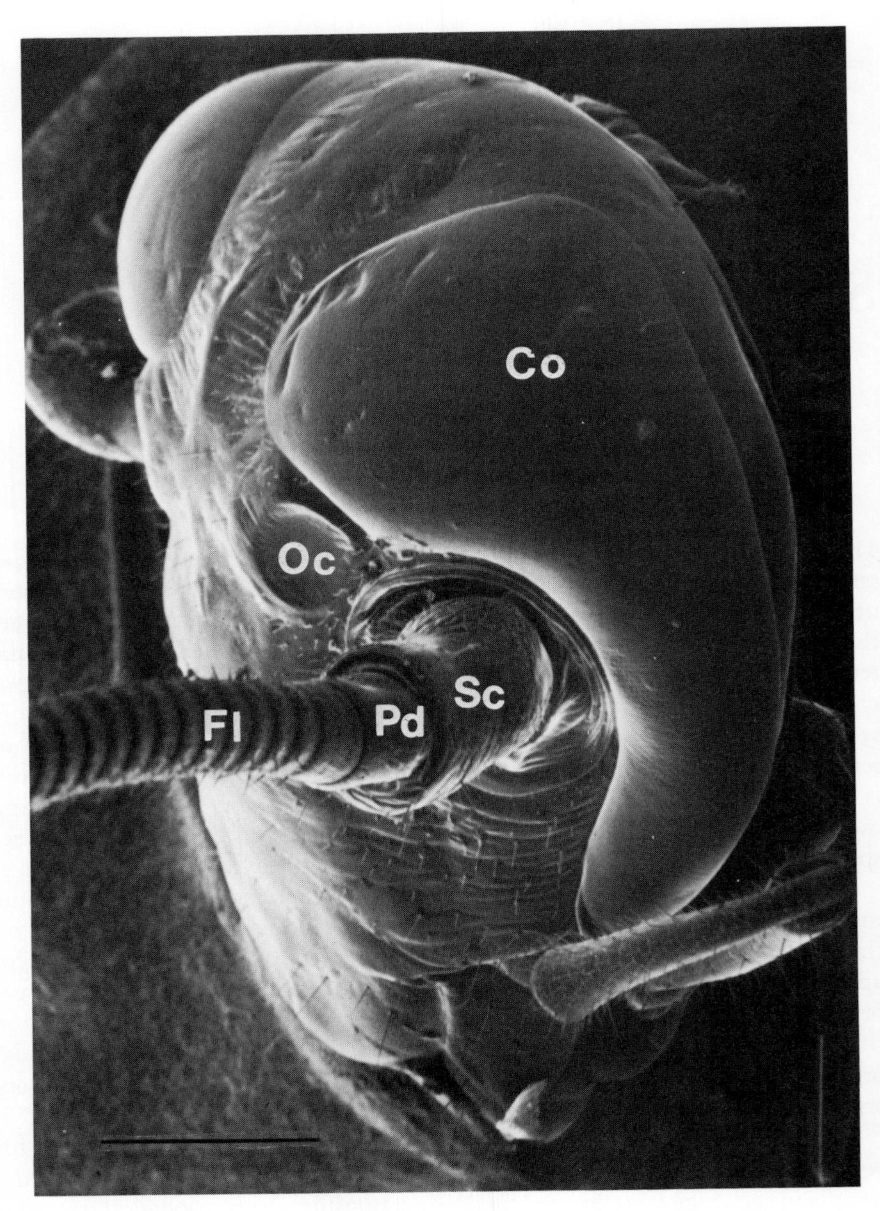

Fig. 9.1 The head of a male *P. americana*. View shows the anterior left portion. Scale bar = 1000 μm. Co, compound eye; Fl, flagellum; Pd, pedicel; Oc, ocellus; Sc, scape.

9.2 Visual system

9.2.1 Structure of the compound eye

The compound eyes of *P. americana* cover both sides of the head capsule (Fig. 9.1). The visual fields overlap to a maximum extent (65°) in the anterior direction, but binocular overlap also occurs in the dorsal and posterior directions. The surface is patterned by the 2000 ommatidia or sensory units which compose each eye. The internal structure of the euconal-type ommatidia consists of a corneal lens, a crystalline cone produced by four cone cells, two primary and additional accessory pigment cells, and eight retinula cells (Butler, 1973a; Fig. 9.2). The retinula cells or sensory cells contain the light-receiving rhabdomeral structures and also

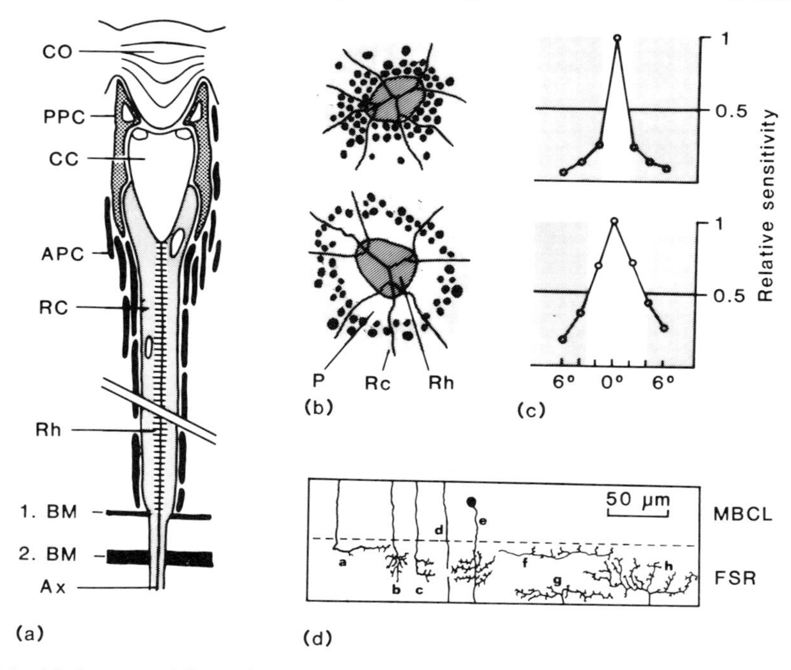

Fig. 9.2 (a) An ommatidium of the compound eye of *P. americana*, shown as a semi-schematic longitudinal section. Total length, 250–350 μm; basal diameter, 5–10 μm; apical diameter 35–40 μm. APC, accessory pigment cells; Ax, axons of the retinula cells; BM, basement membrane; CC, crystalline cone; CO, corneal lens; PPC, primary pigment cells; RC, retinula cells; Rh, rhabdom. (Redrawn from Butler, 1973b.) (b) Adaptation to light by horizontal pigment migration. Cross-sections through the centre of ommatidia in a light-adapted (top), and a dark-adapted (bottom) eye of *P. americana*. P, watery palisade of the endoplasmic reticulum; RC, retinula cells; Rh, rhabdom. (Redrawn from Snyder and Horridge, 1973a.) (c) Increase in the relative angular sensitivity of retinula cells in the dark-adapted (bottom) compared to the light-adapted eye (top). Ordinate: normalized response to a stimulus of constant intensity, given from different angles of incidence (abscissa). The angle of acceptance (white area) is defined by a 50% drop in the response with respect to the optical axis (0°). (Redrawn from Butler and Horridge, 1973a.) (d) Fibre types in the lamina ganglionaris of *P. americana*. MCBL, monopolar cell body layer; FSR, first synaptic region; a,b,c, superficial (a) and deep (b,c) short visual fibres; d, long visual fibres; e, monopolar cell; f,g,h, horizontal cell branchings. (Modified after Ribi, 1977.)

screening pigments. The rhabdomeral microvilli form a fused rhabdom of 2 μm in diameter, with its cup-shaped distal end tightly enclosing the lower part of the crystalline cone. Typical of an apposition eye, the screening pigments of the retinula and pigment cells extend from the cornea to the first basement membrane. A second thicker basement membrane not yet reported in other insects, is located beneath the first (Ribi, 1977). The eight sensory axons of the rentinula cells form a single bundle and extend inward towards the lamina ganglionaris of the brain.

9.2.2 Light–adaptation and visual acuity

Butler and Horridge (1973a,b) investigated the effect of adaptation on the compound eyes using intracelluar recording techniques. In the dark-adapted eye, the sensitivity of a retinula cell ranges only 2 log units of light intensity, which is similar to ranges found in other insects (see Walcott, 1975). Light intensity in most natural environments, however, changes much more from day to night. The difference between bright daylight and a clear moonless night is almost 9 log units. Insects can adjust their eyes to varying intensities of illumination by several mechanisms. First, a fast physiological adaptation involves bleaching of the visual pigment and changes in ionic concentration at the receptor membrane. The contribution of this mechanism to long-term adaptation is unclear in *P. americana* (Butler and Horridge, 1973b).

Secondly, a morphological adaptation in apposition eyes is established by horizontal migration of the screening pigment within the retinula cells. In the dark-adapted *P. americana* eye, the rhabdom is surrounded by a clear, watery palisade of 1 μm thickness formed by cisternae of the endoplasmic reticulum. In response to bright illumination, cytoplasm which contains pigment migrates horizontally towards the rhabdom and can replace the palisade within three minutes (Butler, 1973b; Fig. 9.2b). The absorption coefficient of this screening pigment is 70 times higher than that of the visual pigment (Snyder and Horridge, 1973), and causes a tenfold decrease in sensitivity to axial light (Butler and Horridge, 1973b). Horizontal migration of screening pigment is also reported for apposition and 'neuronal superposition' eyes of diurnal insects, dragonflies, locusts, ants, bees and flies (Walcott, 1975). The sensitivity changes established by this mechanism are relatively moderate (1–2 log units) compared to vertical pigment migration in the superposition eyes of moths (3–5 log units). The smaller range of adaptation in *P. americana* may be related to their association with a strictly nocturnal habit.

Pigment migration also affects the directional sensitivity of the retinula cells. The angle of acceptance is considerably wider in the dark-adapted than in the light-adapted ommatidia (6.7° vs. 2.4°, Butler and Horridge, 1973a) (Fig. 9.2c). The difference can be explained quantitatively by the electromagnetic theory of light guides (Snyder and Horridge, 1973). The refractive index of the rhabdom is 0.010–0.015 higher than that of the watery palisade, hence the rhabdom should function as a light guide in the dark-adapted state. Replacement of the palisade by cytoplasm of higher refractive index narrows the aperture of the rhabdom and consequently narrows the angle of acceptance.

The visual representation of the environment is projected as a mosaic on to the retinula. Each ommatidium provides a single mosaic unit in an apposition eye with fused rhabdoms. Visual acuity is best where inter-ommatidial angles are smallest and overlap between the visual fields of adjacent ommatidia is least. This is found in the anterior binocular field in the *P. americana* eye. During the light-adapted state, the inter-ommatidial angle (2°) almost equals the angle of acceptance (2.4°). Dark-adaptation enhances the degree of overlap and reduces the contrast so that higher absolute sensitivity is gained at the cost of visual acuity. Overall visual acuity is much less in the posterior region of the visual field where the inter-ommatidial angles are much wider (10°) (Butler and Horridge, 1973a).

9.2.3 Colour vision and polarized light sensitivity

The existence of two receptor groups was shown early in electroretinogram recordings by Walther (1958a,b) and Walther and Dodt (1959). The dorsal half of the compound eye showed two sensitivity peaks in green and in u.v. light. The relative height of the peaks could be altered by causing selective adaptation with monochromatic light. Mote and Goldsmith (1970) confirmed the existence of two receptor groups using single cell recordings. A green receptor type is most sensitive at 500–520 nm, and a u.v. type at 365 nm (Fig. 9.3). Both types are about equally sensitive at their optimal wave length.

Butler (1971) used a very different approach to identify the receptor types. He discovered that the receptor cells are arranged in a regular pattern in all ommatidia, and can be mapped by the size and position of the rhabdomeres. After exposure to u.v. light, histological investigation reveals two different categories of receptors: three cells are light-adapted, whereas five cells show well-developed palisades. Adaptation to green light yields the complementary result. This pattern is constant through all the ommatidia of the compound eye. In the dorsal as well as the ventral half of the compound eye, three u.v. and five green receptors exist. This is in contradiction to the physiological finding of Walther (1958b), that the ventral part is insensitive to u.v.

Butler and Horridge (1973b) found the retinula cells to be five times more sensitive to polarized light in one plane of polarization than in the orthogonal plane. Two groups of receptor cells could be separated, with their planes of maximal sensitivity at 90° to each other. Morphologically, two main directions of microvilli orientation are apparent in each ommatidium. It is not clear however if these directions coincide with the physiological preference directions for polarized light, or if they are related to colour vision. Ultraviolet receptors are responsible for polarized light detection in the honeybee, and their central projections are spatially separated from the other types (Sommer and Wehner, 1975). Long visual fibres leading directly into the medulla are also present in

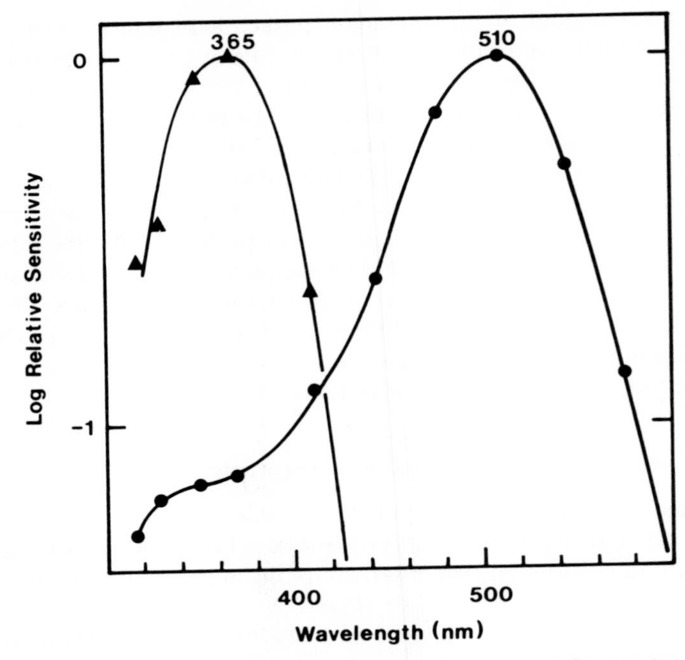

Fig. 9.3 Spectral sensitivity of the two receptor types in the compound eye of *P. americana*. Ordinate: log of reciprocal no. of photons required for a standard response. The optimal wavelength is 365 nm for the u.v.-receptor and 510 nm for the green receptor. (Redrawn from Mote and Goldsmith, 1970.)

P. americana (Ribi, 1977), but no direct evidence has shown them to be u.v. receptors. Polarization sensitivity might be an inherent feature of the insect rhabdom due to dichroism and parallel orientation of the visual pigment molecules. This sensitivity may or may not be used by higher order neurons of the visual system. Still, the degree of polarization sensitivity found by Butler and Horridge (1973b) is as high as in any other insect investigated. From our knowledge of the ecology of *P. americana*, it seems unlikely that polarized light is used for celestial orientation, but it could be used to improve discrimination between light sources and reflecting surfaces such as water.

9.2.4 Central projections: the lamina ganglionaris

The optic ganglia of the brain are located beneath the membrane layers of the compound eye (Section 8.2.2(b)). Only the most distal part, the lamina ganglionaris, has been examined in some detail (Ribi, 1977). As the visual fibres leave the eye they are arranged in regular bundles of eight axons originating from a single ommatidium. Before entering the lamina ganglionaris, the axons regroup into bundles of 6–20 fibres. A long fibre type passes directly through the lamina without branching, but most fibre types branch within the 'first synaptic region' (Fig. 9.2d). Here they have synaptic contacts with either monopolar or horizontal neurons. Monopolar cells extend inward and their cell bodies are located distally from the synaptic region. Several types of horizontal cells project from lateral and proximal regions. A pronounced vertical stratification is lacking. In the neuronal superposition-type eye of the fly, the highly ordered 'cartridge' projection of the retinula on to the lamina is known to contribute substantially to visual acuity by means of lateral inhibition and other neural mechanisms (Braitenberg and Strausfeld, 1973; Laughlin, 1975). The structure of the *P. americana* lamina indicates a comparatively low spatial resolution. On the other hand, a distinct horizontal stratification is apparent. Horizontal and monopolar cells with wide-field arborizations allow convergence of many visual fibres on to one second-order neuron. This arrangement should provide low central thresholds under dim illumination.

9.2.5 Ocelli

Most of our knowledge about the ocelli comes from the extensive work of Ruck (1957, 1958a,b; 1961). These results are reviewed in some detail by Guthrie and Tindall (1968) and are summarized very briefly here. Weber and Renner (1976) give a recent description of the fine structure of the ocellus. The two ocelli of *P. americana* appear as elliptic spots of 500 × 700 μm located dorsomedially to the antennal base (Fig. 9.1). The flattened cornea lacks the brown exocuticle, and the ocelli appear white in colour due to a reflecting tapetum. The space between the tapetum and cornea is almost completely occupied by 10 000 receptor cells, the largest number found in the ocellus of any insect. The club-shaped receptor cells are arranged throughout the lumen in groups of 2–6, which contain rhabdoms of varying cross-sectional shapes. Axon bundles penetrate the tapetum and make contact with numerous dendritic branches of the ocellar nerve. These branches, however, arise from only a few ocellar nerve fibres, some of which are 'giants' (see Section 8.2.2(b)).

Goldsmith and Ruck (1958) showed the existence of only one ocellar receptor type by electroretinogram techniques. The spectral sensitivity of this receptor corresponds to that of the compound eye green receptor. Light adaption reduces the sensitivity of the ocellus by a factor of 5×10^{-5} (Ruck, 1958b), but recovery in the dark is almost completed within one minute. Ruck (1961) analysed three different phases in the electroretinogram of the ocellus. Two phases originated in the receptor layer. The third phase was a fast 'off' spike probably indicating giant fibre activity in the ocellar nerve. Cooter (1973) recorded from spontaneously active neurons in the ventral nerve cord of the thorax, which had been inhibited by illumination of the ocelli. The function of the ocelli is still unknown, but influences on the leg motor neurons have been suggested (Goodman, 1970).

9.3 Mechanoreceptive sensory systems

9.3.1 Cercal thread hair system

The cerci of *P. americana* protrude obliquely out from under the wings at the posterior end of the abdomen. The ventral surface is covered with two types of hair sensilla, bristles and thread hairs. Under SEM, bristle hair sensilla, the most numerous hairs on the cerci, closely resemble the sensilla chaetica found on the antenna. Few details have been reported on these sensilla. Thread hair sensilla are identified as specialized mechanoreceptive hairs which respond with directional sensitivity to movements of air. Previously, the thread hairs were considered auditory hairs, however, they will respond only to the near-field components and not to pressure components of audible sound. The thread hairs have been studied intensively, since they are the receptor organs involved in an escape response to predators (Roeder, 1948).

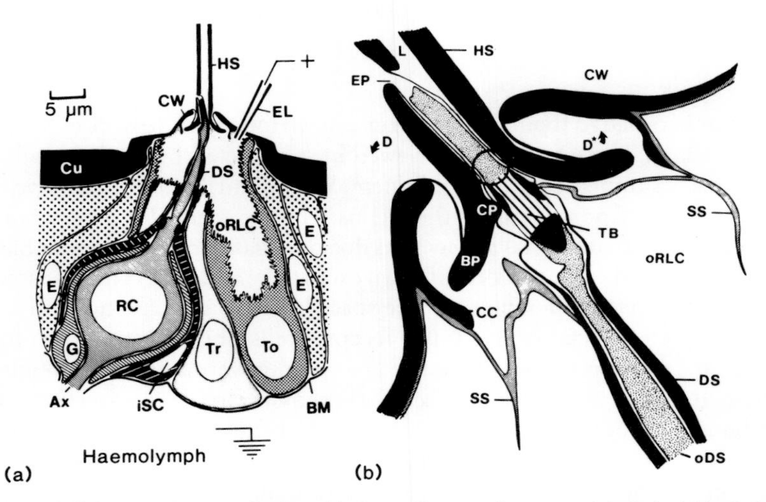

Fig. 9.4 (a) Cellular organization of a thread hair sensillum. Cell types are labelled within their white nuclei. The inner receptor lymph cavity is not labelled; it is the space surrounding the distal receptor cell process and enclosed by the inner sheath cell. The micro-electrode inserted into the outer receptor lymph indicates a positive transepithelial potential with respect to the haemolymph. (Redrawn and modified after Gnatzy, 1976.) (b) The basal structures of a thread hair illustrating the mechanism of stimulus conduction. The circle gives a tentative localization of the pivot. For an explanation see text. The outlines are drawn directly from a TEM photograph (Gnatzy, 1976), except that the broken hair shaft is corrected. Ax, axon; BM, basement membrane; BP, basal plate of the hair shaft, CC, cuticular cylinder; CP, cuticular protrusion; Cu, cuticle; CW, cuticular wall of the socket; D, and D* directions in which the hair shaft and hair base respectively move during a depolarizing deflection of the hair; DS, dendritic sheath; E, epidermal cell; EL, electrode; EP, ecdysial pore; G, glial cell, only one layer is shown; HS, hair shaft; iSC, inner sheath cell; oDS, outer dendritic sheath; oRLC, outer receptor lymph cavity; RC, receptor cell; SS, socket septum; TB, tubular body; To, tormogen cell; Tr, trichogen cell.

The cellular organization of the thread hair sensilla may serve to demonstrate the general insect sensillum scheme (Fig. 9.4). A varying number of receptor cells (only one in each thread hair) are enveloped concentrically by three modified epidermal cells: (1) the inner sheath cell, which produces the cuticular sheath of the receptor process; (2) the trichogen cell, which secretes the hair shaft; and (3) the tormogen cell, which forms the hair socket and the attachment processes of the hair. After the cuticular structures are formed, the enveloping cells withdraw. They leave the hair lumen and a large basal cavity filled with extracelluar fluid, the receptor lymph. Numerous microvilli of the tormogen cell continuously secrete K^+ ions into the receptor lymph. Tight connections close the intercelluar spaces between the sensillar cells. Thus, the receptor lymph is electrically isolated from the haemolymph, and a positive transepithelial potential is maintained (Thurm, 1974). A sensory process of the receptor cell extends distally through the receptor lymph. The outer segment of the dendrite is distinguished by a modified '$9 \times 2 + 0$' cilium. Distally, the cilium usually contains numerous microtubules and may sometimes branch. Mechanoreceptive cilia often have a specialized region, the tubular body, where densely arranged microtubules are embedded in an electron-dense material. It is in this region that the mechanical stimulus is transformed into an electrical event. Stimulation somehow decreases the membrane resistance of the outer dendritic segment and, as a consequence, a current along the cilium is driven by the transepithelial potential. This current depolarizes the receptor membrane and elicits action potentials at the inner dendritic segment. Hansen (1979) and Zacharuk (1980) give further references on the functional organization of insect sensilla.

The structure and function of the cercal thread hairs is described by Nicklaus *et al.* (1967) and Gnatzy (1973, 1976). The slender hair shaft (2.5 μm) projects from the centre of a cuticular cupola and may be up to 700 μm long. The rim of the cupola invaginates inwardly forming an almost circular cylinder. Since the proximal end of the hair shaft widens into an elliptically shaped basal plate, it restricts movement of the hair largely to one plane. Deflection of the hair in one direction (D) along this plane causes depolarization of the receptor potential (Fig. 9.4b). Deflection in the opposite direction causes hyperpolarization. Increased depolarization occurs as the deflection increases in the 'D' direction up to 12°. At larger angles, the cupola rim restricts the movement and the shaft will bend (Nicklaus, 1965). Westin (1979) recorded from the receptor axons and found that the maximum response frequency was elicited when the hair was bent towards only one quadrant (Fig. 9.5b). It is not clear why depolarization occurs by deflection of the hair in one direction. Nicklaus (1965) considered that the hair pivoted about the base of the shaft, where it articulated with the inner rim of the cuticular cylinder. Bending of the tubular body within the dendrite would cause stimulation of the cell. Gaffal *et al.* (1975), however, pointed out a different role of the socket structures in mechanoreceptive hairs. The hair shaft of thread hair sensilla is delicately suspended by membranes. Since the pivot point is located higher than the tubular body, deflection of the hair in direction 'D' would move the hair base in the opposite direction 'D*' (Figs 9.4b and 9.5a). A cuticular protrusion of the hair base would compress the tubular body against the opposing socket septum. Deflection of the hair in the opposite direction will not compress the tubular body, since the socket septum is asymmetrical and does not contact the dendrite on both sides.

The thread hairs respond to constant deflection with long-lasting excitation (Nicklaus, 1965), but, since short wind puffs are more effective in eliciting a high

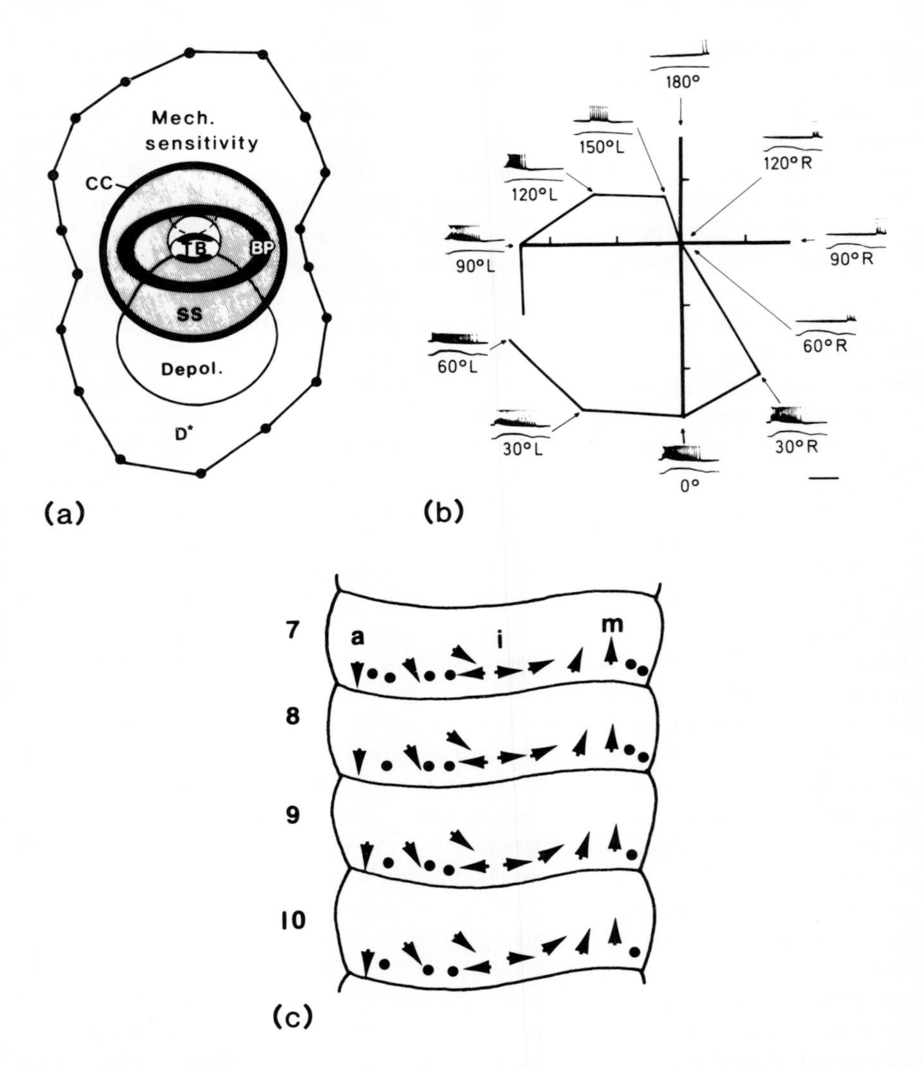

Fig. 9.5 Directional sensitivity of thread hairs. (a) Schematic cross-section of the hair base and cuticular cupola at the level of the dendritic tubular body. The relative mechanical sensitivity indicates the amount of deflection of the hair shaft caused by equal forces in different directions. The relative depolarization area shows the maximum extent of movement of the hair base in different directions which will cause depolarization. The relative mechanical sensitivity and depolarization area are not shown on the same scale or in scale to the base of the shaft. BP, basal plate; CC, cuticular cylinder; D*, direction in which movement of the hair base will cause depolarization and hyperpolarization respectively of the receptor cell; SS, socket septum; TB, tubular body. (b) Responses from a single sensory cell to different angles of the wind. Polar plot of number of action potential (average of 3) vs. the wind angle. 0° is from directly behind the insect. Bar marks units of ten action potentials. Sample recordings are shown for specific angles of the wind. (Top trace) action potentials recorded intracellularly from sensory axon. (Bottom trace) wind speed. Peak speed = 130 ms^{-1} (except 90–180° left = 25 ms^{-1}). Time bar = 50 ms. (From Westin, 1979.) (c) Spatial arrangement of the thread hairs on the cercal segments 7–10 (after Nicklaus, 1965). The arrowheads point in the optimal excitatory wind direction

(after Dagan and Camhi, 1979). Hairs of the same directional sensitivity are arranged in columns on subsequent segments; columns a, i, and m are given as examples. Dark circles = hairs of unknown directional sensitivity.

---

response frequency, Dagan and Camhi (1979) suggest that the receptor may function as a wind-acceleration detector. Westin (1979) identified seven groups of receptors which respond with uniform characteristics to wind puffs from different directions. This finding reflects the highly ordered spatial arrangement of the thread hairs axis found by Nicklaus (1965) (Fig. 9.5c). The hairs occur in rows at the distal edge of each cercal segment. The optimal plane of sensitivity is considerably different within each row, but very similar for the hairs within the parallel columns along the cerci. Wind from any direction will optimally stimulate at least one of these columns, and usually several others to a lesser degree (Dagan and Camhi, 1979). The wind direction is thus coded by an across-fibre comparison of the response pattern of axons within the cercal nerve. The cercal nerves synapse on to numerous interneurons within the last abdominal ganglia, including the seven pairs of giant interneurons which are suspected to mediate the directional information to the leg motoneurons of the thorax (Ritzmann and Camhi, 1978). A wind puff created by the tongue strike of a predatory toad is sufficient to excite the cercal thread hair system and elicite an orientated escape response (Camhi and Tom, 1978; Camhi *et al.*, 1978; see Sections 8.7.3 and 14.2.2).

9.3.2 Proprioreceptors of the leg

The majority of sensory neurons of the leg are mechanoreceptors which function as proprioreceptive organs. Proprioreceptive information concerns contact of the leg with the substrate and surrounding objects, as well as the position, tension, and rate of movement of the leg segments. Peripheral input from some proprioreceptors to higher nervous centers can reflexively modulate motor neuron control (Pringle, 1940; Wong and Pearson, 1976; Zill *et al.*, 1977). Central processing must occur as well, since different classes of receptors may effect the same reflex, and the same sensilla may effect multiple reflexes (Zill *et al.*, 1977; Kramer and Markl, 1978). Four classes of receptors are found: hair sensilla, campaniform sensilla, scolopidia, and multipolar stretch receptors.

(a) *Hair sensilla*

The innervated hairs of the leg closely resemble the mechanoreceptive hairs of the cerci and antennae. A single dendrite inserts on to the base of the fluid-filled lumen of the shaft. Directional sensitivity to movement of the hair is caused by the articulating structure at the hair base and the fine structure of the dendritic process. Unlike the thread hairs of the cerci, the innervated hairs of the leg do not respond to gentle wind puffs or audible vibrations. This insensitivity is due to the larger mass and inertia of the shaft. The hair sensilla are arranged either in large plates located at the intersegmental joints, in groups on the ventral side of the tarsal segments, or scattered about as isolated hairs (Pringle, 1940) (Fig. 9.6).

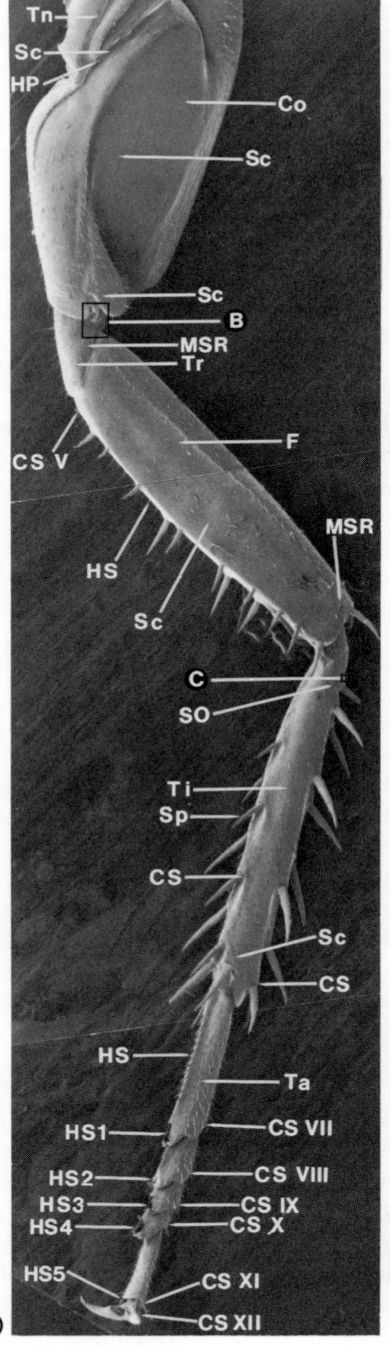
Tn
Sc
HP
Co
Sc
Sc
B
MSR
Tr
F
CS V
HS
MSR
Sc
C
SO
Ti
Sp
CS
Sc
CS
HS
Ta
HS1
CS VII
HS2
CS VIII
HS3
CS IX
HS4
CS X
HS5
CS XI
CS XII
(a)

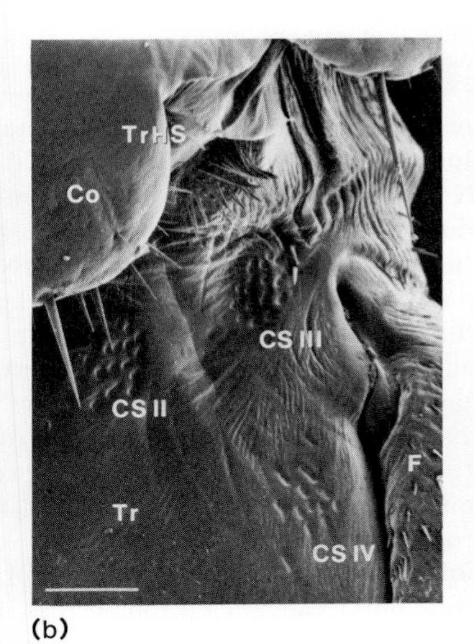
TrHS
Co
CS III
CS II
F
Tr
CS IV
(b)

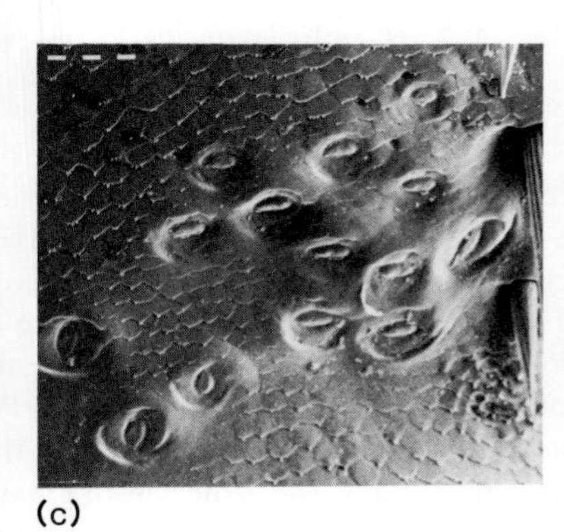
(c)

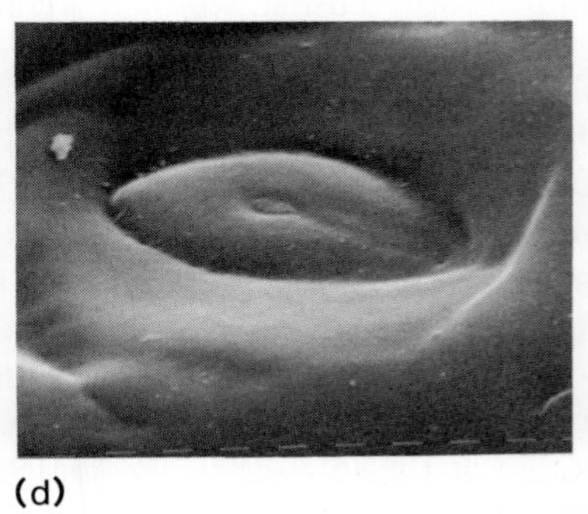
(d)

Fig. 9.6 (a) Distribution of mechanoreceptive sensilla on the left metathoracic leg of *P. americana*. (b) Coxal–trochanteral–femur joint of *P. americana*. A ventral view of the left metathoracic leg, showing the trochanteral hair plate and several fields of campaniform sensilla. Scanning electron micrograph. Scale: Bar = 10 μm. (c) Campaniform sensilla fields located proximally on the tibia of the left metathoracic leg of *P. americana*. Sensilla with a horizontal long axis compose field VIa, vertical long axis, VIb, several isolated mechanoreceptive hairs are on the right. Scanning electron micrograph. Scale: Bar = 10 μm. (d) single campaniform sensillum. Scanning electron micrograph. Co, coxa; CS, campaniform sensilla, fields numbered according to Pringle (1938a); F, femur; HP, hair plates; HS, hair sensilla, groups on tibia numbered according to Kramer and Markl (1978); MSR, multipolar stretch receptors (internal); Sc, scolopidial organs (internal); SO, subgenual organs (internal); Sp, cuticular spine; Ta, tarsi; Ti, tibia; Tn, trochantin of thorax; Tr, trochanter; TrHS, trochanteral hair sensilla.

Three hair plate organs are located on each leg. Two are found at the coxal joint (Pringle, 1938a). Pringle (1938a, 1961) suggested that these receptors are important in the tonic control of leg position. Wong and Pearson (1976) have also demonstrated that, during walking, the trochanteral hair plate functions to limit the extent of femur flexion during each step cycle. The trochanteral hair plate is composed of 50–60 sensilla on the intersegmental membrane of the coxo-trochanteral joint (Fig. 9.6b). About half of the sensilla form a group which rests against a fold in the membrane. The hairs which contact the fold are 30–70 μm-long, and arranged in rows in such a way that flexing the femur progressively bends more rows of sensilla. Electrophysiologically, the sensilla represent two types, of which one responds phasically to the initial deflection of the hair, while the second type adapts more slowly. During walking, the basic rhythmic pattern is centrally generated (see Section 11.7.1). One function of the hair plate is to provide afferent information to stop or limit the flexion of the tibia at the end of the step. Stimulation of the hair plate will produce a transient inhibition of the flexor motor neuron and transiently increase the spontaneous activity of the opposing slow depressor motor neurons (Wong and Pearson, 1976). The slowly adapting type of hair plate sensilla may also provide information about the leg position as suggested by Pringle (1961).

Small groups of 4–10 hair sensilla are located on the ventral side of each tarsal segment. The hair sensilla, as well as numerous campaniform sensilla, are stimulated when the tarsi touch the ground and inhibit flight behaviour (Kramer and Markl, 1978). Contact of only one leg is sufficient, and if the 25 hair sensilla are left intact after destruction of all the campaniform sensilla, the reflex is still effective. Electrophysiological studies of the response characteristics of these hairs have not been reported.

Several very long hair sensilla also occur on the ventral surface near the coxa, trochanter and femur joints. These 1–2 mm-long hair sensilla are distinct from the large spines on the femur. Spencer (1974) reported that the long trochanteral hair has a rapidly adapting phasic response to deflection. The initial velocity of deflection determines the number of impulses produced by a single stimulation. These long hairs are positioned so that contact to the substrate would only be made when the cockroach is in a low crouching position.

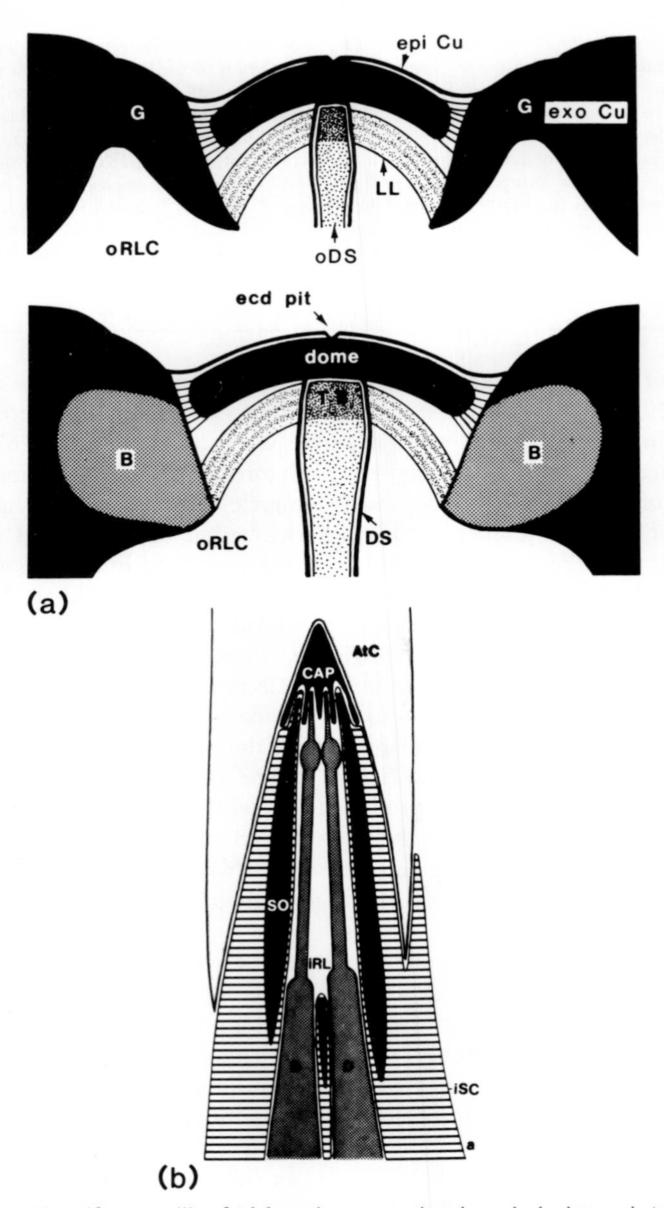

Fig. 9.7 (a) Campaniform sensilla of *Blaberus* in cross-section through the long axis (top) and short axis (bottom) (after Moran and Rowley, 1975). B, buttresses of unknown mechanical properties; dome, cuticular dome of the cap; DS, dendritic sheath; ecd pit, ecdysial pit; epi Cu, epicuticle; exo Cu, exocuticle; G, thin gasket region; LL, laminated layer of the cap; oDS, outer dendritic segment; oRLC, outer receptor lymph cavity; TB, tubular body. (b) Schematic representation of the scolopale region in a scolopidium from the tibio–tarsal chordotanal organ of *P. americana* (after Young, 1970). AtC, attachment cell (= trichogen cell); CAP, cuticular cap (= dendritic sheath); D, dendrite; iRL, inner receptor lymph cavity; iSC, inner sheath cell (= scolopale cell); SO, scolopale.

(b) *Campaniform sensilla*

Campaniform sensilla appear as oval-shaped domes within circular depressions of the cuticle (Fig. 9.6b,c). Although the hair shaft is lacking, the cellular organization of the campaniform sensilla is similar to the mechanoreceptive hairs. The structure of cockroach campaniform sensilla has only been investigated on the antenna of *P. americana* (Toh, 1977), and on the legs of *Blaberus* and *Blattella germanica* (Moran *et al.*, 1971; Moran and Rowley, 1975a). A laminated layer located beneath the cuticular dome is penetrated by the sensory process of the receptor cell (Fig. 9.7a). The tip of the dendrite is modified as a laterally flattened tubular body, with the long axis paralleling the long axis of the dome. The campaniform sensilla are directionally sensitive to cuticular stresses. Compression perpendicular to the long axis stimulates the receptor cell. Since the effect of compression is enhanced when the cuticular dome is simultaneously pushed inwardly by the tip of a fine probe, stimuli seem to displace the dome and the laminated layer towards each other (Spinola and Chapman, 1975). Mechano-electrical transduction probably results from lateral compression of the tubular body by deformation of the laminated layer (Moran and Rowley, 1975a). Further details of how natural stimulating forces affect the movement of the cuticular components are required to understand how directional sensitivity is related to the oval shape of the cap and dendrite.

Campaniform sensilla are specialized for two distinct functions on the leg. Groups of campaniform sensilla are arranged in fields near the joints of each segment (Fig. 9.6b,c), and a single sensillum is located at the base of each tibial spine. Pringle (1938a) showed that, within each field, the long axis and directional sensitivity of the sensilla are orientated identically (Fig. 9.6c). Fields of sensilla with different orientation and directional sensitivity to cuticular stresses contribute information to the postural and locomotory control processes (Pringle, 1940, 1961). Zill *et al.* (1977) reported that stimulation of a single campaniform sensillum reflexively effects the appropriate sets of muscles to compensate for a postural change. Two adjacent fields of campaniform sensilla on the tibia (Fig. 9.6c) are arranged so that one field responds to tibial extension and the other one to flexion. Tibial extension stimulates the proximal field which increases the activity of the slow excitatory axon of the extensor muscle and decreases the activity of excitatory axons of the flexor muscles. Tibial flexion stimulates the antagonistic field and elicits the opposing effect (Zill *et al.*, 1977). Kramer and Markl (1978) identified each of the fields of campaniform sensilla to participate in the inhibition of the flight reflex. Four fields of 12–20 sensilla are located on the trochanter, and fields with fewer sensilla are found on the femur, tibia, and each segment of the tarsus and pretarsus. Leg contact to the substrate needs only to stimulate a small number of these sensilla to inhibit flight even in the absence of the tarsal hairs.

The large tactile spines located on the tibia and femur are not innervated directly, as are the hair sensilla. The large hollow spines are articulated at the base and are surrounded by a flexible membrane. Chapman (1965) reported that each spine contains a single campaniform sensillum in the thick cuticular wall of the spine where it joins the soft cuticle of the socket. A slowly adapting response is elicited by deflection of the spine.

(c) *Proprioreceptive scolopidia*

Proprioreceptive scolopidia or chordotonal organs are internal groups of sensilla responsive to deformations created by the movements of limb segments. Scolopidia are homologous to cuticular sensilla, but they are much more modified than even the campaniform sensilla (Schmidt, 1973). The organization of a scolopidium is shown in Fig. 9.7b (Young, 1970). All of the scolopidia on the leg belong to the mononematic type, in which the cap is connected to the cuticle only by the attachment cell. The short solid cap, homologous to the dendritic sheath of hair sensilla, is produced by the scolopale cell (inner sheath cell of the dendrites). A scolopale formed along the inner surface of the scolopale cell consists of longitudinally arranged microtubules surrounded by electron-dense material. The dendrites of both sensory cells extend the entire length of the rigid tube created by the scolopale. At the distal end of the scolopale, both the dendrites and scolopale insert up into the cap. At the proximal portion, the inner dendritic segment attaches to the scolopale cell by desmosomes. Unlike other mechanoreceptive sensilla, the sensory cilia maintain their '$9 \times 2 + 0$' structure to the tip of the dendrite. Mechano-electrical transduction of longitudinal strain on the scolopidia may involve stretching or compression of the bulge-like region of the cilia close to the cap (Fig. 9.6c) or bending of the ciliary neck (Moran and Rowley, 1975b).

Nijenhus and Dresden (1952) described chordotonal organs associated with all the joints of the leg. Each one is formed by several scolopidial units branching out of the nerve near the distal end of the segment. The proximal end of the scolopidia inserts into either the intersegmental membrane of the joint or the integument of the adjacent segment. The tibio–tarsal organ studied by Young (1970) consists of 12 doubly innervated and two singly innervated scolopidia. The organ responds to backward and downward deflection of the tarsi, but not to the return motion. Functionally, the doubly innervated scolopidia represent two groups. The larger and more heavily sheathed neurons respond phasic-tonically only to extreme flexion, whereas the less well-sheathed neurons respond tonically to the full range of movement. The response of the two small scolopidia was not determined.

(d) *Stretch receptor organs*

Stretch receptor organs formed by multipolar neurons, not associated with a cuticular structure or specialized terminal ending, are found near the leg articulations. Guthrie (1967b) describes five receptors, one of which is a large cell with a centripetal axon possessing twenty branching dendritic processes, which insert into soft cuticle about the anterior ventral condyle of the femoro–tibial articulation. Two smaller cells with 2–3 nonbranching dendrites are located near the postero-dorsal condyle. Two other cells with 3–4 branches, each with numerous fine rami ending in small clubbed terminals, lie against the trochantero-femural articulation. The large stretch receptor in the femur responds to levation of the tibia with very low frequency, slowly adapting, tonic firing. The smaller receptors are believed to give a brief phasic response. As Guthrie (1967b) stresses, the functional contribution of stretch receptors to behavioural or reflex actions is not known. They apparently provide positional information of the joint, supplementing the function of the campaniform and chordotonal organs. The multipolar stretch receptors are heavily sheathed, except near the tips of the dendrites. No ciliary structures are found within the cell, and the process of mechano-electrical transduction is only postulated to involve a local depolarization starting at the dendritic tip when it is bent.

9.3.3 Subgenual organs

The term 'subgenual organs' refers collectively to three groups of scolopidial organs: the subgenual organ, the Nebenorgan (= accessory organ) and the distal organ. They are located at the proximal end of the tibia but, unlike the proprioreceptive scolopidia, their function is not associated with the movement of a joint. The subgenual and Nebenorgan are most sensitive to substrate vibrations (Howse, 1964). The function of the distal organ is not precisely known. It surrounds a haemolymph cavity and is postulated to respond to variations in haemolymph pressure (Schnorbus, 1971).

The mononematic scolopidia of the subgenual organ are singly innervated. The sensory neuron branches off from the 5a nerve bundle at the anterior edge of the tibia. The distal end of the attachment cell is connected to the cuticle on the posterior edge of the tibia just below the fields of campaniform sensilla. The scolopidial unit is arranged in a hook shape. The neuron cell body and axon are vertical in the normal leg position, while the dendrite, scolopale and attachment cells are horizontal (Fig. 9.8). The scolopidial units are organized in a fan-shaped manner, and converge radially towards a single insertion point. Since the scolopales and attachment cell bodies are freely suspended, the haemolymph dampens the movement of the neuron side of the scolopale, while no such dampening occurs to the side anchored to the cuticle. Slight movement or acceleration of the tibia caused by vibrations of the substrate would cause tension changes on the scolopidium. The subgenual organ has a maximum sensitivity between 1000–5000 Hz. The threshold amplitude within this range lies between 10^{-7} and 10^{-10} cm (Schnorbus, 1971).

The scolopidia of the Nebenorgan all lie parallel with one another. The attachment cells insert at the same location as the subgenual organ. The bundle of scolopidia is oriented horizontally, forming the long side of a triangle across the posterior outer corner of the tibia (Fig. 9.8). The unit is stretched perpendicular to the long axis of the tibia so that it can react to tension changes within the scolopale caused by vibrations of the cuticular walls. The Nebenorgan responds most sensitively to low frequency substrate vibrations of 30–500 Hz, with a threshold amplitude of 10^{-6} cm.

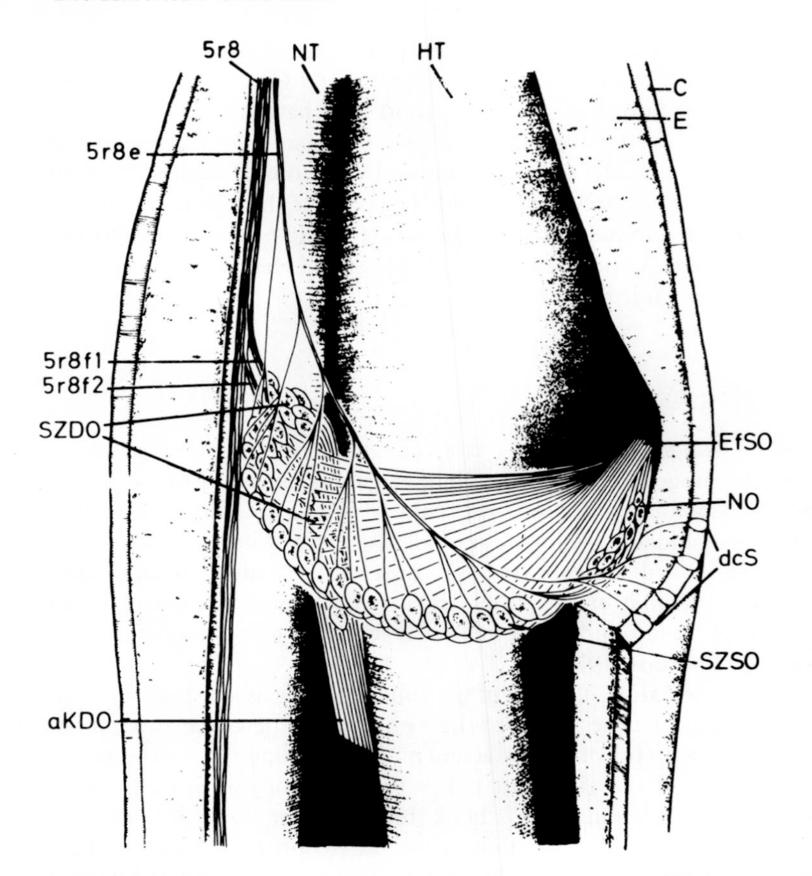

Fig. 9.8 Subgenual organs of *P. americana*. The left tibia opened to show the spatial arrangement of the scolopidial organs. aKDO, accessory cap cells of the distal organ; C, cuticle; dcS, distal campaniform sensilla; E, epidermis; EfSO, terminal fila of the subgenual organ; HT, main trachea; NO, Nebenorgan; NT, small trachea; SZDO, sensory cells of the distal organ; SZSO, sensory cells of the subgenual organ; 5r8, nerve innervating the subgenual organs and the campaniform sensilla. (From Schnorbus, 1971.)

9.4 Antennal sensory system

9.4.1 Structure and classification of antennal sensilla

The antennae of *P. americana* are innervated by chemo-, hygro-, thermo- and mechanoreceptors (Table 9.1). In the adult, the antennae are about 50 mm long and consist of 150–170 segments. As on the leg, a system of mechanoreceptive hairs, campaniform and scolopidial sensilla provide peripheral information on the position of the antenna. These proprioreceptors are predominantly localized on the scape and pedicel. The remaining modalities are perceived by specialized hair sensilla distributed on the flagellar segments. The general cellular organization of

Table 9.1 External antennal sense organs of *P. americana*.

Type of sensillum	*Surface structure*	*Length (μm)*	*Diameter (μm)*	*Number of rec. cells*	*% of male sensilla*	*Function*
Porous-walled						
sw A	smooth	8–12	2–3	2	8	olfaction
sw B	smooth	18–20	3–4	4	54	olfaction (female sex pheromone)
sw C	distal smooth basal grooved	30–40	3	2	6	olfaction
dw A1	16–19 grooves	8–12	2–3	3	8	olfaction
dw A2	22–32 grooves	8–12	2–3	4	rare	olfaction thermoreception
dw B	grooved	20–25		3		olfaction
Porous-tip						
tp	grooved	170	10	5	25	taste mechanoreception
Non-porous						
capped	cap grooved	8	5	3–4	0.2	thermoreception hygroreception
campaniform	dome-like		4	1	0.4	mechanoreception
marginal	pea-like		4	1	0.5	mechanoreception
mechanoreceptive hairs	grooved			1	0.5	mechanoreception

all of the hair sensilla is identical to the cercal thread hairs, but the fine structure of the hairs shows modifications correlated with their specific physiological function.

(a) *Porous-walled hair sensilla*

Porous-walled or olfactory hair sensilla are the thin-walled, non-articulated hairs that possess either pores or channels in the shaft's cuticular surface. The terminology of the porous-walled sensilla is confusing since both the terms sensilla basiconica and sensilla trichodea were originally used to describe the non-articulated hairs. Various authors have attempted to use only one term (see Table 9.2), but recent studies including TEM and high resolution SEM have established structural differences of the wall which are not correlated with the external size and shape of the hair (Altner, 1977). The classification of 'single-' and 'double-wall'

Table 9.2 Terminology used by different authors for *P. americana* sensillum types.

Schaller (1978)	*Toh (1977)*	*Norris and Chu (1974)*	*Schafer and Sanchez (1973)*
Single-walled A Single-walled B	Perforated surface basiconic sensillum	Sensillum basiconicum II Sensillum basiconicum I + III	Sensillum trichodeum
Single-walled C	Trichoid sensillum	Sensillum basiconicum IV	
Double-walled A1 Double-walled A2 Double-walled B	Grooved basiconic	Sensillum basiconicum I + II	
No-pore	Capped sensillum	Smooth cone	Cold-receptor sensillum
Terminal-pore	Chaetic sensillum	Grooved peg I + II	Sensillum chaeticum B

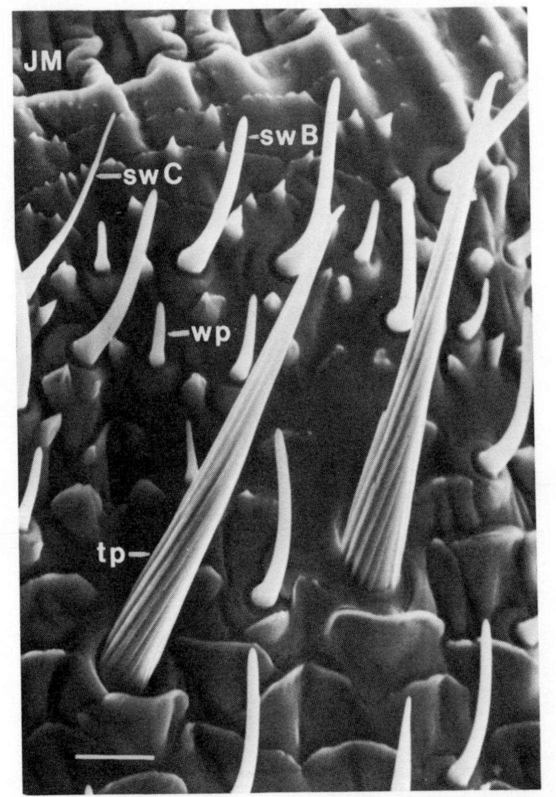

(a)

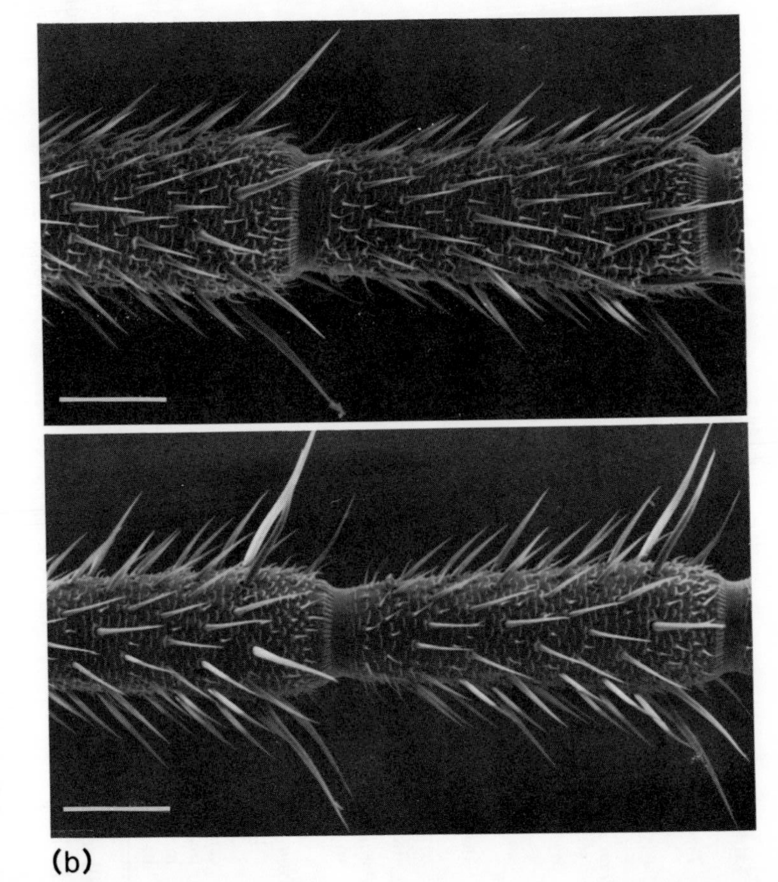

(b)

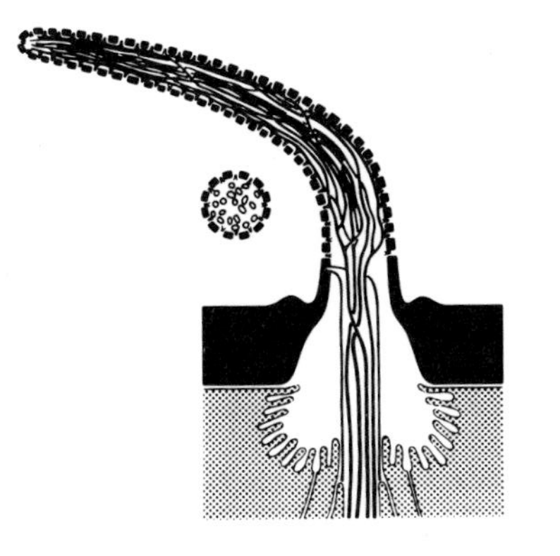

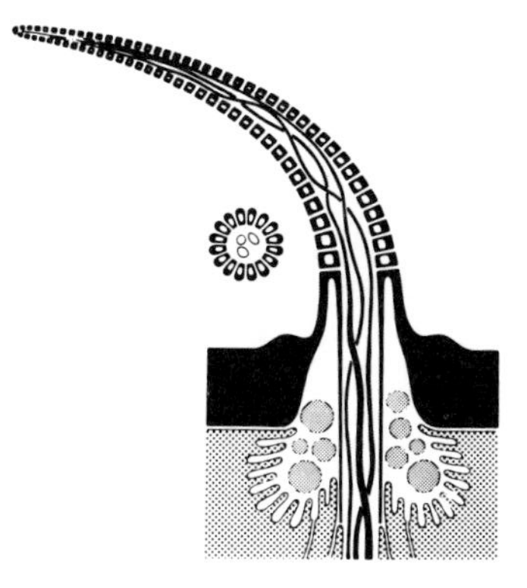

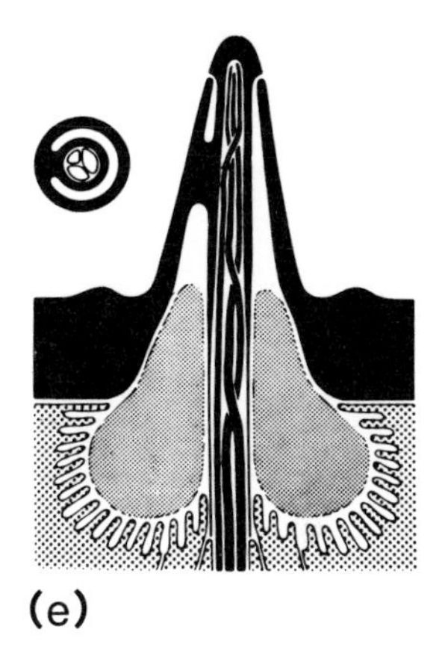

(e)

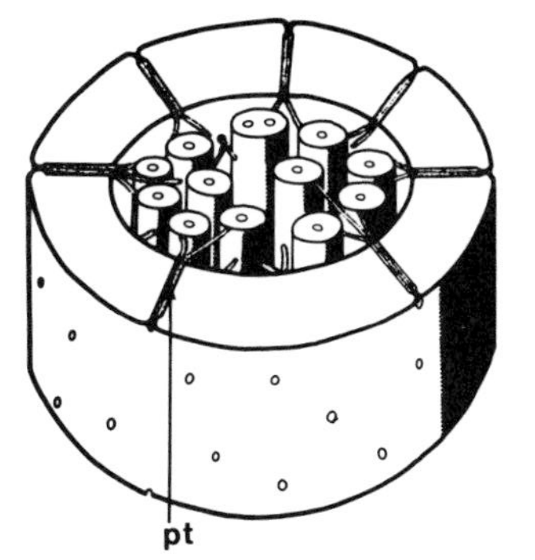

(c)

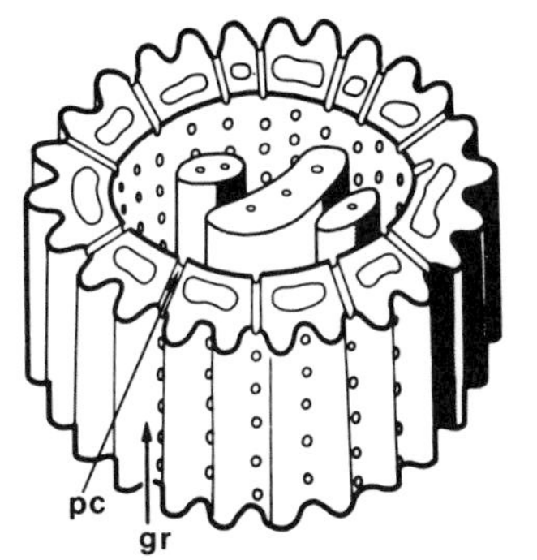

(d)

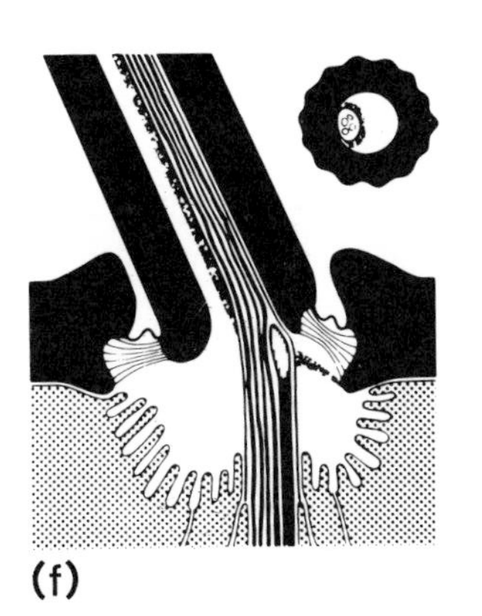

(f)

Fig. 9.9 Antennal sensillar types of *P. americana*. (a) SEM of distal edge of an adult male middle antennal segment. JM, joint membrane between segments; swB, long type B single-wall porous-walled sensillum; sw C, type C single-wall porous-walled sensillum; tp, porous-tip sensillum; wp, porous-walled sensillum (discrimination between the short single-wall (sw A) and double-wall (dw A) porous-walled sensillar types is not possible at this magnification and resolution). Scale: Bar = 10 μm. (b) Lower magnification SEM contrasting two middle antennal segments from an adult male (upper) and an adult female (lower). Scale: Bar = 100 μm. (c–f) Schematic drawings of different sensillar types. (c) single-wall porous-walled sensillum; (d) double-wall porous-walled sensillum; (e) capped non-porous-walled type sensillum; (f) base of porous-tip contact-chemoreceptive sensillum. The longitudinal cross-sections (Schaller, 1978) include an insert of the horizontal section. (The detailed horizontal sections are from Altner *et al.*, 1977 – lower figures c–d.) gr, superficial grooves; pc, pore channels; pt, pore tubules.

---

porous-walled sensilla used by Schaller (1978) to describe *P. americana* sensilla types seems best to fit structural and functional criteria.

The 'single wall' (sw), porous-walled sensilla are innervated by 2–4 neurons, whose dendrites divide into numerous branches as they extend up into the lumen of the shaft (Fig. 9.9c). The wall pores, 150 nm in diameter, contain several pore tubules, which extend into the lumen and often make direct contact with the dendritic membrane. Schaller (1978) discriminated two common subtypes based upon the length of the hair and number of sensory cells. The short 'sw A' are innervated by only two cells, while the longer 'sw B' contain 3–4 sensory cells. The 'sw B' type are highly specific for the female pheromone. They occur only on the adult male antennae, where some are derived from 'sw A' sensilla and some are newly formed during the adult ecdysis (Schaller, 1978). The development of 'sw B' sensilla can be inhibited by topical application of juvenile hormone mimic (Schafer and Sanchez, 1974). A less common third subtype 'sw C' has a smooth distal portion of the hair but, unlike the 'sw A' and 'sw B', has a thicker grooved lower portion.

The 'double wall' (dw), porous-walled sensilla are usually innervated by 3–4 receptor neurons (Fig. 9.9d). Unlike the single wall sensilla the dendrites do not branch. The cuticular sheath is continuous with the inner hair wall separating the dendrites from the basal cavity and the space between the inner and outer walls. Cuticular bridges forming 150–200 nm diameter tubes or channels connect the inner hair lumen with the outside. In cross-sections of the hair, the cuticular bridges have a spoke-wheel appearance. Externally, under high resolution SEM, superficial longitudinal grooves can be seen. Schaller (1978) distinguishes two types of 'dw' sensilla according to length, and characterizes a third subtype which contains a cold receptor cell in addition to olfactory cells.

(b) *Capped hair sensilla*

'Capped' or hygroreceptive hair sensilla are double-walled at the base and possess a distinct cap at the distal end (Yokohari *et al.*, 1975) (Fig. 9.9f). No wall pores are present and the base is not articulated. Each sensillum contains three sense cells with unbranched, but modified, dendrites (Toh, 1977). Electrophysiological work has revealed temperature- and humidity-sensitive receptor cells (Loftus, 1968, 1969; Yokohari and Tateda, 1976).

(c) *Porous-tip sensilla*

Porous-tip or sensilla chaetica, the longest and most rigid hairs on the antenna, are innervated by 2–7 neurons. The base is articulated like mechanoreceptive hair sensilla and the thick wall is superficially carved by deep oblique grooves (Fig. 9.9e). One dendrite forms a tubular body at the basal joint of the hair, like the mechanoreceptive dendrite in the thread hair sensilla. The other dendrites extend unbranched through the hair lumen. They have access to the outside through a single apical pore. The porous-tip sensilla respond to both tactile and chemical stimuli and are representative of a standard insect taste hair. Ruth (1976) has shown, by electrophysiological methods, that sensory dendrites which extend to the terminal pore have specific response spectra and are sensitive to either sugars, salts, fatty acids or alcohols.

(d) *Mechanoreceptive hair sensilla*

Mechanoreceptive hair sensilla are singly innervated articulated hairs located only on the scape and pedicel. They are similar in structure to the mechanoreceptive hairs of the leg, but have been referred to as sensilla chaetica A by Schafer and Sanchez (1973). They differ from the terminal pore sensilla (the sensilla chaetica B of Schafer and Sanchez, 1973) since they are shorter, have smooth rather than grooved walls and lack a porous tip. Over 200 are located on the proximal portion of the scape, and 35 are arranged as three hair plates on the proximal edge of the pedicel. These sensilla undoubtedly respond to flexion of the basal antennal segments (Seelinger, 1979).

(e) *Campaniform sensilla*

Campaniform and marginal sensilla are singly innervated mechanoreceptive organs (Toh, 1977), which closely resemble the campaniform sensilla of the leg (Figs 9.6 and 9.7). Marginal sensilla are distinguishable from the campaniform sensilla by the more convex surface of the cap. Two or three marginal sensilla are located at the distal edges of every other flagellar segment (Campbell, 1972). Isolated campaniform sensilla are located on the proximal edge of each flagellar segment (Schafer and Sanchez, 1973), and approximately 30 'Hick's organs' form a ring around the distal edge of the pedicel. Both types of sensillum probably function to detect movements of the antennal segments.

(f) *Scolopidial organs*

A chordotonal organ is located within both the scape and pedicel. Each contains 12–20 singly innervated mononematic scolopidial units similar to those of the leg (Seelinger, 1979). A Johnston's organ consisting of 200–250 amphinematic scolopidal units is also located within the pedicel. Amphinematic units differ from the mononematic type found in chordotonal and subgenual organs primarily in the fine structure of the cap, which has an attachment process to the cuticle. Each scolopidium is innervated by three sensory cells. One of the scolopidium dendrites extends well past the scolopale region and distally loses the ciliary $9 \times 2 + 0$ arrangement, while the remaining two do not (Seelinger, 1979). The structure of the cockroach Johnston's organ is similar to that in *Chrysopa* (Schmidt, 1969).

9.4.2 Olfaction

Electro-antennogram studies on *P. americana* show that the antennae of both sexes respond to various alcohols, ketones, fatty acids, amines and terpenoids (Nishino and Washio, 1976; Washio and Nishino, 1976). These substances are usually found in fruit, meat, bread and other food sources known to be attractive to cockroaches. They are often referred to as 'general environmental odours', in contrast to more specific pheromonal compounds. Sass (1973) recorded from individual sensilla, and found the receptors for these odours are located in the short-wall pore hair sensilla ('sw A' and 'dw A'). The primary transduction process of olfaction is believed to involve the interaction of the odour molecules with a receptor molecule on the surface of the dendrite (Norris, 1976). The receptor cells can be classified into well-defined physiological types. The response spectrum of each physiological cell type shows that cells will respond to a limited number of odour substances. The specificity of a receptor cell may be largely determined by the specificity of the receptor molecule. All the cells of one type give qualitatively and quantitatively identical responses. Sass (1976) determined the response spectra of eight receptor types, and named the types after the most effective odour substance (Table 9.2). The responses of a pentanol-type receptor cell to several alcohols are shown in Fig. 9.10. The response spectra of some types show considerable overlap in compounds which will stimulate the cells. Since compounds will have quantitatively different effects on each cell type (Fig. 9.10b), a specific compound can be discriminated by the 'across-fibre pattern' it generates in the receptor populations. Theoretically, many odour qualities can be encoded by only a few receptor types using across fibre-comparisons. This system is similar to colour vision, where a certain wavelength is determined by the relative activity levels of spectral cell types. Natural food odours often contain hundreds of different compounds. It is therefore not surprising that fruit or meat odours stimulate several receptor types (Boeckh *et al.*, 1975). The responses elicited in each physiological cell type will be characteristic for a particular fruit. Therefore, a banana could be recognized by an across-fibre comparison of all the olfactory cells.

Each short wall-pore sensillum contains several olfactory cells. One of the receptor cells belongs to a cell type listed in Table 9.2. The response spectra of the additional cells are not yet known in detail, but their specific responses to certain odours suggest that they also represent well-defined receptor types. Standard combinations of different physiological cell types occur within each sensillum. The hexanol type is always accompanied by a cell responding maximally to oil of cinnamon and the hexanoic acid type is always paired with a cold receptor (Sass, 1978).

Altner *et al.* (1977) determined electrophysiologically the receptor cell types within a sensillum, and then used SEM and TEM to determine the morphology of the cell they had recorded from. They found that all the alcohol and terpene cell types occur within single-wall sensilla, 'sw A', whereas fatty acid and amine cell types occur within the double-wall sensilla, 'dw A'. This distribution may be causally related to a specialization in the stimulus-conducting systems. The

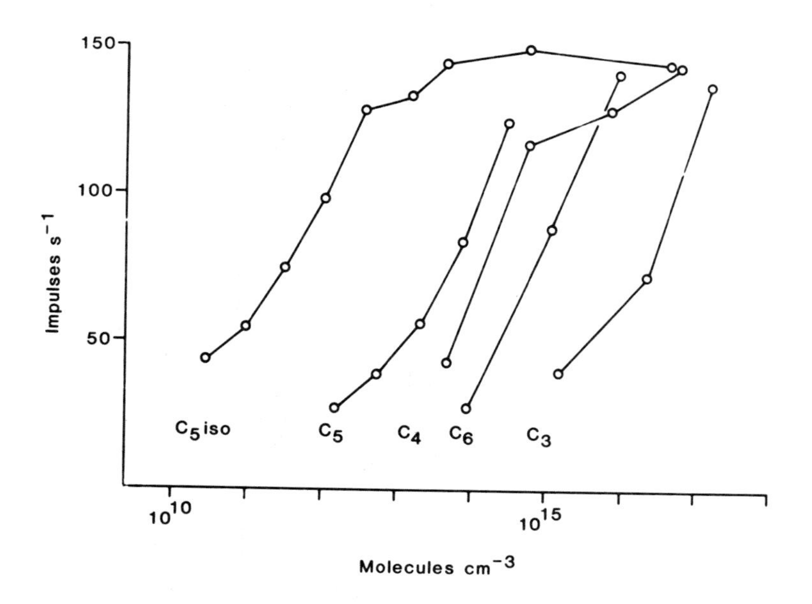

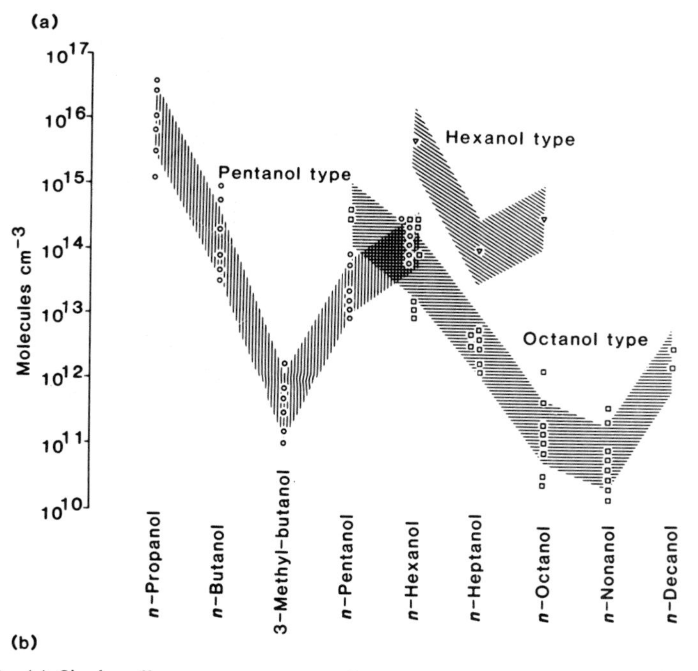

Fig. 9.10 (a) Single cell response spectrum of a 'pentanol type' receptor cell to various alcohol compounds. The number of impulses s^{-1} was recorded during 500–1000 ms after stimulus presentation. C5iso, 3-methyl-butanol; C5, *n*-pentanol; C4, *n*-butanol; C6, *n*-hexanol, C3, *n*-propanol. (From Sass, 1976.) (b) Sensitivity of three alcohol cell types for different alcohols. Each dot represents the concentration needed to elicit 50% of the maximum response frequency in a given cell. (From Sass, 1980.)

tormogen cell of the double-wall sensillum secretes lipid material which penetrates the hair wall at the base, flows distally along the superficial grooves in the hair and enters the pore channels (Altner *et al.*, 1977). No such secretion is found in 'sw A'. Selective diffusion characteristics of the material may limit the transport of odour substances which are not part of the response spectra of the receptor cells.

The axons of the antennal receptors project into the antennal lobe of the deutocerebrum, where they have synaptic contacts on to higher order neurons. Synapses are restricted to approximately 125 ovoid bodies called the glomeruli (Ernst *et al.*, 1977). The ratio of the number of antennal fibres to fibres leaving the deutocerebrum is about 1000 : 1. One consequence of the great reduction in transmission channels is that the deutocerebrum may act as a convergent relay for similar physiological types without additional integrative functions (Weiss, 1974). Summation of the responses of many receptors may be used to enhance the sensitivity of the cell type. Boeckh *et al.* (1975) have recorded from some central neurons which have spatial convergence from different parts of the antennae. Central neurons with integration or convergence of different qualities and modalities have also been found. Since most of the deutocerebral neurons have no processes leaving the deutocerebrum, they are presumably involved in integrative functions. Central neurons have wider response spectra towards single compounds, but often have more specific responses to complex odours than any known receptor cell type. Responses may be of on, off, and on-off patterns depending on quality and concentration of the stimuli (Yamada *et al.*, 1970; Boeckh, 1974; Waldow, 1975, 1977; Selzer, 1979).

9.4.3 Pheromone reception

The male antennae possess many more olfactory sensilla than the female (Schafer and Sanchez, 1973), and give stronger electroantennogram responses to most compounds than the female (Nishino and Washio, 1974). The differential response is most prominent in the adult male's greater electroantennogram response to the female sex pheromone (Boeckh *et al.*, 1970; Nishino *et al.*, 1977; Schafer, 1977a,b). Only the adult male antennae possess long porous-walled sensilla (sw B) (Schaller, 1978), which may total more than 54% of the sensilla on the antennae. Each sw B sensillum contains four receptor cells with very limited response spectra. The female sex pheromone is a mixture of several chemical compounds (Persoons, 1977). One of the 'sw B' cells responds strongly to only the periplanone B component. A second cell responds highly specifically to another pheromonal component (Sass, 1980) (Fig. 9.11). The spectra of the two additional 'sw B' cells have not yet been determined.

A large lenticular mass of neuropil, the 'macroglomerulus' is located within the deutocerebrum of the adult male (Prillinger, in preparation). Neurons from the long 'sw B' sensilla have been traced to the macroglomerulus by filling individual cells with radio-active amino acids locally applied to the surface of removed sensilla. Neurons within the macroglomerulus have been shown to respond to one or both of the purified sex pheromone components (Boeckh *et al.*, in preparation).

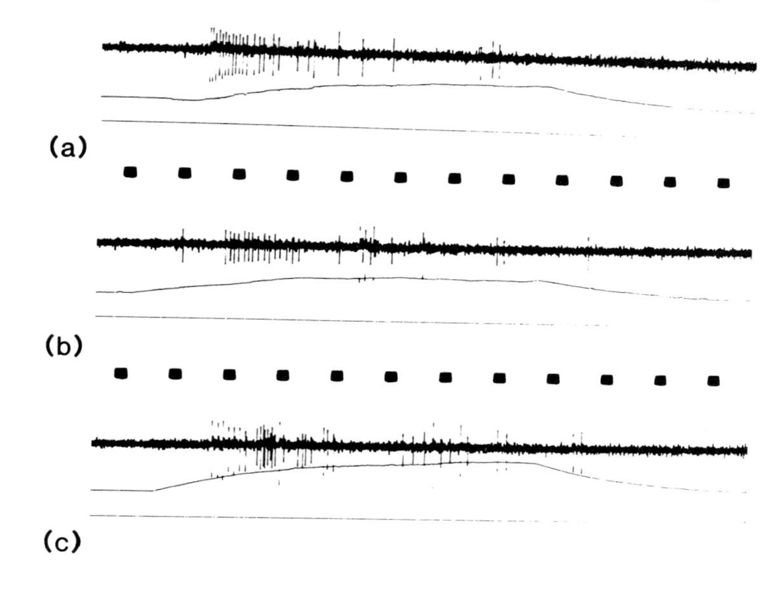

Fig. 9.11 Electrophysiological recordings from a sw B porous-walled sensillum on an adult male antenna. The stimuli are purified natural components of the female sex pheromone. Stimulus duration is indicated by the rise and fall of the lower oscilloscope beam. Time bars = 100 ms. (From Sass, 1980.) (a) Response of receptor with the larger spike to the pheromone component corresponding to periplanone B. (b) Response of receptor with the smaller spike to a second component, probably periplanone A. (c) Response of both receptor cells to a mixture of both components.

In addition to the female sex pheromone, several other plant-derived compounds elicit the behavioural courtship response in males. Some of these compounds like germacrene D (Takahashi *et al.*, 1978) resemble periplanone B, but others are structurally quite different (Nishino *et al.*, 1977). The mode of action of these compounds is unclear but the male electroantennogram will also give a greater sexually specific response to these compounds (Nishino *et al.*, 1977).

9.4.4 Thermoreception

Gunn (1935) showed that *P. americana* has a well-defined temperature preference for 29°C in dry air. Temperatures higher than 33°C cause a drastic increase in desiccation and are avoided. Thermoreceptors could be responsible for this behaviour. Loftus (1966) found a spontaneously active receptor cell in the capped sensillum. The impulse frequency of this cold receptor did not exceed 25 impulses s^{-1} under constant temperature conditions. The cell responds to a temperature drop with peak frequencies up to 250 impulses s^{-1}, but a sudden increase of temperature will result in inhibition. Other stimulus modalities will not effect this receptor. Loftus (1968, 1969) showed that the cold-temperature response is dependent on two components. Under constant temperature conditions, the frequency is negatively correlated to the instantaneous temperature (T). A 0.6 impulse s^{-1} frequency increase would be correlated with a 1° lowering in temperature. The rate of temperature change (dT/dt) will also effect the cell's response. During cooling a −0.01 deg s^{-1} change will result in a 1 impulse s^{-1} increase of frequency.

9.4.5 Hygroreception

The cold receptor of *P. americana* is often associated with two receptor cells responding antagonistically to humidity changes. These units are usually referred to as the 'moist' and 'dry' receptors. They cover the entire range of relative humidity values. However, the response frequency cannot be correlated directly to the humidity of the air stream, when the relative humidity, or the absolute humidity, or the saturation deficit are used as a reference. The temperature will also effect the response frequency (Loftus, 1976). The frequency is considered to be dependent upon both the receptor, and the impulse-generating mechanisms. Yokohari (1978) concluded that the receptor mechanism is temperature-independent while the impulse-generating is temperature-dependent. When they controlled the insect's haemolymph temperature and the temperature of the air stream independently, the thermal component of the response was eliminated. The response frequency was determined by the relative humidity of the air stream. Air velocity would not effect the cell, but mechanical manipulation of the sensillum would effect the moist and dry cells antagonistically. The primary process of stimulus transduction may be non-chemical. The uptake or loss of water by hygroscopic substances in the sensillum is speculated to result in deformation of the dendrites causing excitation of the receptor cell.

9.5 Chemoreceptors on the maxillary palpus

Frings (1946) found the maxillary palpus to be sensitive to sucrose. Contact with a sugar solution elicited feeding behaviour. Low concentrations of acids, added to the sugar water, enhanced the effect. The solution was rejected, however, when higher concentrations of the acid changed the pH to 2 or less. Wieczorek (1978) tested the effect of various sugars on a field of 2500 taste hairs on the lower side of the last segment of the palps. Combined behavioural and biochemical methods gave good evidence for the presence of α-glucosidases as a receptor principle. Sugars eliciting a feeding response were hydrolysed *in vivo*; ineffective sugars were not. About 200 olfactory receptors are present on the upper side of the last segment; they respond to a variety of odours, which are also perceived by antennal sensilla (Altner and Stetter, 1980).

9.6 Conclusions

The present understanding of *P. americana* sensory organs enables us to look within some systems, determine the specificity of response and analyse how information about the stimuli is encoded. Within each modality, the large number of sensory cells belong to a limited number of physiological types which possess distinct, although often overlapping response spectra. The information about the stimulus is encoded within an 'across-fibre pattern' from many sensory cells.

In the future, several areas of research should receive further attention. More information is required on the linking of peripheral inputs to reflexive motor outputs. Very little is known regarding the integration of different sensory modalities and sensory input from different regions of the body. Understanding the functioning of sensory input within higher levels of the nervous system will require further merging of sensory physiology with central co-ordination, muscle output and behavioural studies.

Acknowledgements

The authors greatly appreciate the use of unpublished material contributed by Drs J. Boeckh, H. Sass and L. Prillinger, and the suggestions provided by Drs W. J. Bell and K. Stockhammer.

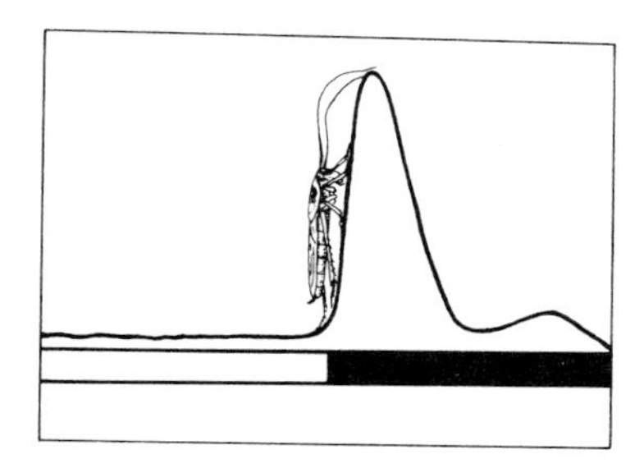

10

Rhythms

Donald J. Sutherland

10.1 Concepts and terminology

Common to all living organisms is the rhythmic nature of their behaviour and the physiological processes supporting that behaviour. Investigations during the last 30 years have confirmed that oscillation is a characteristic of all organisms and levels of organization within an organism, and that many rhythms are inherited and endogenous. Most attention has been given to the circadian rhythms with frequencies of approximately 24 h, which serve to synthronize the organism and its internal processes to the changing environment. There is, however, a wide range of rhythmic frequencies to be detected within any organism (Halberg, 1960; Rapp, 1979), and they have been classified in various ways. On the combined basis of level of organization, regulating mechanisms and frequency, three classes have been proposed (Salanki, 1973): (1) microrhythms of 1000 ms at the cellular level; (2) mesorhythms of 100 ms, minutes, or hours at the organ level; (3) macrorhythms of several hours or more at the organism level. In the latter category would be behavioural rhythms termed *circadian*, *circatidal* (∽12.4 hours), *semilunar* or *circasyzygic* (∽14.7 days), *circalunar* (∽29.4 days) or *circaannual* (∽1 year). An alternate classification (Van der Driessche, 1975) proposes 3 classes of rhythms: (1) high frequency, 1 s–0.5 h; (2) median frequency, 0.5 h–2.5 days; (3) low frequency, more than 2.5 days. Included in class II would be the following rhythms: *circadian* (frequency of about 24 h), *ultradian* (frequency of 0.5–20 h) and *infradian* (frequency of 1–2.5 days).

It is to be assumed that future research will modify such classifications and lead to further precision in terms. Currently the term *circadian* is more precisely used than previously, and it should be reserved for those rhythms which have been shown to be endogenous and persist with a frequency or period (τ) of about 24 h in the absence of an external clue or synchronizer (Zeitgeber). A circadian rhythm can be experimentally entrained to exactly 24 h by a synchronizer such as alternating light and dark periods totalling 24 h. Under entrainment, the rhythm is properly termed an *entrained circadian rhythm*. However, an entrained rhythm, which does not persist in the absence of external synchronizers or has not yet been demonstrated to be circadian, is more correctly termed a *diel* rhythm. Previously, the term *diurnal* has been considered by some writers to be synonymous to *circadian* or to *diel*, but *diurnal* is currently reserved for activity associated with daylight, in contrast to the term *nocturnal* for activity associated with night-time. The study of circadian rhythms, their behavioural and physiological expression and properties, as well as an understanding of their adaptive value, depends on the exact distinction between such terms. Other definitions of terms used in chronobiology are presented elsewhere (Halbert, 1973; Saunders, 1976).

Many insect species have been employed in the study of rhythms of individuals and populations and their governance by the biological clock, pacemaker or oscillator (Bunning, 1973; Brady, 1974; Saunders, 1976). Cockroaches have been particularly suitable for study. Species have included *Periplaneta americana*, *Leucophaea maderae*, *Nauphoeta cinerea*, *Blattella germanica* and *Blaberus* sp. These have not been examined in all aspects, and there is insufficient information on any one species for it to be considered typical of others. It is the objective of the following discussion to bring together information concerning the biological rhythms of *P. americana*. Where details for this species are lacking, or when the contrast between species is known and of value, other species will be discussed. In this way, future areas for study with the American cockroach will become apparent, as well as appropriate areas for comparative studies between species. Thus, the true value of such rhythms to cockroaches in their natural environment can be determined.

10.2 Circadian rhythms of individuals

P. americana is a relatively cosmopolitan species with primary survival requirements of tropical or sub-tropical temperatures, moderate humidity and water. The species is considered omnivorous and negatively phototactic (Guthrie and Tindall, 1968); in situations of natural infestation of high density (Eads *et al.*, 1954), it would appear that the members of the species must interact in many ways. Nevertheless, the species has been reared in various ways in the laboratory, and for experiments on rhythms it has been isolated and subjected to various light : dark regimens. Under such conditions, it is often surprising that rhythms of activity (locomotion), feeding and agonistic behaviour can still be demonstrated.

10.2.1 Locomotion

(a) *Equipment to monitor locomotion*

The size of *P. americana* is sufficient that various methods have been used to monitor its locomotory activity (Miller, 1979b). Since these methods vary in their sensitivity (Brady, 1967b,c) and may even modify behaviour of the individual, a brief discussion on methodology is valuable. There are two major methods: (1) visual observation and trapping as used by Mellanby (1940) in studying the activity of *Blatta orientalis*, and (2) equipment with instruments to actually record movement. This second method can be divided into two groups, based upon the substrate and its mobility. Substrate mobility can be important for its possible feedback to enhance activity and thereby modify natural behaviour patterns. With such equipment, a few individuals may walk almost continuously for more than four hours in mobile running wheels (Lipton and Sutherland, 1970a) whereas on a stationary substrate, an interval of locomotion rarely exceeds two seconds (Hawkins, 1978). However, each method has its advantages and disadvantages. A non-mobile capacitance sensitive circuit, whose space dimension is not much greater than that required for a walking cockroach, often registers grooming activity, particularly notal rubbing against the 'ceiling' of the enclosure. A tilting cage may circumvent this problem, but its dimensions greatly influence the sensitivity of the apparatus, and food and water supply can be a problem. These various methods also can be classified on the basis of data production. Most methods can produce a quantitative measurement of activity per unit time, which allows for comparison with concurrent measurements of physiological parameters as well as the variation between individuals and times. With other methods, namely the running wheel, the activity is recorded as a pen deflection on an event-recorder. As a result, the timing of the event, in this case the onset of a major activity period, is more precise, but the amount of activity is only qualitatively recorded. Most studies on the activity rhythm of cockroaches have been concerned with the phase of the rhythm and the location of physiological clock. For this reason and at times for low cost/unit, the running wheel has been employed more than other devices. Cockroach behavioural differences have also influenced the species chosen for study. *L. maderae* has been most often selected for reasons such as its precise and predictable locomotory activity (Lettau *et al.*, 1977), and the lack of rapid epidermal growth after window insertion over the protocerebrum (Nishiitsutsuji-Uwo and Pittendrigh, 1968). Some studies have included both *L. maderae* and *P. americana* (Roberts, 1960, 1962, 1965a,b, 1966, 1974; Ball, 1971; Roberts *et al.*, 1971) and have indicated a basic similarity in activity rhythms and their control.

(b) *Entrainment by light: dark cycles*

As a nocturnal animal, *P. americana* is active during the dark portion (scotophase) of a 24 h day-night period (L + D = T = 24) and relatively inactive during the light portion (photophase). This behaviour is best displayed by the adult male (Fig. 10.1a), for which the following is most evident. Under conditions of alternating 12 h light and dark (LD 12:12) or 16 h of light and 8 h of dark (LD 16:8), this activity occurs during the first 2–3 h of the scotophase and is thus entrained as rhythm with a period of 24 h (Harker, 1955; Roberts, 1962; Brady, 1967b,c, 1968). There can be considerable variation between individuals (Brady, 1968), with a

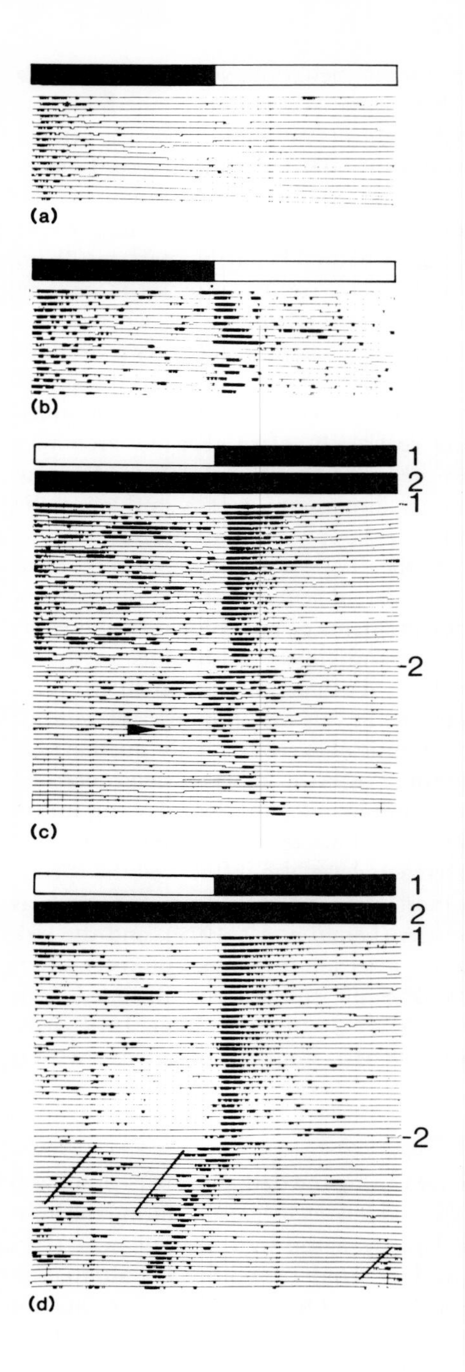
(a)
(b)
1
2
-1
-2
(c)
1
2
-1
-2
(d)

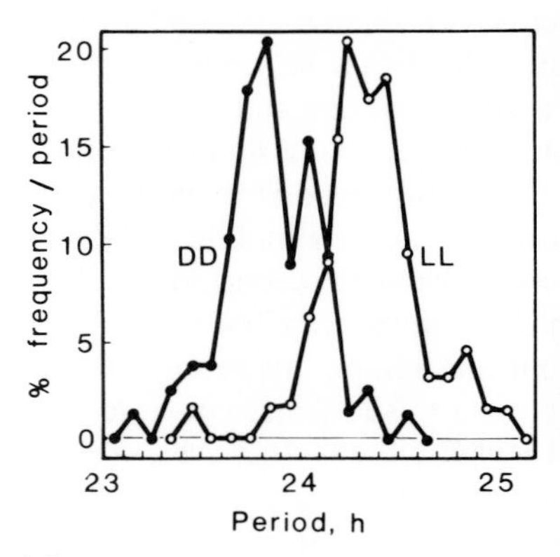
% frequency / period
DD
LL
Period, h
(e)

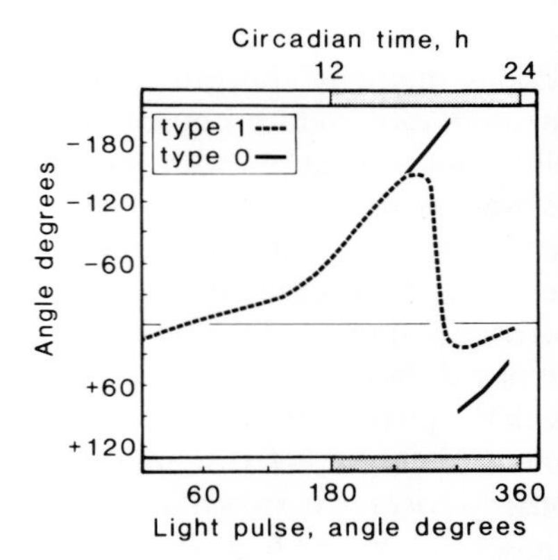
Circadian time, h
type 1
type 0
Angle degrees
Light pulse, angle degrees
(f)

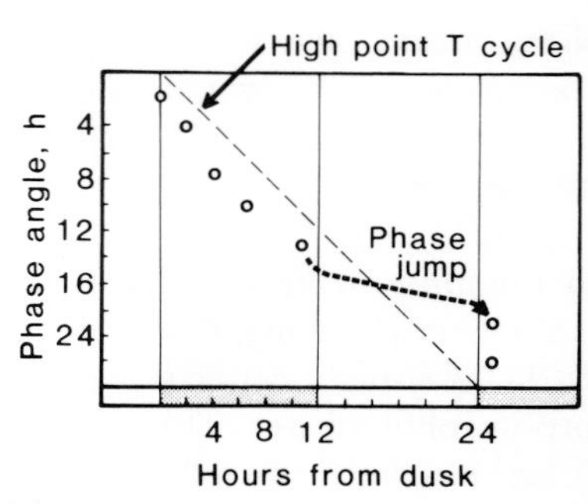
High point T cycle
Phase jump
Phase angle, h
Hours from dusk
(g)

Fig. 10.1 (a) Actogram of an adult male *P. americana* in LD 12 : 12 with activity occurring in early scotophase. (After Lipton and Sutherland, 1970a.) (b) Actogram of an adult male in LD 12 : 12 with secondary activity also in early photophase. (After Lipton and Sutherland, 1970a.) (c) Actogram of an adult male in successive regimens of LD 12 : 12 and DD, with a circadian rhythm (τ = 24.3) developing at dark point after 12 days in DD. (After Lipton, 1969.) (d) Actogram of an adult male in successive regimens of LD 12 : 12 and DD, with a biphasic circadian rhythm (τ = 23.8) developing immediately in DD. (After, Lipton 1969.) (e) Relative frequency distribution of τ for *L. maderae* in DD and LL. (After Lohman, 1967a.) (f) Phase response curves of the activity rhythm of *L. maderae* after high-intensity light pulses. (After Wiedenmann, 1977b.) (g) The onset of activity of *L. maderae* in LD 12 : 12 as a function of the phase angle of the two synchronizers, the high point of the temperature cycle and the LD cycle. (After Pittendrigh, 1960.)

---

range of at least 1 h in the exact time of commencement of activity at the onset of the scotophase; the activity onset of an individual can be as accurate as ± 5 min from day to day. The percentage of individuals displaying the entrainment of the rhythm depends on the researcher, equipment, and possible cockroach strain. Percentage incidence has been reported to be 85–100% (Brady, 1967c) and 95.4% (n = 135) (Lipton and Sutherland, 1970a). Others have also commented on the variation (Pittendrigh, 1960a) or inferred it by discarding those cockroaches which failed to demonstrate clear rhythmicity (Nishiitsutsuji-Uwo *et al.*, 1967). Failure to exhibit an entrained activity rhythm does not mean that an endogenous mechanism does not exist. However, experiments to examine the association between physiological aspects and entrained activity rhythm should confirm the existence of the latter.

Most entrained activity rhythms for *P. americana* in LD 16 : 8 or 12 : 12 have been monophasic. However, a secondary phase of activity, with the onset generally coinciding with the beginning of the photophase (Fig. 10.1b), has been exhibited in almost one third (n = 126) of the males exhibiting a primary activity in the scotophase (Lipton and Sutherland, 1970a). This bimodality has been reported for male *L. maderae* under certain conditions (Roberts, 1960, 1962; Lettau *et al.*, 1977; Wiedenmann, 1977a) and also for male *Blaberus discoidalis* (Shepard and Keeley, 1972), and male *B. germanica* (Sommer, 1975). Although the secondary phase may represent negative phototaxis, the phase generally continues to be expressed under conditions of constant darkness in *P. americana*.

To determine the relative importance of the light and dark periods and also possible mechanisms for photoperiodic time measurements in plants and animals, unnatural period lengths (T = > < 24 h) and resonance cycles (e.g. L = 6 h, D = 18–54 h) have been used (Beck, 1976; Elliot, 1976; Pittendrigh and Daan, 1976b; Saunders, 1976). The periods to which *P. americana* will entrain has not been fully investigated. Early studies (Cloudsley-Thompson, 1953) examined periods of 96, 48 and 18 h for as long as 2 weeks, with the 'most distinct' rhythm resulting from LD 3 : 15, the peak of activity appearing in the photophase. This aspect should be further examined with running wheel cages where onset and span of activity can be better examined. For *L. maderae*, it has been suggested that the range of entrainment is not much larger than the range of the free-running period in DD or LL (Lohmann, 1967a). This species has been entrained to a 25-h cycle (LD 4 : 21) (Page *et al.*, 1977). Harker (1956) reported that in *P. americana* entrainment to a 24-h rhythm can be accomplished, provided that at least 2 h of D or 2 h of L occur during a 24-h cycle. With different equipment, Roberts (1962) entrained *L. maderae* to 24-h cycles with L of 1, 7, 18 and 23 h; however, 8 min of light in every 24 h failed to entrain. Although there may be subtle differences between species and equipment, the general results indicate that the transition of photophase : scotophase is the primary phase setter of activity. *L. maderae* has been entrained by LD cycles whose periods are whole equal submultiples of 24 h (Roberts, 1962); entrainment to LD 2:2 and 4:4 occurred, but not to LD 1:1. Among 20 male *P. americana* examined in LD 2:2,25% entrained to 24 h but the remainder exhibited a circadian period (Sutherland, unpublished; *see* Section 10.2.2). With unequal submultiples of LD 2:6, *P. americana* not only became arrhythmic but also incapable of further entrainment to 24 h and exhibited major pathological changes in the 'gut' (Harker, 1961).

When exposed to significant changes of LD cycles and the onset of scotophase, generally *P. americana* responds with shifts of its major activity to coincide with the onset of the new scotophase. This shift may require several 'transient' cycles before re-entrainment is accomplished and may, as in *B. discoidalis* (Wobus, 1966b; Aschoff *et al.*, 1975), depend on the light intensity. For example, with a complete reversal of LD conditions, the activity rhythm will continue for 2–4 days, but is gradually replaced by a new rhythm, the reversal of the old (Harker, 1956). Less severe changes in entrainment cycles also have been examined (Harker, 1960a). The number of transient cycles necessary to reset the activity phase appeared to depend on the degree to which the onset of the scotophase had been advanced or delayed. These, and other studies relating to the role of the suboesophageal ganglion (Harker, 1960b; Brady, 1967c), have led to a consideration that a phase shift of less than 5 h should be completed within a single cycle, while one of more than 5 h would require at least two cycles for completion. Brady (1967c) did not find the re-entrainment of 12 individuals to an advance of the scotophase to fit this hypothesis; all showed gradual re-entrainment, which in some cases was characterized by shift acceleration followed by deceleration as the activity onset approached the new LD transition time. The shifting of activity phases has been studied in greater detail under free-running conditions of constant darkness, where reset may require as many as 10 cycles or 10 days (Roberts, 1962). Since stress of various forms can modify activity levels (Brady, 1967b), it would seem wise in any experimentation involving change of the photocycle to allow for at least 1–2 weeks for complete entrainment to be achieved.

(c) *Endogenous expression in constant dark (DD) or light (LL)*

In the absence of temporal clues provided by alternating LD cycles, *P. americana* exhibits activity, whose natural period (τ) is not exactly 24 h, but either slightly less or more than this value (Fig. 10.1c) i.e. truly circadian. This has been demonstrated both in constant darkness (DD) and constant light (LL). Some investigators (Wiedenmann, 1977a) have employed continuous red light (RR) which presumably appears as DD to the cockroach but allows the experimenter to check and service equipment. The term 'free-running' has been used to describe the constant conditions (no temporal clues given) as well as the expression of the rhythm and its period (τ). The free-running period (the interval between the onsets of successive circadian activity periods) is especially accessible from running-wheel activity records, and is generally derived from the slope of an eye-fitted line through successive onsets as derived by several people. The reliability of this method has been reported as within 1.73 min (Pittendrigh and Caldarola, 1973).

The demonstration of free-running rhythms in *P. americana* apparently depends on the equipment, the culture and the conditions of LL or DD employed. With the balance-box (Harker, 1956), the species when placed in DD will exhibit for a few days a carryover entrained rhythm from previous LD conditions, but thereafter no rhythm is evident. Similar results occur in LL, although activity is greater than that in the first few days in DD. Under some conditions, the rhythm in DD may be initially absent, but subsequently appear (Harker, 1958a). With running-wheel cages, however, the circadian activity rhythms rapidly become evident in DD (Lipton and Sutherland, 1970a) and LL (Roberts, 1960, 1966) and last for as long as three months.

Not every individual exhibits a circadian activity rhythm. Of 75 males examined in DD, 75% demonstrated a rhythm or a pattern (less exact rhythm) whose periods (τ) were evident to the researchers (Lipton and Sutherland, 1970a). The remainder, who had exhibited entrainment to LD, were classified as arrhythmic in DD. The period of the rhythm or the pattern in DD depends on the individual. Of 56 examined, τ in 34% was < 24 h, in 19% was > 24 h. In 29%, observers derived an τ value very close to 24 h. However, 18% exhibited spontaneous changes in τ, and, therefore, the circadian rhythm is not a fixed value but is labile, as in *L. maderae* and *Byrsotria fumigata* (Roberts, 1960). In some *P. americana* (Lipton and Sutherland, 1970a) and *L. maderae* (Roberts, 1960; Lohmann, 1967c; Wiedenmann, 1977) a secondary activity component appears (Fig. 10.1d) which in many cases is a continuation of activity associated with early photophase in entrainment. Under free-running conditions (LL, DD or RR), this secondary component shifts closer to the main component of activity, with both having the same τ (constant angle). This second component has been induced in *L. maderae* in RR by a variety of disturbances (Wiedenmann, 1977a). The splitting of circadian rhythms into two or more components is evidence of multiple oscillators jointly controlling circadian rhythms (Pittendrigh and Daan, 1976b; Boulos and Terman, 1979).

The values obtained for τ under DD and LL are valuable to judge a species conformance to 'Aschoff's Rule' (Pittendrigh, 1960b). According to this rule, in nocturnal insects, τ lengthens with transfer from DD to LL, or with an increase in light intensity. But in diurnal insects, τ shortens. Not all organisms conform to this rule (Daan and Pittendrigh, 1976); it may not be applicable to tropical animals (Cloudsley-Thompson, 1970) and may not occur in some instances of domestication (Cain and Wilson, 1972). Most cockroach species examined seem to

Table 10.1 The circadian period (h) of various species in constant darkness (DD) and constant light (LL).

Species	*DD*	*LL*	*Reference*
Periplaneta americana	23.8	24.3–24.5	Roberts (1960, 1966)
Byrsotria fumigata	23.9–24.4	24.5–25.5	Roberts (1960)
Blaberus craniifer	24.0–24.7	25.2–25.6	Wobus (1966a); Lohmann (1967b)
Eurycotis floridana	22.5–24.0*		Dreisig (1976)
Blattella germanica	22.8–23.7	Could not be determined	Dreisig and Nielsen (1971)
Leucophaea maderae	23.2–24.7	23.5–25.1	Lohmann (1967a,b,c)
	23.3–24.0	24.0–24.8	Roberts (1960)

* DD inhibitory in this species; conditions ca 0.1 lx with τ_{max} of 27 h.

follow this rule, especially in regard to the transfer from DD to LL (Table 10.1). However, in the case of *P. americana* few values for τ have been reported (one male yielding 2 values at 23.8 h in DD, one male yielding a value of 24.5 in LL, see Roberts, 1960, Figs 4 and 6; 3 male values for LL of 24.3, 24.4 and 24.5 calculated from Roberts, 1966, Fig. 2). Although *L. maderae* (Fig. 10.1e) readily demonstrates longer periods in LL (Lohmann, 1967a,b), the sparse data for *P. americana* prevents a conclusion about this species, especially since the period is labile. Even in *L. maderae*, with repeated changes between DD and LL, the period lengths do not return to their former values under identical conditions, possibly because of 'after-effects' or ageing (Lohmann, 1967b; Caldarola as cited in Pittendrigh and Daan, 1976a). In *B. discoidalis*, a positive or negative $\Delta\tau$ was shown to depend on the timing of brief intervals of entrainment (Harker, 1964).

There is little evidence that in cockroaches the degree of period lengthening is correlated with light intensity (LL). The activity of *Eurycotis floridana* is inhibited and no free-running period is evident unless a light of low intensity (0.1 lx) is provided (Dreisig, 1976); in *B. germanica*, the period in LL could not be suitably determined even in RR (Dreisig and Nielsen, 1971). The range of light intensities which influence period length in *L. maderae* is very small (< 0.5–5 lx) (Lohmann, 1967a); for *B. craniifer* this range is greater, 1–10 lx, with disintegration of the circadian rhythm in 100 or 300 lx (Wobus, 1966a,b; Lohmann, 1967a). For this latter species, it has been reported that the absolute value of τ is dependent on the light intensity in LL (Wobus, 1966b). Further detailed comparative studies are necessary in order to clarify species differences of the pacemaker and the functional meaning of the differences, as has been undertaken for nocturnal rodents (Pittendrigh and Daan, 1976a).

The so-called 'Aschoff's Rule' has been extended and termed the 'circadian rule' to cover 2 more parameters of activity: (1) the ratio of activity-time to rest-time and (2) the total activity per circadian cycle. For diurnal animals these parameters increase with light intensity in LL, but decrease in nocturnal species. This aspect has not been studied in *P. americana*. *L. maderae* and *B. fumigata* do not follow this rule (Roberts, 1959; Lohmann, 1967c), but studies with *B. craniifer* have indicated that some individuals almost conform to the rule (Wobus, 1966a; Aschoff *et al.*, 1975). Possibly, this aspect of the endogenous rhythm has been affected by laboratory domestication of cockroaches.

While τ of the circadian rhythm is susceptible to spontaneous changes in some individuals and influenced by previous experiences, evidence for a general homeostasis for this frequency has been presented for *L. maderae* (Pittendrigh and Caldarola, 1973). The conservation of τ within a narrow range is a functional necessity of the circadian system which is poised to adjust to natural daily periodicity. Although experiments with *P. americana* on this aspect are lacking, it seems plausible to extend the proposition of homeostasis of τ to this species.

Under natural conditions, only a minor phase adjustment is necessary to insure that the onset of the activity component continues to coincide with early scotophase. However, under experimental conditions it is possible to induce great changes or shifts in this phase (termed phase shifts) by a single pulse of light or temperature change and, thereby, to learn the limits of the circadian oscillators' adjustability. Again, *P. americana* has not been examined in this important respect, and it is necessary to cite information about other species for the sake of completeness. *L. maderae* in DD displays a circadian activity rhythm, whose activity phase can be either advanced or delayed by a single light pulse, depending on the timing, intensity and interval of the pulse. By examining the response at all times during the circadian cycle, phase response curves (PRC) can be constructed (Fig. 10.1f). Because of inter-individual variation of the cockroaches, the temporal aspect of the PRC can be standardized for clarity in one of two ways. The circadian cycle (approximately 24 h) can be divided into 'subjective night' (circadian time 12–24 h), whose beginning coincides with the onset of activity, and 'subjective day' (circadian time 0–12 h). The cycle can also be considered as 360°, with the onset of activity assigned to 180°. The pulse timing and its interval, as well as the phase advances or delays, can be specified in terms of circadian h or angle degrees.

In early studies with *L. maderae*, only small phase shifts were observed with light pulses of 2000 lx and 12 h duration (Roberts, 1962). More recently, however, *L. maderae* has been reported (Wiedenmann, 1977b) to demonstrate the two types of PRC's shown by other organisms: type 1 or weak PRC and type 0 or strong PRC (Fig. 10.1f). Type 1, evoked by pulses of 50 000 lx and 90 or 120° (ca. 6 or 8 h), is unsymmetric with the maximal delays (ca. 140° or 9 h) resulting from pulses whose midpoint occurred during the first half of 'subjective night'; small phase advances occurred when timing was during the last half of 'subjective night' and early 'subjective day'. Type 0 PRC resulted from pulses of 80 000 lx and 180° (about 12 h), with maximal delays of 205° (14 h) and maximal advances of 120° (8 h). The phase jump (the time during the cycle when the response changed from one of delay to one of advance) occurred at approximately 300° (around subjective midnight). Pulses given at this time did not cause arrythmicity in *L. maderae* as found in the rhythm of *Drosophila* eclosion, and it has been concluded that either *L. maderae* does not possess a 'point of singularity' or that (1) it is a very unstable state or (2) conditions for inducing arrhythmia must be extremely precise (Wiedenmann, 1977c). *N. cinerea* also exhibits both Type 1 and Type 0 PRCs, the latter having been produced by light pulses of one-half the intensity required for *L. maderae* (Saunders and Thompson, 1977). The transition between Type 1 and Type 0 for *N. cinerea* lies somewhere between 6 and 9 h of 240 μW cm^{-2} (Saunders, unpublished). The differences in intensity requirements as well as the nature and point of phase jump in PRCs for *L. maderae* and *N. cinerea* are strong evidence for differences in circadian mechanisms in cockroaches and the necessity to study *P. americana*.

(d) *Effect of temperature*

Depending on its level and constancy, temperature can have a variable effect on the circadian rhythm. As a daily cycle, temperature change can act as a synchronizer of cockroach activity in the same way as does an LD cycle. In early studies with *P. americana*, activity in DD seemed independent of regularly fluctuating temperatures (Cloudsley-Thompson, 1953). Such cycles were apparently not examined beyond five days, and it remained for Roberts (1962) to show that both *P. americana* and *L. maderae* in DD could be entrained by a 24-h sinusoidal temperature cycle (max 27°C, min 22°C). Such a cycle does not provide the precision of an LD cycle, and the onsets of activity are variable. Even with an activity monitor, such as the running wheel, an extended cycle series is necessary to document entrainment by temperature.

Although entrainable by a temperature cycle, the circadian period (τ) is virtually insensitive to constant temperatures within the normal biological range. The lower point of this range is about 17 or 18°C for *P. americana* (Bunning, 1958), below which point the cockroach does not exhibit outward evidence of circadian activity. However, between 20 and 30°C, τ only varies slightly (between 24 and 27 h for *P. americana*, Bunning, 1958), whereas most metabolic processes would more than double (Q_{10} = 2). The Q_{10} for τ in *L. maderae* has been estimated as 1.03 for one experimental animal (Roberts, 1960) and as a mean of 0.97 for nine individuals (Pittendrigh and Caldarola, 1973). In *B. fumigata* it was greater than 1.0 (Caldarola and Pittendrigh, 1974). Because of inter-individual differences, at least in *L. maderae*, the temperature independence of τ is not simple; instead τ is a non-monotonic function of temperature, where τ is very large at 17°C, falls to a minimum between 20 and 25°C, above which it rises again to a plateau as 30°C is reached. However, such changes in τ are small, and the temperature compensation of τ is evidence of the general homeostatic protection of τ against major change and loss of its functional significance as a time measurement system (Pittendrigh and Caldarola, 1973).

Although τ is temperature-compensated, a brief interval of temperature change can cause a phase-shift of the onset of activity, much as does a pulse of high-intensity light. When *P. americana* in constant dim light was exposed to temperature pulses (+3°C and +5°C) for 5.5–8 h, maximum phase delays resulted from cold treatments initiated towards the end of subjective day and in the first half of subjective night (Bunning, 1973). A similar response has been obtained from *L. maderae* with 12-h pulses of cold (Roberts, 1962). Maximum and minimum phase resets occurred when the end of the temperature pulse coincided with circadian hours 14 and 18 respectively (during subjective night); some resets could not be identified as definite advances or delays. This aspect as well as differences in temperatures and their intervals prevents an exact response comparison between *P. americana* and *L. maderae*. However, for both species chilling for an extended interval (48, 51 h) essentially stops the clock and the phase (onset of activity) is reset by the eventual rise in temperature.

Since light and temperature can both act as synchronizers of the activity rhythm, it has been of interest to consider which of these synchronizers is the strongest (Roberts, 1959; Pittendrigh, 1960b). In the natural environment, the highest temperature generally coincides with late photophase and lowest temperature with late scotophase. Consequently, the phase angle between the two synchronizers is very small. As the phase angle is increased experimentally up to about 12 h, the onset of activity in the scotophase is delayed in order to coincide with the high point of the temperature cycle (Fig. 10.1g). However, beyond this point, where peak temperature falls within the photophase, the onset of activity is delayed until the early portion of the next scotophase. This so-called phase jump indicates that the LD cycle is a stronger synchronizer than the temperature cycle.

(e) *Influence of sex*

Most studies on circadian activity rhythms of *P. americana* and other species have used the adult male since the activity of the female is at least sometimes erratic and a researcher's intent may have been to exclude the possible influence of the reproductive cycle. Sometimes, the investigator neglects to specify sex, and the reader may incorrectly presume that males were used exclusively in experiments.

P. americana females were first shown to display an entrained activity rhythm under LD 12:12, and the production of oöthecae by some females seemed to be associated with an outburst of activity (Cloudsley-Thompson, 1953). However, many periods of increased activity occurred without oöthecal production, and the influence of discontinuous water supply on the events studied cannot be discounted. In a capacitance activity monitor with water and food *ad libitum*, an entrained LD 12:12 activity rhythm was not detected in 7 females, but 4 did appear most active when carrying an oötheca and relatively quiescent after its deposition (Lipton and Sutherland, 1970a). In the running wheel, 9 out of 10 females did exhibit an entrained rhythm, but in only 3 was it similar to that of males. Of the remainder, some were light-active rather than dark-active or biphasic. Virgin adult females also exhibited these types of entrained rhythms. In no case was there a definitive relationship between activity and oötheca production.

These studies have been extended by Faber (1970), wherein some non-virgin and virgin females in running wheels displayed an entrained rhythm (Table 10.2), but none exhibited a circadian activity rhythm in DD. Among 20 mated females and 8 virgin females (not included in Table 10.2) placed in LD 12:12 and observed every

Table 10.2 The presence of entrained and circadian activity rhythms in *P. americana* as influenced by sex and age (after Faber, 1970).

Experi-ment	*Sex*	*Stage**	*LD*†	*n*	*Number of activities judged as*			*n examined in DD*	*% Circadian*
					Rhythm	*Pattern*	*Random*‡		
1	M	A	16:8	20	12	4	4	10	80
2	F	A	12:12	18	4	5	9	9	0
3	F	AV	12:12	11	4	4	3	11	0
4	M	Y	16:8	25	19	0	6	17	0
5	F	Y	16:8I	17	0	6	11	—	—
6	F	Y	16:8	34	18	0	16	10	0

* A adult, Y nymph, V virgin.
† LD, hours of Light : Dark; I, isolated prior to experimentation.
‡ Random, including those inactive.

12 h for extrusion and deposition of oöthecae, no association between these events and activity could be discerned. Since under such experimental conditions the light cycle might not be in synchrony with the reproductive cycle, 9 females with the most regular cycle of oöthecal deposition (3 to 5 days) were placed in DD for extended periods, but it was impossible to correlate the numbers of oöthecae deposited to the numbers of distinct bursts of activity. The amount of activity of female *L. maderae* varies greatly during the reproductive cycle of about 85 days (Leuthold, 1966). In constant darkness, two types of activity could be discerned with a ten-point actograph, (1) a basic activity which was *not* circadian, and (2) a circadian activity which was superimposed on the basic activity but which was most evident during the 8-day interval preceding parturition. Severance of the abdominal nerve cord (immediately anterior to the second abdominal ganglion) prevented transmission of pregnancy information and the circadian activity ceased. The activity of adult *B. germanica* in LD 12:12 is also influenced by the reproductive cycle, with increased activity 2 days before the egg-capsule formation masking the entrained activity displayed at other times (Sommer, 1975). The similarities between the 3 species studied suggest that in female *P. americana* the short reproductive cycle dominates metabolism at the expense of a circadian expression of activity at least as monitored by the running wheel.

(f) *Influence of age*

Few studies have examined how early or late in the life of cockroaches, an activity rhythm, either entrained or circadian, is detectable. In some vertebrates, a rhythm can be detected during embryonic development or at birth (Harker, 1958a, 1964; Bunning, 1973). In *P. americana* possibly the genesis of a rhythm appears during embryonic development, the synchronizer being light which not only can penetrate the oötheca but, if constant, can affect embryonic growth and eventual nymphal development (Sandler and Solomon, 1976; Solomon *et al.*, 1977). Possibly the tactile stimulation between embryos promotes the synchrony of hatch from the oötheca (Provine, 1976a,b) and the synchronizer of the activity rhythm is the actual emergence from the egg (see also Chapter 15).

In the first- and second-instar nymphs, the strong aggregation behaviour (Bell *et al.*, 1972; Burk and Bell, 1973; Roth and Cohen, 1973) may so limit locomotory activity that its rhythm would be imperceptible. The aggregation pheromone is stronger than negative phototaxis at such stages (Burk and Bell, 1973), and behavioural rhythms other than activity such as antennal movement, might be much more evident, especially at LD cycles with low illumination. Possibly such rhythms will fade or not be expressed by isolated individuals. This is suggested by studies with *B. germanica* (Sommer, 1975), where with sensitive radio-actography isolated second- and third-instar nymphs did not consistently exhibit an entrained activity rhythm in LD 12:12. Older nymphs (fifth and sixth instar) did exhibit an entrained rhythm when isolated. However, its true circadian character was not examined in DD.

The activity of last instar *P. americana* nymphs can be entrained by an LD cycle (Cloudsley-Thompson, 1953; Faber, 1970). This is true for the majority of males and females (Table 10.2), but its demonstration depends on pre-test experience of the nymphs. In Experiment 5 (Table 10.2) the individuals were isolated for up to 2 weeks prior to monitoring in wheel cages and did not exhibit an entrained rhythm; those not isolated (Experiment 6) but placed directly into the wheel cages did so. However, the 10 female and 17 male nymphs, which did show entrainment, failed to exhibit a circadian activity rhythm in DD. Fourteen males moulted to adult in DD, after which a circadian rhythm was not evident within ten days. Subsequently, after entrainment for 3–5 days in LD 16:8, 6 did exhibit endogenous activity rhythms when returned to DD. It was concluded that newly moulted males do not spontaneously exhibit the endogenous rhythm (Faber, 1970). Possibly the wheel cages and/or extended separation from siblings have contributed to these results. In addition, the imago may require several days to mature physiologically (Wharton *et al.*, 1965a). Harker (1960c) has noted the low activity just after moulting and considered it an advantage in cuticle hardening. However, the act of moulting in DD or in LL may uncouple activity from the circadian pacemaker, and a brief period of entrainment in an LD cycle or even a pulse of light may be necessary before circadian activity can be demonstrated.

The rearing of *P. americana* from egg hatch in LL or DD is reported to yield arrhythmic individuals (Van Cassel, 1968; Rivault, 1976), but the effect of brief entrainment on the subsequent expression of an endogenous activity rhythm in LL or DD has apparently not been examined. Absence of an endogenous expression does not preclude the presence of a circadian pacemaker. Further cautious experiments are needed to characterize its presence or absence in the immature.

The effect of age on the circadian activity of mature cockroaches has not been examined in detail. The interval of some experiments suggests that the rhythm in DD will continue for up to 3 months in some male *L. maderae* (Roberts, 1960). Harker (1958a) reported circadian rhythms in individual *P. americana* more than a year in the adult stage. In some *P. americana* males, the levels of entrained activity in running wheels markedly declined within 2 months (Faber, 1970) so that an entrained rhythm was not evident. This may have been due more to isolation than to ageing, although the latter would be expected to reduce the amplitude of the rhythm. The period (τ) may decrease with age as shown for nocturnal rodents (Pittendright and Daan, 1976a). A search for such an effect in *P. americana* should separate effects of previous experience, such as light cycle and isolation, from those of ageing, as done recently with *L. maderae* (Page and Block, 1980).

10.2.2 Feeding

Since one purpose of locomotory activity is to locate food and water, it is not surprising to find that *P. americana*, at least adult males, exhibit a daily feeding rhythm with properties similar to the activity rhythm (Lipton and Sutherland, 1970b). As detected by the movement of a food-bearing pendulum, an entrained feeding rhythm (LD 12:12) was exhibited by most males, the onset of feeding occurring within the first hour of scotophase and sometimes intermittently thereafter. Sometimes the latter were concentrated into a definite 'second-meal' rhythm. Some individuals ate also in the photophase, but this did not occur when the cockroaches were provided a small darkened retreat (harbourage), which they occupied during the photophase. The onset of the feeding in the scotophase was advanced about 15 min when a retreat was provided. When the LD cycle was advanced 3 h, the feeding rhythm phase-shifted either immediately or gradually, to coincide with the new onset of the scotophase.

Among the 37 adult male *P. americana* examined in DD, 26 displayed endogenous feeding rhythms or patterns (less exact rhythms). The feeding τ varied between individuals and was labile in some individuals. Feeding and locomotory activity could not be monitored concurrently and the relationship of their period (τ) or event order could not be determined. However, restriction of food supply to the photophase for 3 weeks does not prevent a normal activity rhythm (Harker, 1956). In the total absence of food, the species may seem to become inactive (Olomon *et al.*, 1976). However, in running wheels, the level of activity may increase or decrease with food deprivation depending on the individual, but entrained activity rhythms have continued for at least 19 days (Lipton and Sutherland, 1970b). In DD, circadian patterns of activity were evident in the absence of food, and there were no indications that activity became completely random in a constant search for food.

Sutherland (unpublished) investigated various aspects of the feeding rhythm and its relationship to activity and other physiological factors using running wheels, trophometers, and retreats as described previously (Lipton and Sutherland, 1970b). Males were tested either for activity or feeding rhythms under conditions of frequency demultiplication, i.e. LD whose intervals are submultiples of 24 h. Of 20 males examined in LD 2:2, only 5 exhibited entrained locomotory activity; the remainder were not entrained but free-ran with their activity cycle shifting to the left or right. Of another 66 males assayed for a feeding rhythm in LD 2:2, 35 (53%) were entrained, the others exhibiting a shifting of the feeding phases. The differences in percentage entrainment of activity and feeding may suggest that these behavioural events are governed by two different oscillators. However, the trophograms show types of response not dissimilar to those for activity. Male 1 (Fig. 10.2a) exhibited immediate re-entrainment to a reversal of LD 12 : 12, but was not entrained by LD 2 : 2 . In contrast, male 2 (Fig. 10.2b) only gradually phase-shifted feeding with a reversal of LD 12 : 12, but when LD 2 : 2 was initiated four hours in advance of the developing shift, entrainment of LD 2 : 2 was immediately established at the edge of the shifting phase. In male 3 (Fig. 10.2c) entrainment to LD 2 : 2, with two feeding phases, became circadian in DD, but on return to LD 2 : 2, was entrained again and biphasic for 2 weeks thereafter. In contrast, male 4 (Fig. 10.2d) in LD 2 : 2 was not entrained but shifted to the right, the period of the shift being approximately equal to that of a subsequent interval in DD and a return to LD 2 : 2.

Feeding in LD 2 : 2 did differ from activity in 2 ways: (1) in the absence of entrainment the period of the shift was always greater than 24 h and (2) a phase of feeding always occurred during the scotophases, in contrast to an activity phase which sometimes took place during a photophase (see also *L. maderae* in Fig. 4 Roberts, 1962). In experiments with LD 1.5 : 1.5, activity is not entrainable but is free-running, while feeding is still entrainable in 40% of the individuals. Although different oscillators would seem to be involved, one experimental detail suggests caution in accepting such a conclusion, namely, in the feeding study a retreat was provided for individuals to achieve the goal of negative phototaxis while in activity monitored by a wheel cage no escape from the photophase is provided.

In the absence of water supply, the feeding activity in a trophometer is immediately reduced, but attempts to feed (i.e. handling of the food by mouthparts) continue to coincide with the former entrained feeding rhythm (Sutherland, unpublished) and continue for as long as 2 weeks in the absence of water ($n = 10$). In some cases, when re-offered water in early photophase, the male leaves the retreat to drink and shortly thereafter feeds in the light. However, the feeding rhythm is not shifted but is renewed to coincide with early scotophase. It appears, therefore, that water availability does not control the feeding rhythm but is necessary for its expression. As long as water is available, saliva is not necessary. With salivary duct ligation, feeding was first reduced during a post-operative period but the rhythm returned and continued for several months. In 10 individuals, whose salivary duct and sensory elements to the glands and reservoirs were severed, the feeding rhythm also continued after a short post-operative period. After ligation or severance, feeding intervals seemed longer but this was probably due, not to increased food intake, but to an increased time necessary for ingestion in the absence of saliva.

Severance of the oesophageal nerve, which paralyses the proventriculus and retards food passage, does not greatly affect the feeding rhythm (Sutherland, 1969). In contrast to the hyperphagy resulting in blowflies so treated (Gelperin, 1967), the nocturnal feeding rhythms of 60% of male *P. americana* so treated were not changed but continued even when the crop food level was so abnormally high as to cause lesions in the crop wall. One cannot conclude that crop volume does not

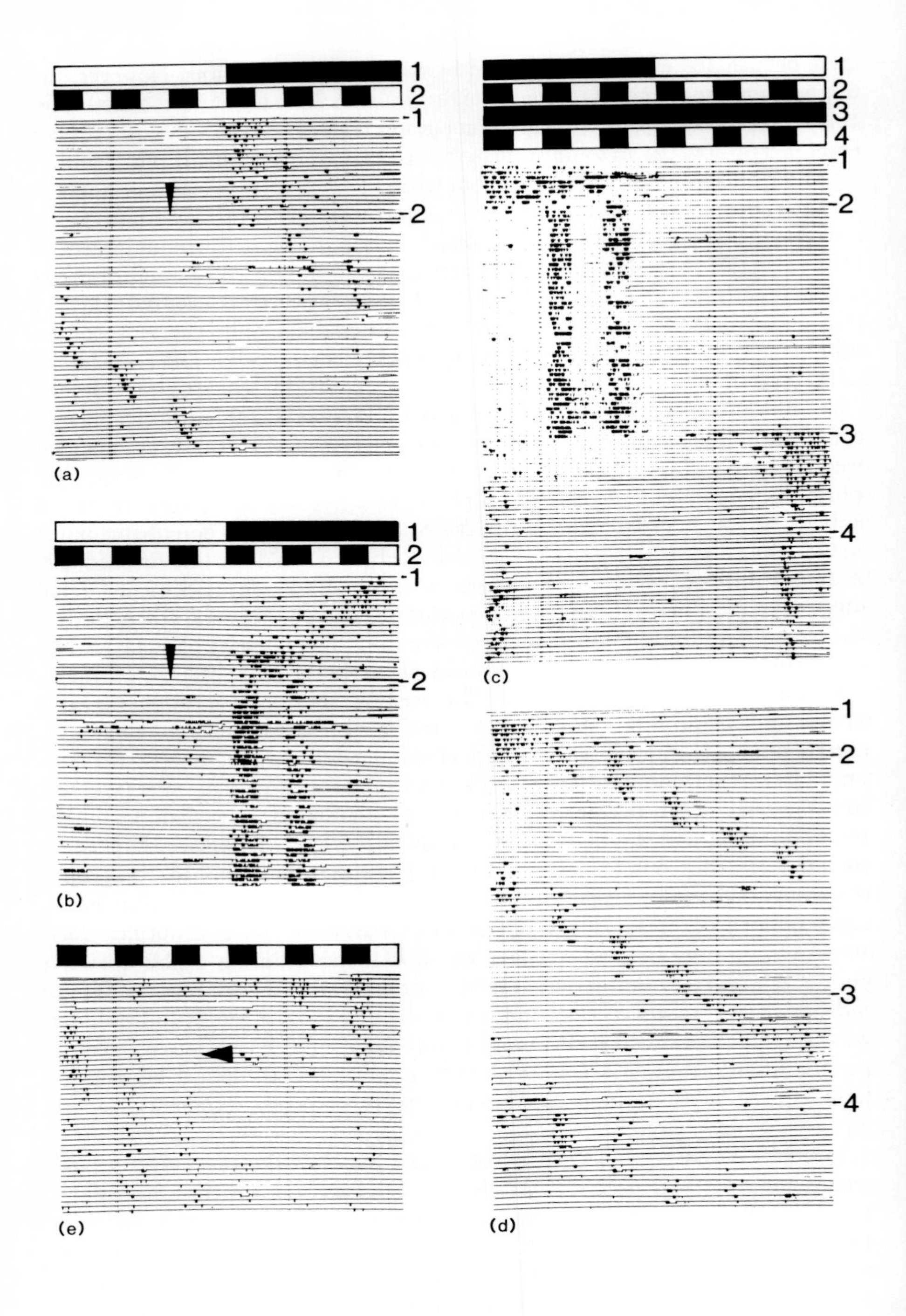
1
2
-1
-2
(a)
1
2
-1
-2
(b)
(e)
1
2
3
4
-1
-2
-3
-4
(c)
-1
-2
-3
-4
(d)

Fig. 10.2 (a and b) Trophograms of two adult male *P. americana*, with reversal of LD 12 : 12 one day prior to day 1 and a subsequent change to DD on day 22 at time of dart point. (c and d) Trophograms of two adult male *P. americana* in successive regimens of LD 12 : 12, LD 2 : 2, DD and LD 2 : 2. (e) Trophogram of an adult male *P. americana* in LD 2 : 2 with a change from laboratory food to sugar at time of dart point. For a better recognition of the slope of onsets through 'doublication' (Lohmann, 1967c; Wiedenmann, 1977b), a photocopy of an actogram or trophogram can be placed immediately to the right of the original, but with upward displacement by one day.

---

influence the feeding cycle, since nerve severance interrupts sensory pathways from crop stretch receptors. A different approach has been employed (Sutherland, unpublished) with cubed sugar replacing laboratory food on the pendulum of the trophometer. With slight moistening and drying, the texture and bitability of the sugar was more closely matched to those of the laboratory food. Since cockroaches ingest less sugar than laboratory food and the nutritive value, osmotic pressure and rate of release from the crop would be expected to differ for the two food types, it is surprising that in LD 2 : 2 a change of food type did not interfere with the shifting of the feeding rhythm (Fig. 10.2e). Extirpation of CA or single injection of trehalose and glucose (1.9 mg and 2.3 mg respectively per male) did not alter the rhythm, and the change from light to dark remains as the dominant stimulus for feeding. Although restriction of food supply to the photophase does not markedly affect the activity rhythm of *P. americana* (Harker, 1956), it should be noted that food availability may be the primary synthronizer of certain cellular rhythms, as shown in higher animals (Nelson *et al.*, 1975; Vilchez *et al.*, 1975).

Food consumption by adult female *P. americana* is greatly influenced by the reproductive cycle (Mills *et al.*, 1966; Bell, 1969b; Verrett and Mills, 1973) (see Fig. 13.6), but females generally exhibit a daily feeding rhythm although it is not as precise as that of the male. This has been shown in studies attempting to detect a simple depletion–repletion system of feeding regulation (Faber, 1975). Recordings of trophometers with retreats were analysed for meal numbers, length of meal and inter-meal intervals. Based on meal frequency per 2-h intervals in LD 16 : 8, males exhibited entrained feeding at temperatures of 35 °C, 30 °C and 25 °C. However, the rhythm of the female was only really evident at 25 °C. For both sexes, at the higher temperatures more food was consumed, but this was accomplished with shorter meal times and with more meals. Therefore, with such interaction, the entrained feeding rhythm would not be precise at the higher temperatures. The influence of the retreat on rhythm precision is not completely due to negative phototropism, since with a transparent retreat made of screen the rhythm remains precise. With LL (about 150–300 lx) and opaque retreats, free-running feeding rhythms were not detected in females but were present in one half of the males (Faber, 1975).

10.2.3 Drinking

Although it is known that water is more important than food for survival of *P. americana* (Faber, 1975) and of *N. cinerea* (Raynieuse *et al.*, 1972), it is not known if a drinking rhythm exists in cockroaches. For adult female *P. americana*, daily imbibition occurs during the reproductive cycle, but the amounts vary as also do the haemolymph volume, total body water and total tissue water (Verrett and Mills, 1973; 1975a). For the male, the daily water intake has not been studied, but some reports suggest that drinking is also a daily event, especially under laboratory conditions. Some evidence of the interdependence of food and water intake is reflected by the volume of the crop and salivary reservoirs under various supply conditions (Sutherland and Chillseyzn, 1968); but the absence of food may reduce the precision of a daily drinking rhythm, much as the absence of water reduces the amount of feeding. Other evidence of the close relationship of the feeding rhythm and a possible drinking rhythm is gained from studies where the effect of water rationing on the feeding rhythm was studied (Faber, 1975). When a ration of 1 drop/day was offered during the photophase, either every day, or every 2nd, 3rd or 4th day, the nocturnal feeding typical with water *ad libitum* disappeared. With rationing, most individuals (4 females, 2 males) rapidly detected and drank the water ration and then fed during the photophase. Two males left the transparent retreats to drink the ration, but waited for 8–10 h before feeding in the scotophase. In these experiments, individuals were exposed to each ration level for 1 month, progressing from water *ad libitum* to water rationed every four days. Therefore, the effect of age and of rationing experience cannot be discounted. Until equipment to monitor the temporal aspects of drinking becomes available, a drinking rhythm with its entrained and endogenous characteristics is only presumptive. With such equipment, this behavioural rhythm and its relationship to feeding can then be studied concurrently, as has been done with higher animals (Boulos and Terman, 1979).

10.2.4 Susceptibility to chemical agents

It is known that circadian rhythms can determine the response of organisms to chemical and physical agents (Reinberg and Halbert, 1971) and this aspect extends to insects and insecticides. This was apparently first shown by Beck (1963, 1968) with *B. germanica* and various types of insecticides. To examine the circadian rather than an entrained response, male adults were first conditioned to LD 12:12 and injected at various times in dim LL, and then observed for mortality 24 h later. Clear sensitivity rhythms were absent for DDT, sodium fluoride, sodium azide and 2,4-dinitrophenol. However, sensitivities to potassium cyanide, dichlorvos and dimetilan were rhythmic, but not in phase with each other. The variation is not unexpected, however, since the above insecticides are of different types. Even among closely related organophosphate insecticides, response differences are to be expected, as shown in studies with the rat (Fatranska *et al.*, 1978). Several factors would influence the response, including the agent solubility, its rate of activation, distribution, metabolism and action.

Other cockroach species (*L. maderae* and *N. cinerea*) have also been shown to exhibit sensitivity rhythms (Sullivan *et al.*, 1970; Halberg *et al.*, 1973, 1974).

Similar phasing for response to pyrethrum aerosols was shown, employing an improved single cosinor method for data analysis. The advantage of this method, over simple inspection of toxicity chronograms, for the detection of rhythms has been well-shown for dichlorvos and *Tribolium confusum* (Vea *et al.*, 1977). These latter authors have suggested that sensitivity is generally related to behavioural activity, i.e. diurnal insects are most susceptible in early photophase and nocturnal insects most susceptible in early scotophase. However, the similarity in sensitivity rhythms for knockdown and mortality of *L. maderae* and *Musca domestica* treated with pyrethrum aerosols (Halberg *et al.*, 1974) does not support this generalization, probably because this insecticide so rapidly stimulates activity. It is also probable that *P. americana* would exhibit a similar rhythm to this insecticide. It should be noted that investigations on sensitivity are based on group, not individual, response, and, except when delivered directly to the insect (Beck, 1963; 1968), the dosage varies between individuals. Also, while response may be closely related to circadian behavioural rhythms, the sensitivity rhythms themselves are only presumed to be circadian. This is understandable since with their examination in LL or DD the period (τ) of free-running behavioural and sensitivity rhythms would differ among individuals.

The influence of noxious agents has also been studied by observing their effects on some behavioural or physiological rhythms of treated individuals. The activity rhythm of *P. americana* is shifted by short doses of irradiation, the degree and permanence of the shift apparently being dependent on the irradiation level (Harker, 1964), the highest of which did cause mortality. Irradiation can also induce changes in susceptibility to insecticides (Shipp and Otton, 1976). Carbon dioxide and nitrogen also cause phase shifts in activity of this species, depending on the timing of the treatment (Ralph, 1959). However, this latter study employed a tilting cage with activity diminishing after 5 days in DD. The phase shifting by these agents or diurnal variation in sensitivity should be re-examined with greater numbers of individuals and for longer intervals under constant conditions. It would also be important to be more specific about the level of sublethality of such agents. This has been done in studies differentiating between the action of DDT and immobilization and their effect on circadian oxygen consumption of *P. americana* nymphs (Patel and Cutkomp, 1967; Patel *et al.*, 1974). During the five-day experiment, the oxygen consumption in all groups decreased, possibly due to the food and water deprivation. In survivors of a topical application of an LD_{50} dosage, the circadian rhythm of oxygen consumption shifted slightly and was less prominent than that of untreated or immobilized individuals. However, the energy-regulating enzyme mitochondrial Mg^{2+} ATPase, whose activity is rhythmic in LD 12:12, apparently does not exhibit a rhythm of sensitivity to DDT (Cutkomp *et al.*, 1977).

L. maderae has served as an experimental animal in an analysis of the effect of D_2O and lithium on circadian rhythms. When this species is deuterated, the period (τ) of the circadian activity rhythm lengthens depending on the temperature (Caldarola and Pittendrigh, 1974). This tends to discount an earlier hypothesis that D_2O exerts its effect by reducing cell temperature, and this agent may only act on one component of the circadian system. When offered lithium chloride the period is increased by approximately 1%, but this effect could not be correlated with internal Li^+ concentration (Hofmann *et al.*, 1978).

10.3 Circadian rhythms of populations

Little is known of the daily activity rhythms of populations or of individuals within a population. However, there are various reports which suggest that the precision of activity rhythms exhibited by isolated individuals is greatly decreased by the presence of and interaction with conspecifics. Such interactions would include sexual as well as agonistic behaviour. Hawkins and Rust (1977) have shown that the entrained activity rhythm of groups of 5 adult males is very shallow, but is influenced by the presence of the sex pheromone. For further details refer to Section 14.3.2.

Agonistic behaviour is also cyclic in LD 12:12 (Bell and Sams, 1973; Olomon *et al.*, 1976). Both nymph and adult *P. americana* fight more intensely during the early scotophase when they are most active. When perturbed by light during the scotophase, activity and agonistic interactions increase unless a retreat provides an escape from the light. The effect of the retreat was found to be approximately proportional to the cockroach density. Whether this escape behaviour (photosensitivity) is variable and cyclic is not known. Since the phase-shifting of the activity of individuals by light pulses depends on timing during the subjective night, a cycle in some of the elements of escape behaviour would not be surprising. However, escape behaviour, as measured by angle of turn away from wind stimulus, is complex and depends on previous experience of the individual (Camhi and Tom, 1978). Cockroach behaviour (*B. orientalis*) may be enhanced or impaired by the presence of conspecifics (Zajonc, 1965; Zajonc *et al.*, 1969), and the cyclic activity of *P. americana* populations might be better studied with very small group size. It has been suggested that *P. americana* exhibits a less advanced type of territoriality than other species (Bell and Sams, 1973). But, since some individuals are more dominant, possibly they serve as the stimulus for activity by passive individuals under entrained conditions. Possibly under constant conditions (DD or LL), the period (τ) and phasing of the circadian activity of the dominant individual synchronizes that of more passive individuals, whose activity rhythms may be weak or non-existent under conditions of isolation. With greater numbers of individuals the interaction between conspecifics may be so great that under constant conditions no circadian population activity rhythms are detectable. This might explain the lack of such a rhythm in *P. americana*, presumably in groups, monitored at the South Pole (Hamner *et al.*, 1962).

One of the first reports on the nocturnal activity rhythm of cockroaches dealt with a natural population of *B. orientalis* (Mellanby, 1940). A study of *P. americana* in sewerage systems (Eads *et al.*, 1954) was not designed to detect daily rhythms, although there were seasonal fluctuations in population levels.

Seasonal activity is also reported for *Periplaneta fuliginosa* (Fleet *et al.*, 1978) but based on mark-recapture techniques, the frequency of female movement was greater than males, as probably influenced by the reproductive cycle and search for suitable oviposition sites. Similar studies on the population ecology of *P. americana* suggest cyclic activity (Faber, unpublished). In a natural population of a roof-top greenhouse, adult females appear at dusk before adult males, who, in turn, remain visible longer during the night. Activity is not biphasic, and neither sex is evident to observers at dawn. The population does not seem to congregate and some individuals have home ranges.

10.4 Circadian rhythms of physiological systems

As more is known of the behavioural rhythms of individuals, greater attention is being given to the rhythms supporting behaviour. Often the physiological fluctuations may be small and may be viewed as variability about an experimental mean. But such fluctuations may be closely related to behaviour, and their exact study can yield information about circadian mechanisms.

10.4.1 Respiration

Apparently, oxygen consumption in *Periplaneta* was the first physiological function shown to be related to the activity rhythm (Janda and Mrciak, 1957). This was subsequently reported to occur also in *B. germanica* (Beck, 1963) and in *Blaberus* sp. (Banks *et al.*, 1975). In the latter species, it was described as circadian, but details are unclear as to the length of pretreatment in DD or LL, and the endogenous expression may be a carryover from entrained conditions of LD. The endogenous nature in *P. americana* in LL for several days has been shown for nymphs (Patel *et al.*, 1974) and male adults (Richards and Halberg, 1964), with ultradian frequencies of 3.4 and 0.86 h also detected in the latter. Such respirometer studies were conducted in the absence of food and water, under which conditions activity and/or basal metabolism decreases and the circadian rhythm can become indistinct in five days (Richards, 1969). The ultradian bands of 3.4 and 0.86 h were detected at 18, 24 and 30°C (Richards and Halberg, 1964) and may occur at lower temperatures. However, while nymphs and adults can show acclimatization of their oxygen consumption to temperature (Dehnel and Segal, 1956), data of Welbers (1976) indicate that an entrained cycle of oxygen consumption (LD 12:12) is essentially imperceptible at 15°C, and possibly ultradian periods might also fade at this temperature. An ultradian rhythm in carbon dioxide output has been detected in *P. americana* at 11°C with a period of 35.8 min, approximately 2.7 times longer than that at 21°C (Wilkins, 1960). These rhythms, detected in the early scotophase (LD 12:12) of entrained individuals, were not affected by order of temperature experiences or high-intensity light. It is not known if there is a circadian rhythm of carbon dioxide output or if the ultradian rhythms persist beyond 5 h.

10.4.2 Circulation

The heartbeat rate of *P. americana* is also circadian in LL for 5 days as detected by the mean cosinor method, and its peak lags approximately 2.5 h behind that of oxygen consumption (Patel *et al.*, 1974). Possibly a periodicity in numbers and categories of haemocytes also occurs, as suggested in work with *Blaberus giganteus* (Arnold, 1969), but successive sampling may interfere in its detection in *P. americana* (Brady, 1968). In *P. americana*, total sugar levels of the haemolymph fluctuate in LD 12:12, but the peak levels occur towards the end of the scotophase and, therefore, are not closely correlated with the activity cycle (Hilliard and Butz, 1969). However, haemolymph sampling at 4°C may have influenced this correlation. In addition, *P. americana* are not highly active insects and the association of feeding, carbohydrate metabolism, and total blood sugar levels is complex. Although not designed to detect rhythms, other studies have indicated that haemolymph glucose levels vary according to activity levels (Matthews *et al.*, 1975). Presumably, diel fluctuations in haemolymph proteins may also occur in *Periplaneta*, although, as has been shown for *L. maderae*, individual variation and successive sampling may interfere with their precise demonstration (Hayes *et al.*, 1970). Such factors did not prevent the demonstration of significant rhythms of uric acid levels in haemolymph of *P. americana* (Hilliard and Butz, 1969); a drop in uric acid in males and females occurred in early scotophase.

Haemolymph hydrocarbons (heptacosadiene, pentacosane and methyl-pentacosane) associated with epicutiluar wax formation exhibit a diel fluctuation, as indicated in an analysis of these components during the photophase of LD 12:12 (Turner and Acree, 1967). After 10 days in constant light these fluctuations decreased. However, since the *P. americana* were grouped and the circadian periods of an individual activity (or even hydrocarbon rhythm) would be expected to differ among individuals in LL, possibly data analysis has concealed the maintainance of a so-called circadian rhythm in the male. These studies did not examine locomotory activity under similar conditions, but the hydrocarbon fluctuations were almost non-existent in the females. This is probably related to the more diffuse nature of the rhythmic activity and feeding of the female during her reproductive cycle.

The activity of *P. americana* is apparently influenced by the ionic content of the haemolymph as modified by food type (Pichon and Boistel, 1963). Conditions of LD were not specified, but a diel fluctuation in activity was more evident among individuals with higher Na^+/K^+ ratios. In experiments to determine whether diel fluctuations in these ions could be more closely associated with activity and whether a reduction in potassium preceded an increase in activity, successive sampling and individual variation prevented consistent detection of the potassium rhythm (Brady, 1968). More recently, the concurrent measurement of potassium activity and locomotory activity in *L. maderae* has circumvented such an experimental problem and indicated a bimodal potassium rhythm in LD 16:8 (Lettau *et al.*, 1977). The peaks of this potassium rhythm appear to be associated with reduced locomotory activity. However, as noted by the authors, a decline in potassium concentration precedes the major activity peak by approximately 1 h, but a secondary activity peak in late scotophase–early photophase precedes the second decline in potassium. The role of this ion in the expression of locomotory activity is indeed complex, but a severe test of the endogenous nature will hopefully allow for the inclusion of food and water.

10.4.3 Digestion

Although a feeding rhythm in *P. americana* has been demonstrated, there are few reports indicating any associated rhythms occurring in the digestive tract. No entrained rhythm in amylase activity could be detected in the glandular portion of male adult salivary glands (Lipton, 1969). Mean amylase activity units immediately prior to feeding, 3–4 h after feeding and 12 h after feeding were 0.48, 0.38 and 0.49 respectively (n = 8 for each time sampled). Although individual variation may have prevented the detection of statistically significant differences, it is more likely that secretory cycle of individual zymogen cells is asynchronous throughout the gland, the salivary reservoir being the component which associates with the feeding rhythm. Other secretory portions of the digestive tract are probably rhythmic, the stimulus being the ingestion of food and the messenger for activity being some blood-borne factor. Rounds (1968), however, has suggested that there may also be a resident factor within the midgut which stimulates cells to release proteinase at 'dusk'. The factor was extracted from female *P. americana* starved for 48 h and tested on the same sex, and this aspect should be re-examined with adult males. In nymphs, a mitotic frequency in the midgut (n = 8) was detected in LD 16:8, with a most significant peak at 2 h after the onset of the scotophase (Choban, 1975). A lesser peak occurred 4 h thereafter and may be associated with the incidence of a second meal. Unfortunately, the feeding rhythm of nymphs has not been examined.

10.4.4 Integument

Although the number and length of instars of *P. americana* are variable depending on the author (Guthrie and Tindall, 1968), it would be as well to always recognize that moulting with its supporting biochemical processes is essentially a macro-rhythm. The rhythm is temporally inexact but could be termed *ecdysial*. The ecdysial rhythm of ecdysone and JH has been revealed for the sixth instar (Shaaya, 1978). Rhythms in tyrosine levels (Wirtz and Hopkins, 1977), and in dopa and tyrosine decarboxylase activity have also been demonstrated (Hopkins and Wirtz, 1976; Wirtz and Hopkins, 1977). Within these ecdysial rhythms, there may be also other rhythms such as the diel growth of the endocuticle. The lamellogenesis in *P. americana* is circadian and continues in LL and DD (Neville, 1967), but its exact relationship to circadian locomotory activity is not known. However, in *Blaberus fuscus*, in which such activity becomes arrhythmic after bilobectomy (excision of pacemaker in optic lobes), the circadian deposition of endocuticle continues, leading to the conclusion that there are at least two self-sustained circadian oscillators (Lukat, 1978).

10.4.5 Endocrine

When early research on the circadian activity rhythms of cockroaches suggested the involvement of haemolymph-borne factors, it stimulated research on the manipulation of the endocrine system as well as its cellular rhythms. The latter aspect was first extensively examined in *P. americana* by Brady (1967d), who monitored cockroach activity and sacrificed animals throughout the LD 12:12 cycle for histochemical analysis of the suboesophageal ganglion. No clear evidence of an entrained cycle in the movement of neurosecretory material was detected, and only the lateral neurosecretory cells were significantly larger at the time of maximum locomotory activity. As discussed by the author, the histological approach is fraught with difficulties, and, in contrast to the histochemistry of the endocrine cycle during the ecdysian macrorhythm, the detection of the precise diel endocrine rhythms and relation of them to behaviour requires more sensitive methodology. This is suggested by other work with *P. americana* (Cymborowski and Flisínska-Bojanowska, 1970), where apparently arrhythmic animals often exhibited greater histological changes. Cymborowsky (1971) briefly reported that both *P. americana* and the house cricket display a diel rhythmicity of the accumulation of material in cells of the pars intercerebralis and suboesophageal ganglion in LD 12:12; synthesis and nuclear volume were also stated to be rhythmic. However, in individuals reared in LL, and apparently presumed to be arrhythmic, no histochemical rhythmicity could be detected. However, this does not mean that the endocrine cycles are not endogenous. This characteristic of the endocrine system might more profitably be studied in DD or dim LL with close monitoring of locomotory activity and time-sampling. It should be noted that rhythmicity within organs bearing endocrine cells may not always be endocrine in nature. Transplantation of suboesophageal ganglia to *P. americana* of identical or reverse activity phases has led to the hypothesis that there are two factors normally secreted rhythmically from the suboesophageal ganglion (Rounds and Riffel, 1974). However, since no other tissues were tested, this distinction cannot be given to just this ganglion, nor can it yet be termed endocrine. Earlier work by Ralph (1962) discovered numerous factors from various tissues of *P. americana* which could accelerate or decelerate the heartbeat, including not only the CC, but also midgut.

10.4.6 Nerve and muscle

Evidence of diel activity in the nervous system was first suggested by Ralph (1962), when neural components were extracted and tested on the heart. Although slight differences existed between the four sampling times for some tissues, a more complete examination of suboesophageal ganglion and the CC–CA complex did not detect significant rhythmicity.

Spontaneous electrical activity of the ventral nerve cord is apparently rhythmical (Rao, 1968; Schulz and Schwarzberg, 1971); however, exact details have not been fully reported. In the first report (Rao, 1968), most details dealt with the scorpion, but maximum electrical activity in *P. americana* apparently extended some 7 h (4 pm–11 pm) during which interval blood and suboesophageal extracts caused a marked increase in such activity. The exact LD cycle or its timing was not given, but, in the second study in LD 12:12 (Schulz and Schwarzberg, 1971), the maximum electrical activity occurred in the last half of the scotophase, a time when activity of the cercal nerve was at its peak. Here the sex and group experience of the cockroaches are not given. Such information would help to explain why individuals raised in constant darkness throughout nymphal development to the adult displayed the 4 h rhythm in both nerve preparations. A recent report of opposite diel rhythms in acetylcholinesterase and acetylcholine (Vijayalakshmi *et al.*, 1977) would also be more significant if the timing of the LD 12:12 cycle had been stated. Such rhythms were not examined in constant conditions, and their circadian nature is not proven. A later report (Vijayalakshmi *et al.*, 1978) has described increased activity levels of aminotransferases of male muscle and nervous system during the night. However, the exact relationship to higher energy requirements during increased locomotory activity cannot be examined, since the locomotory activity of this wild strain was not monitored and the exact LD cycle was not stated.

Since the visual units of the nervous system are an important part of the entrainment of rhythms, it is important to note that, in the housefly, 'eye-nerve' activity can be entrained by an LD cycle, with an increase associated with the approach of the photophase (Gunning and Shipp, 1976). It is endogenous in DD and LL, and remains synchronous among individuals placed in DD but becomes asynchronous among those placed in LL. Possibly similar 'eye-activity' occurs in *P. americana*, and visual rhythmicity may even involve migration of visual pigments.

10.4.7 Reproduction

The apparent effect of the sexual cycle on the locomotory activity of the female cockroach has been discussed (Section 10.2.1(e)), but it is appropriate to emphasize here that the reproductive cycle of the female adult is continuous and rhythmic (Bell, 1969b) and should be considered a low-frequency macrorhythm of 3–6 days involving amounts of feeding (Mills *et al.*, 1966), vitellogenesis (Krolak *et al.*, 1977; Clore *et al.*, 1978) and water balance (Verrett and Mills, 1975b). These are apparently synchronized with a CA cycle, but 2 cycles of CA activity occur during the growth of each wave of oöcytes (Weaver *et al.*, 1975; Weaver and Pratt, 1976) (see also Section 13.2.2 and Fig. 13.6). Production of the female sex attractant is depressed by copulation but can be sporadic thereafter (Wharton and Wharton, 1957). The male *P. americana* can rapidly recover from the pheromone stimulus, and the production of the male spermatophore may be a diel occurrence and even weakly circadian as shown for the cricket (Loher, 1974).

10.5 Location of photoreceptors and circadian pacemakers

In 1954, there began a series of reports by Harker (1954, 1955, 1956, 1958a,b, 1960a,b,c), which led to one of the most interesting controversies about the circadian rhythm and its control in *P. americana* and *L. maderae*. Much of the difficulty has been due to the brevity of reporting, sometimes encouraged by scientific literature and the different nuances detected and expressed by authors reviewing and citing the work of others. In addition, errors (as described by Brady, 1969) have sometimes been made in delicate operations, and probably minor but important details have not always been noted. Species, strains, handling and activity monitors have also differed. The various aspects have been competently considered in recent reviews (Brady, 1974; Saunders, 1976, 1977; Truman, 1976; Kawamura and Ibuka, 1978) and the discussion here will summarize briefly the current view. Some of the original observations of Harker are still unexplained, and recent reports (Rivault, 1976; Lettau *et al.*, 1977) may help to do so.

The photoreceptors by which the endogenous activity rhythm can be entrained are the compound eyes and not the ocelli or non-optic integument. However, the participation of the latter two under some conditions cannot be completely discounted. Rivault (1976) has reported that *P. americana*, raised in LL (88 lx) and with their ocelli and peri-antennal cuticle varnished before placement in LD, are not entrained until the varnish is removed. This does suggest that the ocelli under experimental conditions of apparent arrhythmia in LL can set the rhythm 'the first time'. However, constant LL is not a natural condition, and the non-optic light-sensitive areas may function in the selection of darkened areas in LD; apparently, the compound eyes and ocelli are unnecessary in this selection (Ball, 1965). While light transmission through the cuticle near the terminal abdominal ganglion does not entrain the rhythm (Ball, 1971, 1977), masking of this area possibly 'scatters' a secondary activity phase associated with early photophase.

The location of the pacemaker or circadian oscillator is the optic lobe, as first reported by Nishiitsutsuji-Uwo and Pittendrigh (1968) with mainly *L. maderae* and later confirmed with *P. americana* (Roberts *et al.*, 1971). The lamina apparently serves in the entrainment of the circadian oscillator in LD cycles, but the lobulla and medulla were found to be essential elements for circadian expression in both species (Roberts, 1974). Later experiments with *L. maderae* indicate that the cell bodies in the lobulla are most important (Sokolove, 1975). Although the two optic lobes have connections (Roth and Sokolove, 1975), the right or left lobe can function to allow the animal to free-run in DD (Page *et al.*, 1977). However, excision or surgical isolation of one lobe results in a significant increase in τ over pre-operative values. Recent findings support the hypothesis that (1) the paired pacemakers in the optic lobes are coupled and mutually accelerate each other via a polysynaptic pathway, and (2) the entrainment of each pacemaker by the contralateral eye is mediated by this pathway (Page, 1978). Whether this system governs other behavioural rhythms or is a part in a network of oscillators is not known.

10.6 Rhythms other than circadian

In contrast to the circadian rhythms of *P. americana* little is known of other types of rhythms. At least one physiological system, respiration, exhibits ultradian rhythms (Richards and Halberg, 1964) which may be harmonics of the circadian rhythm. Even ultradian activity rhythms may exist in the species (Harker, 1956) as well as lunar rhythms (Harker, 1958a; Rounds, 1975). The detection and confirmation of such rhythms, and of even circa-annual rhythms, will require very sensitive analysis and complete temporal records. The importance of the latter is demonstrated by Treherne *et al.* (1975) in detecting a seasonal variation in Na^+ in *P. americana*.

10.7 Conclusions

Countless individual cockroaches have co-operated in the experimental demonstration of the cockroach circadian system and its components. So intent has been the aim to test the system, that little consideration has been given to its value to the cockroach. From the cockroach point of view many questions arise. For example, is the circadian system of a laboratory population representative of that of those in man's structures of the temperate region or of the wild populations of tropical areas? Is the circadian system of an individual influenced by population levels? What component of circadian activity (dispersal, ranging, thirst, hunger) is the strongest? What is the significance of arrhythmicity, real or experimentally induced? Many other questions arise when consulting current reports of cycle research in other organisms, often termed chronobiology and biochronometry. Of particular value are the advances in data analysis to detect rhythms. Those researchers continuing research on the rhythms in *P. americana* must employ these advances and more fully define experimental conditions, including pre-experimental rearing conditions. Newer monitoring equipment (Page and Block, 1980) will be important in investigating the antogeny of rhythms. Researchers more interested in other aspects of this species must be aware of the circadian fluctuation in physiological and biochemical variables (Scheving and Pauly, 1974).

Acknowledgements

The author gratefully acknowledges the comments of B. Faber, G. Lipton and S. Roberts in reviews of early drafts, the permission of D. S. Saunders to cite unpublished results, and the valuable assistance of A. Hajek.

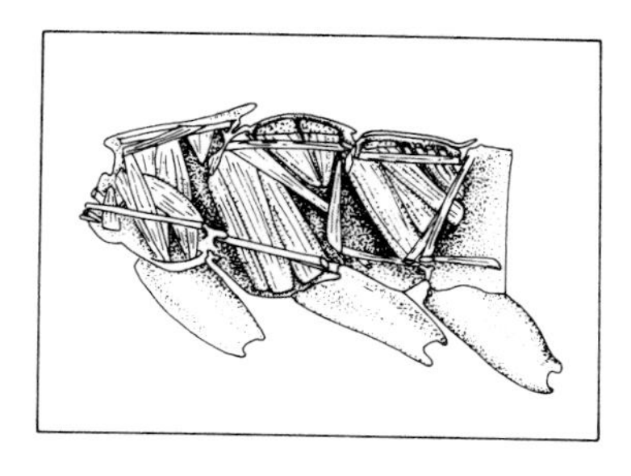

11

Muscles and muscular activity

Fred Delcomyn

11.1 Introduction

By virtue of its contractile properties, the muscular system comprises the main effector system of the body. All movements are due to contractions of muscles, be they the rapid thrust of legs during escape running, the leisurely sway of an antenna, or the churning of food in the gut. Actions of the muscular system are therefore the basis for an animal's behaviour. However, most muscles cannot contract without an excitatory signal from a motoneuron, so one cannot fully understand the functional properties of muscles without also understanding the way motoneurons control them. In this chapter, I will therefore discuss not only the structure and physiology of the muscular system of *P. americana*, but also its neural control. Organization of the nervous system itself, although clearly reflected ultimately in the use of muscles, is considered in Chapter 8.

Since space limitations preclude anything near an exhaustive review of the muscular system, I will refer the reader to appropriate reviews as necessary. The best and most comprehensive treatment of insect muscle is the volume edited by Usherwood (1975). Specific information on cockroach muscle is provided by Guthrie and Tindall (1968).

11.2 Organization and innervation

All muscle in *P. americana*, as in every insect, is striated; that is, a pattern of alternate dark and light bands can be seen in properly prepared living, or fixed and stained muscle tissue. Nevertheless, on both morphological and functional grounds researchers have traditionally identified two types, skeletal and visceral muscle.

Skeletal muscles are the muscles responsible for moving or stabilizing the position of one part of the exoskeleton relative to another; they are therefore the ones which move the animal around in the world. Since that world often contains predators from which rapid escape is essential, these muscles are ordinarily quite fast-acting. They may themselves be subdivided by functional capacity (phasic or tonic) or colour (white or the reddish-brown colour called red or pink). Variations in functional capacity are due to differences in electrical properties (see Section 11.4.2), while colour variations are due to biochemical differences, which also have functional consequences (Section 11.3).

Visceral muscle includes any other muscle, and consists primarily of the muscles of the alimentary canal, although muscles such as those associated with the heart and the malpighian tubules are sometimes also included.

Classifications of muscle which take greater account of its diversity have been proposed (e.g. Mill and Lowe, 1971) but these have not come into general use.

11.2.1 Gross structure and innvervation

(a) *Skeletal muscle*

The gross organization and innervation of the main skeletal muscles in *P. americana* are fairly well known, although unexpected areas of ignorance seem to be exposed from time to time. The main descriptions of gross musculature are those by Carbonell (1947), Dresden and Nijenhuis (1953), Chadwick (1957), Teutsch-Felber (1970) and Alsop (1978) for the legs and thorax, and Shankland (1965) for the abdomen. Guthrie and Tindall (1968) summarize much of this work, and also provide a number of original drawings of parts like the head which have been otherwise neglected. Storch and Chadwick (1967) have surveyed the neck and thoracic muscle structure of nymphs.

From a functional point of view, probably the most important feature of the gross organization of the musculature of *P. americana* is its incredible complexitity. There are literally hundreds of small individual muscles in various parts of the body (e.g. Fig. 11.1). This extreme specialization of muscles makes the task of understanding the role played by each unit very difficult, a task made more so by uncertainties in how to define an individual muscle, especially in places where divisions between them are not clear (Alsop, 1978).

The gross nerve supply of the larger and better-studied muscles is well-known, and has been detailed by Guthrie and Tindall (1968). Readers requiring specific information as to the source of the nerves innervating a particular muscle can also consult references cited in Chapter 8. However, two important points should be made about this information. First, it is quite incomplete for many of the smaller muscles throughout the body, especially those in the abdomen and the ventral part of the thorax. And second, knowing the gross innervation is a long way from knowing how the muscle might be used, since there are one inhibitory and two types of excitatory neurons which can supply it, and the action a muscle may take

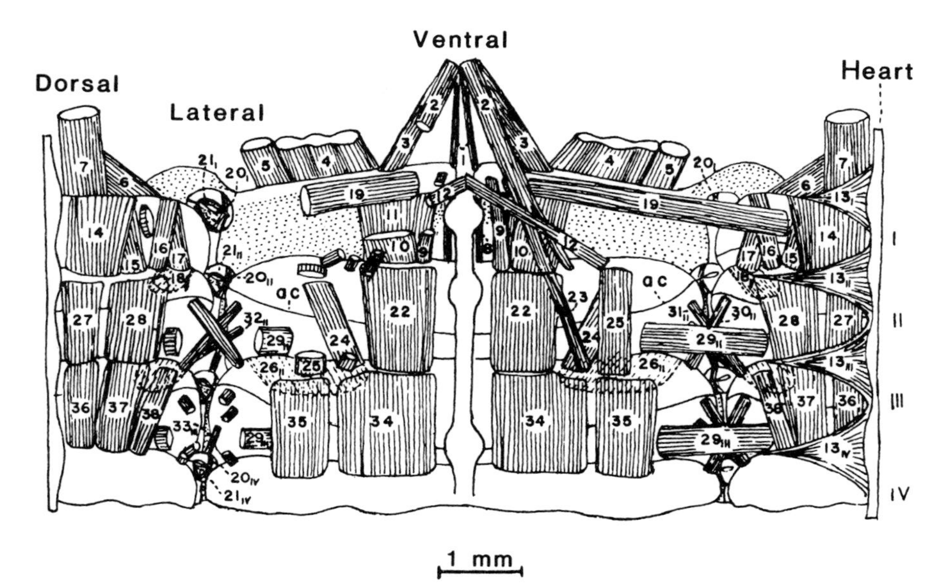

Fig. 11.1 Diagram of the musculature of the first four abdominal segments of *P. americana*. Note the many small individual muscles. (From Shankland, 1965.)

during a behaviour depends strongly on the type of innervation it receives (Sections 11.4 and 11.7).

Patterns of innervation in individual leg muscles have been described in detail for *P. americana*, and are typical for an insect. Each axon makes multiple contacts with each muscle fibre it innervates (*multiterminal* innervation), and each fibre may receive synaptic input from more than one neuron (*polyneuronal* innervation). Dresden and Nijenhuis (1958) have published a detailed nerve by nerve analysis of the distribution of neurons to muscles of the mesothoracic legs, and various other authors have made inferences about specific innervation based on physiological results (e.g. Usherwood 1962a). Guthrie and Tindall (1968) summarize this work.

However, determination of the identity and numbers of axons which synapse on a specific muscle fibre requires careful intracellular recording from the muscle and extracellular monitoring of the motor nerve, and studies in this detail have rarely been done. The most reliable information that we have on *P. americana* comes from the work of Pearson and his collaborators. Pearson and Bergman (1969) showed that some fibres in some femoral flexor muscles of the coxa (m. 182c, d; notation of Carbonnell, 1947) receive two inhibitory, three 'slow' and possibly one 'fast' axon (Section 11.4.3), while Pearson and Iles (1971) showed that some fibres in certain antagonist (extensor) muscles (m. 177d, e, and the mesothoracic homologues, m. 135d, e) receive one 'slow' and three separate inhibitory axons. Other fibres may have only one or two inhibitory axons, and other muscles receive only a single 'fast' or a 'fast' and a 'slow' axon. In many muscles, such a complex

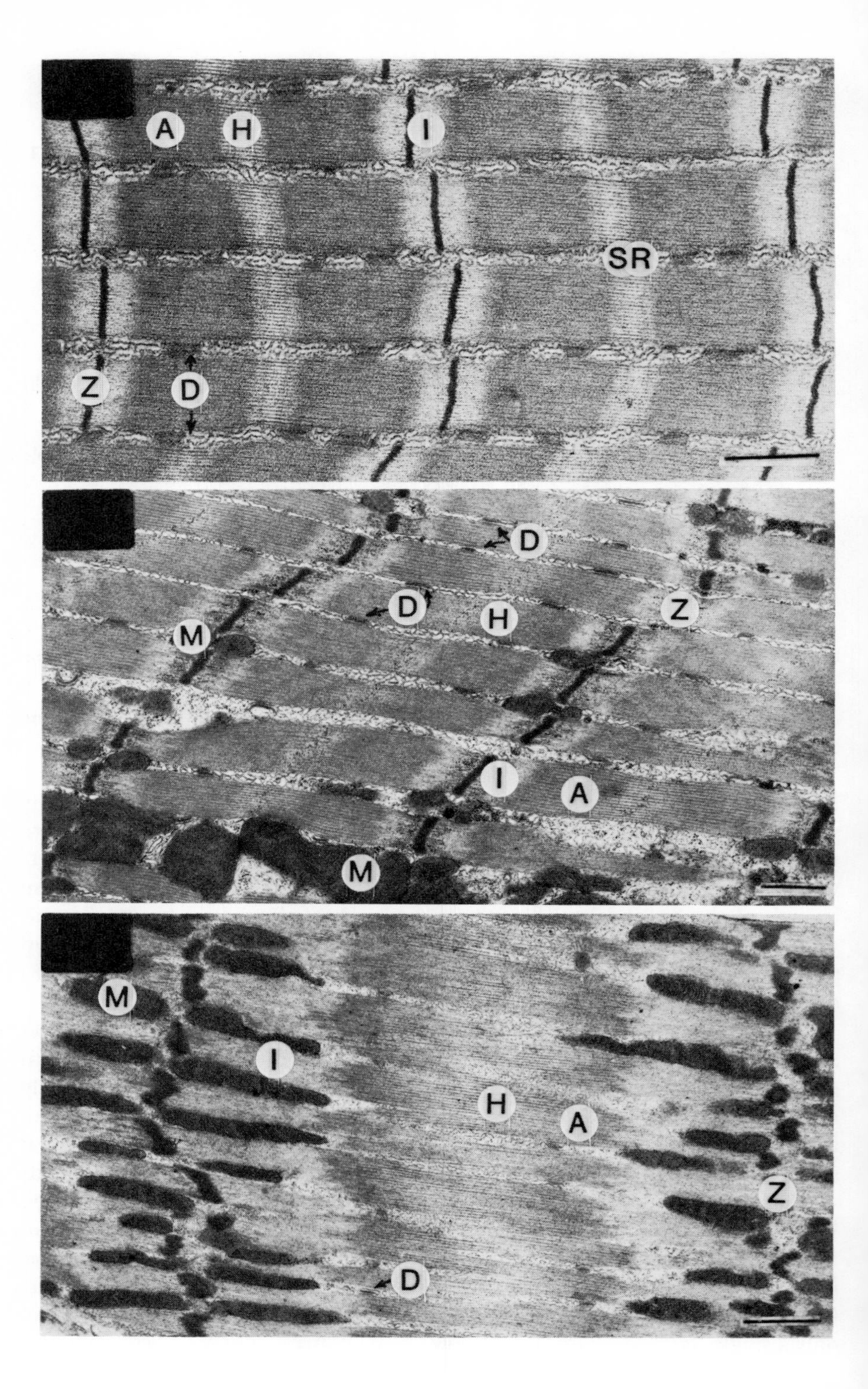
A
H
I
SR
Z
D
D
D
M
H
Z
I
A
M
M
I
H
A
Z
D

Figs. 11.2–11.4 Electron micrographs of longitudinal sections of the three types of muscle in a *P. americana* leg. Fig. 11.2, m. 179 (notation of Carbonell, 1947); Fig. 11.3, m. 177e′; Fig. 11.4, m. 177d. Note the differences in sarcomere length and mitochondrial content between the fast (Fig. 11.2) and slow-contracting muscles (Fig. 11.4). Abbreviations: A, A band; D, diads; H, H zone; I, I band; M, mitochondria; SR, sarcoplasmic reticulum; Z, Z line. Scale: 1.0 μm. (From Fourtner, 1978.)

---

pattern of innervation may be the rule rather than the exception. Tyrer (1971) found fibres in abdominal muscles of the locust receiving up to one 'fast', 2 'slow' and 1 inhibitory axons, and Shepheard (1974) found fibres in locust neck muscles with up to three 'slow' axons and one inhibitory, as well as the combination described by Tyrer (1971).

(b) *Visceral muscle*

The technical difficulties involved in working with the gut have kept to a minimum the number of studies on visceral muscle. The gross morphology of the visceral muscles of the gut has been described in Section 4.2. Innervation of the gut arises primarily from the frontal ganglion and the retrocerebral complex for the foregut (Willey, 1961; Section 8.1.4 and Guthrie and Tindall, 1968, p. 204), and the sixth abdominal ganglion for the hindgut (Brown and Nagai, 1969). The midgut has received considerably less attention from morphologists and physiologists alike, and its innervation is not well-known.

At the level of individual muscle fibres, innervation of the hindgut is multiterminal and polyneuronal (Belton and Brown, 1969). Axons with different post-synaptic effects have not been described.

11.2.2 Fine structure

(a) *Skeletal muscle*

The fine structure of insect muscle, investigated initially for what it might reveal about the generality of the sliding filament hypothesis of muscular contraction, has now also been studied for what it can tell us about insect muscle itself. The fine review by Elder (1975) makes clear that insect muscle shares the basic characteristics of other muscle, such as the presence of an overlapping arrangement of thick and thin filaments. These form the alternating A and I bands which give striated muscle its distinctive banding pattern (see Fig. 11.2). Also present are the sarcoplasmic reticulum (SR) and the system of transverse tubules (TTS) formed by invaginations of the cell membrane (e.g. Fig. 11.5). Other features are more specific to insects.

One such important feature is a non-cellular membrane which surrounds each muscle fibre. This basement membrane, also called the basal lamella (Piek and Njio, 1979), lies just outside the plasma membrane of the cell. It may play an important role in regulating the ionic environment of the neuromuscular junction or of the muscle itself, since it does not invaginate with the plasma membrane in the formation of tubules, and as much as 70% of the plasma membrane may be invaginated in tubules (Piek, 1974).

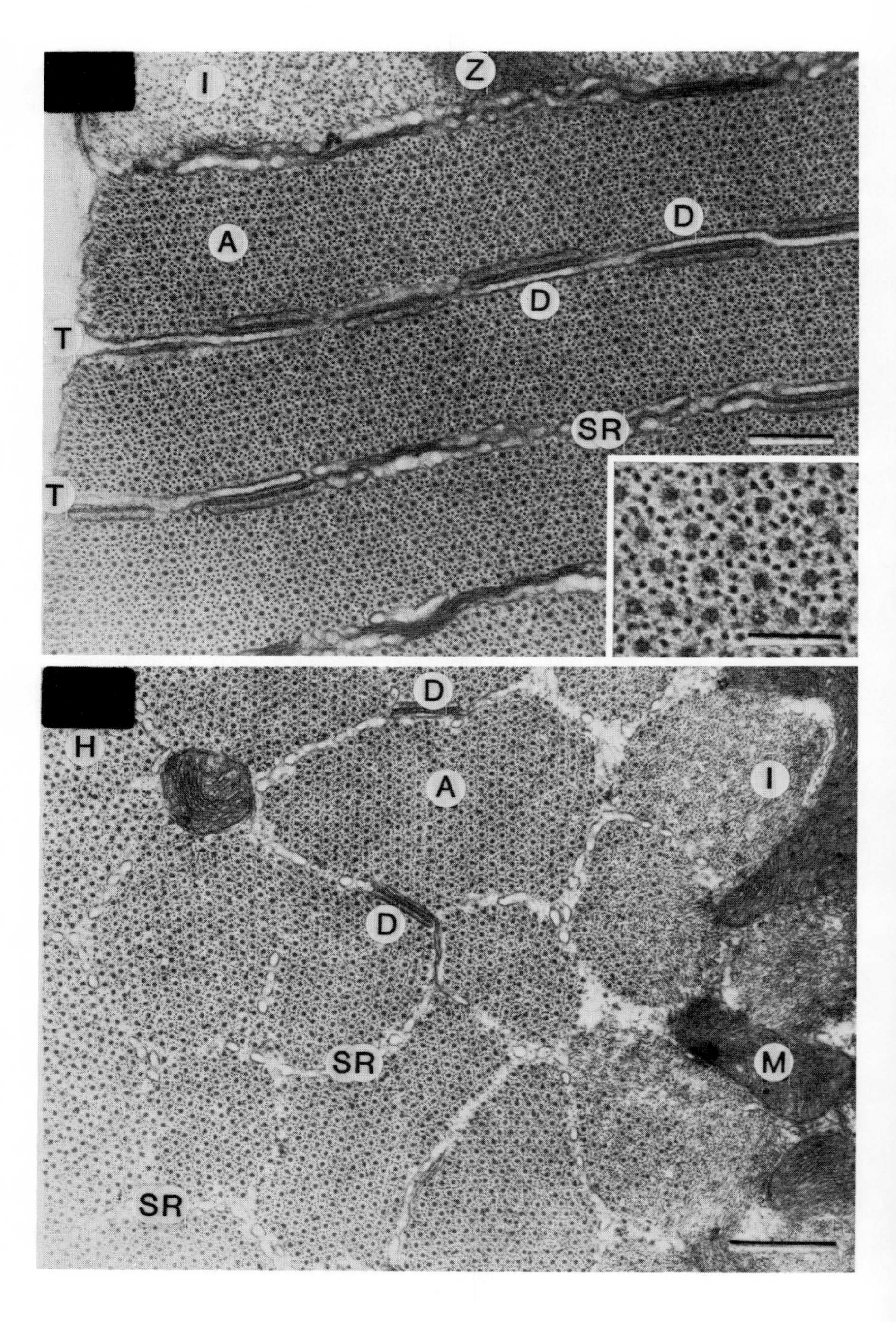
I
Z
D
A
D
T
SR
T
D
H
A
I
D
SR
M
SR

Figs. 11.5–11.6 Electron micrographs of cross-sections of fast-contracting (Fig. 11.5, m. 179) and slow-contracting (Fig. 11.6, m. 177e) muscles in a leg of *P. americana*. Note the layered, strap-like arrangement of fibrils in m. 179 (see also Fig. 11.8) and the polygonal arrangement in m. 177e. Inset: higher power, showing thin to thick arrangement characteristic of all three muscle types. Abbreviations as in Figs. 11.2–4; T, invaginations of the transverse tubular system. Scales: Fig. 11.5, 0.25 μm; inset, 0.1 μm; Fig. 11.6, 0.5 μm. (From Fourtner, 1978.)

Because *P. americana* is not a strong flier and therefore does not share some of the special adaptations for flight (such as the presence of fibrillar muscle), attention in this insect has been focused on leg muscle. Although there have been occasional strong claims, based on physiological results, that the different contractile properties of muscles are due almost entirely to the characteristics of the axons which innervate them (Usherwood, 1962a), there are actually many morphological (and biochemical – see Section 11.3) differences between different muscles. What is not known is the extent to which these differences contribute to the different contractile properties, since they are usually also correlated with different patterns of innervation.

One difference is at the level of organization of the myofibrils in the individual muscle fibres. In many muscles the myofibrils form strap-like arrays. This may be seen in large, fast-acting muscles in the coxa (e.g., m. 179, 137; Jahromi and Atwood, 1969; Fourtner, 1978; Fig. 11.5) or in the tibia (e.g., tibialis extensor: Atwood *et al.*, 1969; Fig. 11.8). In other muscles, the myofibrils appear round or polygonal. This seems to be characteristic of slower, weaker muscles (e.g., m. 177e; Fourtner, 1978; Fig. 11.6).

At the ultrastructural level, Jahromi and Atwood (1969) have shown that muscles innervated by 'fast' axons have a greater content of SR, lower mitochondrial content, shorter sarcomeres (averaging less than 4 μm) and a lower ratio of thin to thick filaments (as low as 3 : 1) (see Fig. 11.2, 11.5) than do muscles innervated by 'slow' axons. The latter show the opposite characteristics: low SR content, many mitochondria, sarcomeres over 8 μm in length and a thin to thick ratio of 6 : 1 or more (Fig. 11.4, 11.6). This general trend seems to hold in other insects as well (e.g. Cochrane *et al.*, 1972). Fourtner (1978) recently studied the ultrastructure of leg muscles differing in their innervation: those with 'fast' only (m. 178, 179), 'fast' and 'slow' together (m. 177d′, e′), and 'slow' only (m. 177d, e). He found striking differences in sarcomere (Figs. 11.2–11.4) and myofilament (Figs. 11.5, 11.6) arrangement, and in other characteristics. He suggested that these muscles, and by extension other skeletal muscles in *P. americana* and other insects, could be grouped into one of three types on the basis of their structural differences, and pointed out that this structural classification correlated well with a functional classification of the same muscles into those used in rapid (running), intermediate (walking) or slow (postural control) movement.

Clear differences can be shown even in muscles with similar innervation if the functional roles of the muscles is different. Smit *et al.* (1967) reported that a flight muscle (m. 135) capable of sustained activity had considerably more mitochondria than leg muscles (m. 136, 137) which fatigued rapidly, even though all three muscles were innervated exclusively by a 'fast' axon.

In addition to this intermuscular heterogeneity, there also appear to be differences between fibres within some muscles. Bhat (1970) has reported that, in *P. americana* flight muscles, some of the fibres are twice the size of others in the same muscle. And at the fine structural level, Hoyle (1967) and Jahromi and Atwood (1969) have pointed out that many leg muscles are not homogeneous. This structural heterogeneity has been studied in leg muscle by Morgan and Stokes (1979), who showed that even within a single anatomical muscle (m. 135d′) one could find groups of muscle fibres with structural characteristics quite different from those of nearby fibres.

It is clear that as more detailed studies are made of skeletal muscles in *P. americana*, more structural (and biochemical, see Section 11.3) diversity and heterogeneity will be found. The challenge now is to bring order to this tremendous variety. The most promising line of attack would seem to be that already suggested by several authors (Jahromi and Atwood, 1969; Morgan and Stokes, 1979): to see whether the *exact* innervation of each muscle fibre in terms of 'fast', 'slow' and inhibitory axons is related to the structural and other variance of those fibres. As pointed out above, all fibres in a single muscle do not receive innervation from all the motor axons which make contact with part of that muscle. It may be, therefore, that the small bundle of fibres in m. 135d′ found by Morgan and Stokes (1979) was missed by workers studying the innervation of the muscle, and that the fibres actually have a different innervation than the rest. The data of Denburg (1978) on innervation-correlated differences in protein content of muscles give us some hope that this line of inquiry will produce fruitful results.

We should also not forget that all the studies cited above have dealt only with leg or flight muscle. The few existing studies of other skeletal muscle in *P. americana* (e.g. Smith, 1966) and other insects suggest some important differences from those muscles we know well. It may be too early to generalize.

(b) *Visceral muscle*

One of the interesting features of the structural organization of skeletal muscles shown in Figs. 11.2–11.4 is that the slower contracting fibres have less clearly defined banding patterns, because the various bands are not as precisely aligned as in faster muscles. Visceral muscle in *P. americana* can be considered as a continuation of this trend (Fig. 11.7). Proctodeal muscle, for example, has extremely slow contraction times (Section 11.6), and rather indistinct A and I bands (Nagai and Graham, 1974). The sarcomere length varies from about 6–7 μm, and the ratio of thin to thick filaments is about 4 : 1 (inset, Fig. 11.7), with the distribution being uneven compared to that seen in skeletal muscle (inset, Fig. 11.5). The SR is not extensive, nor are mitochondria abundant.

I am not aware of any studies of the fine structure of the fore- or midgut muscles of *P. americana*. One would expect the general features described above to be found in those muscles as well. Anderson and Cochrane (1978), for example,

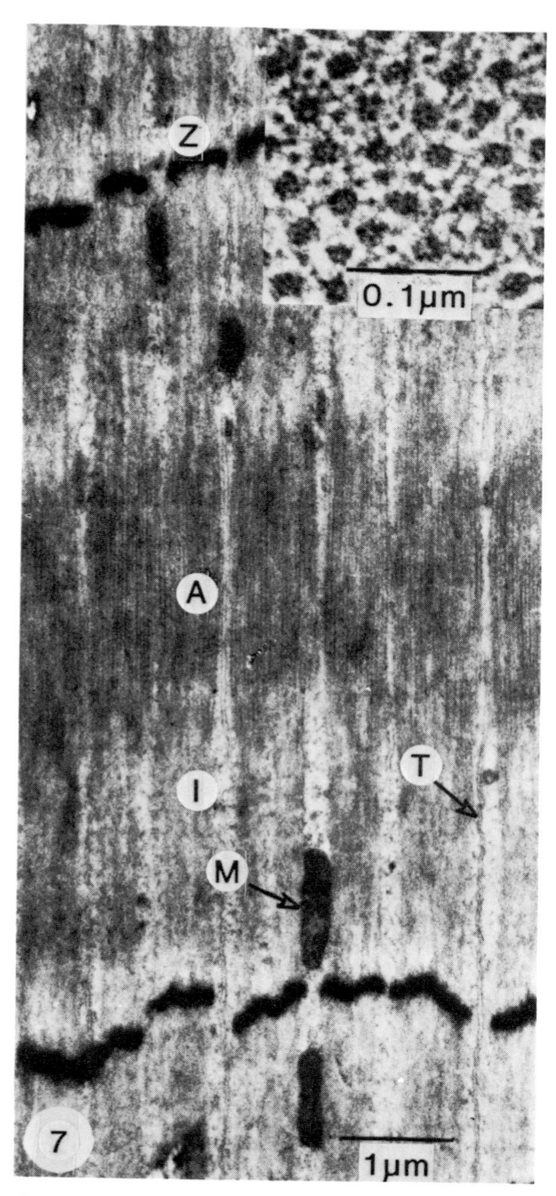

Fig. 11.7 Electron micrograph of a longitudinal and (inset) cross-section of *P. americana* proctodeal (hindgut) muscle. Note the paucity of SR and mitochondria, and the relatively long sarcomeres. Abbreviations as in Figs. 11.2–11.6. Scale: 1.0 μm; inset, 0.1 μm. (After Nagai and Graham, 1974.)

found the same features in the muscles of the mid-gut in the locust *Schistocerca gregaria*, except that sarcomeres tended to be longer (up to about 11 μm).

Miller (1975a) gives a good general review of insect visceral muscle. His article includes reference to the scattered and sparse information on visceral muscles other than those of the gut, which I have not discussed here.

11.2.3 Neuromuscular junctions

In general, neuromuscular junctions in *P. americana* are similar to those of most insects (O'Connor *et al.*, 1965; Smit *et al.*, 1967; Atwood *et al.*, 1969; Hoyle, 1974b; general review by Osborne, 1975). Axons travelling along the muscle are generally invested with a sheath composed of glial cells, variously also referred to as Schwann or lemnoblast cells. Usually, the sheath surrounds the axon on all sides except in the region of the synapse, which shows close apposition between the axon and the nearby muscle, and where a number of synaptic vesicles are visible. In other cases, the axon may lie in a cleft in the muscle, and be without any glial investment (Fig. 11.8). If the axon is not in a cleft, the basal lamella of the muscle fuses with the membrane of the investing glial cell over the synapse (see Piek and Njio, 1979).

There has been some interest in the identification of specific types of axons (i.e., 'fast', 'slow' or inhibitory) in electron micrographs of neuromuscular junctions. Atwood *et al.* (1969) were successful in carrying out a study of this type on 'slow' and 'fast' axon terminals. Their main finding was that 'slow' axons had significantly smaller axon profiles (Fig. 11.8) than did 'fast' axons (Fig. 11.9), and seemed in general to form fewer and smaller synaptic contacts than the latter. These differences might be general in other muscles and other insects. Edwards (1959) found two distinct sizes of axons in polyneuronally innervated abdominal muscle of *Blattella germanica*. Since his was a purely morphological study, he had no evidence that these two sizes might represent 'fast' and 'slow' axons, but at least his observation is consistent with the findings of Atwood *et al.* (1969).

11.3 Biochemistry

The general biochemistry and metabolism of muscle will not be considered here. This is in part because the bulk of the work on the subject has been on flight muscle, either fibrillar or from strong fliers, and *P. americana* has and is neither of these. Readers interested in this topic can consult recent general reviews by Crabtree and Newsholme (1975), Sacktor (1976) and Chippendale (1978), or for specific information about *P. americana* and other cockroaches, the earlier detailed survey by Guthrie and Tindall (1968).

However, we can ask about the relationship between the biochemical characteristics of a muscle, and its functional use by the insect. This problem is clearly related to that of structural specificity discussed above. The main biological differences between 'red' and 'white' muscles has, of course, long been known. Red muscles have a higher content of cytochrome (which confers on them their

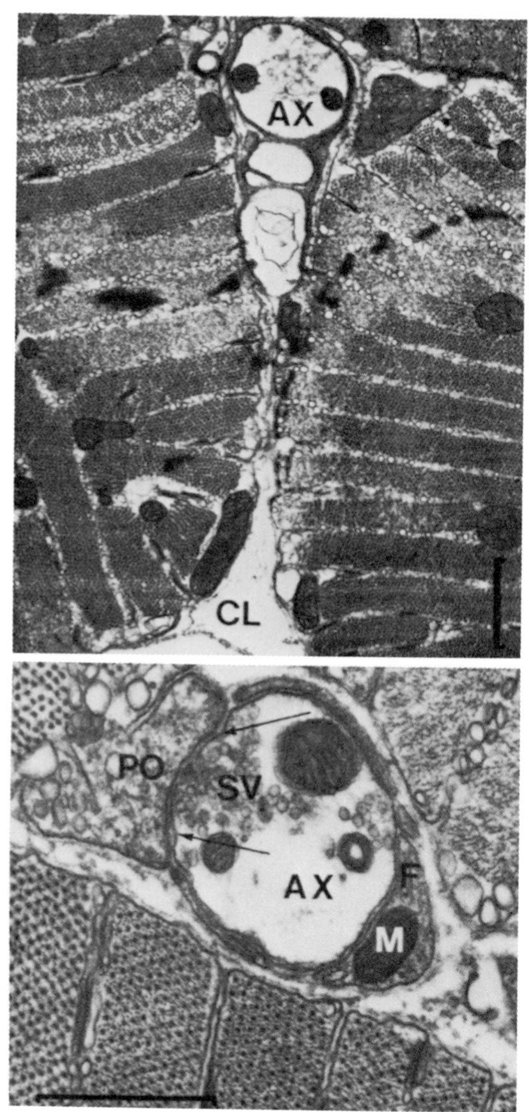

Fig. 11.8 Electron micrographs of 'slow' axon terminals in the extensor tibialis muscle in *P. americana*. Area of synaptic contact is marked by arrows. Abbreviations: AX, axon; CL, muscle cleft; M, mitochondrion; PO, post-synaptic pillar (of the muscle); SV, synaptic vesicles. Scale: 1.0 μm for both parts. (From Atwood *et al.*, 1969.)

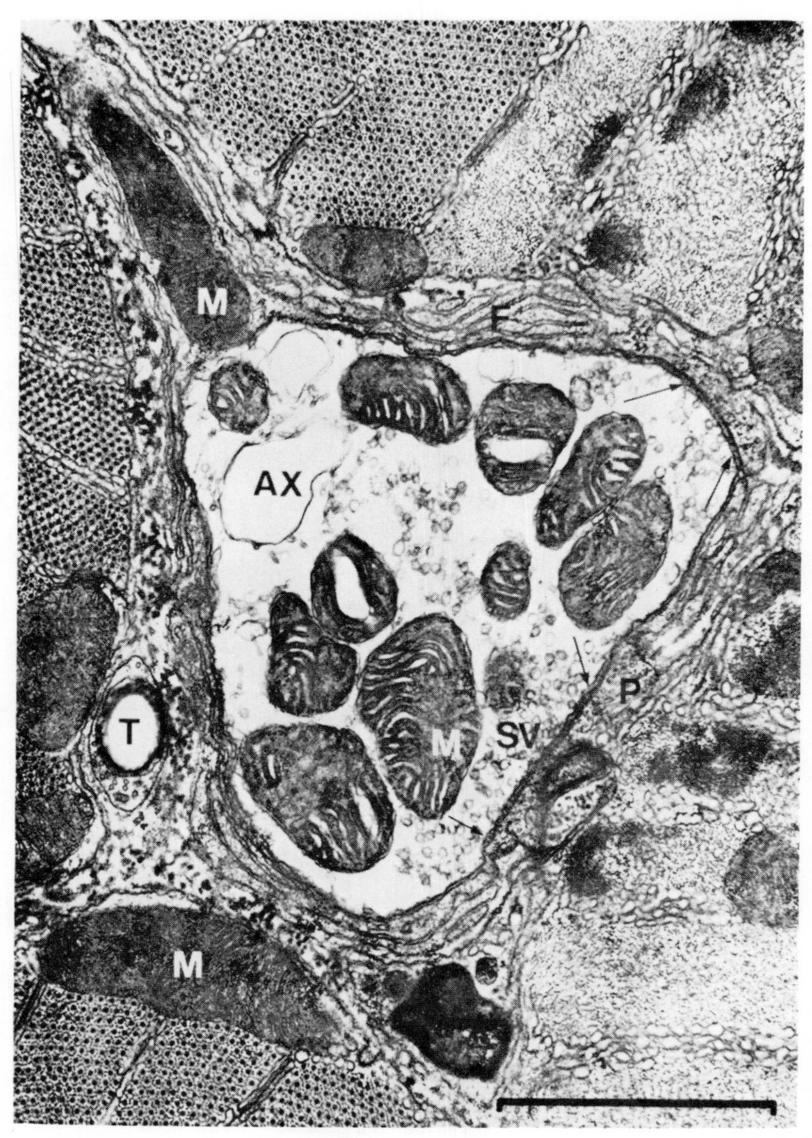

Fig. 11.9 Electron micrograph of 'fast' axon terminal in a cleft of the extensor tibialis muscle of *P. americana*. Two areas of synaptic contact are marked by arrows. Note larger size of terminal and of areas of contact here as compared to 'slow' terminal (Fig. 11.8). Abbreviations as in Fig. 11.8. T, tracheole. Scale: 1.0 μm. (From Atwood *et al.*, 1969.)

characteristic colour), more succinic dehydrogenase (indicative of greater oxidative capacity), more glycogen and more mitochondria, than do white muscles (see reviews by Smit *et al.*, 1967; Elder, 1975). Together, these characteristics confer on the muscle a significantly greater capacity for sustained work. In general, red muscle occurs where the insect requires long-term activity without significant fatigue, as in flight.

But just as there are structural differences between muscles with different functional roles, so there are biochemical ones. For example, Denburg (1975) was able to show for a set of metacoxal muscles in *P. americana* that three specific and characteristic patterns of protein distribution existed. (The muscles studied by Denburg (1975) were the same set later examined ultrastructurally by Fourtner, 1978). Muscles which receive only fast axon innervation (m. 178, 179) have one pattern, muscles with only slow axon innervation (m. 177d, e) a second, and muscles with mixed innervation (m. 177d′, e′) a third, intermediate between the first two. Histochemical analysis of these muscles has also shown that those with fast innervation have significantly less dehydrogenase activity and glycogen than the other two types (Hart and Fourtner, 1979). Since Fourtner (1978) and Denburg (1975) were able to group the same set of muscles the same way on entirely different grounds, we now have good reason to investigate further the later suggestion by Denburg (1978) that the mechanical and contractile properties of a given muscle may be due at least as much to intrinsic properties of the muscle fibres in it as to the excitation properties of its innervating neurons. This proposal has the additional merit, as alluded to by Denburg (1978), of allowing us to make sense out of the fact that two different muscles innervated only by fast axons may nevertheless show striking biochemical (Smit *et al.*, 1967) and morphological (Jahromi and Atwood, 1969) differences.

But biochemical heterogeneity can also exist *within* single anatomical muscles. Stokes *et al.* (1979) have demonstrated such a heterogeneity in fibres within muscle 135d′, just as Morgan and Stokes (1979) demonstrated a structural one (Section 11.2.2). In any case, it appears that, for a complete understanding of the biochemical and structural basis for the different functional roles of various leg muscles, a fibre-by-fibre analysis of muscles will be necessary.

11.4 Electrophysiology

11.4.1 Ionic basis of electrical responses

In general, insect muscle is an electrically excitable tissue, and like other such tissues, its electrical characteristics depend on the distribution and movement of ions across the membranes of its constituent cells. What is the ionic basis of these characteristics? Although this problem is not especially severe in *P. americana*, as we shall see, it is by no means trivial for insects in general. As has been pointed out in several recent reviews (Usherwood, 1969a; Piek and Njio, 1979), some insect muscle contracts with vigour in salines with ionic compositions that would almost instantly knock out a vertebrate muscle.

Traditionally, the approach to this problem has been to assess the concentrations of the major ions in muscle tissue and in the haemolymph to determine the effects of substituting for one or more ions in the external medium, and compare the resulting change in resting potential and active responses (if any) with the changes predicted by electrochemical theory. Early measurements of this sort (Hoyle, 1955; Wood, 1963, 1965) suggested that Na^+ and K^+ were quite important while Cl^- was less so in generating both active responses and resting potentials.

Two matters which have caused some difficulty in interpreting the ionic basis of muscle activity in some insects must, however, be considered (see Usherwood, 1969a; Piek and Njio, 1979). First, it is important that the concentration (more correctly, activity) of the free, available extracellular ion be measured. This will probably not be the same as the total ionic content of the haemolymph bathing the muscle since some of the ions will be bound. Weidler and Sieck (1977), for example, found that about 22% of Na^+ and 10% of Cl^- in *P. americana* haemolymph was in fact bound, and therefore not available to contribute to any electrical event in muscle. Secondly, even the free ions in the haemolymph may not have ready access to the surface of the muscle plasma membrane because of the presence of the basal lamina. Piek (1974) has suggested that the ionic content of the medium just outside the plasma membrane may be regulated by the combined action of the acellular covering and the muscle itself.

The most recent work which takes into account these considerations suggests that in *P. americana*, the resting potential in skeletal muscle, about −60 mV, is based largely on passive distribution of K^+, with a small (around 18 mV) active electrogenic component (Wareham *et al.*, 1974a,b; see also Piek and Njio, 1979). Active responses seem to be based primarily on conventional Na^+ and K^+ movements. However, the question has not been investigated thoroughly in *P. americana*, and one should be aware that in other insects (flies, moths) an important role for calcium has clearly been demonstrated (Piek and Njio, 1979, p. 229).

The situation in visceral muscle is less well known. Investigation of the ionic basis of the resting and action potentials in muscles of the hindgut of *P. americana* (Nagai, 1972) has shown that the usual ions are certainly important. The resting potential is largely dependent on external K^+ concentration, but it is clear that one or more other ions, perhaps Na^+, Cl^- or Ca^{2+}, are also involved. Action potentials seem to be based on influx of Na^+, since reducing external sodium reduces the amplitude and rate of rise of the spike. However, these spikes are not entirely conventional, since TTX (Chapter 8), which blocks Na^+ channels in nerve, fails to block spike production in this muscle. This tissue thus might repay careful study of its electrochemistry.

11.4.2 Electrical properties

(a) *Passive ('cable') properties*

Although the passive electrical properties of skeletal muscle (which determine the rate and extent of spread of post-synaptic potentials along the plasma membrane) have been determined for a number of insects (see Table in Usherwood, 1969a, pp. 212–213), *P. americana* is not among them. The average values which have been determined in other insects range from 0.4–2.8 mm for the length constant, 1.2–5.3 μFcm^{-2} for membrane capacitance and 2.5–16 ms for the time constant. Presumably, those for *P. americana* would fall in these ranges.

The values of these constants have been determined for rectal muscle of *P. americana* by Belton and Brown (1969). These authors reported length constants of from 1.5–4.0 mm, membrane capacities of 0.4–3.5 μFcm^{-2} and a time constant of 70 ms. The much larger time constant in this muscle compared to skeletal muscle is reflected in the longer rise and decay times of its electrical potentials, as discussed below.

(b) *Active properties*

Most insect muscle (referred to as phasic muscle) is electrically excitable, that is, will produce an active electrical response of the membrane when sufficient current is passed through it. However, non-excitable muscle (called tonic muscle) has been reported in cockroaches and other insects (see Usherwood, 1967; Cochrane *et al.*, 1972). Unfortunately, the terms tonic and phasic both have sometimes been applied to electrically excitable muscle, to refer respectively to slow-contracting and fast-contracting fibres. Cochrane *et al.* (1972) suggest that such usage should be abandoned in arthropods since some fibres can contract either quickly or slowly depending on which type of excitatory axon stimulates the contraction (Section 11.4.3).

Phasic (which is to say nearly all) muscle in insects responds in a *graded* fashion to direct stimulation (Cerf *et al.*, 1959). That is, the muscle membrane responds with a potential the amplitude of which is proportional to the strength of the stimulus. This active response is thus not all-or-none. The response is always smaller than the stimulus, so it propagates decrementally (i.e. gets smaller as it travels away from the point of origin). This type of response has sometimes been referred to as an action potential (AP), but Usherwood (1969a) suggests that the term AP should be reserved for all-or-none potentials. Decrementally conducted, graded responses occur in *P. americana* (Becht *et al.*, 1960; Usherwood, 1962a) and many other insects (reviews by Usherwood, 1967, 1969a; Hoyle 1974b).

An interesting feature of insect skeletal muscle is that in the presence of certain drugs or ions it is possible to convert graded responses to all-or-none action potentials. Usherwood (1962b) demonstrated this in *P. americana* using the alkaloid ryanodine, and Werman *et al.* (1961) in grasshoppers using Ba^{2+}, Sr^{2+} and Ca^{2+}. The latter authors have suggested that normally, muscle fibres have a rather high conductance to K^+, and that depolarizing the membrane results in an activation of K^+ conductance which is much faster than in vertebrate muscle or in nerve, resulting in an effective short-circuit of the spike. Ryanodine and the divalent cations may have the effect of reducing this K^+ conductance, allowing a normal AP to develop.

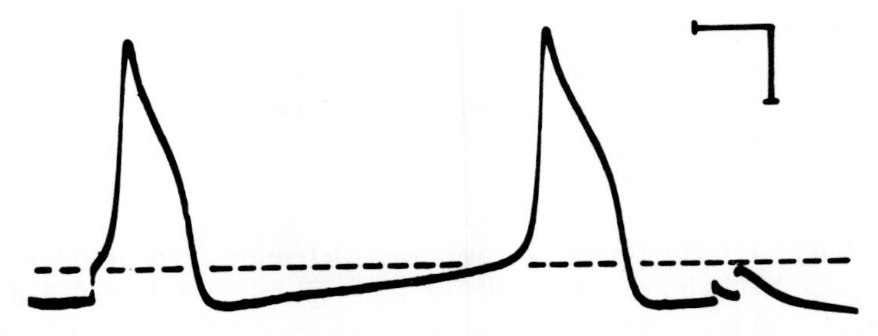

Fig. 11.10 Action potentials in a proctodeal muscle fibre of *P. americana*. The potentials were produced from (left) a spontaneous post-synaptic potential and (right) the depolarizing phase of fluctuating membrane potential (this muscle shows spontaneous rhythmicity of contraction without neural stimulation.) Calibration: 10 mV and 0.5 s. (After Nagai and Brown, 1969.)

In contrast to skeletal muscle, visceral muscle (in *P. americana*) shows all-or-none active responses to stimulation (Nagai and Brown, 1969; see Fig. 11.10). The responses, although rather long lasting (up to 1 s), may properly be referred to as action potentials since they do not vary in height, and show a clear threshold of elicitation.

11.4.3 Electrical responses to neural stimulation

It was pointed out in Sections 11.2.2 and 11.3 that there are important structural and biochemical differences between different muscle fibres. Yet it is also true that different axons may evoke rather different electrical responses, even in a single fibre. The type of innervation a muscle receives therefore has much to do with the way in which that muscle may be used by the insect in day to day activities. Traditionally, two types of excitatory innervation, 'fast' and 'slow', have been identified. Usherwood (1969a, 1974) has pointed out that although these terms originally referred only to the mechanical responses of whole muscles to stimulation of an axon, they have gradually come to be used to describe the electrical responses of individual muscle fibres as well.

Unfortunately, categorizing these electrical responses is not always an easy matter, since they may vary quite considerably in characteristics from one muscle fibre to another, and a suitably varied terminology has not yet come into use. Some authors have occasionally tried to add to or alter the traditional terminology (see, for example, Usherwood, 1969a, 1974), but so far without noticeable impact on the practitioners in the field. The problem is not severe for *P. americana*, probably only because we do not have recordings from a great diversity of fibres. However, readers should expect occasional ambiguities in terminology in some original reports and in reviews.

(a) *'Fast' innervation*

The electrical response of a muscle fibre to stimulation of a 'fast' axon is a large, usually overshooting potential (i.e., one which depolarizes the fibre membrane

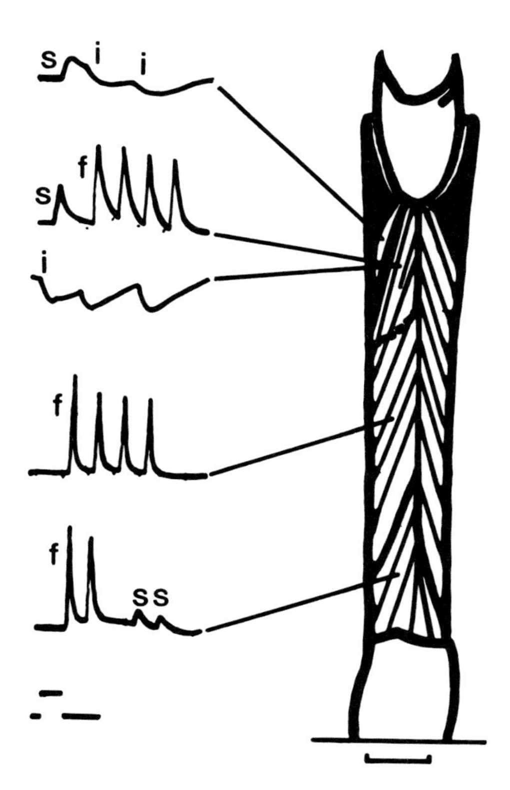

Fig. 11.11 Diagram of *P. americana* femur showing the extensor tibialis muscle and the three main types of post-synaptic potentials which can be recorded from it, 'fast' (f), 'slow' (s) and inhibitory (i). Note variation in amplitude of 'slow' potentials and antifacilitation in 'fast' responses. Calibration: 10 mV and 10 ms. Scale: 1 mm. (From Atwood *et al.*, 1969.)

to zero or a positive value), which does not show facilitation (an increase in size upon repetitive stimulation). 'Fast' axon responses are shown in Fig. 11.11. The potential is the result of the summation of the excitatory post-synaptic potential (EPSP; in muscle sometimes called the excitatory junctional potential or ejp), which is the potential developed at the synapse, and the active electrical response of the rest of the membrane. (An EPSP can of course be recorded throughout the fibre because the fibre is multiterminally innervated.) The active response of the non-synaptic membrane is graded and propagates decrementally.

In *P. americana* coxal muscle, the EPSPs generated by 'fast' axon stimulation are about 40–50 mV (Becht *et al.*, 1960), and the active response of the membrane is around 20–35 mV, giving total responses of about 60–80 mV. This yields a transient membrane potential of from 0 to +20 mV, which is typical of many insect muscles. Muscles in the femur, on the other hand, have smaller total responses, around 45 mV (Atwood *et al.*, 1969; Fig. 11.11), so there is obviously some variation.

(b) *'Slow' innervation*
The electrical response of a muscle fibre to 'slow' axon stimulation is extremely variable. In *P. americana*, measured amplitudes of 'slow' responses in femoral and coxal muscle may vary from a few to perhaps 25 mV (Becht *et al.*, 1960; Atwood *et al.*, 1969; Pearson and Bergman, 1960). These can show considerable facilitation; Pearson and Iles (1971) show facilitation of well over 300%. In the locust *S. gregaria*, which has been more intensively investigated, Hoyle (1957) has shown 'slow' axon responses in leg muscles ranging from less than 10 mV to greater than 60 mV. The latter value was reached after facilitation of about 150%. Potentials produced by 'slow' axon stimulation are often below the threshold for active membrane response, and therefore consist only of an EPSP. Large potentials clearly have an active component, but the exact value of this has not been determined. The EPSPs usually have a rather long time course relative to their amplitude, compared to a 'fast' axon response (Fig. 11.11 and Hoyle, 1957). Most measured durations in *P. americana* are about 5–10 ms (Becht *et al.*, 1960; Atwood *et al.*, 1969), but Pearson and Iles (1971) reported durations of up to 30–40 ms, as did Hoyle (1957) for the locust. The study of innervation by Pearson and Iles (1971) is especially interesting because it revealed EPSPs of considerably different time courses in different muscle fibres which receive input from branches of a single 'slow' excitatory neuron. Since the rate of decay of the EPSP is a function of the passive properties of the muscle membrane, this difference cannot be attributed to a difference in the placement of the recording electrode relative to the synapse, as can some of the variation in height.

(c) *Inhibitory innervation*
Direct peripheral inhibition was demonstrated by Usherwood and Grundfest (1965) in locusts, grasshoppers and two species of cockroach (*Blaberus* sp.). The phenomenon was probably observed by earlier workers (see Hoyle, 1974b), but could not be demonstrated convincingly. The electrical response to stimulation of an inhibitory axon is generally a small hyperpolarizing potential (Usherwood and Grundfest, 1965), although if recordings are made with KCl-filled electrodes this may be converted to a depolarizing one due to the influence of the Cl^- which tends to leak from the electrode tip (Usherwood, 1968).

Post-synaptic responses to inhibitory neuron stimulation in *P. americana* have been studied in detail by Pearson and Bergman (1969) and Pearson and Iles (1971). Both sets of authors found differences in the amplitudes of the IPSPs (inhibitory post-synaptic potential, sometimes called the inhibitory junctional potential, ijp) generated by stimulation of different axons innervating the same muscle fibre. The IPSPs, all of which were hyperpolarizing, varied from less than 1 mV to about 5mV. While some of the variation was clearly a function of electrode position relative to the synapses, it was also clear that some inhibitory axons consistently produce larger post-synaptic responses than do others, just as some 'slow' axons consistently produce larger EPSPs than others.

Pre-synaptic inhibition in which the inhibitory axon causes a reduction in the amount of transmitter released by an excitatory axon, has been reported to occur in *P. americana* (Parnas and Grossman, 1973). The physiology of this inhibition has not been studied in detail.

(d) *Visceral muscle*

The electrical responses of proctodeal muscle fibres in *P. americana* have been studied by Nagai and Brown (1969) and by Nagai (1973). Nerve stimulation results in the generation of one or more EPSPs, ranging from a few millivolts to as much as 12 mV in amplitude (but see below) with a duration of roughly 100 mV. Individual EPSPs can summate, and if the post-synaptic response is large enough, can generate an action potential (Fig. 11.10). Since APs generated in a muscle fibre are all of the same size and appear in an all-or-none fashion, it is presumed that they are fully regenerative and propagated; i.e. they are not conducted decrementally like the active responses evoked in skeletal muscle.

Although EPSPs as large as 12 mV are illustrated by Nagai and Brown (1969), the authors point out that these may be composite potentials built up from summed individual PSPs at several synapses in close proximity to one another. Most of the EPSPs that large show 'humps', more clearly indicating their composite nature. Unitary EPSPs are therefore probably only 2–5 mV, values which correspond to those obtained by Anderson and Cochrane (1977) in locust midgut fibres. EPSPs in the locust muscle show facilitation at frequencies of stimulation up to 4 s^{-1}, but antifacilitation (a decline in amplitude) at higher rates (Anderson and Cochrane, 1977). Nagai and Brown (1969) did not explicitly study the effect of frequency of stimulation on EPSP size, but the data they report tend to show a similar phenomenon at work.

11.5 Neuromuscular transmission and its chemical basis

In some respects, the study of neuromuscular (NM) transmission in insects has produced results with significant impact far beyond the insects alone. This is because there was evidence that the transmitter substance at excitatory NM junctions was glutamate, a common amino acid which is distributed widely throughout the insect body. At the time this evidence was accumulating, nearly all known transmitter substances were materials which seemed to be present almost exclusively at synapses, and to have no other role. Therefore, the suggestion that a substance as common as glutamate might serve as a transmitter met a great deal of resistance. But as the evidence in favour of this view mounted, it had the effect of opening in the minds of other researchers the idea that a common substance could be a transmitter, with consequences for our understanding of vertebrate brain function which we are just beginning to recognize (see Usherwood, 1978, for further discussion of this point).

General aspects of NM transmission have been reviewed thoroughly by Usherwood (1974). Usherwood and Cull-Candy (1975) and Usherwood (1978) have discussed in more detail than can be done here the substantial literature on the identification or detailed action of specific transmitter substances.

11.5.1 Transmission at excitatory neuromuscular junctions

(a) *Glutamate as a transmitter substance*

The transmitter at some insect NM junctions is L-glutamate. This statement can be supported by at least five general lines of evidence: (1) The electrophysiology of post-synaptic responses to axonal stimulation are similar in time course and other characteristics to that of responses to iontophoretically applied glutamate (e.g. Berânek and Miller, 1968; Anwyl and Usherwood, 1974). (2) The pharmacology of synaptic transmission and of glutamate are similar. That is, substances which block glutamate action also block transmission (e.g. Usherwood and Machili, 1968). (3) Glutamate can be collected from active synapses (e.g. Usherwood *et al.*, 1968). (However, the work supporting this statement is not without controversy; see Usherwood and Cull-Candy, 1975, p. 249.) (4) A high-affinity uptake system for glutamate exists (e.g. Faeder and Salpeter, 1970; Salpeter and Faeder, 1971; Faeder *et al.*, 1974; Botham *et al.*, 1978). Since no enzyme capable of degrading glutamate at the NM junction has been found, rapid re-uptake is presumably the mechanism by which the material is inactivated. (5) The effects on glutamate sensitivity of denervating insect muscle are similar to the effects on acetylcholine sensitivity of denervating a vertebrate muscle (Usherwood, 1969b).

Most of the accumulated evidence comes from work on locusts. The specific evidence regarding NM transmission in *P. americana* is almost non-existent. It has been shown that bath-applied L-glutamate will cause depolarizations of leg muscles in this insect (Kerkut *et al.*, 1965b; Kerkut and Walker, 1966, 1967), but iontophoretic application has not been tried. One study (Atwood and Jahromi, 1967) has been carried out on the pharmacology of glutamate and transmission in *P. americana*. The single study on glutamate release from active synapses in this insect (Kerkut *et al.*, 1965a) has been criticized on technical grounds (see Usherwood and Cull-Candy, 1975, p. 249).

One of the main obstacles to the acceptance of glutamate as a transmitter consisted of reports which suggested that glutamate was present in high enough concentrations in the blood of locusts and other insects that it ought to depolarize and desensitize the muscles of the body. Miller (1978) has written an unusually clear review of the issues and arguments involved, and I will not repeat his analysis. The crux of the matter until recently has been whether glutamate might not be in a bound or otherwise inactive form. However, Irving *et al.* (1979) have recently claimed, on the basis of new and more careful chemical analysis, that there is effectively *no* glutamate in the haemolymph of *Periplaneta* or of several other insect species (i.e. the amount was less than the 10^{-5} M they could detect with their techniques). Irving *et al.* (1979) suggest that previous work yielded erroneous results because inadequate precautions were taken to prevent the conversion of glutamine (which is abundant in insect haemolymph) to glutamate, which occurs readily under the conditions of temperature and pH in which most analyses are carried out. This work clearly needs to be repeated in other laboratories, but for the present it gives us some hope that this entire controversy can be laid to rest and most of the doubt which remains about glutamate as a transmitter can be removed.

(b) *Other skeletal muscle transmitters*

Evidence that aspartate may be the transmitter at 'slow' NM junctions in fly larvae has recently been published by Irving and Miller (1980a,b). In this context, it is interesting to recall that Usherwood *et al.* (1968) found a slight increase in the release of aspartate on stimulation in a locust neuromuscular preparation. The fact that the quantity of aspartate detected was much less than the quantity of glutamate, could be a reflection of release from the smaller neurons and fewer synaptic contacts which are characteristic of 'slow' axon innervation (Atwood *et al.*, 1969; Section 11.2.3). However, the possibility that aspartate may act as a transmitter at slow NM junctions in *P. americana* or other orthopteroid insects must be recognized as pure speculation at this point.

(c) *Transmitter release*

The release of transmitter from the NM junction appears to be quantal in insects just as it is in vertebrates (see Usherwood 1974, for review). Certainly spontaneous release in insects produces miniature end plate potentials (mepps), whose frequency of appearance is enhanced by high Ca^{2+} and depressed by high Mg^{2+} concentrations (Usherwood, 1963; Smyth *et al.*, 1973; Washio and Inouye, 1978). However, the appearance of miniatures does not appear to be entirely random. Usherwood (1972) has shown in locusts, and Rees (1974a) in *P. americana*, that there is a tendency for short 'bursts' of miniature potentials to occur. Washio and Inouye (1975) reported that in *P. americana*, if sequences in which these bursts appear are omitted, the remaining mepps occur in a random (Poisson) distribution. However, Rees (1974a), using a more sensitive statistical test, suggested that even if the bursts are left out (which in his case eliminated about 5% of his recordings), the occurrence of mepps is better described by a negative binomial distribution than by a Poisson one. The difference between these two distributions is quite small, but it is nevertheless important that we know which gives the better fit. If it is not a Poisson distribution, we must infer that there is a slight increase in the probability of the release of a packet of transmitter just after one has just been released, and it is clearly important that the mechanism for this be worked out.

In another study, Rees (1974b), has demonstrated how many complex factors may be involved in transmitter release. He showed that in *P. americana*, metabolic *inhibitors* have the short-term effect of increasing the frequency of appearance of mepps. He suggested that the increased release of quanta that this implied may be due to various secondary effects of the inhibitors on factors which control release. There were also important differences between the effects of a particular dose of inhibitor on white and red muscle fibres. Obviously, further studies of this system must be done with careful control of the type and physiological condition of muscle fibre being investigated.

11.5.2 Transmission at inhibitory junctions

The identification of γ-amino butyric acid (GABA) as the transmitter at many inhibitory NM junctions has not been beset with nearly as much controversy as has that of glutamate as the excitatory transmitter. The main reason for this seems to be the lack of wide distribution of GABA except in nerve tissue. In addition, by the time GABA was implicated as a transmitter in insects it was well established in this role in Crustacea (see Usherwood, 1978).

Little detailed work with GABA exists on cockroaches. The original demonstration that GABA would mimic the effects of stimulation of an inhibitory neuron, made by Usherwood and Grundfest (1965) on locusts and grasshoppers, was quickly followed by qualitative studies showing that, in *P. americana*, bath application of the substance would prevent or reduce the response to excitatory stimulation or to application of glutamate (Kerkut *et al.*, 1965b). Smyth *et al.* (1973), on the basis of their recordings of miniature IPSPs, have suggested that the transmitter is released quantally. It is generally assumed that other details, established in Crustacea and locusts, hold for *P. americana* as well. GABA also appears to be the transmitter at pre-synaptic inhibitory junctions (Parnas and Grossman, 1973).

11.5.3 Transmission in visceral muscles

Current evidence suggests that there is at least some neural control over visceral muscle, exerted in the conventional way by the release of transmitter substances at NM junctions (Davey, 1964; Cook and Holman, 1975a,b).

Several possible candidate compounds have been proposed as neurotransmitter substances. In the cockroach *Leucophaea maderae*, Holman and Cook (1970) have suggested that glutamate, and possibly aspartate, might act as transmitters in the hindgut. They cite the following (and other) evidence: L-glutamate is present in the nerves innervating the hindgut; application of 2×10^{-6} gml^{-1} glutamate to the hindgut mimics neural stimulation; and agents which block glutamate action also block the effects of neural stimulation.

Another important candidate transmitter is the pentapeptide molecule called proctolin, the first neuroactive peptide to be identified in insects (see Section 12.4.1). Proctolin has a powerful effect on the activity of muscles of the hindgut and the foregut. A concentration of 10^{-9} M is sufficient to bring about contractions of gut muscle similar to those which follow repetitive nerve stimulation (Brown, 1975). In addition, other physiological and pharmacological tests (similar to those described above for glutamate) strongly suggest proctolin is the natural transmitter at some visceral NM junctions (Brown, 1975). Later work by Brown (1977) indicated that proctolin is widespread throughout many orders of insects.

However, it seems likely that proctolin and glutamate are not the only transmitter substances used at visceral NM junctions. There is some evidence that serotonin (5-hydroxytryptamine), for example, may also be a transmitter in both *P. americana* (Brown, 1975) and *L. maderae* (Cook and Holman, 1978). But further, there is no reason to assume that the same transmitter is employed in all parts of the gut. Indeed, Cook and Holman (1978) have shown marked differences in the sensitivity of fore- and hindgut muscle in *L. maderae* to glutamate and proctolin.

Midgut muscle has, as usual, been neglected entirely with respect to its pharmacology.

11.5.4 Octopamine and neuromodulation

One interesting development over the last decade has been the gradual realization of the extent to which the endocrine and nervous systems overlap in their functions. The neuroendocrine component of the nervous system and its neurosecretory function have, of course, been known for many years (see Chapter 12), but only recently has it been recognized that the distinction between a hormone and a neurotransmitter substance may be so fine as to allow no certain separation.

The role played by octopamine in the control of muscle contraction is a case in point. Crossman *et al.* (1971) described a number of unpaired neurons in the thoracic ganglia of *P. americana* and several grasshopper species. These neurons send branches to several leg muscles, but are apparently neurosecretory, not conventional motoneurons, even though their axons terminate on muscle fibres (Hoyle *et al.*, 1974). Subsequent work, all on locusts or grasshoppers, showed that at least one of these DUM (dorsal unpaired median) neurons releases octopamine at its terminals (Hoyle, 1975; Evans and O'Shea, 1978), modulates the inherent rhythmicity of certain tonic muscle fibres in the leg (Hoyle and O'Shea, 1974; Hoyle, 1974a, 1978a; Evans and O'Shea, 1978) and affects the rate of development of tension and relaxation in other muscle fibres in the leg (O'Shea and Evans, 1979).

Although most studies of the DUM neurons have dealt with their control of inherent rhythmicity (which *P. americana* lacks) of the tonic leg muscles, their effects on other leg muscles would seem to be of more general significance, and more applicable to cockroaches as well. These effects can be summed up in one word: modulation. Neither octopamine nor electrical stimulation of the DUM neuron has any significant direct effect on phasic muscle fibres (O'Shea and Evans, 1979). But when octopamine or stimulation is paired with stimulation of a 'slow' excitatory neuron, the effect is a significant increase in the amplitude of the postsynaptic potential and in the force of contraction of the muscle. O'Shea and Evans (1979) show that these modulatory effects are due to the interaction of octopamine with receptors on both the muscle and the 'slow' axon terminals.

Some substances which are apparently classical neurotransmitters also appear to have modulatory effects on insect muscle. Proctolin can induce rhythmic contractions in locust tonic muscle (Piek *et al.*, 1979), and serotonin can increase contractions in the same muscle (Evans and O'Shea, 1978). It can also induce contractions in *P. americana* malpighian tubule muscle (Crowder and Shankland, 1972b, 1974). The discovery of these hormone-like modulatory effects on skeletal muscle may force drastic revisions in our views on how muscle actions are regulated.

11.5.5 Effects of toxins

No synthetic pesticides have any significant effect on muscles or the neuromuscular junction, presumably because their actions are specific to cholinergic systems or to the axonal membrane (Section 8.6.1). Ryanodine, a plant extract with insecticidal properties, does act directly on insect muscle, but the substance does not have widespread commercial use.

On the other hand, many of the toxins found in animal venoms, especially from animals which use their venoms against arthropods, have dramatic effects on the neuromuscular system. In *P. americana*, the main effect of the venom of the black widow spider, for example, is a massive depletion of transmitter from both excitatory and inhibitory synapses (Griffiths and Smyth, 1973). Electron micrographic study of the depleted terminals in other animals affected by the venom reveals them to be bloated and empty of vesicles, suggesting that the venom acts by completely disrupting the motor terminal endings (see Smyth *et al.*, 1978).

Most studies of the effects of other venoms (e.g., from scorpions, solitary wasps, etc.) on insects have used locusts or Lepidoptera (the prey of many of the solitary wasps, etc.) as subjects. Interestingly, the results suggest that the venoms from the different groups of arthropods contain toxins which have quite different modes of action. The toxic agent of some solitary wasp venoms, for example, appear to act by slowing or stopping the flow of transmitter substance down the motor neurons, thereby leading to local depletion of the transmitter at the synapse, and consequent paralysis of the animal. Some components of scorpion venom, on the other hand, seem to have an excitatory effect on motor neuron endings, causing massive excitation and, in the whole animal, tremors and convulsions. Interested readers should consult the appropriate chapters in Shankland *et al.* (1978) for further information.

11.6 Mechanical properties

11.6.1 Skeletal muscle

Few authors have studied the mechanical properties of non-fibrillar insect skeletal muscle in any detail. Hoyle (1974b) and Aidley (1975) provide recent reviews in which they discuss what is known about elasticity, velocity of shortening, tension as a function of length and other mechanical characteristics of such muscle.

In *P. americana*, skeletal muscle has been most thoroughly investigated by Usherwood (1962a), who carefully examined the isometric mechanical responses of leg muscles to stimulation of their innervating axons. His results verified and expanded the well-known differences between 'slow' and 'fast' axon-innervated muscles. In general, a single shock to a 'slow' axon produces no detectable mechanical response in the whole muscle. Low frequency stimulation produces a slow, weak response, with a long time course (Fig. 11.12). Higher frequencies of stimulation produce increasingly more vigorous (but still slow) contractions, which may be 20 or more times as powerful as those produced by low frequency stimulation.

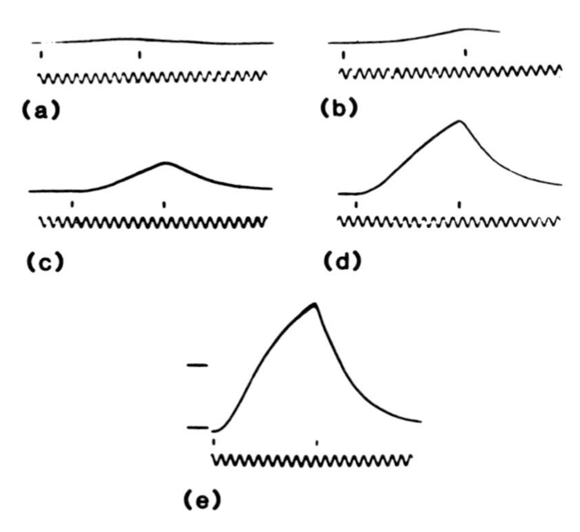

Fig. 11.12 Mechanical responses of m. 135d,e (innervated by a slow but no fast axon) to stimulation of its motor nerve at various frequencies. (a) 10 Hz, (b) 25 Hz, (c) 35 Hz, (d) 80 Hz, (e) 125 Hz. Note the lack of any twitch response even at low frequencies of stimulation, and the tremendous increase in the total tension at higher ones. Calibration: 0.5 g. (After Usherwood, 1962a.)

In contrast, a single shock to a 'fast' axon always yields a vigorous twitch (Fig. 11.13). Low frequencies of stimulation produce a series of twitches, not a stronger contraction. Only at frequencies of stimulation greater than about 30–40 Hz do individual twitches begin to fuse into a single smooth contraction. The fused contraction is always more powerful than individual twitches, but only by a factor of two to five.

This work is a good illustration of the classical differences between 'slow' and 'fast' axon effects of skeletal muscle. Unfortunately, it was done at a time when there was no recognition of the non-uniformity of distribution of axons to individual fibres of a particular muscle. We now recognize not only this heterogeneity of innervation within muscles, but also often a pronounced heterogeneity of structure and physiology as well (Sections 11.2.2 and 11.3). In order to understand how the mechanical properties of muscle contribute to the way in which that muscle is normally used (which is one important goal of such studies), what is really required is a fibre-by-fibre analysis of the mechanical response characteristics of a muscle. This has been provided for a locust tibial extensor

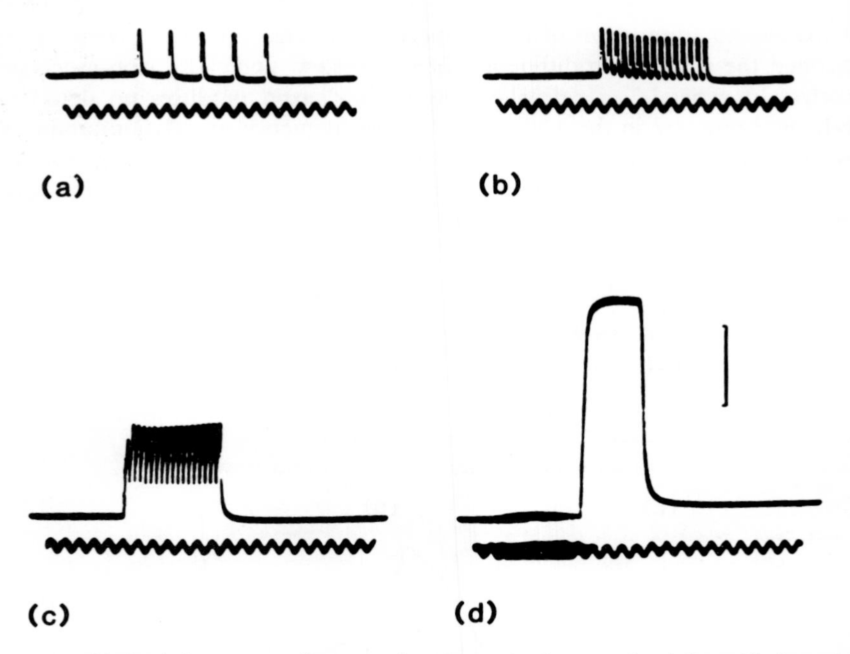

Fig. 11.13 Mechanical responses of m. 139a,b and m. 140 (innervated only by a 'fast' axon) to stimulation of its motor nerve at various frequencies. (a) 7 Hz, (b) 25 Hz, (c) 35 Hz, (d) 50 Hz. Note the strong twitch contractions even at low frequencies of stimulation (cf. Fig. 11.12). Calibration: 2 g. (After Usherwood, 1962a.)

muscle by Hoyle (1978b), who revealed an unexpected complexity of fibre types, innervation and properties; a similar study of the important leg muscles in *P. americana* would undoubtedly yield similarly important information.

11.6.2 Visceral muscle

The mechanical properties of visceral muscle in *P. americana* have been investigated by Nagai and Brown (1969) and Nagai (1970). Two features stand out. First, single action potentials in the muscle produce slow single contractions of proctodeal muscle fibres (Nagai and Brown, 1969) (Fig. 11.14a). If a high enough frequency of stimulation is delivered so that the fibre has not completely relaxed before the next AP is elicited, the contractions sum (Fig. 11.14b). The overall tension of the gut musculature can therefore be quite finely regulated by the frequency of nerve impulses delivered to it.

The other feature, the ability to contract in response to mechanical stretch, is one not shared by most skeletal muscle (Nagai, 1970; Fig. 11.14c). A stretch of the proctodael muscle causes a depolarization of the muscle membrane proportional to the degree of stretch. If sufficient stretch is applied, APs are generated and contraction follows. Stretch-activated tension can be elicited at lengths greater than twice the resting length (Nagai, 1970).

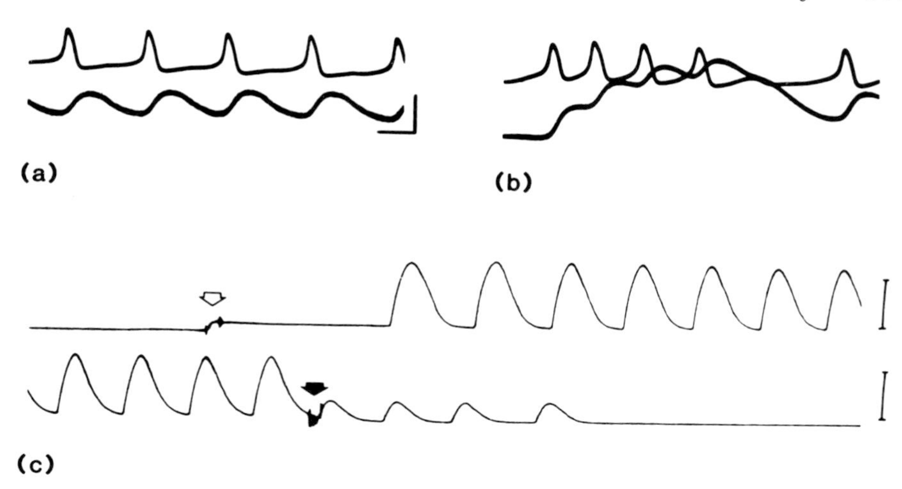

Fig. 11.14 Electrical (a,b top trace) and mechanical (a,b bottom trace, and (c) responses of *P. americana* proctodaeal muscle. (a,b) The effect of a single (a) or trains of (b) action potential(s) on tension. (c) The effect of applied stretch (arrow) on tension. Stretch released at second (black) arrow. Calibration: (a,b) 10 mV and 1 s; (c) 10 s. (a,b after Nagai and Brown, 1969; c after Nagai, 1970.)

11.7 Functional use of muscle

Having discussed the properties of the neuromuscular system, we are now in a position to ask two further questions: how are skeletal muscles used in an insect's day to day activities, and, does this use depend on specific physiological properties of the muscles or their controlling nerves? It is clear that these questions can only be answered by studying muscular and neural activity during normal behaviour (e.g. Runion and Usherwood, 1966; Delcomyn, 1976), and then determining whether the normal pattern of use in any way exploits the properties of the NM system so as to allow it to operate more effectively. Studies of this type have been carried out on several behaviours, but walking is by far the most extensively studied, and will be dealt with here.

11.7.1 Muscle activity during walking

How are the leg muscles driven to produce the movement necessary to walking? Their most basic activity is a reciprocity of action between extensor and flexor muscles (*P. americana*: Pearson, 1972; Delcomyn and Usherwood, 1973; Krauthamer and Fourtner, 1978; *S. gregaria*: Burns and Usherwood, 1979). Two earlier reports (Hoyle, 1964; Ewing and Manning, 1966) described a different pattern, but these reports have not been confirmed and probably resulted from erroneous interpretation of ambiguous recordings (Krauthamer and Fourtner, 1978). The duration of the burst of motoneuron APs driving each flexor and extensor muscle is proportional to the duration of the whole step. One or the other set of muscles is driven at all times, usually reciprocally, but at very low speeds of walking there may be a slight overlap between bursts (Krauthamer and Fourtner, 1978). While the bursts shorten as walking speed increases (because of the decreasing cycle time of each step), the average frequency of firing in each burst increases (Delcomyn and Usherwood, 1973), thereby bringing about a faster, more vigorous contraction in the muscle.

Another factor which increases the vigour of movements at higher walking speeds is the recruitment of more motor axons. At the slowest speeds, only one or two slow axons to each muscle are active. As walking speed increases, however, the 'fast' and more 'slow' axons also become active (Pearson, 1972; Krauthamer and Fourtner, 1978). Since some muscles are innervated primarily or exclusively by fast axons (e.g. the large extensors of the femur located in the coxa, m. 178 and 179 in the metathoracic legs) these muscles begin to add their power only at extremely rapid rates of leg movement. (In *P. americana*, this rate may be over 20 steps s^{-1}; Delcomyn, 1971a.)

It is known that the basic reciprocity of extensor and flexor bursts as well as the properties of the bursts themselves just described are determined by central nervous pathways without the need for sensory feedback (Pearson and Iles, 1970; Pearson, 1972; Section 8.7.4). Nevertheless, sensory feedback can affect both the timing (Wong and Pearson, 1976) and the structure (Pearson, 1972) of individual bursts. For example, at a moderate walking speed, the average rate of firing to m. 177d, a small extensor muscle in the metacoxa, is about 140 impulses s^{-1} (Pearson, 1972; Delcomyn and Usherwood, 1973). When the animal is made to pull a weight, thereby stressing sense organs in the legs, this rate goes up to roughly 170 impulses s^{-1} (Pearson, 1972). If the animal is suspended and allowed to walk holding a lightweight ball (analogous to walking on the ceiling), thereby reducing the stress on the legs, the rate drops to about 100 impulses s^{-1} (Delcomyn, 1973).

Inhibitory axons also appear to play an important role in regulating muscular activity. Iles and Pearson (1971) have shown in pinned-out preparations of *P. americana* that stimulation of an inhibitory axon will help speed the rate of relaxation of a muscle (Fig. 11.15). Burns and Usherwood (1979) found in free-walking locusts that the common inhibitor neuron innervating the metathoracic extensor tibiae tended to fire just at the end of its period of contraction. The importance of inhibition in relaxing muscles seems fairly well-established in the instances cited. However, one must keep in mind that some muscle fibres receive a complex inhibitory innervation, and that not all the axons may serve a relaxation function (Pearson, 1973). Burns and Usherwood (1979) found that, in some legs, the common inhibitor fired at the wrong time to produce functionally significant relaxation.

11.7.2 Muscle activity and muscle properties

How do the properties of specific muscles complement the functional role played by each? It has already been pointed out that the structural (Fourtner, 1978)

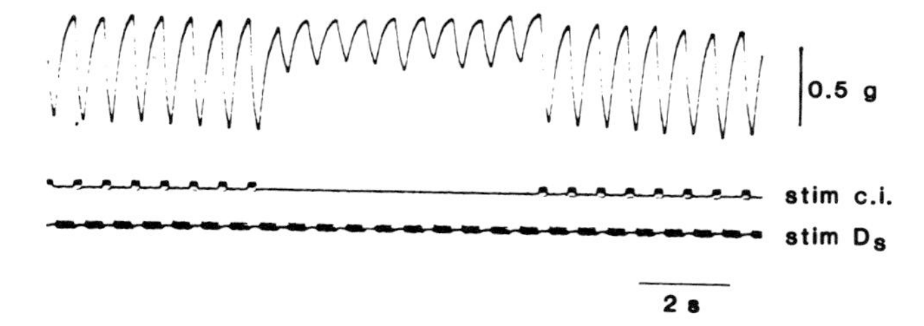

Fig. 11.15 Recording of tension (top trace) in m. 177d,e during repetitive stimulation of its innervating 'slow' neuron (bottom trace, D_s) and one of its inhibitory neurons (middle trace, c.i.). Note that when the inhibitor was stimulated during the intervals between firings of the 'slow' axon (first and last thirds of the traces) the muscle relaxed completely between stimulations, but when it was not (middle third) there was considerable residual tension left in the muscle. (From Pearson, 1973.)

and biochemical (Hart and Fourtner, 1979) differences between muscles tend to parallel functional ones. Muscles specialized for rapid, strong contractions are used in fast movements requiring great power, while muscles specialized for sustained activity are used in postural control.

But further, the innervation of each muscle also seems well-suited to produce contractions with the necessary characteristics. Large muscles like the metathoracic m. 178, 179 (coxal extensors of the femur) receive only a single fast axon. This means that not only are these muscles prepared to give strong, fast contractions, they are innervated by a type of axon which will produce just that kind of response (Usherwood, 1962a) and which will be activated only in circumstances in which the insect will require the powerful thrust the muscles will produce (Pearson, 1972). Small, weak muscles such as metathoracic 177d′,e′ (also extensors of the femur), on the other hand, may receive a complex mixture of 'slow', inhibitory and, in some cases, perhaps even 'fast' axon innervation. This means that these muscles can produce a much greater range of tension than can the large ones, and therefore, are ideally suited for use in slow movements and in maintaining static positions.

It may be possible to bring the match between muscle properties and muscle function down to the level of individual muscle fibres. Hoyle (1978b) has conducted a careful and detailed analysis of the innervation, fine structure and response characteristics of individual fibres of the locust extensor tibiae muscle. In the context of the present discussion, his most important finding was that different muscle fibres tended to be specialized to produce different strengths or rates of contractions. This specialization was due to a combination of structural and innervation differences, so that different fibres could respond differently even to the pattern of excitation delivered by a single axon which innervated them all. The result of this is a single muscle, parts of which can functionally be used in quite different ways. In view of the complexity of both structure and innervation of some leg muscles of *P. americana*, this insect may have similar functional specializations in some of its muscles.

The trend clearly is towards the discovery of more and more complexity in all aspects of muscle physiology and function, especially since there is much we still do not understand about these topics. For example, what is the functional usefulness to the insect of having up to three 'slow' or three inhibitory axons innervating a single muscle fibre (see Hoyle, 1974b)? This situation could be an evolutionary relic of some long-gone, more primitive condition, but such a possibility should be accepted only after all reasonable alternatives have been thoroughly tested. In any event, there seems little doubt that research on *P. americana* will continue to challenge and advance our understanding of insect muscle.

Acknowledgements

I thank Dr C. R. Fourtner for reading part of a draft of this chapter. Preparation of the chapter was aided in part by NIH grant NS 15632.

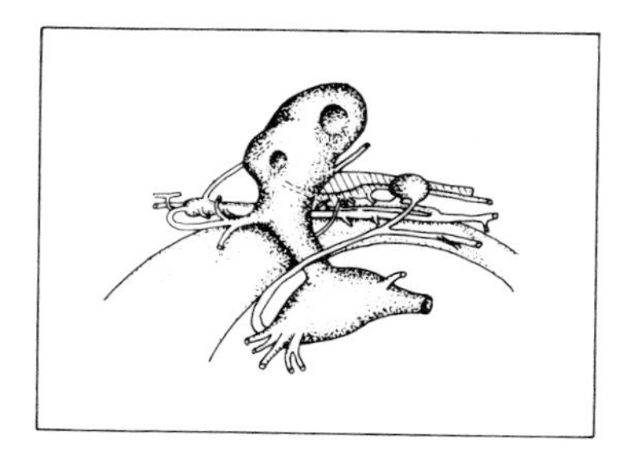

12

Neurosecretions and hormones

Stephen S. Tobe and Barbara Stay

12.1 Introduction

In this chapter we have considered the morphological, physiological and biochemical aspects of the endocrinology of *Periplaneta americana*. Every attempt has been made to refer to relevant literature in which *P. americana* is the experimental animal. In those instances in which useful experimental data have been obtained from cockroach species other than *P. americana*, this information has also been included. Unless otherwise noted, when the species is not specified, the information concerns *P. americana*.

No attempt has been made to include a comprehensive review of the modes of action of the various hormones/neurohormones. These are, however, listed in tables in the text, and for more complete considerations of modes of action, the reader is referred to other relevant chapters dealing with specific target organs.

12.2 Neurosecretory cells (NSC)

Certain cells in the nervous system have, in addition to the ability to transmit nerve impulses, the ability to release neurohormones into the circulatory system and effect a response from target organs. In addition, there are other cells with neurosecretory characteristics which 'innervate' directly specific target organs, usually endocrine glands (Scharrer, 1969). The function of such cells has been established by ablation experiments and by correlation of morphological changes with physiological events. Even without these functional criteria, nerve cells are frequently classified as neurosecretory purely because they contain distinctive secretory granules or neurosecretory material (NSM). Certain staining procedures for light microscopy differentiate NSC but no procedure stains NSC to the exclusion of other neurons (Rowell, 1976). The number of cells stained depends upon the degree of specificity of the light microscopic technique. It should also be noted that haemocytes react with stains used for NSC (Willey, 1961); in addition, lysosomes may also be paraldehyde–fuchsin (PAF)-positive (Steel and Morris, 1977). Different categories of NSC have been distinguished on the basis of their reaction to staining procedures (which include the fixative and prestaining treatment as well as the particular stain used). The stain is generally believed to react with protein (neurophysins) associated with the hormones rather than with the hormones themselves. Cells have been designated type A, B or C according to their affinity for stains (see Rowell, 1976). Since there is lack of uniformity in applying this terminology, we will list cells according to the technique by which they were identified.

The fine structural analysis of NSC reveals membrane-bound granules of varying electron densities and sizes from 100–300 nm in diameter. Distinction of NSC types on the basis of granule size and density has also been attempted. One generalization (Knowles, 1965; Scharrer, 1969) suggests that NSC which secrete peptides (Type A fibres – peptidergic) have dense granules greater than 100 nm in diameter, and those which contain granules less than 100 nm in diameter secrete monoamines (Type B fibres – aminergic). As with light microscopy, the methods used to fix the tissue influence the granule type found. It is not possible to identify cells on the basis of granule type alone.

12.2.1 Brain

(a) *Protocerebrum, pars intercerebralis (PI) medial cells*

The medial cells are the largest, most easily stained group of NSC in the brain (Fig. 12.1); their axons give rise to the nervi corporis cardiaci I (NCC I) which exit the brain on the contralateral side (Fig. 12.3 and 12.4). Pipa (1978a) counted about 250 somata on one side of the brain in preparations filled with cobalt through the nervi corporis allati I (NCA I) (Fig. 12.1). Most of these somata were small, between 10 and 25 μm in diameter. Stains for NSM demonstrate only a fraction of the somata filled with cobalt. By careful study of *in toto* preparations stained with paraldehyde–resorcin–fuchsin (PARF), Pipa (1978a) counted 90 cells, all of small size (10–20 μm in diameter). Other studies have shown even fewer cells (Seshan *et al.* 1974; Khan, 1976; Fraser and Pipa, 1977) and may have been due to difficulty in making accurate counts. Pipa's (1978a) rigorous quantitation still suggests that many of the cells which send axons into the corpora cardiaca (CC) are not stained by methods that demonstrate NSM. It would be of interest to know whether the stained cells are a spatially identifiable subpopulation of the cobalt-filled cells or whether the stainability of the cells changes temporally.

A limited electron microscopical analysis of the medial cells has shown that at least some of them fit the criterion of NSC, in having electron-dense granules

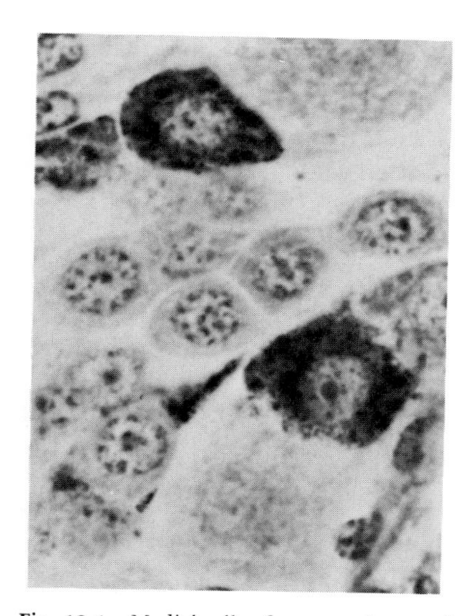

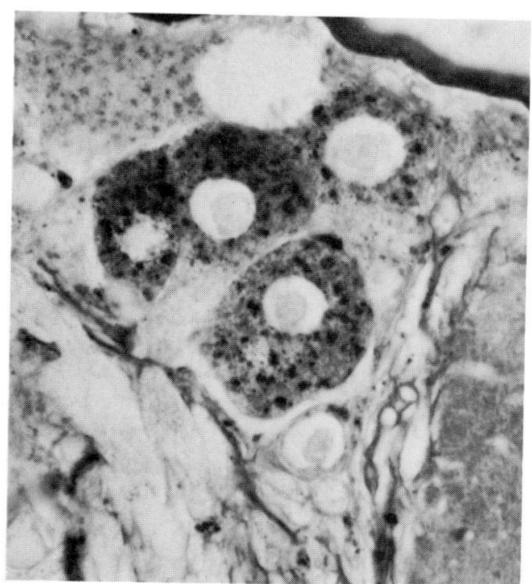

Fig. 12.1 Medial cells of protocerebrum of *P. americana* stained with chrome alum–haematoxylin–phloxin after Helley's fixation. × 1400. (Courtesy of R. Pipa.)
Fig. 12.2 Lateral cells of protocerebrum of starved adult male *P. americana* stained with PAF after Helley's fixation. × 550. (Courtesy of R. Pipa.)

between 100–300 nm in diameter (Seshan *et al.*, 1974). The medial cells can be separated into topologically distinct groups with silver staining methods and neurosecretory stains. Depending on the investigator, there are four or five groups of cells, most of which contribute to the contralateral NCC I (Brousse-Gaury, 1971b). A potentially more instructive method of categorizing the NSC is that of Eckert (1973, 1977) and Eckert and Gersch (1978), who raised antibodies to fractions of CC extracts and used these to localize NSC in the brain and CC. They found that certain of the antisera reacted with brain cells, not all of which were PAF-positive. One antiserum reacted with intrinsic cells of the CC, whereas another reacted with the terminals of brain cells within tracts of the CC (Eckert, 1977). This technique may be useful for identifying specific cells after the functions of the fractions are established.

The NSC of the brain contribute to axons which exit the brain from its posterior face (see Table 12.1 and Figs. 12.3 and 12.4). The axons of the medial cells are readily demonstrable with neurosecretory stains. They form large bundles which decussate at the level of the nervus connectivus (Willey, 1961; Dogra, 1968; Brousse-Gaury, 1971b; Pipa, 1978a). Both pre- and post-chiasmal parts of these axons possess collaterals extending into the neuropile. The collaterals contain NSM demonstrable with performic acid–Victoria Blue (PAVB) (Adiyodi and Bern, 1968), and can be filled with cobalt through the NCA I (Fraser and Pipa, 1977; Pipa, 1978a). It is presumed that these processes provide sites for synaptic communication. From silver-stained preparations, synapses between ocellar nerves and nerve fibres which give rise to the NCC I and NCC II have also been described (Brousse-Gaury, 1975b).

Table 12.1 Origins of nervi corporis cardiaci (NCC) (Somata)

Nerve tract	*Brain region*	*Reference*
NCC I	Protocerebrum Pars intercerebralis (Medial neurosecretory cells)	Willey (1961); Dogra (1968); Brousse-Gaury (1971b); Fraser and Pipa (1977); Pipa (1978a)
NCC II	Protocerebrum Pars lateralis	Willey (1961); Dogra (1968); Brousse-Gaury (1971b); Fraser and Pipa (1977); Pipa (1978a)
NCC III	Tritocerebrum	Brousse-Gaury (1971b)
NCC IV	Deutocerebrum	Brousse-Gaury (1971b)

These pathways do not seem to have been confirmed by cobalt-filling of the ocellar nerve in combination with PAF staining (Bernard, 1977).

The axons from medial cells leave the protocerebrum near the midline on its posterior ventral surface to form the NCC I. This nerve is about 33 μm in diameter in adults and is the largest nerve entering the CC (Brousse-Gaury, 1971b).

(b) *Protocerebrum, lateral cells*

The lateral cells of the protocerebrum which give rise to NCC II are far less conspicuous than the medial cells because they are fewer and not so readily stained for NSM (Fig. 12.2). Nevertheless, under certain conditions, they do contain NSM (Willey, 1961; Brousse-Gaury, 1971b; Fraser and Pipa, 1977; Pipa, 1978a). Using cobalt back-filling, Pipa (1978a) demonstrated 20–45 cells, most of which were 15–25 μm in diameter, located under the dorsal lip of the α lobe of the corpus pedunculatum. Only about 6 large cells of this group stained with PAF in starved adult males, but not with PARF (Pipa, 1978a). Using silver-stained preparations and neurosecretory stains, Brousse-Gaury (1971b) observed other groups of cells which contribute to the NCC II, namely medial cells and cells from the ventral-posterior part of the optic lobe. However, careful cobalt back-filling of one NCC II showed its projection to be only to the ipsilateral lateral group and to some of the tiny (8–15 μm in diameter) globuli cells near the corpus pedunculatum (Pipa, 1978a). Because some globuli cells also stain with PAF, he proposed that they may be neurosecretory in function (Pipa, 1978a).

The most remarkable observation from the back-filling into the lateral cells is the extensive arborizations which they send into the neuropile (Pipa, 1978a) (Section 8.2.2 for details of their distribution). The possibility of synapses which occur at these arborizations requires further investigation.

The NCC II nerve, about 20 μm in diameter in an adult, leaves the protocerebrum lateral and dorsal (at the level of the central body) to the NCC I (Brousse-Gaury, 1971b). NSM has been detected in the NCC II (Brousse-Gaury, 1971b).

(c) *Optic lobes*

Beattie (1971) studied the small group of NSC on the posterior side of optic lobes in living *P. americana* brain. About 120 monopolar ovoid cells (length 16 μm) occur in each lobe near the lamina ganglionaris. They are distinguishable in the light microscope by the Azan method and are stained by light green, and hence might be classed as 'C cells'. Fine structural analysis has shown a population of membrane-bound granules 100–170 nm in diameter of variable electron density. The axons form synapses with axons containing smaller neurosecretory particles ($\sim$70 nm in diameter). Beattie (1971) speculated that the population of peptidergic (large granules) NSC in optic lobes are co-ordinated by aminergic (small granules) cells and suggested that these NSC are associated with the clock which controls activity rhythms, although Brady (1974) considers the latter unlikely (see also Section 10.5).

(d) *Deutocerebrum*

Brousse-Gaury (1968, 1971b) described 5 groups of cells in the deutocerebrum (4 in the external cortex, 1 in the internal cortex) which contribute axons to a small nerve, about 15 μm in diameter in adults, leaving the posterior face of the dorsal lobe of the deutocerebrum. This nerve, NCC IV, joins the CC at the dorsal cardiac commissure (Brousse-Gaury, 1971b). NSM was found in this nerve fibre and its perikarya under certain experimental conditions in *P. americana* and *Blaberus fuscus* (i.e. *Blaberus craniifer*) (Brousse-Gaury, 1971b). Neither Willey (1961) nor Fraser and Pipa (1977) observed this small NCC IV.

(e) *Tritocerebrum*

A group of lateral-external cells in the tritocerebrum was described by Brousse-Gaury (1968, 1971b) as the origin of the NCC III. She found NSM (PAF- or Alcian blue-positive) in these cells under certain experimental conditions in *B. craniifer* (Brousse-Gaury, 1971b).

The NCC III, a nerve of 15 μm diameter in adults, leaves the tritocerebrum at the level of the dorsal substomodeal commissure (Brousse-Gaury, 1971b) and enters the cardiaca-commissural organ (CCO). Just before it joins the CC, it is joined by many nerves (Willey, 1961) one of which is from the anterior branch of the NCA II (Fraser and Pipa, 1977). Within the CCO, the NCC III forms a chiasma (Willey, 1961). Fraser and Pipa (1977) were unable to fill the NCC III nerve and thus could not confirm its cells of origin.

12.2.2 Biogenic amines in the brain

Since biogenic amines have been shown to have neurosecretory as well as transmitter function (Knowles, 1965), it is appropriate to mention their distribution in the brain.

The quantity of dopamine, noradrenaline and octopamine in the various parts of the nervous system of *P. americana* was determined by Evans (1978a) and Dymond and Evans (1979). These authors observed that octopamine is conspicuously concentrated in the CC, in larger quantities than either dopamine or noradrenaline. Octopamine is not demonstrable by fluorescence histochemistry but it has been localized in the globuli cells and calyx by microdissection and assay of these areas (Dymond and Evans, 1979). The antennal lobes and tritocerebrum were also shown to be rich in octopamine (Evans, 1978a). The cells in these regions could be contributing to the octopamine found in the CC.

Fluorescence histochemistry has been used to localize catecholamines in two dense synaptic areas of the brain, the β-lobes of the corpora pedunculata and the central body (Mancini and Frontali, 1970). Electron microscopic analysis of these areas after glutaraldehyde–formaldehyde–osmium fixation showed two types of fibres distinguished on the basis of granule content and size (all less than about 100 nm in diameter). Incubation of brain slices in noradrenaline or α-methylnoradrenaline followed by $KMnO_4$ fixation showed a class of small (32 nm in diameter) vesicles with dense cores not seen in control tissue slices; these were cautiously interpreted as catecholamine granules which can be demonstrated only after these manoeuvres and were not seen after glutaraldehyde–formaldehyde–osmium fixation (Mancini and Frontali, 1970).

12.2.3 Functions of the neurosecretory cells of the brain

The designation of functions for NSC of the brain is far more elusive than their histological localization. Since the CC contains intrinsic cells and their release sites, as well as the release sites of cells from the brain, it is not clear that all factors extracted from the CC or obtained from them by stimulation arise from brain cells. Factors which have been isolated from the CC are given in Table 12.2. Many have been extracted also from the brain, for example, neurohormone D and prothoracic gland (PG) activation factors (Gersch and Stürzebecher, 1968); these can also be released from the CC *in vitro* by stimulating the NCC I and II respectively (Kater, 1967; Gersch *et al.*, 1970; Gersch, 1972). However, the cells of origin of NCC I and II may not be the source of the brain hormones released by stimulation of these nerves because there could be interactions between the axons in these two nerves such that stimulating NCC I would release substances arriving via axons of NCC II or vice versa. In short, the specific sites of origin of the numerous brain hormones of demonstrated function are unknown in cockroaches.

Hyperglycaemic hormone (HGH) can be released from the brain by stimulation (Gersch, 1974a), and it appears that the lateral cells of the brain are the source of this hormone, because: (1) starvation of the cockroach results in more pronounced staining of NSM in the lateral cells (Fraser and Pipa, 1977); (2) stimulation of NCC II results in release of HGH (Gersch, 1972); and (3) section of NCC II results in a reduction of trehalose level in the haemolymph (retention of HGH presumably) whereas section of NCC I has no effect (Gersch, 1974b).

Hindgut stimulating neurohormone (HSN) activity was released via NCC I and CC from cultured brains of *Leucophaea maderae*, whereas severance of the CC from the brain inhibited release of the substance (Holman and Marks, 1974). Thus the hormone probably originates in the medial cells. Similarly, Kater (1968) has suggested that a neural pathway between the brain and CC (via NCC I) mediates release of a cardioaccelerator.

(a) *Control of the corpora allata (CA) by the brain*
No brain factor has been isolated to date which acts on the CA; yet there is circumstantial evidence that there is such a substance(s) (Section 12.5.4).

(b) *Control of the ovary*
A brain hormone extracted from *P. americana* larvae and adults has been demonstrated by Hagedorn *et al.* (1979) to stimulate ecdysone production by the ovaries of the mosquito. This hormone may have a similar function in the cockroach in view of the fact that adult cockroach haemolymph and ovaries contain ecdysteroids (Bullière *et al.*, 1979) (see Section 12.7).

12.2.4 Control of NSC of brain

(a) *Neural activity*
NSC appear to be spontaneously active. Extracellular recordings of cultured medial cells showed constant low rates of discharge (1 s^{-1}) (Seshan *et al.*, 1974). *In situ* intracellular recordings from the soma of medial cells also showed spontaneous discharge which lasted 4–12 ms (Gosbee *et al.*, 1968; Krauthamer, 1978). The long duration of discharge is a characteristic of NSC (Maddrell, 1974; Berlind, 1977). Cook and Milligan (1972) also recorded spontaneous nerve impulses in these cells *in situ* and, in addition, found post-synaptic potentials, the majority of which were excitatory. Krauthamer (1978) noted from *in situ* recording that action potentials were often driven by excitatory potentials. The presence of post-synaptic potentials indicates that NSC receive synaptic input which could influence the function of the cells. Indeed, light affected both resting and impulse potentials of these MNSC (Cook and Milligan, 1972). Experimental stimulation of the brain was found to deplete NSM (Gosbee *et al.*, 1968) and to cause release of neurohormone (Kater, 1968; Gersch, 1974a). The release of NSM can be controlled by electrical inputs under experimental conditions.

Pathways of neural input to the NSC have been studied by observing changes in NSM in cells after experimentally altering the neural input from sensory receptors (Brousse-Gaury, 1971b). It is difficult to establish the meaning of these alterations in NSM.

(b) *Biogenic amines*
Biogenic amines have been postulated by Hentschel (1972) to influence the release of NSM. He has demonstrated that reserpine-treated females, in which brain amines are presumably depleted (Frontali, 1968), have reduced production of oöthecae. Inhibitor studies suggest that dopamine is responsible for the presumed suppression in CA activity (Hentschel, 1975). There is no evidence in *P. americana*

Table 12.2 CC factors with demonstrated action from cockroaches

General action	*Species*	*Names*	*Actions*	*References*
Cardioacceleration	*P. americana*	Cardioaccelerator	Synthesis of indolalkylamines by pericardial cells	Davey (1961a,b, 1963); Ralph (1962)
		Neurohormone D	(1) Increased excitability of myocardial cells (2) Increased rate of firing of neurosecretory cells	Unger (1957); Gersch *et al.* (1960, 1963); Gersch and Stürzebecher (1967); Gersch (1975)
		Cardioaccelerator peak factors 1a, 1b, 2a, 2b		Natalizi *et al.* (1970); Traina *et al.* (1976)
		P_1 and P_3		Brown (1965)
Hindgut stimulation	*P. americana*	Hindgut-stimulating factor	Increased tonus, frequency and amplitude of hindgut muscle contraction	Cameron (1953); Davey (1962b)
		P_1 and P_2	Increased tonus, frequency and amplitude of hindgut muscle contraction	Brown (1965)
	L. maderae	Hindgut-stimulating neurohormone (HSN)	Increased excitability of proctodeal muscles at non-synaptic membrane sites	Holman and Cook (1972, 1979)
Hyperglycemia	*P. americana*	Hyperglycemic hormone (HGH)	Elevation of fat body glycogen phosphorylase	Steele (1961, 1963); Brown (1965)
			Activation of adenylate cyclase and cAMP-dependent protein kinase	Hanoaka and Takahashi (1976, 1977); Gäde (1977)
		Adipokinetic hormone (AKH)		Jones *et al.* (1977); Holwerda *et al.* (1977)
Hypolipemia	*P. americana*	Hypolipemic hormone	Mobilization of di- and triglycerides from fat body	Downer and Steele (1969, 1972); Downer (1972)
Neural activity	*P. americana*		Depression of spontaneous firing rate of neurons	Ozbas and Hodgson (1958)
			Stimulation of spontaneous firing rate of neurons	Milburn *et al.* (1960); Milburn and Roeder (1962); Brown (1965); Natalizi *et al.* (1970)
Respiratory metabolism	*B. discoidalis*		Stimulation of fat body respiration	Keeley and Friedman (1969)
	L. maderae		Stimulation of cytochrome synthesis	Keeley (1978)
			Stimulation of fat body respiration	Lüscher and Leuthold (1965); Müller and Englemann (1968)
	N. cinerea		Stimulation of fat body respiration	Lüscher (1968a)
Activation of prothoracic glands	*P. americana*	Activation factor I	Increased RNA synthesis in prothoracic glands	Gersch and Stürzbecher (1970); Gersch and
		Activation factor II	Increased membrane potential of prothoracic gland cells of *Galleria*	Bräuer (1974); Gersch *et al.* (1973)
Protection of prothoracic glands	*P. americana*			Bodenstein (1953c); Lanzrein (1975)
Protection of oocytes	*N. cinerea*			Lüscher (1968b)
Diuresis	*P. americana*		Increased rectal water absorption	Wall (1967)

Table 12.3 Factors from ventral nerve cord of *P. americana*

Name or action	*Location**	*Reference*
Hyperglycemic hormone	S	Gersch (1974a)
Cardio-accelerators	S, T1–3, A2–4	Rounds and Gardner (1968); Gardner and Rounds (1969)
	A6	Krolak *et al.* (1977)
(Neurohormone D)	A6	Gersch (1974a)
Cardio-decelerator	T1–3	Gersch (1974a)
Prothoracicotropic (Lepidopteran development)	A	Gersch and Stürzebecher (1968)
Diuretic hormone	A6	Mills (1967); Goldbard *et al.* (1970)
Antidiuretic hormone	S, T1,3 A3,6	Wall and Ralph (1964; 1965)
	T3, A6	Wall (1967)
Bursicon	A6	Mills *et al.* (1965)

* S, suboesophageal ganglion; T1–3, thoracic ganglia; A1–6, abdominal ganglia.

that this is a direct effect on the NSC, but it is an intriguing possibility. Reserpine may also act on CA that have been implanted into allatectomized females. Serotonin (5-hydroxytryptamine, 5HT) is effective in causing the release of NS granules from the CC (Scharrer and Wurzelmann, 1978). Gersch (1972) suggested that catecholamines control the release of HGH (when NCC II is stimulated), whereas cholinergic control is implicated in the release of Neurohormone D (when NCC I is stimulated).

(c) *Ecdysterone*

The medial NSC in the brain of *L. maderae* may be regulated by ecdysterone (Marks *et al.*, 1972). Cells which had accumulated NSM in culture over several days were observed to have less NSM within 24 h of ecdysterone treatment, thus suggesting that the ecdysterone facilitates release of NSM (Marks *et al.*, 1972).

12.2.5 Suboesophageal and ventral ganglia; perisympathetic organs; other NSC terminals

(a) *Suboesophageal and ventral ganglia*

NSC occur in these ganglia and factors have been obtained from them which have similar physiological activity to some of the factors isolated from the brain and CC (Tables 12.2 and 12.3). The release sites for the NSC of the ventral ganglia are presumed to be segmental neurohaemal organs called perisympathetic organs (see below) (de Bessé, 1967).

The location, soma size and staining characteristics of the NSC have been described (Füller, 1960; de Bessé, 1967). Small, Azan-positive cells (20–30 μm in diameter) predominate; they are located in the midline and in the periphery of the ganglia, especially the last abdominal ganglion. The PAF and chrome-haematoxylin–phloxin(CHP)-positive cells are evenly divided between large

(40–50 μm) and small (25 μm) cells, all of which are located at the periphery of the ganglia (de Bessé, 1967).

More detailed studies of various ganglia have added to these earlier descriptions. Pipa and Novak (1979) using cobalt back-filling of the suboesophageal ganglion, have demonstrated two groups of cells which send axons to the NCA II. Electron microscopial analysis of this nerve shows the presence of NS granules in the constituent fibres and a neurohaemal type structure at the periphery of the nerve. It is probable that the axons arise in the suboesophageal ganglion and that their release sites are at the nerve surface.

In the abdominal ganglia (1 through 5), Smalley (1970) studied histologically three groups of 20–30 cells and traced their axons into the neurohaemal organs. The presence of electron-transparent vesicles, 110–160 nm in diameter, in the median nerve leading to the perisympathetic organ adds support to their presumed neurohaemal function. Brady and Maddrell (1967) reported similar transparent vesicles and a few dense granules, presumably peptidergic, in the perisympathetic organs. Ali and Pipa (1978) confirmed by cobalt filling from the neurohaemal organ that axons in this organ arise from cells in the midsagittal plane of the 2nd and 3rd abdominal ganglia. Cells which stain with PAF and Azocarmine are located in the same region as those which could be demonstrated with cobalt; the perisympathetic organs were similarly stained (Ali and Pipa, 1978).

Dymond and Evans (1979) isolated dorsal medial soma from both the last abdominal and the metathoracic ganglion and found them to contain octopamine. The medial neurohaemal organ also contained octopamine (Evans, 1978a) although dopamine and noradrenaline are more prominent constituents of these organs (Dymond and Evans, 1979).

(c) *Perisympathetic organs*

The median nerves from the ventral nerve cord are thickened and stain for NSM. In the thorax, these occur posterior to the ganglia and are inconspicuous. In the abdomen, they are anterior to the ganglia, of variable size, but more conspicuous than in the thorax, especially at the 2nd and 3rd (of 7) ganglia (de Bessé, 1967).

(c) *Other NSC terminals*

A neurohaemal organ with single or small groups of terminals occurs in the wall of the pulsatile organ at the base of each antenna (Beattie, 1976). The origin and function of these axons are unknown.

Neurosecretory terminals also occur in other organs. To be considered functional, Maddrell (1974) suggests that these terminals be free of a glial coat, penetrate the basement lamella of the organ and show morphological features consistent with release of NSM. Axons with all of these characteristics have been reported for *P. americana* heart (Johnson, 1966), and for *L. maderae* prothoracic glands (Scharrer, 1964b) and corpora allata (Scharrer and Wurzelmann, 1974). In the spermatheca (Gupta and Smith, 1969) and hindgut muscles (Oschman and

Wall, 1969), the nerves containing NS granules showed only the first of these criteria. However, in the case of the hindgut, it seems clear from physiological evidence that proctodaeal nerves contain and release proctolin (Section 12.4.1) (Brown, 1967, 1975).

Peripheral cells, presumably sensory in origin, in the abdomen of stick insects, have been found to contain NS granules (Finlayson and Osborne, 1968). No such cells have been described in *P. americana* but it is likely that future work may demonstrate similar cells.

12.3 Corpora cardiaca (CC)

12.3.1 General structure

The CC are paired structures which form the anterior portion of the retrocerebral complex (Fig. 12.3 and 12.4). They are located dorsal to the hypocerebral ganglion and each CC is connected with the other by a single dorsal commissure (Willey, 1961; Adiyodi, 1974; Fraser and Pipa, 1977). The anterior portion of the CC and the CCO form the latero-ventral wall of the aorta. The posterior portion of the CC is not associated directly with the aorta (Willey, 1961). According to Willey (1961), the CC develop from the same anlage as the hypocerebral ganglion.

The CC can generally be divided into two components, the axons of the NSC of the brain which store and release their products here (neurohaemal organ) and the intrinsic NSC bodies and their release sites. In *P. americana*, the division between these two areas is not as distinct as in other orthopteroid species. The CC are the possible sites of synthesis and release of several neurohormones (see Section 12.3.3). The two posterior ventral lobes of the CC are joined to the hypocerebral ganglion by two separate connectives and give rise to the stout NCA I which innervate the CA (Willey, 1961).

12.3.2 Morphology and ultrastructure of the CC

(a) *Intrinsic cells (parenchymal cells)*

The intrinsic cells of the CC and storage axons are neurosecretory (Fig. 12.5). Scharrer (1963) describes them from *L. maderae* as 'neuroglandular' elements, possessing characteristics of both neural cells and secretory cells. According to Willey (1961), these cells do not stain with PAF. The large (20 μm in diameter) intrinsic cells are located in the anterior portion of the CC and are associated with the aorta (Willey, 1961), although few are found in the commissural organ. The intrinsic cells of *L. maderae* are pear-shaped and measure up to 22 by 13 μm (Fig. 12.5). They possess typical NS granules closely associated with the prominent Golgi complexes (Figs. 12.5 and 12.6). No distinction between different functional intrinsic cell types can be made (Scharrer, 1963).

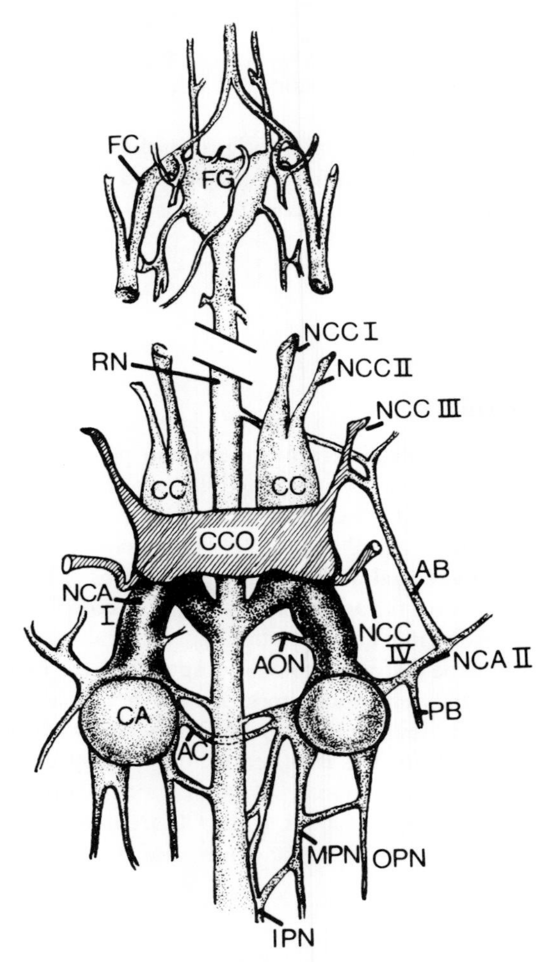

Fig. 12.3 Dorsal view of *P. americana* retrocerebral complex and frontal ganglion. The posterior section of the CCO has been removed. × 40. AB, anterior branch of NCA II; AC, allatal commissure; AON, aortic nerve; CA, corpus allatum; CC, corpus cardiacum; CCO, cardiaca-commissural organ; FC, frontal connective; FG, frontal ganglion; IPN, inner post-allatal nerve; MPN, middle post-allatal nerve; NCA I, II, nervi corporis allati; NCC I, II, III and IV, nervi corporis cardiaci; OPN, outer post-allatal nerve; PB, posterior branch of NCA II; RN, recurrent nerve. (After Willey, 1961; Fraser and Pipa, 1977.)

Ultrastructural studies of the CC of *P. americana* and other blattids have revealed that the intrinsic cells contain putative NS granules that are morphologically similar to the granules of the storage axons from the brain (Figs. 12.5 and 12.6) (Scharrer, 1963, 1968; Scharrer and Kater, 1969). Differences between intrinsic cells and storage axons of brain NSC cannot be made either on the basis of structure or function (see, however, Eckert and Gersch, 1978).

Scharrer and Kater (1969) observed signs of enhanced release of neurosecretion

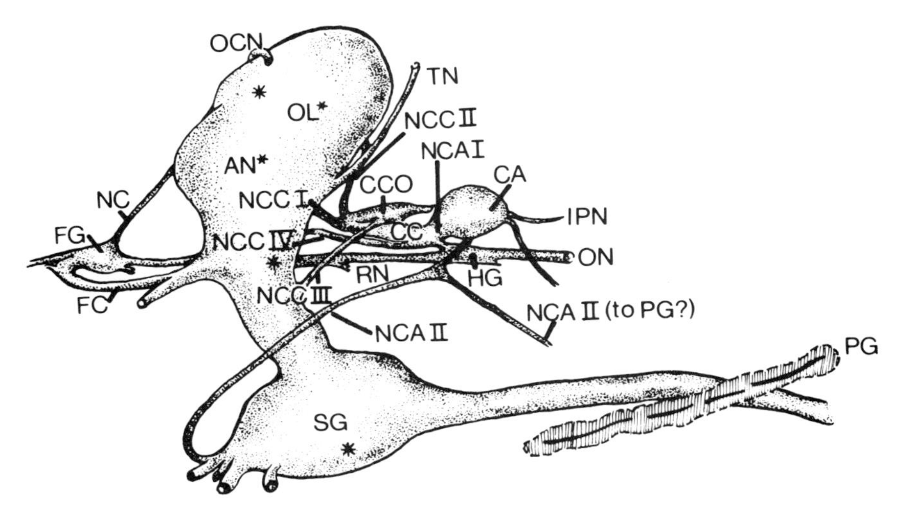

Fig. 12.4 Nervous system of head plus retrocerebral complex of *P. americana*, left lateral view. Abbreviations as for Fig. 12.3 and, in addition: AN, antennal nerve; HG, hypocerebral ganglion; NC, nervus connectivus; OCN, ocellar nerve; OL, optic lobe; ON, oesophageal nerve; PG, prothoracic gland; SG, suboesophageal ganglion; TN, tegumental nerve. Asterisks denote defined areas of neurosecretion. (After Willey, 1961; Brousse-Gaury, 1971b.)

in electrically stimulated preparations of CC, although they were unable to identify the cells which released the specific neurohormone (cardioaccelerator – see Section 12.3.3) on the basis of depletion of NS granules. Subsequently, Scharrer and Wurzelmann (1978) were able to show rapid and profound changes in the ultrastructure of the CC of *L. maderae* after release induced by serotonin. In particular, they observed numerous 'synaptoid' vesicles (see below), omega profiles and extracellular electron-dense granules within 3 min of treatment (Fig. 12.7). These morphological traits can be associated with the process of exocytosis and lend support to the hypothesis that release of NS material from the CC occurs by this process (Fig. 12.7).

(b) *Interstitial cells ('glial' cells)*

In *L. maderae*, the CC also possess a second complement of readily distinguishable cells, the interstitial cells, which surround, ensheath and separate the intrinsic cells (Fig. 12.5). These polymorphic cells have prominent nuclei but little cytoplasm, and appear to adapt to the space within the CC, surrounding and separating the perikarya and 'axonal' processes of the intrinsic cells (Fig. 12.5) (Scharrer, 1963). They are also associated with the axons of extrinsic cells originating in the brain (Scharrer, 1963). Scharrer (1963) suggests that these cells exhibit all the structural properties of glial cells.

(c) *Connective tissue sheath*

The entire CC is covered by an acellular connective tissue sheath, which serves to separate the organ from the surrounding haemolymph. The sheath contains

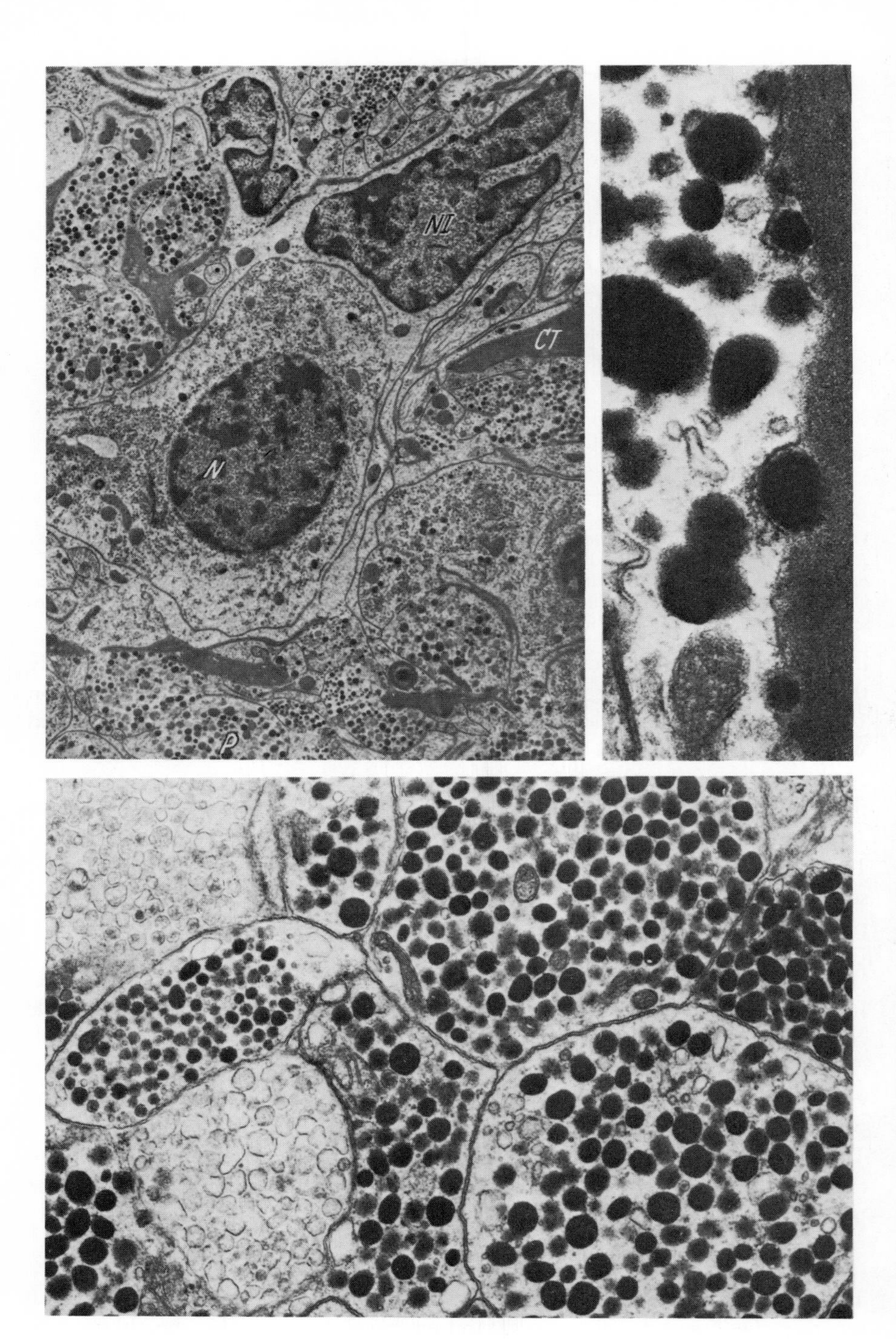

NI
CT
N
P

Fig. 12.5 Low power electron micrograph of the corpus cardiacum of female *L. maderae*. Two types of cell bodies are shown: intrinsic (parenchymal) cells with ovoid nuclei, N, and glia-like interstitial cells with polymorphic nuclei, NI. CT, connective tissue. × 5000. (Reproduced by permission, B. Scharrer and Springer-Verlag; Scharrer, 1963; Fig. 2.)

Fig. 12.6 Electron micrograph of the corpus cardiacum of male *L. maderae* showing profiles of neurosecretory neurons. Note variations in size and density of granules. Large granules are presumably peptidergic, whereas small granules (left) are possibly aminergic. Axon with small granules shows 'synaptoid' contact with axon containing large granules. × 17 600. (Reproduced by permission, B. Scharrer and Springer-Verlag; Scharrer, 1968; Fig. 1.)

Fig. 12.7 Electron micrograph of the corpus cardiacum of *L. maderae*, 25 min after injection of 50 μl of serotonin creatine sulphate (0.27 μg) to stimulate release of neurosecretory material. Neurosecretory granules can be seen within axon (left). Exocytotic figures associated with release of neurosecretory material into stroma (right) are shown. × 50 000. (Reproduced by permission, B. Scharrer and Springer-Verlag; Scharrer and Wurzelmann, 1978; Fig. 1.)

---

filaments with collagenous characteristics (Scharrer, 1963), and processes of the sheath penetrate throughout the CC, in association with tracheoles and elements of the haemolymph (Fig. 12.5) (Scharrer, 1963). This interdigitation provides a system of channels throughout the CC through which the secretory products gain access to the haemolymph and *vice versa*.

(d) *Regulation of release from the CC*

The axons within the CC of *P. americana* contain mostly peptidergic NS granules and far fewer vesicles presumed to contain monoamines (Scharrer, 1963, 1968, 1969). Scharrer (1968) has proposed that aminergic axons may regulate release of neurohormones from the adjacent peptidergic axons. Scharrer (1968) also described a second type of contact, that between peptidergic axons, in which pre-synaptic elements were clear but post-synaptic ones were not (Fig. 12.6). She notes that it is incorrect to describe these structures as 'synapses' and that they may be more properly regarded as 'synaptoid' (i.e. having characteristics similar but not identical to synapses) (Scharrer, 1969). These axo-axonal associations may function in intercellular communication and regulate the release of peptidergic granules.

12.3.3 Neurohormones of the CC (see Table 12.2)

(a) *Cardioaccelerators*

It is now well established that substances from the CC are able to accelerate the rate of beating of the cockroach heart (hence the term *cardioacceleration*). But, as is clear from recent reviews (Miller, 1975b, 1979; Gersch, 1975; Frontali and Gainer, 1977), there is little agreement as to the nature or mode of action of these compounds. We present here a brief summary of the different points of view.

There appears to be a fundamental disagreement over the existence of cardio-accelerators. Although the work of Davey (1961a,b, 1963) and of Gersch and co-workers (see Gersch, 1975) has demonstrated factors in the CC which are able to accelerate the heart beat rate in *P. americana*, a real physiological function for them has not been proven (Miller, 1975b), and the existence of a neurohormone which regulates heart beat rate in the intact insect remains to be demonstrated.

It is perhaps indicative of the problem that there is considerable variability in the responses which *P. americana* shows to injections of CC extracts (Davey, 1961b, 1962a,b, 1963; Hertel, 1971), and there is variability in the amount of putative hormone required to elicit a response (Gersch *et al.*, 1970; Hertel, 1971; Mordue (cited in Miller, 1975b); Gersch, 1975). The complexity of *P. americana* heart innervation may account to some extent for the variability observed by many authors (Miller, 1975b, 1979).

Caution should be exercised in interpreting the results on cardioaccelerators. Much of the early work with these substances involved the use of whole extracts of CC. Such extracts have been found to contain other pharmacological substances (such as 5-HT) which are known to be potent cardioaccelerators (Natalizi *et al.*, 1970; Miller, 1979). Accordingly, experiments using unpurified extracts must be interpreted with caution. Studies which have used chromatographic procedures to separate, partially at least, the active substances in the CC are to be preferred. Cardioaccelerator activity does remain after such purification and provides stronger support for the existence of such substances (Brown, 1965; Natalizi *et al.*, 1970; Traina *et al.*, 1976).

Neurohormone D was originally described by Unger (1957) and was isolated from the CNS of *P. americana*. Subsequently, Gersch *et al.* (1960) isolated a factor with similar characteristics from the CC. Neurohormone D was only one of several isolated compounds with cardioaccelerator activity, but it is the substance which has received most attention. Neurohormone D appears to be a peptide of approximately 2000 mol wt (Gersch and Stürzebecher, 1967), although the peptide nature of the factor remains open to question. The loss of cardioaccelerator activity after treatment with proteolytic enzymes such as trypsin, chymotrypsin and pepsin have been used by numerous authors including Gersch *et al.* (1963) (see also Davey, 1961a,b; Brown, 1965; Kater, 1968; Natalizi *et al.*, 1970) to support the hypothesis that the CC cardioaccelerators are peptides. In fact, Traina *et al.* (1976) have utilized the differential sensitivity of CC fractions of *P. americana* to proteolytic enzymes, as well as chromatographic techniques, to identify in part the various cardioaccelerators.

Neurohormone D has been hypothesized to exert its cardioaccelerator effect in two different ways: firstly, it is believed to increase directly the membrane excitability of the myocardial cells and the neurons innervating the alary muscles; secondly, Neurohormone D stimulates the rate of firing in NS axons that are believed to synapse with the lateral cardiac nerves (Richter, 1973; Gersch, 1975). It must be noted, however, that high concentrations of CC extract are necessary to evoke these changes in the heart system. Dose-response curves suggest that extract of at least two pairs of CC are necessary for a minimum response (see for example, Richter and Gersch, 1974). The physiological significance of these high doses remains open to question.

Davey (1964) also hypothesized an indirect effect of the putative cardioaccelerator, via the pericardial cells. However, Gersch *et al.* (1974) observed, on the basis of histochemistry and electron microscopy, that the lateral cardiac nerves of

B. craniifer contain both aminergic and peptidergic substances. The presence of peptidergic substances suggests that neurosecretory axons may in fact directly innervate the heart. Indeed innervation of this kind has been observed previously in *P. americana* (Johnson, 1966). Such observations led Frontali and Gainer (1977) to suggest that there is a direct neuronal transport of neurohormones to the target tissues.

Factors outside the CC have also been observed to possess cardioaccelerator activity. In particular, extracts of various portions of the CNS, including the ventral nerve cord and associated ganglia, show potent cardioaccelerator activity (Gardner and Rounds, 1969; Rounds and Gardner, 1968). Gardner and Rounds (1969) in fact suggested that their extracts contained both cholinergic and non-cholinergic substances, on the basis of experiments using agonists and antagonists of cholinergic neurotransmitters. Partial purifiction by column chromatography of extracts of the terminal abdominal ganglion has revealed at least two fractions with cardio-accelerator activity, one presumably a peptide and the other a low molecular weight compound suggested to be acetylcholine (Krolak *et al.*, 1977). The putative peptide was suggested by these authors to be Neurohormone D of Gersch, although it is tempting to speculate that it is the pentapeptide *proctolin* originally described by Brown and Starratt (1975) and Starratt and Brown (1975). In this context, it should be noted that the proctodaeal nerves appear to possess substantial quantities of proctolin (Brown, 1967; 1975) and because these nerves originate in the abdominal ventral nerve cord, Krolak *et al.* (1977) may in fact have extracted this peptide.

(b) *Hyperglycemic-adipokinetic factors* (see Table 12.2)

A marked hyperglycemia was observed to occur in *P. americana* after injection of extracts of the CC (Steele, 1961). In particular, the haemolymph concentration of the non-reducing disaccharide trehalose increased dramatically within 30 min of injection and reached a maximum of over twice the control value by 5 h (Steele, 1961). Subsequent observations by Steele (1963) and Ralph and McCarthy (1964) revealed that the activity of fat body glycogen phosphorylase was enhanced after injection of CC extracts and that a concomitant decrease in fat body glycogen reserves (but not muscle) occurred. On the basis of its stability after heating to 100 °C and its inactivation by chymotrypsin, Steele (1961, 1963) suggested that this HGH factor was a peptide. Subsequent studies by Brown (1965), Natalizi and Frontali (1966) and Traina *et al.* (1976), supported Steele's suggestion and, despite the fact that the hyperglycemic hormone (HGH) of *P. americana* has not been characterized, there seems little doubt that it is a peptide. HGH is an extremely potent neurohormone, with a threshold of response of 0.002 gland equivalents (Steele, 1976).

The mode of action of HGH is similar to that of glucagon on vertebrate liver, with the exception that trehalose, rather than glucose, is ultimately released into the haemolymph (Section 7.6). Thus, Downer (1979a,c) more correctly refers to the CC hormone as the *hypertrehalosemic* factor. The CC factor causes an elevation in the levels of cAMP and adenylate cyclase in the fat body (Hanaoka and Takahashi, 1976, 1977, 1978; Gäde, 1977). cAMP-dependent protein kinases in the fat body have been reported (Takahashi and Hanaoka, cited in Hanaoka and Takahashi, 1977), and it is likely that activation of these kinases by cAMP, after treatment with CC extract, is responsible for the activation of the glycogen phosphorylase system. The action of CC extract on the glycogen phosphorylase system can be mimicked by cAMP or phosphodiesterase inhibitors (Steele, 1963, 1964, 1976).

In addition to the hyperglycemic effect, injection of crude aqueous extracts of the CC causes a decrease in the lipid content of the haemolymph (Downer and Steele, 1969; Downer, 1972). This *hypolipemic* effect appears to be confined to the di- and triglycerides (Downer, 1972) with the effect on the latter especially pronounced (Downer and Steele, 1972). Elevation in fat body triglycerides has been observed in association with haemolymph hypolipemia (Downer and Steele, 1969, 1972).

Interestingly, injections of *P. americana* CC extracts into locusts cause *hyperlipemia*, whereas *Locusta migratoria* CC extracts injected into *P. americana* cause *hypolipemia* (Downer, 1972; Goldsworthy *et al.*, 1972). The CC factor from locusts is AKH, usually associated with the mobilization of glycerides during flight. Locust AKH has been isolated and characterized; it is a decapeptide and has the following structure: PCA-Leu-Asn-Phe-Thr-Pro-Asn-Trp-Gly-Thr-NH_2 (Stone *et al.*, 1976). *P. americana* HGH has not been purified; thus it is not known whether AKH and HGH have the same structure. Their effects in the two organisms are related but not the same: (i) AKH injected into *P. americana* causes a marked hyperglycemia (Mordue and Stone, 1977; Jones *et al.*, 1977), but in *L. migratoria* it causes a hyperlipemia; (ii) fat body is a major target organ in both species, but locust fat body is one to two orders of magnitude more sensitive to AKH (Mordue and Stone, 1977; Jones *et al.*, 1977); (iii) CC extract of *L. migratoria* causes a decrease in haemolymph lipid in *P. americana* (Downer, 1972; Goldsworthy *et al.*, 1972), whereas AKH increases haemolymph lipid in *L. migratoria*. If the effects in *P. americana* alone are considered, it can be seen that CC extracts from both organisms cause a drop in haemolymph lipids; and synthetic AKH, when injected into *P. americana*, causes an increase in haemolymph trehalose, as does *P. americana* CC extract. Thus, the differences in the actions of the hormones noted previously may be related to differences in the response of target organs in the two organisms. There is evidence, however, for more than one factor with HGH and/or AKH activity in cockroaches (Brown, 1965; Natalizi and Frontali, 1966; Holwerda *et al.*, 1977). Holwerda and co-workers (1977) have, in fact, found two factors in CC extracts, one of which causes hyperglycemia in *P. americana* and one which causes hyperlipemia in locusts. Hence, the existence of an AKH in *P. americana* remains to be determined.

12.4 Neurohormones of the ventral nerve cord

12.4.1 Proctolin and hindgut-stimulating neurohormones

Different parts of the nervous system are sources of factors that stimulate the muscular activity of the hindgut (Section 4.5.1). Their precise source, identity and mode of action is still under debate (Brown, 1975, 1977; Holman and Cook, 1979), but it is clear that the proctodeal nerves of *P. americana* possess a substance that is able to increase the frequency and amplitude of contraction as well as the tonus of the muscles of the hindgut, or proctodeum (Brown, 1967). This factor, named proctolin, was isolated from whole body extracts (Brown and Starratt, 1975; Starratt and Brown, 1975) and found to be a pentapeptide with the following structure: Arg-Tyr-Leu-Pro-Thr. A peptide with similar pharmacological, chromatographic and electrophoretic properties has been found in other insects and has been suggested to occur virtually 'universally' (Brown, 1977). Brown (1977) speculated that this compound may serve a secretomotor function in insects in general.

Proctolin has a molecular weight of 648 and the synthetic product shows an identical dose response curve in the hindgut bioassay to the naturally occurring substance (Starratt and Brown, 1975). It is active at a threshold concentration of 10^{-9} M. Removal of either of the terminal amino acids threonine or arginine results in a substantial loss of activity (Brown, 1975; Starratt and Brown, 1975). Starratt (1979) has estimated that individual adult *P. americana* contain about 12 ng proctolin.

Crude extracts of CC (Cameron, 1953; Davey, 1962b) or purified fractions of CC (Brown, 1965) also appear to possess hindgut stimulating activity. Brown (1965) utilized paper chromatography and chymotrypsin digestion to isolate three fractions, P_1, P_2 and P_3. P_1 was observed to be active both on the hindgut and on the heart (as a cardioaccelerator), whereas P_2 and P_3 were specifically active on the hindgut and on the heart, respectively. Brown (1965) concluded that these substances were peptides, but he was unable to determine whether they were products of brain NSC or of intrinsic cells of the CC.

Subsequent work by Brown (1967) revealed that the hindgut itself, as well as the foregut, possesses a factor which stimulated contractions of the hindgut. It was also suggested to be a peptide, but differs from hindgut-stimulating substances P_1 and P_2 in that it is not chymotrypsin-sensitive. This substance was present in particularly high concentrations in the proctodaeal nerves, and Brown (1967) suggested that it acted as a neuromuscular transmitter, not only on the hindgut but on visceral muscles in general. This substance was subsequently identified as proctolin. The site of synthesis of the gut factor was not identified, but section of the proctodeal nerve results in a gradual loss of activity in the hindgut (Brown, 1967). However, it appears that, in *L. maderae*, the hindgut-stimulating factor isolated from the gut is not present in the head and may be different from that isolated from the CC (Holman and Cook, 1979). It is possible that Brown's (1965) P_1 and P_2 fractions from CC of *P. americana* may be similar to the hindgut-stimulating neurohormone (HSN) from the head of *L. maderae* described by Holman and Cook (1979).

There has been considerable debate over the similarity between proctolin in *P. americana* and the HSN in *L. maderae*. For example, it has been proposed that proctolin and HSN are identical (Brown and Starratt, 1975; Starratt and Brown, 1975), whereas Holman and Cook (1979) have suggested that these substances are distinct. Holman and Cook (1979) argue that HSN is found in heads of *L. maderae* (Holman and Cook, 1972), is apparently stored in the brain and released from the CC *in vitro* (Marks *et al.*, 1973; Marks and Holman, 1974; Holman and Marks, 1974), and it is distinct from proctolin. In addition, HSN-activity can be obtained in putative neurosecretory granules, isolated from brains of *L. maderae* (Sowa and Borg, 1975). Studies such as that of Holman and Cook (1979) should be extended to *P. americana* to demonstrate conclusively the presence or absence of a distinct hindgut-stimultory peptide such as that found in *L. maderae*.

The modes of action of proctolin and HSN also appear to differ, providing further evidence for the existence of at least two distinct peptides. Brown (1975) proposed that proctolin acts as an excitatory mediator of visceral muscle activity – in essence, as a peptide neurotransmitter. Brown (1975) based this conclusion on the similarities between responses evoked by nerve stimulation and proctolin and on studies using a variety of known neurotransmitters and antagonists. However, as Holman and Cook (1979) correctly point out, the precise sites of action of proctolin are not known. According to these authors, Brown's (1975) results may be interpreted equally well as indicating that proctolin functions as a neuromodulator rather than a neurotransmitter. Peptides have been demonstrated to be modulators of neural function in many systems (for example, see Barker, 1977) and accordingly, this suggestion of Holman and Cook (1979) requires further investigation. In addition, Cook and Holman (1975a,b) and Cook *et al.* (1975) have shown that HSN functions at non-synaptic sites on the hindgut muscle, presumably at the level of the cell membrane of the muscle cells, to alter their excitability. Brown (1975) and Starratt (1979), on the other hand, suggest that their pharmacological evidence indicates that proctolin acts at the *post-synaptic* receptor on the muscle cell membrane. Thus the mode of action of HSN may be very different from that of proctolin.

Finally, proctolin has been observed to exert effects on other visceral muscles, as suggested by Brown (1967, 1977). For example, Miller (1979) has demonstrated that proctolin is able to increase the rate and amplitude of contraction of semi-isolated heart preparations (see cardioaccelerators, Section 12.3.3) and this had led Miller (1979) to speculate that proctolin may be similar to Neurohormone D of Gersch. However, the fact that proctolin is not present in the head extracts of *P. americana* argues against this possibility.

12.4.2 Bursicon

Bursicon, the hormone which regulates cuticular sclerotization at ecdysis, is found in the ventral nerve cord, particularly in the last abdominal ganglion (Mills *et al.*, 1965; Vandenberg and Mills, 1974). Bursicon of *P. americana* appears to be a protein with a molecular weight of approximately 40 000 (Mills and Lake, 1966; Mills and Nielsen, 1967). Although it has been partially purified, the mode of action of bursicon remains uncertain. Studies utilizing haemocyte preparations revealed that bursicon stimulated the conversion of tyrosine to dopamine and *N*-acetyl dopamine (Mills and Whitehead, 1970). These authors suggested that bursicon acts to increase the permeability of the haemocytes to tyrosine. Because haemocytes appear to contain the enzymes responsible for the conversion of tyrosine to *N*-acetyl dopamine, substrate availability may be the rate-limiting factor and hence, regulation of substrate concentration by bursicon could provide an effective system for controlling sclerotization (see also Chapter 2).

Further studies have suggested that cAMP may be the mediator in the conversion of tyrosine to *N*-acetyl dopamine; incorporation of radio labelled metabolites (after injection of ^{14}C-tyrosine) into the cuticle is greatly enhanced by the injection of cAMP, which thus appears to mimic the action of bursicon (Vandenberg and Mills, 1974). These authors suggest that bursicon may activate an adenylate cyclase in the haemocyte membrane, with increased synthesis of cAMP altering the permeability of the membrane. Although adenylate cyclase activity has been subsequently demonstrated in haemocyte membranes (Vandenberg and Mills, 1975), it remains to be shown that they are the primary target cells for bursicon *in vivo*.

Bursicon is released at the time of ecdysis although the precise timing seems to differ among species. In *P. americana*, bursicon was not detected in the haemolymph until 10–20 min after ecdysis and activity was highest 90 min after ecdysis (Mills, 1966). On the other hand, in *L. maderae*, bursicon activity in the haemolymph was significant during ecdysis and remained so for 90 min; it was virtually undetectable after 180 min (Srivastava and Hopkins, 1975). It is not clear whether this difference in timing of bursicon release is real or the result of different bioassay techniques.

12.4.3 Antidiuretic hormone (see Section 6.5.2)

12.5 Corpora allata (CA)

12.5.1 Fine structure of the CA

Fine structural analysis is not available for *P. americana* but has been carried out for *L. maderae* (Scharrer, 1964a, 1971, 1978) and *Blaberus fuscus* (i.e. *craniifer*) (Brousse-Gaury *et al.*, 1973; Brousse-Gaury and Cassier, 1975). A micrograph of *Diploptera punctata* CA is shown in Fig. 12.8. Most observations have been on glands from adult females during the gonotrophic cycle.

A general fine structural characterization of the glands is as follows: A conspicuous sheath with 'collagenous' fibrils covers the glands. Gland cells are stellate in shape, even when cells are isolated (Scharrer, 1964a); this characteristic is manifested in electron micrographs by extensive interdigitation of processes from adjacent cells. Hemidesmosomes are seen at the interface with the sheath; the junctions between cells include simple and septate desmosomes and maculae adhaerentes (Brousse-Gaury *et al.*, 1973). Pinocytotic vesicles occur on the outer

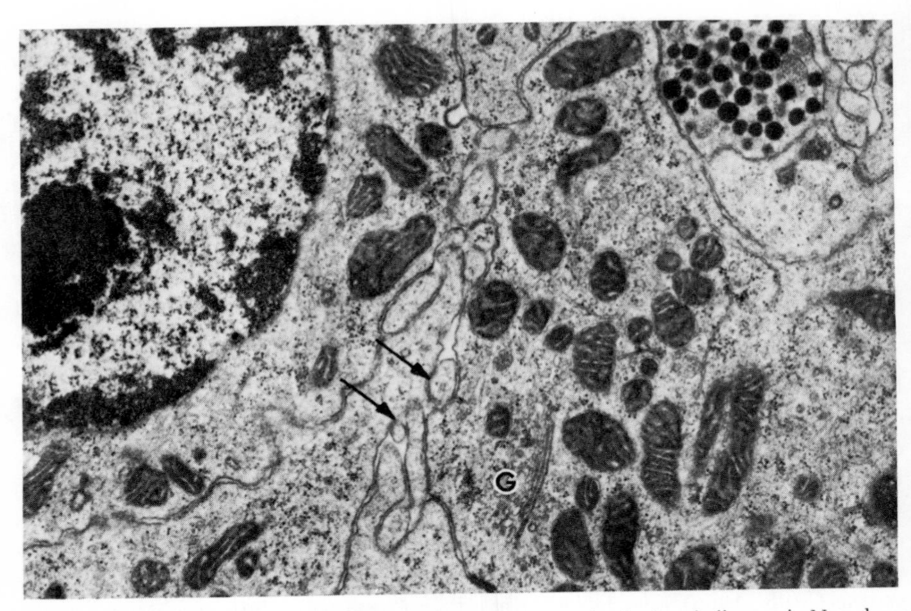

Fig. 12.8 Electron micrograph of corpus allatum of *D. punctata*, beginning vitellogenesis. Note dense mitochondria, ribosomes, smooth and rough ER and flattened Golgi apparatus, G. Cell processes interdigitate (arrows). Neurosecretory axon, upper right, lies adjacent to gland cells. × 17 500. (Micrograph courtesy of G. Johnson.)

and lateral borders of the cells (Brousse-Gaury *et al.*, 1973). The most consistently conspicuous cellular organelles in the cytoplasm are the numerous mitochondria; there are also smooth-(SER) and rough-surfaced endoplasmic reticulum (RER), free ribosomes, relatively small Golgi bodies, and more or less electron-dense inclusion bodies of various sizes and shapes which are presumably lysosomes (Scharrer, 1964a). Brousse-Gaury *et al.* (1973) described a compartmentalization of some organelles in the cytoplasm: mitochondria are grouped at one pole of the nucleus, the RER is essentially perinuclear (Scharrer, 1964a also found whorls of RER near the nucleus), whereas the SER is more likely to be in the periphery of the cells.

The characteristics of an active gland, using the period of rapid oöcyte growth as an indicator, emerge from the description of *L. maderae* and *B. craniifer*. The numerous mitochondria are long and may be variously shaped (e.g. cup, dumbbell), the RER is sometimes in long profiles and in concentric whorls and many free ribosomes are present (Scharrer, 1964a and Brousse-Gaury *et al.*, 1973). Scharrer (1964a) noted vesicles and tubules of SER; Brousse-Gaury *et al.* (1973) found that SER is abundant in the periphery of presumably active cells. Dense bodies, presumably residual bodies of lysosomal activity, are also observed. These increase in number as the activity cycle progresses (Scharrer, 1964a; Brousse-Gaury *et al.*, 1973). Golgi bodies show no signs of massive sequestration of synthetic product in active glands (Scharrer, 1964a). If increase in volume of the cells is a major criterion for increased synthetic activity of the cells, it is noteworthy that peripheral cells seem to enlarge before the central cells (Scharrer, 1964a). At any given time in the cycle, not all cells appear similar (Scharrer, 1964a; Brousse-Gaury and Cassier, 1975). Brousse-Gaury and Cassier (1975) referred to light and dark cells in activating glands of male *B. craniifer*. They attributed the darkness to the greater density of RER and ribosomes and presumed that the dark cells are the most 'active'. This is contrary to Dorn's (1973) supposition for the CA of *Oncopeltus fasciatus*, in which he considered the 'light cells' to be active. However, the 'light cells' of *O. fasciatus* do have extraordinarily large quantities of free ribosomes as do the dark cells of *B. craniifer*.

In contrast, the picture of an inactive gland, either before the period of rapid oöcyte growth (Brousse-Gaury *et al.*, 1973) or in newly emerged or pregnant females (Scharrer, 1964a), is unimpressive; the RER is in few, small saccules; the globular or bilobed mitochondria are smaller than in active glands; some SER is present (Brouse-Gaury *et al.*, 1973). There is a decrease in amount of cytoplasm, but not much change in distribution of organelles. Thus, there must have been a reduction in numbers of organelles (Scharrer, 1964a), presumably by autolysis. No whorls of RER are found in these inactive glands; centrioles are frequently seen, and Golgi bodies are conspicuous (Scharrer, 1964a). Scharrer (1971) has also described, for glands presumed to be inactive, a dense material which appears to arise in the Golgi bodies and migrates to extracellular spaces. The cribriform pattern evident after OsO_4 fixation gave rise to the designation 'C body material'. The relationship of C body material with JH synthesis has not been established.

After prolonged ovariectomy, Scharrer (1978) observed a striking abundance of SER and possibly less RER compared to control animals. This hypertrophy of the SER suggests that it plays a prominent role in JH synthesis and that the glands of castrates are highly active. In *L. maderae*, long-term ovariectomy has been reported to result in the accumulation of vitellogenin and its continued synthesis by the fat body (Engelmann, 1978). On this basis, Engelmann (1979) suggested that the CA of ovariectomized females remain active. However, short-term ovariectomy results in inactive CA in *D. punctata* (Stay and Tobe, 1978).

12.5.2 Innervation

The CA of *P. americana* are innervated by axon tracts orginating in the brain and in the suboesophageal ganglion. This innervation appears to be complex and variable, as evidenced in the various diagrams which have been constructed for the retrocerebral complex (Willey, 1961; Dogra, 1968; Brousse-Gaury, 1971b; Adiyodi, 1974; Fraser and Pipa, 1977). Fig. 12.3 shows a composite, simplified diagram of CA innervation – this figure includes most of the common features of the innervation and can be summarized as follows: The major tract arriving at the CA from the cephalic perspective is the NCA I. The NCA I is continuous with tracts in the CC which contain axons whose cell bodies are located in the contralateral pars intercerebralis and in the ipsilateral pars lateralis, as demonstrated by retrograde diffusion and precipitation of cobalt (Fraser and Pipa, 1977; Pipa, 1978a).

Which of these cells continue on to actually innervate the CA remains to be determined. Nor is it known how many of the NCA I axons are 'non-neurosecretory'. This is an important question because such neurons may regulate or modulate the activity of the NS axons. It appears that at least some of the NCA I axons traverse the CA and enter the NCA II (Adiyodi, 1974; Fraser and Pipa, 1977). However, cobalt diffusion through the NCA II towards the CA did not reveal any axons entering the NCA I (Pipa and Novak, 1979), although the variable number of small axons may indicate that some of these axons descend from the NCA I. Their small size may also preclude their filling or visualization with cobalt.

The CA are also 'innervated' by a second major tract – the NCA II (Fraser and Pipa, 1977) '(Figs. 12.3 and 12.4). The NCA II shows at least two branches, designated by Fraser and Pipa (1977) as the anterior and posterior branches. The cell bodies of these axons are probably located in the suboesophageal ganglion and, because of the presence of putative NS granules, are believed to be NSC (Pipa and Novak, 1979). On the basis of their morphology, Pipa and Novak (1979) have suggested that the peripheral region of the NCA II may be a neurohaemal organ (Section 8.2.4).

At the junction with the CA, many of the axons of the NCA II enter the transverse allatal tract (Fig. 12.3). Some of these axons cross the CA, enter the allatal commissure (Fig. 12.3) and extend to the opposite CA (Adiyodi, 1974; Fraser and Pipa, 1977). Branches of the NCA II axons also enter the 'cap' at the base of the NCA I; this cap has been suggested to be another neurohaemal area (Pipa and Novak, 1979).

The direct innervation of CA cells by axons of the NCA II remains to be demonstrated. Although Pipa and Novak (1979) did observe occasional axons branching within the CA, release sites for these particular axons within the CA have yet to be shown in cockroaches. Scharrer and Wurzelmann (1974) have reported release sites within the CA of *P. americana*, but the origin of the axons is not known. Thus, the relative importance of the NCA II to the regulation of the CA remains unclear.

The only other major axon tracts associated with the CA are the post-allatal nerves, divided by Fraser and Pipa (1977) into inner, middle and outer branches (Fig. 12.3) some of which extend to foregut musculature (Section 8.2.4).

12.5.3 Juvenile hormones

The juvenile hormone (JH) of adult female *P. americana*, the released product of the CA, has been identified as C_{16}JH (JH-III; methyl 10R, 11-epoxy-3,7,11-trimethyl-2E,6E-dodecadienoate) (Pratt and Weaver, 1975; Müller *et al.*, 1975; Hamnett and Pratt, 1978). The identification of this hormone was made possible by the development of an *in vitro* radiochemical assay for JH synthesis which utilizes the incorporation of the radiolabelled methyl moiety of methionine into JH by CA in a defined tissue culture medium (Pratt and Tobe, 1974; Tobe and Pratt, 1974a). This radiochemical assay has been used subsequently to identify the JH of other cockroach species (*D. punctata* – Tobe and Stay, 1977; *Nauphoeta cinerea*

– Lanzrein *et al.*, 1978). Only C_{16}JH appears to be synthesized by isolated adult CA of these species. Lanzrein *et al.* (1975) have reported the presence of C_{17}JH and C_{18}JH (JH II and I respectively) in the haemolymph of adult and larval *N. cinerea*, in addition to C_{16}JH. However, it must be noted that Hamnett and Pratt (1978) and Pratt *et al.* (1978), using high specific radioactivity methionine, were unable to detect the synthesis of either C_{17}JH or C_{18}JH by isolated CA of adult *P. americana* by radio gas–liquid chromatography, in spite of a sensitivity ratio of $1:10^{-4}$. Thus, C_{16}JH appears to be the major, if not exclusive, JH of those Dictyoptera studied.

The three identified JHs show different biological activities in many insects studied, including cockroaches (Kunkel, 1973; Lanzrein, 1979) when injected or topically applied. It has been suggested that, in *N. cinerea*, C_{16}JH is the gonadotrophic hormone whereas C_{17}JH and C_{18}JH are 'status quo' hormones, responsible for the retention of larval characteristics (Lanzrein *et al.*, 1975; Lanzrein, 1979). However, it must be stressed that such a model may not be applicable to *P. americana*. This area requires further clarification.

(a) *Biosynthetic pathway*

Little is known of the biosynthetic pathway for JH in cockroaches, with the exception of the final two stages (see below). It is reasonable to assume that the intermediates in the biosynthetic pathway are similar, if not identical, to those reported for C_{16}JH synthesis in *Manduca sexta* (Schooley *et al.*, 1973). Such a pathway probably utilizes acetate as a precursor and appears to have the same intermediates as the well-known pathway for the synthesis of isopentenyl pyrophosphate. Polymerization of these C_5 units produces first geranyl pyrophosphate (C_{10}), then farnesyl pyrophosphate (C_{15}). The points of rate limitation in JH biosynthesis are unknown at present although it is known that rate limitation occurs prior to the final two biosynthetic stages (see below). A likely biosynthetic stage for rate limitation is: 3-hydroxy-3-methylglutaryl-Coenzyme A $\rightarrow$ mevalonic acid, by regulation of the activity of the enzyme 3-hydroxy-3-methylglutaryl-Coenzyme A reductase (HMG CoA reductase). Rate limitation at this stage is known to occur in vertebrate cholesterol biosynthesis, although the compounds which regulate the activity of the enzyme remain open to question (Kandutsch *et al.*, 1978).

The points of rate limitation in JH biosynthesis remain an important area for future research, although such studies have been hampered by the fact that the complement of intermediates and associated enzymes have not been isolated in any CA system to date. An alternate method for the study of rate limitation makes use of the ability of CA to utilize exogenous precursors or presumed intermediates as substrates for JH biosynthesis *in vitro*. Under such conditions, rates of JH biosynthesis can be stimulated appreciably and as described below, this characteristic may provide important information on rate limitation.

(b) *Stimulated synthesis of JH*

Addition of the exogenous precursor farnesenic acid to the CA incubation medium results in a significant stimulation in the rates of JH synthesis by CA (Pratt *et al.*,

1975a, 1976). It appears that this compound easily penetrates the CA, just as in *Schistocerca gregaria* (Tobe and Pratt, 1974a), and is efficiently incorporated into C_{16}JH. The stimulation of C_{16}JH synthesis by farnesenic acid, at this late stage in the biosynthetic pathway, suggests strongly that rate limitation in JH biosynthesis occurs before this step (see Tobe and Pratt, 1976). The conversion of farnesenic acid to C_{16}JH appears to occur in the following sequence:

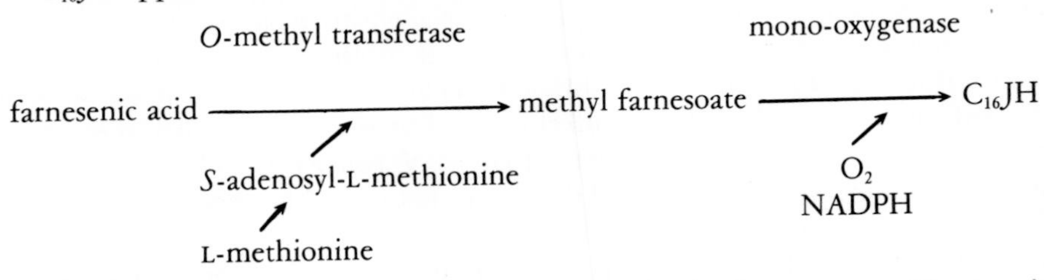

This pathway is based in part on the observation that in the presence of farnesenic acid, methyl farnesoate accumulates *within* the CA (but is not released into the incubation medium) (Pratt *et al.*, 1975a). Up to 150 pmol of methyl farnesoate can be found within a pair of CA, but rates of conversion of this compound to C_{16}JH are not stimulated over the entire range of methyl farnesoate intraglandular content. This indicates that the mono-oxygenase system is saturable under stimulated conditions. However, under conditions of *de novo* JH synthesis (i.e. *spontaneous JH synthesis* – no farnesenic acid added), no accumulation of methyl farnesoate is observed, demonstrating that under normal conditions, the rate of conversion of the putative farnesyl precursor to methyl farnesoate does not exceed the rate of conversion of methyl farnesoate to C_{16}JH (Pratt *et al.*, 1975a). Thus, under conditions of *de novo* C_{16}JH biosynthesis, the mono-oxygenase is not saturated and so rate limitation in C_{16}JH biosynthesis does *not* occur at this stage in the biosynthetic pathway. Feedback of the product at this stage may regulate the rate of synthesis of the farnesyl precursor so as to avoid saturation of the mono-oxygenase system. Hammock (1975) who also used precursors of JH to study the nature and specificity of the enzymes responsible for JH synthesis, observed a conversion of radiolabelled methyl farnesoate to C_{16}JH in CA homogenates of *Blaberus giganteus*. Mono-oxygenase activity was associated with the microsomal fraction (100 000 g precipitate) and this is consistent with the idea that at least the terminal stage of JH synthesis occurs in the ER (Tobe and Saleuddin, 1977). CA homogenates may prove valuable for the study of the enzymes involed in JH synthesis, provided that esterase activity in these homogenates can be inhibited (to prevent esterolysis of biosynthesized JH).

Addition of the higher homologues of arnesenic acid to incubation media containing CA also results in the biosynthesis of the corresponding olefinic esters. For example, addition of the C_{18} homologue of farnesenic acid results in the synthesis of C_{18}JH (Pratt *et al.*, 1978). Other analogues of farnesenic acid are also esterified and epoxidized, provided that they possess an olefinic bond at the position corresponding to the 10,11 position in farnesenic acid (Pratt and Weaver, 1978;

Pratt *et al.*, 1978). These results show that both the *O*-methyl transferase and the mono-oxygenase have low substrate specificity in *P. americana*. Hammock and Mumby (1978) have also studied the specificity of the mono-oxygenase in CA homogenates of *B. giganteus*. These authors observed that the unnatural 2Z geometric isomer of methyl farnesoate is epoxidized considerably faster than the natural 2E isomer; again this suggests low substrate specificity for the mono-oxygenase.

Several authors have incorrectly associated this low substrate specificity with the ability of the CA to synthesize the higher JH homologues *de novo*. As noted above, the higher homologues of JH have not been reported in adult *P. americana* and *only* in the presence of the C_{18} homologue of farnesenic acid will synthesis of C_{18}JH be observed.

(c) *Release of JH and its regulation*

In endocrine systems, either the release or the biosynthesis of hormone may be regulated. It has now been demonstrated in several species, including *P. americana* and *D. punctata* that it is JH biosynthesis rather than release which ultimately regulates the amount of JH which enters the haemolymph (Tobe and Pratt, 1974b; Pratt *et al.*, 1975b; Tobe and Stay, 1977; Weaver and Pratt, 1977). JH is released as soon as it is biosynthesized and does not accumulate within the glands. The rate of release of JH is directly proportional to intraglandular content; thus high intraglandular levels of JH can be associated with high rates of release (Pratt *et al.*, 1975b; Tobe and Stay, 1977).

(d) *Changes in rates of JH biosynthesis*

Using the radiochemical assay described above, the changes in rates of JH biosynthesis have been followed in females of several cockroach species (*P. americana* – Weaver *et al.*, 1975; Weaver and Pratt, 1977; *D. punctata* – Tobe and Stay, 1977; *N. cinerea* – Lanzrein *et al.*, 1978). These studies have followed JH biosynthesis in relation to the female gonotrophic cycle; only in *D. punctata* has JH biosynthesis been determined in the male (Tobe *et al.*, 1979).

Large and predictable changes occur in JH biosynthesis during the female gonotrophic cycle. High levels of synthesis are associated with periods of vitellogenesis, whereas lower levels occur before the onset of vitellogenesis and at the completion of oöcyte maturation. Depending on the species, this correlation can be very precise; in *D. punctata*, length of the basal oöcytes can be predictors of JH synthesis in the *normal* gonotrophic cycle (Tobe, 1980).

Fig. 12.9 shows the rates of JH synthesis by CA of mated female *P. americana* during the first 17 days of adult life (Weaver and Pratt, 1977). It is apparent from this figure that the CA show cyclic peaks in JH synthesis which can be associated with waves of oöcyte growth. Thus, the peaks in CA activity are synchronized with oöcyte development, with successive peaks at intervals of 4–5 days. This corresponds well with the successive cycles of oöthecal formation (3–4 days). However, because two oöcytes in any given ovariole are vitellogenic at any given time, it is

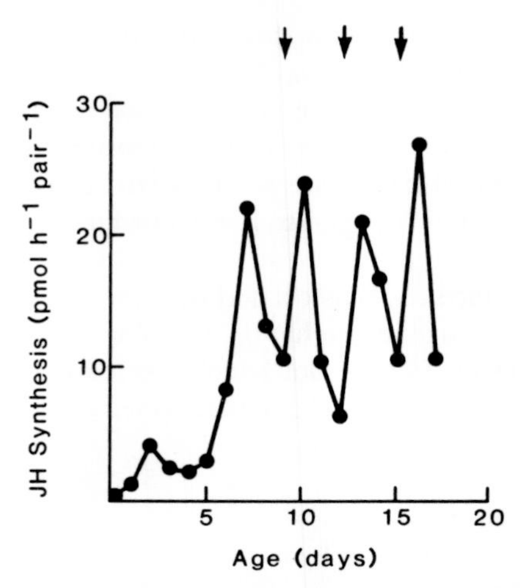

Fig. 12.9 C_{16}JH synthesis by corpora allata of adult mated females of *P. americana* as a function of adult age. JH synthesis was determined using the *in vitro* radiochemical assay of Pratt and Tobe (1974) and Tobe and Pratt (1974a). Arrows indicate times of oötheca formation. (Redrawn from Weaver and Pratt, 1977.)

impossible to ascertain which peak in JH synthesis functionally corresponds to a given wave of oöcyte growth and maturation (Weaver and Pratt, 1977).

The first gonotrophic cycle differs from subsequent ones in that (1) the first peak in JH synthesis (on day 2) is quantitatively much smaller and (2) there are two peaks in JH synthesis prior to the formation of the first oötheca (Fig. 12.9). The first small peak may be associated with stimulation of previtellogenic growth in the basal oöcytes and other developmental events, associated with reproduction, as has been suggested in other species (Tobe and Pratt, 1975; Weaver and Pratt, 1977), whereas the second peak is probably associated with the early vitellogenic growth period in the penultimate oöcytes. It is striking that growth of oöcytes of the first gonotrophic cycle occurs along with other JH-dependent events, including synthesis of vitellogenin (Bell, 1969a) and growth of colleterial glands (Bodenstein and Shaaya, 1968; Shaaya and Bodenstein, 1969) at these low levels of CA activity. Further studies will be necessary to establish if JH titres are lower during this period than in subsequent gonotrophic cycles.

Although measurement of CA activity by direct radiochemical assay is a quantitatively accurate method for determining JH synthesis, it must be noted that previous authors have estimated CA activity in *P. americana* by other methods, in particular, by nucleo-cytoplasmic ratio and/or CA volume (Adiyodi and Nayar, 1965; Adiyodi, R. G., 1969; Brousse-Gaury, 1977, 1978a). Comparisons of CA volume (not shown) and JH synthesis (Fig. 12.9) suggest a relationship between CA volume/nucleo–cytoplasmic ratio and JH synthesis.

The changes which occur in JH biosynthesis, as described above, pertain to a normal gonotrophic cycle. Numerous factors may influence JH biosynthesis including *mating* and *starvation*. In virgin female *P. americana*, rates of JH biosynthesis are similar to those of mated females during the first six days of adult life; thereafter rates of JH synthesis are much lower than those of mated females and remain so for three weeks (Weaver and Pratt, 1977). Significantly, the cyclical changes in JH biosynthesis associated with the gonotrophic cycles of mated females are virtually abolished in virgin females, although maturation of some basal oöcytes may occur (Weaver and Pratt, 1977). The effect which mating has on the endocrine activity of the CA in *P. americana* is not known. In *D. punctata*, CA of virgin females show low rates of JH biosynthesis compared to mated female CA during the gonotrophic cycle (Stay and Tobe, 1977; Tobe and Stay, 1977). However, mating can be mimicked by denervation of CA of virgin females and from this evidence it appears that mating in cockroaches results in the lifting of inhibitory signals from the CNS (Stay and Tobe, 1977, 1978; Weaver and Pratt, 1977 – see also Engelmann, 1959a, 1960).

Starvation of mated female *P. americana* dramatically reduces the nucleo-cytoplasmic ratio of the CA and the rate of oöcyte growth (Brousse-Gaury, 1977). Subsequent feeding results in an increase in nucleo-cytoplasmic ratio and the stimulation of basal oöcyte growth (Brousse-Gaury, 1977). These results suggest that starvation suppresses the activity of the CA (see Scharrer, 1952). However, the mechanism of this suppression remains unknown. Engelmann (1965) has suggested that this effect of starvation on CA activity is not mediated through the CNS, but rather that nutritional milieu in starved animals directly influences the CA. This hypothesis should now be directly tested, using the radiochemical assay for CA activity.

12.5.4 Regulation of the CA

From the foregoing discussion, it is apparent that the CA of adult cockroaches are precisely regulated. Although the precise mechanisms whereby CA are regulated remain to be elucidated, it is proposed that both *allatostatic* (CA-inhibitory) and *allatotropic* (CA-stimulatory) factors must operate.

Allatostatins may act to inhibit the CA prior to mating or feeding, at the completion of oöcyte development and during gestation in ovoviviparous and viviparous species; they also may suppress CA activity during metamorphosis: (1) At the end of the gonotrophic cycle, JH synthesis declines to low levels (Weaver and Pratt, 1977; Tobe and Stay, 1977; Lanzrein *et al.*, 1978). In *D. punctata*, this decline is observed in both normal innervated CA and in denervated CA (Stay and Tobe, 1977), indicating that allatostatins operate by a humoral pathway. The ovary appears to be the source of an allatostatin (Stay *et al.*, 1980b). Engelmann (1959a) showed that in *L. maderae* injections of ecdysone or implantation of active prothoracic glands results in an inhibition of oöcyte growth and high nucleo-cytoplasmic ratios in CA (presumably indicating low levels of JH synthesis). Ecdysterone has been detected in the haemolymph and ovaries of cockroaches (*B. craniifer* – Bullière *et al.*, 1979; *N. cinerea* – Imboden *et al.*, 1978; *D. punctata* – Stay *et al.*, 1980a), and it is likely that it will be found in other species. In *D. punctata*, ecdysterone inhibits JH synthesis in a dose-dependent fashion *in vivo* (Stay *et al.*, 1980a; Friedel *et al.*, 1980), but not *in vitro* (Friedel *et al.*, 1980), suggesting that its allatostatic activity is mediated by the CNS. (2) In *D. punctata*, the activity of the CA is depressed in the presence of elevated JH titres (Tobe and Stay, 1980) and in the presence of a functional mimic of JH, the JH analogue ZR512 (Tobe and Stay, 1979). (3) Allatostatic factors also appear to reach the CA by the nervous tracts. Severance of the NCA I in *P. americana* last instar nymphs results in a supernumerary moult (Fraser and Pipa, 1977). In *D. punctata*, denervation of the CA of virgin adult females leads to a cycle of JH synthesis quantitatively similar to that of normal mated females (Stay and Tobe, 1977). In *L. maderae*, section of the NCC I results in an increase in size of the ipsilateral CA in both nymphs and adults and in supernumerary moults of the nymphs (Scharrer, 1952; Engelmann and Lüscher, 1957). Severance of NCC II does not appear to result in CA 'activation' (Engelmann and Lüscher, 1957).

Allatotropins act to stimulate JH synthesis by the CA. They probably operate during periods of high JH requirements, for example, during the gonotrophic cycle and may operate either humorally or via the nervous tracts: (1) Treatment of *D. punctata* females with a functional mimic of JH, ZR512, stimulates JH synthesis significantly at low doses (Tobe and Stay, 1979). This stimulation suggests the existence of a positive feedback loop, which may operate by causing the release of an allatotropin from the CNS. (2) Ovariectomy of *D. punctata* results in dramatic decline in JH synthesis (Stay and Tobe, 1977). Implantation of ovaries into ovariectomized females evokes an apparently normal cycle of JH synthesis associated with the gonotrophic cycle (Stay and Tobe, in preparation). The ovary probably releases an allatotropic factor which operates either directly on the CA or through the CNS. (3) JH synthesis by right and left CA of adult females shows a high degree of symmetry in *P. americana* (Weaver, 1979) at all levels of JH synthesis. A humorally acting allatotropin could cause the same degree of stimulation in both CA. (4) The brain is needed for a short time after adult emergence for CA-dependent oöcyte growth in *N. cinerea* (Barth and Sroka, 1975) and has been suggested to be the source of an allatotropin in *D. punctata* (Stay and Tobe, 1977).

12.5.5 Regulation of JH titre

The interaction between JH biosynthesis and its degradation and/or excretion, mainly by esterolysis of the methyl function and/or hydration of the epoxide function, ultimately determines the amount of JH in the haemolymph. However, it is unclear whether biosynthesis or degradation (and/or excretion) plays the more important role in regulation of JH titres. In the Lepidoptera, esterolysis of JH has been proposed to be the mechanism regulating the titre of JH at the time of metamorphosis (Akamatsu *et al.*, 1975). Esterolysis may be carried out by either general esterases or by esterases specific for JH; in the Lepidoptera, these esterases are insensitive to diisopropylfluorophosphate (DFP) (Sanburg *et al.*, 1975).

Table 12.4 Actions of juvenile hormone in cockroaches

Target organ	*Actions*	*Species*	*Reference*
Left colleterial gland	Production of protocatechuic acid glucoside	*P. americana*	Bodenstein and Shaaya (1968); Willis and Brunet (1966)
	Production of glucoside	*B. fumigata*	Bell and Barth (1970); Barth and Bell (1970)
	Production of glucoside	*B. germanica*	Zalokar (1968)
Ovary	Uptake of vitellogenin	*P. americana*	Bell (1969a, 1970)
	Uptake of vitellogenin	*N. cinerea*	Wilhelm and Lüscher (1974)
	Uptake of vitellogenin	*Eublaberus posticus*	Bell and Barth (1971)
Fat body	Synthesis of vitellogenin	*P. americana*	Pan, Bell and Telfer (1969)
	Synthesis of vitellogenin	*N. cinerea*	Lüscher (1968a)
	Synthesis of vitellogenin	*L. maderae*	Brookes (1969); Engelmann (1969)
	Decrease in fat body respiration		Müller and Engelmann (1968)
	Stimulation of fat body respiration		Lüscher and Leuthold (1965)
Spermathecal accessory gland	Release of spermatophore	*L. maderae*	Engelmann (1960)
		D. punctata	Engelmann (1960)
Genital tract	Stimulation of pheromone production (low doses)	*B. fumigata*	Bell and Barth (1970); Barth and Bell (1970)
	Inhibition of pheromone production (high doses)		
Numerous body tissues	Prevention of metamorphosis	*P. americana*	Bodenstein (1953c,d)
Antennae (male)	Proliferation of pheromonal sensory receptors	*P. americana*	Schafer and Sanchez (1976)

Because C_{16}JH appears to be the exclusive JH of adult *P. americana* (Müller *et al.*, 1975; Pratt and Weaver, 1975), studies on specific JH esterases must utilize this substrate. However, several studies have utilized the substrate C_{18}JH and it is thus difficult to evaluate the significance of the results (Weirich and Wren, 1976; de Kort *et al.*, 1979). de Kort *et al.* (1979) also utilized C_{16}JH as substrate for haemolymph esterases of *P. americana* adults and reported rates of hydrolysis of 0.8 nmol min^{-1} ml^{-1} haemolymph. No effect of DFP on rates of hydrolysis of either C_{16}JH or C_{18}JH was observed. These authors suggested that little degradation of C_{16}JH occurs in the haemolymph of adults, but it is unfortunate that a more quantitative, temporal analysis of the haemolymph esterases was not performed. However, in the light of the rates of hydrolysis of C_{16}JH in the haemolymph, de Kort *et al.* (1979) suggested that most degradation of C_{16}JH occurs in the tissues.

JH-specific binding proteins, which serve to transport JH and to protect it from esterolysis in the haemolymph in Lepidoptera, have not been demonstrated in the Dictyoptera. It should be stressed that future studies should focus, not on the ability of specific haemolymph proteins to bind JH, but rather on the total binding capacity of the haemolymph. To this end, equilibrium dialysis and/or gel filtration chromatography of esterase-inhibited haemolymph may provide important information.

The activity of the CA and the rate of JH synthesis in particular may be an important determinant in regulating JH titre. This view has been put forward by Tobe and Stay (1977, 1980), Tobe (1980) and by de Kort and co-workers (for example de Kort *et al.*, 1979). Nonetheless, it should be noted that other authors believe that JH-specific esterases play the major role in regulating JH titres (Akamatsu *et al.*, 1975; Rüfenacht and Lüscher, 1976). For a more complete discussion of the importance of changing rates of JH synthesis in the determination of JH titres, see Tobe (1980).

12.5.6 Actions of JH

Table 12.4 shows a compilation of established actions of JH in cockroaches. For a more complete discussion, refer to the relevant chapter.

12.6 Prothoracic glands (PG)

The PG are a pair of transparent tissue bands which lie in the anterior ventral thorax near the prothoracic ganglion (Fig. 12.10). Each gland is a discrete flat band, 1 or 2 cells thick and 4–12 cells wide, surrounding a thin strand of muscle. Each band extends from the first cervical sclerite to the contralateral basal region of the procoxa crossing just anterior to the prothoracic ganglion (Chadwick, 1956). The bands taper at the ends such that the tips are muscle fibre only (Bodenstein, 1953c). Because the PG of *P. americana* are less branched and have fewer tracheal attachments to other tissues than do those of other cockroach species, they are easier to extirpate completely (Chadwick, 1956).

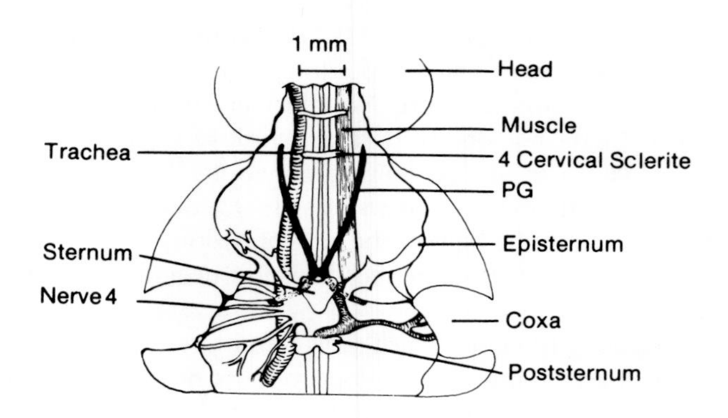

Fig. 12.10 Partial dissection of cervical and prothoracic region of larval *P. americana* showing pertinent anatomical features and prothoracic glands. (After Chadwick, 1956.)

12.6.1 Tracheation and innervation

Tracheae and nerves accompany the central muscle fibre. The innervation arrives at the coxal end as a narrow branch from the 4th segmental nerve of the prothoracic ganglion (Chadwick, 1956). A posterior branch from the NCA II also passes near the PG but does not appear to innervate it (Fraser and Pipa, 1977). Also several small branches from NCA II pass to what may be a cephalic part of the PG, but again it is not certain that they innervate the PG (Fraser and Pipa, 1977).

12.6.2 Fine structure

The fine structure of the PG of *P. americana* has not been studied. The description which follows comes from Scharrer's (1964b) study of nymphal glands of *L. maderae* at various times in the moult cycle.

A sheath containing collagenous fibres covers the PG; it is thinner than that covering the CA. Elements of the sheath penetrate into and cover the muscular core. Sheath elements also cover the extensively infolded processes of the gland cells.

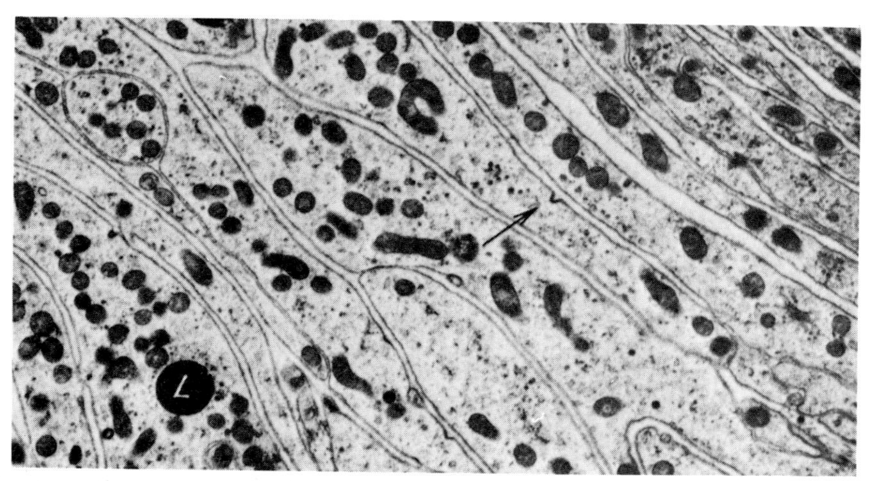

Fig. 12.11 Electron micrograph of peripheral area of prothoracic gland of *L. maderae* (female 8 days after last larval moult) shows extracellular channels penetrating the gland and numerous mitochondria. Micropinocytotic pit (arrow), lysosome-like body (L.). × 800. (Reproduced by permission, B. Scharrer and Springer-Verlag; Scharrer 1964b; Fig. 6.)

The gland cells are densely packed and a most conspicuous characteristic of their morphology is the extensive surface infoldings which Scharrer (1964b) describes as slender cytoplasmic processes extending towards the surface (Fig. 12.11). The extracellular spaces between these processes are regular straight channels about 0.05 μm wide with irregular wider spaces up to 0.5 μm. Coated micropinocytotic vesicles, from 100–200 nm in diameter, occur adjacent to the plasma membrane and mitochondria are prominent organelles in these processes (Fig. 12.11).

The cytoplasm contains little ER: SER is vesicular and tubular and RER occurs in short profiles. Some flattened vesicles of ER are adjacent to the plasma membrane.

The Golgi apparatus is relatively inconspicuous and shows no signs of cyclic changes. Only small inconspicuous vesicles occur adjacent to the Golgi saccules; no secretory granules are evident in the cytoplasm. Other cytoplasmic organelles include free ribosomes, lysosomes (see Osinchak, 1966), multivesicular bodies, filaments and glycogen. When ribosomes are abundant, the cells appear electron dense.

Scharrer (1964b) observed variations in fine structural features with the cyclic activity of the glands which include larger diameter of the mitchondria in the first half of the moult cycle, an increase in size and irregularity of nucleolar outline with increasing interval after moult, and a greater abundance of lysosomes and mitotic figures around the middle of the intermoult. However, she did not find conspicuous changes in gland morphology or changes in nucleo–cytoplasmic ratio such as those which are associated with the cyclic activity of adult CA (Scharrer, 1964a).

12.6.3 Growth and regression of the PG

The PG of *P. americana* grow by mitosis during each stadium and reach maximal size at the end of the last stage. After the adult moult, the glandular cells degenerate and by day 12–14, only the muscular core is apparent (Bodenstein, 1953c).

A detailed study of PG diameter in the last two larval stadia and early adult life of *N. cinerea* showed that diameter increases during the first half of the larval stadia, followed by a decrease several days before the moult; by three days after the moult, the gland had degenerated (Lanzrein, 1975). Mosconi-Bernadini (1966) found that the PG is female *L. maderae* degenerated less rapidly than those of males and suggested some interaction between CA activity and PG degeneration. More information is needed before the point can be established.

The fine structural features of regressing PG is adult males and females are known for *L. maderae* and *B. craniifer*. At metamorphosis the glands are similar to nymphal glands; within 5 days they show features of regression (Scharrer, 1966). The organelles of glandular cells disappear in an orderly process, starting with autophagy of cytoplasmic organelles, and pyknosis of the nuclei. Subsequently, the plasma membranes disappear and the cellular debris is invaded and engulfed by haemocytes. The autophagy involves lysosomal enzymes (Scharrer, 1966; Osinchak, 1966). At the completion of the degeneration, the haemocytes depart leaving only the central axis of muscle, trachea and nerve surrounded by a consolidated sheath (Scharrer, 1966).

12.6.4 Control of regression of the PG in adult

Lanzrein (1975) has studied the factors which contribute to degeneration of the PG in *N. cinerea*. Using transplantation experiments, application of JH analogue, decapitation and implantation of CC, she has concluded: (1) Only PG which have been activated can be induced to degenerate. (2) The induction of degeneration occurs over about a two-day period, six to seven days prior to adult ecdysis. Since apolysis occurs at this time, ecdysterone may be the induction factor. (3) The PG can respond to induction factor only if JH has been absent for six to ten days prior to induction. After this time, the presence of JH does not prevent degeneration. (4) The PG can be protected from degeneration after induction by a 'protecting factor' which is released from the CC and may be produced by the brain. The protecting factor need be present for only a day to inhibit degeneration of PG after adult ecdysis.

12.6.5 Function of the PG

Bodenstein (1953c,d) induced moulting in old adults implanted with PG from last-instar nymphs. That PG do synthesize a moulting hormone was demonstrated *in vitro* with PG of *L. maderae* larvae (King and Marks, 1974; Borst and Engelmann, 1974). Ecdysone was identified as the secretory product using high resolution liquid chromatography and mass spectrometry by the former authors and GLC by the latter. Gersch (1980) has shown with radio-immunoassay (RIA) that *P. americana* PG also synthesize ecdysteroids (ECD) *in vitro*. Although the structure of the ECD remains to be determined it seems likely that it will be shown to be ecdysone.

12.6.6 Cyclic activity of PG

Direct evidence for cyclic activity of the PG during the last instar exists from measurements of *in vitro* synthesis of *P. americana* glands by RIA (Gersch, 1980). ECD synthesis is very low for two-thirds of the stadium after which production increases about 10-fold just before ecdysis. An *in vitro* study of Borst and Engelmann (1974) in *L. maderae* showed that PG of mid last-instar produced 10-fold more hormone than glands from early last-instar or one day old adults. Cyclic changes in diameter of PG in *N. cinerea* larvae also suggests a cyclic synthetic activity (Lanzrein, 1975).

12.6.7 Actions of ecdysone

Table 12.5 shows a compilation of established actions of ecdysone/ecdysterone in cockroaches. For a more complete discussion, refer to the relevant chapter.

12.6.8 Titres of ecdysteroids

Gersch (1980) has shown that haemolymph ECD parallel the production of hormone by the PG. Maximal levels were found shortly before ecdysis. Ecdysterone appears to be the principal component at that time (Gersch, 1980). In the

Table 12.5 Actions of ecdysone in cockroaches

Target organ	*Actions*	*Species*	*Reference*
Epidermis (larval regenerates, *in vitro*)	Cuticle formation	*L. maderae*	Marks and Leopold (1971); Marks (1972, 1973a)
	Differentiation, moult-linked	*L. maderae*	Marks (1973b)
Epidermis (embryonic leg, *in vitro*)	RNA and cuticle synthesis	*B. craniifer*	Bullière (1977)
	DNA synthesis, differentiation-linked	*B. craniifer*	Bullière and Bullière (1977a)
	Regeneration, inhibition of	*B. craniifer*	Bullière and Bullière (1977b)
Brain (larval *in vitro*)	Release of NSM	*L. maderae*	Marks *et al.* (1972)
Corpora allata (adult)	Inhibition of CA activity	*L. maderae*	Engelmann (1959b)
		D. punctata	Stay *et al.* (1980b)
Fat body (adult)	RNA and protein synthesis	*Gromphadorhina portentosa*	Bar-Zev and Kaulenas (1975)

penultimate larval stadium of *P. americana*, moulting hormone activity, quantitated by *Calliphora* bioassay, was undetectable in the first half of the instar, increased gradually in the second half to a peak and then declined sharply to an undetectable level again at apolysis (Shaaya, 1978).

There have been no measurements of ECD titres in the adults of *P. americana*, but measurements exist for the pre-oviposition period in *B. craniifer* (Bullière *et al.*, 1979) and the pregnancy period in *N. cinerea* (Imboden *et al.*, 1978). ECD would not be expected in the haemolymph of such adults since the PG begin regression shortly after metamorphosis in both of these species (Scharrer, 1966; Lanzrein, 1975). However, RIA of haemolymph from *B. craniifer* revealed a small peak just after copulation and a large peak just before ovulation (Bullière *et al.*, 1979). In *N. cinerea*, small amounts of ECD were present in female haemolymph prior to dorsal closure of the embryos as determined by RIA (Imboden *et al.*, 1978). Thus there is obviously an alternative source of ecdysteroids in the adult female and this seems very likely to be the ovary (Section 12.7).

The titres of ECD in embryos have also been determined by RIA for *N. cinerea* (Imboden *et al.*, 1978) and *B. craniifer* (Bullière *et al.*, 1979). In both species, there are several peaks, the largest just before hatching. Bullière *et al.* (1979) related some of the other peaks to developmental events known to require ECD, such as first and second cuticle formation; other peaks have no such obvious role.

12.6.9 Alternative sources of moulting hormone

Extirpation of PG is difficult but not impossible in cockroaches and the results leave little doubt that there must be an alternative source of moulting hormone. Chadwick (1956), Nutting (1955) and Gersch (1977) have observed moults in *P. americana* nymphs deprived of PG. Furthermore, Gersch and Eibisch (1977) found radiolabelled compounds co-chromatographing with ecdysone after injection of ^{14}C-cholesterol into larvae lacking PG. The source of the ECD is unknown, but ligature experiments with *B. craniifer* embryos indicate that tissues posterior to the mesothoracic segment can produce moulting hormone (Bullière and Bullière, 1974). The oenocytes are possible candidates.

12.6.10 Activation of the PG

Fraction I, extractable from brain or CC, was shown to activate PG RNA synthesis *in vivo* (Gersch and Stürzebecher, 1970) and *in vitro* (Gersch and Bräuer, 1974). This was thought to be indicative of PG activation because it did not stimulate similar RNA synthesis in other tissues. However, much RNA synthesis occurs in the absence of ECD synthesis (Gersch, 1980). Thus, an increase in RNA synthesis is not a conclusive indicator of hormone synthesis and Factor I cannot be considered the prothoracicotropic factor because it does not stimulate ECD synthesis (Gersch, 1980). Nonetheless, whole brain homogenates do stimulate ECD synthesis *in vitro* and, consequently, Gersch (1980) postulates different activating factors.

The period during which the brain is necessary in order for a moult to occur has not yet been determined in *P. americana*. Kunkel (1975a) found in the fourth larval instar of *Blattella germanica* that the brain is no longer necessary after 77 h of the 130-h intermoult period. The critical period during which limb amputation could cause a delay in moulting is similar to the critical period for the presence of the brain; hence Kunkel (1975a) suggests that regeneration causes a delay in brain activation of the PG.

There has been no experimental demonstration that ECD have a positive feedback on the PG to stimulate the synthesis of ecdysone as has been demonstrated in other insects (e.g. Agui and Hiruma, 1977). However, there is evidence that ecdysone promotes the release of NSM from the brain, at least *in vitro* (Marks *et al.*, 1972). This is in contrast to Steel's (1973) assertion that, in *Rhodnius prolixus* larvae, ecdysone stimulates synthesis and inhibits release of NSM.

12.7 Ovary as an endocrine organ

There are several pieces of evidence which suggest an endocrine function for the ovary. In *L. maderae*, ovariectomy results in a pronounced change in the appearance of NS cells of the suboesophageal ganglion (Scharrer, 1952). In *N. cinerea*, ovariectomy prevents an increase in size of CA (Wilhelm and Lüscher, 1974). In *D. punctata*, JH synthesis is suppressed in the absence of the ovary (Stay and Tobe, 1978). However, at the end of vitellogenesis, the ovary also appears to trigger the decline in CA activity and this can be mimicked by injections of ecdysterone (Tobe and Stay, 1980; Stay *et al.*, 1980; Friedel *et al.*, 1980). As noted above, ECD have been found in the haemolymph of adult females of *N. cinerea*, *L. maderae* and *B. craniifer* (King and Marks, 1974; Imboden *et al.*, 1978; Bullière *et al.*, 1979) in spite of the degeneration of the PG shortly after adult ecdysis (Scharrer, 1948; Bodenstein, 1953c,d; Lanzrein, 1975). ECD have also been found in the ovary. During the gonotrophic cycle of *B. craniifer*, the quantities of ECD in ovary and haemolymph are parallel, but 10 times higher in the ovary with the peak occurring just before chorion formation (Bullière *et al.*, 1979). Nijhout and Koeppe (1978) have also found ECD in the ovary of *Leucophaea maderae* and their observation that the ovary can synthesize ECD *in vitro* suggests that the ovary is the source of ovarian and haemolymph ecdysteroids.

It is not yet clear which ECD are present in the ovary and haemolymph. Only ecdysterone was detected in adult female haemolymph of *L. maderae* (King and Marks, 1974), whereas ecdysone and ecdysterone occur in *N. cinerea* haemolymph (Imboden *et al.*, 1978). Not only do the identities of the ECD in the haemolymph and ovaries need clarification, but also a precise quantitation throughout the reproductive cycle, together with an elucidation of the biosynthetic pathway and its regulation before a hypothesis of humoral function for the ovary can be substantiated. None of this information is available as yet for *P. americana*.

Acknowledgements

We thank Professor S. R. Pipa and Drs R. Feyereisen, T. Friedel, E. Mundall, B. Scharrer and G. Johnson, for their suggestions and criticisms of the manuscript. Supported by grants from the Natural Sciences and Engineering Research Council of Canada and the US Public Health Service.

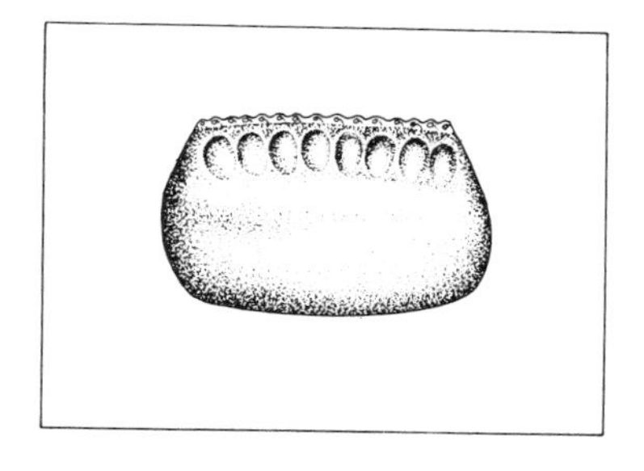

13

Reproduction

William J. Bell and K. G. Adiyodi

13.1 Introduction

This chapter deals almost exclusively with reproductive physiology of *Periplaneta americana*. Substantially more space would be required for a comparative approach, including other cockroach species – *Leucophaea maderea*, *Nauphoeta cinerea*, *Brysotria fumigata*, *Blattella germanica* and *Diploptera punctata* – that have also been subjects of intensive reproductive studies. For more comprehensive reviews, the reader should consult *The Physiology of Insect Reproduction* by Engelmann (1970) and recent reviews by DeWilde and DeLoof (1973), Doane (1973), Adiyodi and Adiyodi (1974), Engelmann (1979) and Roosen-Runge (1977). Chapter 12 of this volume discusses hormones and the control of hormone secretion as related to reproductive processes in cockroaches.

13.2 Female reproductive system

The reproductive organs of *P. americana* consist of paired ovaries, colleterial (accessory) glands and a single spermatheca. Each ovary has eight ovarioles connected by a short pedicel to the oviduct. Paired oviducts join to form a common oviduct opening into the genital chamber close to the opening of the spermatheca. Colleterial glands enter the vestibulum of the genital pouch at the base of the ovipositor valves.

13.2.1 Morphology of the ovary

The panoistic ovariole of *P. americana* is covered by an inner relatively thick membrane, the tunica propria, and an outer sheath of connective tissue

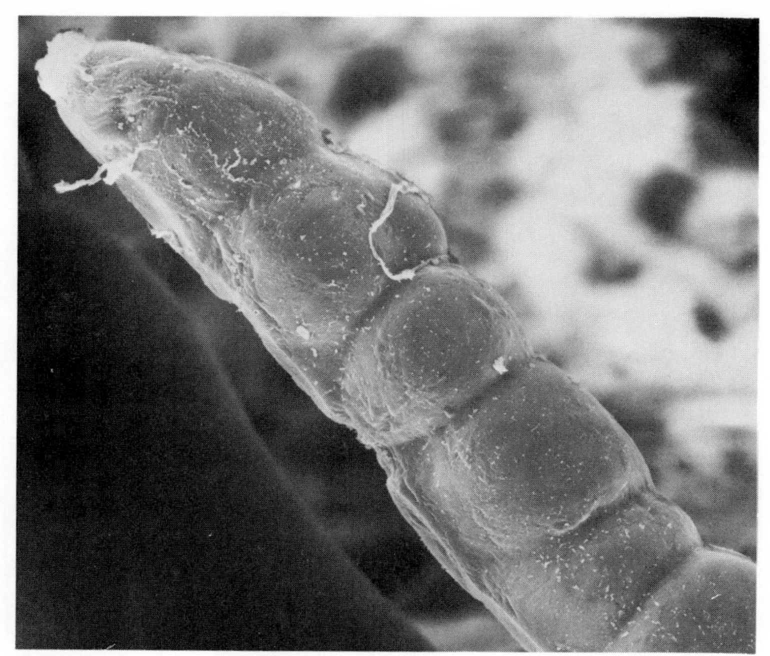

(a)

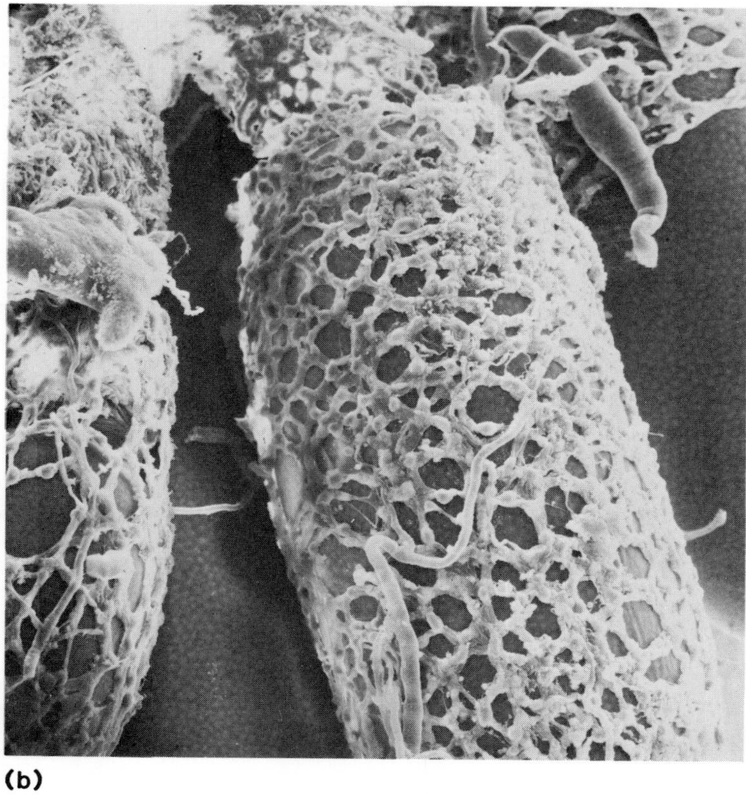

(b)

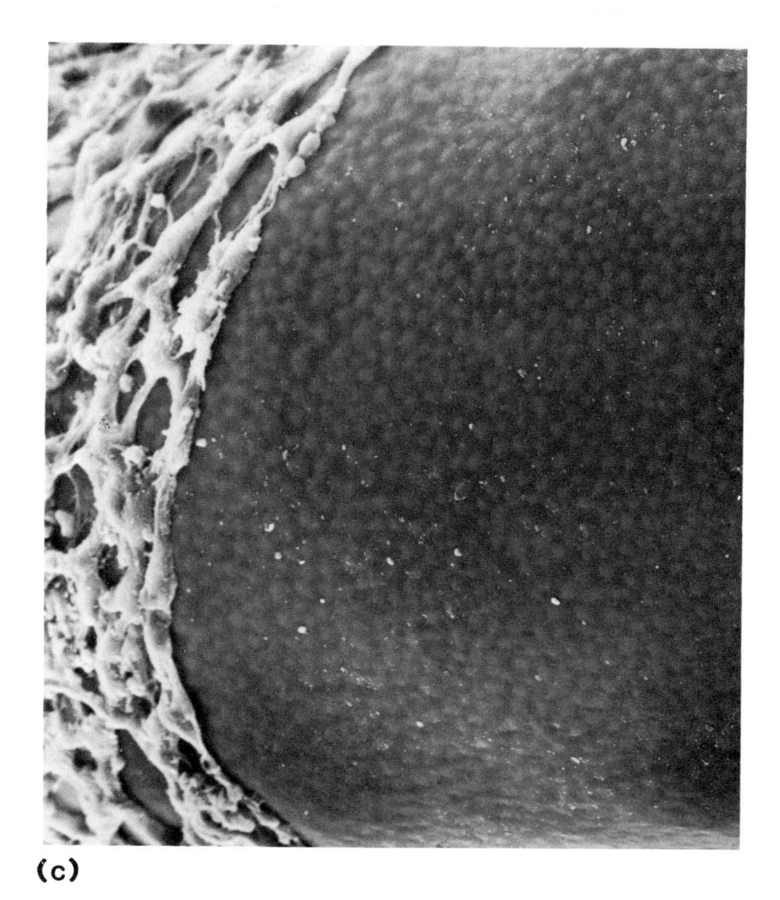

Fig. 13.1 Scanning electron micrographs of the ovariole of *P. americana*. (a) Germarium and previtellogenic follicles (anterior tip of the ovariole). × 520. (b) Basal and penultimate follicles × 110. (c) Enlargement of basal follicle revealing follicle cells through the tunica propria and extracellular ovarial sheath. × 320.

(Bonhag and Arnold, 1961). Bonhag (1959) and Anderson (1964, 1969) divided the ovariole into functional stages proceeding posteriorly from the germarium to the zone of chorionated oöcytes (Fig. 13.1). Light and ultrastructural studies of oogenesis and vitellogenesis by these workers are extensive, and only a summary can be presented here.

The germarium contains oögonia, prefollicular cells and young oöcytes. The first maturation division is initiated and proceeds to the diplotene stage of prophase; meiosis terminates at the time of fertilization. In the germarium, each oöcyte becomes completely surrounded by a single layer of follicle cells; the unit of an oöcyte and its complement of follicle cells is referred to here as a follicle. Each follicle is separated from adjacent ones by layers of interfollicular tissue.

13.2.2 Oogenesis and vitellogenesis

(a) *Previtellogenic follicles*

Prior to the initiation of yolk deposition there is uptake of material by pinocytosis, forming vesicles of approximately 85–125 μm in diameter. The plasma membrane of the follicle cells adheres closely to the oölemma and the follicle cells are cuboidal (Anderson, 1964). Growth of previtellogenic follicles proceeds slowly, accounting for an increase in volume from 0.01–0.03 mm^3 (Bell, 1969a). It is probable that material sequestered at this stage is utilized solely for nutrition of the oöcytes and not for storage as yolk.

The nucleoli of previtellogenic oöcytes produce emission bodies, composed of RNA, that move through pores in the nuclear membrane into the oöplasm (Montgomery, 1898; Bonhag, 1959; Anderson, 1964); whether this population of RNA is necessary for vitellogenesis or embryogenesis has not been established.

(b) *Vitellogenic follicles*

Yolk deposition is signalled by the retraction of follicle cell membranes from the oölemma and enlargement of spaces between the follicle cells (Figs. 13.2 and 13.3) (Anderson, 1964, 1969; Sams and Bell, 1977; Sams, 1977; Wasilenko *et al.*, unpublished). These spaces or channels between the follicle cells, the formation of which is termed patency (Davey and Huebner, 1974), differ in size in any given section observed owing to septate desmosomes that make intercellular supportive connections (Anderson, 1964; Sams, 1977). The oölemma becomes folded and endowed with microvilli to a greater extent than in previtellogenic follicles (Fig. 13.4). Anderson (1969) demonstrated the movement of horse-radish peroxidase through interfollicular spaces to the oölemma and its incorporation into vitellogenic yolk spheres. No such incorporation occurred in previtellogenic follicles.

Yolk deposition proceeds at a rapid rate, the oöcyte increasing in volume 100-fold in approximately 12 days (Bell, 1969a). The process, as revealed by electron microscopy, consists of pinosome formation at membrane sites coated externally with 'fuzzy' material and coated internally with descrete spikes or projections (Fig. 13.5), the internalization of pinosomes, and finally the coalescence of these bodies into large (1 μm diameter) yolk spheres (Anderson, 1969; Sams, 1977).

Bell and Sams (1974) and Krolak *et al.* (1977) demonstrated that proteins synthesized and secreted by the follicle cells enter the oöcyte along with vitellogenin. It is not certain if follicle cell proteins of *P. americana* are actively involved in the process of selective haemolymph protein uptake as in the *Cecropia* moth (Anderson and Telfer, 1969; Anderson, 1971).

Lipids (Nath *et al.*, 1958), but not glycogen (Bonhag, 1959), enter the oöcyte, primarily as haemolymph-borne triglyceride (Nath *et al.*, 1958; Vroman *et al.*, 1965). The final increase in oöcyte volume occurs just prior to chorion formation and entails uptake of water by the oöcyte (Verrett and Mills, 1975a,b).

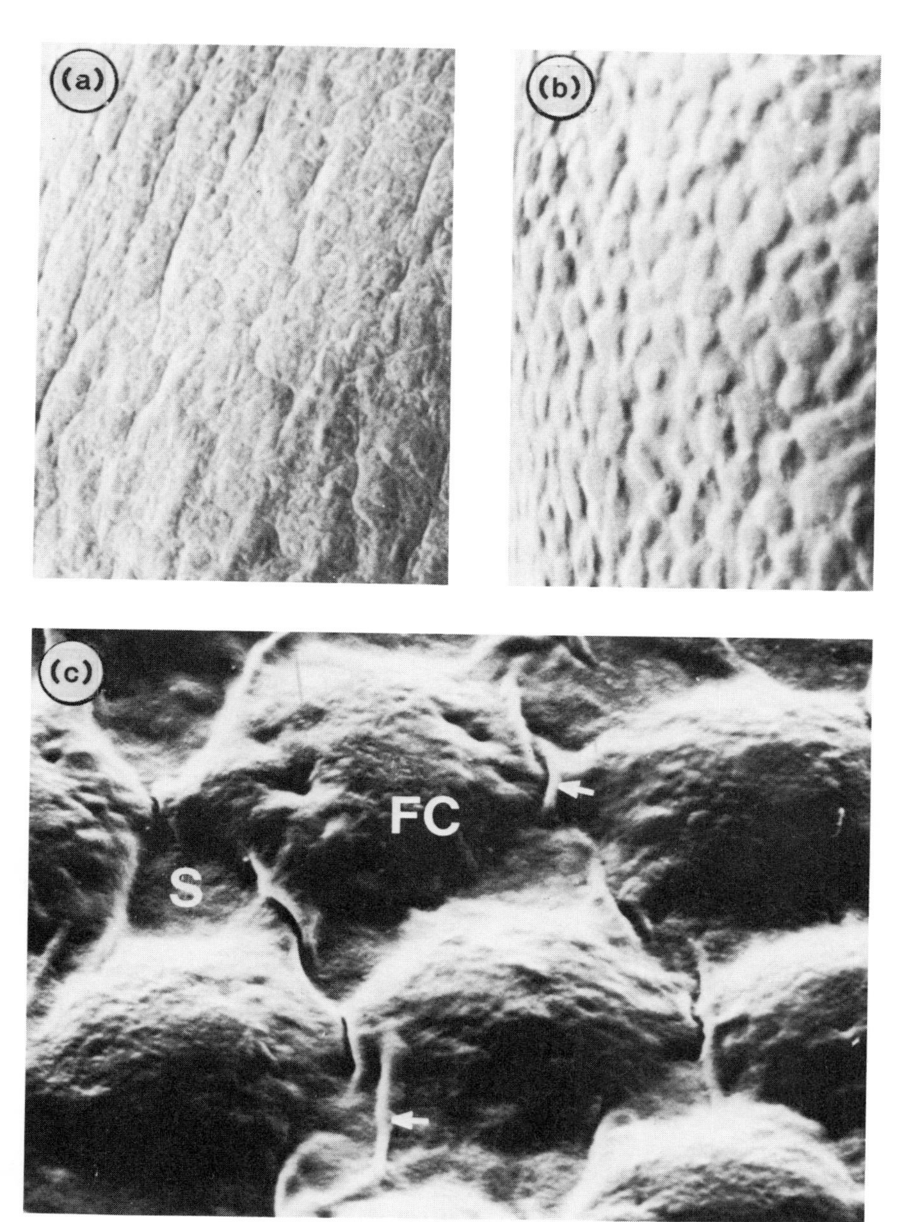

Fig. 13.2 Scanning electron micrographs of the surface of basal follicles of (a) 2-day, (b) 5-day and (c) 8-day post-ecdysis female *P. americana*. Tunica propria becomes sculptured by the underlying follicle cells as patency progresses. Pseudopodia (arrows) extend from one cell to another. S, interfollicular space beneath tunica propria; FC, follicle cell. × 1500. (Wasilenko *et al.*, unpublished.)

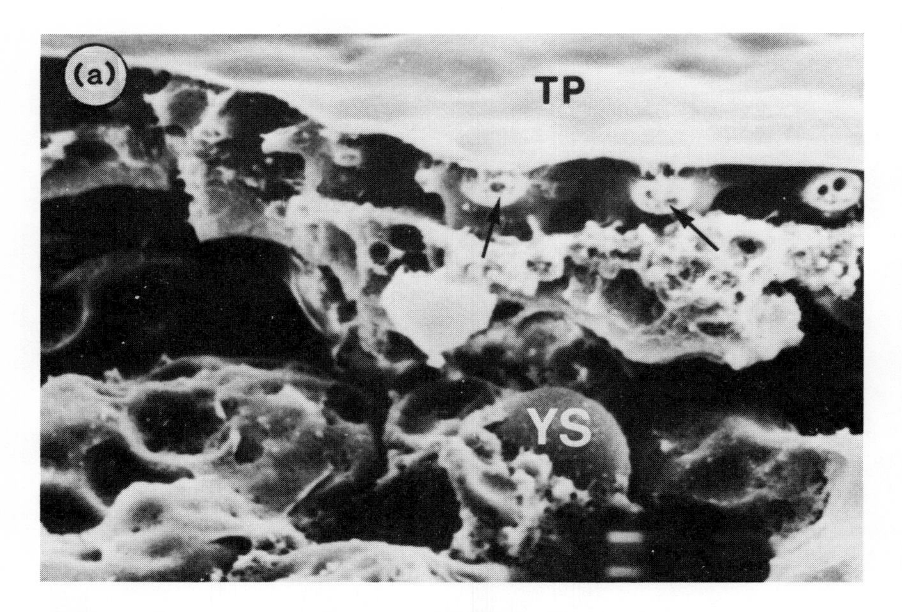

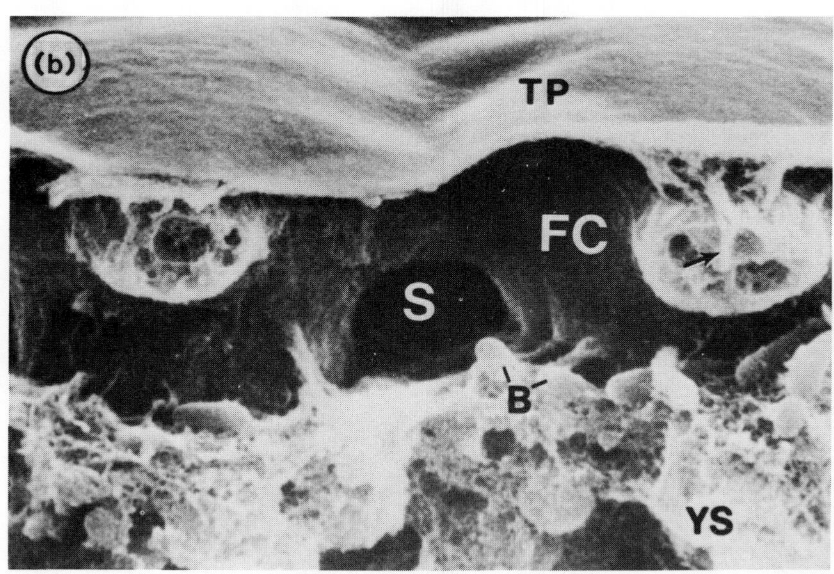

Fig. 13.3 Scanning electron micrograph of *P. americana* vitellogenic follicle fractured after critical point drying. (a) Below the tunica propria (TP) is a layer of follicle cells from which pseudopodia extend (arrows). Beneath the follicle cells are the oölemma and bacteroids (see (b)). Large yolk spheres (YS) are visible within the oöcyte as well as indentations where lipid yolk spheres were lost during preparation of the tissue. × 1500. (b) Two follicle cells are shown, each with pseudopodial extensions (arrows). Tunnel-like zone between the follicle cells is an interfollicular space (S). Bacteroids (B) are visible within a meshwork of material that is presumably the zone of microvilli of the oölemma. × 6000. (Wasilenko *et al.*, unpublished.)

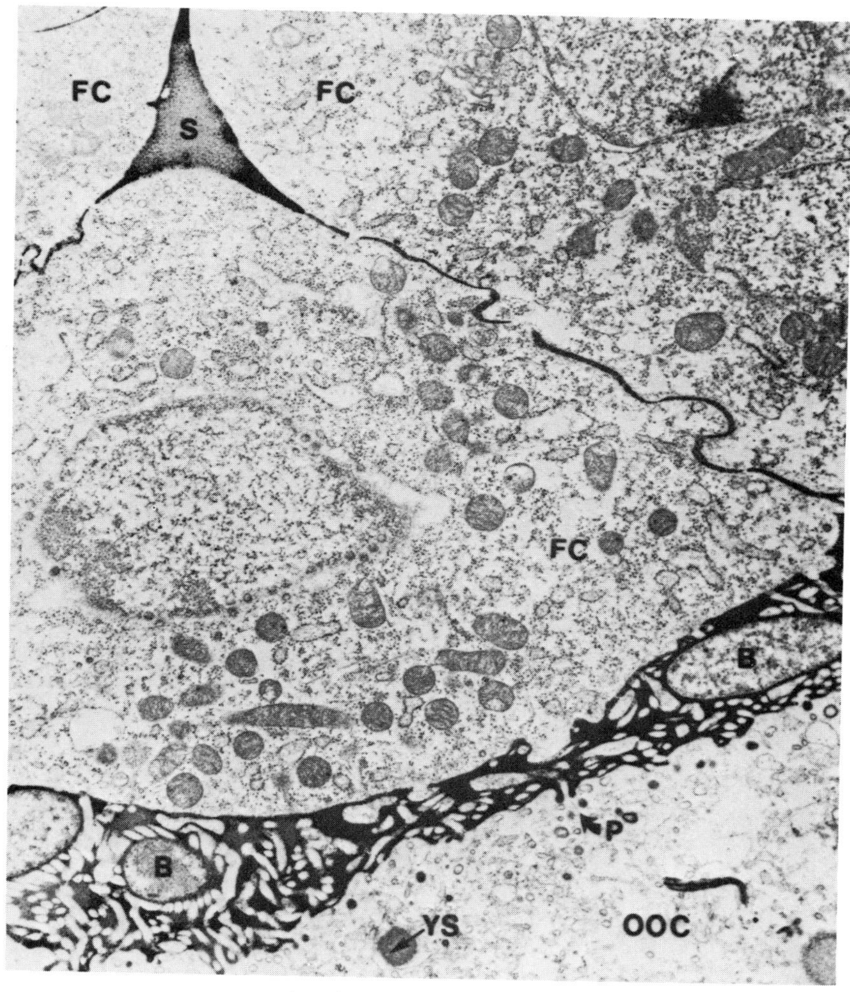

Fig. 13.4 Transmission electron micrograph of vitellogenic follicle of *P. americana*. Lanthanum treatment demonstrates localization of extracellular glycoproteins, possibly follicle cell products or vitellogenin within the spaces between follicle cells, S, and between the oöcyte, OOC, and follicle cells, FC. Electron-dense lanthanum deposits are found in pinosomes (P) in the process of being internalized by the oöcyte. B, bacteroids; YS, yolk sphere. × 20 000. (Sams, 1977.)

(c) *Chorion formation*

Follicle cells become flattened as though stretched by the vast increase in volume of the oöcyte that they contain. They then secrete a chorion or 'egg shell'. During ovulation the chorionated oöcyte slips out of the follicular epithelium at the calyx and moves down the oviduct. The follicle cells gradually shrink to small yellowish bodies sometimes referred to as the 'corpus luteum'.

(d) *Vitellogenic cycle*

Among species of cockroaches there are two basic types of vitellogenic cycles that correlate with the mechanism of ovulation and parturition (Roth and Willis, 1958). Oviparous cockroaches (e.g. *P. americana*, *Blatta orientalis*) have continuous vitellogenic cycles in which the penultimate and basal follicles sequester yolk protein concurrently; when the basal oöcytes are ovulated, the penultimate set

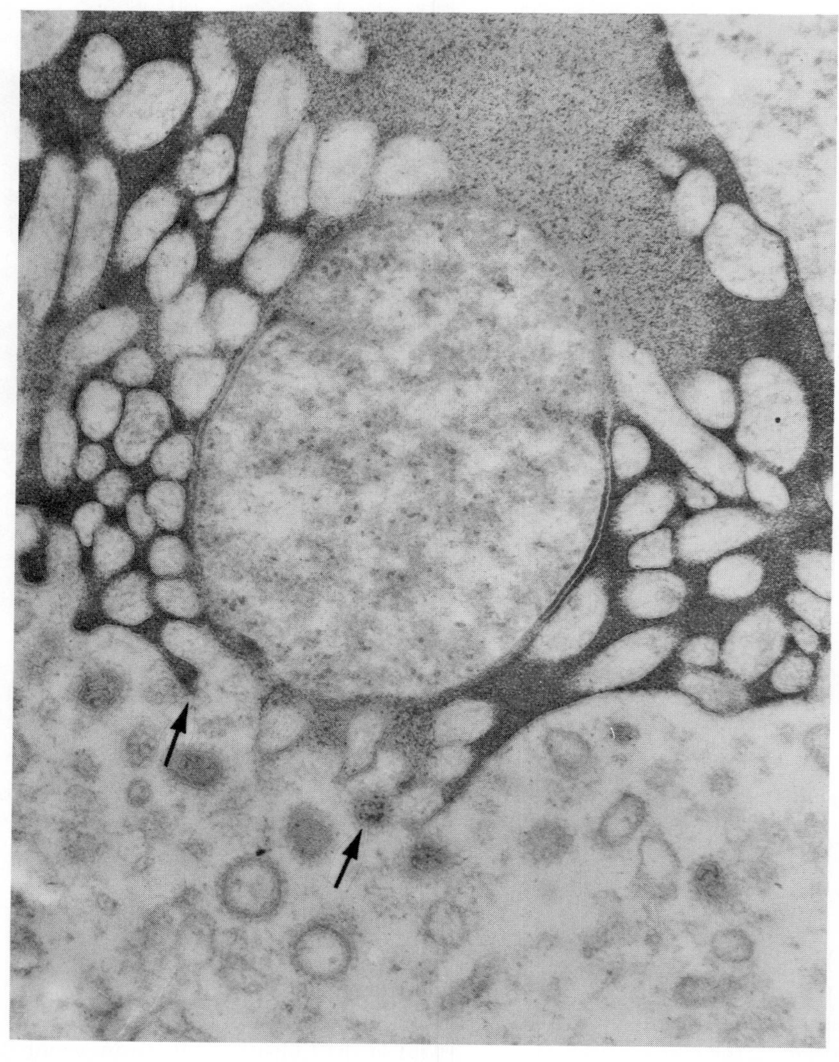

Fig. 13.5 Enlargement of the zone of microvilli in Fig. 13.4. Formation and internalization of lanthanum-bound pinosomes is visible (arrows) as well as coated vesicles within the oöcyte. × 55 000. (Sams, 1977.)

assumes the basal position and the third set becomes penultimate (Fig. 13.6a). Ovoviviparous (*L. maderae*, *B. fumigata*) and viviparous species (*D. punctata*) have periodic vitellogenic cycles in which the basal oöcytes are ovulated and then incubated in the uterus; during the period of pregnancy there is little if any yolk deposition in the follicles.

Control of vitellogenic cycles in ovoviviparous cockroach species is through feedback to the brain from the bursa (during or shortly after mating) and from the uterus or mature follicles (during pregnancy); thus juvenile hormone (JH) is secreted from the corpora allata (CA), and vitellogenesis is stimulated when females are mated and are not pregnant (Stay and Tobe, 1978; Tobe and Stay, 1979; see also earlier reviews by Cornwell, 1968; Engelmann, 1970; and Roth, 1970). Oviparous species, such as *P. americana*, do not incubate their eggs (the uterus is not present) and therefore a mechanism involving feedback during pregnancy is not necessary. Feedback from mature follicles to the brain in *P. americana* is possible, but no evidence for such a mechanism is available.

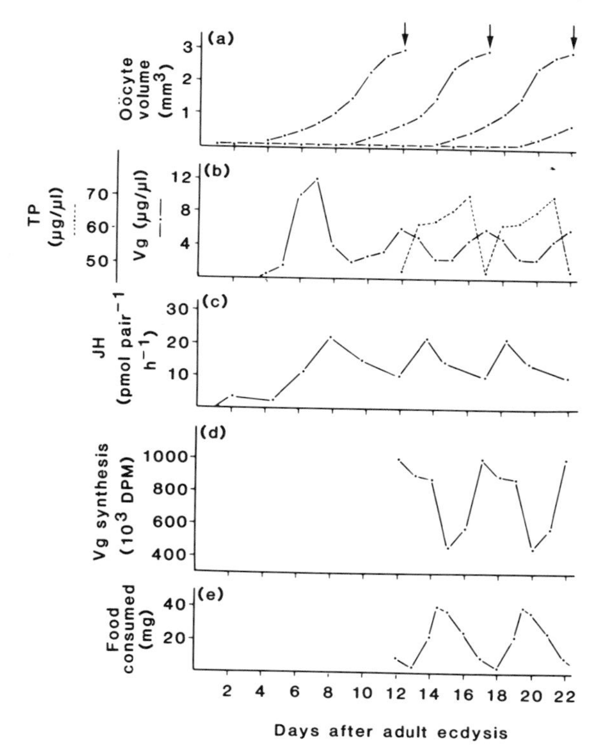

Fig. 13.6 Correlation of factors that may be involved in vitellogenesis of *P. americana*. (a) Oöcyte volume (Bell, 1969a,b). (b) Haemolymph vitellogenin (Vg) and total haemolymph protein (TP) concentration (Bell, 1969a,b; Mills *et al.*, 1966). (c) JH synthesis by pairs of CA *in vitro* (Weaver and Pratt, 1977). (d) Vitellogenin synthesis by fat body *in vitro* (Pan *et al.*, 1969; Bell, unpublished). (e) Food consumed by individual females (Mills *et al.*, 1966; Bell, 1969b). Arrows at top show period of oötheca formation (ovulation).

That mating is important to support continuous vitellogenesis was shown by Weaver and Pratt (1977); JH synthesis by the CA and yolk deposition in the ovaries decreased markedly in females under conditions of enforced virginity. Feeding also seems essential for continuous vitellogenesis, as the process is abruptly curtailed by starvation (Bell, 1971; Brousse-Gaury, 1977) and restimulated by feeding.

Fig. 13.6 summarizes the events that occur during the vitellogenic cycle of *P. americana* from adult ecdysis through three oviposition periods. The data of Weaver and Pratt (1977) on JH synthesis (Fig. 13.6c) are based on a 3-day vitellogenic cycle, but it was possible, using ovulation as a marker, to convert their values into the 5- or 6-day cycle of Bell (1969a) and Mills *et al.* (1966). In addition to the parameters illustrated in the figure, Adiyodi (1967) and Brousse-Gaury (1977) found that stainable material is released from the pars intercerebralis–corpus cardiacum complex just prior to ovulation and immediately following oviposition. These periods of secretion presumably occur, for example, on days 11 and 13 in the time frame of Fig. 13.6. Rhythmic changes in the structure and cell–nuclear volume ratio (Girardie, 1962; Adiyodi and Nayar, 1965, 1966a; R. G. Adiyodi, 1969; Brousse-Gaury, 1978c) confirm the more direct measurements of CA activity (Weaver and Pratt, 1977) (Chapter 12).

Based on the data in Fig. 13.6, it is tempting to suggest that the first JH pulse on about day 2 after adult ecdysis stimulates the initiation of processes involved in vitellogenesis (see Section 13.2.2.f). After the first pre-oviposition period, however, it is difficult to correlate JH secretion or neurosecretion exactly with the cycle of yolk deposition or vitellogenin levels. One hypothesis that could explain the phenomena in Fig. 13.6 is that the cycle of vitellogenin levels in the haemolymph results from differential rates of vitellogenin uptake by the ovaries and that the CA secrete JH in a cyclic manner because of stimuli supplied during feeding that are mediated through the brain (Brousse-Gaury, 1978b) or internal controls related to circadian rhythm.

Recently, 20-hydroxy ecdysone has been implicated in the control of JH synthesis in *D. punctata* (Stay *et al.*, 1980b). If mature follicles of *P. americana* secrete 20-hydroxy ecdysone which reduces the rate of JH synthesis, as suggested for *D. punctata*, then ecdysone secretion might be responsible for the cycle of JH synthesis observed by Weaver and Pratt (1977), and possibly play a role in the control of the vitellogenic cycle in *P. americana*.

One component of the control system in *P. americana* appears to be endogenous, in that tandem follicles of an ovariole are developmentally programmed to initiate vitellogenesis at 5-day intervals. Transplantation studies have shown that the normal 5-day cycle of *P. americana* is preserved when the ovary is transplanted into a closely related species having a longer or shorter vitellogenic cycle (Bell, 1972b). The possibility that follicles within an ovariole communicate with each other was explored by Maa and Bell (1977). They found, for example, that when the 4 posterior follicles were removed and the remainder of the ovariole implanted into a host at the same stage of vitellogenesis as the donor, the new basal follicle did not initiate vitellogenesis until about 20 days later; that the fifth follicle 'waits' until the time at which vitellogenesis would normally be initiated suggests that vitellogenesis is not controlled by inhibitory factors flowing from more advanced to less advanced follicles. These experiments indicate that there is a signal that marks each 5-day cycle in some way, possibly peak JH titre or release of neurosecretory substances, and that these cues determine the time of developmental competence for each follicle.

(e) *Vitellogenin: the major yolk precursor*

The origin of yolk in *P. americana* and related species has been traced to the follicle cells (Nath *et al.*, 1958; Aggarwal, 1960), oöcyte mitochondria (Ranade, 1933), oöcyte nucleus (Hogben, 1920; Nath and Mohan, 1929; Gresson, 1931) and the midgut (Mills *et al.*, 1966). Direct immunological evidence, however, revealed that the major yolk protein, vitellogenin, comprising more than 85% of the protein in *P. americana* yolk, is synthesized and secreted (Pan *et al.*, 1969; Bell, 1970), but not stored (Subramoniam, 1973), by the fat body. *P. americana* vitellogenin, which was observed electrophoretically by Menon (1963), Thomas and Nation (1966) and Nielsen and Mills (1968), has now been characterized as a single fraction with a molecular weight of 600 000 and consisting of three polypeptide subunits in a configuration of A2B2C2 (Sams *et al.*, 1981) (see also Section 7.6.4). It is certain that the two vitellogenin fractions isolated by various researchers, including Bell (1970), consisted of the vitellogenin and a breakdown product. Clore *et al.* (1978) identified only two major polypeptides in an analysis of *P. americana* vitellogenin, but it is probable that their peptide A is composed of A and B peptides observed by Sams *et al.* (1981).

Vitellogenin is female-specific in all cockroaches studied thus far; in *P. americana* the fat body of males cannot be induced to synthesize vitellogenin by ovary transplants (Bell, 1972a) nor by adult female blood or JH *in vitro* or *in vivo* (Sudarsanam, 1980). Mundall *et al.* (1979) showed that vitellogenin synthesis can be elicited in male *D. punctata* by treating with relatively high dosages of a JH analogue or implanting female CA. It is possible therefore that higher JH dosages would stimulate vitellogenin synthesis in male *P. americana* or perhaps that there is a basic difference in the male potential for this process among various species of cockroaches.

Vitellogenin, first secreted on about the 4th day after adult ecdysis (Bell, 1969b), occurs in the haemolymph at a concentration of 1.5–3.0 $\mu g\ \mu l^{-1}$, which is approximately 3% of the total haemolymph protein (Bell, 1969b; 1970). The concentration of vitellogenin in the yolk is at least 180 $\mu g\ \mu l^{-1}$, indicating an active, selective uptake process of vitellogenin from the haemolymph into the oöcyte (Bell, 1969a; 1970). The magnitude of protein uptake is revealed if one considers the entire ovary in which 1.5–3.0 mg of protein are sequestered each day.

(f) *Control of yolk deposition*

Juvenile hormone secreted by the corpora allata, CA, directly or indirectly stimulates the synthesis of yolk proteins by the fat body (Menon, 1963;

Thomas and Nation, 1966; Adiyodi and Nayar, 1966b, 1967; Bell, 1960b, 1970), and the uptake of these proteins by the ovary of *P. americana* (Chen *et al.*, 1962; Girardie, 1962; Mills *et al.*, 1966; Bell, 1969a).

Two stages are evident in the control of yolk deposition. First, ovaries gain competence during the last moult to respond to the gonadotropic signal. Second, yolk deposition is initiated in the adult female.

Scharrer (1946) showed in cockroaches, that to achieve adulthood, JH must be absent and moulting hormones present during the final moult. Bell and Sams (1975) found that a similar regime is necessary to promote ovarian competence. By manipulating JH and ecdysone in last instar nymphs they found that the ovary must be exposed to ecdysone without JH in order to become vitellogenic when exposed to JH at a later time. Since DNA synthesis was observed in follicle cells at the time of ecdysone exposure, the authors speculated that DNA replication may be instrumental in permitting the cells to reprogram for adult functions.

JH seems to affect the ovary of *P. americana* directly, i.e. the hormone is a gonadotropin. Sams and Bell (1977) were able to stimulate low levels of vitellogenin uptake *in vitro* using a synthetic medium containing JH III and *Manduca sexta* binding protein. More substantial levels of yolk deposition were achieved using transplantation experiments with controls for possible hormonal intermediates (Sudarsanam, 1980). When two ovarioles, one incubated with JH and the other without JH, were implanted into a ligated newly emerged female, the JH-treated ovariole exhibited normal rates of uptake of vitellogenin (injected into ovariole recipients), whereas neither the control implant nor the host's own ovaries became vitellogenic. If JH had stimulated a secondary intermediate in this experiment, then both the control implant and host's ovary would also have been expected to respond to the intermediate.

The action of JH on the insect ovary was clarified recently by Davey and Huebner (1974) when they discovered that treating *Rhodnius prolixus* ovaries *in vitro* with JH stimulated an almost immediate formation of interfollicular spaces (increased patency). When vitellogenic ovaries were deprived of JH *in vitro* the spaces closed once again as in the previtellogenic condition. An interesting aspect of their results is that only those follicles previously exposed to JH could be stimulated by JH treatment to enlarge their inter-follicular spaces (Abu-Hakima and Davey, 1975). The theory developed by Davey and co-workers (Abu-Hakima and Davey, 1977a,b,c) is that the hormone stimulates vitellogenesis by reducing follicle cell volume and/or shape, thereby providing access for haemolymph vitellogenins moving through interfollicular spaces to the oöcyte. Wasilenko *et al.* (unpublished) repeated these experiments with *P. americana* and failed to find such immediate induction of patency. Among several different experiments, previtellogenic ovaries from day 4.5 (post-adult ecdysis) females, which presumably had been exposed to the day 2 JH peak observed by Weaver and Pratt (1977), failed to exhibit patency when incubated up to 3 days in medium containing JH. Transplantation of 4.5 day ovaries to day 8 females resulted in patency and yolk deposition after 48 h, but this is the period normally required *in situ*. Thus JH may indeed control patency in

P. americana, but if so the effect in the cockroach is not so dramatic nor immediate as in *R. prolixus*. Placing vitellogenic ovarioles into medium or haemolymph devoid of JH failed to close the interfollicular spaces until 36 h in culture. Transplantation of ovaries from vitellogenic females into day 1 post-ecdysis females likewise resulted in a decrease in patency but only after 2.5 days. Thus the control mechanism that operates in *R. prolixus* cannot be entirely extrapolated to cockroaches. The evidence may suggest, however, that in blood feeders the process of yolk deposition must be triggered quickly after the feeding stimulus to ensure utilization of available nutrients, whereas in insects that have vitellogenic cycles at least partially controlled internally, the effect of JH can be less immediate.

The possibility that, of the haemolymph constituents, vitellogenin alone can stimulate yolk deposition in *P. americana* females has been rejected on the following grounds. First, injections of vitellogenin into allatectomized (Bell, 1969b) or starved (Bell, 1971) females failed to trigger yolk deposition. Secondly, it is possible to stimulate yolk deposition in ovaries transplanted into males; the proteins sequestered are those synthesized by the male fat body and do not include vitellogenin (Bell, 1972a). The presence of vitellogenin in male haemolymph (added by injection) significantly enhances yolk deposition owing to the process of selective protein uptake (Bell and Barth, 1971; Bell, 1972a).

(g) *Control of vitellogenin synthesis*

The fat body becomes competent to synthesize and secrete vitellogenin during adult ecdysis and then must be exposed to female haemolymph for four days before it is capable of responding to JH, the stimulus for vitellogenin synthesis and secretion (Sudarsanam, 1980). These conclusions are based on transplantation of fat body from day-0 to day-10 post-ecdysis adult females into adult males that received corpora allata of vitellogenic females. Only fat body from day-4 or older females secreted vitellogenin into the male haemolymph. The factor or factors present in female haemolymph after the fourth post-ecdysial day may be specific titres of JH. Two lines of evidence support this hypothesis: Davey and co-workers found that early exposure to JH (and possibly other factors) is essential for later patency responses to JH in *R. prolixus*, and Weaver and Pratt (1977) observed a pulse of low titre JH at about day 2 after adult ecdysis in *P. americana*.

That the factor or factors required to stimulate vitellogenin synthesis by the cockroach fat body are still incompletely understood is indicated by an absence of reports documenting JH-stimulation of vitellogenin synthesis *in vitro* (Engelmann, 1979). To date there is evidence for vitellogenin synthesis by insect fat body only when the female was treated with JH, prior to removal and culture *in vitro* (e.g. Engelmann, 1969b; Pan and Wyatt, 1971; Koeppe and Ofengand, 1976).

The ovary of the cockroach does not feedback to the fat body to regulate vitellogenin synthesis in *P. americana*. Females ovariectomized as nymphs accumulate up to 220 $\mu g\,\mu l^{-1}$ vitellogenin in the haemolymph, showing that absence of the ovary does not curtail vitellogenin production (Bell, 1969b). The same is true for many other insect species (review: Engelmann, 1979). Recently, Stay and

Tobe (1978) reported that ovariectomy causes a decrease in JH synthesis in *D. punctata*; thus, in some species, the low JH level that probably exists in ovariectomized females is sufficient to support vitellogenin synthesis. Females of several cockroach species were able to support yolk deposition in one or two additional implanted ovaries, suggesting, at least indirectly, that vitellogenin synthesis was enhanced by the implanted ovaries (Bell, 1972b).

13.2.3 Oöcyte resorption

Oöcyte resorption occurs in many species of insects under conditions of nutritional deficiency, virginity, social pressure, lack of ovipositional sites, parasitism and, to a certain extent, in seemingly normal females (Bell and Bohm, 1975). Resorption in *P. americana* entails the breakdown of oöcytes and follicle cells of vitellogenic follicles (Bonhag, 1959); vitellogenins of resorbing oöcytes are salvaged intact and accumulate in the haemolymph (Bell, 1971). Synthesis and secretion of vitellogenin by the fat body is curtailed abruptly in starved females and does not resume unless feeding occurs (Bell, 1971). When starved females are allowed to feed, yolk deposition begins within a few days (Bell, 1971; Brousse-Gaury, 1977); similar results were obtained by injecting starved females with JH (Bell, 1971). Similarly, enforced virginity of *P. americana* females leads to a reduction in levels of JH synthesis and to oöcyte resorption; mating was followed by renewed JH synthesis and yolk deposition (Weaver and Pratt, 1977). Because vitellogenin is available to the follicles of starving females, owing to salvage of yolk vitellogenin, and given the evidence of Weaver and Pratt (1977) on JH synthesis in virgin females, it appears that JH is the primary limiting factor blocking vitellogenesis during starvation and enforced virginity of females.

The relative developmental sequence of each follicle established during successive 5-day cycles is retained during the starvation-induced vitellogenic hiatus. This suggests that the developmental sequence is not merely established in young females and then maintained through ensuing oviposition cycles.

13.2.4 Transmission of bacteroids

The importance of bacteroids to the nutrition of *P. americana* is discussed in Chapters 4 and 7. Transmission of these organisms from generation to generation is accomplished by their entry into developing oöcytes (Gier, 1936; Anderson, 1964). Bacteroids reside in mycetocytes in the connective tissue surrounding the germarium. Their route and mechanism of entry is uncertain, but they come to lie between the oöcyte and follicle cells as the follicles differentiate. There they increase in number through fission and remain extracellular, surrounded by the oölemma, throughout vitellogenesis (see Fig. 13.4). Gier (1936) suggested that the bacteroids enter the cytoplasm of the oöcyte when the oölemma breaks down just prior to the time of ovulation.

13.2.5 Ovulation and oviposition

Ovulation and oviposition in cockroaches are temporally separated owing to the grouping of eggs following ovulation into an egg mass of two rows of eggs that are enveloped by secretions of colleterial glands to form the oötheca; deposition of the oötheca constitutes oviposition. During oötheca formation, the secretions of the colleterial glands flow over the inner surface of the vestibulum. This becomes filled with eggs, one passing into the developing oötheca from each ovariole. As the process continues, the eggs are surrounded by the secretion and move posteriorly. The shape of the oötheca is dependent on shaping by the walls of the genital chamber. The oötheca is constructed such that a keel is produced dorsally. It is along this line of weakness that emerging nymphs break out of the oötheca at the time of hatching (Lawson, 1951).

Control of ovulation and oviposition is poorly understood in cockroaches, although in *P. americana* ovulation is known to be accompanied by a steep increase in the size of the CA (Adiyodi, 1969) and a release of PI-CC neurosecretory materials (Adiyodi and Nayar, 1965). In other insect species it has been shown that haemolymph-borne factors are responsible for causing oviduct contraction during the process of ovulation (Davey, 1964; Highnam, 1964; Thomas and Mesnier, 1973).

P. americana has been the subject of extensive research concerning the control of oötheca formation and the biochemical events that lead to the construction of this structure (detailed in Chapter 2). The left colleterial gland produces a structural protein, β-glucoside of protocatechuic acid, and a polyphenol oxidase (laccase) (Brunet and Kent, 1955). The right gland secretes β-glucosidase. When the contents of the glands mix at the time of ovulation the glucosidase hydrolyses the glucoside and liberates the phenol, protocatechuic acid. This is oxidized to give a *O*-quinone which is believed to cross-link the protein to produce sclerotization or the hardening of the protein (Brunet and Kent, 1955; Kent and Brunet, 1959; Whitehead *et al.*, 1960).

The characteristics of the colleterial gland cells and the products of these cells have been studied in *Periplaneta* and *Blatta* (Bordas, 1909; Pryor, 1940; Pryor *et al.*, 1946; Brunet, 1951, 1952; Mercer and Brunet, 1959). Especially interesting are the calcium oxalate crystals secreted by the left colleterial gland into the oötheca; it is possible that the crystals are involved in water balance of the oötheca. Production of structural protein, glucoside and calcium oxalate is controlled by JH (Bodenstein and Sprague, 1959; Stay *et al.*, 1960; Willis and Brunet, 1966; Bodenstein and Shaaya, 1968; Shaaya and Bodenstein, 1969). The system has been used effectively as a bioassay for JH (e.g. Bodenstein and Shaaya, 1968; Barth and Bell, 1970; Shaaya, 1978).

13.3 Male reproductive system

Aspects of reproductive physiology of the male have been much less investigated than those of the female, probably because testicular activity and spermatophore

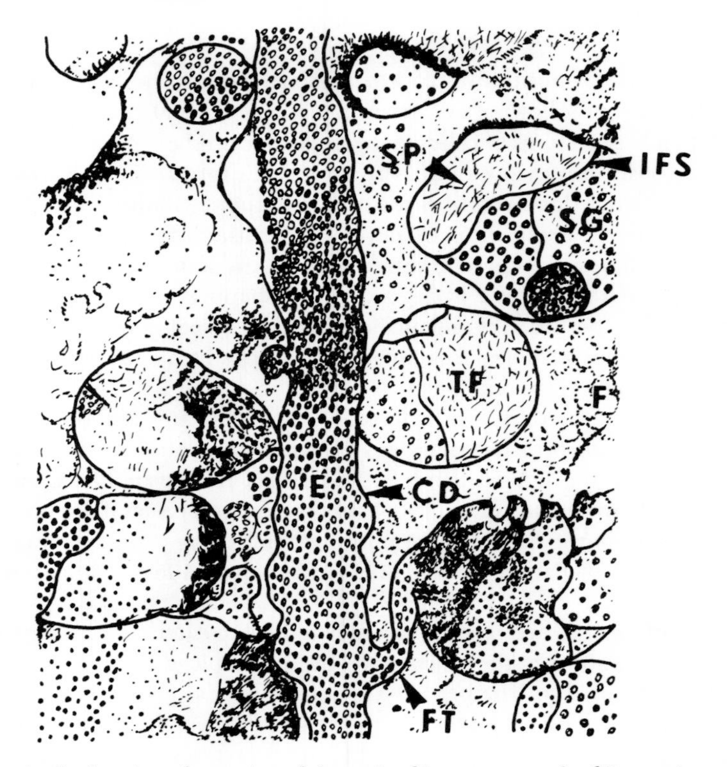

Fig. 13.7 Longitudinal section of a portion of the testis of last-stage nymph of *P. americana*. CD, intratesticular vas deferens; E, spherical cells; F, fatty tissue; FT, vas efferens; IFS, intrafollicular septum; SG, spermatogonia; SP, spermatozoa; TF, testicular follicle. (From Jaiswal and Naidu, 1972.)

formation, two key processes in male reproductive physiology, are not as spectacular as is oötheca production. The testis of *P. americana* becomes mature in the last nymphal instar, if not earlier, and produces what look like healthy spermatozoa. The last male nymphs are, however, not able to mate successfully: the accessory sex gland apparatus which supplies the material needed for spermatophore production does not mature until a few days after the final moult. Spermatozoa have been observed by Jaiswal and Naidu (1972) in the 'seminal lobe' at the base of the vesiculae seminales, 5–6 h after the nymphal–adult moult, and by Vijayakelshmi (1976) in the vesiculae seminales themselves 2 days after emergence as adult.

13.3.1 Morphology of the testis and sperm ducts

(a) *Testis*

The male reproductive system of *P. americana* consists of a pair of testis in segments 4 and 5 of the abdomen, a pair of vasa deferentia, vesiculae seminales and

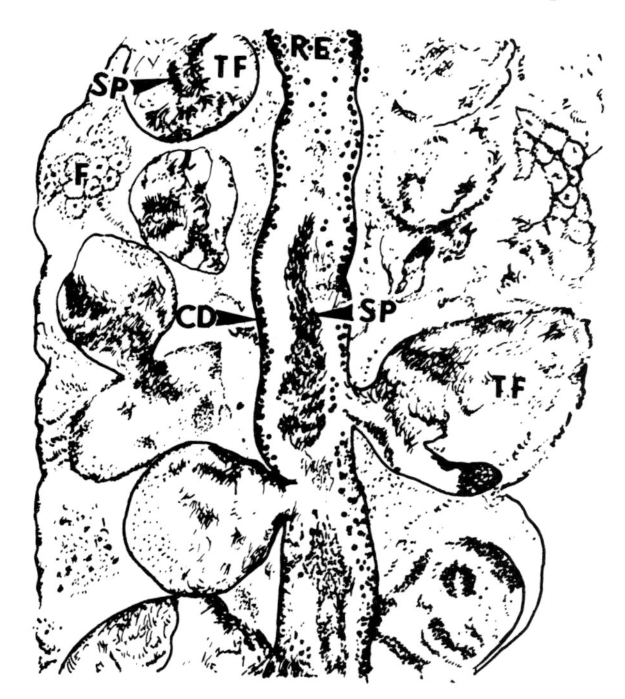

Fig. 13.8 Longitudinal section of a portion of the testis of 24-h-old adult male *P. americana*. Abbreviations as for Fig. 13.7. In addition, RE, remnants of cells. (From Jaiswal and Naidu, 1972.)

ductus ejaculatorius. Accessory reproductive organs associated with the male tract include a mushroom gland covering the anterior region of the ductus ejaculatorius, and a club-shaped conglobate (phallic) gland that opens out separately. The testis, well-developed in last-instar male nymphs (in which it may be 8 mm long and 4 mm broad) is regressed in adults and buried in fatty tissue, and can be detected only with difficulty. Information on ultrastructure of the testis of *P. americana* is unfortunately not available, but light microscopic data on testis of last nymphs have been provided by Jaiswal and Naidu (1972). The testis consists of 25–32 round, membraneous follicles arranged in three or four beadlike linear groups of 8–10 follicles each. A thin peritoneal membrane is said to be present enveloping the testis. The space between adjoining testicular follicles as well as that between the follicles and peritoneal membrane is filled with a 'fatty tissue' with spherical nuclei and vacuolated cytoplasm. The testicular follicles open either directly or through vasa efferentia into the vas deferens running down the centre of the testis (Fig. 13.7). Certain spherical cells with prominent nuclei and non-grannular cytoplasm fill the vasa efferentia and the lumen of the intratesticular regions of the vas deferens.

Testicular follicles are divided by 3–5 membraneous septa placed at random within the follicle; the more developed spermatozoa occur in follicular compartments closer to the intra-testicular part of the vas deferens. In last-instar nymphs,

Jaiswal and Naidu (1972) could not find any spermatozoa among the spherical cells in the main duct within the testis, but in 24-h-old adult males, nearly all the spherical cells had disappeared, and the vasa efferentia and the intratesticular vasa deferentia were filled with migrating spermatozoa (Fig. 13.8). There is thus reason to suspect that the spherical cells may function as a plug blocking precocious spermiation. The suggestion of Jaiswal and Naidu (1972) that the spherical cells on disruption might form a medium for sperm transport and also function as nutrient for spermatozoa migrating to the vesiculae seminales is of some interest, but requires more rigorous proof: *P. americana* spermatozoa are themselves motile; furthermore, the migration of sperm masses and individual spermatozoa could be assisted by the action of circular muscles in the wall of the vas deferens.

The testicular follicles shrink in size with the adult moult, some disappear, and interfollicular septa become indistinguishable, but this does not preclude the possibility that small peaks of spermatophore production might nevertheless occur in the reduced testis of adult *P. americana*. Raichoudhuri and Mitra (1941) observed active spermatozoa in the testis of adults; Hughes and Davey (1969) also observed developing spermatids and mature spermatozoa in the testis of adults, though no indication of the age has been given. In last instar nymphs, the intra-follicular septa are broken during spermiation, but the actual mechanism of spermiation is not understood.

(b) *Vas deferens*

The walls of the vasa efferentia and the intratesticular portions of the vasa deferentia are formed of a single layer of small cuboidal cells with prominent nuclei and nucleoli (Jaiswal and Naidu, 1972). The epithelium of the main duct rests on a basement membrane which in turn is covered by circular muscles and a peritoneal sheath.

(c) *Vesiculae seminales*

The vesiculae seminales, situated antero-ventrally within the mushroom gland at the proximal end of the ductus ejaculatorius are comprised of 12–14 'spongy' club-shaped tubules of uneven length (Vijayalekshmi and Adiyodi, 1973b) (Table 13.1). The spermatozoa which reach the paired seminal lobes from the vasa deferentia pass to the vesiculae seminales where they are stored for variable periods of time. The secretory epithelium of vesiculae seminales is covered by a thin layer of what looks like smooth muscles.

13.3.2 Spermatogenesis

Smear preparations of vesiculae seminales of adult males are by far the best material for studies on sperm morphology. The normal mature spermatozoon of *P. americana* resembles rather closely the vertebrate type. Nath *et al.* (1957) found using phase contrast microscopy that in spermatozoa taken from the male, the acrosome at the tip of the head is complex consisting of a large swollen leaf which

Table 13.1 Morphology and histology of the mushroom gland complex of adult male *P. americana*. (From Vijayalekshmi and Adiyodi, 1973b.)

Type of tubule	*Position*	*Shape*	*Nature*	*Approx. no.*	*Length and diameter μm*	*Breadth of epithelium**, *μm*	*Solubility in water*	*Histological nature of secretion*
Utriculi breviores	Anterior and dorsal	Elongated with a stalk like portion which is narrow	Opaque, but not milky	90	1350, 144	11.9	Very little fragment	Brownish-red with brownish-black granules
Vesiculae seminales	Anterior and ventral	Club-shaped with small stalk	Spongy	12–14	1062, 270	8.0	Sediment almost nil	Seminal plasma with large bluish-black droplets
Utriculi intermedia	Anterior and lateral	Elongated with a narrow stalk	Opaque, but not milky	200	1782, 202	9.5	Moderate amount of coagulum	Discrete disc-like or granular secretion with or without black granules enmeshed in a ground substance to give a crimson-red appearance
Utriculi majores	Ventro-lateral	Elongated with uniform width throughout	Milky white	25	2718, 324	5.1	Large clumps	Light violet and dark violet secretions
Utriculi translucentes	Posterior and dorsal	Flat with uniform width	Translucent	12–16	2070, 223	13.6	Coagulum almost nil	Magenta-coloured with some dark granules

* Measured in paraffin sections.

in turn encloses a cup-like structure covering a dense axial rod. These observations have been electron microscopically confirmed by Hughes and Davey (1969). The vesiculae seminales of adults contain an insignificantly small number of giant spermatozoa – 0.6% of the total on day 3 (Vijayalekshmi and Adiyodi, 1973b); they are either polyploid or the products of multinucleate spermatids. Giants form up to 30% of the total spermatozoa and spermatids in the testis of last-instar nymphs suggesting that most of them disintegrate before reaching the vesiculae seminales (Richards, 1963b).

There is very little to add to the classical description of spermatogenesis given by Nath *et al.* (1957) with special reference to the fate of mitochondria and Golgi bodies in *P. americana*. Secondary spermatocytes, like the first, have thick, filamentous mitochondria; Golgi bodies are in the form of granules, rodlets or spheroids. During spermatidogenesis, spheroid Golgi bodies fuse to form an acroblast, which in turn gives rise to a proacrosome; the complicated acrosome described above originates from the proacrosome. Within the spermatid, both the nebenkern which is formed by condensation of the mitochondria, and the axial filament which passes through the nebenkern elongate more quickly to begin with than the spermatid. As a result, the nebenkern and axial filament are for a time found coiled within the spermatid. This is soon corrected, and spermiogenesis is complete with the sloughing off of the blebs of cytoplasm attached to the tail.

The behaviour of chromosomes in the prophase of reduction division of the primary spermatocyte has been controversial in *P. americana* (Guthrie and Tindall, 1968 for references), for it is difficult to find diplotene and diakinesis stages. Most authors failed to detect chiasmata formation, but Rajasekarasetty and Ramanamurthy (1963) claim to have succeeded in observing the chiasmata, very briefly though, during meiosis.

13.3.3 Sperm storage

In the male, spermatozoa are stored in an inactive state in the vesiculae seminales to conserve energy and prolong the life span of the spermatozoa. Activity, judged by tail beat frequency, diffuses inwards when the vesiculae seminales are ruptured. The seminal plasma, a product of the secretory epithelium of vesiculae seminales, mixes with groups of spermatozoa within the vesicle. It is rich in saliva-non-labile, PAS-positive, substances not appreciably destroyed by acetylation and in phospholipids (Table 13.2). Moderate amounts of tyrosine-deficient protein and neutral and bound lipids are present (Vijayalekshmi and Adiyodi, 1973c); further, small amounts of glucose have been chromatographically detected (Vijayalekshmi and Adiyodi, 1973a). The spermatozoa of *P. americana* seem to rely on glucose, saliva-non-labile polysaccharides, trehalose, phospholipids, fatty acids and free amino acids to meet their basic energy demands in the vesiculae seminales (Vijayalekshmi, 1976).

In *P. americana*, as in other cockroaches, the spermatozoa are packed into spermatophores and transferred to the female during copulation; the spermatophore

Table 13.2 Histochemistry of the mushroom (MG) and conglobate glands (CG) of *P. americana*. (From Vijayalekshmi and Adiyodi, 1973c.)

Stain or histochemical technique employed	*MG*										*CG*
	UB		*VS*		*UI*		*UM*		*UT*		
	Secretion	*Epithelia*	*Seminal plasma*	*Epithelia*	*Secretion*	*Epithelia*	*Secretion*	*Epithelia*	*Secretion*	*Epithelia*	*Epithelia*
Feulgen		+ +		+ +		+ +		+ +		+ +	+ +
Mercury bromophenol blue	+ +	+	±	−	+ +	±	+ +	±	+ + ±	+ +	±
Million's reaction	+ +	+	−	− −	+ ±	±	+ + ±	+	+ + ±	+ ±	±
Oil red O	+ ±	ND	±	ND	+ +	ND	+ + +	ND	±	ND	−
Sudan III & IV	+ ±	ND	±	ND	+ +	ND	+ + +	ND	±	ND	−
Acetone Sudan Black	+ +	+ +	+ ±	±	+ +	+ +	+ ±	+	+ +	+ +	+ ±
Nile Blue	+ + +	+ + ±	+ +	+ +	+ +	+ + ±	+ + ±	+ + ±	+ +	+ +	+ + +
Best's carmine	+ +	+ +	+ + ±	+ + +	+ ±	+ +	?	+ ±	+ +	+ +	+
Best's carmine after digestion in saliva at 37°C for 3 h	+ +	+ +	+ + ±	+ + +	+ ±	+ +	?	+ ±	+ +	+ +	+
Alcian blue (after Steedman)	− −	−	ND	ND	+	−	+	−	+	−	+
Alcian blue 1.0 procedure	− −	−	ND	ND	+	−	+ +	−	+	−	+
Periodic acid–Schiff	+ +	+ +	+ +	+ +	+ +	+	+	+	+ + +	+ +	+ + ±
Alcian blue followed by PAS											
Red	ND	ND	ND	ND	ND	ND	ND	ND	ND	ND	ND
Bluish red	+	+	+	+	+	+	+	+	+ +	+ +	+
Blue	− −	ND	ND	ND	±	ND	+	ND	ND	ND	ND

− −, traces; −, low; ±, medium; +, fairly strong; + +, strong; + + +, very strong; ?, doubtful; ND, not detected.

UB, utriculi breviores; VS, vesiculae seminales; UI, utriculi intermedia; UM, utriculi majores; UT, utriculi translucentes.

gets attached to the spermathecal opening (Gupta, 1947 for details) (Fig. 13.10). According to Hughes and Davey (1969), the spermatheca has two branches: a plain branch, forked and tightly coiled (spermathecal gland?), and an S-shaped capsular branch which terminates in a large swelling lined with prominent cuticular intima within. Following mating, the capsule and its duct, and the plain branch in particular, are full of spermatozoa. The plain branch seems to release the speramatozoa more slowly, or only after the capsule has discharged them. However, little is known of the histology of spermatheca in *P. americana* or the nature of secretions produced by it. From comparison with other cockroaches, it appears likely that spermatophore envelope might undergo partial enzymatic lysis and disintegration by spermathecal secretions; the remnants may be thrown off by the female within 1 or 2 days after copulation to facilitate the passage of eggs.

In *P. americana*, speramatozoa seem to undergo a part of the 'maturation' (capacitation?) process within the spermatheca as borne out by changes in morphology of the spermatozoa: the acrosomal region becomes compact and homogeneous by a reduction in the size of the sac around the acrosome between 5 and 48 h after copulation. Hughes and Davey (1969) also observed an almost two-fold increase in tail beat frequency of the spermatozoa taken from spermatheca and isolated in Ringer solution (approximately 1050 beats min^{-1}) compared to that of the spermatozoa taken from the vesiculae seminales (approximately 600 beats min^{-1}).

13.3.4 Sperm motility

Rate of motility of the spermatozoa of *P. americana*, like many other physiological processes, is a function of temperature as shown by Richards (1963a) (16 $\mu m\ s^{-1}$ at 15–16°C and 54 $\mu m\ s^{-1}$ at 37–39°C in saline). Studies by Davey (1968) (cited in Hughes and Davey, 1969) show that dilution has only slight effect at densities between 10^6 and 10^8 spermatozoa ml^{-1}; optimal pH for activity appears to be 7. Though saline solutions (>400 mosmol) give satisfactory results in studies on sperm motility, a good suspending medium is the Gouldin's solution used by Richards (1963a), but with the pH adjusted to 7 after Hughes and Davey (1969).

There is reason to believe that in female *P. americana*, activation of the spermatozoa occurs in the spermatophore itself. Activation takes some time to manifest itself probably because the spermathecal secretions released under the stimulus of coitus have to find their way into the spermatophore. Extracts of spermathecae, however, fail to activate the spermatozoa *in vitro*. The vesiculae seminales do not seem to produce any inhibitor of sperm activation; extracts of mushroom and conglobate (phallic) glands are also stated to be without effect (Hughes and Davey, 1969). The time course in which morphological changes and activation occur in spermatozoa in the female has not been precisely determined, though the authors tend to think that activation precedes morphological changes.

13.3.5 Accessory sex glands and their functions

(a) *Mushroom gland*

Based on morphological parameters of the tubules (size, position, shape and nature) and features such as solubility properties, tinctorial behaviour and textural patterns of their secretory products, four types of tubules have been distinguished by Vijayalekshmi and Adiyodi (1973b,c) in the mushroom gland of *P. americana*: utriculi breviores (minores), intermedia, majores and translucentes (Table 10.1) besides the vesiculae seminales (Section 13.3.1.c). Tinctorial differences between tubules are not always absolute: utriculi intermedia and majores may grade into each other. The majores have the thinnest epithelium, and the translucentes, the thickest. Ultrastructural observations by Adiyodi and Adiyodi (1974) on utriculi majores show that they have prominent RER and Golgi complex, typical of glands synthesizing protein for export. There is a thin layer of slow-acting muscle over the majores, mostly monomyofibrillar in nature with an ill-developed T-system, sarcoplasmic reticulum and striations. Possibly, other utricles in the mushroom gland also have a similar muscular coat, though it is difficult to distinguish the same with optical microscopy. Conceivably, the products of utriculi majores (which form the chief component of spermatophore) are exuded through slow muscle action. A cuticular intima is not apparent in any of the utricules of the mushroom gland, suggesting that the mushroom gland may be of mesodermal origin. Secretion is apparently of the merocrine type, and storage mostly extracellular, i.e. in the lumen. Stocks of secretory products depleted during mating are soon replenished.

The time course of maturation of the mushroom gland of *P. americana* has been followed independently by Dixon and Blaine (1973) and Vijayalekshmi and Adiyodi (1973b). The primordium of the gland is no more than the size of a pin head in early nymphal stages; some increase in size occurs with each nymphal instar. After about 45 days into the tenth and final instar, the gland becomes triangular and cell clusters begin to be evident; by 53 days, the utriculi majores and breviores can be distinguished by their outlines. Tubules are internally formed by day 55, and two days later, the tubules protrude from the surface. The nymphal–adult moult takes place on the average on day 60.

Though small amounts of secretion may be produced by late last nymphs, the tinctorial integrity of different secretions of the gland becomes clear in adult males only by day 2 in postecdysis adult males. During early adulthood there is some rearrangement of epithelial cells of the utricles, more secretion is produced in each tubule, and the secretions acquire the configuration and texture, characteristic of those of mature adults.

The secretions of the mushroom gland are predominantly proteinaceous; total protein amounted to 4.16 ± 0.11 mg (mean ± S.E.) per gland (Vijayalekshmi, 1976). One of the major carbohydrate reserves of the gland is the non-reducing polyol, inositol. Small amounts of glucose are present in the vesiculae seminales and also some acidic sulphated and non-sulphated mucopolysaccharides. Levels of free amino acids in adult glands were 65 ± 1.1 μg, lipids 467 ± 1.00 μg and trehalose 4.1 ± 0.11 μg per gland (Vijayalekshmi, 1976). Phospholipids observed include phosphatidylcholine followed by plasmalogens and traces of phosphatidylethanolamine and *lyso*-phosphatidylethanolamine (Adiyodi and Adiyodi, 1972; Vijayalekshmi and Adiyodi, 1973a). Table 13.2 suggests that the histochemically different utricles of the mushroom gland may have different roles to play in sperm maintenance and/or in spermatophore formation.

(b) *Conglobate (phallic) gland*

The wall of the conglobate gland is formed of large secretory cells with spherical nuclei and small epidermal cells interspersed between secretory cells at the luminar end. Golgi bodies are not abundant in gland cells; the ergastoplasm contains vesicular elements and rosettes or aggregations of small ribosomes. The lumen of the gland is lined with cuticle composed of epi-, exo- and endocuticles. Beams *et al.* (1962) found a system of ductules piercing the cuticular layer and establishing continuity with ductules within the gland cells. Microvilli abut on the reticulate layer formed by the epicuticle on entering the gland cell; rows of small vesicles which apparently form a part of the transport system are found in the luminar border of the cell. Mitochondria are distributed below and close to the base of the microvilli. The microvilli–ductile system reported by Beams *et al.* (1962) in the epithelial cells of the conglobate gland of *P. americana* is reminiscent of the 'end apparatus' described by Mercer and Brunet (1959) in type II and type III cells of the left colleterial glands of female *P. americana*.

Secretions of the conglobate gland are stored for the most part intracellularly as in the right colleterial gland of females: the lumina are narrow and practically devoid of secretory products. Secretions of the gland are released apparently under copulatory stimuli. Histochemical reactions of the epithelium of conglobate gland are shown in Table 13.2. One of the prominent carbohydrate reserves is inositol as in the mushroom gland; major phospholipids are plasmalogens and phosphatidylcholine (Vijayalekshmi and Adiyodi, 1973a). Total proteins amounted to 0.62 ± 0.01 mg, lipids 367 ± 13 μg, trehalose 4.4 ± 0.2 μg and free amino acids 10 ± 0.01 μg per gland (Vijayalekshmi, 1976). Polyacrylamide gel electrophoresis yielded two peroxidase isozymes; the relatively small glands of day 0 adults had only feeble activity. The activity increased with adult age, day 7 males having nearly as much activity as adults (Adiyodi and Adiyodi, 1970).

The time course of maturation of the conglobate gland of *P. americana* has been worked out by Vijayalekshmi (1976). The gland, distinct in early last stage nymphs, increases in size towards the end of the last instar. In newly emerged adult males, the cells are small and the nuclei ovoid and closely packed. By day 2, the cells have become larger, and their borders distinct; epidermal cells with narrow, elongated nuclei can be distinguished at this stage from the larger secretory cells. By day 3, secretory protein rich in tyrosine residues can be detected in the intertubular spaces in some regions of the gland. Epidermal cells have arranged themselves as a layer bordering the lumen by day 5. The gland cells show increased basophilia and

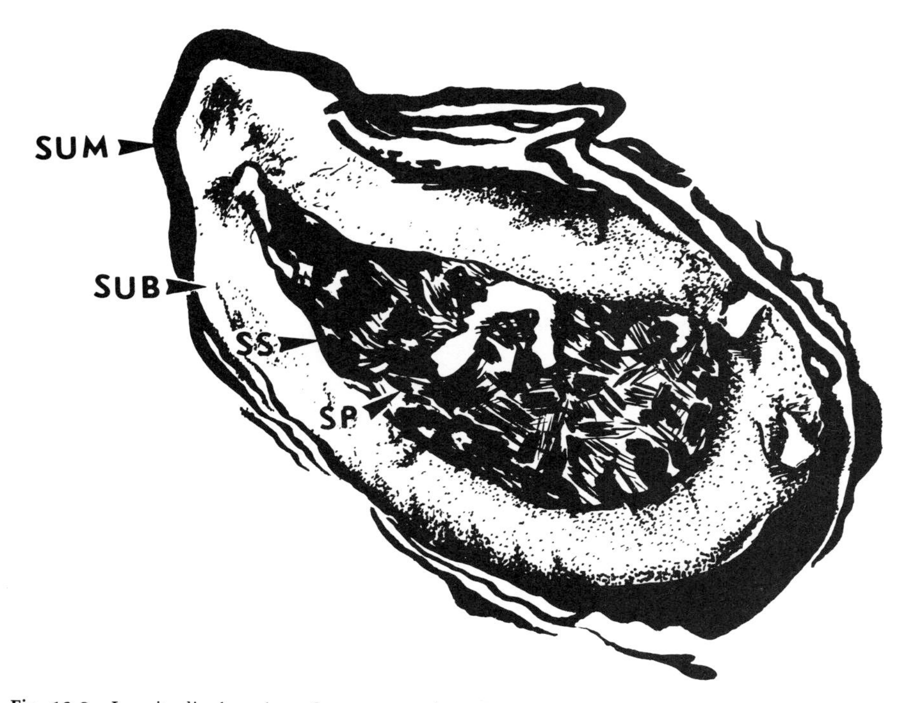

Fig. 13.9 Longitudinal section of a spermatophore deposited by a male from which the conglobate had been removed. SUB, secretion of utriculi breviores; SUM, secretion of utriculi majores; SP, spermatozoa; SS, spermatophore sac. (From Jaiswal and Naidu, 1976.)

stainability with mercury bromophenol blue by days 6 and 7. Between 8 and 14 days after emergence, the conglobate gland acquires the cytomorphological and cytochemical characteristics of the gland of adult males.

(c) *Spermatophore*

Spermatophores of *P. americana* are roughly ovoid whitish bodies slightly larger than the head of a pin. Reports on the composition and mode of formation of the spermatophore are conflicting. According to Gupta (1946), the spermatophore wall is three-layered. Paired gelatinous masses from the peripheral glands (utriculi majores) appear at the anterior end of the ductus ejaculatorius, and fuse to form the inner layer of the spermatophore. At about this time, secretions from the utriculi breviores together with the spermatozoa enter into its cavity. The middle layer composed of secretions from ductus ejaculatorius is added next. The third and outermost layer is formed by the secretions of the conglobate gland during copulation, within the genital atrium of the female. Gupta's (1946) opinion that the formation of the spermatophore may begin before copulation is supported by the findings of Vijayalekshmi and Adiyodi (1973b) and Jaiswal and Naidu (1976). There is no evidence to show that the epithelium of the ductus ejaculatorius of *P. americana* produces the middle layer of the spermatophore as suggested by

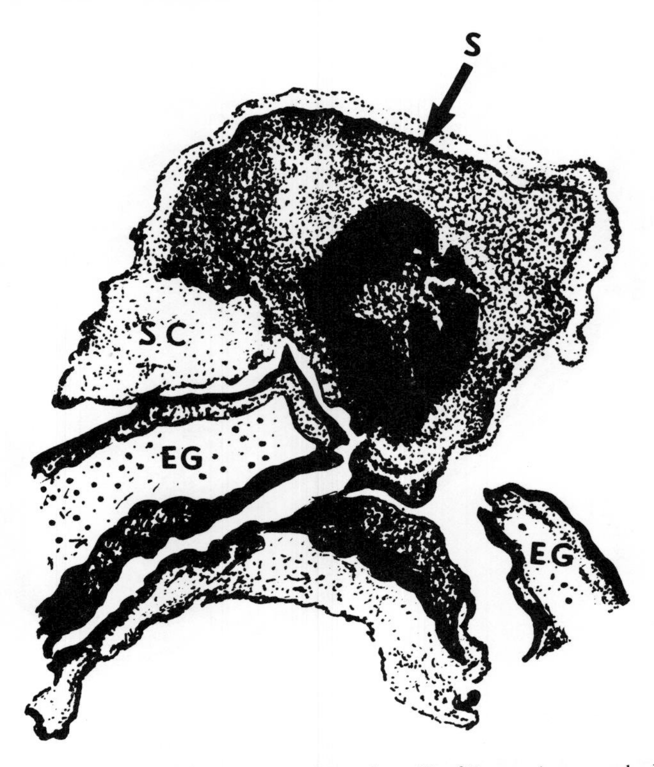

Fig. 13.10 Longitudinal section of a normal spermatophore (S) of *P. americana* attached to the female genitalia. EG, part of external genitalia of the female; SC, secretion of the conglobate gland. (From Jaiswal and Naidu, 1976.)

Gupta (1946); the observations of Vijayalekshmi and Adiyodi (1973b), however, confirm that secretions of the utriculi majores reaching the bulbus from the utricular sacs are kneaded into the spermatophore envelope by the cuticular intima of the bulbus before copulation. They found a mass of spermatozoa together with seminal plasma and also secretions of utriculi translucentes at the mouth of the spermatophore being shaped out. At least some of the utriculi breviores open into the seminal lobe as do vesiculae seminales; spermatozoa have occasionally been found to enter the utriculi breviores. It is, therefore, conceivable that the secretions of utriculi breviores also enter the spermatophore, as was originally proposed by Gupta (1946) and confirmed by Jaiswal and Naidu (1976). The small amount of seminal plasma produced by the vesiculae seminales can hardly hold the spermatozoa together or supply them with nourishment; secretions of utriculi translucentes and breviores, rich in organic substances (Table 13.2), are therefore pressed into service (Vijayalekshmi and Adiyodi, 1973c). Jaiswal and Naidu (1976) maintain, based on similarity of staining behaviour, that the innermost layer of the spermatophore is formed of secretions of the utriculi breviores, and that the middle layer is built with the products of utriculi majores (Fig. 13.9). The third and outermost layer which is added on to the spermatophore during mating is composed of secretions of the conglobate gland (Fig. 13.10). This layer was missing from spermatophores produced by males from which the conglobate gland had been surgically removed prior to copulation (Jaiswal and Naidu, 1976). The reduction observed by them in the size of the conglobate gland following mating, together with the fact that the gland opens separately to the outside and not into the ductus ejaculatorius, suggests that its secretions are poured over the spermatophore after the spermatophore has left the ductus ejaculatorius. Conglobate gland secretions are believed to serve as a glue to fix the spermatophore in the right position in the genital chamber of the female so that the spermatozoa can migrate directly into the spermatheca (Fig. 13.10). It is doubtful whether this is the only function of the secretions of conglobate gland, rich as they are in certain organic substances (Table 13.2) (see Adiyodi and Adiyodi, 1975 for discussion). Conceivably, not only spermatozoa, but some components contributed by the male with the spermatophore, are also used by the female. There is apparently some reinforcement of the spermatophore wall. It is not known whether phenolic groups abundantly present in mushroom and conglobate glands and the peroxidase present in the conglobate gland are involved.

13.3.6 Mechanisms controlling sexual maturation and reproduction in the male

Studies by Vijayalekshmi (1976) have shown that content of neurosecretory cells (NSC) of the thoracic ganglia of male *P. americana* shows clear fluctuations during the first 15 days after imaginal moult. There is a progressive accumulation of neurosecretory material (NSM) in the perikarya of A_1 and A_2 cells during the first week in postecdysis adult males. NSM accumulation reaches a peak on day 6 in the prothoracic ganglion, and on day 8 and 9 in the meso- and metathoracic ganglia respectively. Release occurs from the prothoracic ganglion between days 7–10, and from the meso- and metathoracic ganglia after day 9. Some accumulation of NSM occurs again in the prothoracic ganglion around day 10, but this is soon followed by a release. In the meso- and metathoracic ganglia also, there is a second accumulation of NSM in the perikarya. Though the quantity of NSM present in Group I NSC in the pars intercerebralis (PI) is not particularly high, the pattern of accumulation and release of NSM in these cells is strikingly comparable to that of the prothoracic ganglion. The massive cyclic accumulation and release of NSM observed during the first 15 days after imaginal moult is not observed later in adult life. Vijayalekshmi (1976) is therefore of opinion that secretions from the NSC of the thoracic ganglia and Group I cells of the PI may have some role in sexual maturation, particularly in maturation of the accessory sex glands. The possibility that the cyclic release of NSM may be related to other post-emergence processes cannot be discarded.

In *P. americana*, the testis is programmed to be functional in last-stage male nymphs, but the mushroom and conglobate glands are, by comparison, only marginally active in the last nymphal instar. This may be taken to suggest that hormonal 'priming' and or genetic programming needed for testicular activity might be qualitatively and temporally different from that (or those) needed for accessory gland activity. If the stimulus is hormonal, a longer 'priming' under the influence of appropriate hormonal milieu is required for accessory glands to be activated. Brain hormone, ecdysone, and JH suggest themselves as likely candidates. Studies by Blaine and Dixon (1970) seem to show that NSC and ecdysial glands have no effect on retarding or accelerating the development of testis in *P. americana*; testis would continue its development in the absence of CA and ecdysterone, but CA retard and ecdysterone accelerates the development of the testis (Blaine and Dixon, 1975). More critical work using pure hormone preparations in *in vitro* cultures and inter-instar transplantations of testis is necessary to solve the problem of endocrine influences in testicular development.

Corpora allata are apparently not required for development of the nymphal mushroom gland of *P. americana* (Dixon and Blaine, 1973): eighth-instar nymphs which had been allatectomized at emergence and had become adults after two moults had fully developed mushroom glands in about the same time required for normal 8th instar nymphs to moult twice and become adults. Ninth-instar mushroom gland transplants are claimed to develop better in the last-stage nymph than do eighth-instar transplants, more so if the mushroom gland of the host has been surgically removed. It appears that imaginal differentiation of the mushroom gland requires something more than the mere absence of JH: tenth-instar glands which normally take about 60 days to differentiate, fail to do so, if transplanted into adult males allatectomized within 6 h after emergence.

Blaine and Dixon (1973) claim that, in adult male *P. americana*, the CA or JH are essential for mushroom gland activity and protein production for spermatophore. They have not, however, estimated the actual quantity of protein present in the gland under different experimental conditions and have based their conclusions on the assumption that the dry weight of the gland represents its total protein. In general, this might be, but it remains to be studied whether the production of other organic constituents is also JH-dependent. Notwithstanding the ability of the utricular epithelia themselves to secrete protein, there is a need to investigate the possibility that some of the protein found in the mushroom gland may be incorporated from without as in growing oöcytes.

Acknowledgements

W. J. B. is indebted to Drs Barbara Stay and Stephen S. Tobe for many helpful suggestions and criticisms, and to former students, Drs G. R. Sams, W. Wasilenko and B. Sudarsanam for providing electron micrographs and unpublished data discussed in this review. K. G. A. thanks Dr V. R. Vijayalekshmi for permitting him to cite from her unpublished thesis.

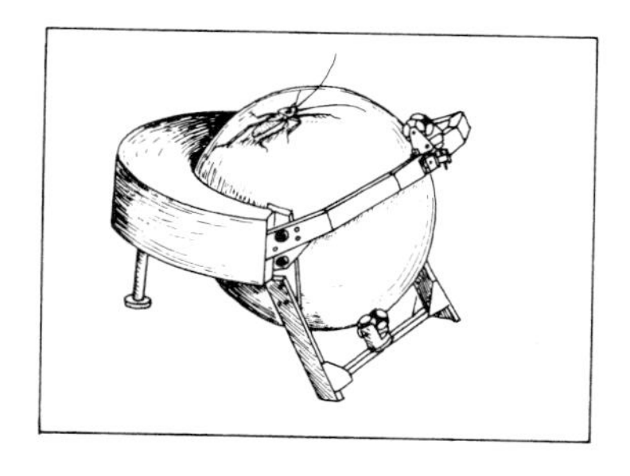

14

Pheromones and behaviour

William J. Bell

14.1 Introduction

Coakroaches, especially *Periplaneta americana*, have been subjects of behavioural experimentation in which the underlying neural bases for behavioural actions are explored relative to motor systems (see Chapter 11) and sensory information processing (see Chapter 9). The active movement of cockroaches in time and space is a most important aspect of their behaviour because these activities lead to and are components of all other types of more complex behaviour. This chapter begins, therefore, with spatial orientation and then proceeds to behaviours that involve interactions among individuals and groups of individuals.

14.2 Orientation

Orientation is defined as self-controlled maintenance or change of body position relative to environmental space (Jander, 1970). Two types of orientation can be distinguished: *positional orientation*, whereby the position of the body is maintained in space relative to virtually constant stimuli such as light and gravity, and *object orientation*, the change in body position (through locomotion) with respect to the spatial patterns of resources and stress sources in the environment.

14.2.1 Positional orientation

When a cockroach is turned on its back a highly stereotyped, though complex, series of righting movements occurs (*Blattella germanica*–Wille, 1920; *Blatta orientalis*–Hoffman, 1933; *Gromphadorhina portentosa*–Camhi, 1974, 1977).

In *P. americana*, the righting response, in which methathoracic legs kick outward rhythmically against the substratum, can be evoked by inverting a tethered animal (Reingold and Camhi, 1977). *P. americana* also exhibits a behaviour called 'tarsal reflex' which involves the initiation of flight when the legs lose contact with the substrate (Diakonoff, 1936; Krämer and Markl, 1978). Optomotor studies of cockroaches are totally lacking; such experiments have been performed with other Orthoptera (e.g. locusts–Kien, 1976), where the animal turns without translatory movements in response to a stimulus such as a vertical stripe pattern that is moved around it.

14.2.2 Object orientation

The function of object orientation is to facilitate the localization of resource objects in the environment (mates, food, shelters) and to avoid predators or stressful conditions. One way to functionally delineate object orientation (Jander, 1975) is to distinguish between two orientation phases – *search* and *approach* (or *avoidance*). 'They are based on a motile organism having on the one hand, insufficient, and on the other hand, precise, information about the spatial location of resource (or stress source) objects'. Search orientation enhances the probability of an organism locating discrete objects in its environment without directional or distance information about those objects. The two major components of search are *ranging*, where an animal has no knowledge of the location of resource objects and therefore attempts to cover the maximum areas possible, and *local search* (= area-restricted search), where an animal perceives non-directional information about the proximity of an object and can then limit the area of its searching. Two major categories of information that control object orientation are internally stored (= idiothetic) information and external (= allothetic) information that is available through the sensory organs (Mittelstaedt and Mittelstaedt, 1973; Mittelstaedt-Burger, 1973); both sources of information are referred to in the following discussion.

(a) *Ranging*

Ranging of cockroaches was monitored by Bell and Kramer (1979) using a servosphere device that permits the animal to run freely while a locomotion compensator moves the sphere and maintains the position of the animal at the upper pole (Kramer, 1976). Movements of the cockroach are recorded over time by electronically storing the turning increments of the sphere with respect to the *x* and *y* axes of movement. An attempt was made to minimize orientation cues such as lights, air currents and vibrations so that idiothetically controlled ranging behaviour would be measured. Three cockroach species, *P. americana*, *Blaberus craniifer* and *B. germanica*, range in patterns that reduce repetitive search of the same area through a combination of circling and relatively straight movements (Fig. 14.1). Similar experiments were performed in a 2.4 m circular arena that differed from the servosphere principally because the arena has an edge. *P. americana*, as with other

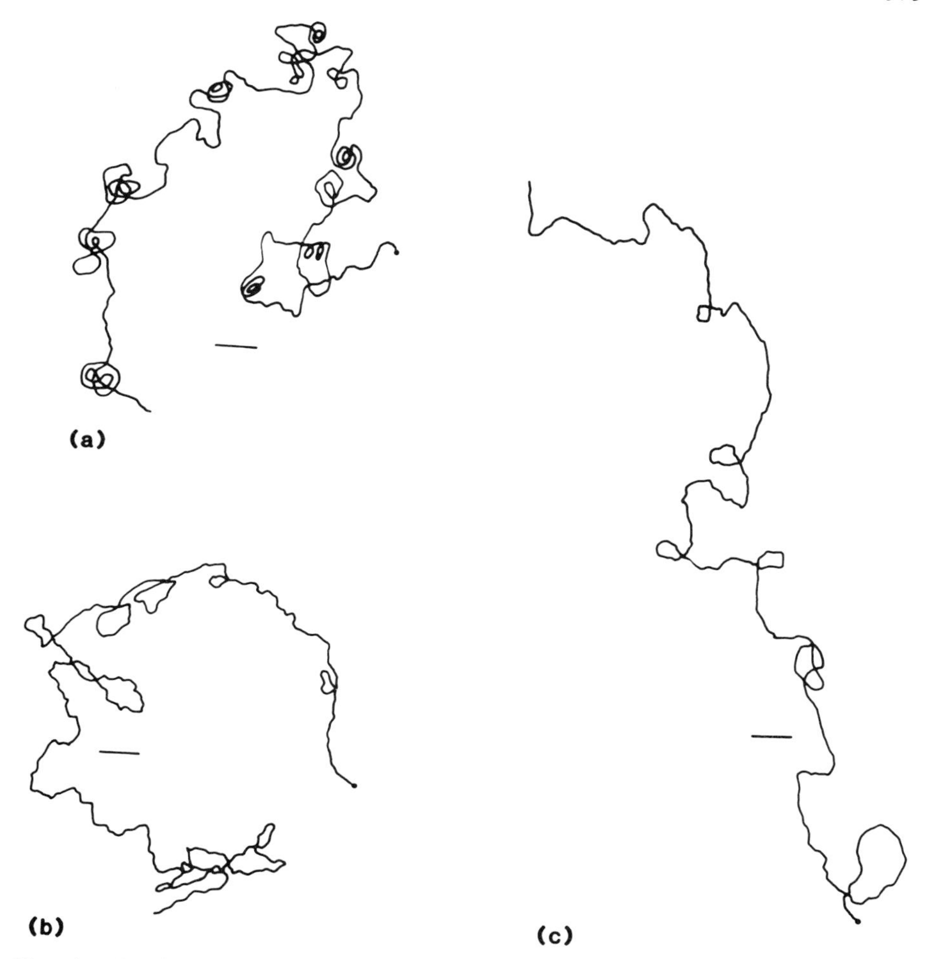

Fig. 14.1 Ranging orientation of three cockroach species (a) *B. craniifer*, (b) *P. americana*, (c) *B. germanica*) recorded on a servosphere device in the absence of wind currents and other external stimuli. Time duration is approximately 1000 s. Solid dot indicates starting position. Scale: Bar = 1 m. (Bell and Kramer, 1979.)

cockroach species, is strongly thigmotactic, and 70% of its time in the arena was spent near or within antennal contact of the edge. When running in the central portions of the arena, patterns similar to those recorded on the servosphere were observed (Hawkins, 1978; Bell and Tobin, in press). These experiments suggest that *P. americana* (and probably most animals) can revert to the use of internally stored (idiothetic) orientation information when external sensory cues are below a biologically relevant level.

Ranging under natural conditions, however, is probably controlled by a combination of idiothetic and allothetic information. For example, in attempting to eliminate sensory cues around the servosphere apparatus for ranging experiments, it was observed that *P. americana* oriented in a relatively straight line if even a pinpoint of directional light or a slight wind current could be perceived by the animal. Sensory cues to guide ranging orientation (e.g. light, gravity, wind currents) are nearly always present in the environment and are utilized by cockroaches as described below.

Early studies showed clearly that cockroaches are negatively phototactic (e.g. Szymanski, 1912; Eldering, 1919) but no data are available as to the extent and mechanisms by which cockroaches employ light in their orientation, a topic well investigated in ants (e.g. Schneirla, 1953) and other animals. In servosphere experiments, *P. americana* and *B. germanica* males ran primarily at one specific angle relative to a horizontal light beam (Bell, unpublished; Mittelstaedt-Burger, 1972), although the particular angle differed for an individual from trial to trial. Too often, perhaps, researchers focus on tactile and olfactory cues because most cockroaches are nocturnal, although directional sources of light, even at night, are probably important for cockroach orientation. That many species of cockroaches including *P. americana* have well-developed eyes and visual acuity attests to this hypothesis (see Chapter 9).

Cockroaches also employ gravity in control of course direction. Jander *et al.* (1970) tested *Leucophaea maderae* on a sloping substrate to determine the mechanism of gravity perception. When the slope was reduced from 30° to 14.5° the standard deviation of upward pathway directions roughly doubled. Adding a weight equal to that of the cockroach reduced the variation in pathway directions on a 14.5° slope, suggesting that geotaxis, is mediated by gravity receptors in the legs. In many higher insects, such as Diptera and Hymenoptera, other joints in the neck, petiolus and antennae also contribute to gravity perception.

Orientation of cockroaches in wind (= anemotaxis) has been investigated more thoroughly than with light or gravity. Rust and Bell (1976) found that male *P. americana* moved consistently downwind when they were dorsally tethered and running on a styrofoam Y-maze. Such responses were further investigated by Bell and Kramer (1979, 1980) using a servosphere device. Cockroaches were tested in still air and in air currents of 1.5, 3, 6, 12, 24 and 48 cm s^{-1} (Fig. 14.2). The lower threshold for the perception of wind direction by *P. americana* is between 1.5–3.0 cm s^{-1}. Positive anemotaxis is exhibited with low velocity air currents (1.5 to 6 cm s^{-1}), random orientation occurs at 12 cm s^{-1}, and negative anemotaxis occurs at 24–48 cm s^{-1}. In addition to changes in the direction of orientation, there is an increase in the straightness of locomotory movements as air speed is increased (Fig. 14.2). These data suggest that wind alone provides adequate information for the cockroach to maintain a relatively straight course. The pathways are neither entirely straight nor directly up- or downwind, but generally conform to a zig-zag pattern at an angle that is about 20–30° relative to the wind.

P. americana, *B. craniifer* and *B. germanica* perceive low velocity, continuous air currents with the antennae (Silverman, 1978; Bell and Kramer, 1979); shorter-duration 'wind puffs' are perceived through sense organs of the cerci (see Section 14.2.2(c)). When both antennae are ablated or prevented from moving, cockroaches

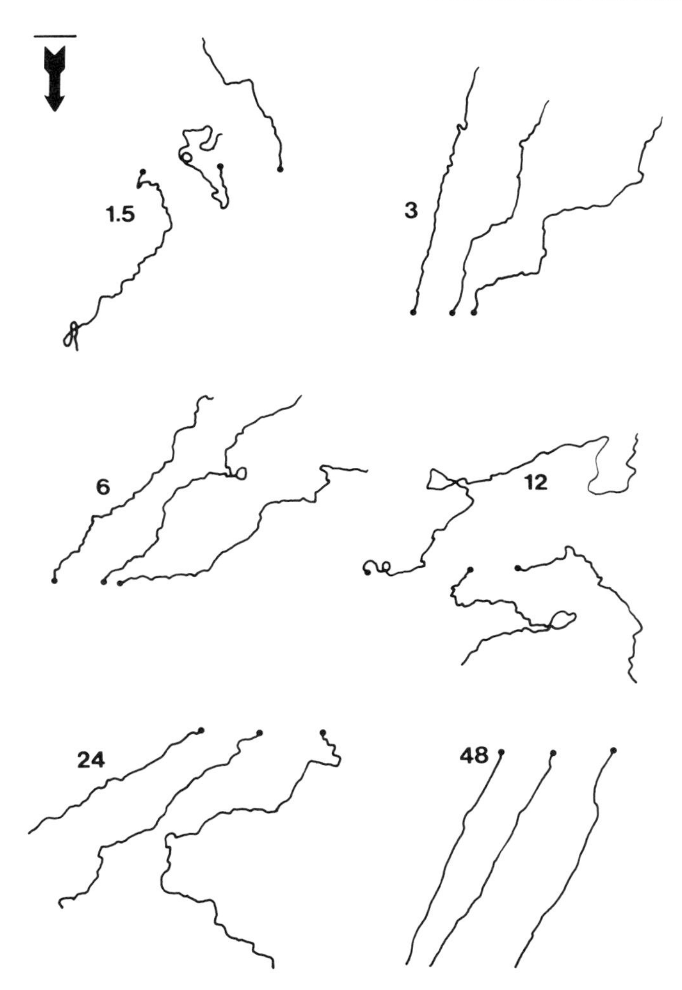

Fig. 14.2 Pathways of male *P. americana* in an air current recorded on a servosphere device. Three representative paths are depicted for each wind speed tested (values in cm s^{-1}). Vertical arrow indicates wind direction. Scale: Bar = 1 m. (Bell and Kramer, 1979.)

fail to perceive wind and move in circling, looping patterns similar to those recorded when external sensory information was removed. Analysis of orientation pathways of intact cockroaches in an air stream suggests that the animals run at angles (usually 20–30°) relative to the wind, stop and continue to run upwind. Often they turn after stopping and run approximately perpendicular to the wind before turning upwind once again (Bell and Kramer, 1979). Cockroaches probably detect wind through the mechanoreceptors of the antennal pedicellus: Johnston's organs, the campaniform sensilla of Hick's organ and a chordotonal organ (Section 9.4). As shown by Gewecke (1974) and Heinzel and Gewecke (1979), the campaniform sensilla of the Hick's organ of the locust can perceive both the directional movements of the antennae and the intensity of the thrust during flight.

(b) *Local search*

When an animal detects the proximity of a resource object, but cannot derive sufficient information to orient in a directional manner (i.e. at one point in time a differential cannot be perceived), it switches from ranging to local search. To simulate conditions where local search would be expected to occur, female sex pheromone was 'pulsed' from above *P. americana* males in an arena (Hawkins, unpublished) or on a servosphere (Bell, unpublished). The orientation of the cockroach was recorded after it was subjected to a 10-s pulse of sex pheromone. The rate of locomotion increased (controls, $\bar{X} = 5$ cm s^{-1}; sex pheromone, $\bar{X} = 11$ cm s^{-1}), wall-seeking behaviour was reduced, and the insect changed its gait from a relatively steady walk to short fast locomotory bursts punctuated by stop periods of one to three seconds. On the servosphere, the male cockroach ran quickly in circles, seldom moving more than two metres from the place at which it was stimulated. The switch from ranging to local search in this context thus entails behaviour that would seem to function to bring the animal, by chance, close enough to the stimulus source that directional information can be obtained. In *P. americana*, an increase in the frequency and amplitude of antennal movements is associated with the sex pheromone-stimulated switch from ranging to local search (Rust *et al.*, 1976; Tourtellot and Franklin, 1978). Antennal sweeping, along with head movements, probably functions to move the antennae to positions where they are maximally separated, thus facilitating a comparison between the inputs of the two antennae. To date there is no obvious explanation for the vibration of these sense organs, although similar movements are observed in many insects and other arthropods that encounter an odour (Heymons and Lengerken, 1929; Murr, 1930; Bogenschütz, 1978; Suzuki, 1975). This may be analogous to sniffing in mammals.

(c) *Approach and avoidance*

When search orientation leads an animal close enough to an object to detect directional information about that object, then either approach or avoidance occurs based upon the information obtained. The following discussion deals with types of information known to be utilized by cockroaches to locate resource objects (e.g. mates) or to avoid stress source objects (e.g. predators).

Several studies have been performed on directional orientation that is stimulated by tactile cues. Dominant male *Nauphoeta cinerea* accurately turn the body towards a tactile stimulus and then lunge in that direction (Bell, 1978). *B. craniifer* and *G. portentosa* lower their pronotum or abdomen on the side receiving tactile stimulation, apparently as a defensive manoeuvre (Gautier, 1974a; Camhi, 1974).

A combination of tactile and chemical stimuli controls certain types of orientation. *B. germanica* males execute courtship turns when their antennae are stimulated by an object containing female sex pheromone (Roth and Willis, 1952).

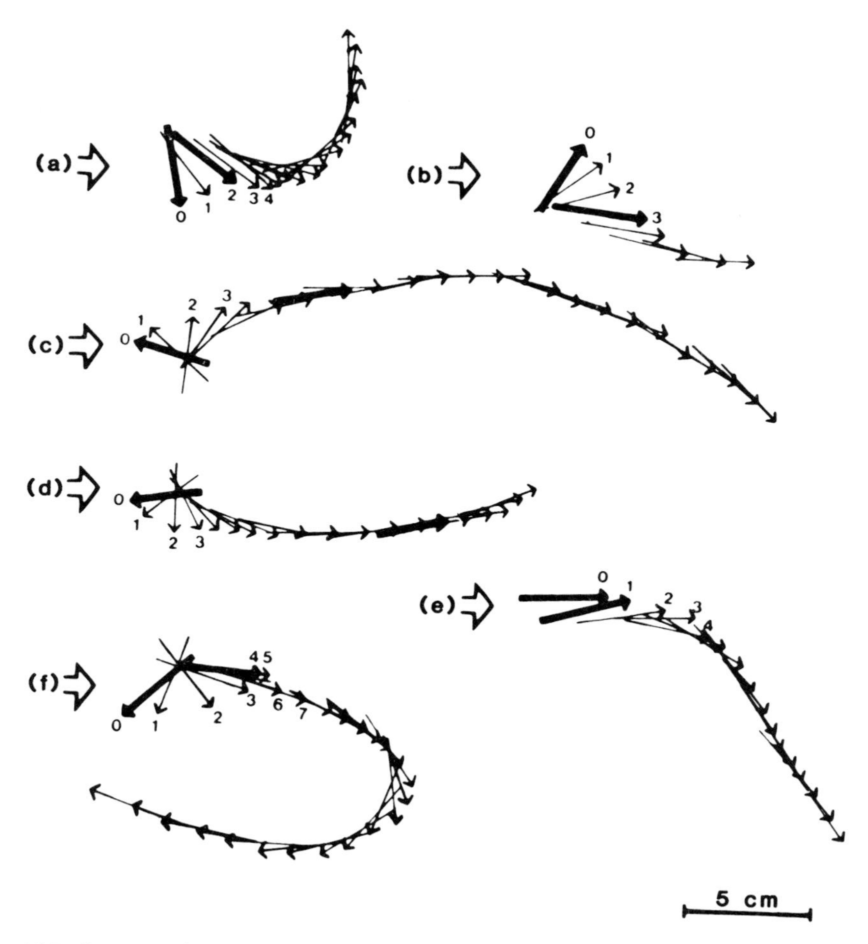

Fig. 14.3 Representative turning responses to wind puffs by free-ranging *P. americana*. (a–f) are 6 different trials. For each trial the sequence of arrows, 0–1–2 . . . represents the positions of the body (arrows point anteriorly) with respect to the wind tunnel (open arrow) on successive cine frames, 16 ms apart. Arrow '0' shows the insect's position on the cine frame just prior to the onset of the turning response. The angle between the two thick arrows for each trial is defined as the 'angle of turn'. (Camhi and Tom, 1978.)

The angle of turn, which normally positions the abdomen of the male towards the female to facilitate mounting, is specified by the position of the antennae when the male is stimulated (Bell *et al.*, 1978b; Bell and Schal, 1980; Franklin, Bell and Jander, 1981).

Turning responses of *P. americana* to wind puffs that might be created by a predator were investigated by Camhi and Tom (1978) and Camhi *et al.* (1978). A wind puff was aimed at a cockroach in the plane of movement of the animal from various directions relative to the longitudinal body axis. Cockroaches turned away from the stimulus as shown in Fig. 14.3. The direction of the initial escape turn is

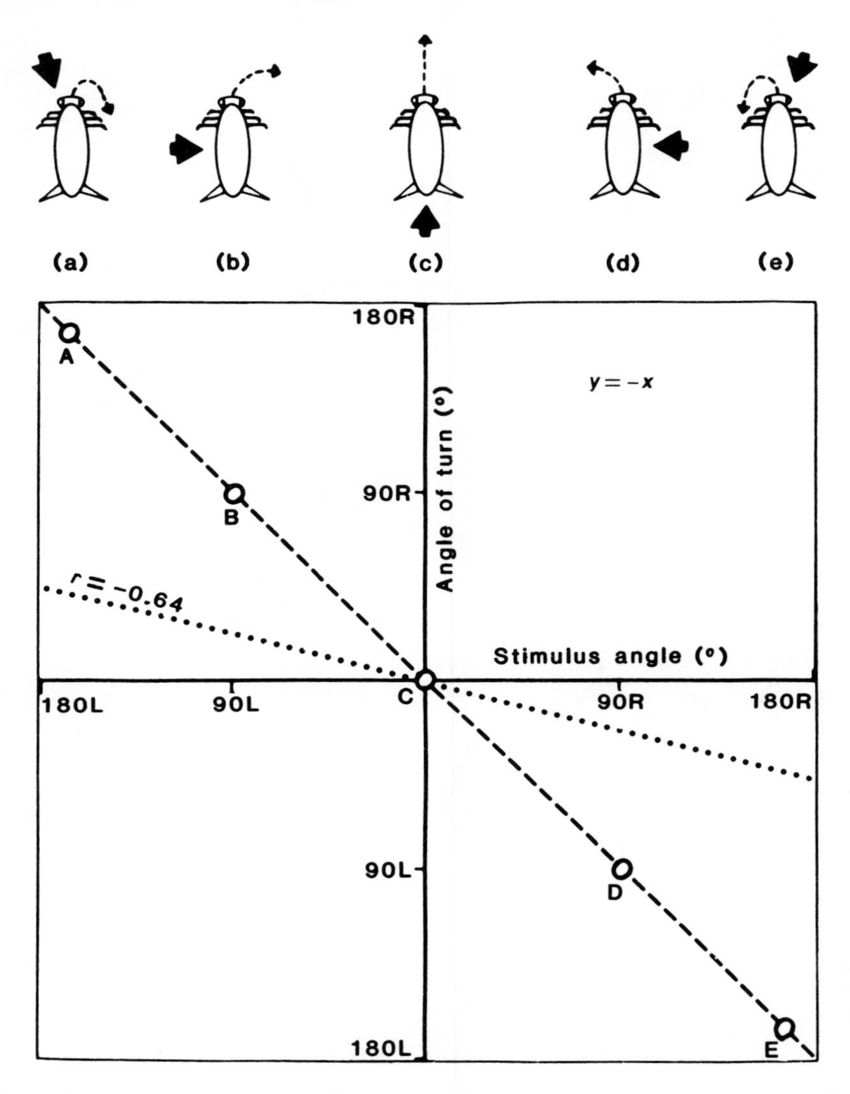

Fig. 14.4 Angle of turn by *P. americana* exposed to a wind puff versus the stimulus angle relative to the longitudinal axis of the body. Small figures (a–b) show idealized turns (small arrow) exactly away from the wind (large arrow), e.g. a stimulus angle of 170° left leads to a turn of 170° right. The dashed line plots these idealized data and has a formula $y = -x$. Dotted line represents turning responses of cockroaches maintained under inverted cups prior to stimulation by wind puffs ($n = 110$ trials; $r = -0.64$; $p < 0.01$; formula $y = -0.24x - 3.1$). (Modified from Camhi and Tom, 1978.)

consistent for a given stimulus angle (Fig. 14.4), although subsequent turning is more variable. That these responses are actually employed in escape behaviour was suggested by another set of experiments using a toad, *Bufo marinus* (Fig. 14.5). Cockroaches escaped the strikes of the toad although the mean wind speed

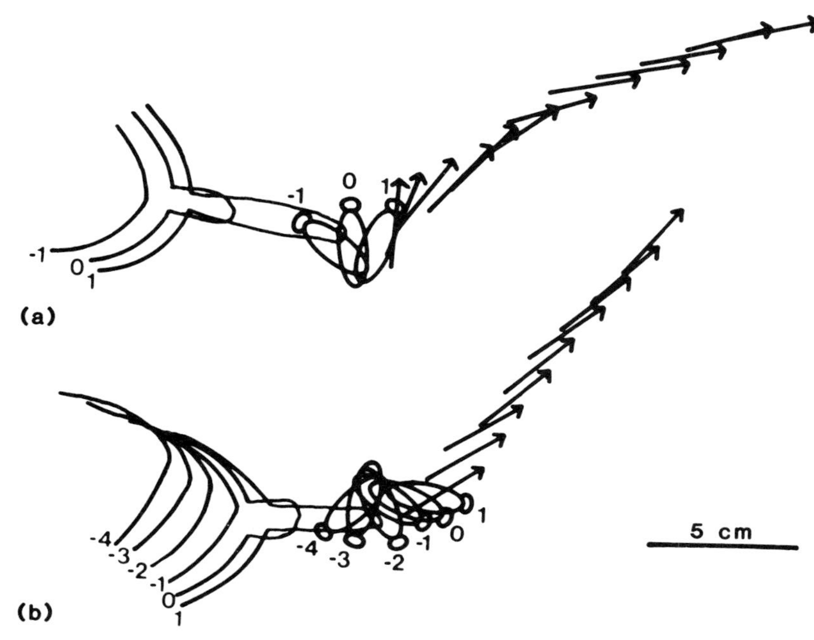

Fig. 14.5 Cine sequences of two toad–*P. americana* interactions. For each interaction the front of toad's head and tongue are traced from a series of cine frames. Tracings are numbered with reference to the frame on which toad's tongue first appeared (frame 0). The earliest frame shown in (a) and (b) is that just before the cockroach began to move (thus the initial part of the toad's tongue is not shown). On the first several frames, until the tongue reaches its most extended position, the cockroach's outline is drawn; subsequently, the positions of the cockroach's body are indicated by arrows. Numbers beside the outline of each cockroach indicate cine frames, corresponding to the numbers on the tracings of the toad. (Camhi *et al.*, 1978.)

generated by the striking toad was only 22 mm s^{-1} (Fig. 14.6). Turning and escape reactions are probably 'open-loop', because correct turning responses occurred during the first 16 ms of movement, a time span which seems insufficient to allow use of feedback information during turning. The appropriateness of the toad as a potential predator is documented by predation of cockroaches by toads in caves of Trinidad (Deleporte, 1976; Gautier, 1974a,b; Brossut, unpublished).

Silverman and Bell (1979) applied wind puffs to cockroaches on horizontal or vertical surfaces. The wind was directed from 'above', meaning perpendicular to the plane of movement of the cockroach. On a horizontal surface, the cockroach either ran straight ahead or ran to the left or right at an angle of 90° or 45°. Individual cockroaches showed a tendency to use different angles in sequential testing. On a vertical surface, males and females ran upwards when exposed to a wind puff; this may be an adaptation for fleeing from predators that climb less effectively than cockroaches. In both wind puff studies, the cerci were shown to be the receivers of short wind bursts, as discovered by Roeder (1948). The function of cerci is discussed further in Section 9.3.1.

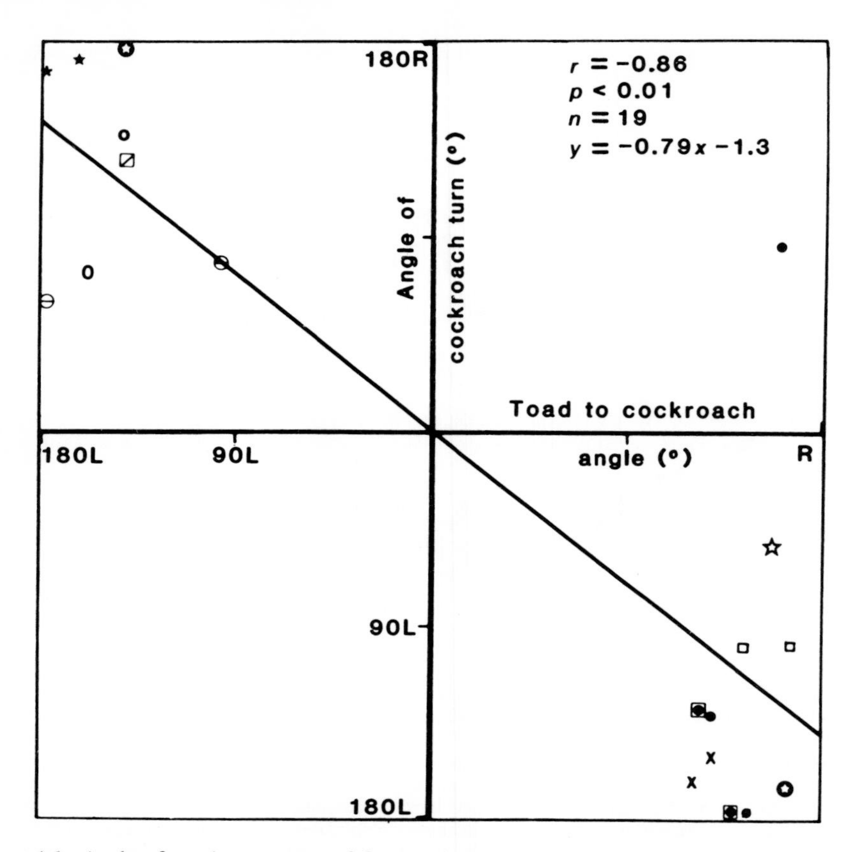

Fig. 14.6 Angle of turning response of free-ranging *P. americana* to predatory strikes of toads. Definitions: angle of toad to cockroach (abscissa), 0° = toad's strike from behind cockroach; 180° = head-on encounters. Angle of cockroach's turn (ordinate), 0° = forward locomotion; 180°R = about face to right; 180°L = about face to left (see also Fig. 14.4). Each type of symbol represents the response of a different animal. All data points are for relatively large angles of strike because most of the encounters occurred when an insect walked towards a toad. Slope of regression is larger than for responses to wind puffs (Fig. 14.4) because only turns of cockroaches which escaped from the toads are plotted here; the captured cockroaches might have been those making smaller turns. (Camhi *et al.*, 1978.)

The predators of cockroaches include warm-blooded animals as well as amphibians, reptiles and arthropods (Roth and Willis, 1960; Gautier, 1974a,b) (see also Chapter 1). Silverman (1978) found that wood rats, mice and laboratory rats captured and fed upon *P. americana*. He tested the response of cockroaches on a vertical surface to puffs of air with or without 10% CO_2 at different temperatures and humidities. Response latencies were recorded from the time when an air puff was delivered to the time when the subject responded by moving (upwards). The shortest response times were recorded with air carrying CO_2, at the highest temperature tested (29°) and at the highest humidity tested (64% R.H.) (mean, 2.2 s). Presence of CO_2 was found to be the most important factor; temperature and humidity were less important, but had an additive effect in reducing response times. These results suggest that responses to wind puffs can be modulated by additional factors such as temperatures and odours.

P. americana males utilize wind direction in orienting to objects releasing odours, as do many other insects (Kennedy, 1977). This type of orientation, chemo-anemotaxis, was documented for *P. americana* males walking on a styrofoam Y-maze where the animal had a choice of 'attempting' to turn up- or downwind (Rust *et al.*, 1976; Rust and Bell, 1976; Seelinger, in preparation); upwind orientation occurred when the wind carried female sex pheromone or aggregation pheromone. More recently, Bell and Kramer (1980) showed that *P. americana* running freely on a servosphere moved upwind when the wind carried sex pheromone. In addition to stimulating upwind orientation, the pathways of the cockroach became straighter as the concentration of pheromone was increased (Fig. 14.7). Tobin (unpublished) tested males in a 2.4 m wind tunnel similar in design to that used for investigating chemo-anemotaxis in moths (e.g. Miller and Roelofs, 1978; Cardé and Hagaman, 1979). As shown in Fig. 14.8, cockroaches in a pheromone plume run upwind in zig-zag movements that are very similar to those recorded for other insects walking or flying in an odour plume.

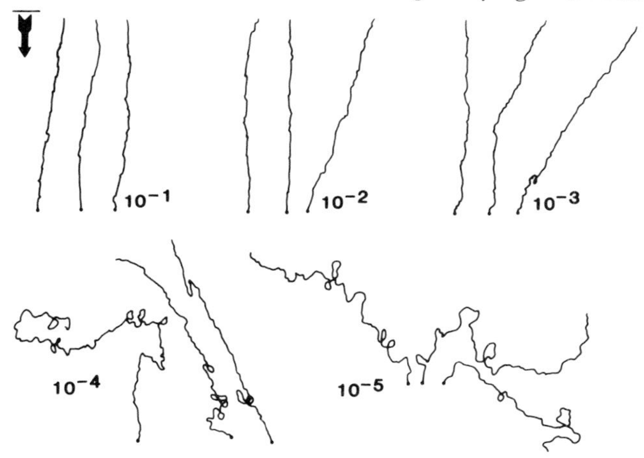

Fig. 14.7 Pathways of male *P. americana* in an air current carrying female sex pheromone extract as recorded on a servosphere device. Vertical arrow indicates wind direction (24 cm s^{-1}). Values are concentrations in μg. Scale: Bar = 1 m. (Bell and Kramer, 1980.)

Gravity orientation is also employed to locate females releasing sex pheromone. When male *P. americana* in a vertical arena were stimulated with sex pheromone in a puff of air, they ran downward, where as a puff of clean air stimulated upward movements (Silverman and Bell, 1979). Because males are usually located above females when housed in a 3-dimensional arena (Silverman, 1978), it is likely that sex pheromone-stimulated males move downward to locate females.

In addition to using wind and gravity orientation for locating an odour source, *P. americana* is also capable of direct chemo-orientation, whereby a chemical

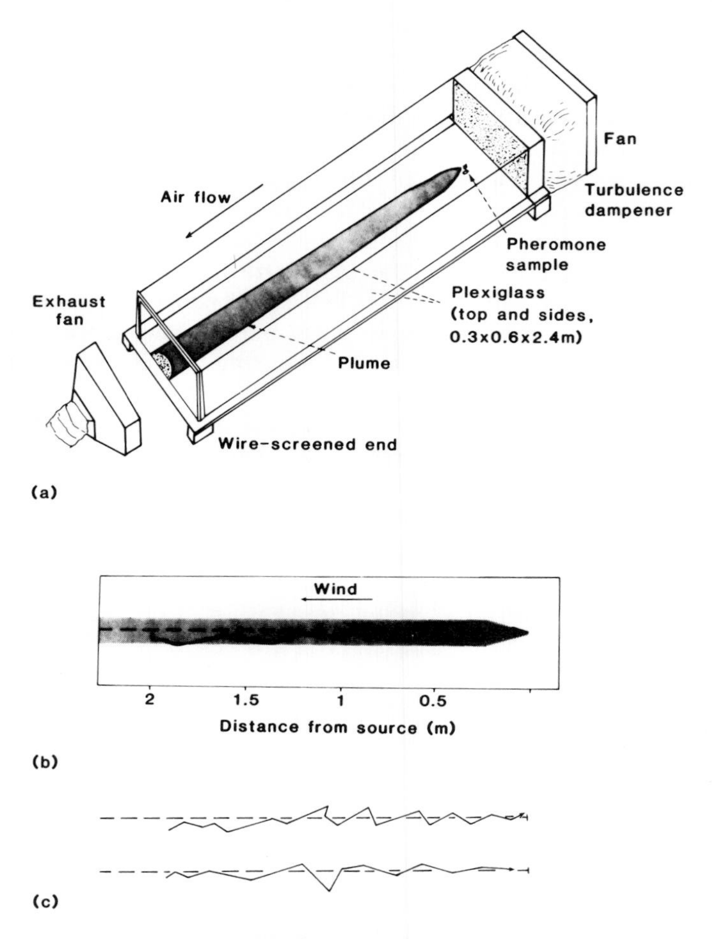

Fig. 14.8 Anemotactic orientation of male *P. americana* photographed in a 2.4 m wind tunnel. (a) Wind tunnel to scale showing dimensions of a periplanone B plume; (b) Cockroach pathway superimposed on a periplanone B plume (17.8 s, mean time duration); (c) Representative pathways showing zig-zag form of orientation path. (Tobin, unpublished.)

gradient is deciphered to determine the direction to the odour source. Individuals tested in a T-maze turned into the tube impregnated with aggregation pheromone (Bell *et al.*, 1973) or into the tube of a Y-maze impregnated with 'marking pheromone' (Brousse-Gaury, 1975a). Rust *et al.* (1976) showed that male *P. americana* orientated towards a source of aggregation pheromone or sex pheromone located within 20 cm of an animal walking on a styrofoam Y-maze (Fig. 14.9). Chemo-orientation was also studied in a 2.4 m-diameter circular arena (Hawkins, 1977, 1978; Bell *et al.*, 1977; Bell and Tobin, in press). A sex pheromone source or a tethered virgin female was placed in the centre of the arena and a male cockroach was released at the periphery. The orientation pathway was recorded using

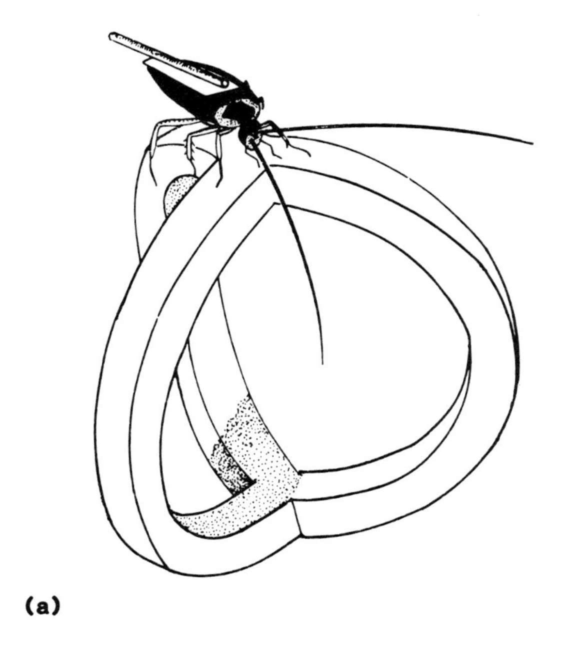

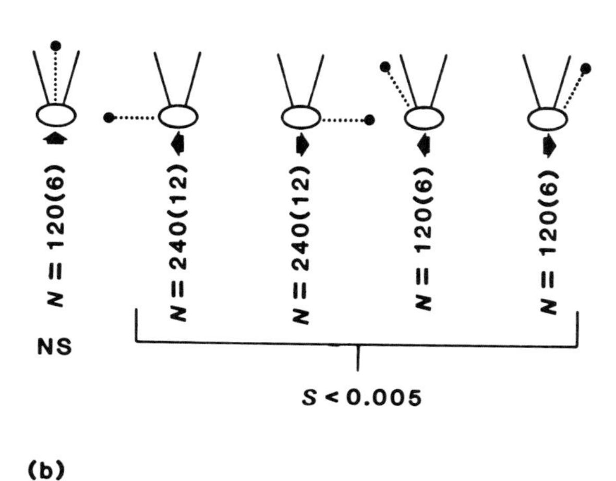

Fig. 14.9 (a) Diagram of a dorsally tethered cockroach on a styrofoam Y-maze; (b) Female sex pheromone (extract) stimulated turning in male *P. americana*. Diagrams show the cockroach head and antennae with the position of the pheromone source depicted by a solid dot in front of or beside the head. Arrows indicate direction of turning. (NS) no significant turning preference as compared with controls (Wilcoxon matched pairs signed rank test); (*n*) number of trials (number of animals tested in parentheses). (Redrawn from Rust *et al.*, 1976.)

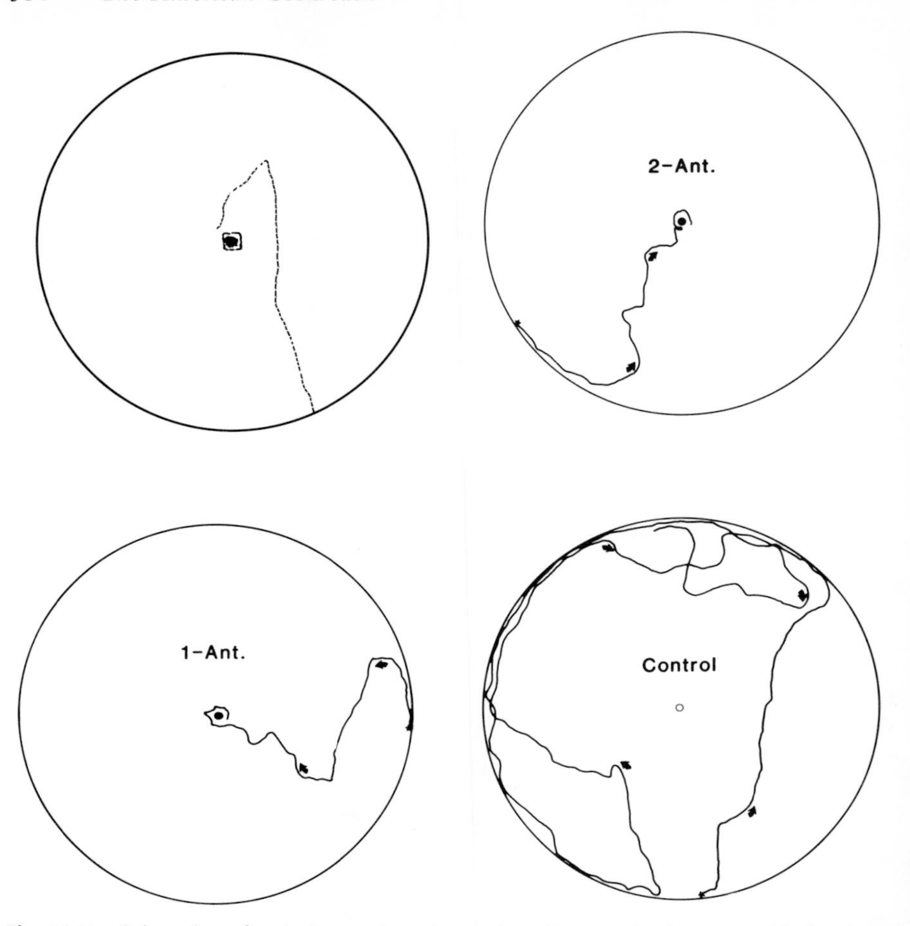

Fig. 14.10 Orientation of male *P. americana* in a 2.4 m-diameter circular arena with female sex pheromone source in the centre. Upper left shows a diagram redrawn from a photographic print of actual 20 s-duration slotted-disc-camera time exposures. Other figures are representative approach pathways of intact (2-ANT) and unilaterally antennectomized (1-ANT) cockroaches compiled from sequences of 20-s exposures (30–55 s duration). Circle in centre of arena indicates position of sex pheromone source (solid circle), or solvent control (open circle). (Bell and Tobin, in press.)

time-lapse photography. A slotted disc moving beneath the camera caused a spot of reflecting paint on the cockroach to appear as a sequence of dots on the films (Fig. 14.10, top left). Within 20 seconds after being released, a male cockroach exhibited the same changes in locomotion that were observed with a non-directional sex pheromone source: increased locomotion, reduced wall-seeking tendency and stop-start gait. As shown in Fig. 14.10, a cockroach runs towards the pheromone source in a zig-zag or looping manner; 90% of animals tested were able to locate and contact the source within two minutes.

As the cockroach moved closer to the centre of the arena its ability to 'home in' on the pheromone source increased. Hawkins (1977) assessed this tendency using a vector analysis of the angle of movement relative to a straight line towards the source. The analysis was performed on trails 35, 70 and 105 cm from the centre of the arena. With a solvent control in the centre of the arena, the resulting vectors were significantly smaller (0.34, 35 cm; 0.21, 70 cm; 0.22, 105 cm) than with sex pheromone (0.97, 35 cm; 0.66, 70 cm; 0.41, 105 cm), indicating significant levels of unidirectionality when sex pheromone was present in the centre of the arena.

Cockroaches, as well as other insects, probably decipher a chemical gradient by one of two possible means: (1) the concentration perceived by each antenna is compared, and the cockroach turns in the direction of the antenna receiving the greatest stimulation (chemotropotaxis), (2) the cockroach sequentially samples the concentration at one place and then another, and by comparing the two values it determines whether the direction taken is towards or away from the odour source (longitudinal klinotaxis). Results obtained thus far for *P. americana* are somewhat ambiguous and suggest that both mechanisms are employed. In an arena with a sex pheromone source in the centre, the orientation pathway of a cockroach with one antenna removed is nearly identical to that of an intact cockroach (Fig. 14.10) (Bell and Tobin, in press). This suggests that the mechanism employed by unilaterally antennectomized cockroaches and intact cockroaches is the same. Since cockroaches with one antenna cannot, by definition, employ tropotaxis, it would seem that both use longitudinal klinotaxis. On the other hand, a fixed cockroach walking on a movable sytrofoam Y-maze (Rust *et al.*, 1976) cannot detect differences in odour concentration over time because the odour source and the cockroach are stationary; this suggests that tropotaxis is employed under these circumstances. It is possible, therefore, that the experimental apparatus influences the chemo-orientation mechanism employed by an insect, and that cockroaches are capable of switching among these mechanisms depending on the spatial distribution of the orientation cue.

14.3 Courtship

14.3.1 Analysis of courtship behaviour

Detailed analysis of the courtship of *P. americana* and evidence for a volatile female sex pheromone were first reported by Roth and Willis (1952). The sequence of courtship acts of males and females was described by Wharton *et al.* (1954a), Barth (1970), and further elaborated upon by Simon and Barth (1977a). Fig. 14.11 is a summary of courtship behaviour based primarily on Barth's (1970) work. The approach of the male is directed (see above) by sex pheromone-stimulated chemo-orientation. Moreover, Schal and Tobin (unpublished) have recently discovered that females of many cockroach species, including *P. americana*, exhibit a 'calling pose'. The female assumes a stance with the abdomen slightly lowered and with the 8th and 9th terga spread apart so that the underlying tissues are exposed. When the male antennates the female, the female assumes a motionless stance,

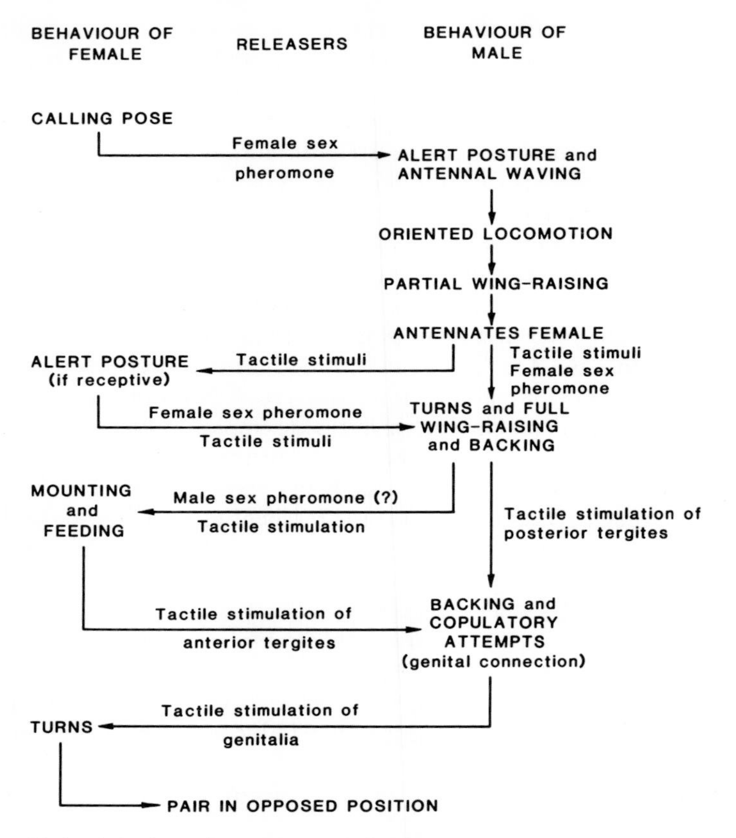

Fig. 14.11 Mating behaviour of *P. americana*, indicating the probable releasers for each step in the sequence. (Primarily from Barth, 1970, with additions from Simon and Barth, 1977a and Schal and Tobin, in preparation.)

and the male turns approximately 180°, raises and flutters its wings and moves backwards. In some encounters, the female mounts and then the male attempts copulation; in others, the male attempts copulation without female stimulation. After copulation is attained the female turns so that the pair are in an opposed position.

Simon and Barth (1977a,b) point out the importance of female receptivity for successful courtship. Unreceptive females typically decamp or exhibit aggressive behaviour towards courting males. The behaviour of the male is largely defensive, but prolonged bouts may ensue. Simon and Barth (1977b,c) consider courtship in *P. americana* to be a conflict situation in which the potential for aggressive responses is initially high: '. . . Certain aspects of courtship behaviour may have evolved either to suppress the release of agonistic behaviour or to channel or divert it in some way'.

Males of many cockroach species, including some species of *P. americana*, have modified dermal moulting glands (DMG) (also called tergal glands or dermal sexual glands, DSG) on the first abdominal terga (Roth, 1969; Brousse-Gaury, 1973; Brossut *et al.*, 1975; Brossut and Roth, 1977). Females feed upon glandular secretions while mounting the male (Barth, 1970; Simon and Barth, 1977a). In *P. americana*, the dermal moulting glands are almost evenly distributed beneath the cuticle in both sexes throughout life, and are probably involved in secretion of cuticular cement at moulting. In the males the DMG are supplemented by a large number of anatomically similar DSG, particularly beneath the anterior abdominal terga. DSG are believed to produce the arrestant pheromone described by Barth (1970). In sexually mature adult males, DSG are rich in arylsulphatase, whereas DMG are rich in arylsulphatase only during the moult (Brunet, Adiyodi and Adiyodi, 1980, unpublished).

Tactile stimuli are important releasers for both the male and female. For example, when stimulated by sex pheromone the males 'court' any object encountered (Barth, 1970; Roth and Willis, 1952). Tactile stimulation of the genitalia seems to be required for the male to correctly orientate its copulation attempt. Air currents produced by wing fluttering stimulate mounting by females, although it is not known whether the air currents also carry a male pheromone (Simon and Barth, 1977a).

14.3.2 Bioassay of female sex pheromone

The courtship response of *P. americana* males to sex pheromone has been used extensively for bioassay (e.g. Roth and Willis, 1952; Wharton *et al.*, 1954a,b; Chen, 1974; Takahashi and Kitamura, 1972, 1976). Quantitative assays were devised by Wharton *et al.* (1954a,b) and Block and Bell (1974). The Wharton method quantifies the percentage of males (in a group) responding to a given quantity of pheromone; responses are proportional to the log pheromone concentration. Rust (1976) showed that the rate of locomotion of male cockroaches increases proportionally with the log pheromone concentration. Similar results have been obtained with other cockroaches species (Bell *et al.*, 1974; Schafer, 1977a,b). Locomotion, which can be assessed automatically with an activity meter, is therefore an appropriate behaviour to assay for sex pheromone. Counting the number of times each male exhibits wing-raising is a less powerful tool because its release requires higher concentrations of pheromone than does running (Rust, 1976; Tobin *et al.*, in press), and depends in part on tactile stimulation from other individuals or objects in the test arena. Recently, a chemical structure of the sex pheromone of *P. americana* was suggested by Persoons *et al.* (1976, 1979) and Talman *et al.* (1978). The two major components, periplanone A and B, are sesquiterpenoids and a minor component has not yet been elucidated. Periplanone B has now been synthesized (Still, 1979; Adams *et al.*, 1979). Tobin *et al.* (in press) compared the responses of males periplanone B and a sex pheromone extract and found that periplanone B released all of the behaviours normally observed after

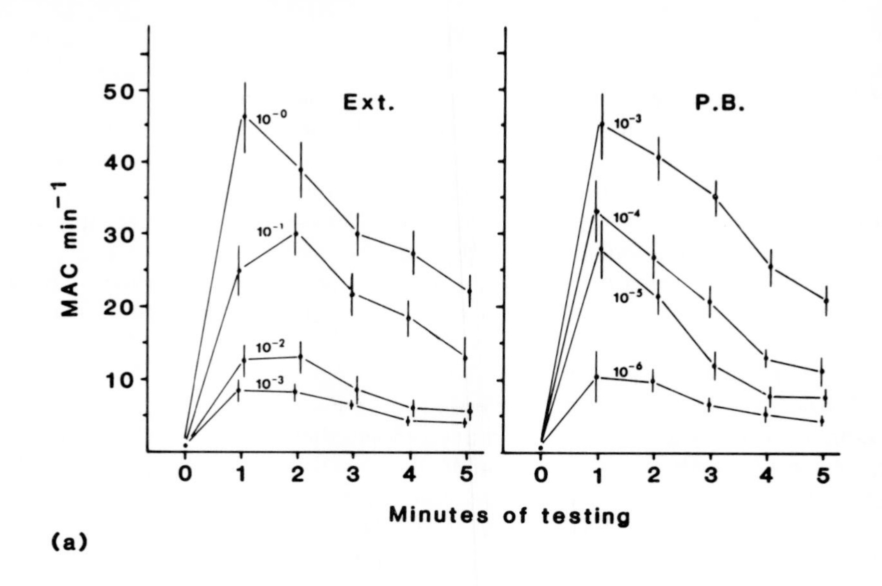

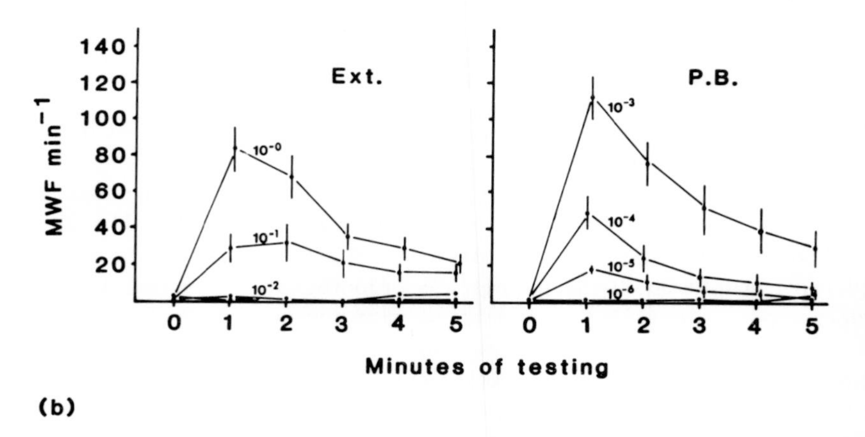

Fig. 14.12 Responses of *P. americana* males to female sex pheromone extract (Ext.) and periplanone B (P.B.). Pheromone concentrations expressed in μg (a) Locomotory activity recorded as mean activity counts min^{-1} (MAC min^{-1}) over five min of exposure; (b) Wing-fluttering activity recorded as mean wing flutters min^{-1} (MWF min^{-1}) over five min of exposure. Vertical lines are standard deviations of the mean for five animals. (Tobin *et al.*, in press).

stimulation by a sex pheromone extract (Fig. 14.12), including chemo-anaemotactic and chemotactic orientation (Tobin, 1980; Bell, unpublished). These studies showed that 1 pg of periplanone B equals the activity of one female equivalent of extract. Seelinger (in preparation) tested two fractions of *P. americana* female sex pheromone isolated by chromatography (Sass, unpublished), and observed that both fractions elicited courtship and orientation behaviour in male cockroaches. That the two fractions are indeed different compounds is suggested by adaptation studies. Individuals adapted by the first fraction to a condition of unresponsiveness would then respond to the second fraction; analogous results were obtained using single-cell electrophysiological recording of antennal chemoreceptors (Chapter 9). Mixtures of both fractions are not more effective in eliciting a behavioural response than an equal quantity of either single fraction, eliminating the possibility of a synergistic effect. At present it is not certain which fraction, if either, is periplanone B.

Physiological and environmental factors affect male responsiveness to sex pheromone. Hawkins and Rust (1977) showed that the pheromone concentration required to stimulate responses during the light phase of the photocycle was 1000 times greater than the effective concentration during the dark phase (Fig. 14.13).

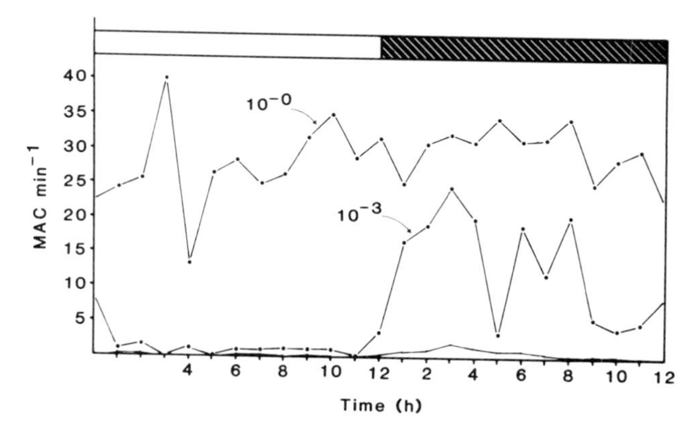

Fig. 14.13 Relationship between photocycle and level of male *P. americana* locomotory response to two concentrations of sex pheromone extract. 10^{-1} = undiluted extract (equal to 10^{-3} μg periplanone B); 10^{-3} = 1000-fold dilution of extract (equal to 10^{-6} μg periplanone B). Light period on left (12 h); dark period on right (12 h). Control activity (no pheromone) shown as line with small dots. MAC min^{-1} = mean activity counts min^{-1}. Each point is the mean of 5 groups of 5 males. Standard errors of means less than 20%. (Hawkins and Rust, 1977.)

Males adapt or habituate with constant exposure to sex pheromone (Block and Bell, 1974; Hawkins and Rust, 1977), and become unresponsive after about 25 min. Given 30 s pulses, however, males continue to exhibit peak responsiveness every 8 min (Fig. 14.14). Recovery is, therefore, relatively fast.

The age of males and females determines when they begin to participate in sexual behaviour. Females of *P. americana* begin to secrete sex pheromone about 8 days after adult ecdysis (Hawkins and Rust, 1977) as a result of juvenile hormone secretion (Roufa and Barth, unpublished, cited in Barth and Lester, 1973). They continue to emit sex pheromone until they are mated (Wharton and Wharton, 1957). Juvenile hormone secretion, stimulated by feeding, is also necessary for sexual receptivity of females (Brousse-Gaury, 1978a,c). Behavioural responses of males to sex pheromone develop sequentially beginning at adult ecdysis (Silverman, 1977). *Rapid attennation* is induced by sex pheromone even on the day

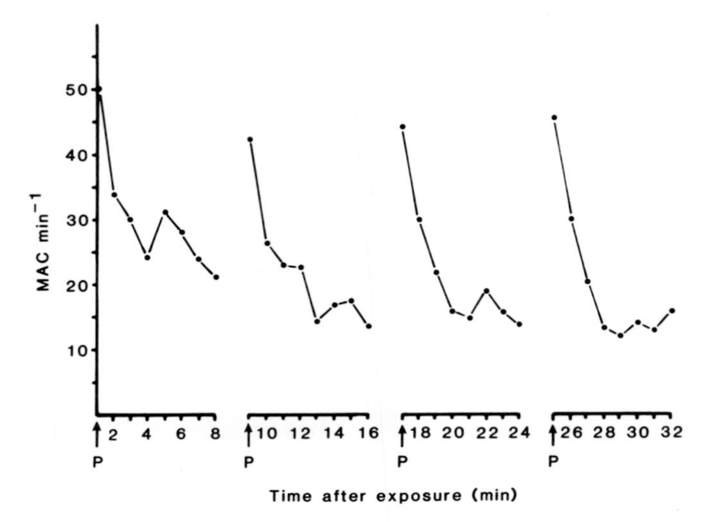

Fig. 14.14 Male *P. americana* locomotory response to four consecutive 20-s pulses of 10^{-3} sex pheromone extract. Symbols are the same as in Fig. 14.13. (Hawkins and Rust, 1977.)

after ecdysis, although other odours do not have this effect. On day three or four *erect body posture* and increased locomotory response were observed. *Wing-raising* and *abdominal extension* were not released until days eight and ten, respectively. Interestingly, the ontogenetic, temporally dependent, sequence observed by Silverman (1977) is the same sequence observed by Rust (1976) when he recorded responses of adults to step-wise increases in sex pheromone concentration. This suggests that the same 'sequencer' that operates during ontogeny, orders responses to different pheromone concentration in adult animals.

14.4 Agonism

Agonistic behaviour of *P. americana* was first described in detail by Bell and Sams (1973), although Barth (1970) discussed certain aspects of agonism during courtship and Wharton *et al.* (1968) provided an account of contact and agonistic display behaviour among nymphs. Bell *et al.* (1979) defined the individual acts and the probability of each during an encounter (Table 14.1) and the frequencies of two-act sequences or dyads (Fig. 14.15). Different agonistic acts are not 'selected' randomly by aggressive partners. For example, the animal that is challenged *kicks* more often than it *bites*. The animal initiating an encounter 'selects' certain acts more commonly than the responding animal and vice versa. *Approach* by the initiator often leads to *retreat* by the responder; however, the responder commonly employs *kicking* in defence. Some acts release more predictable responses than others. For example, *lunge* by the initiator elicits *retreat* in nearly all instances; on the other hand *stilt-posture* by either initiator or responder elicits any of the possible responses with nearly equal frequency. Bell *et al.* (1979), using information

Table 14.1 Agonistic acts of male *P. americana* and the probability of each act occurring during an encounter. Probability values, *p*, are designated for *Initiator* and *Responder* (I/R). (Adapted from Bell *et al.*, 1979.)

Agonistic act	*p (I/R)*	*Description*
Approach	—	Moving directly towards another individual, initiating an encounter
Does nothing	0.05/0.23	No behavioural act occurs (within the limitations of experimental observation)
Antennate	0.29/0.15	Contacting another individual with the antennae
Stilt-posture	0.05/0.14	Extending the legs and lifting the body off the substrate
Jerk	0.05/0.14	Forward/backward or sideways vibration of low frequency, performed while in stilt-posture
Climb	0.08/0.01	Mounting the abdomen of another individual
Bite	0.07/0.02	Using the mandibles against another individual
Kick	0.02/0.08	Thrusting a leg towards another individual
Lunge	0.05/0.02	Moving quickly forward towards another individual, terminating in contact
Wing-raise	0.01/0.00	Raising the wings and presenting the abdomen towards another individual
Retreat	0.32/0.21	Running or walking from the site of an encounter, terminating the encounter.

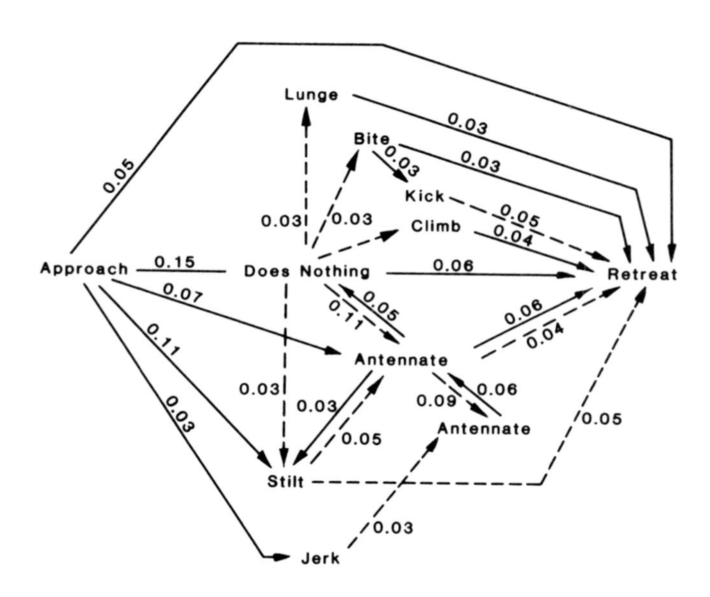

Fig. 14.15 Frequency of two-act sequences (dyads) in agonistic encounters of male *P. americana*. Dyads of relatively low frequency are not included. Solid lines are sequences progressing from the initiator of an encounter to the responder; dashed lines are the reverse. (Bell *et al.*, 1979.)

theory analysis, suggest that the act of *lunge* transmits more information than *stilt-posture*, since the *lunge* limits the kind of responses elicited to a greater extent than *stilt-posture*.

Zanforlin *et al.* (1973) and Deleporte (1974, 1978) are justly critical of conclusions suggesting that an aggressive act of one individual actually causes another individual to respond in a certain manner. Indeed there is no direct evidence as to exactly how individuals perceive an agonistic act such as *stilt-posture*, whereas tactile information is certainly perceived in acts such as *antennation* and substratum vibrations are probably perceived in other behavioural acts. Further analysis is required to dissect out the stimuli provided by one animal that affect the other.

Breed and Rasmussen (1980) extended the analysis of agonistic encounters among male *P. americana*. Four principal conclusions are drawn from their results. First, there is variability in the way in which encounters begin, and this heterogeneity is in part associated with differences in encounter length (number of acts). Second, the probability of escalation along the ranked behavioural scale (see below) during an encounter increases with the length of an encounter. Third, the length of encounters (i.e. number of acts) is non-randomly distributed, so that short and long encounters occur in less than expected frequency. Fourth, in short encounters, the initiator usually wins.

These results support the conclusion that the outcome of many encounters are decided non-randomly and early, probably because of the behavioural acts or other signals performed during the first dyad. These encounters are won by the initiator. Other encounters that have a different distribution of initial dyads and which include significantly more acts than the mean, are escalated more often by the non-initiator, and, if the non-initiator escalates first, are won significantly more frequently than expected by chance, by the responder.

Assignments of levels of aggressiveness that are thought to correspond to the likelihood of inflicting physical damage have been suggested for agonistic bouts of male *P. americana*. The classification of Bell and Sams (1973) is virtually identical to that of Simon and Barth (1977c), except that Bell and Sams (1973) subdivide the most intense level (4) into two levels according to duration of the encounter. Most (62%) male–male encounters are level 3, in which one animal attacks and the other defends or retreats. Fewer encounters are devoid of aggressive acts (level 1, 15%), or consist solely of threat and evasion (level 2, 14%), or consists of prolonged bouts of aggression (level 4, 6%) (Bell and Sams, 1973; Simon and Barth, 1977c). The classification system of Deleporte (1976) is more complicated than that described above. The dominance status of an agonistic participant is first determined by the type of acts displayed, i.e. (I) acts tending to separate individuals, either because one animal exhibits an aggressive act or in the absence of such acts; (II) acts that facilitate information exchange; and (III) acts that provoke a sudden increase in the intensity of the encounter. The sequences may terminate without exchanges or with exchange of non-aggressive acts (type a and b) or after aggressive acts by one or both participants (types c–g).

Both starvation and water deprivation enhance frequency of agonistic encounters (Bell, unpublished) and level of agonism (Simon and Barth, 1977c). When introduced to novel surroundings (e.g. a clean cage) the frequency of agonism increased significantly (Simon and Barth, 1977c). 'Foreign males' introduced to another colony were attacked by 'resident males', suggesting that colony odours are relevant for agonistic behaviour (Bell and Sams, 1973; Raisbeck, 1976). Female sex pheromone increased the frequency of encounters and the level of agonism among males (Bell and Sams, 1973). Lastly, crowding enhanced the incidence of agonism; agonistic encounters ceased when each animal had sufficient space to avoid contacting other individuals (Bell and Sams, 1973; Takagi, 1977). Cannibalism among crowded nymphs was reported by Wharton *et al.* (1967), although wounded or ecdysing individuals were primarily susceptible.

Social hierarchies are known to occur among individuals in groups of some cockroach species. *N. cinerea*, *Eublaberus posticus* and *B. craniifer*, for example, maintain fairly stable dominance hierarchies among males through ritualized dominance and submissive behaviour (Ewing, 1972, 1973; Gautier, 1976, 1979; Bell and Gorton, 1978; Bell *et al.*, 1978a, 1979; Breed *et al.*, 1980). Whether or not *P. americana* establishes dominance hierarchies is a controversial issue. Deleporte (1974, 1976, 1978) observed that stable, linear hierarchies form within 10 days after a group of males is placed together, whereas Bell and Sams (1973), Zanforlin *et al.* (1973) and Breed and Rasmussen (unpublished) found only transient episodes of dominance behaviour and hierarchy formation. Experiments performed by Deleporte (1974, 1978) were carried out over longer time periods (months) than those of Bell and Sams (1973) (days), and since Deleporte (1976) observed *P. americana* agonism in 'natural' habitats, e.g. caves in Trinidad, as well as in the laboratory, he is in a better position to assess hierarchy formation in this species. Indeed, Deleporte (1978) contends that even first-instar nymphs exhibit agonistic behaviour, can be separated into dominant and submissive individuals and that dominant nymphs often become dominant adults. Olomon *et al.* (1976) found that *P. americana* nymphs are not as aggressive as adults and that the level of aggression increases gradually from first to last instars to adults.

A very basic question that has intrigued cockroach behaviourists is why do male cockroaches fight? More specifically, what resources are they fighting for? Hierarchies, or simply dominance behaviour, could facilitate the relatively more effective exploration of the home range and/or territory by dominant individuals, e.g. more effective access to food, mates or shelters. Agonistic encounters among *P. americana* males have never been observed in the context of mate protection or competition for mates, although such behaviour may occur under particular circumstances, and has been reported for other cockroach species (e.g. *Cryptocercus punctualatus* – Ritter, 1964). Obviously, a paucity of field data hampers our ability to understand how agonistic behaviour operates in nature. Simon and Barth (1977c) suggest that agonistic behaviour simply serves to increase the distance between individuals, while other kinds of behaviours serve to decrease this distance. That cockroaches require 'individual space' is implied by observations of movement and agonistic behaviour of *P. americana* males under crowded conditions until contact between individuals is minimal (Bell and Sams, 1973). It seems likely that dominance/subordinance behaviour is a primitive character in *P. americana*, and is much more involved to include ritualized behaviour in cockroaches such as *N. cinerea*. If this is true then the agonistic behaviour and hierarchy formation of *P. americana* males placed in certain experimental situations may serve as an excellent model for the transitional evolutionary stage between the more primitive and more advanced levels of behavioural interactions for cockroach species.

14.5 Aggregation

To form an aggregate, animals must locate one another or a common site through orientation and then must remain together. *P. americana* aggregates during the light phase of the photocycle when individuals are inactive. Sites for aggregation are usually concealed, dark and offer thigmotactic stimuli. An aggregation pheromone is secreted in the faeces of several species in the Blattidae and Blattellidae (Ishii and Kuwahara, 1967, 1968; Ishii, 1970; Bell *et al.*, 1972, 1973; Burk and Bell, 1973; Roth and Cohen, 1973; Sommer, 1974, 1975; Metzger and Trier, 1975). Ishii and Kuwahara (1967) suggest that the aggregation pheromone of *B. germanica* is produced in rectal pad cells and passed with the faeces. *B. craniifer*, and probably other blaberids, secrete an aggregation pheromone from the mandibular glands (Brossut, 1970; Brossut *et al.*, 1974). Only in *B. craniifer* has an aggregation pheromone been characterized and tested as a synthetic (Brossut *et al.*, 1974).

P. americana adults and nymphs are attracted to filter papers contaminated with faeces (Bell *et al.*, 1972; Roth and Cohen, 1973), and nymphs stop when they encounter a contaminated paper (Burk and Bell, 1973). Interactions among individuals are also important in promoting aggregations. A nymph is more likely to stop on a faeces-contaminated filter paper if one or more other nymphs are already in residence (Burk, unpublished). Wharton *et al.* (1968) describe mutual antennation and agonistic acts prior to 'huddling together'. Presumably these behaviours provide nymphs with information about the conspecificity of other individuals. Bell *et al.* (1972) and Roth and Cohen (1973) investigated the species specificity of aggregation pheromones. Both studies provided choices of contaminated papers of three or more species and then recorded the distributions of nymphs on these papers. Both studies conclude that aggregation pheromones from faeces of closely related species are not strictly species-specific, although individuals of a species prefer the conspecific odour.

Different species occupy separate resting places in close approximation to each other (Ledoux, 1945; Lederer, 1952), although some species are repelled by the odours of others (Roth and Cohen, 1973). Further investigation is needed to determine the role of aggregation pheromones as interspecific isolation agents.

Aggregation has been reported to accelerate growth of nymphs of *B. germanica*

(Izutsu *et al.*, 1970) and *P. americana* (Wharton *et al.*, 1968) and to enhance the ability of nymphs to learn simple tasks (Gates and Allee, 1933). Aggregation pheromones probably function to bring individuals of both sexes together, especially in species that lack volatile sex pheromones (Brossut and Roth, 1977; Brossut *et al.*, 1975).

14.6 Learning

Cockroaches have been subjects in two main types of learning experiments. Their ability to learn a maze has been shown using both positive (Gates and Allee, 1933; Chauvin, 1947) and negative (Szymanski, 1912; Turner, 1912; Eldering, 1919; Hunter, 1932) reinforcement. *P. americana* also learned to avoid electric shocks by choosing correct left or right turns into a light box instead of a normally selected dark box (Eldering, 1919). The number of errors was reduced over a period of 7 days to a level of 10% of trials conducted. Retention spans of about 2 days were deemed to be poor (Eldering, 1919; Hunter, 1932), however, as compared to animals such as rodents.

Lovell and Eisenstein (1973) trained *P. americana* to avoid the dark side of a box by using an electric shock as a negative reinforcement. Most learning occurred within the first minute of training. Retention lasted at least two hours after training. When CO_2 (known to cause amnesia – Freckleton and Wahlsten, 1968) was administered immediately after training, no retention was observed two hours later. When CO_2 was given one hour after training some retention was observed two hours after training. These results indicate that memory phases with different susceptibilities to disruption, occur in cockroaches as in other animals.

Extensive studies of Chauvin (1947) with *B. germanica* led to the thesis that cockroaches are unable to generalize from one learning task to another. That cockroaches have relatively poor learning retention and cannot generalize from one learning task to another seems in one respect to be a deficiency, relegating the cockroach to a lesser stature than the rat. When viewed simply as one tool for survival, however, learning may have little advantage over the mechanisms employed by cockroaches in mate and food finding and in predator avoidance. In fact, since learning for a cockroach requires practice trials, the advantage gained through learning must be weighed against potential risks incurred during the learning process. The plasticity of cockroaches is of interest when viewed in perspective with rats. If the rat is trained to turn right in a T-maze, and then is trained to go left, it typically makes many more errors on the reversal problem than in the original testing. Longo (1964) trained cockroaches in a Y-maze, with 10 trials per day. Some were reversed daily, others at the end of 4 days; the latter made fewer errors daily than the former, but neither group showed the drastic increase in errors after the first reversal typical of rats or fish.

Given that cockroaches are poor learners as compared to bees or mammals, it is ironic that they have provided an excellent system to investigate learning at the level of a single ganglion. Horridge (1962) demonstrated that the headless cockroach is capable of learning to avoid electric shocks by holding its leg above a certain level, below which the leg received shocks. The technique is potentially a convenient one for studying acquisition and retention of information as well as for the investigation of the neural mechanisms involved.

Eisenstein and Cohen (1965) went one step further in testing the ability discovered by Horridge (1962). They used a preparation consisting essentially of one ganglion isolated from the rest of the CNS. Learning by a single ganglion was later confirmed by Pritchatt (1968) and Aranda and Luco (1969). These studies show that an association between the position of a leg and the electric shocks, resulting in an avoidance reaction, can occur without the head and without other portions of the CNS.

Chen *et al*. (1970) used intact animals, headless animals or isolated ganglia of *P. americana* in the kind of experiment outlined above to determine if the head or other ganglia influence learning or retention. They demonstrated that intact cockroaches learnt to avoid shocks in about ten sessions of ten trials per session. Retention of the learned behaviour, up to seven days, was not affected by severance of the head nor even by the isolation of the single ganglion that learned the task. Headless animals and isolated ganglia, by contrast, never reached 100% avoidance, as did intact cockroaches. Moreover, retention was less than one day in duration for isolated ganglia and less than three days for headless animals. The authors concluded that to establish long-term memory an intact nervous system is required, but once accomplished the acquired learning can be retained by the lower portions of the CNS.

Recently, Willner (1978) showed that speed of learning the leg-lifting response is dependent on task difficulty, as defined by the height above the saline at which the leg is held at the start of training. He suggests that differences in height might explain certain discrepancies in the literature. For example, time to achieve asymptotic performance has been reported at 1.5 min (Disterhoft, 1972) to 1 h (Horridge, 1962). Similarly, headless animals were found to learn better than intact animals by Eisenstein (1967), Kerkut *et al*. (1970) and Rick *et al*. (1972), but not by Horridge (1965). Such differences in results might be expected if headless animals hold the leg lower than intact animals. Willner's (1978) leg-lifting experiments also challenge the idea of retention of learned behaviour in headless cockroaches. His preparation did not remember the correct avoidance response for as long as 15 min, suggesting that the headless cockroach is unlikely to be of value in elucidating the physiological bases of memory.

Plasticity of a behavioural pattern is illustrated by experiments on grooming (Luco and Aranda, 1964). The forelimb of *B. orientalis* normally guides labial grooming of the contralateral antenna, but when the forelimb is amputated the cockroach immediately switches to using the remaining foreleg. When both forelegs are removed, the cockroach learns to use its middle leg for grooming. Camhi, Moran and Eisner (unpublished) showed that *P. americana* grooms the abdomen with an ipsilateral hindleg; when the leg is amputated the movements are performed by the ipsilateral middle leg. Loci on the abdomen too far posteriorly for the middle to reach are groomed by the contralateral hind leg. Thus the grooming field for the hindleg is fractionated when that leg is lost.

14.7 Conclusion

Our understanding of the behaviour of *P. americana* is greatly enhanced by excellent studies of sense organ functions (Chapter 9), locomotion and muscle actions (Chapter 11), the CNS (Chapter 8) and rhythms (Chapter 10). Hormonal influences on the behaviour of *P. americana* or other cockroach species are not complete and in some cases not understood at all (e.g. agonism). Studies of cockroach behaviour suffer mainly, however, from a paucity of ecological data. It is especially ironic that few eco-behavioural correlations can be drawn for *P. americana*, a species that for centuries has lived in close association with man. There are few organisms that could be the subject of a book without a chapter on ecology.

Acknowledgements

The author appreciates the suggestions and comments of Drs L. M. Roth, R. Jander, C. D. Michener, G. Seelinger and S. Jones. Studies from my laboratory were conducted in association with former students, Drs M. K. Rust, J. M. Silverman, M. D. Breed, M. K. Tourtellot, E. Block, T. Tobin, W. A. Hawkins and C. Schal; to these colleagues the author is deeply indebted. Supported in part by grants from the National Science Foundation (BNS 75-21019 and BNS 77-24898) and the National Institutes of Health (NS 13798 and K04-NS00190).

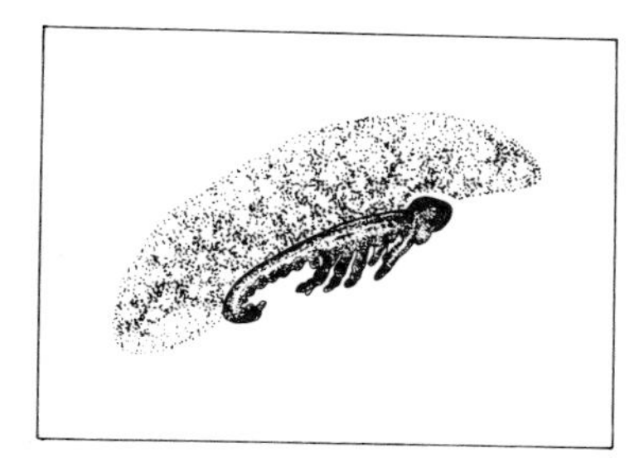

15

Embryonic and post-embryonic development

Robert R. Provine

15.1 Introduction

The modern era of insect developmental study began almost a century ago with the outstanding descriptive work of Wheeler (1889, 1893) in the United States and Heymons (1895) in Germany. It is ironic that the developmental analysis of cockroaches, among the most studied insects, has advanced so little since. This bleak state of affairs is reflected in the absence of a section on embryogenesis in a book on the biology of the cockroach which was published only 13 years ago (Guthrie and Tindall, 1968). Important research has been done on fertilization, post-embryonic development and the life cycle (Chapter 1), the development of specific subsystems and on related topics such as reproduction (Chapter 13) and regeneration (Chapter 16). What is missing are detailed descriptions of normal embryonic development and experimental analyses which reveal the mechanisms of the developmental process.

15.2 Embryogenesis

The present account of cockroach embryonic development is necessarily general because it is a synthesis of data from various related orthopteran and hemimetabolous species. The scholarly and comprehensive reviews by Johannsen and Butt (1941) and Anderson (1972a,b, 1973) were useful in preparing this description as were the original research papers of Wheeler (1889) on *Blattella germanica* and Heymons (1895) on *Periplaneta americana*.

The egg of *P. americana* is very rich in yolk as are the eggs of other Orthoptera. The egg is bounded by two membranes, a thin, inner vitelline membrane and a thicker, outer chorion. Improved views of embryos may be obtained by clearing the opaque chorion by immersing the egg in a 3% hypochlorite solution (Bentley *et al.*, 1979).

Some openings in the chorion serve as micropyles which are apertures through which sperm enter the egg; others are involved in respiratory processes. The zygote nucleus is embedded in a small island of cytoplasm in the yolk of the eggs' central region. The zygote nucleus undergoes successive mitotic divisions producing cleavage energids. The energids later contribute to the embryo, embryonic membranes and yolk cells (Anderson, 1972a,b). The energids initially migrate to the surface of the posterior pole of the egg where they initiate blastoderm formation (Fig. 15.1a). Other energids stay behind forming primary vitellophages. These are nuclei with surrounding cytoplasm which become involved in yolk digestion. Other vitellophages, termed secondary vitellophages, later migrate inward from the proliferating surface energids to join the primary vitellophages (Heymons, 1895).

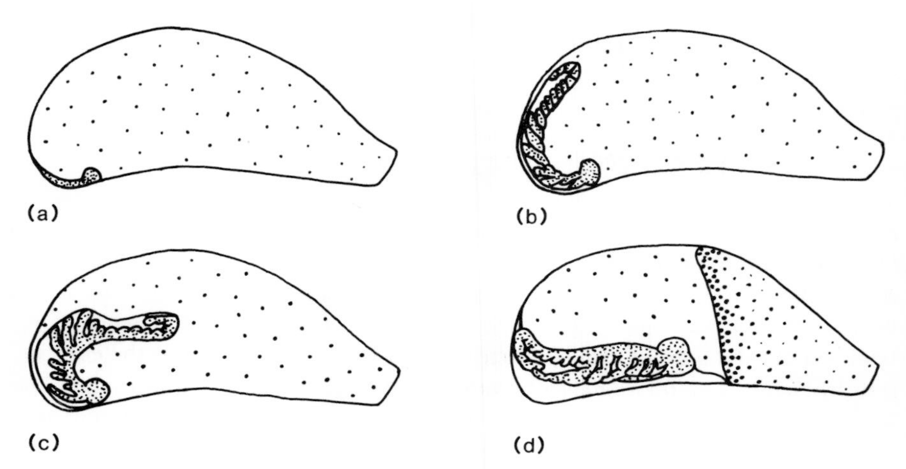

Fig. 15.1 Development of body segmentation and shifts in body position of the *P. americana* embryo during early incubation stages. (After Heymons, 1895.)

The blastoderm consists of a tightly packed monolayer of cells on the ventral surface of the yolk mass which gives rise to the embryonic primordium and a less dense, surrounding, extra-embryonic ectoderm which produces the extra-embryonic membranes. When the embryonic primordium starts to increase in length, but before the onset of external segmentation, it is called the germ band (Fig. 15.1a). Discrete body regions are later blocked out, forming a segmented germ band (Fig. 15.1b). Eventually, the lateral edges of the embryonic ectoderm extend laterally and dorsally around the periphery of the yolk mass, and meet and fuse at the dorsal surface of the egg, an event termed dorsal closure. The position of the embryo within the egg changes during development as a result of two types of movement that Wheeler (1893) collectively termed blastokinesis. The first type of movement is anatrepsis, the emersion of the embryo in the yolk (Fig. 15.1c). The second is katatrepsis, the emergence of the embryo from the yolk (Fig. 15.1d). As a result of katatrepsis, the embryo assumes the approximate position within the egg which will be maintained until hatching.

Gastrulation is the complex process through which cells from the monolayered embryonic primordium immigrate inwards and form an inner cellular layer. The details of this process are not well understood in any insect. The inward migration (infolding) of cells occurs during the germ band stage and produces the gastral groove which runs along the ventral midline of the embryonic primordium. The infolded cells produce an inner embryonic layer, the hypoblast (entomesoderm) which gives rise primarily to mesoderm and secondarily to entoderm. Entodermal structures are thought to have multiple origins, although the details of their genesis are open to question. The gastral groove disappears when adjacent ectodermal cells, which form the shoulders of the groove, move together joining at the midline.

The central nervous system (CNS) has its origin in neuroblasts which differentiate in the ventral ectoderm. The proliferation of neuroblasts produces a neural ridge on each side of a mid-ventral neural groove. The division of neuroblasts also produces rows of preganglion cells (ganglion mother cells) which divide and differentiate into the neurons which constitute the ganglia. Older cells are displaced inward by the most recently generated cells. Thus, a column of daughter cells is produced with the oldest cells at the head of the column farthest from the neuroblast which maintains its initial peripheral position. As the cellular population increases, the rows of daughter cells (neurons) on either side of the neural groove merge. The segmentation of the ganglia along the embryonic rostrocaudal axis may be the result of decreased intersegmental neuronal proliferation. Glia are probably produced by the preganglion cells which produce neurons.

The cells which will constitute the ganglia move to the interior of the embryo. There they produce the neurites which form the neuropile. Gradually, the adult ganglion configuration is assumed which is a cortex of cell bodies surrounding a central neuropile (Cohen and Jacklet, 1967). Later, intersegmental ganglionic connectives and intraganglionic commissural processes are established. An analysis of interganglionic connective formation *in vitro* is presented in Section 15.7. For a recent account of brain development in *P. americana*, consult Malzacher (1968). Schafer and Sanchez (1973) provide a description of the post-embryonic development and morphology of the antennal sensory organs. Contemporary research in the general area of insect neurogenesis is considered in several publications (Edwards, 1969; Young, 1973; Bate, 1976a,b; Goodman *et al.*, 1979).

15.2.1 Determination and regulation

The phylogenetically primitive insects including the Orthoptera are thought to possess considerable regulative capacities in contrast to the more advanced forms such as the higher Diptera and Lepidoptera which have determined, mosaic eggs. Regulative capacity is evaluated by destroying selected anlagen or separating regions of the egg and observing the extent to which the missing parts influence subsequent embryogenesis. Highly regulative eggs can replace the missing parts while determined 'mosaic' eggs cannot. Therefore, the cell fates in mosaic eggs are irreversibly determined at the time of the experiment while the cells in regulating eggs are not because the cells alter their fate to compensate for the missing parts. The regulative capacity of all eggs and embryos diminishes with age and the regions capable of self-regulation shrink in size.

The regulative capacity of embryonic cockroaches has not been experimentally tested but, in the following discussion, it is probably much more similar to that of the regulative dragonfly than the phylogenetically more distant mosaic fruit fly. For more comprehensive discussions of regulation in insect eggs and phylogenetic comparisons, refer to Bodenstein (1953a,b, 1955a), Agrell (1964), Kume and Dan (1968), Counce (1973) and Anderson (1972a,b).

Eggs from the dragonfly (*Platycnemis penipes*) have great regulatory powers. For example, smaller than normal but complete twin embryos can be experimentally produced by dividing the germ band longitudinally during early cleavage (Seidel, 1936). In contrast, the determination of presumptive embryonic parts of dipterans (e.g. *Drosophila*) is complete at the time of fertilization (Seidel *et al.*, 1940). In mosaic eggs of the latter type, body regions seem to be mapped out in the egg's cortical plasma (periplasm) before the nucleus has begun to divide. The fate of the cleavage energids seems to be determined by the regional nature of the periplasm, because elimination of some of the migrating energids by ultraviolet irradiation produces no structural deficits in the embryo; other energids replace them (Seidel, 1932). This suggests that before the energids reach the cortical region, the periplasm is chemically differentiated and the arriving energids are specified by this influence. However, in the eggs of hemimetabolous insects such as the dragonfly, determination of the periplasm does not occur until after the arrival of the energids. This is consistent with the finding that ligation of an anterior egg region results in an intact but smaller than normal embryo in the residual posterior egg region (Seidel, 1936). Recent discussions of these phenomena are given in reviews by Counce (1973) and Sander (1976). The considerable research concerning the determinative, regulative and inductive influences on regeneration are reviewed in Chapter 16.

15.2.2 Induction

Induction is the process through which cytoplasmic differentiation is initiated in material which has received no previous information as to fate. Much of our knowledge of embryonic induction in insects comes from experiments by Bock (Seidel *et al.*, 1940; Bock, 1941) on the *Chrysopa perla* egg. The experiments were performed at the stage of germ-layer formation. If the whole of the presumptive hypoblast (entomesoderm) was removed, the ectodermal derivatives of the mesoderm-free embryo differentiated in a nearly normal manner. This suggests that the ectoderm has the capacity for self-differentiation. If only part of the hypoblast was destroyed, the residual mesoderm moved beneath the ectoderm and formed nearly normal mesodermal derivative organs. However, if the mesoderm moved into a region lacking ectoderm, it degenerated. This result suggests that the ectoderm induces differentiation of the underlying mesoderm. In vertebrates, the induction acts in the opposite direction with the mesoderm inducing ectodermal differentiation.

15.3 The oötheca

The oötheca, the communal egg case, resembles a small reddish-brown bean with a notched ridge (the crista or keel) on one side (Fig. 15.2a,b). The hard oötheca provides the embryo with protection from mechanical trauma, moisture damage, desiccation and many types of microbial invasion. The oötheca contains sufficient water to permit the eggs to develop in the absence of additional water from the environment (see Chapter 1). It is made of quinone-tanned protein (Brunet, 1951) and contains no chitin (Campbell, 1929) (see Section 2.4 for details). Within the oötheca, 12–16 translucent white embryos (Fig. 15.2d) are grouped in two rows of 6–8, with the embryos in each row lined up side by side. The interior of an oötheca which was opened at a late incubation stage is shown in Fig. 15.2c. This preparation is termed 'cockroaches on the half-shell'. In the intact oötheca, the embryos of both rows face inward towards each other. Each embryo is ensheathed in an individual chorionic membrane. The heads of all embryos are oriented towards the notched edge of the oötheca which contains vents through which the embryos probably respire (Wigglesworth and Beament, 1950).

During hatching, all embryos simultaneously pry open the clam-like oötheca along a seam which runs along the midline of the notched edge and emerge head-first (Fig. 15.3a,b,c). Hatching and eclosion, the shedding of the embryonic cuticle, occur simultaneously using identical movements which are described later in Section 15.6. The oötheca has a 'spring-loaded' hinge, the resistance of which must be overcome by the embryos. The resistance to opening has at least two important consequences. If insufficient force is generated by the embryos attempting to hatch, due to the death, lack of synchronized hatching or weakness of littermates, all embryos will be trapped in the oötheca and die. (Mechanisms involved in the synchronization of hatching and eclosion are considered in Section 15.6.) A second consequence of the resistance to opening is that hatching embryos scrape off their embryonic cuticles on the mouth of the oötheca when exiting. This occurrence is apparently adaptive because higher rates of incomplete eclosion are observed in embryos reared *ex ovo* or removed from the oötheca immediately after the initiation of eclosion movements. Fig. 15.3d shows shed cuticles in the mouth of an oötheca which returned to the natural, closed position after the embryos hatched.

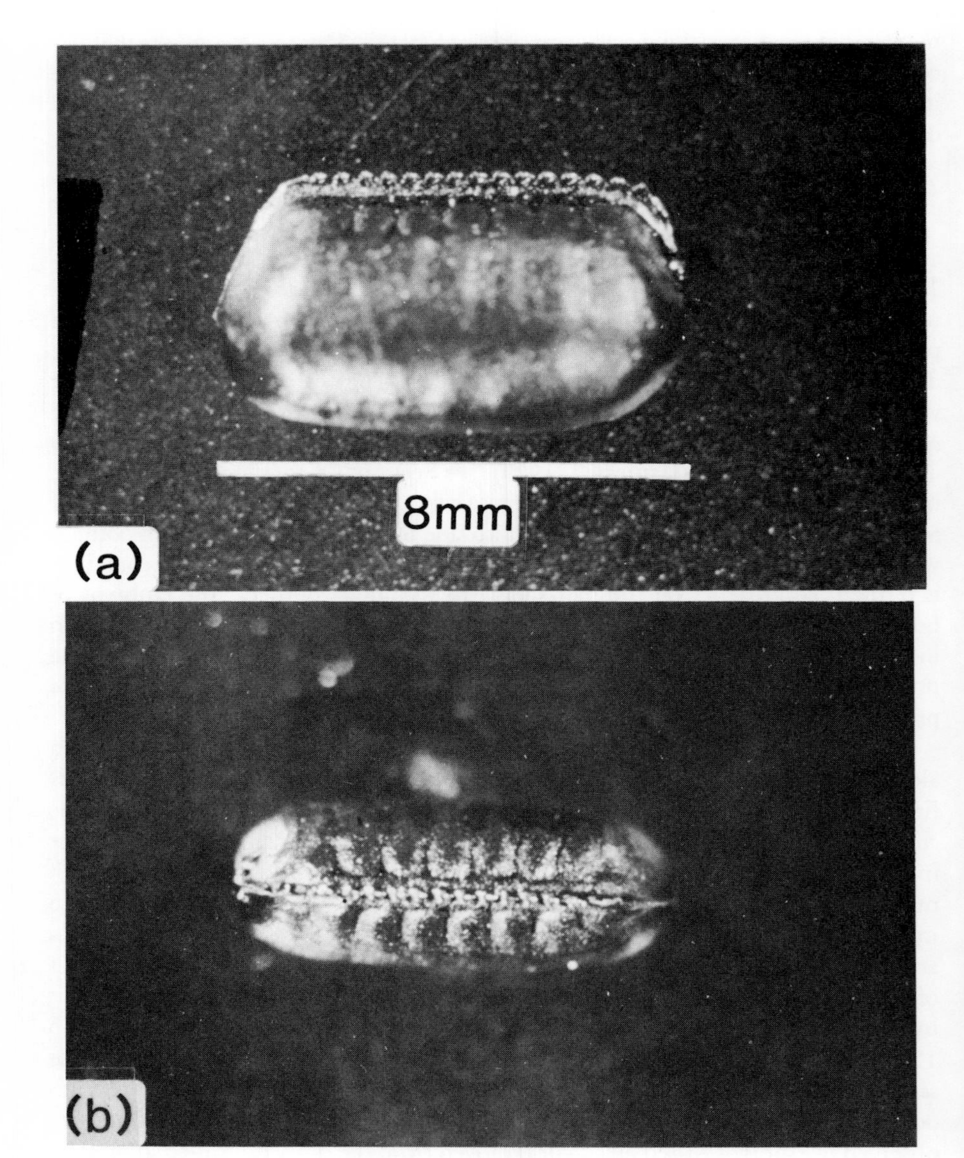
8mm
(a)
(b)

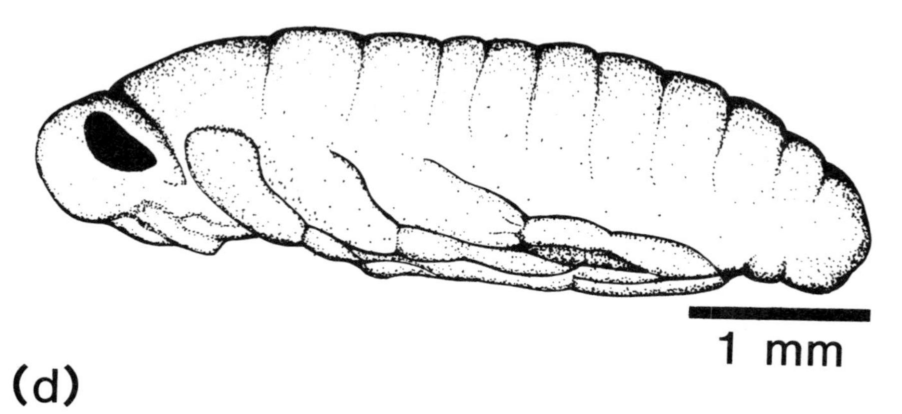

Fig. 15.2 Oötheca of *P. americana*. (a) Intact oötheca lying on side which shows the notched ridges (cristae) on the dorsal edge; (b) Oötheca viewed from the dorsal edge; (c) Oötheca in (b) after spreading open to reveal embryos whose heads are oriented towards the top and bottom of figure; (d) A sketch of an embryo at a late incubation stage which is reproduced from Provine, 1977, by permission of Pergamon Press.

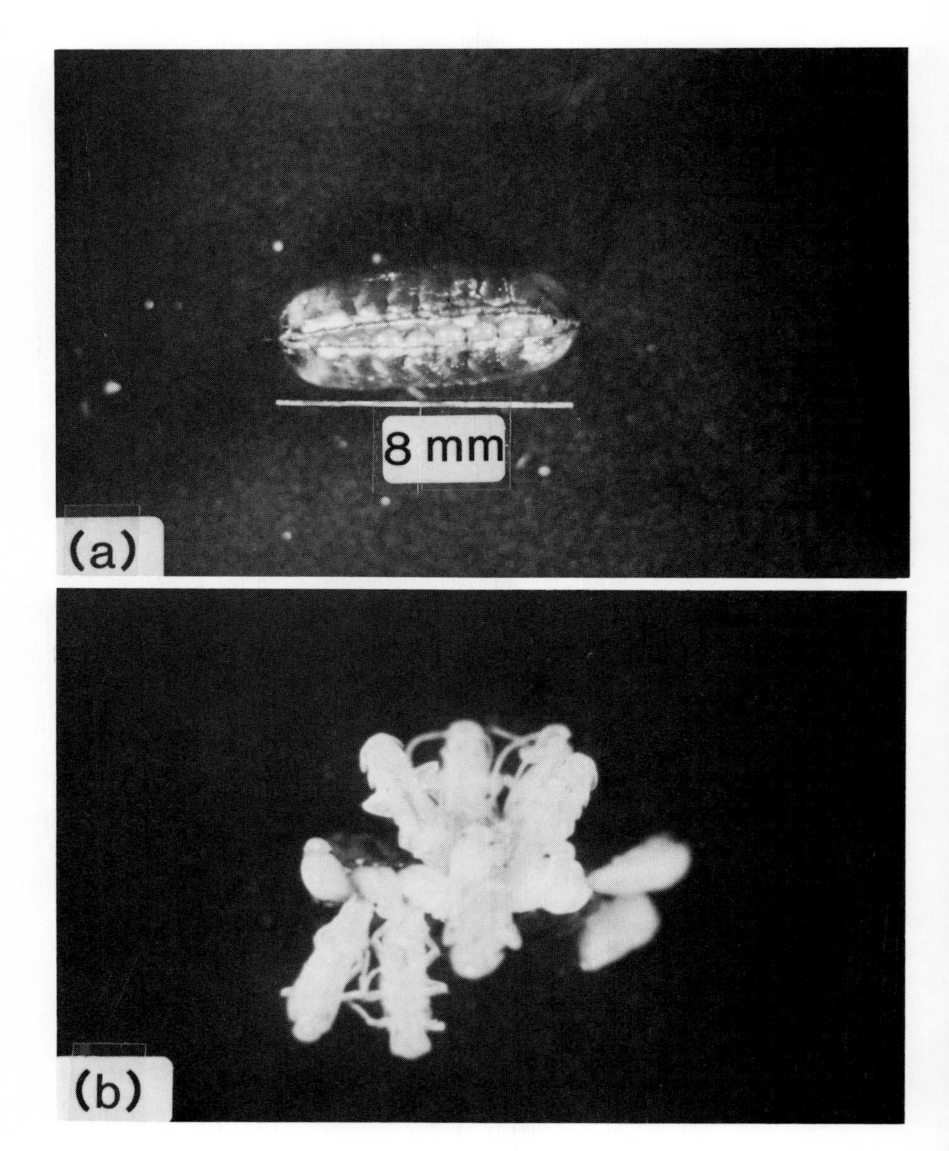
8 mm
(a)
(b)

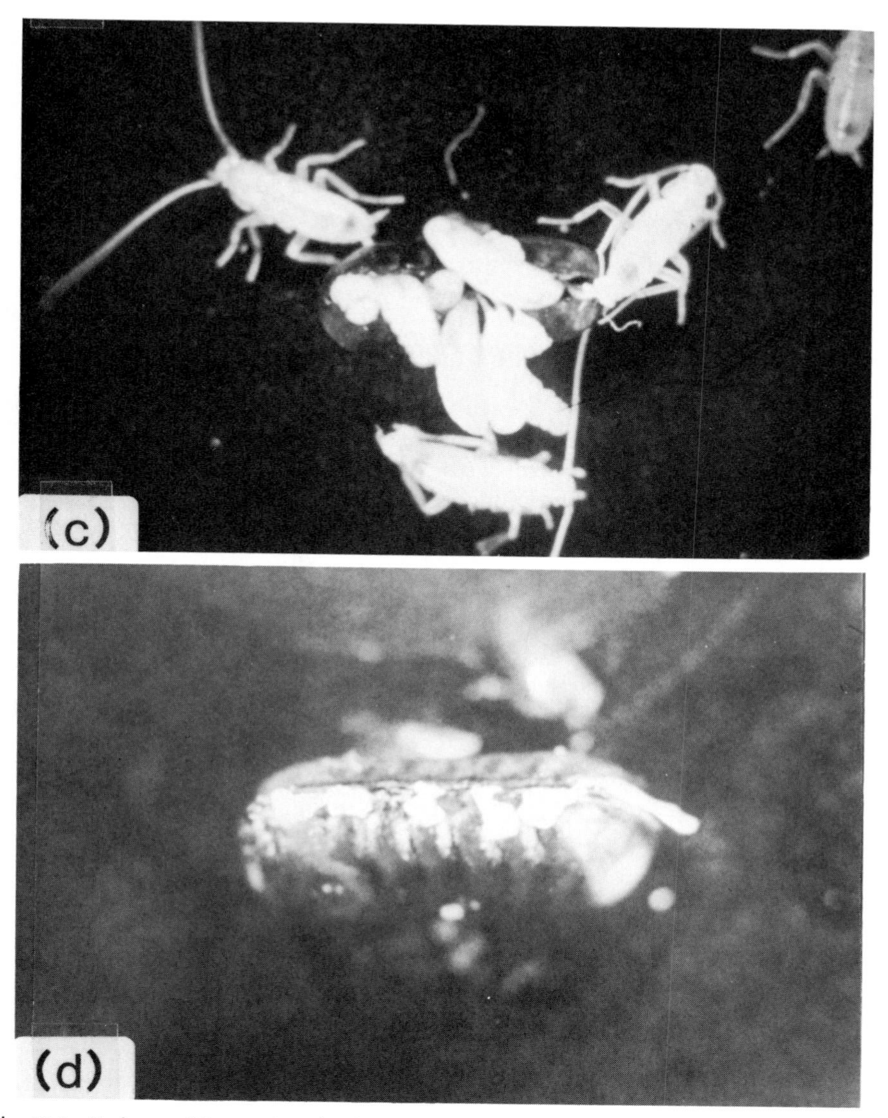

Fig. 15.3 Embryos of *P. americana* hatching from an oötheca. (a) The heads of embryos are shown emerging from an oötheca in an early hatching stage; (b) Embryos are surging head-first out of the oötheca and shedding the embryonic cuticle which frees the legs; (c) Some embryos have hatched and eclosed and are running away from the oötheca; (d) Shed embryonic cuticles are in the mouth of the oötheca which closed after escape of the embryos.

The extent to which the complex environment of the oötheca influences embryonic development is not well understood. However, the *ex ovo* preparation described below provides a useful preparation for analysis of the problem. If precautions are taken against microbial infection and desiccation, embryos removed from the oötheca during the last third of incubation and stripped of their chorions may be successfully incubated through eclosion and 'hatching'. This demonstrates that the oötheca and the chorion do not provide unique conditions for normal development during the last part of the embryonic period.

15.4 Culture of intact embryos *ex ovo*

A major recommendation for *P. americana* as a subject for developmental analysis is the ease with which embryos can be removed from the oötheca and chorion and incubated *ex ovo* during the last third of the incubation period (Provine, 1976a, 1977). The capacity of embryos to survive extraction from the oötheca and the removal of the chorion enables investigators to evaluate the relative impact of various intrinsic and extrinsic factors on development. This attribute is exploited in an analysis of behaviour development (Section 15.5), eclosion and hatching (Section 15.6). Since individual embryos can be incubated *ex ovo*, investigators can observe or experimentally manipulate the embryo with relative ease. The *ex ovo* incubation procedures do not seriously disrupt normal development as indicated by the performance of successful eclosion – 'hatching' movements by embryos at the usual time. This would not be true of viviparous cockroaches such as *Diploptera punctata* which require nourishment from the mother.

The initial step in embryo removal is the sterilization of the oötheca by sequentially dipping it in solutions of 2% iodine, 95% ethyl alcohol, 95% ethyl alcohol and finally in sterile distilled water. The oötheca is pryed open with sterile forceps forming the 'half-shell' configuration as in Fig. 15.2c. Individual embryos still encased in their chorions are placed on the sterile stage of a dissecting microscope under which the chorions are removed. Successful development occurs most often if embryos are removed from the oötheca during the last third of the 30-day incubation period at 29°C. Sixteen to 18-day embryos have been incubated *ex ovo* but younger embryos are like fragile yolk-filled bags with weak body walls that flatten out and often rupture when stripped of the supporting chorion.

Embryos removed from the oötheca are placed on the moistened filter paper floor of a Petri dish. Organ culture dishes (Falcon Plastics, Style 3010) which have absorbent paper rings at their bottoms and a central well are especially convenient for *ex ovo* incubation. The embryos are kept moist by drops of sterile distilled water which are periodically added to the paper ring and the central well. The culture dishes are covered with their plastic lids and placed in sealed desiccator jars, the bottoms of which are filled with wet cotton. The jars are kept in a darkened incubator which is maintained at 29°C.

15.5 Embryonic behaviour development

Behavioural analyses of *P. americana* embryos are facilitated because individual embryos can be removed from the opaque oötheca and chorion and incubated *ex ovo* in the simplified and controlled environment of the culture dish. The *ex ovo* technique permits normally developing embryos to be observed continuously during the last third of incubation through the clear plastic lids of the dishes. Most of the following observations are from Provine (1976a, 1977) who used the *ex ovo* preparation.

During the second half of the 30-day incubation period at 29°C, the embryo performs spontaneous movements which are clearly the product of muscular contractions. The movements range from barely perceptible twitches of 16–17-day embryos to the relatively powerful hatching and eclosion movements which occur around day 30. The behavioural repertoire of the embryo is restricted because its body surface is soft and flexible, providing little exoskeletal support. Tanning and hardening of the cuticle do not occur until after hatching and eclosion. Until then, the legs and other body parts are capable of little physical displacement. Muscle contractions produce a twitch or depression in the cuticular surface, probably at the point where the muscle is anchored. During the last half of embryonic development, body proportions remain stable, the body segmentation is clearly defined and the legs, antennae and cerci are differentiated and lie folded beneath the body (Fig. 15.2d). Body bristles appear at 26–27 days beneath the embryonic cuticle, but remain pinned in a horizontal position until they are freed during eclosion and hatching. The first-instar nymph which emerges after hatching has the general appearance and behaviour of a miniature adult except that it has no wings.

The earliest spontaneous movements are isolated, low-amplitude, slow depressions in the cuticle (*dimpling*) which most often occur in the upper leg region of the prothoracic segment of 16–17-day embryos. Movements may be present even earlier but observations have not been made because of embryonic fragility. (Minute movements of the cardiovascular apparatus are not considered in this behavioural analysis.) At 18 days, dimpling is present in the upper region of all legs, the lateral thoracic wall and less often in the mid-abdominal segments and head. The variety, frequency (Fig. 15.4) and amplitude of movements increase with age as does the number of body parts showing activity. By 20 days, occasional, weak, synchronous dorso-ventral *flattening* movements of the abdominal segments (they resemble postnatal ventilation movements) are present as are rapid, localized *twitches* in many body regions. The flattening movements are the first clear evidence of multi-segmental movement co-ordination. The flattenings are more vigorous and frequent by day 22 when they are usually performed in series having regular intermovement intervals. The interval between flattening movements remains essentially unchanged during embryonic development, being 6.4 ± 1.5 s at 22 days and 5.3 ± 0.7 s at 29 days, one day before hatching-eclosion. Another type of multisegmental movement, the wave, appears at 22 days. Waves are ripples of muscle contraction which pass smoothly from segment to segment through part

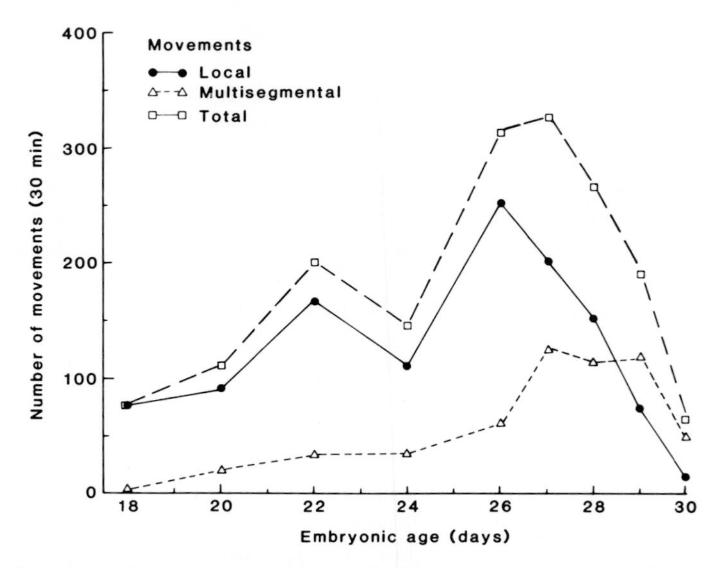

Fig. 15.4 Frequency of spontaneous movements of 18-to-30 day *P. americana* embryos during a 30 min observation period. Local (segmental) movements are isolated muscle contractions within a single segment. Multisegmental movements are co-ordinated contractions involving several segments. These include simultaneous dorsal-ventral abdominal flattening movements and peristaltic waves. Total movements are the summed local and multisegmental movements at a given stage. (Reproduced from Provine, 1977; by permission of Pergamon Press.)

or all of the body. They may begin in thoracic or abdominal segments and sweep rostrally or caudally or begin in mid-body regions and sweep in both directions. Occasionally, waves are simultaneously initiated in different body parts and collide in intermediate regions. By 24 days, waves, flattening, dimpling and twitching movements are joined by *shudders* which are violent twitching movements that occur simultaneously in all body segments. Between 24 and 29 days, the proportion of multisegmental relative to local movements increases (Fig. 15.4). Also, during this interval multisegmental movements become more vigorous, show more precise intersegmental synchronization and are more likely to be performed in series. Very few movements are observed on days 29 to 30 (Fig. 15.4) except for flattenings and occasional *eclosion* movements. Eclosion movements are vigorous, caudal-to-rostral peristaltic waves and ventral thoracic flexures which usually occur in series with regular intermovement intervals of 5–6 s. Early appearing eclosion movements are less vigorous and not repeated to the extent of the movements of the actual eclosion-hatching sequence.

The continuity between many embryonic and postnatal movements is difficult to establish because equivalent patterns of neuromuscular events would look different before and after hatching. Also, embryonic behaviour has not been shown to be neurogenic, the product of neuromuscular processes. However, multisegmental movements such as flattening and eclosion which have postnatal equivalents and require intersegmental co-ordination are almost certainly neurogenic.

The dorsal-ventral abdominal flattening is probably an early appearing ventilation movement (Miller, 1966; Chapter 5), although its participation in gas exchange has not been demonstrated at embryonic stages. An informative and simple experiment would be to challenge *ex ovo* embryos with heightened levels of atmospheric CO_2; an increased frequency of flattening movements would suggest the presence of respiratory control mechanisms similar to those of the adult. If the flattening movements are shown to have a respiratory function, they would be the only movements other than eclosion-hatching that have definite adaptive significance for the embryo. Other movements may be epiphenomena of neuromuscular development. The eclosion-hatching movements have the distinction of being periodically recapitulated postnatally on the occasion of each nymphal moult.

The determination of the postnatal equivalents of various cuticle depressions and twitches which constitute much of the embryonic behavioural repertoire must await electrophysiological analysis. This would require a comparison of patterns of electromyographic activity of the embryos with known patterns in the adult such as those of walking. The availability of the *ex ovo* preparation and the relatively large size of the *P. americana* embryo make it an attractive candidate for such an analysis. An added attraction of the cockroach embryo to students of behavioural embryology is that it represents a 'natural experiment' on the effects of movement deprivation on the development of movement. Preliminary behavioural observations indicate that walking and other behaviours are performed immediately after hatching and eclosion, the first occasion on which the limbs can normally respond to neuromuscular commands. Thus, practice and experience are not necessary for the development of locomotor behaviour, a finding in many invertebrate and vertebrate species (Davis, 1973; Kammer and Kinnamon, 1970; Provine, 1976b, 1979). The finding that limb displacement is not necessary for the normal development of the skeletomuscular system may explain the low level of spontaneous motility in cockroach embryos. In the much more active vertebrate embryos, motility may be important in the sculpting of joints and the development of muscles (Provine, 1976a).

15.6 Eclosion and hatching

Eclosion (moulting) refers to a series of stereotyped caudal-to-rostral peristaltic waves and thoracic ventral flexures that enable *P. americana* embryos and nymphs to shed their cuticles. Embryonic eclosion (the first ecdysis) resembles and may be identical to the numerous ($\sim$10) nymphal moults.

Embryonic eclosion transforms the immature looking, soft-white embryos to first-instar nymphs which resemble miniature adults; it is also the behaviour used for hatching. Newly hatched nymphs are white except for dark eyes, and mandibles, but quickly tan, becoming the normal reddish-brown colour. Embryonic hatching and eclosion usually occur on day 30 at 29°C. Since the 12–16 embryos that reside in the oötheca hatch simultaneously (Fig. 15.3), hatching is a

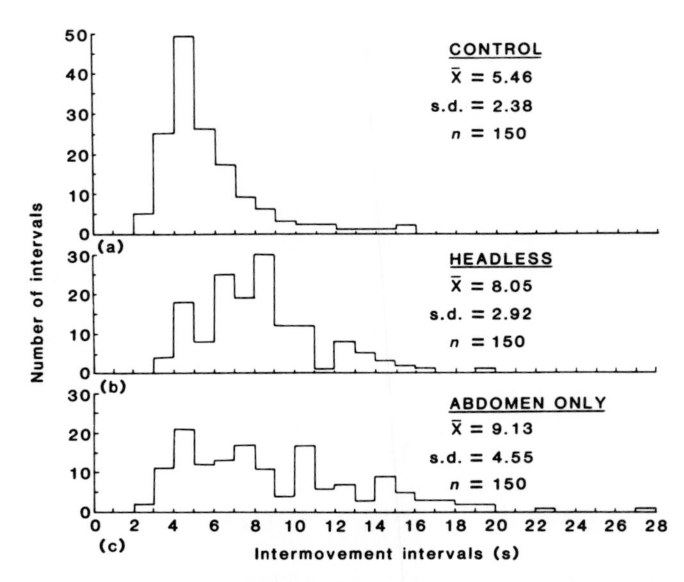

Fig. 15.5 Intermovement interval histograms of peristaltic eclosion movements in the abdomens of intact and decapitated *P. americana* embryos and isolated abdomens. The eclosion movements were present in all conditions but intermovement intervals were longer and more variable in the headless embryos and isolated abdomens than in intact embryos. (Reprinted from Provine, 1977; by permission of Pergamon Press.)

'social affair'. The synchronization of hatching is probably necessary because as discussed previously, a group effort is required by embryos to force open the mouth of the oötheca and hatch.

Individual embryos stripped of their chorions and raised for the last several days of incubation *ex ovo* in tissue culture dishes were observed to provide much of the following description of eclosion (Provine, 1976a, 1977). Twenty-nine- and 30-day embryos are almost inactive except for occasional multisegmental flattening and eclosion movements (Fig. 15.4). The eclosion movements are vigorous, smooth, rostrally propagated, peristaltic waves that pass through the abdomen and are translated as segment-to-segment ventral flexures when they reach the thorax. Dorsal thoracic flexures and abdominal rotary movements are occasionally performed. Some embryos perform only a few eclosion movements and halt while others continue until eclosion is complete. The eclosion movements are performed at regular intervals of 5.5 s (Fig. 15.5). Air-swallowing movements are performed concurrently with the eclosion movements. The swallowed bubbles may be seen inside the translucent bodies of the embryos. As eclosion proceeds, the originally sleek embryonic cuticle becomes slack and wrinkled which is probably the result of the combined effects of stretching produced by the eclosion movements and the increase in body volume (inflation) brought about by air swallowing. The cuticle tears along the dorsal surface of the thorax, probably as a consequence of the ventral bending of the thorax, and is passed over the head and moved caudally along the remainder of the body by the caudal-to-rostral peristaltic waves. The removal of the cuticle transforms both appearance (Figs. 15.2d, 15.3c) and behaviour. Newly eclosed first-instar nymphs immediately cease eclosion movements when the cuticle clears the last abdominal segment and walk or run away.

The eclosion-'hatching' behaviour sequence is performed by single embryos removed from the oötheca, stripped of their chorions and incubated in organ culture dishes. Therefore, the initiation and performance of the behaviour does not depend upon environmental factors associated with the oötheca, chorion and/or littermates. This does not exclude all environmental influences on the triggering of eclosion-hatching behaviour. For example, some form of tactile, mutual stimulation between embryos is probably responsible for the synchronization of hatching of the embryos in each oötheca. It has been shown that gentle stroking of 30-day-old embryonic littermates reared *ex ovo* produced eclosion after an average latency of 7.5 min as compared with a day or more for unstimulated littermate controls (Provine, 1976a). This suggests that once eclosion has been initiated in one or several embryos in an oötheca, they rub against each other and trigger eclosion-hatching in their neighbours, producing the mass effort necessary to pry open the oötheca and hatch. However, the finding that eclosion spontaneously occurs in isolated embryos indicates that tactile stimulation is neither necessary nor sufficient for the initiation of eclosion, an event that is probably hormonally mediated as in silkmoths (Truman and Sokolove, 1972).

Other sensory factors play a role in the control of the eclosion movements once they are initiated (Provine, 1976a). The finding that eclosion behaviour was immediately terminated after removal of the cuticle from the terminal abdominal segment suggests that cuticle-related factors or the cerci are involved in the maintenance or termination of ongoing eclosion movements. This possibility was tested by removing the cuticles from embryos immediately after the initiation of eclosion. This procedure resulted in the immediate cessation of the behaviour and the switching to the adult locomotor pattern. Conversely, by glueing the cuticles to the bodies of eclosing embryos, the eclosion behaviour which is usually completed within 2–3 min can be extended up to at least 10–15 min. These results indicate that, if we are dealing with a hormonally triggered readout of an endogenous central programme (Truman and Sokolove, 1972), the termination and thus the time that the programme runs is under cuticle-related sensory control. The embryonic cuticle has not been removed prior to eclosion to test whether the presence of the cuticle is necessary for the initiation of the eclosion behaviour sequence.

The neural origin of eclosion behaviour within the segmental ganglia has been evaluated by ligating various body segments of eclosing embryos and observing the behavioural effects (Provine, 1977). Eclosion movements were usually initiated in 30-day-old *ex ovo* embryos by gentle tactile stimulation.

Decapitation of eclosing embryos did not disrupt the basic caudal to rostral eclosion movements (Fig. 15.6a,b), but significantly increased the duration of the mean intermovement intervals (Fig. 15.5a,b). Abdomens that were isolated by a

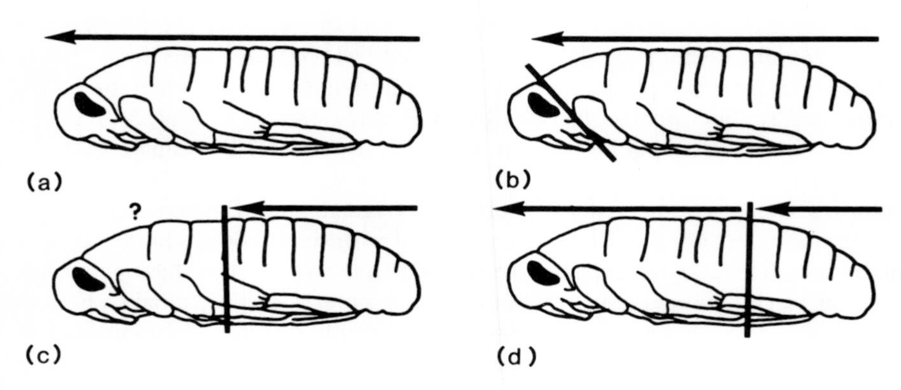

Fig. 15.6 Effects of various abdominal ligations on eclosion movements in *P. americana*. (a) The arrow shows the direction of propagation of the peristaltic eclosion movements in intact control embryos. When the wave reaches the thoracic segments, they are propagated as ventral flexures; (b) Decapitated embryos show caudal to rostral waves typical of intact control embryos; (c) Embryos with ligations between the last thoracic and first abdominal segments show independent movements on both sides of ligation. Eclosion waves are maintained in the abdomen. The question mark indicates uncertainty about the relation to eclosion of post-ligation movements in the thorax; (d) Embryos with mid-abdominal ligations show independent (asynchronous) caudal-to-rostral eclosion-type movements on both sides of the ligation. (After Provine, 1977; by permission of Pergamon Press.)

ligation placed between the metathoracic and first abdominal segments had much longer than normal mean intermovement intervals (Fig. 15.5c), but performed the basic eclosion movements (Fig. 15.6c). Relatively normal eclosion waves were even maintained in embryos with mid-abdominal ligations, although the waves on the two sides of the ligation were asynchronous (Fig. 15.6d). These results indicate that the neural circuitry within isolated chains of abdominal ganglia is sufficient to maintain eclosion behaviour once it is initiated. This conclusion accords with the demonstration by Truman and Sokolove (1972) that the silkmoth eclosion motor programme resides in the abdominal ganglia. Since the eclosion wave in *P. americana* embryos is propagated in a caudal-to-rostral direction, the most caudal abdominal ganglion in a chain probably triggers the activity in its next most rostral neighbour. This sequence of ganglion to ganglion excitation progresses rostrally. The finding of independent eclosion movements on both sides of embryos with mid-abdominal ligations indicates further that any abdominal ganglion is capable of driving the eclosion pattern of the next most rostral ganglion. The absence of rostral-to-caudral eclosion waves suggests that a ganglion may inhibit activity in the adjacent, caudal ganglion. However, a rostrally propagated excitatory wave initiated in the terminal abdominal ganglion and coupled with a post-excitatory refractory period of ganglia could produce a similar result. The role of the thoracic and other more rostral ganglia in eclosion is uncertain because the behaviour of the isolated thorax is difficult to interpret. In intact embryos, the most anterior ganglion participating in the eclosion pattern may reset the pattern and trigger the next wave in the most caudal abdominal ganglion. However, the behaviour of isolated abdomens indicates either that such resetting does not occur or that any abdominal ganglion is capable of resetting an eclosion sequence if it is the most rostral member of a chain.

Work needs to be done to establish the role of hormones in triggering embryonic and nymphal eclosion (moulting) in *P. americana*. Electrophysiological studies are also required to determine the 'fine structure' of the ganglionic motor output to the abdominal and thoracic muscles and the nature of interganglionic co-ordination. Recent behavioural and electrophysiological work with the silkmoth, an organism that possesses an eclosion pattern similar to that of *P. americana*, suggests possible plans of action (Truman and Sokolove, 1972; Truman, 1979).

Embryonic eclosion and hatching occur simultaneously using the same movements. Therefore, the above consideration of eclosion is an analysis of 'hatching' examined *ex ovo*. The rostrally propagated peristaltic waves of eclosion which propel the cuticle caudally also drive the embryo forward out of the oötheca. The ventral thoracic flexures of the two opposed rows of embryos coupled with the increased body volume caused by air swallowing, provide added lateral force useful in prying open the oötheca. If embryos are oriented wrong end up in the oötheca, the rostrally propagated waves drive the embryos away from the mouth of the oötheca and they fail to hatch. This observation argues against the involvement of passive (non-behavioural) processes in the hatching of individual embryos.

A technique useful in hatching studies is to break off the notched edge of a 29- or 30-day oötheca with forceps. Hatching usually follows within minutes. This technique may facilitate the escape of embryos which are already performing eclosion-hatching movements by weakening the oötheca. It may also induce eclosion-hatching of quiescent embryos.

15.7 Tissue culture of the embryonic nervous system

In recent years, important strides have been made in the development of insect tissue culture media and procedures. Organ and cell cultures have been prepared from *P. americana*, *Blaberus fuscus*, *Leucophaea maderae*, and *B. germanica* (Marks and Reinecke, 1965; Landureau, 1966; Marks *et al.*, 1968; Chen and Levi-Montalcini, 1969) using media of very different composition (Levi-Montalcini *et al.*, 1973). Tissue culture preparations of the embryonic nervous system by Rita Levi-Montalcini and colleagues will be emphasized here because they use *P. americana*, produce long-term organotypic cultures and contribute to the neurobehavioural orientation of this Chapter. Other virtues of the preparation are that ganglia are explanted on clean glass coverslips and maintained in a chemically defined liquid growth medium. Since the coverlips provide a smooth, flat growth surface, the configurations assumed by outgrowing nerve fibres are an expression of the inherent growth tendencies of the cultured tissue. This is an important consideration in studies examining directional outgrowth of nerve fibres. Many *in vitro* studies of nerve outgrowth are carried out on solid (collagen) or semi-solid (plasma clot) substrates which possess various amounts of inherent orientation which potentially confound attempts to segregate intrinsic and extrinsic (environmental) influences on directional outgrowth. The present studies document the remarkable capacity of the cockroach embryonic nervous system to self-differentiate and form interganglionic connectives in the absence of normal environmental 'instructions' and constraints. The appreciation of this capacity is helpful in understanding the nature and extent of environmental influences on the developing nervous system *in situ*.

Cultures are usually prepared using ganglionic explants of 16–18-day embryos because of the good survival and outgrowth that they produce *in vitro*. A whole mount preparation of the 16-day CNS is shown in Fig. 15.7a. Proliferative activity in the CNS is essentially complete at this age and the major features of the ganglionic chain resemble those of a miniature adult. At this stage, the embryonic CNS consists of a ganglionic chain consisting of a brain (B), a suboesophageal ganglion (SE), three thoracic ganglia (T1–3), five abdominal ganglia (A1–5) and a terminal abdominal or cercal ganglion (C). Embryos were removed from oöthecae using the procedures previously described for the *ex ovo* preparation and placed on Maximow depression slides filled with sterile culture medium. Parts of the nervous system were dissected away using sterile tools and placed on cover slides immersed in small culture dishes (18 mm-inside diameter) with liquid nutrient medium. The cultures were incubated at 29°C in the dark in sealed desiccator jars filled with 5% CO_2 in 95% air. The interior of the jars was humidified by wet cotton. The culture medium was replaced every 4–7 days.

A suitable culture medium for cockroach CNS explants was determined empirically by Chen and Levi-Montalcini (1969). It was a mixture of 4 parts commercial Eagle basal medium and 5 parts commercial Schneiders *Drosophila* solution to which was added 100 ml^{-1} pencillin, 100 g ml^{-1} streptomycin and 0.25 g ml^{-1} fungizone to prevent microbial growth. The two solutions were mixed immediately before use. The frequent appearance of precipitates in cultures using this mixture of commercial media made it desirable to compound a similar but completely defined chemical medium that was used in most of the work reported here. The formula for the solution is provided by Levi-Montalcini *et al*. (1973, p. 8–9).

Every type of CNS explant tested, the brain (B), suboesophageal ganglion (SE), thoracic ganglion (T), abdominal ganglion (A) and terminal abdominal or cercal ganglion (C) produced nerve fibre outgrowth. The first outgrowth appeared after 2–4 days *in vitro* and was distributed in a radial pattern around the explant (Fig. 15.8a). Outgrowth was most vigorous between 1–3 weeks, after which it tapered off. The internal structure of a ganglion, the central neuropile surrounded by a rind of cell bodies, and much of the peripheral fibre outgrowth remained intact for periods of 3 months or more. Generally, the fibre outgrowth was more vigorous and was maintained longer if numerous explants shared a culture dish. This suggests that the explants 'conditioned' the medium in some yet undefined manner.

Isolated ganglia that do not form contacts with other tissues usually produce only thin fascicles and radial, apparently random, nerve fibre outgrowth. After 2–6 weeks *in vitro*, many fibres that do not form contacts with other tissues either

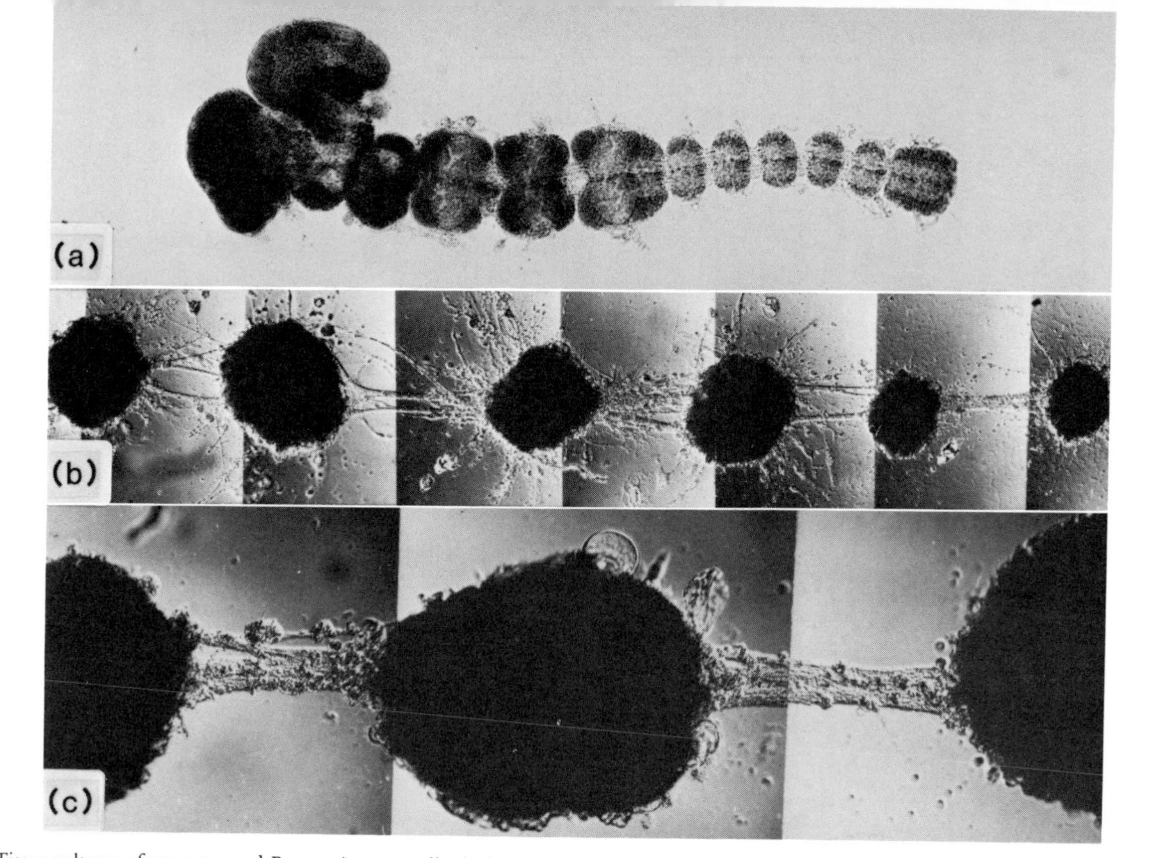

Fig. 15.7 Tissue cultures of reconstructed *P. americana* ganglionic chains. (a) A methylene blue stained whole-mount preparation of the ganglionic chain of a 16-day *P. americana* embryo. From left to right, the structures are the B, SE, T1–2–3, A1–2–3–4–5 and C. (b and c) Nomarski microphotographs of living cultures showing the substantial interganglion connectives which formed between rows of separated ganglia of 16-day embryos after three weeks *in vitro*. The ganglia in (b) are from left to right, SE, T1, T2, T3, A and A2. The ganglia in (c) are T1, T2, T3. ((a) and (b) are reproduced from Provine, 1976; by permission of the Wistar Press.)

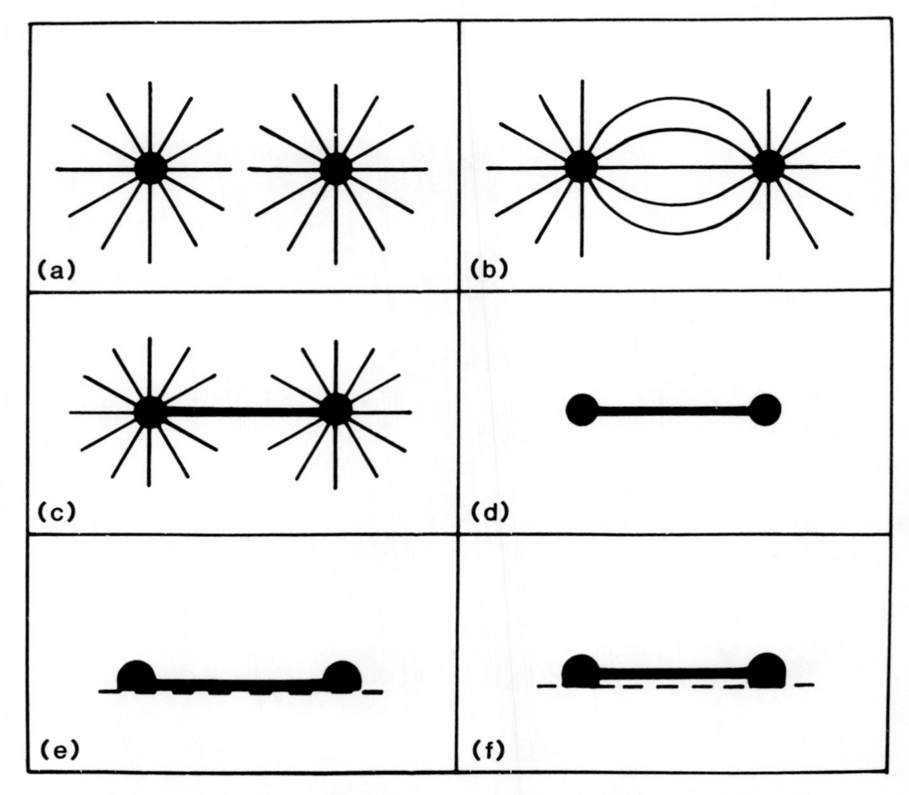

Fig. 15.8 Sequence of events involved in the formation of interganglionic connectives *in vitro* in *P. americana*. (a) The initial pattern of neurite outgrowth is radial; (b) Outgrowing neurites from the two ganglia intersect, adhere, and grow towards each other's perikarya in the adjacent ganglion; (c) An increase in fibre tension (shortening) is correlated with the pulling in of initially bowed connectives and the formation of substantial straight fasicles; (d) Some radial fibres are retracted and others degenerate, leaving only the straight interganglionic connectives; (e) A side view of a chick fasicle before and after (f) straightening (shortening) which results in an increase in fibre tension and a lifting of the fasicle off the glass substrate.

degenerate or are retracted as described below. The most luxuriant outgrowth was produced by abdominal and thoracic ganglia being followed by cercal ganglia, suboesophageal ganglia and the brain.

The radial nerve fibre growth pattern of isolated ganglia contrasts with the straight, stout connectives that form between ganglia dissected from the chain, separated and arranged *in vitro* in their usual sequence. Explanted ganglia were placed further apart than *in vivo* to prevent fusion of adjacent ganglia. The substantial connectives which form after 1–3 weeks are often double as *in vivo* (Fig. 15.7b). Frequently, lateral processes were observed to grow away from the rostral–caudal axis of the chain. These laterally growing fibres may correspond to lateral nerve roots that are present *in vivo*.

The formation of the interganglionic connectives appears to be relatively non-specific as suggested by the growth of substantial connections between ganglia in various biologically nonsensical configurations. Stout connectives were formed between multiple homologous ganglia (T1–T1–T1–T1, A1–A1–A1–A1, B–B–B–B, SE–SE–SE–SE, C–C–C–C, etc.), ganglia in abnormal order (T1–T3–T2–, A1–A5–A2–A4–A3, B–A1–B–A1–B–A1, etc.) (Provine *et al.*, 1976) and mixed ganglia from *P. americana* and chick. Substantial connectives were also formed between ganglia with abnormal axial relationships (Provine *et al.*, 1976). Therefore, the interganglionic connectives formed *in vitro* are not the result of outgrowth from ganglia preprogrammed to send out fibres in specific directions; adjacent ganglia are free to form connections even when placed in non-naturally occurring rectangular, triangular and circular configurations (Provine *et al.*, 1974).

Daily observations of living cultures made with Normarski optics suggest the developmental processes responsible for the formation of interganglionic connectives (Provine *et al.*, 1976). The first fibres grow out of ganglia after 2–4 days of incubation. The initial outgrowth is usually in a radial pattern like that of isolated ganglia (Fig. 15.8a). A few thin fibres growing from adjacent ganglia contact each other at 4–8 days (Fig. 15.8b). Intersecting fibres from adjacent ganglia tend to adhere to each other and grow along each others axes. Thus, the initial radial fibre arrays (fibre catchers) serve as guidelines which channel intersecting fibres towards soma in the ganglia. New outgrowing fibres tend to grow along existing fibre pathways, enlarging established fasicles. After 2–3 weeks *in vitro*, the usually bowed fibres which form the initial interganglionic contacts (Fig. 15.8b) are straightened (Fig. 15.8c). The straightening process is striking when it occurs over short 1–3 day periods. Straightened fibres are under tension and lift off the glass growth surface and span the liquid medium between the miniature hills of ganglionic nerve cells (Fig. 15.8e,f). After several weeks in culture, nerve fibres which do not establish firm contacts with the glass growth surface or other tissues are absorbed (Fig. 15.8d). The absorbtions may be associated with the increased fibre tension described above. Some fibres, particularly those of older cultures, degenerate, leaving cellular debris. The present account argues against the involvement in our cultures of the 'two-centre effect' which Weiss (1955, 1959) offered as an explanation of interganglionic connective formation in vertebrate cultures (Provine *et al.*, 1976).

The fibre-straightening (shortening) process which is a prominent step in connective formation ensures that fasicles take the shortest route between two points and that fibres which have a common origin and termination coalesce into a common tract even though their original paths to the destination were different. Similar processes may be involved in faciculation and tract formation in vertebrates. The present observations are a warning that straight fibre bundles (Figs. 15.7a,b, 15.8d) need not be the product of direction-specific outgrowth.

The fibre-straightening (shortening) process of *P. americana* may be similar to the shortening of the ganglionic chain of lepidopterans (Pipa, 1963, 1973; Heywood, 1965). The mechanisms are unclear in all cases. Robertson and Pipa (1973) suggest that axons of *Galleria* are coiled within the neurilemma as a result of ecdysone-induced glial migration or contraction. However, in *P. americana*, ecdysone is probably not necessary because shortening occurs *in vitro* in a chemically defined medium containing tissues which are not known to produce ecdysone. Heywood (1965) suggests that shortening in *Pieris* is accomplished by absorption of axonal material.

The present account stresses the importance of non-directed fibre outgrowth, fibre–fibre interactions and fibre shortening in the production of connectives, but it does not rule out the existence of selective, target-specific outgrowth. Selective outgrowth of some fibres may be overwhelmed by the random outgrowth of surrounding fibres. Also, specific synaptic connections are probably formed once a fibre reaches a neighbouring ganglion. The *in vitro* preparations of ganglia described above are not well-suited to detect target specific outgrowth. The problem may be attacked by examining the behaviour of known populations of outgrowing fibres towards specific target tissues or substances applied electrophoretically through micropipettes placed near neurite growth cones. However, evidence of target-specific nerve fibre outgrowth has been provided by Levi-Montalcini and Chen (1971). They showed that fibres growing out of thoracic ganglia which normally innervate the legs *in vivo* often grew towards explanted legs *in vitro*, while fibres from abdominal ganglia which never innervate the leg *in vivo* did not.

Thoracic and abdominal ganglia produce spontaneous action potentials and polyneuronal bursts which are propagated to adjacent ganglia by connectives that form *in vitro* (Provine *et al.*, 1973; Seshan *et al.*, 1974; Provine, 1976b). These spontaneous bioelectrical phenomena may be involved in the production of the embryonic motility described previously.

Good long-term survival and fibre outgrowth have been obtained from the frontal ganglion (Aloe and Levi-Montalcini, 1972a,b), the ingluvial ganglion (Aloe and Levi-Montalcini, 1972a,b), the corpora allata (Seshan and Levi-Montalcini, 1971), the corpora cardica (Seshan and Levi-Montalcini, 1971; Seshan *et al.*, 1974) and the medial neurosecretory cells (MNSC) of the protocerebral nuclei (Seshan and Levi-Montalcini, 1973; Seshan *et al.*, 1974) of nymphal cockroaches. Unique patterns of nerve fibre outgrowth (Levi-Montalcini *et al.*, 1973) and unit bioelectrical activity (Seshan *et al.*, 1974) were observed in many of these preparations. Explants of the protocerebral MNSC maintained their characteristic high level of neurosecretory granules for many weeks *in vitro* (Seshan *et al.*, 1974). These results suggest that the explants retained much of their structural and functional integrity *in vitro*.

Neurite outgrowth and survival has also been produced from isolated, mechanically dissociated neurons *in vitro* (Chen and Levi-Montalcini, 1970; Seshan *et al.*, 1974). The fibre processes of the neurons in Fig. 15.9 grew out entirely *in vitro*. Such isolated neurons survive best when cultured in combinations with foregut or other tissues. The single cell cultures are probably glia-free because the fragile glia do not survive the process of dissociation. This indicates that glia are not necessary

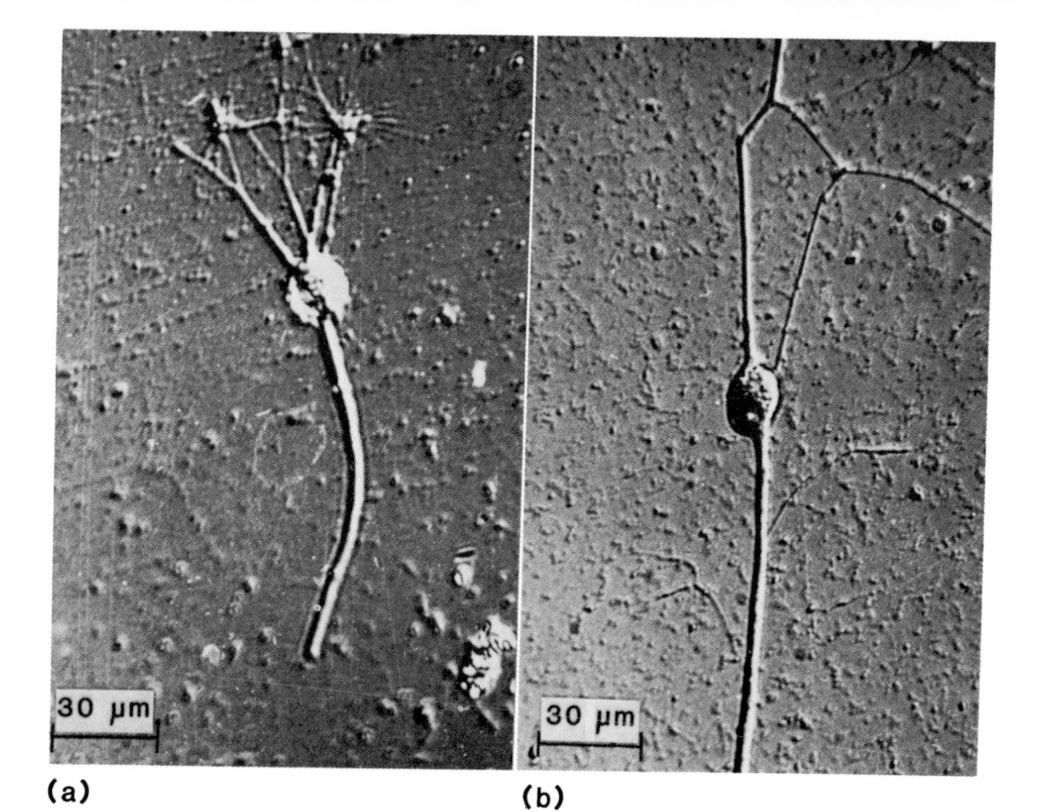

Fig. 15.9 Nomarski microphotographs of single living median protocerebral neurosecretory cells of *P. americana*. (a) 33-day culture from the seventh-instar nymph; (b) 20-day culture from a fifth-instar nymph. The cells were spherical on the day of explantation. All fibre processes grew out *in vitro*. (Reproduced from Seshan *et al*., 1974; with permission of Elsevier Press.)

for neuron survival *in vitro* (Chen and Levi-Montalcini, 1970). However, Schlapfer *et al.* (1972) report that isolated neurons maintained *in vitro* for two weeks are deficient in cholinesterase and choline acetyltransferase while whole brain and ganglion explants show increases in these enzymes. This suggests that isolated neurons may require the enriched surround of the intact ganglia for normal biochemical maintenance and development.

15.8 Conclusion

The study of cockroach development is at a critical period. If a concerted and carefully orchestrated effort to understand cockroach embryology is not launched soon, some of the attractiveness of the cockroach as a model system for hemimetabolous insect development will diminish. Researchers anxious to gain new insights into developmental processes will be discouraged by the great amount of preliminary groundwork which must be done and turn to other better understood embryonic preparations such as the grasshopper or cricket. Consensus about a 'target species' must be reached. Entomology can no longer afford the scattering of efforts and species chauvinism that pervaded much past developmental research.

Given the paucity of modern cockroach developmental research, it is appropriate to conclude with some suggestions for future research. Foremost is the need for a detailed description of embryonic development using the best modern techniques. A descriptive analysis using sectioned whole-mount embryos would be a useful beginning. Little work of this kind has been done since the last century. Macrophotographic analysis of intact, living embryos is a promising complementary procedure that does not shrink or otherwise distort the embryo (Bentley *et al.*, 1979). The latter study would suggest developmental landmarks which could be used to produce a development stage series. The series could describe development in terms of discrete numbered steps or as a percentage of total time for embryogenesis (Bentley *et al.*, 1979). Such descriptive accounts would also help establish the time of occurrence of developmental events which could be evaluated in subsequent experimental analyses.

Much of insect experimental embryology needs to be redone using *P. americana*. At present, too many assumptions about *P. americana* development are based upon generalizations from other species. The traditional procedures of extirpation, transplantation and tissue culture would help specify the inductive and regulative capacities of the embryo. Fate maps based upon the history of single labelled cells or small clusters of cells would provide especially useful data concerning the developmental mechanics of processes such as gastrulation.

The neuro-embryologist should attempt to trace the fate of individual, intracellularly marked neuroblasts (Goodman and Spitzer, 1979; Goodman *et al.*, 1979). The procedure of back-filling cut sensory and motor rootlets with light and electron opaque dyes to study the emergence of neuronal branching patterns may be useful at later developmental stages. Conditions are favourable for the establishment of neurobehavioural correlations, because electromyographic activity may be recorded simultaneously from a variety of easily accessible muscles of large, immobile, hearty *ex ovo* embryos. A promising tissue culture preparation also exists although much normative research of the intact embryo needs to be done before the adequacy of the *in vitro* 'model' may be evaluated. These many positive features suggest that the cockroach can be as effective a subject for the study of neuro-embryological and neurobehavioural development as it has been for the analysis of the neural mechanisms of behaviour in the adult.

Acknowledgements

I thank Drs D. T. Anderson, R. W. Oppenheim, L. M. Roth, Paula Sullivan and John Edwards who commented upon early drafts of this article. Much of the research by the author was conducted in collaboration with Dr K. R. Seshan and Luigi Aloe in the laboratory of Dr Rita Levi-Montalcini. The research was supported in part by grants from the NIH (NS-03777) and from the NSF (GB-16330X; GB-37142).

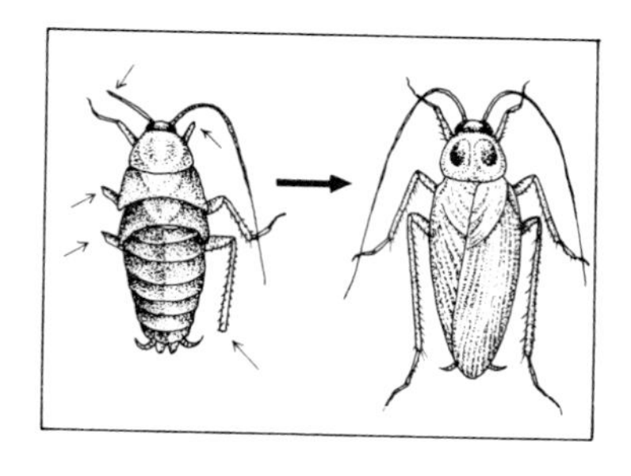

16

Regeneration

J. G. Kunkel

16.1 Introduction

Regeneration is found to varying extents throughout the animal kingdom (Goss, 1969). The first recorded discussions of the phenomenon in cockroaches occurred in the 1840s at meetings of the Royal Society of London. At that time, a purely philosophical debate ensued on whether cockroaches could replace lost limbs. The first productive experimentation confirming that leg regeneration does occur in cockroaches did not take place until a half century later (Brindley, 1897). Since then, the phenomenon has been studied in considerable detail, particularly in cockroaches, with the majority of work concentrated on leg regeneration. The reason for this preoccupation with legs is the special nature of the leg regeneration process in cockroaches. As in other insects, many tissues of the cockroach have the capacity to regenerate themselves relatively slowly over the course of a number of larval instars. Cockroach legs, however, will regenerate in an 'all-or-none' fashion in a single instar (O'Farrell and Stock, 1953). This all-or-none response has not been described for insect appendages other than the legs of cockroaches and the wing imaginal discs of Lepidoptera (Pohley, 1965).

The ability of cockroach legs to regenerate their form and function within a single moulting cycle has made them an attractive object of research in a number of areas including studies of pattern formation (Bohn, 1976; Bryant *et al.*, 1977, 1981), endocrine regulation (Bodenstein, 1959; O'Farrell *et al.*, 1960; Penzlin, 1965; Bullière and Bullière, 1977b; Kunkel, 1977), neural specificity (Bodenstein, 1957; Young, 1973; Cohen, 1974; Guthrie, 1975), *in vitro* cuticle synthesis (Marks and Leopold, 1971) and endocrine (Marks, 1973a,b) and insecticide (Sowa and Marks, 1975) action.

16.2 Tissues with regenerative potential

16.2.1 Epidermal structures

While only legs in cockroaches regenerate in an all-or-none fashion, other appendages, such as eyes (Hyde, 1972; Shelton *et al.*, 1977), antennae (Haas, 1955; Pohley, 1959; Drescher, 1960; Schafer, 1973), cerci (O'Farrell and Stock, 1956a), and exoskeletal features, such as ecdysial lines (Shelton, 1979), regenerate more or less gradually over a number of moulting cycles depending on the severity of loss. The failure to regenerate completely in a single moulting cycle can be due to the inability of appendages, other than legs, to delay the moulting cycle. Antennal amputation, while it does not delay individual moulting cycles, has been observed to increase the number of instars which *Periplaneta americana* and other species of cockroach take to reach the adult stage (Pohley, 1959; Ishii, 1971).

In order for tissues involving epidermal structures to regenerate, they must go through a series of stages which have been described by various authors (Penzlin, 1963; Bullière and Bullière, 1977b). After an initial wound healing phase, a period of embryological dedifferentiation and redifferentiation occurs under the old cuticle during which the pattern of the lost structure is in some way re-established. Next, a growth phase allows for the growth of the structure to approximate the size of the eventual regenerate. Finally, the pharate-regenerated epidermis secretes a cuticle and awaits ecdysis to reveal its form. The timing of the regenerative programme corresponds to the normal cyclical pattern of determination, proliferation and differentiation described for epidermis (Kunkel, 1975a). The initial regression and redetermination of the pattern of a limb can only occur in the intermoult phase of the moult/intermoult cycle. Any mitoses necessary to re-establish bristles or glands in the regenerate must occur during the interphase period, the normal timing of this type of mitosis for the general epidermis. If regeneration is not initiated during this phase, no regeneration occurs in the current stadium. If a structure is lost during the intermoult phase it can only regenerate in proportion to how much of the intermoult phase is left for pattern re-establishment and differentiative mitoses. If it is a leg that is lost, a mechanism exists for extending the intermoult phase and allowing a functionally complete pattern to be re-established.

16.2.2 Internal tissues

The extent of regeneration of internal organs other than leg-related tissues is treated briefly in the literature. Of particular interest to endocrine research is the apparent ability of the prothoracic glands of *P. americana* to regenerate after extirpation (Bodenstein, 1955b, 1956). Neuro-endocrine cell bodies have no capacity to regenerate (Drescher, 1960), however, the neuropile and commisures of the brain show substantial ability to reform after ablation and section experiments. Two categories of neurons have been proposed: Category one neurons of early embryological origin, which cannot regenerate, and Category two neurons of later origin, which retain a higher growth rate and capacity for regeneration (Guthrie, 1975; Jacobson, 1978).

Leg-related internal tissues including muscle and nervous tissue have been shown to have extensive regenerative potential. Muscle (Cowden and Bodenstein, 1961) and neuronal regeneration (Young, 1973) have also been studied in the absence of leg regeneration. Following nerve and muscle regeneration in the adult also allows the experimenter to isolate the phenomenon from the confines of a moulting cycle.

16.2.3 Regenerative fields

A morphological structure is surrounded by a space called its field. The field is a circumscribed area of tissue from which the original structure can regenerate if at least a portion of the field is left remaining. Regenerative fields have been observed for eye, antenna, cercus and leg of cockroaches. Of particular interest to the general study of fields is the demonstration (Bohn, 1974b, 1976) of a two-part epidermal field for leg regeneration. The leg field includes both sclerites anterior to the coxa and a membraneous epidermal region posterior to the coxa. Both must be present for regeneration to occur and each contributes a longitudinal half of the eventual regenerate. This supports the intercalation rule of Bryant *et al*. (Section 16.3.4d).

A tissue might not necessarily have to have been a part of a structural field in order to be incorporated into and contribute to its structure. Pronotal cuticle transplanted to the head adjacent to the eye was reported to incorporate into the advancing margin of the compound eye and contribute to facet development (Hyde, 1972). However, this result has not so far been repeatable in other workers hands (Shelton *et al*., 1977) which might argue that only cells within an eye field, as embryonically determined, are competent to form eye cells.

16.3 Phenomenology of limb regeneration

16.3.1 Faithfulness of regeneration

(a) *Gross morphology of regenerate*

That the leg regenerate is not a faithful copy of the original cockroach leg is one of the oldest facts to be reported in the literature (Brindley, 1898). The most obvious difference is that the regenerated tarsus has four segments instead of the normal five (Fig. 16.1). This artifact allows a simple tarsal segment count to establish whether a cockroach has ever regenerated a limb. Since regeneration can affect the rate of development (O'Farrell and Stock, 1956b), this artifact can be a valuable aid to identifying and eliminating animals with a history of regeneration from a developmental study.

Differences between the original and regenerated limb of *P. americana* were

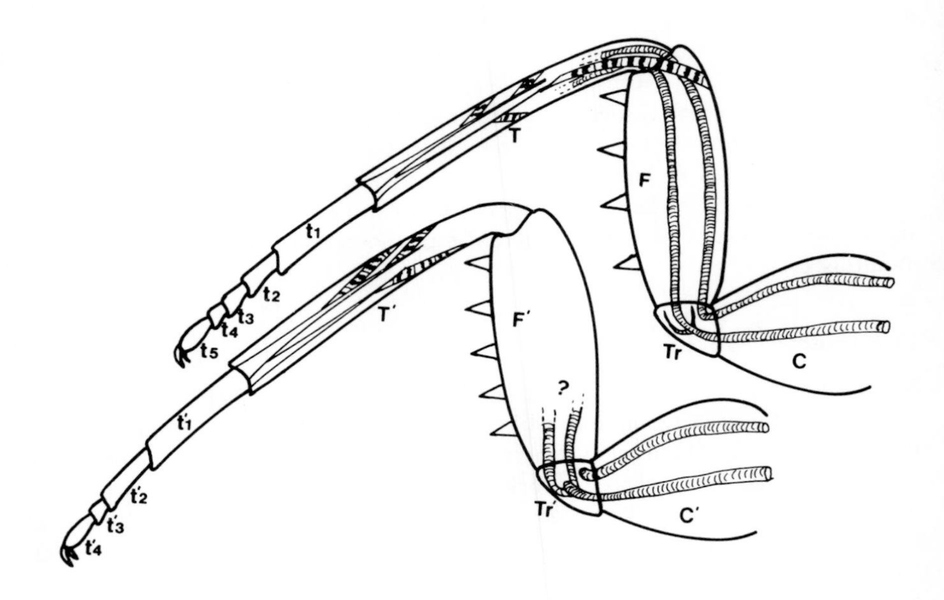

Fig. 16.1 Normal versus regenerated cockroach legs. Composite qualitative differences observed in *P. americana* (Penzlin, 1963) and *B. germanica* (O'Farrell, 1959; Kunkel, 1968). The major differences involve the tracheal supply, an absence of a tarsal muscle inserted in the femur in normal legs, femoral spine number and tarsal segment number. The regenerate exemplifies the form and size attained by at least three instars after the time of autotomy by which time all size regeneration has been completed (Roberts, 1973). The question mark in the regenerated femur indicates a highly variable tracheal branching pattern in the regenerate.

catalogued more completely by Penzlin (1963). He noted differences in internal morphology including differences in musculature and a highly variable tracheal supply to the regenerated femur, tibia and tarsus. The differences in tracheation could reflect the *ad hoc* active role of tissues in directing the distribution of tracheation in other systems (Wigglesworth, 1959).

(b) *Size of regenerates*

The size of initial regenerates is highly dependent on the time in the moulting cycle at which autotomy occurs (Fig. 16.2a) as well as the time after autotomy at which the delayed ecdysis occurs (Fig. 16.2b) (Kunkel, 1977). The new regenerate, although qualitatively functional, is never equivalent in size to the contralateral normal leg. However, through subsequent moulting cycles the regenerate rapidly approaches the size of a normal leg (Roberts, 1973). This approach to normality in subsequent moulting cycles includes such details as tibial spine number (Kunkel, 1968). But, the traits mentioned in the previous section (tarsal segment number and tracheal branching pattern) never return to normal. In the regenerated tarsus, the reduction in tarsal segment number is compensated by an increase in the lengths of the first two segments, but primarily tarsal segment two (Roberts, 1973).

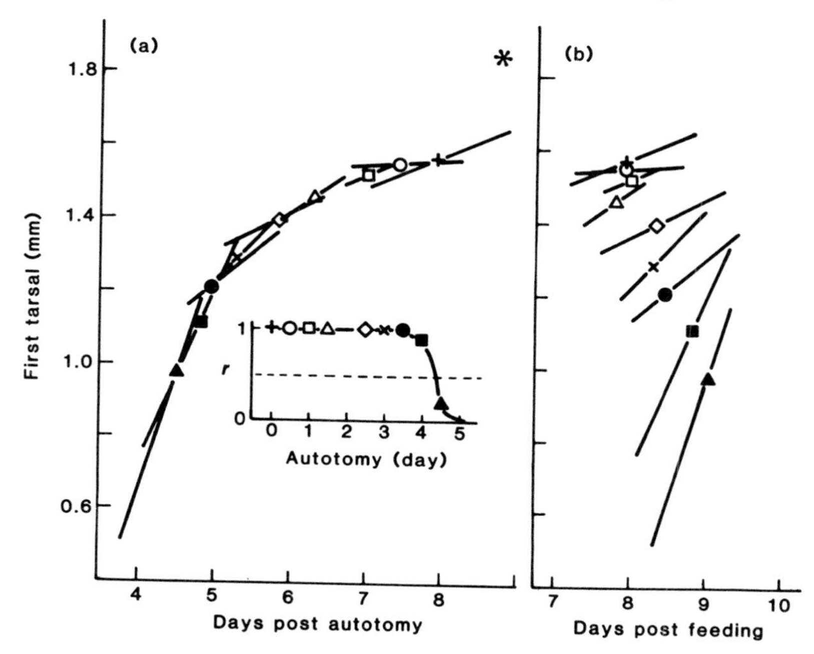

Fig. 16.2 Tarsal size regenerated at the first moult after tarsal autotomy. Size of the first tarsal segment at ecdysis is plotted versus (a) time from feeding to ecdysis and (b) time from autotomy to ecdysis. Groups of 25 animals had a metathoracic tarsus removed at different times after feeding in the VI (metamorphic) instar. An inset to (a) identifies the time of autotomy as well as the proportion regenerating for each group, *r*. The 50% point of this inset curve represents the midpoint of the critical period for regeneration for this stage animal. Each regression line in (a) and (b) passes through the mean size of regenerates as well as the mean time of ecdysis for that group. The regression line spans the times of ecdysis for that group. An asterisk in (a) indicates the length of the normal first tarsal segment (modified from Kunkel, 1977).

After three or four moulting cycles, regenerated segments three and four are indistinguishable in size from segments four and five of the contralateral normal leg. Segment two has also been shown to be the critical tarsal segment that must be lost if a four-segmented regenerate is going to be formed. Amputation at tarsal segment three, four or five results in a normal five-segmented tarsus (Bohn, 1965; cf. Penzlin, 1963).

Another departure from faithfulness only becomes obvious in the regenerate after a number of instars. The posterior border of the femur is usually spined. While the femur approaches normal size within a few moulting cycles, it consistently has one fourth more spines on its border than a normal leg (Kunkel, 1968). The addition of each spine occurs at a position formerly occupied by a bristle in the previous instar (Kunkel, unpublished). Studying the spacing of spines along the posterior edge of the femur in regenerated versus normal legs provides a one-dimensional system in which to study bristle pattern formation as opposed to the usual two-dimensional models (Lawrence, 1973).

The difference in pattern between normal and regenerate limbs may represent the effects of different conditions existing in the embryo compared to the regenerating limb during the critical phases when the pattern of the limb is being determined. It is clear that the pattern of bristles and the segment number and length are controlled by gradient phenomena in the epidermal tissue. It is not clear how these natural aberrations in pattern, observed in regenerating limbs, relate to the experimentally induced alterations in pattern which will be discussed later (Section 16.3.4d).

16.3.2 Moulting delay

(a) *Time of single leg autotomy*

A critical period for leg regeneration was originally described for *Blattella germanica* (O'Farrell and Stock, 1953), but is known to be a general phenomenon in all families of cockroaches, including *P. americana* (Penzlin, 1963). Prior to the critical period an animal will delay its moulting cycle to regenerate an autotomized limb. After the critical period, there is no moulting delay and no regeneration in response to an autotomized limb. Only a healing phase occurs when the moulting cycle is not delayed. This regeneration critical period was shown to occur simultaneously with the brain critical period (Kunkel, 1975a).

(b) *Severity of loss*

The length of delay of the moulting cycle fits the severity of the loss of tissue by autotomy. Thus, within each thoracic segment, loss of a tarsus will delay less than the loss of a femur–tibia–tarsus. Also loss of a prothoracic femur–tibia–tarsus causes less delay than loss of the same segments of the larger metathoracic limb. Loss of two limbs of the same thoracic segment at the same time will cause additional delay compared to loss of one limb (Stock and O'Farrell, 1954).

These correlations of the amount of delay with amount of tissue to be regenerated were construed by a number of workers to mean that the amount of regenerating tissue was controlling the delay. However, evidence to the contrary suggests rather that the length of delay determines how much regeneration can occur (see Kunkel, 1977 for references and discussion). Exactly what mechanism fashions the length of delay so that it is appropriate for the amount of tissue to be regenerated is not yet understood.

(c) *Autotomy versus amputation*

While leg regeneration will initiate from non-autotomy points, the delay of the moulting cycle depends upon limb loss at either of two autotomy points, the trochanter–femur or the tibia–tarsus joints. Loss at any other joint or mid-segment is rare since the two autotomy points yield so easily to tension. The distinction between autotomy and amputation is most clearly seen in large groups of synchronized animals which have a limb either autotomized or amputated shortly before their critical period for regeneration (Fig. 16.3). In such contrasts, the amputated animals do not delay the moulting cycle and, as a result, do not regenerate the

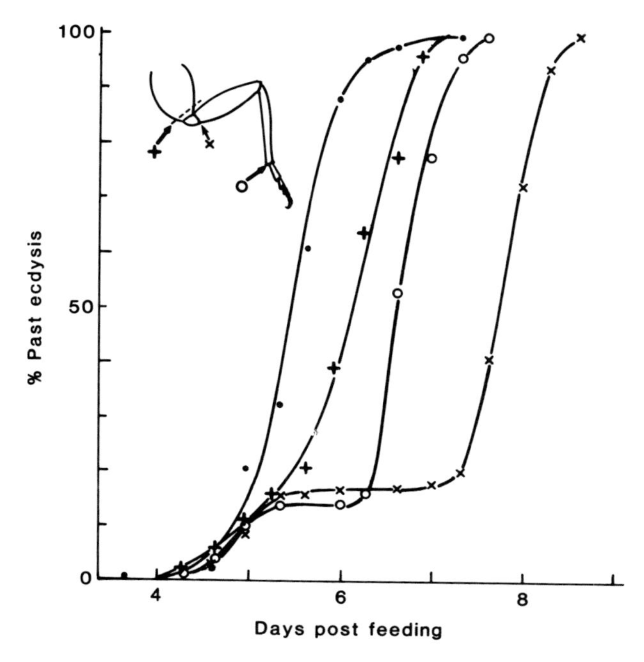

Fig. 16.3 Differences in effect of autotomy and amputation on the cumulative moulting curves of fourth instar *B. germanica*. Autotomy of either a tarsus (o) or femur–tibia–tarsus (×) results in a biphasic curve for cumulative ecdyses. Early non-regenerates are separated by a plateau from the delayed regenerating animals. The amputation at the distal end of the coxa (+) produces a largely monophasic curve somewhat delayed compared to the unoperated controls (●) but resulting in no regeneration (modified from Kunkel, 1977).

lost segment (Kunkel, 1975a). The same contrast with the limb removed earlier in the stadium results in regeneration in both cases without any substantial delay of the moulting cycle.

(d) *Roles of nerves*

Experiments with feeding-synchronized *P. americana* have shown that innervation of the leg is essential for establishing a delay of moulting. If the nerve trunks innervating the leg are cut, autotomy of the leg close to the critical period for regeneration does not result in regeneration or delay of the moulting cycle. Simultaneous autotomy of a contralateral leg does result in delay and allows time for some of the non-innervated legs to regenerate also (Kunkel, 1977). Furthermore, an intact connection between the brain and the ganglion innervating the regenerating limb in *P. americana* was found necessary for moulting delay and regeneration to be established.

(c) *Role of hormones*

Circumstantial evidence suggests a role of hormones in the delay of the moulting process; however, most of this evidence bears on the effects of hormones on the regeneration process itself (Section 16.4.2a) rather than on the delay. The size of the corpora allata changes during the regenerative process (O'Farrell *et al.*, 1960). Ligating the head shortly after autotomy prevents regeneration but also removes the source of both juvenile hormone (JH) and brain hormone (Penzlin, 1965). Since JH is known to delay brain hormone secretion in other insects (Nijhout and Williams, 1974) and possibly in cockroaches (Kunkel, 1981), the exact route of feedback in delaying moulting is obscure at this point. Beyond this, it is seen that the regeneration critical period is coincident with the brain critical period which suggests that delay of moulting could be caused by a delay in brain hormone secretion (Kunkel, 1975a). Direct measurement of ecdysteroid (ECD) titres in normal and limb-regenerating animals has demonstrated that moulting delay is paralleled by a delay in the major peak of (ECD) in *Blatta orientalis* (Kunkel, 1977). The delay of moulting could thereby be caused by an insufficient titre of brain hormone to turn on the prothoracic glands secretion of ecdysone.

16.3.3 Synchronization of moulting

(a) *Single leg autotomy*

Species of cockroach from all major families, including *P. americana*, have their moulting cycles synchronized by being forced to regenerate a limb (Kunkel, 1977). Autotomy shuts a moulting gate. Animals which delay their moulting cycle to regenerate an autotomized limb queue up at this gate. When some as yet unidentified regenerative process is completed, the gate is opened and all the animals initiate a moulting cycle in a relatively synchronous fashion.

(b) *Independence of delays due to multiple autotomies*

Two similar types of evidence suggest that the delays imposed by multiple simultaneous autotomies are independent of one another. The first evidence is the predictable extra delay and synchrony of moulting observed when two legs rather than one are autotomized (Fig. 16.4). The extra delay and synchrony derive precisely from the independence of the two delay processes. If a model of a six-legged hypothetical cockroach is considered, the moulting behaviour after autotomy could be easily predicted if all six legs had identical, but independent, delay properties. Autotomy of six legs would have closed six identical and independent gates. Since all gates must be opened for moulting to be initiated, the distribution of moulting initiation then could be predicted from the distribution of the last of six independent and identically distributed events.

However, since pro-, meso- and metathoracic legs have substantially different mean delay times associated with their autotomies, there is no chance for their gate-opening distributions to interact to create a greater delay than an individual leg would cause. In such cases, all the gates for the lesser delay (smaller) limbs will have opened by the time that the frequency distribution of the longer delay gates starts, and thus the delay-of-moulting distribution follows the characteristic of the larger leg. This is

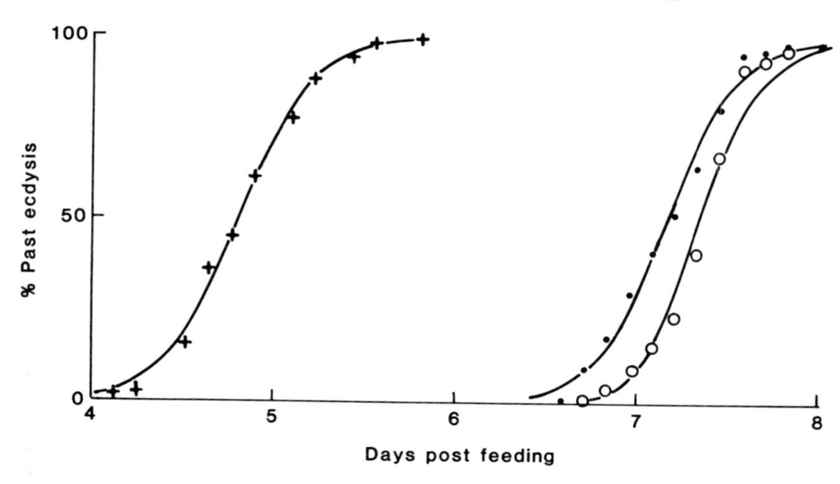

Fig. 16.4 Delay of moulting due to autotomy of 2 versus 1 metathoracic femur–tibia–tarsi. Cumulative percent past ecdysis is plotted against time of ecdysis measured from the time of feeding. The curves for control and single leg regenerates were fitted using a logistic curve. The curve for the double leg regenerates was computed by squaring the curve for the single leg regenerates. The data fit the curve according to a Kolmogorov–Smirnov goodness-of-fit test (Kunkel, 1977).

precisely the pattern observed when two dissimilar legs are autotomized simultaneously.

When two identical legs are autotomized sequentially at a sufficiently long interval apart, the last leg to be autotomized controls the timing and synchrony of the delayed moult. Such sequentially autotomized legs can act in concert to extend the intermoult phase considerably because each autotomized limb defines a new, secondary, critical period which is the time before which any subsequent autotomy must occur to further delay the moulting phase and ensure regeneration in the current moulting cycle. Such delays have their obvious limits since the cockroach has but six legs to lose in each instar.

16.3.4 The regenerative process

(a) *Wound healing*

Whether regeneration occurs after an autotomy or amputation, the first response is that of a healing process. This process has been described at the light microscope level in *P. americana* (Bodenstein, 1955b; Penzlin, 1963) and in *Blaberus discoidalis* (Bohn, 1976). Initially, a clot is formed involving haemocytes plugging the opening of the wound. Within a few days, the epidermis extends from its ruptured borders closing the gap spanned by the clot. The closure is an active process. The epidermis is not passively guided by the clot, but must actively pinch through the clot to become continuous. Besides their role in controlling infections attendant to loss of the limb, the haemocytes also play an active role in the healing, since epidermal cells are reluctant to grow out from their cuticular sites of attachment if haemocytes are not present or if the clotting process is inhibited (Bohn, 1976, 1977a). A haemocyte 'conditioning factor' may play a role in the growth behaviour of the epidermis (Bohn, 1977b). This factor has been shown to be a protein and to be antigenic.

(b) *Dedifferentiation*

The regeneration of the metathoracic limb autotomized at the trochanter–femur autotomy plane is the most convenient process to follow since the process occurs largely within the coxal segment, and the progress can be monitored visually through the broad flat ventral face of the metathoracic coxa (Penzlin, 1963). After wound healing, the tissues of the stump undergo a period of dedifferentiation and regression. The epidermis detaches from the cuticle and retreats within the coxa (Penzlin, 1963). The underlying muscles release their attachments to apodemes and also lose substantial portions of their structural detail, including the fibrillar components of their contractile apparatus (Bullière, 1968). The regression of at least some of the muscles is under the control of the nerves innervating them in *P. americana* (Shapiro, 1976). In autotomized limbs which have also been experimentally axotomized, the muscles of the coxa undergo a more drastic regression. In normally innervated legs, neuromuscular synapses maintain their structure during the regression; and the muscle cells maintain their insertions into epidermal cells, even though the epidermis has released from the apodemes. In the majority of muscle, regression involves loss of mass, including sarcoplasmic reticulum, mitochondria and myofilaments, rather than loss of cells (Shapiro, 1976). This regression involves all the muscles of the coxa, but the distal muscle is involved sooner than the proximal. After losing 40% of their weight, the coxal muscles undergo no further dedifferentiation. This limited and orderly regression is characterized by Shapiro (1976) as the reversal of normal muscle development and is contrasted to the degeneration and eventual autolysis of muscles observed during metamorphosis of holometabolous insects (cf. Lockshin and Beauleton, 1974).

How coxal muscle regression is induced is not known. Although there is a general reduction in activity of the nerves innervating the coxal muscle after autotomy, mini-endplate potentials continue to be recordable during the regression period and muscle spike potentials are recordable from coxal muscles during walking, though in a generally unpredictable pattern compared to the normal phasing of contractions of intact legs. If this innervation of the regressing muscles is removed, severe autolysis of the muscle occurs (Shapiro, 1976).

(c) *Blastema formation*

As in many regenerating systems (Goss, 1969) the form of the regenerate emerges from an apparently undifferentiated cluster of cells, the blastema. The origin of the blastema tissue is of considerable theoretical interest. Does pre-existing coxal muscle dedifferentiate into a pluripotential state, a blastemal cell, from which it can redifferentiate into a different muscle cell type at some other morphological location? Does all the blastemal tissue arise at the wound site or does it come from some pool of cells held in reserve in another part of the body?

Some of these questions have been answered by transplant experiments between species in which the donor and host cuticular products are clearly distinguishable. It is clear, for instance, that the regenerated epidermis is not clonally derived and that it does originate from tissues adjacent to the wound site (Bohn, 1972). Tracheal regeneration does not originate from an extension of the prior tracheal trunks. Rather it derives from growth of small tracheal branches previously serving local needs (O'Farrell, 1959; Penzlin, 1963). Motor axons regenerate from their central origins (Guthrie, 1975). The only tissue whose origin has not been traced is muscle. During limb regeneration new myotube and muscle differentiation is occurring continuously up to the time of ecdysis (Cowden and Bodenstein, 1961). Initially, these cells, which form the new muscles of the lost limb segments, were thought to be derived from the regressing and dedifferentiating coxal muscle (Penzlin, 1963). However, since current studies at the electron microscope level (Schapiro, 1976) suggest that coxal muscle remains essentially intact, except for an atrophy of their internal machinery, the immediate source of the muscle blastematous tissue is conjectural. Perhaps, like the epidermis, it is derived from a very localized set of cells immediately adjacent to the wound site. The solution of this problem is of general interest since it may shed some light on the general phenomenon of muscle proliferation in insects.

During the blastemal stage the segmental and intrasegmental patterns of the limb are being established. The small distances involved at the blastemal stage may afford the geometry within which the necessary events for pattern formation occur. From the rare occasions when animals seem to have initiated moulting when they had barely passed the blastemal stage, it is obvious to this observer that a primitive segmentation is already set up in the late blastema. Clearly, this stage is the most critical and least understood of the regeneration process. The explanation of the phenomena occurring during blastema development would be an unveiling of the most basic mysteries of developmental biology.

(d) *Epidermal pattern restoration*

The regenerating limb exhibits many of the phenomena associated with pattern formation in vertebrates and other invertebrate systems (Bryant *et al.*, 1977; Palka, 1979). In particular, the segments of the regenerating limb are a set of repeating gradients (Bohn, 1966) similar to the repeating gradients of the abdominal sclerites (Locke, 1959; Lawrence, 1973) (Fig. 16.5). Gradients in adjacent segments of the cockroach limb have been shown to be similar by virtue of compatibility of equivalent levels when intersegmental grafts are performed (Fig. 16.6b) (Bohn, 1966).

Incompatibilities of tissues in grafts can be explained with reference to two types of gradient operative in each segment of an appendage: an axial gradient and a circumferential gradient (Bohn, 1965, 1966; French, 1976, 1978). The length and longitudinal pattern of each segment is controlled by the axial gradients. When the proximal end of a segment (nominal high end of a gradient) is grafted on to the

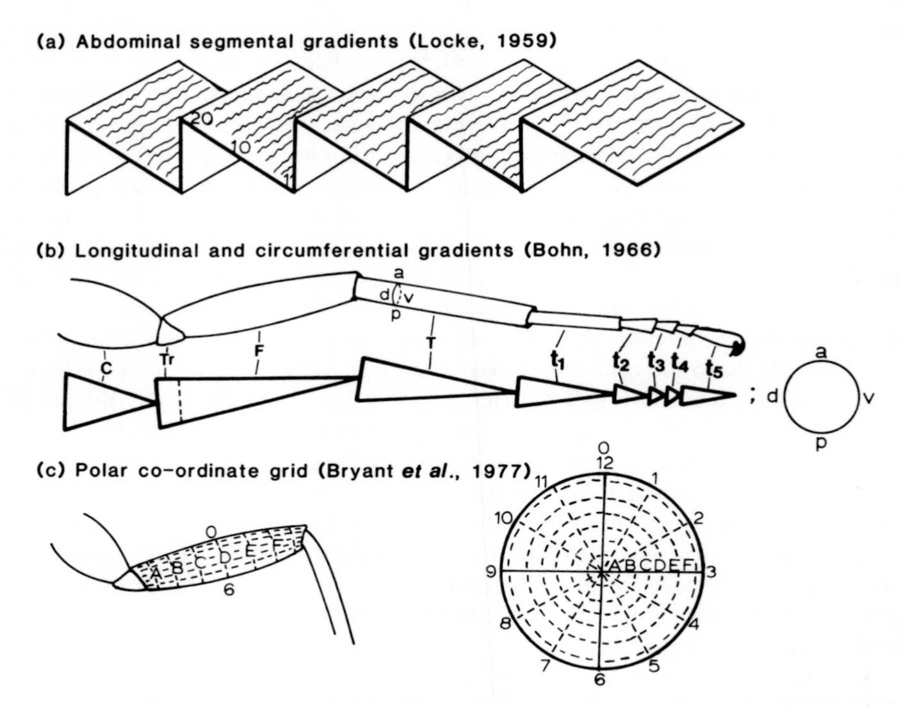

Fig. 16.5 Gradient models of abdominal and leg segments. (a) The earliest gradient model used to describe the behaviour of the adult transverse tergal ripple pattern of adult *Rhodnius prolixus* when grafts are performed during the previous larval stages. Hypothetical gradient levels (1–20) in each host segment were required to be in alignment with graft gradient levels in order for the ripple pattern of the adult to be undisturbed. (b) Application of gradient concepts to the cockroach leg. Each leg segment is suggested to contain a gradient (note that the trochanter–femur comprises one gradient). A separate circumferential gradient was postulated to explain results of contralateral grafts. (c) The polar co-ordinate grid as it is envisioned to apply to each cockroach limb segment. Each point on the surface of a limb segment can be mapped to a point on the polar co-ordinate grid.

distal end of an identical segment (low end of a gradient) in a host (Fig. 16.6c), an incompatibility of levels of the opposed gradients is set up, which results in the generation of superfluous tissues intervening in gradient level between the two opposed ends. Generation of superfluous tissue can be interpreted as a key to understanding the nature of gradients and how they control tissue growth and pattern.

Epidermal cells grow, divide and orient themselves and their cuticular secretions in relation to the gradient position they find themselves in. This inferential statement is a direct outcome of the observations on the amounts and orientation of regenerated superfluous tissues that result from experiments such as that described in Fig. 16.6c. The active role of axial gradients in orienting cells was vividly demonstrated by Bohn (1974a) using interspecific grafts of identifiable tissue rotated 90° from the host segmental axis. The orientation of the bristles of the transplant gradually rotated in a step-wise manner in subsequent moulting cycles

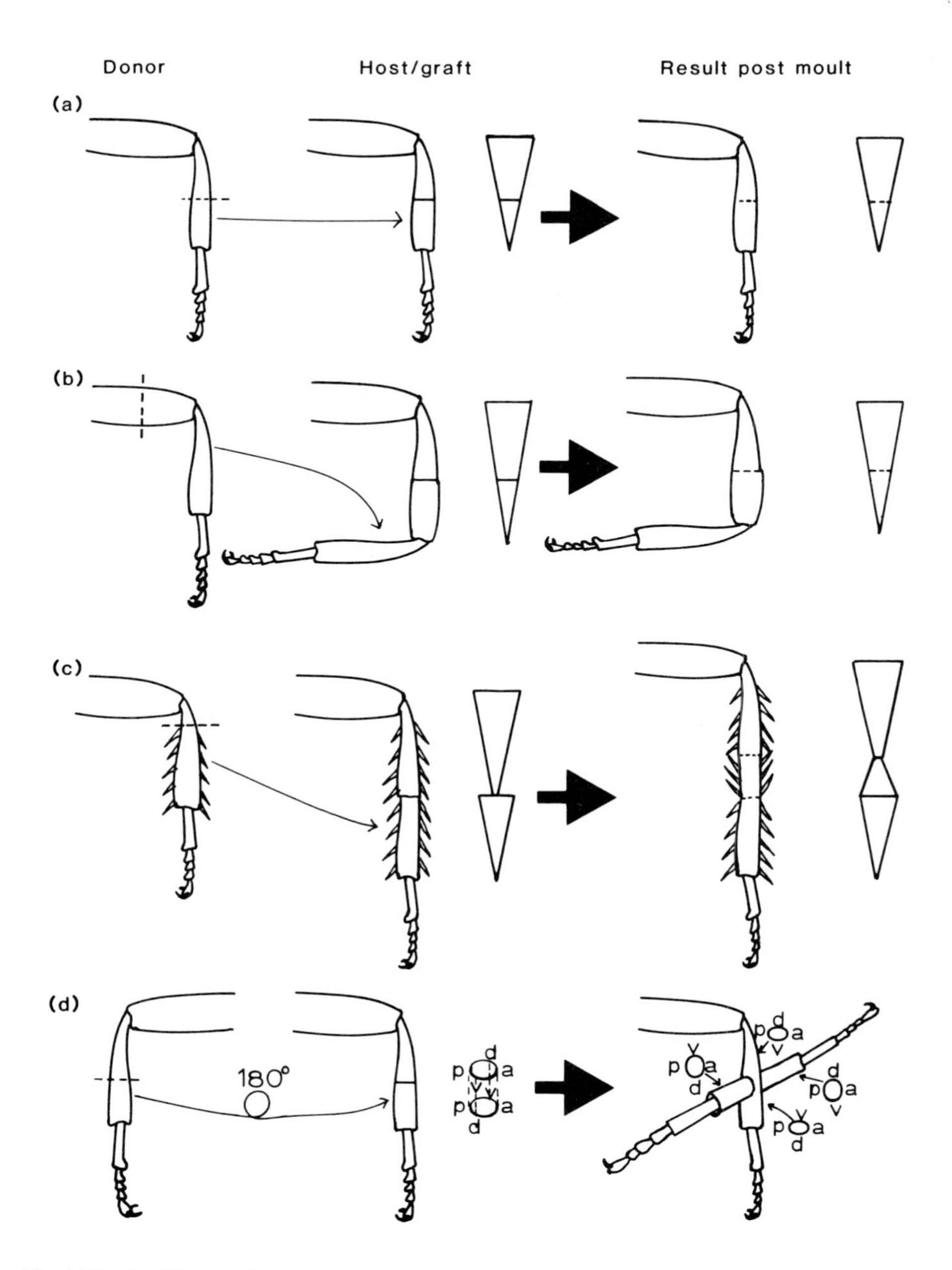

Fig. 16.6 Grafting experiments. (a) Homologous ipsilateral tibial graft (same leg; same segment; same level). (b) Inter-segmental graft (same leg; femur graft on to tibial host; same level). (c) Intra-segmental graft (same leg; same segment; proximal level graft on to distal level host tibia). (d) Contralateral tibial graft (contralateral leg; same segment; same linear gradient level but incompatible circumferential gradients). The outcome of each experiment can be predicted by referring to gradient compatibilities. Orientation of the tibial spines is used to predict the nominal direction of the gradient.

to eventually assume the orientation of the host epidermal cells adjacent to the graft.

While circumferential gradients perpendicular to the axial gradient have not been described in abdominal segments (Locke, 1959; cf. Shelton, 1979), in limbs, such gradients control the pattern and increase in girth of limb segments (French, 1978). These circumferential gradients are most graphically demonstrated by grafts of donor limbs to host limbs at a similar axial gradient level, but on a contralateral side of the animal (Fig. 16.6d). Such grafts result in incompatibilities of the circumferential gradients of apposed tissues which cannot be satisfied by a simple rotation of the tissues relative to one another. The net result is often supernumerary limbs generated at the sites of incompatibility.

A mathematically descriptive and predictive model encompassing both the axial and circumferential gradients has been proposed (French, 1976; Bryant *et al.*, 1977), which is consistent with most of the phenomena so far observed. This model is capable of predicting the outcome of grafting experiments in which incongruities of the apposed gradients result. This extends to predicting the amount of superfluous tissue (Fig. 16.6c), the placement of supernumerary limbs (Fig. 16.6d) and even the handedness of the supernumerary regenerates. This model describes the limb gradients as a polar co-ordinate grid in which the axial gradient is represented by the centripetal rays of a circle (Fig. 16.5c), while the circumferential gradients at each point along the axis of the limb are represented by concentric rings in the model. Any point on the leg segment can be mapped to a gradient point within the polar co-ordinate grid.

Along with this fine resolution description of the gradients go rules for the behaviour of tissue when the integrity of the system is destroyed by amputation or grafting. When a portion of a limb is ablated, the resulting healing brings together epidermal cells which differ more or less in their gradient values. This incompatibility stimulates growth to replace the intervening tissue and gradient values. An additional rule states that the shortest route between the two apposed gradient levels will be taken in intercalating a complete pattern.

This model is partially an outgrowth of a more disjointed earlier model of Bohn (1965) in which the poles of the limb circumference were defined more crudely in terms of dorsal versus ventral and anterior versus posterior borders, plus a distinct longitudinal axial gradient. This earlier model was based almost entirely on cockroach experiments and predicts most of the same results as the finer resolution model of Bryant *et al.* (1977). Neither of the models deals effectively with the transitions from segment to segment of the limb nor with what happens in the terminal segment, which would seem to have some trouble rounding itself off at the tip. Neither approach the fundamental problem of the nature of the gradients involved: Are they gradients of substance or of cell adhesivity or of cell compatibility?

The fundamental contribution of the gradient models is their provision of a mechanism for recognition of wholeness at the tissue level. When, due to autotomy or amputation or the artificial introduction of a transplant, cells with different gradient level properties come to abut one another, the incompatabilities of gradient level result in a spontaneous correction of the problem. The result for most of the situations an animal will normally encounter (i.e. autotomy or amputation) is normal regeneration; the result of grafting contralateral legs is supernumerary regeneration. By understanding the artificial situations created by grafting, we may at some later date be able to understand the normal generation and regeneration of pattern.

(e) *Neural specificity*

Another aspect of pattern restoration exhibited by the regenerating limb is the specificity of nerve cell – muscle cell recognition and re-innervation. This topic has been abundantly reviewed in the past decade (Young, 1973; Guthrie, 1975; Anderson *et al.* 1980; Pipa, 1978b). This aspect of regeneration is rarely studied in the normal regenerating limb. As demonstrated by Shapiro (1976), the muscles of the coxa never lose their innervation during coxal regression attendant on limb autotomy. Despite this, the majority of experiments on neural specificity in the cockroach leg concentrate on the specificity of coxal muscle re-innervation after nerve section in adults (Young, 1972; Pearson and Bradley, 1972).

From experiments on artificial nerve section and leg transplantation it has been demonstrated that specific afferent and efferent re-innervation of coxal muscles occurs (Pearson and Bradley, 1972; Fourtner *et al.*, 1978). Innervation cues are equivalent on right and left sides of the body (Bate, 1976a). Adjacent thoracic segments use the same cues for motor axon recognition of homologous muscles (Young, 1972). The initial finding of the appropriate muscle by a motor neuron is perhaps a trial-and-error search process with inappropriate connections made at first (Young, 1973; Whitington, 1977; Denburg *et al.*, 1977). Subsequently, the appropriate connections are singled out to survive.

If this later suggestion is correct, then more weight is placed on a recognition of appropriateness and inappropriateness of cell junctions. Identifiable muscle fibres of *P. americana* differ from one another in the pattern of their polypeptides (Denburg, 1975, 1978). Whether any of the so-far catalogued differences in *P. americana* are sufficient to serve as a decisive factor in stabilizing an appropriate re-innervation has been questioned (Tyrer and Johnson, 1977). Further study of what happens at initially inappropriate innervation sites is needed (cf. Urvoy, 1970).

16.4 Regulation of regeneration

16.4.1 Roles of nerves

(a) *Direct trophic effects of nerves*

The earliest work on the role of nerves in cockroach leg regeneration (Bodenstein, 1957; Penzlin, 1964) demonstrated that nerves were not essential for regeneration to be initiated nor for the maintenance of the process once started. These experiments were done on animals autotomized shortly after ecdysis to allow ample time for the regenerative process to be established before a moulting cycle was initiated. While the epidermal components of regenerates from denervated legs appear qualitatively complete, they are in general smaller than normally innervated regenerates and are 'hollow', lacking a well-developed musculature (Penzlin, 1964). A general trophic role of nerves seems to consist of maintaining muscle size and integrity (Guthrie, 1967a; Lockshin and Beauleton, 1974). A more specialized trophic role for nerves is an important part of the normal coxal muscle regression which attends autotomy (Shapiro, 1976; Section 16.3.4d). The effects of nerves on the size of the epidermal components of a regenerate are most likely secondary to neuromuscular trophic effects (though this point is somewhat mute due to rapid regeneration of nerves, cf. Edwards and Palka, 1976).

(b) *Indirect role of nerves*

Without the proper sensory innervation of a limb and without the proper sensory stimulation by autotomy, no delay of the moulting cycle will occur (Section 16.3.2b,c). In this sense, an intact nervous sytem is essential to a consistent all-or-none regeneration of a leg. A leg lost early in the stadium will regenerate its epidermal form at least, irrespective of whether it is innervated. However, if a leg is lost close to the time of initiation of a moulting cycle and is denervated, an incompletely formed regenerate results, since the moulting cycle is not delayed and the critical phase of morpholaxis is interrupted.

16.4.2 Role of hormones

(a) *In vivo results*

Regeneration is not an entirely spontaneous activity of the tissue remaining after autotomy or amputation of a structure. This is evident from the observation that adult *P. americana* legs do not go through any of the steps of regeneration other than healing (Bodenstein, 1953c), unless transplanted to or parabiosed to a larval animal. In addition, in larvae of *P. americana*, the steps of the leg regenerative process beyond healing can be prevented by ligating the head from the body within the first two days after autotomy (Penzlin, 1965). While it is clear from *in vivo* experiments that large doses of ECD will terminate the autotomy-induced regenerative process (Bullière, 1972; Kunkel, 1975b, 1977) by precipitating a moult cycle, it is unclear from these studies what hormones might be involved in promoting regeneration in any positive sense (O'Farrell *et al.*, 1960; Penzlin, 1965). One suggestion that ecdysteroids might play an enabling role in regeneration derived from the observation that *P. americana* adults would initiate regeneration if a leg was taken off shortly after ecdysis, a time prior to prothoracic gland involution (Bodenstein, 1953c) (see also Section 7.6.13).

(b) *In vitro effects*
The original suggestion (Marks and Reinecke, 1964) that an organ culture approach to studying limb regeneration phenomena might be fruitful, has paid off handsomely (Marks, 1980). Two dose-dependent functions of ECD in controlling leg regeneration have been demonstrated *in vitro*. Early morphogenetic phases require low doses of ECD to continue while the later formalization of the regenerate by cuticle production is dependent on higher titres of the same hormones (Marks, 1973b). It is still unclear what specific phases of the morphogenetic process require ECD and whether there are any other hormones involved.

16.5 Unified model

16.5.1 Systems involved

The process of leg regeneration requires the involvement and interfacing of three systems in the domain of the leg itself as well as one major system extrinsic to the leg. The three leg systems are: (1) a neuromuscular servo-system involved in walking, (2) an epidermal system involved in cuticle maintenance and moulting and (3) a morphogenetic system involved in pattern generation and restoration. The outside system with which the three leg systems interact is the neuro-endocrine system controlling the moulting process. Each of these systems can be considered separately as a black box with inputs and outputs (Fig. 16.7).

16.5.2 Interactions of the systems

(a) *Neuro-endocrine system*
The inputs of the neuro-endocrine system are feeding, which controls the initiation of the moulting process (Kunkel, 1966), and the delay status input from each of the six independent leg sensory-neuromuscular systems. As long as there are no autotomized legs, the neuro-endocrine system is free to respond to food availability by a timed production of the moulting hormone ecdysone. Ecdysone titre is the output of this system. The ecdysone output impinges on two leg systems, the epidermal moulting system and the epidermal morphogenetic system.

(b) *Epidermal moulting system*
This system requires a high titre of ecdysone to undergo its cycle of cuticle deposition. In carrying out its role in the moulting process, this system may cut short any regeneration that is taking place. This behavioural response is a fundamental property of all epidermal tissue.

(c) *Epidermal morphogenetic system*
In the early part of the stadium, the low ecdysone titre stimulates the epidermal morphogenetic system to operate. This self-contained system responds to local

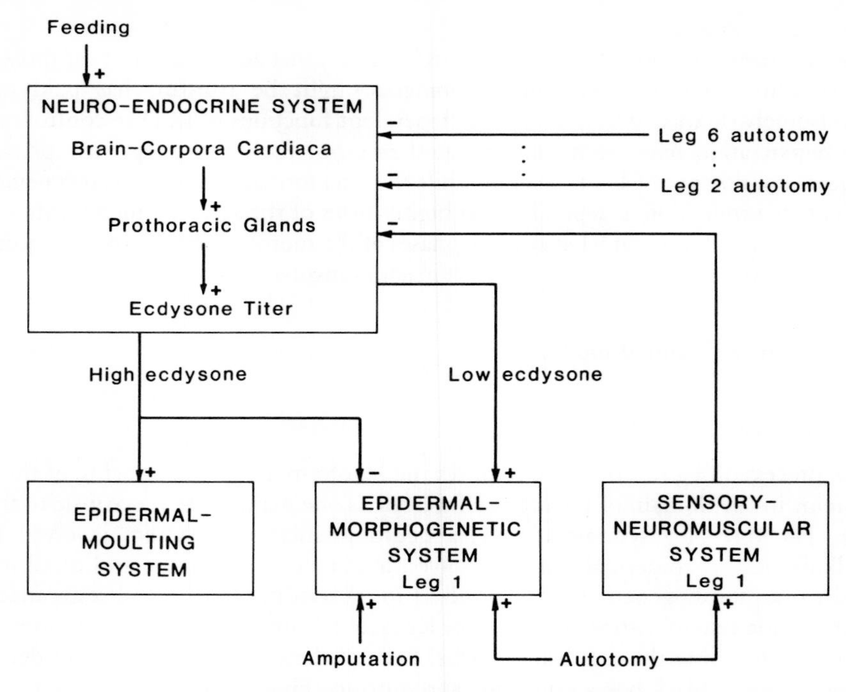

Fig. 16.7 Flow diagram of regeneration regulation. Arrows reflect information flow between systems contained in each box. Signs (+ and −) refer to stimulation or inhibition of one system on another.

needs for new structures such as new glands or sensory bristles by differentiative mitoses of epidermal cells (Kunkel, 1975a). When an autotomy or amputation has occurred, this system responds to the resultant incompatibilities of adjacent epidermal cells by stimulating growth and differentiation until the intervening gradient values are regenerated. This process can continue to completion as long as the epidermal moulting system has not been activated. When moulting is activated, all regenerative activities cease and the epidermal structures so far regenerated go through a cycle of cuticle production. The completeness of regeneration depends on the appropriateness of the delay provided by the sensory neuromuscular system.

(d) *Sensory neuromuscular system*

This system is made up of the muscles of the leg, the sensory neurons, motor neurons and interneurons involved in locomotion behaviour. Critically relevant to regeneration is the ability of the nervous system to perceive and react behaviourally to the loss of a limb. The perception of autotomy of a limb is reflected in an immediate change in locomotory pattern (Hughes, 1957; Delcomyn, 1971b). This change in pattern depends on loss of the limb at the autotomy point. Artificial limb loss at the femur–tibia joint by amputation does not result in an altered stepping pattern. It is likely that it is the leg sensory system that is involved in the perception of a need for a delay of the moulting cycle, the hemiganglion serving the autotomized limb relaying this information to the brain (Kunkel, 1977; Section 16.3.2d). An earlier hypothesis suggested that the bulk of the regenerating muscle (Cowden and Bodenstein, 1961) or the growing regenerate (O'Farrell *et al.*, 1960) was involved in metabolizing the moulting hormone, as a result delaying moulting.

The perception of loss at the trochanter–femur autotomy plane most likely involves the fields of campaniform sensillae found just pre- and post- the autotomy plane. The pattern of input from these sensillae are likely essential in the instruction of what type of delay is needed. Simple removal of all sensory and motor communication between the hemiganglion and the leg does not result in any delay of moulting whether or not the leg is autotomized. All evidence points to an informed decision within the leg or its hemiganglion as to how much delay to create in order to accomplish the desired amount of regeneration. Exactly how the delay length is meted out is not known. The neuronal regenerative process, initially observed in *P. americana* when the nerve axons to a leg are cut (Cohen and Jacklet, 1965; Jacklet and Cohen, 1967), is an obvious candidate for supplying the control of delay length. The nature and timing of this process in the neuron cell body has been difficult to establish (Young *et al.*, 1970; Denburg and Hood, 1977) and, as a result, it may be hard to correlate the cytologically observed changes with the physiological delay process.

16.6 Conclusion

Clearly, much more work is needed to detail what is happening within the sphere of each of the subsystems involved in regeneration. The phenomenon in cockroaches is a mixture of neuronal, hormonal and developmental processes which can be orchestrated by the investigator to occur using the cues of feeding and autotomy. Among models of regeneration, the cockroach leg model presents a unique opportunity to study the process from a variety of perspectives. Leg autotomy is also a way of reproducibly perturbing the endocrine system in studies aimed at the physiology of the moulting process.

Acknowledgement

I am indebted to John Edwards for helpful comments on the manuscript.

References

Abu-Hakima, R. and Davey, K. G. (1975) Two actions of juvenile hormone on the follicle cells of *Rhodnius prolixus* Stal. *Can. J. Zool.* 53, 1187–1188.

Abu-Hakima, R. and Davey, K. G. (1977a) Effects of hormones and inhibitors of macromolecular synthesis on the follicle cells of *Rhodnius. J. Insect Physiol.* 23, 913–917.

Abu-Hakima, R. and Davey, K. G. (1977b) The action of juvenile hormone on the follicle cells of *Rhodnius prolixus*: the importance of volume changes. *J. exp. Biol.* 69, 33–44.

Abu-Hakima, R. and Davey, K. G. (1977c). The action of juvenile hormone on follicle cells of *Rhodnius prolixus in vitro*: The effect of colchicine and cytochalasin B. *Gen. comp. Endocrinol.* 32, 360–370.

Acholonu, A. D. and Finn, O. J. (1974) *Moniliformis moniliformis* (Acanthocephala: Moniliformidae) in the cockroach *Periplaneta americana* in Puerto Rico. *Trans. Am. Micros. Soc.* 93, 141–142.

Adair, E. W. (1923) Notes sur *Periplaneta americana* L. et *Blatta orientalis* L. (Orthop.). *Bull. Soc. ent. Egypt* 7, 18–38.

Adams, M. A., Nakanishi, K., Still, W. C., Arnold, E. V., Clardy, J. and Persoons, C. J. (1979) Sex pheromone of the American cockroach: absolute configuration of Periplanone B. *J. Am. Chem. Soc.* 101, 2495–2498.

Adiyodi, K. G. (1967) Abstracts of papers presented at the seminar of neuroendocrinology: Neuroendocrine mechanisms of reproduction in two cockroaches. *J. Animal Morphol. Physiol.* 14, 140–154.

Adiyodi, K. G. (1974) Extracerebral cephalic neuroendocrine complex of the blattids, *Periplaneta americana* (L.) and *Neostylopyga rhombifolia* (Stoll): an *in situ* study. *J. Morph.* 144, 469–484.

Adiyodi, K. G. and Adiyodi, R. G. (1970) Peroxidase isozymes in the conglobate gland of male cockroaches, *Periplaneta americana* L. *Ind. J. exp. Biol.* 8, 55–56.

Adiyodi, K. G. and Adiyodi, R. G. (1972) Lipids in the accessory sex gland complex of adult male American cockroaches. *Proc. IV Int. Cong. Histochem. Cytochem.*, p. 255.

Adiyodi, K. G. and Adiyodi, R. G. (1975) Morphology and cytology of the accessory sex glands in invertebrates. *Int. Rev. Cytol.* 43, 353–398.

Adiyodi, K. G. and Bern, H. A. (1968) Neuronal appearance of neurosecretory cells in the pars intercerebralis of *Periplaneta americana* (L.). *Gen. comp. Endocrinol.* 11, 88–91.

Adiyodi, K. G. and Nayar, K. K. (1965) Effects of administration of clomiphene on reproduction in female cockroaches. *Zool. Jahrb. Abt. Physiol.* 71, 669–676.

Adiyodi, K. G. and Nayar, K. K. (1966a) Some neuroendocrine aspects of reproduction in the viviparous cockroach, *Tucho blatta sericea.* (Saussure). *Zool. Jahrb. Abt. Physiol.* 72, 453–462.

Adiyodi, K. G. and Nayar, K. K. (1966b) Haemolymph proteins and reproduction in *Periplaneta americana. Curr. Sci.* 23, 587–588.

Adiyodi, K. G. and Nayar, K. K. (1967) Haemolymph proteins and reproduction in *Periplaneta americana*: The nature of conjugated proteins and the effect of cardiac allatectomy on protein metabolism. *Biol. Bull.* 133, 271–286.

Adiyodi, K. G. and Nayar, K. K. (1968) The conjugated plasma proteins in adult females of *Periplaneta americana* (L.) under starvation and other stress. *Comp. Biochem. Physiol.* 27, 95–104.

Adiyodi, R. G. (1969) Morphometric relations between corpus allatum and ovary in *Periplaneta americana* (L.). *J. Kerala Acad. Biol.* 1, 36–39.

Adiyodi, R. G. and Adiyodi, K. G. (1974) Ultrastructure of the utriculi majores in the mushroom-shaped male accessory gland of *Periplaneta americana* (L.). *Z. Zellforsch.* 147, 433–440.

Agarwal, H. C. and Casida, J. E. (1960) Nature of housefly sterols. *Biochem. biophys. Res. Commun.* 3, 508–512.

Aggarwal, S. K. (1960) Histochemical studies of the carbohydrates, proteins and nucleic acids in the oogenesis of *Periplaneta americana* and *Chrotogonus rachypterus. Res. Bull. Panjab Univ. Sci.* 11, 147–153.

Agrell, I. (1964) Physiological and biochemical changes during insect development. In: *The Physiology of Insecta*, Rockstein, M., Ed., Vol. I., pp. 91–148. Academic Press, New York.

Agui, N. and Hiruma, K. (1977) Ecdysone as a feedback regulator for the neurosecretory brain cells in *Mamestra brassicae. J. Insect Physiol.* 23, 1393–1396.

Aidley, D. J. (1975) Excitation–contraction coupling and mechanical properties. In: *Insect Muscle*, Usherwood, P. N. R., Ed., pp. 337–356. Academic Press, London.

Akamatsu, Y., Dunn, P. E., Kezdy, F. J., Kramer, K. J., Law, J. H., Reibstein, D. and Sanburg, L. L. (1975) Biochemical aspects of juvenile hormone action in insects. In: *Control Mechanisms in Development*, Meints, R. and Davis, E., Eds., pp. 123–149. Plenum Press, New York.

Ali, Z. I. and Pipa, R. (1978) The abdominal perisympathetic neurohemal organs of the cockroach *Periplaneta americana*: innervation revealed by cobalt chloride diffusion. *Gen. comp. Endocrinol.* 36, 396–401.

Aloe, L. and Levi-Montalcini, R. (1972a) Interrelation and dynamic activity of visceral muscle and nerve cells from insect embryos in long-term cultures. *J. Neurobiol.* 3, 3–23.

Aloe, L. and Levi-Montalcini, R. (1972b) *In vitro* analysis of the frontal and ingluvial ganglia from nymphal specimens of the cockroach *Periplaneta americana. Brain Res.* 44, 147–163.

Alsop, D. W. (1978) Comparative analysis of the intrinsic leg musculature of the American cockroach, *Periplaneta americana* (L.). *J. Morph.* 158, 199–242.

Altner, H. (1977) Insect sensillum specificity and structure: an approach to a new typology. In: *Olfaction and Taste*, Vol. VI., McLeod, P., Ed., pp. 295–303. Academic Press, London.

Altner, H., Sass, H. and Altner, I. (1977) Relationship between structure and function of antennal chemo-, hygro-, and thermoreceptive sensilla in *Periplaneta americana. Cell Tiss. Res.* 176, 389–405.

Altner, H. and Stetter, H. (1980) Olfactory input from the maxillary palps in the cockroach as compared with the antennal input. *Joint Cong. Chemocreception ECRO IV/ISOT VII*, Noordwiijkeshout, Holland.

Andersen, S. O. (1970) Isolation of arterenone (2-amino-3′, 4′-dihydroxy-acetophenone) from hydrolysates of sclerotized insect cuticle. *J. Insect Physiol.* 16, 1951–1959.

Andersen, S. O. (1971) Phenolic compounds isolated from insect hard cuticle and their relationship to the sclerotization process. *Insect Biochem.* 1, 157–170.

Andersen, S. O. (1972) An enzyme from locust cuticle involved in the formation of crosslinks from *N*-acetyldopamine. *J. Insect Physiol.* 18, 527–540.

Andersen, S. O. (1974) Evidence for two mechanisms of sclerotization in insect cuticle. *Nature* 251, 507–508.

Andersen, S. O. (1979) Biochemistry of insect cuticle. *A. Rev. Ent.* 24, 29–61.

Andersen, S. O. and Barrett, F. M. (1971) The isolation of ketocatechols from insect cuticle and their possible role in sclerotization. *J. Insect Physiol.* 17, 69–83.

Andersen, S. O. and Weis-Fogh, T. (1964) Resilin, a rubber-like protein in arthropod cuticle. *Adv. Insect Physiol.* 2, 1–65.

Anderson, A. D. and March, R. B. (1956) Inhibitors of carbonic anhydrase in the American cockroach, *Periplaneta americana* (L.). *Can. J. Zool.* 34, 68–74.

Anderson, A. D. and Patton, R. L. (1955) *In vitro* studies of uric acid synthesis in insects. *J. exp. Zool.* 128, 443–451.

Anderson, D. T. (1972a) The development of hemimetabolous insects. In: *Developmental Systems: Insects*, Vol. 1, Counce, S. J. and Waddington, C. H., Eds., pp. 96–165. Academic Press, New York.

Anderson, D. T. (1972b) The development of holometabolous insects. In: *Developmental Systems: Insects*, Vol. 1, Counce, S. J. and Waddington, C. H., Eds., pp. 166–241. Academic Press, New York.

Anderson, D. T., (1973) *Embryology and Phylogeny in Annelids and Arthropods*. Pergamon Press, Oxford.

Anderson, E. (1964) Oocyte differentiation and vitellogenesis in the roach, *Periplaneta americana*. *J. Cell Biol.* 20, 131–155.

Anderson, E. (1969) Oogenesis in the cockroach, *Periplaneta americana*, with special reference to the specialization of the oolemma and the fate of coated vesicles. *J. Microscopie* 8, 721–238.

Anderson, H., Edwards, J. S. and Palka, J. (1980) Developmental Neurobiology of invertebrates. *A. Rev. Neurosci.* 3, 97–139.

Anderson, L. M. (1971) Protein synthesis and uptake by isolated *Cecropia* oocytes. *J. Cell Sci.* 8, 735–750.

Anderson, L. M. and Telfer, W. J. (1969) A follicle cell contribution to the yolk spheres of moth oocytes. *Tissue Cell* 1, 633–644.

Anderson, M. and Cochrane, D. G. (1977) Studies on the mid-gut of the desert locust, *Schistocerca gregaria*. I. Morphology and electrophysiology of the muscle coat. *Physiol. Ent.* 2, 247–253.

Anderson, M. and Cochrane, D. G. (1978) Studies on the mid-gut of the desert locust, *Schistocerca gregaria*. II. Ultrastructure of the muscle coat and its innervation. *J. Morph.* 156, 257–278.

Annandale, N. (1910) Cockroaches as predatory insects. *Rec. Ind. Mus.* 3, 201–202.

Anonymous (1967) Up a tree for American roach control. *Pest Control* 35, 16.

Anwyl, R. and Usherwood, P. N. R. (1974) Voltage clamp studies of glutamate synapse. *Nature* 252, 591–593.

Applebaum, S. W. and Schlesinger, H. M. (1973) Regulation of locust fat-body phosphorylase. *Biochem. J.* 135, 37–41.

Aranda, L. C. and Luco, J. V. (1969) Further studies of an electrical correlate to learning: Experiments in an isolated insect ganglion. *Physiol. Behav.* 4, 133–137.

Arlian, L. G. and Veselica, M. M. (1979) Water balance in insects and mites. *Comp. Biochem. Physiol.* 64, 191–200.

Arnold, J. W. (1969) Periodicity in the proportion of hemocyte categories in the giant cockroach, *Blaberus giganteus. Can. Ent.* 101, 68–77.

Arnold, J. W. (1972) A comparative study of the haemocytes (blood cells) of cockroaches (Insecta: Dictyoptera: Blattaria) with a view of their significance in taxonomy. *Can. Ent.* 104, 309–326.

Arnold, W. J. (1960). *The Anatomy of the Cephalic Nervous System of the Cockroach, Periplaneta americana* (L.), *with histological and histochemical studies of the brain.* Ph.D. Thesis, University of California, Berkeley.

Aschoff, J., Hoffman, K., Hermann, P. and Wever, R. (1975) Re-entrainment of circadian rhythms after phase-shifts of the Zeitgeber. *Chronobiologia* 2, 23–78.

Ashhurst, D. E. (1961a) A histochemical study of the connective-tissue sheath of the nervous system of *Periplaneta americana. Q. J. Micros. Sci.* 102, 455–461.

Ashhurst, D. E. (1961b) An acid mucopolysaccharide in cockroach ganglia. *Nature* 191, 1224–1225.

Ashhurst, D. E. and Costin, N. M. (1971) Insect mucosubstances. II. The mucosubstances of the central nervous system. *Histochem. J.* 3, 297–310.

Ashhurst, D. E. and Patel, N. G. (1963) Hyaluronic acid in cockroach ganglia. *Ann. ent. Soc. Am.* 56, 182–184.

Aston, R. J. (1975) The role of adenosine-3′; 5′-cyclic monophosphate in relation to the diuretic hormone of *Rhodnius prolixus. J. Insect Physiol.* 21, 1873–1877.

Atwood, H. L. and Jahromi, S. S. (1967) Strychnine and neuromuscular transmission in the cockroach. *J. Insect. Physiol.* 13, 1065–1073.

Atwood, H. L., Smyth, Jr., T. and Johnston, H. S. (1969) Neuromuscular synapses in the cockroach extensor tibiae muscle. *J. Insect Physiol.* 15, 529–535.

Aubele, E. and Klemm, N. (1977) Origin, destination and mapping of tritocerebral neurons of locust. *Cell Tissue Res.* 178, 199–219.

Auclair, J. L. (1959) Amino acid oxidase activity in the fat body and Malpighian tubes of some insects. *J. Insect Physiol.* 3, 57–62.

Bade, M. L. (1964) Biosynthesis of fatty acids in the roach *Eurycotis floridana. J. Insect Physiol.* 10, 333–341.

Bailey, E. (1975) Biochemistry of insect flight. Part 2: Fuel supply. In: *Insect Biochemistry and Function*, Candy, D. and Kilby, B. A., Eds., pp. 91–176. Chapman and Hall, London.

Baker, E. W., Evans, T. M., Gould, D. J., Hull, W. B. and Keegan, H. L. (1956) *A Manual of Parasitic Mites of Medical or Economic Importance.* Nat. Pest Control Ass., New York.

Baker, G. L., Vroman, H. E. and Padmore, J. (1963) Hydrocarbons of the American cockroach. *Biochem. biophys. Res. Commun.* 13, 360–365.

Ball, H. J. (1965) Photosensitivity in the terminal ganglion of *Periplaneta americana* (L.). *J. Insect Physiol.* 11, 1311–1315.

Ball, H. J. (1971) The receptor site for photic entrainment of circadian activity rhythms in the cockroach *Periplaneta americana. Ann. ent. Soc. Am.* 64, 1010–1015.

Ball, H. J. (1977) Spectral transmission through the cuticle of the American cockroach, *Periplaneta americana. J. Insect Physiol.* 23, 1–4.

Ballan-Dufrançais, C. (1972) Ultrastructure de l'iléon de *Blattella germanica* (L.) (Dictyoptère). Localisation, genèse et composition des concretions minerales intra-cytoplasmiques. *Z. Zellforsch.* 133, 163–179.

Banks, W. M., Bruce, A. S. and Peart, H. T. (1975) The effects of temperature, sex and circadian rhythm on oxygen consumption in two species of cockroaches. *Comp. Biochem. Physiol.* 52A, 223–228.

Barker, J. L. (1977) Physiological roles of peptides in the nervous system. In: *Peptides in Neurobiology*, Gainer, H., Ed., pp. 295–343. Plenum Press, New York and London.

Barth, R. H. (1970) The mating behavior of *Periplaneta americana* (Linnaeus) and *Blatta*

orientalis Linnaeus (Blattaria-Blattinae), with notes on three additional species of *Periplaneta* and interspecific action of female sex pheromones. *Z. Tierpsychol.* 27, 722–748.
Barth, R. H., Jr. and Bell, W. J. (1970) Physiology of the reproductive cycle in the cockroach *Byrsotria fumigata* (Guérin). *Biol. Bull.* 139, 447–460.
Barth, R. H. and Lester, L. J. (1973) Neuro-hormonal control of sexual behavior in insects. *A. Rev. Ent.* 18, 445–472.
Barth, R. H. and Sroka, P. (1975) Initiation and regulation of oöcyte growth by the brain and corpora allata of the cockroach, *Nauphoeta cinerea. J. Insect Physiol.* 21, 321–330.
Barton-Browne, L. (1975) Regulatory mechanisms in insect feeding. *Adv. Insect Physiol.* 11, 1–116.
Bar-Zev, A. and Kaulenas, M. S. (1975) The effect of β-ecdysone on *Gromphadorhina* adult female fat body at the transcriptional and translational levels. *Comp. Biochem. Physiol.* 51B, 355–361.
Bate, C. M. (1976a) Nerve growth in cockroaches (*Periplaneta americana*) with rotated ganglia. *Experientia* 32, 451–452.
Bate, C. M. (1976b) Embryogenesis of an insect nervous system. 1. A map of the thoracic and abdominal neuroblasts in *Locusta migratoria. J. Embryol. exp. Morph.* 35, 107–123.
Baudet, J. L. (1974a) Recherches sur l'appareil respiratoire des Blattes. I. Étude du plan des trachées à partir d'un exemple: *Nauphoeta cinerea* (Dict. Blaberidae). *Ann. Soc. Ent. France* 10, 903–916.
Baudet, J. L. (1974b) Morphologie et ultrastructure du positif trachéen permettant l'émission sonore chez *Gromphadorhina portentosa* Schaum (Dictyoptères, Blaberidae). *C. R. Acad. Sci. Paris* Ser. D, 279, 1175–1178.
Baudet, J. L. and Sellier, E. (1975) Recherches sur l'appareil respiratoire des Blattes. II. Les vésiculations trachéenes et leur évolution dans le sous-ordre de Blattaria. *Ann. Soc. Ent. France* 11, 481–489.
Beament, J. W. L. (1955) Wax secretion in the cockroach. *J. exp. Biol.* 32, 514–538.
Beament, J. W. L. (1958a) A paralyzing agent in the blood of cockroaches. *J. Insect Physiol.* 2, 199–214.
Beament, J. W. L. (1958b) The effect of temperature on the waterproofing mechanisms of an insect. *J. exp. Biol.* 35, 494–519.
Beams, H. W., Anderson, E. and Kessel, R. G. (1962) Electron microscope observations on the phallic (conglobate) gland of the cockroach *Periplaneta americana. J. R. micros. Soc.* 81, 85–89.
Beattie, T. M. (1971) Histology, histochemistry and ultrastructure of neurosecretory cells in the optic lobe of the cockroach, *Periplaneta americana. J. Insect Physiol.* 17, 1843–1855.
Beattie, T. M. (1976) Autolysis in axon terminals of a new neurohaemal organ in the cockroach *Periplaneta americana. Tissue Cell* 8, 305–310.
Becht, G. Hoyle, G. and Usherwood, P. N. R. (1960) Neuromuscular transmission in the coxal muscles of the cockroach. *J. Insect. Physiol.* 4, 191–201.
Beck, S. D. (1963) Physiology and ecology of photoperiodism. *Bull. ent. Soc. Am.* 9, 8–16.
Beck, S. D. (1968) *Insect Photoperiodism*. Academic Press, New York.
Beck, S. D. (1976) Photoperiodic determination of insect development and diapause. V. Diapause, circadian rhythms and phase response curves, according to the dual system theory. *J. comp. Physiol.* 107, 97–111.
Beckel, W. E. (1958) The morphology, histology and physiology of the spiracular regulatory apparatus of *Hyalophora cecropia* L. (Lep.). *Proc. 10th Int. Cong. Ent., Montreal* Vol. 2, pp. 87–115.
Bell, W. J. (1969a) Dual role of juvenile hormone in the control of yolk formation in *Periplaneta americana. J. Insect Physiol.* 15, 1270–1290.
Bell, W. J. (1969b) Continuous and rhythmic reproductive cycle observed in *Periplaneta americana* L. *Biol. Bull.* 137, 239–249.
Bell, W. J. (1970) Demonstration and characterization of two vitellogenic blood proteins

in *Periplaneta americana*: an immunochemical analysis. *J. Insect Physiol.* 16, 291–299.
Bell, W. J. (1971) Starvation-induced oöcyte resorption and yolk protein salvage in *Periplaneta americana. J. Insect Physiol.* 17, 1099–1111.
Bell, W. J. (1972a) Transplantation of ovaries into male *Periplaneta americana*: effects of vitellogenesis and vitellogenin secretion. *J. Insect Physiol.* 18, 851–855.
Bell, W. J. (1972b) Yolk formation by transplanted cockroach oöcytes. *J. exp. Zool.* 181, 41–48.
Bell, W. J. (1978) Directional information in tactile behavioral acts involved in agonistic encounters of cockroaches. *Physiol. Ent.* 3, 1–6.
Bell, W. J. and Barth, Jr, R. H. (1970) Quantitative effects of juvenile hormone on reproduction in the cockroach *Byrsotria fumigata. J. Insect Physiol.* 16, 2303–2313.
Bell, W. J. and Barth, Jr, R. H. (1971) Initiation of yolk deposition by juvenile hormone. *Nature* 230, 220–221.
Bell, W. J. and Bohm, M. (1975) Oosorption in insects. *Biol. Rev. Camb. Philosoph. Soc.* 50, 373–396.
Bell, W. J., Burk, T. and Sams, G. R. (1973) Cockroach aggregation pheromone: directional orientation. *Behav. Biol.* 9, 251–255.
Bell, W. J., Burns, R. E. and Barth, R. H. (1974) Quantitative aspects of the male courting response in the cockroach *Byrsotria fumigata* (Guérin) (Blattaria). *Behav. Biol.* 10, 419–433.
Bell, W. J. and Gorton, Jr, R. E. (1978) Informational analysis of agonistic behaviour and dominance hierarchy formation in a cockroach, *Nauphoeta cinerea. Behaviour* 67, 217–235.
Bell, W. J., Gorton, R. E., Tourtellot, M. K. and Breed, M. D. (1979) Comparison of agonistic behaviour in five species of cockroaches. *Insectes Sociaux* 26, 252–263.
Bell, W. J. and Kramer, E. (1979) Search and anemotaxis in cockroaches. *J. Insect Physiol.* 25, 631–640.
Bell, W. J. and Kramer, E. (1980) Sex pheromone-stimulated orientation responses by the American cockroach on a servosphere apparatus. *J. chem. Ecol.* 6, 287–295.
Bell, W. J., Parsons, C. and Martinko, E. A. (1972) Cockroach aggregation pheromones: analysis of aggregation tendency and species specificity. *J. Kansas Ent. Soc.* 45, 414–420.
Bell, W. J., Robinson, S., Tourtellot, M. K. and Breed, M. D. (1978a) An ethometric analysis of agonistic behavior and social hierarchies in the cockroach, *Eublaberus posticus. Z. Tierpsychol.* 48, 203–218.
Bell, W. J. and Sams, G. R. (1973) Aggressiveness in the cockroach *Periplaneta americana* (Orthoptera–Blattidea). *Behav. Biol.* 9, 581–593.
Bell, W. J. and Sams, G. R. (1974) Factors promoting vitellogenic competence and yolk deposition in the post-ecdysis female. *J. Insect Physiol.* 20, 2475–2485.
Bell, W. J. and Sams, G. R. (1975) Factors promoting vitellogenic competence and yolk deposition in the cockroach ovary: transitions during adult ecdysis. *J. Insect Physiol.* 21, 173–180.
Bell, W. J. and Schal, C. (1980) Patterns of turning in courtship orientation of the male German cockroach. *Animal Behav.* 28, 86–94.
Bell, W. J. and Tobin, T. (1981) Orientation to sex pheromone in the American cockroach: analysis of chemotactic mechanisms. *J. Insect Physiol.* (In press.)
Bell, W. J., Vuturo, S. B. and Bennett, M. (1978b) Endokinetic turning and programmed courtship acts of the male German cockroach. *J. Insect Physiol.* 24, 369–374.
Bell, W. J., Vuturo, S. B., Robinson, S. and Hawkins, W. A. (1977) Attractancy of the American cockroach sex pheromone. *J. Kansas Ent. Soc.* 50, 503–507.
Belton, P. and Brown, B. E. (1969) The electrical activity of cockroach visceral muscle fibers. *Comp. Biochem. Physiol.* 28, 853–863.
Benke, G. W., Wilkinson, C. F. and Telford, J. N. (1972). Microsomal oxidases in a cockroach, *Gromphadorhina portentosa. J. econ. Ent.* 65, 1221–1229.

Bentley, D., Keshishian, H., Shankland, M. and Toroian-Raymond, A. (1979) Quantitative staging of embryonic development of the grasshopper, *Schistocerca nitens. J. Embryol. exp. Morph.* 54, 47–74.

Beránek, R. and Miller, P. L. (1968) The action of iontophoretically applied glutamate on insect muscle fibres. *J. exp. Biol.* 49, 83–93.

Berlind, A. (1977) Cellular dynamics in invertebrate neurosecretory systems. *Int. Rev. Cytol.* 49, 172–251.

Bernard, A. (1976) Étude topographique des inteneurones ocellaires et de quelques uns de leurs prolongements chez *Periplaneta americana. J. Insect Physiol.* 22, 569–577.

Bernard, A. (1977) Mise en évidence simultanée des neurones ocellaires et des cellules neuro-sécrétrices de la pars sur montage *in toto* chez la Blatte. *C.R. Acad. Sci. Paris* 284, 1807–1810.

Bernton, H. S. and Brown, H. (1964) Insect allergy – preliminary studies of the cockroach. *J. Allergy* 35, 506–513.

Bernton, H. S. and Brown, H. (1969) Insect allergy: the allergenic potentials of the cockroach. *South. med. J. Nashville* 62, 1207–1210.

Berridge, M. J. (1970a) A structural analysis of intestinal absorption. *Symp. R. ent. Soc. Lond.* 5, 135–151.

Berridge, M. J. (1970b) Osmoregulation in terrestrial arthropods. In: *Chemical Zoology*, Florkin, M. and Scheer, B. T., Eds., pp. 287–319. Academic Press, New York.

de Bessé, N. (1967) Neurosecretion dans la chaine neurveuse ventrale de deux blattes, *Leucophaea maderae* (F.) et *Periplaneta americana* (L.). *Bull. Soc. Zool. France* 92, 73–86.

Bhakthan, N. M. G. and Gilbert, L. I. (1970) An autoradiographic and biochemical analysis of palmitate incorporation into fat body lipid. *J. Insect Physiol.* 16, 1783–1796.

Bhat, U. K. M. (1970) Heterogeneity in the fibre composition in the flight muscles of *Periplaneta americana. Experientia* 26, 995–997.

Bhatia, S. S. and Tonapi, G. T. (1968) Effects of fumigants on the spiracles and the possible role of a neurohumor on the spiracular activity of *Periplaneta americana* (L.). *Experientia* 24, 1224–1225.

Biellman, G. (1960) Étude du cycle des mues chez *Periplaneta americana. Bull. Soc. Zool. France* 84, 340–351.

Bignell, D. E. (1976) Gnawing activity, dietary carbohydrate deficiency and oothecal production in the American cockroach (*Periplaneta americana*). *Experientia* 32, 1405–1406.

Bignell, D. E. (1977a) Some observations on the distribution of gut flora in the American cockroach, *Periplaneta americana. J. Invert. Pathol.* 29, 338–343.

Bignell, D. E. (1977b) An experimental study of cellulose and hemicellulose degradation in the alimentary canal of the American cockroach. *Can. J. Zool.* 55, 579–589.

Bignell, D. E. (1978) Effects of cellulose in the diet of cockroaches. *Ent. exp. appl.* 24, 54–57.

Bignell, D. E. (1980) An ultrastructural study and stereological analysis of the colon wall in the cockroach *Periplaneta americana. Tissue Cell* 12, 153–164.

Bignell, D. E. and Mullins, D. E. (1977) A preliminary investigation of the effects of diets on lesion formation in the hindgut of adult female American cockroaches. *Can. J. Zool.* 55, 1100–1109.

Blaine, W. D. and Dixon, S. E. (1970) Hormonal control of spermatogenesis in the cockroach, *Periplaneta americana* (L.). *Can. J. Zool.* 48, 283–287.

Blaine, W. D. and Dixon, S. E. (1973) The effect of juvenile hormone on the function of the accessory gland of the adult male cockroach *Periplaneta americana* (Orthoptera: Blattidea). *Can. Ent.* 105, 1275–1280.

Blaine, W. D. and Dixon, S. E. (1975) Testicular development in the cockroach, *Periplaneta americana* (L.). *Proc. ent. Soc. Ont.* 106, 60–65.

Bland, K. P. and House, C. R. (1971) Function of the salivary glands of the cockroach *Nauphoeta cinerea*. *J. Insect Physiol.* 17, 2069–2084.
Blankemeyer, J. T. and Harvey, W. R. (1977) Insect midgut as a model epithelium. In: *Water Relations in Membrane Transport in Plants and Animals*. Jungreis, A. M., Hodges, T. K., Kleinzeller, A. and Schultz, S. G., Eds., pp. 161–182. Academic Press, New York.
Blaschko, H., Colhoun, E. H. and Frontali, N. (1961) Occurrence of amine oxidase in an insect, *Periplaneta americana* L. *J. Physiol.* 156.
Blochmann, F. (1887) Vorkommen bakterienahnliche Korperchen in den Geweben und Eiern verschiedener Insekten. *Biol. Zentralbl.* 7, 606–608.
Block, E. F. IV and Bell, W. J. (1974) Ethometric analysis of pheromone receptor function in cockroaches, *Periplaneta americana*. *J. Insect Physiol.* 20, 993–1003.
Bock, E. (1941) Wechselbeziehungen zwischen den Keimblättern bei der Organbildung von *Chrysopa perla* (L.) 1: Die Entwicklung des Ektoderms in mesodermdefekten keimteilen. *Roux Arch.* 141, 159–247.
Bodenstein, D. (1953a) Embryonic development. In: *Insect Physiology*, Roeder, K. D., Ed., Ch. 29, pp. 780–821. Wiley, New York.
Bodenstein, D. (1953b) Postembryonic development. In: *Insect Physiology*, Roeder, K. D., Ed., Ch. 30, pp. 822–865.
Bodenstein, D. (1953c) Studies on the humoral mechanisms in growth and metamorphosis of the cockroach, *Periplaneta americana*. II. The function of the prothoracic gland and the corpus cardiacum. *J. exp. Zool.* 123, 413–433.
Bodenstein, D. (1953d) Studies on the humoral mechanisms in growth and metamorphosis of the cockroach, *Periplaneta americana*. III. Humoral effects on metabolism. *J. exp. Zool.* 124, 105–115.
Bodenstein, D. (1955a) Insects. In: *Analysis of Development*, Willier, B., Weiss, P. and Hamburger, V., Eds., pp. 337–345. Saunders, New York.
Bodenstein, D. (1955b) Contributions to the problem of regeneration in insects. *J. exp. Zool.* 129, 209–224.
Bodenstein, D. (1956) Humoral aspects of insect regeneration. *Proc. X Int. Cong. Ent.* Vol. 2, p. 43.
Bodenstein, D. (1957) Studies on nerve regeneration in *Periplaneta americana*. *J. exp. Zool.* 136, 89–115.
Bodenstein, D. (1959) The role of hormones in the regeneration of insect organs. *Scientia (Milan)* 94, 19–23.
Bodenstein, D. and Shaaya, E. (1968) The function of the accessory sex glands in *Periplaneta americana* (L.). 1. A quantitative bioassay for the juvenile hormone. *Proc. natn. Acad. Sci. U.S.A.*, 59, 1223–1230.
Bodenstein, D. and Sprague, I. B. (1959) The developmental capacities of the accessory sex glands in *Periplaneta americana*. *J. exp. Zool.* 142, 177–202.
Bodnaryk, R. P., Brunet, P. C. J. and Koeppe, J. K. (1974) On the metabolism of *N*-acetyldopamine in *Periplaneta americana*. *J. Insect Physiol.* 20, 911–923.
Boeckh, J. (1974) Die Reaktionen olfaktorischer Neurone im Deutocerebrum von Insekten im Vergleich zu den Antwortmustern der Geruchssinneszellen. *J. comp. Physiol.* 90, 183–205.
Boeckh, J., Ernst, K. D., Sass, H. and Waldow, U. (1975) Coding of odour quality in the insect olfactory pathway. In: *Olfaction and Taste*, Vol. V Denton, D. A. and Coghlan, J. P., Eds., Academic Press, New York, pp. 239–245.
Boeckh, J., Sass, H. and Wharton, D. R. A. (1970) Antennal receptor reactions to female sex attractant in *Periplaneta americana*. *Science* 168, 589.
Bogenschütz, H. (1978) Fuhler-Bewegungen wirtesuchender Weibchen der Schlupfwespe *Coccygomimus turionellae* (Hymenoptera: Ichneumonidea). *Ent. Germ.* 4, 122–132.
Bohn, H. (1965) Analyse der Regenerationsfehigkeit der insektenextremität durch Amputations-und Transplantationsversuche an larven der Afrikanische Schabe

(*Leucophaea maderae* Fabr.) I. Regeneratspotenzen. II. Achsendetermination. *Roux Arch.* 156, 49–74, 449–503.

Bohn, H. (1966) Transplantationsexperimente mit interkalerer Regeneration zum Nachweis eines sich segmental widerholenden Gradienten im bein von *Leucophaea maderae* (Blattaria.) *Verhandl. Deutsch. Zool. Gesel., Gottingen* 30, suppl. bd. 499–508.

Bohn, H. (1972) The origin of the epidermis in the supernumerary regenerates of triple legs in cockroaches (Blattaria). *J. Embryol. exp. Morph.* 28, 185–208.

Bohn, H. (1974a) Pattern reconstitution in abdominal segments of *Leucophaea maderae*, Blattaria. *Nature* 248, 608–609.

Bohn, H. (1974b) Extent and properties of the regeneration field in the larval legs of cockroaches (*Leucophaea maderae*) I. Extirpation experiments. *J. Embryol. exp. Morphol.* 31, 557–572.

Bohn, H. (1976) Tissue interactions in the regenerating cockroach leg. *Symp. R. ent. Soc.* 8, 170–185.

Bohn, H. (1977a) Conditioning of a glass surface for the outgrowth of insect epidermis (*Leucophaea maderae*, Blattaria). *In Vitro* 13, 100–107.

Bohn, H. (1977b) Enzymatic and immunological characterization of the conditioning factor for epidermal outgrowth in the cockroach *Leucophaea maderae. J. Insect Physiol.* 23, 1063–1073.

Boistel, J. (1959) Quelques caractéristiques électriques de la membrane de la fibre nerveuse au repos d'un insecte (*Periplaneta americana*). *C. R. Soc. Biol. Paris* 153, 1009–1013.

Bollade, D. and Boucrot, P. (1973) Incorporation d'acides gras radioactifs alimentaires dans les lipides chez *Periplaneta americana. Insect Biochem.* 3, 123–137.

Bollade, D., Paris, R. and Moulins, M. (1970) Origine et mode d'action de la lipase intestinale chez les blattes. *J. Insect Physiol.* 16, 45–53.

Bonhag, P. F. (1959) Histological and histochemical studies on the ovary of the American cockroach *Periplaneta americana* (L.). *Univ. Calif. Publ. Ent.* 16, 81–124.

Bonhag, P. F. and Arnold, W. J. (1961) Histology, histochemistry and tracheation of the ovariole sheaths in the American cockroach *Periplaneta americana* (L.). *J. Morphol.* 108, 107–129.

Bordas, L. (1909) Recherches anatomiques, histologiques et physiologiques sur les organes appendiculaires de l'appareil reproducteur femelles des Blattes (*Periplaneta orientalis*). *Ann. Sci. nat.* 9, 71–121.

Borst, D. W. and Engelmann, F. (1974). *In vitro* secretion of α-ecdysone by prothoracic glands of a hemimetabolous insect, *Leucophaea maderae* (Blattaria). *J. exp. Zool.* 189, 413–419.

Botham, R. P., Beadle, D. J., Hart, R. J., Potter, C. and Wilson, R. G. (1978) Synaptic vesicle depletion and glutamate uptake in a nerve–muscle preparation of the locust, *Locusta migratoria* L. *Experientia* 34, 209–210.

Bouligand, Y. (1965) Sur une architecture torsadee repandue dans de nombreuses cuticules d'arthropodes. *C.R. Acad. Sci., Paris* 261, 3665–3668.

Boulos, Z. and Terman, M. (1979) Splitting of circadian rhythms in the rat. *J. comp. Physiol.* 134, 75–83.

Bowser-Riley, F. and House, C. R. (1976) The actions of some putative neurotransmitters on the cockroach salivary gland. *J. exp. Biol.* 64, 665–676.

Boyer, A. C. (1975) Sorption of tetrachlorvinphos insecticide (Gardona) to the hemolymph of *Periplaneta americana. Pestic. Biochem. Physiol.* 5, 135–141.

Bracke, J. W., Cruden, D. L. and Markovetz, A. J. (1978) Effect of metranidazole on the intestinal microflora of the American cockroach, *Periplaneta americana* L. *Antimicrob. Agents Chemother.* 13, 115–120.

Bracke, J. W., Cruden, D. L. and Markovetz, A. J. (1979) Intestinal microbial flora of the American cockroach, *Periplaneta americana*, L. *Appl. environ. Microbiol.* 38, 945–955.

Bracke, J. W. and Markovetz, A. J. (1980) Transport of bacterial end products from the colon of *Periplaneta americana. J. Insect Physiol.* 26, 85–90.
Brady, J. (1967a) The relationship between blood ions and blood cell density in insects. *J. exp. Biol.* 48, 31–38.
Brady, J. (1967b) Control of the circadian rhythm of activity in the cockroach. I. The role of the corpora cardiaca, brain and stress. *J. exp. Biol.* 47, 153–163.
Brady, J. (1967c) Control of circadian rhythm of activity in the cockroach. II. The role of the suboesophageal ganglion and ventral nerve cord. *J. exp. Biol.* 47, 165–178.
Brady, J. (1967d) Histological observations on circadian changes in the neurosecretory cells of cockroach suboesophageal ganglia. *J. Insect Physiol.* 13, 201–213.
Brady, J. (1968) Control of the circadian rhythm of activity in the cockroach. III. A possible role of the blood-electrolytes. *J. exp. Biol.* 49, 39–47.
Brady, J. (1969) How are insect circadian rhythms controlled? *Nature* 233, 781–784.
Brady, J. (1974) The physiology of insect circadian rhythms. *Adv. Insect Physiol.* 10, 1–116.
Brady, J. and Maddrell, S. H. P. (1967) Neurohaemal organs in the medial nervous system of insects. *Z. Zellforsch.* 76, 389–404.
Braitenberg, V. and Strausfeld, N. J. (1973) Principles of the mosaic organisation in the visual system's neuropil of *Musca domestica* L. In: *Handbook of Sensory Physiology*, Vol. VII/3, Central processing of visual information, Part A. pp. 631–660. Springer-Verlag, New York.
Breed, M. D. and Rasmussen, C. D. (1980) Behavioral strategies during inter-male agonistic encounters in a cockroach. *Anim. Behav.* 28, 1063–1069.
Breed, M. D., Smith, S. K. and Gall, B. G. (1980) Systems of mate selection in a cockroach species with male dominance hierarchies. *Anim. Behav.* 28, 130–134.
Bretschneider, F. (1913) Der Centralkörper und die pilzförmigen Körper im Gehirn der Insekten. *Zool. Anz.* 41, 560–569.
Bretschneider, F. (1914) Über die Gehirne der Küchenschabe und des Mehlkäfers. *Jena Z. Naturw.* 52, 269–362.
Brindley, H. H. (1897) On the regeneration of the legs in the Blattidae. *Proc. Zool. Soc.* 1897, 903–916.
Brindley, H. H. (1898) On the reproduced appendages in arthropoda. *Proc. Zool. Soc.* 1898, 924–958.
Brindley, W. A. and Dahm, P. A. (1970) Microscopic examination of parathion-activating microsomes from American cockroach fat body. *J. econ. Ent.* 63, 31–38.
Brookes, V. J. (1969) The maturation of the oocytes in the isolated abdomen of *Leucophaea maderae. J. Insect Physiol.* 15, 621–631.
Brooks, M. A. (1960) Some dietary factors that affect ovarial transmission of symbiotes. *Proc. Helm. Soc., Wash.* 27, 212–220.
Brooks, M. A. (1970) Comments on the classification of intracellular symbiotes of cockroaches and a description of the species. *J. Invert. Pathol.* 16, 249–258.
Brooks, M. A. and Kringen, W. B. (1972) Polypeptides and proteins as growth factors for aposymbiotic *Blattella germanica* (L.). In: *Insect and Mite Nutrition*, Rodrigues, J. G., Ed., pp. 353–364. North Holland Publ. Co., Amsterdam.
Brooks, M. A. and Richards, A. G. (1955) Intracellular symbiosis in cockroaches. I. production of aposymbiotic cockroaches. *Biol. Bull.* 109, 22–39.
Brooks, M. A. and Richards, A. G. (1956) Intracellular symbiosis in cockroaches. III. Reinfection of aposymbiotic cockroaches with symbiotes. *J. exp. Zool.* 132, 447–465.
Brooks, M. A. and Richards, K. (1966) On the *in vitro* culture of intracellular symbiotes of cockroaches. *J. Invert. Pathol.* 8, 150–157.
Brossut, R. (1970) L'interattraction chez *Blabera craniifer* Burm. (Insecta, dictyoptera): Secretion d'une pheromone par les glandes mandibulaires. *C.R. Acad. Sci., Paris Ser. D.* 270, 714–716.
Brossut, R., Dubois, P. and Rigaud, J. (1974) Le gregarisme chez *Blaberus craniifer*: isolement et identification de la pheromone. *J. Insect Physiol.* 20, 529–543.

Brossut, R., Dubois, P., Rigaud, J. and Sreng, L. (1975) Étude biochemique de la secretion des glandes tergales des Blattaria *Insect Biochem.* 5, 719–732.
Brossut, R. and Roth, L. M. (1977) Tergal modifications associated with abdominal glandular cells in the Blattaria. *J. Morph.* 151, 259–298.
Brousse-Gaury, P. (1968) Localisation des noyaux-origines des nerfs paracardiaques do Dictyoptères Blaberidae et Blattidae. *C.R. Acad. Sci, Paris* 266, 1972–1975.
Brousse-Gaury, P. (1971a) Influence de stimuli externes sur le comportement neuro-endocrinien de blattes. *Ann. Sci. nat. Zool.* 13, 181–450.
Brousse-Gaury, P. (1971b) Description d'arcs reflexes neuro-endocriniens partant des ocelles chez quelque orthopteres. *Bull. Biol. France Belg.* 105, 84–93.
Brousse-Gaury, P. (1973) L'apparente specificite des glandes tergales chez les males de *Periplaneta brunnea* Burmeister. *C.R. Acad. Sci., Paris* 276, 2175–2177.
Brousse-Gaury, P. (1975a) Mise en evidence d'une pheromone de marquage de piste chez *Periplaneta americana. C.R. Acad. Sci., Paris* 280, 319–322.
Brousse-Gaury, P. (1975b) Projection des informations ocellaires dans le cerveau des blattes: analyse et control. *Bull. Biol. France Belg.* 109, 125–163.
Brousse-Gaury, P. (1977) Starvation and reproduction in *Periplaneta americana* L.: Control of mating behaviour in the female. *Adv. Invert. Reprod.* 1, 328–343.
Brousse-Gaury, P. (1978a) Activité sexuelle des femelles de *Periplaneta americana* L. *Bull. Biol. France Belg.* 112, 129–165.
Brousse-Gaury, P. (1978b) Pieces buccales et reactivite endocrinienne chez *Periplaneta americana* L. *C.R. Acad. Sci., Paris* 280, 1261–1264.
Brousse-Gaury, P. and Cassier, P. (1975) Contribution a l'étude de la dynamique des corpora allata chez *Blabera fusca*. Influence du jeune, du groupement et du régime photoperiodique. *Bull. Biol. France Belg.* 109, 253–277.
Brousse-Gaury, P., Cassier, P. and Fain-Maurel, M. A. (1973) Contribution expérimentale et infrastructural a l'étude de la dynamique des corpora allata chez *Blabera fusca*. Influence du groupement visuel, des afférences ocellaires et antennaires. *Bull. Biol. France Belg.* 107, 143–169.
Brown, B. E. (1965) Pharmacologically active constituents of the cockroach corpus cardiacum: Resolution and some characteristics. *Gen. comp. Endocrinol.* 5, 387–401.
Brown, B. E. (1967) Neuromuscular transmitter substance in insect visceral muscle. *Science* 155, 595–597.
Brown, B. E. (1975) Proctolin: a peptide transmitter candidate in insects. *Life Sci.* 17, 1241–1252.
Brown, B. E. (1977) Occurrence of proctolin in six orders of insects. *J. Insect Physiol.* 23, 861–864.
Brown, B. E. and Nagai, T. (1969) Insect visceral muscle: neural relations of the proctodaeal muscles of the cockroach. *J. Insect Physiol.* 15, 1767–1783.
Brown, B. E. and Starratt, A. N. (1975) Isolation of proctolin, a myotropic peptide from *Periplaneta americana. J. Insect Physiol.* 21, 1879–1881.
Brück, E. Stocken, W. (1972) Morphologische Untersuchungen an der Cuticula von Insekten. II. Die Feinstruktur der larvalen Cuticula von *Periplaneta americana* (L.). *Z. Zellforsch.* 132, 417–430.
Brunet, P. C. J. (1951) The formation of the oötheca by *Periplaneta americana*. I. The micro-anatomy and histology of the posterior part of the abdomen. *Q. J. Micros. Sci.* 92, 113–127.
Brunet, P. C. J. (1952) The formation of the oötheca by *Periplaneta americana*. II. The structure and function of the left colleterial gland. *Q. J. Micros. Sci.* 93, 47–69.
Brunet, P. C. J. (1963) Tyrosine metabolism in insects. *Ann. N.Y. Acad. Sci.* 100, 1020–1034.
Brunet, P. C. J. and Kent, P. W. (1955) Observations on the mechanism of tanning reaction in *Periplaneta* and *Blatta*. *Proc. R. Soc. Lond.* Ser. B 144, 259–274.

Bruno, C. F. Cochran, D. G. (1965) Enzymes from insect tissues which catabolize pyrimidine compounds. *Comp. Biochem. Physiol.* 15, 113–124.
Bryan, Jr, E. H. (1926) Insects of Hawaii, Johnston Island and Wake Island. *Bull. Bernice P. Bishop Mus.* 31, 1–94.
Bryant, P. J., Bryant, S. V. and French, V. (1977) Biological regeneration and pattern formation. *Sci. Am.* 237, 66–81.
Buck, J. (1958) Cyclic CO_2 release in insects. IV. A theory of mechanism. *Biol. Bull.* 114, 118–140.
Buck, J. (1962) Some physical aspects of insect respiration. *A. Rev. Ent.* 7, 27–56.
Bullière, D. (1968) Dedifferentiatin of the muscle fibers of an insect in the course of regeneration. *J. Micros.* 7, 647–652.
Bullière, D. (1972) Action of ecdysone and inokosterone on the regeneration of an appendage in the larva of *Blabera cranifer* (Dictyoptere, Insecta) *C.R. Hebd. Seances Acad. Sci. Nat.* 274, 1349–1352.
Bullière, F. (1977) Effects of moulting hormone on RNA and cuticle synthesis in the epidermis of cockroach embryos cultured *in vitro. J. Insect Physiol.* 23, 393–401.
Bullière, D. and Bullière, F. (1974) Recherche des sites de synthèse de l'hormone de mue chez les embryons de *Blaberus craniifer* Burm., Insecte Dictyoptère. *C.R. Acad. Sci., Paris* 278, 377–380.
Bullière, F. and Bullière, D. (1977a) DNA synthesis and epidermal differentiation in the cockroach embryo and pharate first instar larva: Moulting hormone and mitomycin. *J. Insect Physiol.* 23, 1475–1489.
Bullière, F. and Bullière, D. (1977b) Régénération différenciation et hormones de mues chez l'embryon de blatte en culture *in vitro. Wilhelm Roux's Arch.* 182, 255–275.
Bullière, D., Bullière, F. and de Reggi, M. (1979) Ecdysteroid titres during ovarian and embryonic development in *Blaberus craniifer. Wilhelm Roux's Arch.* 186, 103–114.
Bunning, E. (1958) Über den Temperatureinflusz auf die endogene Tagesrhythmik, besonders bei *Periplaneta americana. Biol. Zentralbl.* 77, 141–152.
Bunning, E. (1973) *The Physiological Clock. Circadian Rhythms and Biological Chronometry*, 3rd edn. Springer-Verlag, New York.
Burgess, N. R. H. (1979) The cockroach – a hazard to health. *Proc. 5th British Pest Control Conference (Stratford)*, Third Session, Paper no. 8.
Burk, T. and Bell, W. J. (1973) Cockroach aggregation pheromone: inhibition of locomotion (Orthoptera: Blattidae). *J. Kansas Ent. Soc.* 46, 36–41.
Burkett, B. N. and Schneiderman, H. A. (1974) Roles of oxygen and carbon dioxide in the control of spiracular function in *Cecropia* pupae. *Biol. bull.* 147, 274–293.
Burns, M. D. and Usherwood, P. N. R. (1979) The control of walking in Orthoptera. II. Motor neurone activity in normal free-walking animals. *J. exp. Biol.* 79, 69–98.
Burrows, M. (1979) Graded synaptic interactions between local premotor interneurons of the locust. *J. Neurophysiol.* 42, 1108–1123.
Burrows, M. (1980) The control of sets of motoneurones by local interneurones in the locust. *J. Physiol.* 298, 213–233.
Burrows, M. and Siegler, M. V. S. (1978) Graded synaptic transmission between local interneurones and motor neurones in the metathoracic ganglion of the locust. *J. Physiol.* 285, 231–255.
Bursell, E. (1967) The excretion of nitrogen in insects. *Adv. Insect Physiol.* 4, 33–67.
Bursell, E. (1974a) Environmental aspects–temperature. In: The Physiology of Insecta, 2nd edn., Rockstein, M., Ed., Vol. II, Ch. 1, pp. 1–41. Academic Press, New York.
Bursell, E. (1974b) Environmental aspects–humidity. In: *The Physiology of Insecta.* 2nd edn., Rockstein, M., Ed., Vol. II, Ch. 2, pp. 44–84. Academic Press, New York.
Butler, R. (1971) The identification and mapping of spectral cell types in the retina of *Periplaneta americana. Z. vergl. Physiol.* 72, 67–80.
Butler, R. (1973a) The anatomy of the compound eye of *Periplaneta americana* L. I. General

features. *J. comp. Physiol.* 83, 223–238.

Butler, R. (1973b) The anatomy of the compound eye of *Periplaneta americana* L. II. Fine structure. *J. comp. Physiol.* 83, 239–262.

Butler, R. and Horridge, G. A. (1973a) The electrophysiology of the retina of *Periplaneta americana*. I. Changes in receptor acuity upon light/dark adaptation. *J. comp. Physiol.* 83, 263–278.

Butler, R. and Horridge, G. A. (1973b) The electrophysiology of the retina of *Periplaneta americana*. II. Receptor sensitivity and polarized light sensivity. *J. comp. Physiol.* 83, 279–288.

Cain, J. R. and Wilson, W. O. (1972) A test of the circadian rule of Aschoff with chicken hens. *J. Interdiscipl. Cycle Res.* 3, 77–85.

Caldarola, P. C. and Pittendrigh, C. S. (1974) A test of the hypothesis that D_2O affects circadian oscillations by diminishing the apparent temperature. *Proc. natn. Acad. Sci. U.S.A.* 71, 4386–4388.

Callec, J.-J. (1974) Synaptic transmission in the central nervous system of insects. In: *Insect Neurobiology*, Treherne J. E., Ed., pp. 119–185. Elsevier, New York.

Cameron, E. (1955) On the parasites and predators of the cockroach. I. *Tetrastichus hagenowii* (Ratz.). *Bull. ent. Res.* 46, 137–147.

Cameron, M. L. (1953a) Some Pharmacologically Active Substances Found in Insects. PhD Thesis, University of Cambridge, ii and 104 pp.

Cameron, M. L. (1953b) Secretion of an ortho-diphenol in the corpus caridacum of the insect. *Nature* 172, 349–350.

Camhi, J. M. (1974) Neural mechanisms of response modification in insects. In: *Experimental Analysis of Insect Behaviour*, Barton Browne, L., Ed., pp. 60–86. Springer-Verlag, New York.

Camhi, J. M. (1976) Non-rhythmic sensory inputs: influence of locomotory outputs in arthropods. In: *Neural Control of Locomotion*, Herman, R. M., Grillner, S., Stein, P. S. G. and Stuart, D. G., Eds., pp. 561–586. Plenum Press, New York.

Camhi, J. M. (1977) Behavioral switching in cockroaches: Transformations of tactile reflexes during righting behavior. *J. comp. Physiol.* 113, 283–301.

Camhi, J. M. and Hinkle, M. (1974) Response modification by the central flight oscillator of locusts. *J. exp. Biol.* 60, 477–492.

Camhi, J. M. and Tom, W. (1978). The escape behavior of the cockroach *Periplaneta americana*. I. Turning response to wind puffs. *J. comp. Physiol.* 128, 193–201.

Camhi, J. M., Tom, W. and Volman, S. (1978) The escape behavior of the cockroach *Periplaneta americana*. II. Detection of natural predators by air displacement. *J. comp. Physiol.* 128, 203–212.

Campbell, F. L. (1929) The detection and estimation of insect chitin, and the irrelation of 'chitinisation' to hardness and pigmentation of the cuticla of the American cockroach, *Periplaneta americana. Ann. ent. Soc. Am.* 22, 401–426.

Campbell, F. L. (1972) A new antennal sensillum of *Blattella germanica* (Dictyoptera: Blattidae) and its presence in other Blattaria. *Ann. ent. Soc. Am.* 65, 888–892.

Campos, M. and Vargas, M. (1977) Biología de *Protospirura muricola* Gedoelst, 1916 y *Mastophorus muris* (Gmelin, 1790) (Nematoda: Spiruridae), en Costa Rica. Huéspedes intermediarios. *Rev. Biol. Trop.* 25, 191–207.

Carbonell, C. S. (1947) The thoracic muscles of the cockroach *Periplaneta americana* (L.). *Smithson. Misc. Coll.* 107, 1–23.

Cardé, R. T. and Hagaman, T. E. (1979) Behavioral responses of the gypsy moth in a wind tunnel to air-borne enantiomers of disparlure. *Environ. Ent.* 8, 475–484.

Carlstrom, D. (1957) The crystal structure of α-chitin (poly-*N*-acetyl-D-glucosamine). *J. biophys. biochem. Cytol.* 3, 669–683.

Case, J. F. (1956) Spontaneous activity in denervated insect muscle. *Science* 124, 1079–1080.

Case, J. F. (1957) Median nerves and cockroach spiracular function. *J. Insect Physiol.* 1, 85–94.
Case, J. F. (1961) Organization of the cockroach respiratory centre. *Biol. Bull.* 121, 385.
Casida, J. E., Beck, S. D. and Cole, M. J. (1957) Sterol metabolism in the American cockroach. *J. biol. Chem.* 224, 365–371.
Cerf, J. A., Grundfest, H., Hoyle, G. and McCann, F. V. (1959) The mechanism of dual responsiveness in muscle fibers of the grasshopper *Romalea microptera. J. gen. Physiol.* 43, 377–395.
Cervenkova, E. (1960) Metabolismus svala *Periplaneta americana* za hladoveni. *Ceskoslovenska zoologicka spolecnost Vestnik* 24, 183–193 (cited in Tucker, 1977a).
Chadwick, L. E. (1956) Removal of prothoracic glands from the nymphal cockroach. *J. exp. Zool.* 131, 291–306.
Chadwick, L. E. (1957) The ventral intersegmental muscles of cockroaches. *Smithson. Misc. Coll.* 131, 1–30.
Chambille, I., Masson, C. and Rospars, J. P. (1980) The deutocerebrum of the cockroach *Blaberus craniifer* Burm. Spatial organization of the sensory glomeruli. *J. Neurobiol.* 11, 135–157.
Chang, F. and Friedman, S. (1971) A developmental analysis of the uptake and release of lipids by the fat body of the tobacco hornworm, *Manduca sexta. Insect Biochem.* 1, 63–80.
Chanussot, B. and Pentreath, V. W. (1973) Étude des monoamines du système nerveux stomatogastrique extracéphalique de *Periplaneta americana. C. R. Seances Soc. Biol. (France)* 167, 1405.
Chapman, K. M. (1965) Campaniform sensilla on the tactile spines of the legs of the cockroach. *J. exp. Biol.* 42, 191–203.
Chauvin, R. (1947) Études sur le comportement de *Blattella germanica* dans divers types de labyrinthes. *Bull. Biol. France Belg.* 61, 92–128.
Chefurka, W. (1965) Intermediary metabolism of carbohydrates in insects. In: *The Physiology of Insecta*, Rockstein, M., Ed., Vol. II, pp. 581–667. Academic Press, New York.
Chefurka, W., Horie, Y and Robinson, J. R. (1970) Contribution of the pentose cycle to glucose metabolism by insects. *Comp. Biochem. Physiol.* 37, 143–165.
Chen, D. H., Robbins, W. E. and Monroe, R. E. (1962) The gonadotropic action of cecropia extracts in allatectomized American cockroaches. *Experientia* 18, 577–578.
Chen, J. S. and Levi-Montalcini, R. (1969) Axonal outgrowth and cell migration *in vitro* from nervous system of cockroach embryos. *Science* 166, 631–632.
Chen, J. S. and Levi-Montalcini, R. (1970) Axonal growth from insect neurons in glia-free cultures. *Proc. natn. Acad. Sci. U.S.A.* 66, 32–39.
Chen, P. S. (1978) Protein synthesis in relation to cellular activation and deactivation. In: *Biochemistry of Insects*, Rockstein, M., Ed., pp. 145–203. Academic Press, New York.
Chen, S. M. L. (1974 *Sex pheromone of American cockroach, Periplaneta americana, Isolation and some Structural Features.* Ph.D. Thesis, Columbia University.
Chen, W., Aranda, L. C. and Luco, J. V. (1970) Nerve structures required for conditioning of varying complexity. *Arch. biol. med. Exp.* 7, 58.
Chino, H. and Downer, R. G. H. (1979) The role of diacylglycerol in absorption of dietary glyceride in the American cockroach, *Periplaneta americana* L. *Insect Biochem.* 9, 379–382.
Chino, H. and Gilbert, L. I. (1964) Diglyceride release from insect fat body: a possible means of lipid transport. *Science* 143, 359–361.
Chino, H. and Gilbert, L. I. (1965) Lipid release and transport in insects. *Biochim. biophys. Acta* 98, 94–110.
Chino, H., Katase, H., Downer, R. G. H. and Takahashi, K. (1981) Diacylglycerol-carrying lipoprotein of hemolymph of the American cockroach: purification, characterization, and function. *J. Lipid Res.* 22, 7–15.

Chino, H., Murakami, T. and Harashima, K. (1969) Diglyceride-carrying lipoproteins in insect hemolymph: isolation, purification and properties. *Biochim. biophys. Acta* 176, 1–26.
Chino, H., Sudo, A. and Harashima, K. (1965) Isolation of diglyceride-bound lipoprotein from insect hemolymph. *Biochim. biophys. Acta* 144, 177–179.
Chippendale, G. M. (1978) The functions of carbohydrates in insect life processes. In: *Biochemistry of Insects*, Rockstein, M., Ed., pp. 1–55. Academic Press, New York.
Choban, R. G. (1975) *Hyperplasia and Associated Responses of the Ventriculus following Cobalt-60 Gamma Irradiation of Periplaneta americana* (L.). Ph.D. Thesis, Rutgers University.
Clark, A. J. and Bloch, K. (1959) Conversion of ergosterol to 22-dehydrocholesterol in *Blattella germanica*. *J. biol. chem.* 234, 2589–2593.
Clarke, K. U. (1958) Studies on the relationship between changes in the volume of the tracheal system and growth in *Locusta migratoria* L. *Proc. 10th Int. Cong. Ent., Montreal* 2, 205–211.
Clayton, R. B. (1960) The role of intestinal symbionts in the sterol metabolism of *Blattella germanica*. *J. biol. Chem.* 235, 3421–3425.
Clayton, R. B. (1964) The utilization of sterols by insects. *J. Lipid Res.* 5, 3–19.
Clayton, R. B., Hinkle, P. C., Smith, D. A. and Edwards, A. M. (1964) The intestinal absorption of cholesterol, its esters and some related sterols and analogues in the roach *Eurycotis floridana*. *Comp. Biochem. Physiol.* 11, 333–350.
Clore, J. N., Petrovitch, E., Koeppe, J. K. and Mills, R. R. (1978) Vitellogenesis by the American cockroach: electrophoretic and antigenic characterization of hemolymph and oocyte proteins. *J. Insect Physiol.* 24, 45–52.
Cloudsley-Thompson, J. L. (1953) Studies in diurnal rhythms. 3. Photoperiodism in the cockroach *Periplaneta americana* (L.). *An. Mag. Nat. Hist.* 6, 705–712.
Cloudsley-Thompson, J. L. (1970) Recent work on the adaptive functions of circadian and seasonal rhythms in animals. *J. interdiscipl. Cycle Res.* 1, 5–19.
Cochrane, D. G. (1975) Excretion in insects. In: *Insect Biochemistry and Function*, Candy, D. J. and Kilby, B. A., Eds. pp. 177–282. Chapman and Hall, London.
Cochrane, D. G. (1976) Kynurenine formamidase activity in the American cockroach. *Insect Biochem.* 6, 267–272.
Cochrane, D. G. (1977) Cytology of urate storate in *Periplaneta americana* (L.) (Dictyoptera: Blattidae). *J.N.Y. ent. Soc.* 85, 170.
Cochrane, D. G., Elder, H. Y. and Usherwood, P. N. R. (1972) Physiology and ultrastructure of phasic and tonic skeletal muscle fibres in the locust, *Schistocerca gregaria*. *J. Cell Sci.* 10, 419–441.
Cochrane, D. G., Mullins, D. E. and Mullins, K. J. (1979) Cytological changes in the fat body of the American cockroach *Periplaneta americana* in relation to dietary nitrogen levels. *Ann. ent. Soc. Am.* 72, 197–205.
Coenen-Stass, D. (1976) Vorzugstemperatur und Vorzugsluftfeuchtigkeit der beiden Schabenarten *Periplaneta americana* und *Blaberus trapezoideus* (Insecta: Blattaria). *Ent. exp. appl.* 20, 143–153.
Coenen-Stass, D. and Kloft, W. (1976a) Sorptionmessungen an den beiden Schabenarten *Periplaneta americana* und *Blaberus trapezoideus*. *J. Insect Physiol.* 22, 1127–1133.
Coenen-Stass, D. and Kloft, W. (1976b) Transpirationsmessungen und den beiden Schabenarten *Periplaneta americana* und *Blaberus trapezoideus*. *J. Insect Physiol.* 22, 945–950.
Coenen-Stass, D. and Kloft, W. (1977) Auswirkungen der Verdunstungskulung und der Stoffwechselwarme auf der Korpertemptur der Schabenarten *Periplaneta americana* und *Blaberus trapezoideus*. *J. Insect Physiol.* 23, 1397–1406.
Cohen, M. J. (1974) Trophic interactions in excitable systems of invertebrates. *Ann. N.Y. Acad. Sci.* 228, 364–380.
Cohen, M. J. and Jacklet, J. W. (1965) Neurons in insects: RNA changes during injury and regeneration. *Science* 148, 1237–1239.

Cohen, M. J. and Jacklett, J. W. (1967) The functional organization of motor neurons in an insect ganglion. *Phil. Trans. R. Soc. Lond. Ser. B.* 252, 561–569.

Colhoun, E. H. (1963) Synthesis of 5-hydroxytryptamine in the American cockroach. *Experientia* 19, 9–10.

Compton, D. A. and Mills, R. R. (1981) Bursicon-induced adenyl cyclase activity in the haemocytes of the American cockroach. *J. Insect Physiol.* (In press.).

Conrad, C. W. and Jackson, L. L. (1971) Hydrocarbon biosynthesis in *Periplaneta americana. J. Insect Physiol.* 17, 1907–1916.

Cook, B. J. (1967) An investigation of factor S, a neuromuscular excitatory substance from insects and crustacea. *Biol. Bull.* 133, 526–538.

Cook, B. J. and Eddington, L. C. (1967) The release of triglycerides and free fatty acids from the fat body of the cockroach, *Periplaneta americana. J. Insect Physiol.* 13, 1361–1372.

Cook, B. J., Eraker, J. and Anderson, G. R. (1969a) The effect of various biogenic amines on the activity of the foregut of the cockroach, *Blaberus giganteus. J. Insect Physiol.* 15, 445–455.

Cook, B. J. and Holman, G. M. (1975a) Sites of action of a peptide neurohormone that controls hindgut muscle activity in the cockroach *Leucophaea maderae. J. Insect Physiol.* 21, 1187–1192.

Cook, B. J. and Holman, G. M. (1975b) The neural control of muscular activity in the hindgut of the cockroach *Leucophaea maderae*: prospects of its chemical mediation. *Comp. Biochem. Physiol.* 50C, 137–146.

Cook, B. J. and Holman, G. M. (1978) Comparative pharmacological properties of muscle function in the foregut and the hindgut of the cockroach *Leucophaea maderae. Comp. Biochem. Physiol.* 61C, 291–295.

Cook, B. J. and Holman, G. M. (1979a) The pharmacology of insect visceral muscle. *Comp. Biochem. Physiol.* 64C, 183–190.

Cook, B. J. and Holman, G. M. (1979b) The action of proctolin and L-glutamic acid on the visceral muscles of the hindgut of the cockroach *Leucophaea maderae. Comp. Biochem. Physiol.* 64C, 21–28.

Cook, B. J., Holman, G. M. and Marks, E. P. (1975) Calcium and cyclic AMP as possible mediators of neurohormone action in the hindgut of the cockroach, *Leucophaea maderae. J. Insect Physiol.* 21, 1807–1814.

Cook, B. J. and Holt, G. G. (1974) Neurophysiological changes associated with paralysis arising from body stress in the cockroach, *Periplaneta americana. J. Insect Physiol.* 20, 21–40.

Cook, B. J. and Milligan, J. V. (1972) Electrophysiology and histology of the medial neurosecretory cells in adult male cockroaches, *Periplaneta americana. J. Insect Physiol.* 18, 1197–1214.

Cook, B. J., Nelson, D. R. and Hipps, P. (1969b) Esterases and phosphatases in the gastric secretion of the cockroach *Periplaneta americana. J. Insect Physiol.* 15, 581–589.

Cook, B. J. and Reinecke, J. P. (1973) Visceral muscles and myogenic activity in the hindgut of the cockroach *Leucophaea maderae. J. comp. Physiol.* 84, 95–118.

Cooter, R. J. (1973) Visual and multimodal interneurons in the ventral nerve cord of the cockroach, *Periplaneta americana. J. exp. Biol.* 59, 675–696.

Cooter, R. J. (1975) Ocellus and ocellar nerves of *Periplaneta americana* L. (Orthoptera: Dictyoptera). *Int. J. Insect Morph. Embryol.* 4, 273–288.

Corbett, J. R. (1974) *The Biochemical Mode of Action of Pesticides.* Academic Press, New York.

Cordero, S. M. and Ludwig, D. (1963) Purification and activities of purine enzymes from various tissues of the American cockroach, *Periplaneta americana. J.N.Y. ent. Soc.* 71, 66–73.

Cornwell, P. B. (1968) *The Cockroach, Vol. I. A Laboratory Insect and an Industrial Pest.* Hutchinson Press, London.

Cornwell, P. B. (1976) *The Cockroach, Vol. II. Insecticides and Cockroach Control.* Assoc. Business Prog. Ltd, London.

Couch, E. F. and Mills, R. R. (1968) The midgut epithelium of the American cockroach: acid phosphomonoesterase activity during the formation of autophagic vacuoles. *J. Insect Physiol.* 14, 55–62.

Counce, S. J. (1973) The causal analysis of insect embryogenesis. In: *Developmental Systems: Insects*, Counce, S. J. and Waddington, C. H., Eds., Vol. II, Ch. 1, pp. 1–156. Academic Press, New York.

Cowden, R. and Bodenstein, D. (1961) A cytochemical investigation of striated muscle differentiation in regenerating limbs of the roach, *Periplaneta americana. Embryologia (Nagoya)* 6, 36–50.

Crabtree, B. and Newsholme, E. A. (1975) Comparative aspects of fuel utilization and metabolism by muscle. In: *Insect Muscle*, Usherwood, P. N. R., Ed., pp. 405–500. Academic Press, London.

Crossley, A. C. (1975) The cytophysiology of insect blood. *Adv. Insect Physiol.* 11, 117–222.

Crossman, A. R., Kerkut, G. A., Pitman, R. M. and Walker, R. J. (1971) Electrically excitable nerve cell bodies in the central ganglia of two insect species, *Periplaneta americana* and *Schistocerca gregaria*. Investigation of cell geometry and morphology by intracellular dye injection. *Comp. Biochem. Physiol.* 40A, 579–594.

Crowder, L. A. and Shankland, D. L. (1972a) Structure of the Malpighian tubule muscle of the American cockroach, *Periplaneta americana. Ann. ent. Soc. Am.* 65, 614–619.

Crowder, L. A. and Shankland, D. L. (1972b) Pharmacology of the malpighian tubule muscle of the American cockroach *Periplaneta americana. J. Insect Physiol.* 18, 929–936.

Crowder, L. A. and Shankland, D. L. (1974) Response to 5-hydroxytryptamine and electrophysiology of the Malpighian tubule muscle of the American cockroach, *Periplaneta americana. Ann. ent. Soc. Am.* 67, 281–284.

Cruden, D. L. and Markovetz, A. J. (1979) Carboxymethylcellulose decomposition by the intestinal bacteria of cockroaches. *Appl. environ. Microbiol.* 38, 369–372.

Cuenot, L. (1896) Études physiologiques sur les Orthopteres. *Arch. Biol.* 14, 293–341.

Cuenot, L. (1899) les pretendus organs phagocytaries decrits par Koulvetch chez la blatte. *Arch. Zool. exp. Gen.* 7, 62–75.

Cunliffe, F. (1952) Biology of the cockroach parasite, *Pimeliaphilus podapolipophagus* Trägardh, with a discussion of the genera *Pimeliaphilus* and *Hirstiella*. (Acarina, Pterygosomidae). *Proc. ent. Soc., Wash.* 54, 153–169.

Cutkomp, L. K., Khan, H. and Sudershan, P. (1977) Rhythmicity of the activity of the ATPase enzyme system in cockroaches and sensitivity to DDT. *Chronobiologia* 4, 172.

Cymborowski, B. (1971) Circadian histochemical changes in the neurosecretory cells of the insect nervous system. *Folia histochem. cytochem.* 9, 343–344.

Cymborowski, B. and Flisínska-Bojanowska, A. (1970) The effect of light on the locomotory activity and structure of neurosecretory cells of the brain and subesophageal ganglion of *Periplaneta americana* L. *Zool. Pol.* 20, 387–399.

Daan, S. and Pittendrigh, C. S. (1976) A functional analysis of circadian pacemakers in nocturnal rodents. III. Heavy water and constant light: Homeostasis or frequency? *J. comp. Physiol.* 106, 267–290.

Dadd, R. H. (1970a) Arthropod nutrition. In: *Chemical Zoology*, Florkin, M. and Scheer, B. T., Eds., Vol. 5, pp. 35–95. Academic Press, New York.

Dadd, R. H. (1970b) Digestion in insects. In: *Chemical Zoology*, Florkin, M. and Scheer, B. T., Eds., Vol. 5, pp. 117–145. Academic Press, New York.

Dagan, D. and Camhi, J. M. (1979) Responses to wind recorded from the cercal nerve of the cockroach *Periplaneta americana*, II. Directional sensitivity of the sensory neurons innervating single columns of filiform hairs. *J. comp. Physiol.* 133, 103–110.

Dagan, D. and Parnas, I. (1970) Giant fibre and small fibre pathways involved in the evasive response of the cockroach, *Periplaneta americana. J. exp. Biol.* 52, 313–324.

Daley, D. L. and Delcomyn, F. (1980a) Modulation of excitability of cockroach giant interneurons during walking. I. Simultaneous excitation and inhibition. *J. comp. Physiol.* 138, 231–239.

Daley, D. L. and Delcomyn, F. (1980b) Modulation of excitability of cockroach giant interneurons during walking. II. Central and peripheral components. *J. comp. Physiol.* 138, 241–251.

Daley, D. L., Vardi, N., Appignani, B. and Camhi, J. M. (1981) Morphology of the giant interneurons and cercal nerve projections of the American cockroach. *J. comp. Neurol.* 196, 41–52.

Daniel, R. S. and Brooks, M. A. (1972) Intracellular bacteroids: electron microscopy of *Periplaneta americana* injected with lysozyme. *Exp. Parasitol.* 31, 232–246.

Davey, K. G. (1961a) The mode of action of the heart accelerating factor from the corpus cardiacum of insects. *Gen. comp. Endocrinol.* 1, 24–29.

Davey, K. G. (1961b) Substances controlling the rate of beating of the heart of *Periplaneta. Nature* 192, 284.

Davey, K. G. (1962a) The release by feeding of a pharmacologically active factor from the corpus cardiacum of *Periplaneta americana. J. Insect Physiol.* 8, 205–208.

Davey, K. G. (1962b) The mode of action of the corpus cardiacum on the hindgut in *Periplaneta americana. J. exp. Biol.* 39, 319–324.

Davey, K. G. (1962c) The nervous pathway involved in the release by feeding of a pharmacologically active factor from the corpus cardiacum of *Periplaneta. J. Insect Physiol.* 8, 579–583.

Davey, K. G. (1963) The release by enforced activity of the cardiac accelerator from the corpus cardiacum of *Periplaneta americana. J. Insect Physiol.* 9, 375–381.

Davey, K. G. (1964) The control of visceral muscles in insects. *Adv. Insect Physiol.* 2, 219–245.

Davey, K. G. and Huebner, E. (1974) The response of the follicle cells of *Rhodnius prolixus* to juvenile hormone and antigonadotropin *in vitro. Can. J. Zool.* 52, 1407–1412.

Davey, K. G. and Treherne, J. E. (1963a) Studies on crop function in the cockroach *Periplaneta americana*. I. The mechanism of crop-emptying. *J. exp. Biol.* 40, 763–773.

Davey, K. G. and Treherne, J. E. (1963b) Studies on crop function in the cockroach *Periplaneta americana*. II. The nervous control of crop-emptying. *J. exp. Biol.* 40, 775–780.

Davey, K. G. and Treherne, J. E. (1964) Studies on crop function in the cockroach *Periplaneta americana*. III. Pressure changes during feeding and crop-emptying. *J. exp. Biol.* 41, 513–524.

Davis, B. J. (1964) Disc electrophoresis II. Method and application to human serum proteins. *Ann. N.Y. Acad. Sci.* 121, 404.

Davis, W. J. (1973) Development of locomotor patterns in the absence of peripheral sense organs and muscles. *Proc. natn. Acad. Sci. U.S.A.* 70, 954–958.

Day, M. F. (1951) The mechanism of secretion by the salivary gland of the cockroach, *Periplaneta americana. Aust. J. sci. Res.* (*B*) 4, 136–143.

Day, M. F. (1952) Wound healing in the gut of the cockroach *Periplaneta. Aust. J. sci. Res.* 5, 283–289.

Day, M. F. and Powning, R. F. (1949) A study of the processes of digestion in certain insects. *Aust. J. sci. Res.* (*B*) 2, 175–215.

Dehnel, P. A. and Segal, E. (1956) Acclimation of oxygen consumption to temperature in the American cockroach (*Periplaneta americana*). *Biol. Bull.* 111, 53–61.

Dejmal, R. K. and Brookes, V. J. (1972) Insect lipovitellin: chemical and physical characteristics of a yolk protein from the ovaries of *Leucophaea. J. biol. Chem.* 247, 869–874.

Delachambre, J., Delbecque J. P., Provansal, A., de Reggi, M. L. and Cailla, H. (1979) Induction of epidermal cyclic AMP by bursicon in mealworm, *Tenebrio molitor. Experientia* 35, 701–702.

Delbecque, J. P., Hirn, M., Delachambre, J. and de Reggi, M. (1978) Cuticular cycle and molting hormone levels during the metamorphosis of *Tenebrio molitor* (Insecta Coleoptera). *Dev. Biol.* 64, 11–30.

Delcomyn, F. (1971a) The locomotion of the cockroach *Periplaneta americana. J. exp. Biol.* 54, 443–452.

Delcomyn, F. (1971b) The effect of limb amputation on locomotion in the cockroach, *Periplaneta americana. J. exp. Biol.* 54, 453–469.

Delcomyn, F. (1971c) Computer-aided analysis of a locomotor leg reflex in the cockroach *Periplaneta americana. Z. vergl. Physiol.* 74, 427–445.

Delcomyn, F. (1973) Motor activity during walking in the cockroach *Periplaneta americana*. II. Tethered walking. *J. exp. Biol.* 59, 643–654.

Delcomyn, F. (1976) An approach to the study of neural activity during behaviour in insects. *J. Insect Physiol.* 22, 1223–1227.

Delcomyn, F. (1977) Corollary discharge to cockroach giant interneurons. *Nature* 269, 160–162.

Delcomyn, F. (1980) Neural basis of rhythmic behavior in animals. *Science*. 210, 492–498.

Delcomyn, F. and Daley, D. L. (1979) Central excitation of cockroach giant interneurons during walking. *J. comp. Physiol.* 139, 39–48.

Delcomyn, F. and Usherwood, P. N. R. (1973) Motor activity during walking in the cockroach *Periplaneta americana*. I. Free walking. *J. exp. Biol.* 59, 629–642.

Deleporte, P. (1974) Contribution a l'etude de l'organization sociale chez *Periplaneta americana* (Dictyoptères), Evolution des relations interindividuelles chez les larves et adultes males. D.E.A., Univ. Rennes.

Deleporte, P. (1976) *L'Organisation Sociale chez Periplaneta americana* (Dictyoptères). Aspects Eco-ethologiques – Ontogenese des Relations inter-individuelles. These Doctorat, 3° cycle, Univ. Rennes.

Deleporte, P. (1978) Ontogenese des relations inter-individuelles chez *Periplaneta americana*. I. Étude longitudinale par confrontation de males aux differents stades de developement *Biol. du comp.* 3, 259–272.

Denburg, J. L. (1975) Possible biochemical explanation for specific formation of synapses between muscle and regenerating motoneurons in cockroach. *Nature* 258, 535–537.

Denburg, J. L. (1978) Protein composition of cockroach muscles: identification of candidate recognition molecules. *J. Neurobiol.* 9, 93–110.

Denburg, J. L. and Hood, N. A. (1977) Protein synthesis in regenerating motor neurons in the cockroach. *Brain Res.* 125, 227–239.

Denburg, J. L., Seecof, R. L. and Horridge, G. A. (1977) The path and rate of growth of regenerating motor neurons in the cockroach. *Brain Res.* 125, 213–226.

Desportes, I. (1966) L'ultrastructure de la jonction entre le primite et le satellite des associations de *Gregarina blattarum* Sieb. (Eugrégarines, Gregarinidae). *C. R. Acad. Sci. Ser. D, Paris* 262, 1869–1870.

DeWilde, J. and DeLoof, A. (1973) Reproduction. In: *The Physiology of Insecta*, Vol. 1. Rockstein, M., Ed., pp. 11–95. Academic Press, New York.

Diakonoff, A. (1936) Contribution to the knowledge of the fly reflexes and the static sense in *Periplaneta americana* L. *Arch. neerl. Physiol.* 21, 104–129.

Disterhoft, J. F. (1972) Learning in the intact cockroach (*Periplaneta americana*) when placed in a punishment situation. *J. comp. physiol. Psych.* 79, 1–7.

Dixon, S. E. and Blaine, W. D. (1973) Hormonal control of male accessory gland development in the cockroach, *Periplaneta americana* (L.). *Proc. ent. Soc. Ont.* 103, 97–103.

Doane, W. (1973) Role of hormones in insect development. In: *Developmental Systems: Insects*, Vol. 2, Counce, S. J. and Waddington, C. H., Eds., pp. 291–497. Academic Press, New York.

Dogra, G. S. (1968) The study of the neurosecretory system of *Periplaneta americana* (L.) *in situ* using a technique specific for cystine and/or cysteine. *Acta anat.* 70, 288–303.

Donnellan, J. F. and Kilby, B. A. (1967) Uric acid metabolism by symbiotic bacteria from the fat body of *Periplaneta americana. Comp. Biochem. Physiol.* 22, 235–252.

Dorn, A. (1973) Electron microscopic study on the larval and adult corpus allatum of *Oncopeltus fasciatus* Dallas (Insecta, heteroptera). *Z. Zellforsch.* 145, 447–458.
Downer, R. G. H. (1972) Interspecificity of lipid-regulating factors from insect corpus cardiacum. *Can. J. Zool.* 50, 63–65.
Downer, R. G. H. (1978a) Functional role of lipids in insects. In: *Biochemistry of Insects.* Rockstein, M., Ed., pp. 58–92. Academic Press, New York.
Downer, R. G. H. (1978b) Regulation of carbohydrate metabolism in excited cockroaches. In: *Comparative Endorcrinology*, Gaillard P. J. and Boer, H. H., Eds., pp. 465. Elsevier/North Holland, Amsterdam.
Downer, R. G. H. (1979a) Induction of hypertrehalosemia by excitation in *Periplaneta americana. J. Insect Physiol.* 25, 59–63.
Downer, R. G. H. (1979b) Trehalose production in isolated fat body of the American cockroach, *Periplaneta americana. Comp. Biochem. Physiol.* 62C, 31–34.
Downer, R. G. H. (1980) Short-term hypertrehalosemia induced by octopamine in the American cockroach, *Periplaneta americana* L. In: *Proc. Int. Symp. Insect Neurobiol. Pesticide Action.* pp. 335–339. Society of Chemical Industry, London.
Downer, R. G. H. (1981) Physiological and environmental considerations in insect bioenergetics. In: *Energy metabolism in insects.* Downer, R. G. H., Ed., pp. 1–17. Plenum Press, New York.
Downer, R. G. H. and Chino, H. (1979) Cholesterol and cholesterol ester in haemolymph of the American cockroach, *Periplaneta americana* L. *Can. J. Zool.* 57, 1333–1336.
Downer, R. G. H. and Matthews, J. R. (1976a) Patterns of lipid storage and utilisation in insects. *Am. Zool.* 16, 733–745.
Downer, R. G. H. and Matthews, J. R. (1976b) Glycogen depletion of thoracic musculature during flight in the American cockroach, *Periplaneta americana* L. *Comp. Biochem. Physiol.* 55B, 501–502.
Downer, R. G. H. and Parker, G. H. (1979) Glycogen utilisation during flight in the American cockroach, *Periplaneta americana* L. *Comp. Biochem. Physiol.* 64A, 29–32.
Downer, R. G. H. and Steele, J. E. (1969) Hormonal control of lipid concentration in fat body and haemolymph of the American cockroach, *Periplaneta americana. Proc. ent. Soc. Ontario.* 100, 113–116.
Downer, R. G. H. and Steele, J. E. (1972) Hormonal stimulation of lipid transport in the American cockroach, *Periplaneta americana. Gen. comp. endocrinol.* 19, 259–265.
Downer, R. G. H. and Steele, J. E. (1973) Haemolymph lipase activity in the American cockroach, *Periplaneta americana. J. Insect Physiol.* 19, 523–532.
Dreisig, H. (1976) Phase shifting the circadian rhythms of nocturnal insects by temperature changes. *Physiol. Ent.* 1, 123–129.
Dreisig, H. and Nielsen, E. T. (1971) Circadian rhythm of locomotion and its temperature dependence in *Blattella germanica. J. exp. Biol.* 54, 187–198.
Drescher, W. (1960) Regenerationsversuche am Gehirn von *Periplaneta americana. Z. Morphol. Oekol. Tiere* 48, 576–649.
Dresden, D. and Nijenhuis, E. D. (1953) On the anatomy and mechanism of motion of the mesothoracic leg of *Periplaneta americana. Proc. Kon. Ned. Akad. Wetensch. C.* 56, 39–47.
Dresden, D. and Nijenhuis, E. D. (1958) Fibre analysis of the nerves of the second thoracic leg in *Periplaneta americana. Proc. Kon. Ned. Akad. Wetensch. C.* 61, 213–223.
Dubowsky, N. and Pierre, L. L. 1967 Activity of isocitric dehyrogenase in the fat bodies of the cockroach, *Leucophaea maderae* (F.) *Nature* 213, 209–210.
Dujardin, F. (1850) Mémoire sur le système nerveux des Insectes. *Ann. Sci. Nat., Zool.* (3) 14, 195–206.
Dumbleton, L. J. (1957) Parasites and predators introduced into the Pacific Islands for the biological control of insects and other pests. *South Pac. Comm. Tech. Pap. No. 101*, 40 pp. [Appeared erroneously under the name of C. P. Hoyt.]

Dumortier, B. (1963) Morphology of sound emission apparatus in arthropods. In: *Acoustic Behaviour of Animals*, Busnel, R. G., Ed., pp. 277–373. Elsevier, Amsterdam.
Dumortier, B. (1965) L'émission sonore dans le genre *Gromphadorhina brunneri* (Blattodea, Perisphaeriidae), étude morphologique et biologique. *Bull. Soc. Zool. France* 90, 89–101.
Dutton, G. J. (1962) The mechanism of *o*-aminophenyl glucoside formation in *Periplaneta americana. Comp. Biochem. Physiol.* 7, 39–46.
Dymond, G. R. and Evans, P. D. (1979) Biogenic amines in the nervous system of the cockroach, *Periplaneta americana*: Association of octopamine with mushroom bodies and dorsal unpaired median (DUM) neurons. *Insect Biochem.* 9, 535–545.
Eads, R. B., VonZuben, F. J., Bennett, S. E. and Walker, O. L. (1954) Studies on cockroaches in a municipal sewerage system. *Am. J. Trop. Med. Hyg.* 3, 1092–1098.
Eckert, M. (1973) Immunologische Untersuchungen des neuroendokrinen Systems von Insekten. III. Immunhistochemische Markierung des neuroendokrinen Systems von *Periplaneta americana* mit durch Fraktionierung von Retrocerebralkomplex-extrakten gewonnenon Anti-Seren. *Zool. Jb. Physiol.* 77, 50–59.
Eckert, M. (1977) Immunologische Untersuchungen des neuroendokrinen Systems von Insekten. IV. Differenzierte immunhistochemische Darstellung von Neurosekreten des Gehirns und der Corpora cardiaca bei der Schabe *Periplaneta americana. Zool. Zb. Physiol.* 81, 25–41.
Eckert, M. and Gersch, M. (1978) Immunochemical investigation on the neuroendocrine system of the cockroach, *Periplaneta americana* L. In: *Neurosecretion and Neuroendocrine Activity, Proc. VII Int. Symp. Neurosec. Leningrad, 1976*, Bargmann, W. *et al.*, Ed., pp. 365–369. Springer-Verlag, Berlin.
Edmunds, L. R. (1953) Some notes on the Evaniidae as household pests and as a factor in the control of roaches. *Ohio J. Sci.* 53, 121–122.
Edney, E. B. (1966) Absorption of water vapour from unsaturated air by *Arenivaga* sp. (Polyphagidae: Dictyoptera). *Comp. Biochem. Physiol.* 91, 387–408.
Edney, E. B. (1968) The effect of water loss on the haemolymph of *Arenivaga* sp. and *Periplaneta americana. Comp. Biochem. Physiol.* 25, 149–158.
Edney, E. B. (1977) *Water Balance in Land Arthropods*. Springer-Verlag, Berlin.
Edwards, G. A. (1959) The fine structure of a multiterminal innervation of an insect muscle. *J. biophys. biochem. Cytol.* 5, 241–244.
Edwards, J. S. (1969) Postembryonic development and regeneration of the insect nervous system. *Adv. Insect Physiol.* 6, 97–137.
Edwards, J. S. and Palka, J. (1976) Neural generation and regeneration in insects. In: *Simpler Networks and Behavior*, Fentress, J. C., Ed., pp. 167–185. Sinauer Ass., Sunderland, Mass.
Edwards, L. J. (1970) A thermal conductivity device for detecting the discontinuous release of carbon dioxide from insects. *Ann. ent. Soc. Am.* 63, 627–629.
Edwards, L. J. and Patton, R. L. (1967) Carbonic anhydrase in the house cricket *Acheta domesticus. J. Insect Physiol.* 13, 1333–1341.
Eisenstein, E. M. (1967) The effects of CNS lesions on the demonstration of shock avoidance learning by the roach, *Periplaneta americana. Physiologist* 10, 160.
Eisenstein, E. M. (1972) Learning and memory in isolated insect ganglia. *Adv. Insect Physiol.* 9, 111–181.
Eisenstein, E. M. and Cohen, M. J. (1965) Learning in an isolated prothoracic insect ganglion. *Animal Behav.* 13, 104–108.
Eisner, T. (1955) The digestion and absorption of fats in the fore-gut of the cockroach *Periplaneta americana. J. exp. Zool.* 130, 159–182.
Ela, R., Chefurka, W. and Robinson, J. R. (1970) *In vivo* glucose metabolism in the normal and poisoned cockroach, *Periplaneta americana. J. Insect Physiol.* 16, 2137–2156.
Elder, H. Y. (1975) Muscle structure. In: *Insect Muscle*, Usherwood, P. N. R., Ed., pp. 1–74. Academic Press, London.

Eldering, F. (1919) Acquisition d'habitudes par les insectes. *Arch. neerl. Physiol.* 3, 469–490.
Elliot, J. A. (1976) Circadian rhythms and photoperiodic time measurements in mammals. *Fedn Proc. fedn Socs exp. Biol. Am.* 35, 2339–2349.
Emmerich, H. and Hartmann, R. (1973) A carrier lipoprotein for juvenile hormone in the haemolymph of *Locusta migratoria. J. Insect Physiol.* 19, 1663–1675.
Engelmann, F. (1959a) The control of reproduction in *Diploptera punctata* (Blattaria). *Biol. Bull.* 116, 406–419.
Engelmann, F. (1959b) Über die Wirkung implantierter Prothoraxdrüsen im adulten Weibchen von *Leucophaea maderae* (Blattaria). *Z. vergl. Physiol.* 41, 455–470.
Engelmann, F. (1960) Mechanisms controlling reproduction in two viviparous cockroaches (Blattaria). *Ann. N.Y. Acad. Sci.* 89, 516–536.
Engelmann, F. (1963) Die Innervation der Genital- und Postgenitalsegmente bei Weibchen der Schabe *Leucophaea maderae. Zool. Jb. Anat.* 81, 1–16.
Engelmann, F. (1965) The mode of regulation of the corpus allatum in adult insects. *Arch. Anat. Micros.* 54, 387–404.
Engelmann, F. (1966) Control of intestinal proteolytic enzymes in a cockroach. *Naturwissenschaften* 53, 113–114.
Engelmann, F. (1969a) Food-stimulated synthesis of intestinal proteolytic enzymes in the cockroach *Leucophaea maderae. J. Insect Physiol.* 15, 217–235.
Engelmann, F. (1969b) Female specific protein: Biosynthesis controlled by corpus allatum in *Leucophaea maderae. Science* 165, 407–409.
Engelmann, F. (1970) *The Physiology of Insect Reproduction.* Pergamon Press, New York.
Engelmann, F. (1972) Juvenile hormone-induced RNA and specific protein synthesis in an adult insect. *Gen. comp Endocrinol.* 3 (suppl.), 168–173.
Engelmann, F. (1974a) Polyribosomal and microsomal profiles of fat body homogenates from reproductively active and inactive females of the cockroach, *Leucophaea maderae. Insect Biochem.* 4, 345–354.
Engelmann, F. (1974b) Juvenile hormone induction of the insect yolk protein precursor. *Am. Zool.* 14, 1195–1206.
Engelmann, F. (1977a) Vitellogenin polysomes from fat bodies of the cockroach *Leucophaea maderae. Am. Zool.* 17, 914.
Engelmann, F. (1977b) Undegraded vitellogenin polysomes from female insect fat bodies. *Biochem. biophys. Res. Commun.* 78, 641–647.
Engelmann, F. (1978) Synthesis of vitellogenin after long-term ovariectomy in a cockroach. *Insect Biochem.* 8, 149–154.
Engelmann, F. (1979) Insect vitellogenin: Identification, biosynthesis and role in vitellogenesis. *Adv. Insect Physiol.* 14, 49–108.
Engelmann, F. and Lüscher, M. (1957) Die hemmende Wirkung des Gehirns auf die Corpora allata bei *Leucophaea maderae* (Orthoptera) *Ver. Deut. Zool. Ges. Hamburg* (1956), 215–220.
Ernst, K.-D., Boeckh, J. and Boeckh, V. (1977) A neuroanatomical study on the organization of the central antennal pathways in insects. II. Deutocerebral connections in *Locusta migratoria* and *Periplaneta americana. Cell Tissue Res.* 176, 285–308.
Evans, P. D. (1978a) Octopamine distribution in the insect nervous system. *J. Neurochem.* 30, 1009–1013.
Evans, P. D. (1978b) Octopamine: a high affinity uptake mechanism in the nervous system of the cockroach. *J. Neurochem.* 30, 1015–1022.
Evans, P. D. and O'Shea, M. (1978) The identification of an octopaminergic neuron and the modulation of a myogenic rhythm in the locust. *J. exp. Biol.* 73, 235–260.
Ewing, A. W. and Manning, A. (1966) Some aspects of the efferent control of walking in three cockroach species. *J. Insect Physiol.* 12, 1115–1118.
Ewing, L. S. (1972) Hierarchy and its relation to territory in the cockroach *Nauphoeta cinerea. Behaviour* 42, 152–174.

Ewing, L. S. (1973) Territoriality and the influence of females on the spacing of males in the cockroach, *Nauphoeta cinerea. Behaviour* 45, 287–304.

Faber, B. L. (1970) *Activity Patterns, Endogenous and under Light–Dark Cycles, of last-instar male and female nymphs and adult American cockroaches. M. S. Thesis, Rutgers University.*

Faber, B. (1975) *The Effects of Several Environmental Factors on Feeding Behavior in the American Cockroach, Periplaneta americana* (L.). PhD Thesis, Rutgers University.

Faeder, I. R., Matthews, J. A. and Salpeter, M. M. (1974) [^{3}H]-Glutamate uptake at insect neuromuscular junctions: effect of chlorpromazine. *Brain Res.* 80, 53–70.

Faeder, I. R. and Salpeter, M. M. (1970) Glutamate uptake by a stimulated insect nerve–muscle preparation. *J. Cell Biol.* 46, 300–307.

Farley, R. D. and Case, J. F. (1968) Sensory modulation of ventilative pacemaker output in the cockroach, *Periplaneta americana. J. Insect Physiol.* 14, 591–601.

Farley, R. D., Case, J. F. and Roeder, K. D. (1967) Pacemaker for tracheal ventilation in the cockroach, *Periplaneta americana. J. Insect Physiol.* 13, 1713–1728.

Farley, R. D. and Milburn, N. S. (1969) Structure and function of the giant fibre system in the cockroach, *Periplaneta americana. J. Insect Physiol.* 15, 457–476.

Farnworth, E. G. (1972) Effect of ambient temperature and humidity on internal temperature and wing-beat frequency of *Periplaneta americana. J. Insect Physiol.* 18, 359–370.

Fast, P. G. (1964) Insect lipids: a review. *Mem. ent. Soc. Canada* #37.

Fast, P. G. (1970) Insect lipids. *Prog. Chem. Fats other Lipids* 11, 181–242.

Fatranska, M., Vargova, M., Rosival, L., Batora, V. and Janekora, D. (1978) Circadian susceptibility rhythms to some organophosphate compounds in the rat. *Chronobiologica* 5, 39–44.

Faucheux, M. J. and Sellier, R. (1971) L'ultrastructure de l'intima articulaire des sacs aériens chez les insectes. *C. R. Acad. Sci., Paris D* 272, 2197–2200.

Feldman, M. R. (1972) Fine structural studies of the intestinal system of the nematode *Leidynema appendiculata* (Leidy, 1850). *Trans. Am. Micros. Soc.* 91, 337–347.

Fibiger, J. A. G. and Ditlevsen, H. (1914) Contributions to the biology and morphology of *Spiroptera* (*Gongylonema*) *neoplastica* n. sp. *Mindeskr. Japetus Steenstrups Fødsel, Copenhagen,* 25.

Fifield, S. M. and Finlayson, L. H. (1978) Peripheral neurons and peripheral neurosecretion in the stick insect, *Carausius morosus. Proc. R. Soc. Lond. Ser. B.* 200, 63–85.

Finlayson, L. H. (1976) Abdominal and thoracic receptors in insects, centipedes and scorpions. In: *Structure and Function of Proprioceptors in the Invertebrates*. Mill, P. J., Ed., pp. 153–212. Chapman and Hall, London.

Finlayson, L. H. and Lowenstein, O. (1958) Structure and function of abdominal stretch receptors in insects. *Proc. R. Soc. Lond. Ser. B.* 148, 433–449.

Finlayson, L. H. and Osborne, M. P. (1968) Peripheral neurosecretory cells in the stick insect (*Carausius morosus*) and the blowfly larva (*Phormia terrae-novae*). *J. Insect Physiol.* 14, 1793–1801.

Fischer, O., von (1928) Die Entwicklung von *Periplaneta americana. Mitt. Naturforsch. Ges., Bern* 1927, pp. V–VII.

Fisher, Jr, F. M. and Sanborn, R. C. (1964) *Nosema* as a source of juvenile hormone in parasitized insects. *Biol. Bull.* 126, 235–252.

Fisk, F. W. (1951) Use of a specific mite control in roach and mouse cultures. *J. econ. Ent.* 44, 1016.

Flattum, R. F., Watkinson, I. A. and Crowder, L. A. (1973) The effect of insect 'autoneurotoxin' on *Periplaneta americana* (L.) and *Schistocerca gregaria* (Forskal) Malpighian tubules. *Pest. Biochem. Physiol.* 3, 237–242.

Fleet, R. R. and Frankie, G. W. (1974) Habits of two household cockroaches in outdoor environments. *Texas Agric. exp. St. Misc. Pub.* 1153, 1–8.

Fleet, R. R. and Frankie, G. W. (1975) Behavioral and ecological characteristics of a eulophid egg parasite of two species of domiciliary cockroaches. *Environ. Ent.* 4, 282–284.

Fleet, R. R., Piper, G. L. and Frankie, G. W. (1978) Studies on the population ecology of the smoky-brown cockroach *Periplaneta fuliginosa* in a Texas USA outdoor urban environment. *Environ. Ent.* 7, 807–814.

Flint, R. A. and Patton, R. L. (1959) Relation of eye color to molting in *Periplaneta americana. L. Bull. Brooklyn ent. Soc.* 54, 140.

Foldesi, L. and Mills, R. R. (1981) Hormonally induced uptake of protein-bound phenols by the cockroach cuticle. (In preparation.)

Forgash, A. J. and Moore, R. F. (1960) Dietary inositol requirement of *Periplaneta americana. Ann. ent. Soc. Am.* 53, 91–94.

Fourtner, C. R. (1976) Central nervous control of cockroach walking. In: *Neural Control of Locomotion*. Herman, R. M., Grillner, S., Stein, P. S. G. and Stuart, D. G., Eds., pp. 401–418. Plenum Press, New York.

Fourtner, C. R. (1978) The ultrastructure of the metathoracic femoral extensors of the cockroach, *Periplaneta americana. J. Morph.* 156, 127–139.

Fourtner, C. R. and Drewes, C. D. (1977) Excitation of the common inhibitory motor neuron: a possible role in the startle reflex of the cockroach, *Periplaneta americana. J. Neurobiol.* 8, 477–489.

Fourtner, C. R. Drewes, C. D. and Holzman, T. W. (1978) Specificity of afferent and efferent regeneration in the cockroach: Establishment of a reflex pathway between contralateral homologous target cells. *J. Neurophysiol.* (*Bethesda*) 41, 885–895.

Fourtner, C. R. and Pearson, K. G. (1977) Morphological and physiological properties of motor neurons innervating insect leg muscles. In: *Identified Neurons and Behavior of Arthropods*, Hoyle, G., Ed., pp. 87–99. Plenum Press, New York.

Fox, F. R. and Mills, R. R. (1969) Changes in haemolymph and cuticle proteins during the molting process in the American cockroach. *Comp. Biochem. Physiol.* 29, 1187–1195.

Fox, F. R., Seed, J. R. and Mills, R. R. (1972) Cuticle sclerotization by the American cockroach: Immunological evidence for the incorporation of blood proteins into the cuticle. *J. Insect Physiol.* 18, 2065–2070.

Fraenkel, G. (1932) Untersuchungen über die Koordination von Reflexen und Automatisch-nervösen Rhythmen bei Insekten. II. Die Nervöse Regulierung der Atmung wahrend des Fluges. *Z. vergl. Physiol.* 16, 394–417.

Fraenkel, G. and Rudall, K. M. (1940) A study of the physical and chemical properties of the insect cuticle. *Proc. R. Soc., Lond. Ser. B.* 129, 1–35.

Fraenkel, G. and Rudall, K. M. (1947) The structure of insect cuticles. *Proc. R. Soc., Lond. Ser. B.* 134, 111–43.

Francois, J. (1978) The ultrastructure and histochemistry of the mesenteric connective tissue of the cockroach *Periplaneta americana* L. (Insecta, Dictyoptera). *Cell Tissue Res.* 189, 91–107.

Franklin, R., Bell, W. J. and Jander, R. (1981) Rotational locomotion by the cockroach *Blattella germanica. J. Insect Physiol.* (In press.)

Fraser, P. J. (1977) Cercal ablation modifies tethered flight behaviour of cockroach. *Nature* 268, 523–524.

Fraser, J. and Pipa, R. (1977) Corpus allatum regulation during the metamorphosis of *Periplaneta americana*: axon pathways. *J. Insect Physiol.* 23, 975–984.

Freckleton, Jr, W. C. and Wahlsten, D. (1968) Carbon dioxide-induced amnesia in the cockroach *Periplaneta americana. Psychon. Sci.* 12, 179–180.

French, V. (1976) Leg regeneration in the cockroach, *Blattella germanica*. II. Regeneration from a non-congruent tibial graft/host junction. *J. Embryol. exp. Morph.* 35, 267–301.

French, V. (1978) Intercalary regeneration around the circumference of the cockroach leg. *J. Embryol. exp. Morph.* 47, 53–84.

Friedel, T., Feyereisen, R., Mundall, E. C. and Tobe, S. S. (1980) The allatostatic effect of

20-hydroxyecdysone on the adult viviparous cockroach, *Diploptera punctata*. *J. Insect Physiol.* 26, 665–670

Friedman, S. (1978) Trehalose regulation, one aspect of metabolic homeostasis. *A. Rev. Ent.* 23, 389–407.

Friedman, S. and Hsueh, T.-F. (1979) Insect trehalose-6-phosphatase: the unactivated type, as illustrated in *Periplaneta americana*, and a survey of the ordinal distribution of the two presently known types. *Comp. Biochem. Physiol.* 64B, 339–344.

Frings, H. (1946) Gustatory thresholds for sucrose and electrolytes for the cockroach, *Periplaneta americana* (L.). *J. exp. Zool.* 102, 23–50.

Frishman, A. M. and Alcamo, I. E. (1977) Domestic cockroaches and human bacterial disease. *Pest Control*, June, 16.

Frontali, N. (1968) Histochemical localization of catecholamines in the brain of normal and drug-treated cockroaches. *J. Insect Physiol.* 14, 881–886.

Frontali, N. and Gainer, H. (1977) Peptides in invertebrate nervous systems. In: *Peptides in Neurobiology*, Gainer, H., Ed., pp. 259–292. Plenum Press, New York.

Frontali, N. and Mancini, G. (1970) Studies on the neuronal organization of cockroach corpora pedunculata. *J. Insect Physiol.* 16, 2293–2301.

Füller, H. B. (1960) Morphologische und experimentelle Untersuchungen über die neurosekretorischen Verhältnesse im Zentralnervesystem von Blattiden und Culiciden. *Zool. Jahrb. Physiol.* 69, 223–250.

Füller, H. B. and Vent, H. O. (1976) Iontophoretische Untersuchungen zu Topographie und Struktur der Riesenneurone des 6 Abdominalganglions von *Periplaneta americana* L. *Zool. Jahrb. Anat.* 96, 438–447.

Fukuto, T. R. (1979) Effect of structure on the interaction of organophosphorus and carbamate esters with acetylcholinesterase. In: *Neurotoxicology of Insecticides and Pheromones*, Narahishi, T., Ed., pp. 277–295. Plenum Press, New York.

Gäde, G. (1977) Effect of corpus cardiacum extract on cyclic AMP concentration in the fat body of *Periplaneta americana*. *Zool. Jahrb. Physiol.* 81, 245–249.

Gaffal, K. P., Tichy, H., Theiss, J. and Seelinger, G. (1975) Structural polarities in mechanosensitive sensilla and their influence on stimulus transmission (Arthropoda). *Zoomorphologie* 82, 79–103.

Gammon, D. W. (1978a) Neural effects of allethrin on the free walking cockroach, *Periplaneta americana*: an investigation using defined doses at 15° and 32°C. *Pest. Sci.* 9, 79–91.

Gammon, D. W. (1978b) Effects of DDT on the cockroach nervous system at three temperatures. *Pest. Sci.* 9, 95–104.

Gammon, D. W. (1979) An analysis of the temperature-dependence of the toxicity of allethrin to the cockroach. In: *Neurotoxicology of Insecticides and Pheromones*, Narahashi, T., Ed., pp. 97–117. Plenum Press, New York.

Gardner, Jr, F. E. and Rounds, H. D. (1969) The pharmacology of cardio-accelerators in the central nervous system of *Periplaneta americana* (L.). *Comp. Biochem. Physiol.* 29, 1071–1078.

Gates, M. F. and Allee, W. C. (1933) Conditioned behavior of isolated and grouped cockroaches on a simple maze. *J. comp. Psychol.* 15, 331–358.

Gautier, J.-Y. (1974a) *Processus de differentiation de l'Organisation Sociale chez quelques especes de Blattes du genre Blaberus*: Aspects Ecologiques et Ethologiques. These de Doctorat d'Etat, Univ. Rennes.

Gautier, J.-Y. (1974b) Études comparee de la distribution spatiale et temporelle des adultes de *Blaberus atropos* et *Blaberus colosseus* (Dictyopteres) dans quatre grottes de Trinidad. *Rev. comp. Animal* 9, 237–258.

Gautier, J.-Y. (1976) Contribution a l'étude de phenomène d'ordre de dominance chez les blattes. I. Affrontement par paires de *Blaberus craniifer* males. *Biol. du Comp.* 1, 353–365.

Gautier, J.-Y. (1979) Contribution a l'étude du phenomène d'ordre de dominance chez les blattes. II. Analyse qualitative des relations males/males chez *Blaberus craniifer*. *Biol. du Comp.* 4, 61–74.

Gee, J. D. (1977) The hormonal control of excretion. In: *Transport of Ions and Water in Animals*. Gupta, B. L., Moreton, R. B., Oschman, J. L. and Wall, B. J., Eds., pp. 265–281. Academic Press, London.

Geiger, J. G., Krolak, J. M. and Mills, R. R. (1977) Possible involvement of cockroach haemocytes in the storage and synthesis of cuticle proteins. *J. Insect Physiol.* 23, 227–230.

Gelperin, A. (1967) Stretch receptors in the foregut of the blowfly. *Science* 157, 208–210.

Gerolt, P. (1969) Mode of entry of contact insecticides. *J. Insect Physiol.* 15, 563–573.

Gerolt, P. (1972) Mode of entry of oxime insecticides into insects. *Pest. Sci.* 3, 43.

Gersch, M. (1972) Experimentelle Untersuchungen zum Freisetzungsmechanismus von Neurohormonen nach elektrischer Reizung der Corpora cardiaca von *Periplaneta americana in vitro. J. Insect Physiol.* 18, 2425–2439.

Gersch, M. (1974a) Experimentelle Untersuchungen zur Ausschüttung von Neurohormonen aus Ganglien des Bauchmarks von *Periplaneta americana* nach elektrischer Reizung *in vitro. Zool. Jahrb. Physiol.* 78, 138–149.

Gersch, M. (1974b) Selektive Freisetzung des hyperglykäemischen Faktors aus den Corpora cardiaca von *Periplaneta americana in vivo. Experientia* 30, 767.

Gersch, M. (1975) *Prinzipien neurohormonaler und neurohumoralor Steuerung physiologischer Prozesse*, Friedrich-Schiller-Universität, Jena.

Gersch, M. (1977) Moulting of insects without moulting gland: Results with larvae of *Periplaneta americana. Experientia* 33, 228–230.

Gersch, M. (1980) Control of activity of the prothoracic glands in insects at the cellular level. In: *Progress in Ecdysone Research*, Hoffman, J. A., Ed., pp. 111–123. Elsevier/North-Holland, Amsterdam.

Gersch, M. and Bräuer, R. (1974) *In vitro*-Stimulation der Prothorakaldrüsen von Insekten als Testsystem (Prothorakaldrüsentest). *J. Insect Physiol.* 20, 735–741.

Gersch, M., Bräuer, R. and Birkenbeil, H. (1973) Experimentelle Untersuchungen zum Wirkungsmechanismus der beiden entwicklungsphysiologisch aktiven Fraktionen des 'Gehirnhormons' der Insekten (Aktivationsfaktor I und II) auf die Prothorakaldrüse. *Experientia* 29, 425–427.

Gersch, M. and Eibisch, H. (1977) Synthesis of ecdysone-^{14}C and ecdysterone-^{14}C from cholesterol-^{14}C in cockroaches (*Periplaneta americana*) without molting glands. *Experientia* 33, 468.

Gersch, M., Fischer, F., Unger, H. and Koch, H. (1960) Die Isolierung neurohormonaler Faktoren aus dem Nervensystem der Küchenschabe *Periplaneta americana. Z. Naturforsch.* 15b, 319–322.

Gersch, M., Hentschel, E. and Ude, J. (1974) Aminerge Substanzen im lateralen Herznerven und im stomatogastrischen Nervensystem der Schabe *Blaberus craniifer* Burm. *Zool. Jahrb. Physiol.* 78, 1–15.

Gersch, M., Richter, K., Böhm, G.-A and Stürzebecher, J. (1970) Selektive Ausschüttung von Neurohormonen nach elektrischer Reizung der Corpora cardiaca von *Periplaneta americana in vitro. J. Insect Physiol.* 16, 1991–2013.

Gersch, M. and Stürzebecher, J. (1967) Zur Frage der Identität und des Vorkommens von Neurohormon D in verschiedenen Bereichen des Zentralnervensystems von *Periplaneta americana. Z. Naturforsch.* 22b, 563.

Gersch, M. and Stürzebecher, J. (1968) Weitere Untersuchungen zur Kennzeichnung des Aktivationshormons der Insektenhäutung. *J. Insect Physiol.* 14, 87–96.

Gersch, M. and Stürzebecher, J. (1970) Experimentelle Stimulierung der zellulären Aktivität der Prothorakaldrüsen von *Periplaneta americana* durch den Aktivationsfaktor. *J. Insect Physiol.* 16, 1813–1826.

Gersch, M. and Unger, H. (1957) Nachweis von Neurohormonen- aus dem Nervensystem von *Dixippus morosus* mit Hilfe papierchromatographischer Trennung. *Naturwiss.* 44, 117.

Gersch, M., Unger, H., Fischer, F. and Kapitza, W. (1963) Identifizierung einiger Wirkstoffe aus dem Nervensystem der Crustaceen und Insekten. *Z. Naturforsch.* 18b, 587–588.

Gewecke, M. (1974) The antennae of insects as air-current sense organs and their relationship to the control of flight. In: *Experimental Analysis of Insect Behaviour*, Barton-Browne, L., Ed., pp. 100–113. Springer-Verlag, New York.

Gharagozlou, I. D. (1972) Localisation d'activite lytique dans le tissu adipeaux de *Periplaneta americana. J. Micros.* 73, 281–284.

Gier, H. T. (1936) The morphology and behaviour of the intracellular bacteroids of roaches. *Biol. bull.* 71, 433–452.

Gier, H. T. (1947a) Intracellular bacteroids in the cockroach *Periplaneta americana* (Linn.). *J. Bact.* 53, 173–189.

Gier, H. T. (1947b) Growth rate in the cockroach *Periplaneta americana* (Linn.). *Ann. ent. Soc. Am.* 40, 303–317.

Gilbert, L. I. (1967a) Lipid metabolism and function in insects. *Adv. Insect Physiol.* 4, 69–211.

Gilbert, L. I. (1967b) Changes in lipid content during the reproductive cycle of *Leucophaea maderae* and effects of the juvenile hormone on lipid metabolism *in viro. Comp. Biochem. Physiol.* 21, 237–257.

Gilbert, L. I. (Ed.) (1976) *The Juvenile Hormones*. Plenum Press, New York.

Gilbert, L. I., Chino, H. and Domroese, K. A. (1965) Lipolytic activity of insect tissues and its significance in lipid transport. *J. Insect Physiol.* 11, 1057–1070.

Gilby, A. R. (1962) The absence of natural volatile solvents in cockroach grease. *Nature* 195, 729.

Gilby, A. R. and Cox, M. E. (1963) The cuticular lipids of the cockroach *Periplaneta americana* (L.). *J. Insect Physiol.* 9, 671–681.

Girardie, A. (1962) Étude biometrique de la croissance ovarienne apres ablation et implantation de corpora allata chez *Periplaneta americana. J. Insect Physiol.* 8, 199–214.

Glaser, R. W. (1946) The intracellular bacteria of the cockroach in relation to symbiosis. *J. Parasitol.* 32, 483–489.

Gnatzy, W. (1973) Die Feinstruktur der Fadenhaare auf den Cerci von *Periplaneta americana* L. *Verh. Dt. Zool. Ges.* 66, 37–42.

Gnatzy, W. (1976) The ultrastructure of the thread-hairs on the cerci of the cockroach *Periplaneta americana* L.: The intermoult phase. *J. ultrastruct. Res.* 54, 124–134.

Goldbard, G. A., Sauer, J. R. and Mills, R. R. (1970) Hormonal control of excretion in the American cockroach, II. Preliminary purification of a diuretic and antidiuretic hormone. *Comp. gen. Pharmacol.* 1, 82–86.

Goldsmith, T. H. and Ruck, P. (1958) The spectral sensitivities of the dorsal ocelli of cockroaches and honeybees. *J. gen. Physiol.* 41, 1171–1185.

Goldsworthy, G. J., Mordue, W. and Guthkelch, J. (1972) Studies on insect adipokinetic hormones. *Gen. comp. Endocrinol.* 18, 545–551.

Gole, J. W. D. and Downer, R. G. H. (1979) Elevation of adenosine 3:5′-monophosphate by octopamine in fat body of the American cockroach. *Periplaneta americana* L. *Comp. Biochem. Physiol.* 64C, 223–236.

Goodman, C. S. (1974) Anatomy of locust ocellar interneurons: constancy and variability. *J. comp. Physiol.* 95, 185–201.

Goodman, C. S. (1976) Anatomy of the ocellar interneurons of acridid grasshoppers. I. The large interneurons. *Cell Tissue Res.* 175, 183–202.

Goodman, C. S., O'Shea, M., McCaman, R. and Spitzer, N. C. (1979) Embryonic development of identified neurons: Temporal pattern of morphological and biochemical differentiation. *Science* 204, 1219–1222.

Goodman, C. S. and Spitzer, N. C. (1979) Embryonic development of identified neurons: Differentiation from neuroblast to neuron. *Nature* 280, 208–214.

Goodman, C. S. and Williams, J. L. D. (1976) anatomy of the ocellar interneurons of acridid grasshoppers. II. The small interneurons. *Cell Tissue Res.* 175, 203–225.
Goodman, L. J. (1970) The structure and function of the insect dorsal ocellus. *Adv. Insect Physiol.* 7, 97–195.
Gordon, H. T. (1959) Minimal nutritional requirements of the German roach, *Blattella germanica. Ann. N.Y. Acad. Sci.* 77, 290–351.
Gordon, H. T. (1968) Intake rates of various solid carbohydrates by male German cockroaches. *J. Insect Physiol.* 14, 41–52.
Gosbee, J. L., Milligan, J. V. and Smallman, B. N. (1968) Neuronal properties of the protocerebral neurosecretory cells of the adult cockroach *Periplaneta americana. J. Insect Physiol.* 14, 1785–1792.
Goss, R. J. (1969) *Principles of Regeneration*. Academic Press, New York.
Gould, G. E. (1941) The effect of temperature upon the development of cockroaches. *Proc. Indiana Acad. Sci.* 50, 242–248.
Gould, G. E. and Deay, H. O. (1938) The biology of the American cockroach. *Ann. ent. Soc. Am.* 31, 489–498.
Gould, G. E. and Deay, H. O. (1940) The biology of six species of cockroaches which inhabit buildings. *Purdue Univ. Agric. exp. St. Bull.* 451, 2–31.
Granett, J. and Leeling, N. C. (1971) Trehalose and glycogen depletion during DDT poisoning of American cockroaches, *Periplaneta americana. Ann. ent. Soc. Am.* 64, 784–789.
Greenberg, B., Kowalski, J. and Karpus, J. (1970) Micro-potentiometric pH determinations of the gut of *Periplaneta americana* fed three different diets. *J. econ. ent.* 63, 1795–1797.
Gregory, G. E. (1974) Neuroanatomy of the mesothoracic ganglion of the cockroach *Periplaneta americana* (L.). I. The roots of the peripheral nerves. *Phil. Trans. R. Soc. Lond. Ser. B.* 267, 421–465.
Gresson, R. A. R. (1931) Yolk-formation in *Periplaneta orientalis. Q. J. Micros. Sci.* 74, 257–274.
Gresson, R. A. R. (1934) The cytology of the mid gut and hepatic caeca of *Periplaneta orientalis. Q. J. Micros. Sci.* 77, 317–334.
Griffiths, D. J. G. and Smyth, Jr, T. (1973) Action of black widow spider venom at insect neuromuscular junctions. *Toxicon* 11, 369–374.
Griffiths, J. T. and Tauber, O. E. (1942a) Fecundity, longevity and parthenogenesis of the American roach, *Periplaneta americana* L. *Physiol. Zool.* 15, 196–209.
Griffiths, J. T. and Tauber, O. E. (1942b) The nymphal development for the roach, *Periplaneta americana* L. *J. N. Y. ent. Soc.* 50, 263–272.
Grossman, Y. and Parnas, I. (1973) Control mechanisms involved in the regulation of the phallic neuromuscular system of the cockroach *Periplaneta americana. J. comp. Physiol.* 82, 1–21.
Gulati, A. N. (1930) Do cockroaches eat bed bugs? *Nature* 125, 858.
Gundel, M. and Penzlin, H. (1978) The neuronal connections of the frontal ganglion of the cockroach *Periplaneta americana*. A histological and iontophoretical study. *Cell Tissue Res.* 193, 353–371.
Gunn, D. L. (1935) The temperature and humidity relations of the cockroach. III. A comparison of temperature preference, and rates of desiccation and respiration of *Periplaneta americana*, *Blatta orientalis* and *Blattella germanica. J. exp. Biol.* 12, 185–190.
Gunn, D. L. and Cosway, C. A. (1938) The temperature and humidity relations of the cockroach. V. Humidity preference. *J. exp. Biol.* 15, 555–563.
Gunn, D. L. and Cosway, C. A. (1942) The temperature and humidity relations of the cockroach. VI. Oxygen consumption. *J. exp. Biol.* 19, 124–132.
Gunn, D. L. and Notley, F. B. (1936) The temperature and humidity relations of the

cockroach. IV. Thermal death-point. *J. exp. Biol.* 13, 28–34.
Gunning, R. and Shipp, E. (1976) Circadian rhythm in endogenous nerve activity in the eye of *Musca domestica* L. *Physiol. Ent.* 1, 241–248.
Gupta, A. P. (Ed.) (1979) *Insect Hemocytes: Development, forms, functions, and techniques*. Cambridge University Press, Cambridge.
Gupta, A. P. and Sutherland, D. J. (1966) *In vitro* transformations of the insect plasmatocyte in some insects. *J. Insect Physiol.* 12, 1369–1375.
Gupta, A. P. and Sutherland, D. J. (1967) Phase contrast and histochemical studies of spherule cells in cockroaches (Dictyoptera). *Ann. ent. Soc. Am.* 60, 557–565.
Gupta, B. J. and Smith, D. S. (1969) Fine structural organization of the spermatheca in the cockroach, *Periplaneta americana. Tissue Cell* 1, 295–324.
Gupta, P. D. (1946) On the structure and function of the spermatophore in the cockroach *Periplaneta americana. Ind. J. Ent.* 8, 79–84.
Gupta, P. D. (1947) On copulation and insemination in the cockroach *Periplaneta americana. Proc. nat. Inst. Sci., India* 13, 65–71.
Guthrie, D. M. (1967a) The regeneration of motor axons in an insect. *J. Insect Physiol.* 13, 1593–1611.
Guthrie, D. M. (1967b) Multipolar stretch receptors and the insect leg reflex. *J. Insect Physiol.* 13, 1637–1644.
Guthrie, D. M. (1975) Regeneration and neural specificity – the contribution of invertebrate studies. In: *'Simple' Nervous Systems*, Usherwood, P. N. R. and Newth, D. R., Eds. Edward Arnold, London.
Guthrie, D. M. and Tindall, A. R. (1968) *The Biology of the Cockroach*. Edward Arnold, London.
Gyure, W. L. (1975) Characterization of isoxanthopterin deaminase isolated from the cockroach, *Periplaneta americana. Insect Biochem.* 5, 813–819.
Haas, H. (1955) Untersuchungen zur segmentbildung an der Antenne von *Periplaneta americana* L. *Roux Arch. Entwickmech.* 147, 434–473.
Haber, V. R. (1919) Cockroach pests of Minnesota. *Univ. Minn. Bull.* 186, 16 pp.
Haber, V. R. (1920) Oviposition by a cockroach *Periplaneta americana* Linn. (Orth.). *Ent. News* 31, 190–193.
Haber, V. R. (1926) The blood of insects with special reference to that of the common household German or croton cockroach. *Bull. Brooklyn ent. Soc.* 21, 61–100.
Hackman, R. H. and Goldberg, M. (1977) Molecular crosslinks in cuticles. *Insect Biochem.* 7, 175–84.
Hagedorn, H. H., Shapiro, J. P. and Hanaoka, K. (1979) Ovarian ecdysone secretion is controlled by a brain hormone in an adult mosquito. *Nature* 282, 92–94.
Halberg, F. (1960) Temporal organization of physiological function. *Cold Spring Harbor Symp. Quant. Biol.* 25, 289–310.
Halberg, F. (1973) Chronobiologic glossary. *Int. J. Chronobiol.* 1, 31–63.
Halberg, J., Cutcomp, L. K., Lee, J., Halberg, F., Sullivan, W. N., Hayes, D. K., Cawley, B. and Rosenthal, J. (1973) Single cosinor technique improves resolution and extends scope of knock-down and mortality-gauged susceptibility rhythms to insecticide. *Int. J. Chronobiol.* 1, 328–329.
Halberg, J., Halberg, F., Lee, J. K., Cutkomp, L., Sullivan, W. N., Hayes, D. K., Cawley, B. M. and Rosenthal, J. (1974) Similar timing of circadian rhythms in sensitivity to pyrethrum of several insects. *Int. J. Chronobiol.* 2, 291–296.
Hammel, H. T. (1976) Colligative properties of a solution. *Science* 192, 748–756.
Hammock, B. D. (1975) NADPH-dependent expoxidation of methyl farnesoate to juvenile hormone in the cockroach *Blaberus giganteus* L. *Life Sci.* 17, 323–328.
Hammock, B. D. and Mumby, S. M. (1978) Inhibition of epoxidation of methyl farnesoate to juvenile hormone III by cockroach corpus allatum homogenates. *Pest. Biochem. Physiol.* 9, 39–47.

Hamner, K. C., Flinn, Jr, J. C., Sirohi, G. S., Hoshizaki, T. and Carpenter, B. H. (1962) Studies of the biological clock at the South Pole. *Nature* 195, 476–480.

Hamnett, A. F. and Pratt, G. E. (1978) Use of automated capillary column radio-gas-chromatography in the identification of insect juvenile hormones. *J. Chromat.* 158, 387–399.

Hanaoka, K. and Takahashi, S. Y. (1976) Effect of hyperglycaemic factor on haemolymph trehalose and fat body carbohydrates in the American cockroach. *Insect Biochem.* 6, 621–625.

Hanaoka, K. and Takahashi, S. Y. (1977) Adenylate cyclase system and the hyperglycemic factor in the cockroach, *Periplaneta americana. Insect Biochem.* 7, 95–99.

Hanaoka, K. and Takashashi, S. Y. (1978) Endocrine control of carbohydrate metabolism including the mechanism of action of the hyperglycemic hormone in insect. In: *Comparative Endocrinology*, Gaillard, P. J. and Boer, H. H., Eds., pp. 455–458. Elsevier/North Holland Press, Amsterdam.

Hansen, K. (1979) Insect chemoreception. In: *Taxis and Behavior*, Hazelbauer, G. L., Ed., pp. 233–292. Chapman and Hall, London.

Hanström, B. (1928) Vergleichende Anatomie des Nervensystems der Wirbellosen Tiere. J. Springer, Berlin.

Hanström, B. (1940) Inkretorische Organe, Sinnesorgane und Nervensystem des Kopfes einiger niederer Insektenordnungen. *Kungl. Svenksa Vetenskapsakad. Handlingar.* Ser 3B 18, 1–266.

Harker, J. E. (1954) Diurnal rhythm in *Periplaneta americana* L. *Nature* 173, 689–690.

Harker, J. E. (1955) Control of diurnal rhythms of activity in *Periplaneta americana* L. *Nature* 175, 733.

Harker, J. E. (1956) Factors controlling the diurnal rhythm of activity in *Periplaneta americana* L. *J. exp. Biol.* 33, 224–234.

Harker, J. E. (1958a) Diurnal rhythms in the animal kingdom. *Biol. Rev.* 33, 1–52.

Harker, J. E. (1958b) Experimental production of midgut tumours in *Periplaneta americana* L. *J. exp. Biol.* 35, 251–259.

Harker, J. E. (1960a) The effect of perturbations in the environmental cycle of *Periplaneta americana* L. *J. exp. Biol.* 37, 154–163.

Harker, J. E. (1960b) Internal factors controlling the suboesophageal ganglion neuro-secretory cycle in *Periplaneta americana* L. *J. exp. Biol.* 37, 164–170.

Harker, J. E. (1960c) Endocrine and nervous factors in insect circadian rhythms. *Cold Spring Harbor Symp. Quant. Biol.* 25, 279–287.

Harker, J. E. (1961) Diurnal rhythms. *A. Rev. Ent.* 6, 131–146.

Harker, J. E. (1964) *The Physiology of Diurnal Rhythms.* Cambridge University Press, Cambridge.

Harper, E., Seifter, S., and Scharrer, B. (1967) Electron microscopic and biochemical characterization of collagen in blattarian insects. *J. Cell Biol.* 33, 385–394.

Harris, C. L. (1977) Giant interneurons of the cockroach neither trigger escape nor 'clear all stations'. *Comp. Biochem. Physiol.* 56A, 333–335.

Harris, C. L. and Garrison, W. (1976) Electrotonic coupling between cercal afferents and giant interneurons in the American cockroach. *J. Insect Physiol.* 22, 31–40.

Harris, C. L. and Smyth, T. (1971) Structural details of cockroach giant axons revealed by injected dye. *Comp. Biochem. Physiol.* 40A, 295–303.

Harrow, I. D., Hue, B., Pelhate, M. and Sattelle, D. B. (1980) Cockroach giant interneurons stained by cobalt-backfilling of dissected axons. *J. exp. Biol.* 84, 341–343.

Harshbarger, J. C. and Forgash, A. J. (1964) Effect of lindane on the intracellular micro-organisms of the American cockroach, *Periplaneta americana. J. econ. Ent.* 57, 994–995.

Harshbarger, J. C. and Taylor, R. L. (1968) Neoplasms of insects. *A. Rev. Ent.* 13, 159–190.

Hart, T. F. and Fourtner, C. R. (1979) Histochemical analysis of physiologically and morphologically identified muscles in an insect leg. *Comp. Biochem. Physiol.* 64A, 437–440.

Hawkins, W. A. (1977) *Locomotion and Sex Pheromone-Stimulated Chemolocation in the Male American Cockroach, Periplaneta americana.* Ph.D. Thesis, University of Kansas.
Hawkins, W. A. (1978) Effects of sex pheromone on locomotion in the male American cockroach *Periplaneta americana. J. chem. Ecol.* 4, 149–160.
Hawkins, W. A. and Rust, M. K. (1977) Factors influencing male sexual response in the American cockroach *Periplaneta americana. J. chem. Ecol.* 3, 85–99.
Hawkins, W. B. and Sternburg, J. (1964) Some chemical characteristics of a DDT-induced neuroactive substance from cockroaches and crayfish. *J. Econ. ent.* 57, 241–247.
Haydak, M. H. (1953) Influence of protein level on the longevity of cockroaches. *Ann. ent. Soc. Am.* 46, 547–560.
Hayes, D. K., Mensing, E. and Schechter, M. S. (1970) Electrophoretic patterns of proteins in hemolymph obtained from the adult Maderia cockroach, *Leucophaea maderae* (F.), during a twenty-four hour period. *Comp. Biochem. Physiol.* 34, 733–737.
Hazelhoff, E. H. (1926) Regeling der Ademhaling bij Insecten en Spinnen. Proefschrift, Rijks-Universiteit de Utrecht (translation kindly loaned by Dr John Buck).
Heit, M., Sauer, J. R. and Mills, R. R. (1973) The effects of high concentrations of sodium in the drinking medium of the American cockroach, *Periplaneta americana* (L.). *Comp. Biochem. Physiol.* 45A, 363–370.
Heinzel, H.-G. and Gewecke, M. (1979) Directional sensitivity of the antennal campaniform sensilla in insects. *Naturwissenschaften* 66, 212–213.
Henry, S. M. and Block, R. J. (1960) The sulphur metabolism of insects. IV. The conversion of inorganic sulphate to organic sulphur compounds in cockroaches. The role of intracellular symbionts. *Cont. Boyce Thompson Inst.* 20, 317–329.
Henry, S. M. and Block, R. J. (1962) Amino acid synthesis, a ruminant-like effect of the intracellular symbionts of the German cockroach. *Fedn Proc. fedn Socs exp. Biol. Am.* 21, 9.
Hentschel, E. (1972) Ovulation and aminerges neurosekretorisches System bei *Periplaneta americana* (L.) (Blattoidea, Insecta). *Zool. Jahrb. Physiol.* 76, 356–367.
Hentschel, E. (1975) Die Workung von 6-Hydroxydopamin und *p*-Chlorophenylalanin auf die Ovulation bein *Periplaneta americana* (L.) (Blattoidea, Insecta). *Zool. Jahrb. Physiol.* 79, 506–512.
Herford, G. M. (1938) Tracheal pulsations in the flea. *J. exp. Biol.* 15, 327–338.
Hertel, W. (1971) Untersuchungen zur neurohormonalen Steuerung des Herzens der Amerikanischen Schabe *Periplaneta americana* (L.). *Zool. Jahrb. Physiol.* 76, 152–184.
Hertel, W., Koch, J. and Penzlin, H. (1978) Electrophysiologische Untersuchungen an Frontalganglion von *Periplaneta americana* L. *J. Insect Physiol.* 24, 721–735.
Heslop, J. P. and Ray, J. W. (1959) The reaction of the cockroach *Periplaneta americana* L. to bodily stress and DDT. *J. Insect Physiol.* 3, 395–401.
Hess, A. (1958a) The fine structure of nerve cells and fibers, neuroglia and sheaths of the ganglion chain in the cockroach (*Periplaneta americana*). *J. biophys biochem. Cytol.* 4, 731–742.
Hess, A. (1958b) Experimental anatomical studies of pathways in the severed central nerve cord of the cockroach. *J. Morph.* 103, 479–502.
Heymons, R. (1895) *Die Embryonalentwicklung von Dermapteren und Orthopteren unter besonderer Berucksichtigung der Keimblatterbildung*. Gustav Fischer, Jena.
Heymons, R. and Lengerken, H. von (1929) Biologische Untersuchungen an copraphagen Lamellicorniern. *Z. morph. Okol. Tiere* 14, 531–613.
Heywood, R. B. (1965) Changes occurring in the central nervous system of *Pieris brassicae*, L. (Lepidoptera) during metamorphosis. *J. Insect Physiol.* 11, 413–430.
Highnam, K. C. (1961) Induced changes in the amounts of material in the neurosecretory system of the desert locust. *Nature* 191, 199–200.
Highnam, K. C. (1964) Endocrine relationships in insect reproduction. *Symp. R. ent. Soc. Lond.* 2, 26–42.

Hillerton, J. E. (1978) changes in the structure and composition of the extensible cuticle of *Rhodnius prolixus* through the 5th larval instar. *J. Insect Physiol.* 24, 399–412.

Hillerton, J. E. and Vincent, J. F. (1979) The stabilisation of insect cuticles. *J. Insect Physiol.* 25, 957–963.

Hilliard, S. D. and Butz, A. (1969) Daily fluctuations in the concentrations of total sugar and uric acid in the hemolymph of *Periplaneta americana*. *Ann. ent. Soc. Am.* 62, 71–74.

Hipps, P. P., Holland, W. H. and Sherman, W. R. (1972) Identification and measurement of *chiro*-inositol in the American cockroach, *Periplaneta americana* L. *Biochem. biophys. Res. Commun.* 46, 1903–1908.

Hipps, P. P. and Nelson, D. R. (1974) Esterases from the midgut and gastric caecum of the American cockroach *Periplaneta americana* (L.), isolation and characterization. *Biochim. biophys. Acta* 341, 421–436.

Hipps, P. P., Sehgal, R. K., Holland, W. H. and Sherman, W. R. (1973) Identification and partial characterization of inositol: NAD^+ epimerase and inosose: NAD(P)H Reductase from the fat body of the American cockroach, *Periplaneta americana* L. *Biochemistry* 12, 4705–4712.

Hitchcock, C. R. and Bell, E. T. (1952) Studies on the nematode parasite, *Gongylonema neoplasticum* (*Spiroptera neoplasticum*), and avitaminosis A in the fore-stomach of rats: Comparison with Fibiger's results. *J. natn. Cancer Inst.* 12, 1345–1387.

Hoffman, A. G. D. and Downer, R. G. H. (1974) Evolution of $^{14}CO_2$ from 1-^{14}C-acetate in the American cockroach, *Periplaneta americana*. *Comp. Biochem. Physiol.* 48B, 199–204.

Hoffman, A. G. D. and Downer, R. G. H. (1976) The crop as an organ of glyceride absorption in the American cockroach, *Periplaneta americana* L. *Can. J. Zool.* 54, 1165–1171.

Hoffman, A. G. D. and Downer, R. G. H. (1977) Diacylglycerols as major end products of triacylglycerol hydrolysis by tissue lipases of the cockroach. *Am. Zool.* 17, 477.

Hoffman, A. G. D. and Downer, R. G. H. (1979a) End product specificity of triacylglycerol lipases from intestine, fat body, muscle and haemolymph of the American cockroach, *Periplaneta americana* L. *Lipids* 14, 893–899.

Hoffman, A. G. D. and Downer, R. G. H. (1979b) Synthesis of diacylglycerols by monoacylglycerol acyltransferase from crop, midgut and fat body tissues of the American cockroach, *Periplaneta americana* L. *Insect Biochem.* 9, 129–134.

Hoffman, R. W. (1933) Zur Analyse des Reflexgeschehens bei *Blatta orientalis* L. *Z. vergl. Physiol.* 18, 740–795.

Hofmann, K., Guenderoth-Palmowsky, M., Wiedenmann, G. and Engelmann, W. (1978) Further evidence for period lengthening effect of L^+ on circadian rhythms. *Z. Naturforsch. Sect. C.* 33, 231–234.

Hofmanová, O., Čerkasaková, A., Foustka, M. and Kubišta, V. (1966) Metabolism of the thoracic musculature of *Periplaneta americana* during flight. *Acta Univ. Carol.* 1966, 183–189.

Hogben, L. (1920) Studies on synapsis. II. Parallel conjugation and the prophase complex in *Periplaneta* with special reference to the premeiotic telophase. *Proc. R. Soc., Lond. Ser. B.* 91, 305–329.

Holan, G. (1969) New halocyclopropane insecticides and the mode of action of DDT. *Nature* 221, 1025–1029.

Hollande, A. C. and Favre, R. (1931) La structure cytologique de *Blattabacterium cuenoti* (Mercier) N. G., symbiote du tissue adipeaux des Blattides. *C.R. Soc. Biol., Paris* 107, 752–754.

Holman, G. M. and Cook, B. J. (1970) Pharmacological properties of excitatory neuromuscular transmission in the hindgut of the cockroach, *Leucophaea maderae*. *J. Insect Physiol.* 16, 1891–1907.

Holman, G. M. and Cook, B. J. (1972) Isolation, partial purification and characterization of a peptide which stimulates the hindgut of the cockroach, *Leucophaea maderae* (Fabr.). *Biol. Bull.* 142, 446–460.

Holman, G. M. and Cook, B. J. (1979) Evidence for proctolin and a second myotropic peptide in the cockroach, *Leucophaea maderae*, determined by bioassay and HPLC analysis. *Insect Biochem.* 9, 149–154.

Holman, G. M. and Marks, E. P. (1974) Synthesis, transport and release of a neurohormone by cultured neuroendocrine glands from the cockroach, *Leucophaea maderae. J. Insect Physiol.* 20, 479–484.

Holwerda, D. A., Weeda, E. and van Doorn, J. M. (1977) Separation of the hyperglycemic and adipokinetic factors from the cockroach corpus cardiacum. *Insect Biochem.* 7, 477–481.

Hominick, W. M. and Davey, K. G. (1972) The influence of host stage and sex upon the size and composition of the population of two species of thelastomatids parasitic in the hindgut of *Periplaneta americana. Can. J. Zool.* 50, 947–954.

Hominick, W. M. and Davey, K. G. (1973) Food and spatial distribution of adult female pinworms parasitic in the hindgut of *Periplaneta americana* L. *Int. J. Parasitol.* 3, 759–771.

Hominick, W. M. and Davey, K. G. (1975) The effect of nutritional level of the host on space and food available to pinworms in the colon of *Periplaneta americana* L. *Comp. Biochem. Physiol.* 51, 83–88.

Honegger, H.-W. and Schürmann, F. W. (1975) Cobalt sulphide staining of optic fibres in the brain of the cricket, *Gryllus campestris. Cell Tissue Res.* 159, 213–225.

Hopkins, T. L. and Lofgren, P. A. (1968) Adenine metabolism and urate storage in the cockroach, *Leucophaea maderae. J. Insect Physiol.* 14, 1803–1814.

Hopkins, T. L., Murdock, L. L. and Wirtz, R. A. (1971a) Tyrosine side-chain metabolism in larval and adult cockroaches, *Periplaneta americana*: post-ecdysial patterns. *Insect Biochem.* 1, 97–101.

Hopkins, T. L. and Srivastava, B. B. L. (1972) Rectal water transport in the cockroach *Leucophaea maderae*: effects of lumen cations, carbon dioxide and ouabain. *J. Insect Physiol.* 18, 2293–2298.

Hopkins, T. L., Srivastava, B. B. L. and Bahadur, J. (1971b) Rectal water transport in the cockroach *Leucophaea maderae*: tritiated water uptake and solute effects. *J. Insect Physiol.* 17, 1857–1864.

Hopkins, T. L. and Wirtz, R. A. (1976) Dopa and tyrosine decarboxylase activity in tissues of *Periplaneta americana* in relation to cuticle formation and ecdysis. *J. Insect Physiol.* 22, 1167–1171.

Horridge, G. A. (1962) Learning of leg position by the ventral nerve cord in headless insects. *Proc. R. Soc. Lond. Ser. B.* 157, 33–52.

Horridge, G. A. (1965) The electrophysiological approach to learning in an isolatable ganglia. *Animal Behav.* (Suppl) 1, 163–182.

House, C. R. (1973) An electrophysiological study of neuroglandular transmission in the isolated salivary glands of the cockroach. *J. exp. Biol.* 58, 29–43.

House, C. R. (1975) Intracellular recording of secretory potentials in a 'mixed' salivary gland. *Experientia* 31, 904–906.

House, C. R. (1977) Cockroach salivary gland: a secretory epithelium with a dopaminergic innervation. In: *Transport of Ions and Water in Animals*, Gupta, B. L., Moreton, R. B., Oschman, J. L. and Wall, B. J., Eds., pp. 403–425. Academic Press, London.

House, H. L. (1974a) Nutrition. In: *The Physiology of Insecta*, Rockstein, M., Ed., 2nd edition, Vol. 5, pp. 1–62. Academic Press, New York.

House, H. L. (1974b) Digestion. In: *The Physiology of Insecta*, Rockstein, M., Ed., 2nd edition, Vol. 5, pp. 63–117. Academic Press, New York.

Howse, P. E. (1964) An investigation into the mode of action of the subgenual organ in the termite, *Zootermopsis angusticollis* Emerson, and in the cockroach, *Periplaneta americana* (L.). *J. Insect Physiol.* 10, 409–424.

Hoyle, G. (1955) The effects of some common cations on neuromuscular transmission in insects. *J. Physiol.* 127, 90–103.
Hoyle, G. (1957) The nervous control of insect muscle. In: *Recent Advances in Invertebrate Physiology*, Scheer, B. T., Ed., pp. 73–98. University of Oregon Press, Eugene.
Hoyle, G. (1960) The action of carbon dioxide gas on an insect spiracular muscle. *J. Insect Physiol.* 4, 63–79.
Hoyle, G. (1961) Functional contracture in a spiracular muscle. *J. Insect Physiol.* 7, 305–314.
Hoyle, G. (1964) Exploration of neuronal mechanisms underlying behavior in insects. In: *Neural Theory and Modeling*, Reiss, R. R., Ed., pp. 346–376. Standford University Press, Stanford.
Hoyle, G. (1967) Diversity of straited muscle. *Am. Zool.* 7, 435–449.
Hoyle, G. (1970) Cellular mechanisms underlying behavior – Neuroethology. *Adv. Insect Physiol.* 7, 349–444.
Hoyle, G. (1974a) A function for neurons (DUM) neurosecretory on skeletal muscle of insects. *J. exp. Zool.* 189, 401–406.
Hoyle, G. (1974b) Neural control of skeletal muscle. In: *The Physiology of Insecta*, 2nd edition, Vol. 4, Rockstein, M., Ed., pp. 175–236. Academic Press, New York.
Hoyle, G. (1975) Evidence that insect dorsal unpaired median (DUM) neurons are octopaminergic. *J. exp. Zool.* 193, 425–431.
Hoyle, G. (1978a) Intrinsic rhythm and basic tonus in insect skeletal muscle. *J. exp. Biol.* 73, 173–203.
Hoyle, G. (1978b) Distribution of nerve and muscle fibre types in locust jumping muscle. *J. exp. Biol.* 73, 205–233.
Hoyle, G. and Burrows, M. (1973) Neural mechanisms underlying behavior in the locust *Schistocerca gregaria*. II. Integrative activity in metathoracic neurons. *J. Neurobiol.* 4, 43–67.
Hoyle, G., Dagan, D., Moberly, B. and Colquhoun, W. (1974) Dorsal unpaired median insect neurons make neurosecretory endings on skeletal muscle. *J. exp. Zool.* 187, 159–165.
Hoyle, G. and O'Shea, M. (1974) Intrinsic rhythmic contractions in insect skeletal muscle. *J. exp. Zool.* 189, 407–412.
Hoyte, H. M. D. (1961a) The protozoa occurring in the hind-gut of cockroaches. I. Response to changes in environment. *Parasitology* 51, 415–436.
Hoyte, H. M. D. (1961b) The protozoa occurring in the hind-gut of cockroaches. II. Morphology of *Nyctotherus ovalis*. *Parasitology* 51, 437–463.
Hoyte, H. M. D. (1961c) The protozoa occurring in the hind-gut of cockroaches. III. Factors affecting the dispersion of *Nyctotherus ovalis*. *Parasitology* 51, 465–495.
Hughes, G. M. (1957) The co-ordination of insect movements. II. The effect of limb amputation and the cutting of commissures in the cockroach *Blatta orientalis*. *J. exp. Biol.* 34, 306–333.
Hughes, M. and Davey, K. G. (1969) The activity of spermatozoa of *Periplaneta*. *J. Insect Physiol.* 15, 1607–1616.
Hunter, W. A. (1932) The effect of inactivity produced by cold upon learning and retention in the cockroach *Blattella germanica*. *J. Genet. Psychol.* 41, 253–266.
Hwang-Hsu, K., Reddy, G., Kumaran, A. K., Bollenbacher, W. E. and Gilbert, L. I. (1979) Correlations between juvenile hormone esterase activity, ecdysone titre and cellular reprogramming in *Galleria mellonella*. *J. Insect Physiol.* 25, 105–112.
Hyatt, A. D. and Marshall, A. T. (1977) Sequestration of haemolymph sodium and potassium by fat body in the water-stressed cockroach, *Periplaneta americana*. *J. Insect Physiol.* 23, 1437–1441.
Hyde, C. A. T. (1972) Regeneration, postembrionic induction and cellular interaction in the eye of *Periplaneta americana*. *J. Embryol. exp. Morph.* 27, 367–379.
Iles, J. F. (1972) Structure and synaptic activation of the fast coxal depressor motoneuron of the cockroach, *Periplaneta americana*. *J. exp. Biol.* 56, 647–656.

Iles, J. F. (1976) Organization of motoneurons in the prothoracic ganglion of the cockroach *Periplaneta americana* (L.). *Phil. Trans. R. Soc. Lond. Ser. B.* 276, 205–219.

Iles, J. F. and Pearson, K. G. (1971) Coxal depressor muscles of the cockroach and the role of peripheral inhibition. *J. exp. Biol.* 55, 151–164.

Imboden, H. Lanzrein, B., Delbecque, J. P. and Lüscher, M. (1978) Ecdysteroids and juvenile hormone during embryogenesis in the ovoviviparous cockroach *Nauphoeta cinerea. Gen. comp. Endocrinol.* 36, 628–635.

Irving, S. N. and Miller, T. A. (1980a) Ionic differences in 'fast' and 'slow' neuromuscular transmission in body wall muscles of *Musca domestica* larvae. *J. comp. Physiol.* 135, 291–298.

Irving, S. N. and Miller, T. A. (1980b) Aspartate and glutamate as possible transmitters at the 'slow' and 'fast' neuromuscular junctions of the body wall muscles of *Musca* larvae. *J. comp. Physiol.* 135, 299–314.

Irving, S. N., Wilson, R. G. and Osborne, M. P. (1979) Studies on L-glutamate in insect haemolymph. 3. Amino acid analyses of the haemolymph of various arthropods. *Physiol. Ent.* 4, 231–240.

Ishii, S. (1970) An aggregation pheromone of the German cockroach *Blattella germanica* (L.). II. Species specificity of the pheromone. *Appl. ent. Zool.* 5, 33–41.

Ishii, S. (1971) Structure and function of the antenna of the German Cockroach, *Blattella germanica* (L.) (Orthoptera: Blattellidae). *Appl. ent. Zool.* 6, 192–197.

Ishii, S., Kaplanis, J. N. and Robbins, W. E. (1963) Distribution and fate of 4-^{14}C-cholesterol in the adult male American cockroach. *Ann. ent. Soc. Am.* 56, 115–119.

Ishii, S. and Kuwahara, Y. (1967) an aggregation pheromone of the German cockroach *Blattella germanica* (L.) (Orth. Blattellidae). I. Site of the pheromone production. *Appl. Ent. Zool.* 2, 203–217.

Ishii, S. and Kuwahara, Y. (1968) Aggregation of German cockroach (*Blattella germanica*) nymphs. *Experientia* 24, 88–89.

Izutsu, M., Ueda, S. O. and Ishii, S. (1970) Aggregation effects on the growth of the German cockroach, *Blattella germanica* (L.) (Blattaria, Blattellidae). *Appl. ent. Zool.* 5, 159–171.

Jacklet, J. W. and Cohen, M. J. (1967) Nerve regeneration: Correlation of electrical, histological and behavioral events. *Science* 156, 1640–1643.

Jacobson, M. (1978) *Developmental Neurobiology*, 2nd Edn. Plenum Press, New York.

Jahromi, S. S. and Atwood, H. L. (1969) Structural features of muscle fibres in the cockroach leg. *J. Insect Physiol.* 15, 2255–2262.

Jaiswal, A. K. and Naidu, M. B. (1972) Studies on the reproductive system of the cockroach *Periplaneta americana* L. (Male reproductive system – Part I). *J. Animal Morph. Physiol.* 19, 1–7.

Jaiswal, A. K. and Naidu, M. B. (1976) Studies on the reproductive system of the cockroach *Periplaneta americana* L. (Male reproductive system – Part II). *J. Animal Morph. Physiol.* 23, 176–184.

Janda, V. and Mrciak, M. (1957) Gesamtstoffwechsel der Insekten. VI. Die Bewegungsaktivät der Schabe *Periplaneta americana* L. während des Tages and ihre Beziehung zum Sauerstoffverbranch. *Acta Soc. zool. Bohem.* 21, 244–255.

Jander, R. (1970) Ein Ansatz fur die moderne Elementarbeschreibung der Orientierungshandlung. *Z. Tierpsychol.* 27, 771–778.

Jander, R. (1975) Ecological aspects of spatial orientation. *A. Rev. Ecol. Syst.* 6, 171–182.

Jander, R., Horn, E. and Hoffman, M. (1970) Die Bedeutung von Gelenkreceptoren in den Beinen fur die Geotaxis der hoheren Insekten (Pterygota). *Z. vergl. Physiol.* 66, 326–342.

Jarry, D. M. and Jarry, D. T. (1963) Quelques Thelastomatidae (Nematoda: Oxyuroidea) parasites des Blattides à l'institut Pasteur de Tunis. *Arch. Inst. Pasteur, Tunis* 40, 229–234.

Jawlowski, H. (1963) On the origin of corpora pedunculata and the structure of the tuberculum opticum (Insecta). *Acta anat.* 53, 346–359.

Jenkin, P. and Hinton, H. E. (1966) Apolysis in arthropod moulting cycles. *Nature* 211, 871.

Johannsen, O. A. and Butt, F. H. (1941) *Embryology of Insects and Myriapods*. McGraw-Hill, New York.

Johnson, B. (1966) Fine structure of the lateral cardiac nerves of the cockroach *Periplaneta americana* (L.). *J. Insect Physiol.* 12, 645–653.

Johnston, J. W. and Jungreis, A. M. (1979) Comparative properties of mammalian and insect carbonic anhydrases: effects of potassium and chloride on the rate of carbon dioxide hydration. *Comp. Biochem. Physiol.* 62B, 465–469.

Jones, J., Stone, J. V. and Mordue, W. (1977) The hyperglycaemic activity of locust adipokinetic hormone. *Physiol. Ent.* 2, 185–187.

Jones, J. C. (1957) A phase contrast study of the blood cells of the adult cockroach, *Periplaneta americana. Anat. Rec.* 128, 571.

Jones, J. C. (1964) The circulatory system of insects. In: *Physiology of Insecta*, Rockstein, M., Ed., Vol. III, pp. 1–107. Academic Press, New York and London.

Jones, J. C. (1974) Factors affecting heart rates in insects. In: *The Physiology of Insecta*, Rockstein, M., Ed., 2nd edition, Vol. 5, pp. 119–167. Academic Press, New York.

Jones, J. C. (1977) *The Circulatory System of Insects*. Charles C. Thomas, Springfield.

Joshi, M. and Agarwal, H. C. (1976) Cholesterol absorption in the roach, *Periplaneta americana. Entomol.* 1, 93–100.

Kaars, C. (1979) Neural control of homologous behaviour patterns in two blaberid cockroaches. *J. Insect Physiol.* 25, 209–218.

Kallapur, V. L., Downer, R. G. H. and George, J. C. (1980) Conversion of [4-^{14}C] glucose to lipid in the American cockroach, *Periplaneta americana* L. *Archs. int. Physiol. Biochim.* 88, 363–369.

Kammer, A. E. (1976) Respiration and the generation of rhythmic outputs in insects. *Fedn Proc. fedn Socs exp. Biol. Am.* 35, 1992–1999.

Kammer, A. E. and Heinrich, B. (1978) Insect flight metabolism. *Adv. Insect Physiol.* 13, 133–228.

Kammer, A. E. and Kinnamon, S. C. (1979) Maturation of the flight motor pattern without movements in *Mandura sexta. J. comp. Physiol.* 130, 29–37.

Kandutsch, A. A., Chen, H. W. and Heiniger, H.-J. (1978) Biological activity of some oxygenated sterols. *Science* 201, 498–501.

Katagiri, C. (1977) Localization of trehalase in the haemolymph of the American cockroach, *Periplaneta americana. Insect Biochem.* 7, 351–354.

Kater, S. B. (1967) Release of a cardio-accelerator substance by stimulation of nerves to the corpora cardiaca in *Periplaneta americana. Am. Zool.* 7, 722.

Kater, S. B. (1968a) Cardio-accelerator release in *Periplaneta americana* (L.). *Science* 160, 765–767.

Kater, S. B. (1968b) Studies on Neurosecretion in the Roach *Periplaneta americana* (L.). PhD Thesis, University of Virginia, Charlottesville, vii and 108 pp.

Kawamura, H. and Ibuka, N. (1978) The search for circadian rhythm pacemakers in the light of lesion experiments. *Chronobiologia* 5, 69–88.

Keeley, L. L. (1970) Insect fat body mitochondria: endocrine and age effects of respiratory and electron transport activities. *Life Sci.* 9, 1003–1011.

Keeley, L. L. (1971) Endocrine effects on the biochemical properties of fat body mitochondria from the cockroach, *Blaberus discoidalis. J. Insect Physiol.* 17, 1501–1515.

Keeley, L. L. (1972) Biogenesis of mitochondria: neuroendocrine effects on the development of respiratory functions in fat body mitochondria of the cockroach, *Blaberus discoidalis. Arch. Biochem. Biophys.* 153, 8–15.

Keeley, L. L. (1975) Neuroendocrine deficiency effects on tropic metabolism and water balance in the cockroach *Blaberus discoidalis. J. Insect Physiol.* 21, 501–510.

Keeley, L. L. (1977) Development and endocrine regulation of cytochrome levels in fat body mitochondria of the cockroach, *Blaberus discoidalis. Insect Biochem.* 7, 297–301.

Keeley, L. L. (1978) Endocrine regulation of fat body development and function. *A. Rev. Ent.* 23, 329–352.

Keeley, L. L. and Friedman, S. (1969) Effects of long-term cardiacectomy–allatectomy on mitochondrial respiration in the cockroach, *Blaberus discoidalis. J. Insect Physiol.* 15, 509–518.

Keister, M. and Buck, J. (1974) Respiration: some exogenous and endogenous effects on rate of respiration. In: *Physiology of Insecta*, Rockstein, M., Ed., Vol. 6, pp. 469–509. Academic Press, New York.

Kennedy, J. S. (1977) Olfactory responses to distant plants and other odor sources. In: *Chemical Control of Insect Behavior*, Shorey, H. H. and McKelvey, Jr, J. J., Eds, pp. 67–91. Wiley-Interscience, New York.

Kent, P. W. and Brunet, P. J. C. (1959) Occurrence of protocatechuic acid and its 4-*o*-β-D-glucoside in *Blatta* and *Periplaneta. Tetrahedron Letters* 7, 252–256.

Kerkut, G. A., Emson, P. C. and Beesley, P. W. (1972) Effect of leg-raising learning on protein synthesis and ChE activity in the cockroach CNS. *Comp. Biochem. Physiol.* 41B., 635–646.

Kerkut, G. A., Leake, L. D., Shapira, A., Cowan, S. and Walker, R. J. (1965a) The presence of glutamate in nerve-muscle perfusates of *Helix, Carcinus* and *Periplaneta. Comp. Biochem. Physiol.* 15, 485–502.

Kerkut, G. A., Oliver, G. W. O., Rick, J. T. and Walker, R. J. (1970) The effects of drugs on learning in a simple preparation. *Comp. gen. Pharmacol.* 1, 437–483.

Kerkut, G. A., Pitman, R. M. and Walker, R. J. (1969) Iontophoretic application of acetylcholine and GABA on to insect central neurons. *Comp. Biochem. Physiol.* 31, 611–633.

Kerkut, G. A., Shapira, A. and Walker, R. J. (1965b) The effect of acetylcholine, glutamic acid and GABA on the contractions of the perfused cockroach leg. *Comp. Biochem. Physiol.* 16, 37–48.

Kerkut, G. A. and Taylor, B. J. R. (1957) A temperature receptor in the tarsus of the cockroach, *Periplaneta americana. J. exp. Biol.* 34, 486–493.

Kerkut, G. A. and Walker, R. J. (1966) The effect of L-glutamate, acetylcholine and γ-aminobutyric acid on the miniature end-plate potentials and contractures of the coxal muscles of the cockroach, *Periplaneta americana. Comp. Biochem. Physiol.* 17, 435–454.

Kerkut, G. A. and Walker, R. J. (1967) The effect of iontophoretic injection of L-glutamic acid and γ-amino-*N*-butyric acid on the miniature end-plate potentials and contractures of the coxal muscles of the cockroach *Periplaneta americana* L. *Comp. Biochem. Physiol.* 20, 999–1003.

Kessel, R. G. and Beams, H. W. (1963) Electron microscope observations on the salivary gland of the cockroach *Periplaneta americana. Z. Zellforsch.* 59, 857–877.

Kestler, P. (1971) *Die Diskontinuierliche Ventilation bei Periplaneta americana* L. *und anderen Insekten*. Dissertation, Würzburg, 1971.

Kestler, P. (1978a) Gas balance of external respiration by electronic weighing in water-saturated air. *Deutsche Physiol. Ges. Abst.* 49 Mtg. R36, No. 113.

Kestler, P. (1978b) Atembewungen und Gasaustausch bei der Ruheatmung adulter terrestrischer Insekten. *Verh. Dtsch. Zool. Ges.* 1978, 269.

Kevan, D. K. McE. (1979) Personal communication.

Khan, T. R. (1976) Neurosecretory cells in the brain and frontal ganglion of the cockroaches, *Periplaneta americana* (L.) and *Blatta orientalis* (L.). *Zool. Anz.* 197, 117–124.

Kien, J. (1976) Arousal changes in the locust optomotor system. *J. Insect Physiol.* 22, 393–395.

Kilby, B. A. (1963) The biochemistry of the insect fat body. *Adv. Insect Physiol.* 1, 112–165.

King, D. S. and Marks, E. P. (1974) The secretion and metabolism of α-ecdysone by cockroach (*Leucophaea maderae*) tissues *in vitro. Life Sci.* 15, 147–154.

Kitchel, R. L. and Hoskins, W. M. (1935) Respiratory ventilation in the cockroach in air, in carbon dioxide and in nicotine atmospheres. *J. econ. Ent.* 28, 924–933.

Klein, H. Z. (1933) Zur Biologie der amerikanischen Schabe (*Periplaneta americana* L.) *Z. Wiss. Zool.* 144, 102–122.

Klemm, N. (1976) Histochemistry of putative transmitter substances in the insect brain. *Prog. Neurobiol.* 7, 99–169.

Kloot, K. G. van der (1963) The electrophysiology and the nervous control of the spiracular muscle of pupae of giant silkmoths. *Comp. Biochem. Physiol.* 9, 317–333.

Kloss, G. R. (1966) Review of the nematodes from cockroaches in Brazil. *Pap. Avulsos Dep. Zool., São Paulo* 18, 147–188. [In Portuguese.]

Klowden, M. J. and Greenberg, B. (1974) Development of *Periplaneta americana* cell cultures and their function with enteroviruses. *J. med. Ent.* 11, 173–178.

Klowden, M. J. and Greenberg, B. (1976) *Salmonella* in the American cockroach: evaluation of vector potential through dosed feeding experiments. *J. Hyg., Cam.* 77, 105–111.

Klowden, M. J. and Greenberg, B. (1977a) *Salmonella* in the American cockroach: outcome of natural invasion of the hemocoele. *J. med. Ent.* 14, 362–366.

Klowden, M. J. and Greenberg, B. (1977b) Effects of antibiotics on the survival of *Salmonella* in the American cockroach. *J. Hyg., Camb.* 79, 339–345.

von Knorre, V. E., Gersch, M. and Kusch, T. (1972) Zur Frage der Beeinflussung des tanning Phanomens durch zyklishes 3′, 5′-AMP. *Zool. Jahrb. Physiol.* 76, 434–440.

Knowles, F. G. W. (1965) Neuroendocrine correlations at the level of ultrastructure. *Arch. Anat. Micros.* 54, 343–357.

Koeppe, J. K. and Mills, R. R. (1972) Hormonal control by the American cockroach: Probable bursicon-mediated translocation of protein-bound phenols. *J. Insect Physiol.* 18, 465–469.

Koeppe, J. K. and Mills, R. R. (1975) Metabolism of noradrenalin and dopamine during ecdysis by the American cockroach. *Insect Biochem.* 5, 399–408.

Koeppe, J. and Ofengand, J. (1976) Juvenile hormone-induced biosynthesis of vitellogenin in *Leucophaea maderae*. *Archs. Biochem. Biophys.* 173, 100–113.

Kok, G. C. and Walop, J. N. (1954) Conversion of *0,0*-diethyl-*0-p*-nitrophenyl thiophosphate (parathion) into an acetylcholinesterase inhibitor by the insect fat body. *Biochem. Biophys. Acta* 13, 510–515.

de Kort, C. A. D., Wieten, M. and Kramer, S. J. (1979) The occurrence of juvenile hormone specific esterases in insects. A comparative study. *Proc. K. Ned. Akad. Wet. C* 82, 325–331.

Kramer, E. (1976) The orientation of walking honeybees in odour fields with small concentration gradients. *Physiol. Ent.* 1, 27–37.

Krämer, K. and Markl, H. (1978) Flight-inhibition on ground contact in the American cockroach, *Periplaneta americana*: I. Contact receptors and a model for their central connections. *J. Insect Physiol.* 24, 577–586.

Krämer, K. A., Sanburg, L. L., Kezdy, E. J. and Law, J. H. (1974) The juvenile hormone binding protein in the hemolymph of *Manduca sexta* Johannson (Lepidoptera: Sphingidae). *Proc. natn. Acad. Sci., U.S.A.* 71, 493–497.

Krauthamer, V. (1978) Electrical properties of neurons of the pars intercerebralis of *Periplaneta americana*. *Am. Zool.* 18, 579.

Krauthamer, V. and Fourtner, C. R. (1978) Locomotory activity in the extensor and flexor tibiae of the cockroach, *Periplaneta americana*. *J. Insect Physiol.* 24, 813–819.

Krijgsman, B. J. (1952) Contractile and pacemaker mechanisms of the heart of arthropods. *Biol. Rev.* 27, 320–346.

Krolak, J. M., Clore, J. N., Petrovitch, E. and Mills, R. R. (1977) Vitellogenesis by the American cockroach: haemolymph and follicle protein patterns during vitellogenin synthesis. *J. Insect Physiol.* 23, 381–385.

Krolak, J. M. and Mills, R. R. (1981a) Post-ecdysial changes in cuticle dehydration and total body water by the American cockroach. *J. Insect Physiol.* (Submitted.)

Krolak, J. M. and Mills (1981b) Correlation of structural changes in the cockroach integument with the synthesis of *N*-acetyldopamine. *J. Insect Physiol.* (Submitted.)

Krolak, J. M., Shafer, S. C. and Mills, R. R. (1981) Haemolymph proteins and changes in tyrosine metabolism during the ecdysial cycle by the American cockroach. (In preparation.)
Krolak, J. M., Zimmerman, M. L. and Mills, R. R. (1977) Cardio-accelerating factors from the terminal abdominal ganglion of the American cockroach. *J. Insect Physiol.* 23, 1343–1347.
Kubišta, V. (1966) Preparations of isolated insect musculature. *Acta Univ. Carol.* 1966, 197–208.
Kumar, S. S. and Hodgson, E. (1970). Partial purification and properties of choline kinase from the cockroach, *Periplaneta americana. Comp. Biochem. Physiol.* 33, 73–84.
Kume, M. and Dan, K. (1968) *Invertebrate Embryology*. Nolit, Belgrade.
Kunkel, J. G. (1966) Development and the availability of food in the German cockroach, *Blattella germanica* (L.). *J. Insect Physiol.* 12, 227–235.
Kunkel, J. G. (1968) *The Control of Cockroach Development*, PhD. Dissertation, Case Western Reserve University, *Diss. Abst.* 29B, 4532.
Kunkel, J. G. (1973) Gonadotrophic effect of juvenile hormone in *Blattella germanica*: A rapid, simple quantitative bioassay. *J. Insect Physiol.* 19, 1285–1297.
Kunkel, J. G. (1975a) Cockroach Molting. I. Temporal organization of events during molting cycle of *Blattella germanica* (L.). *Biol. Bull.* 148, 259–273.
Kunkel, J. G. (1975b) Larval-specific protein in the order Dictyoptera. II. Antagonistic effects of ecdysone and regeneration on LSP concentration in the hemolymph of the oriental cockroach, *Blatta orientalis. Comp. Biochem. Physiol.* 51B, 177–180.
Kunkel, J. G. (1977) Cockroach molting. II. The nature of regeneration induced delay of molting hormone secretion. *Biol. Bull.* 153, 145–162.
Kunkel, J. G. (1981) A minimal model of metamorphosis: fat body competence to respond to juvenile hormone. In: *Current Topics in Insect Endocrinology and Nutrition*, Bhaskarah, G., Friedman, S. and Rodriguez, J. G., Eds., pp. 107–129. Plenum Press, New York.
Kunkel, J. G. and Lawler, D. M. (1974) Larval-specific serum protein in the order Dictyoptera. I. Immunologic characterization in larval *Blattella germanica* and cross-reaction through the order. *Comp. Biochem. Physiol.* 47B, 697–710.
Kuo, J. F., Wyatt, G. R. and Greengard, P. (1971) Cyclic nucleotide-dependent protein kinases. IX. Partial purification and some properties of guanosine 3 : 5′-monophosphate-dependent and adenosine 3 : 5′-monophosphate-dependent protein kinases from various tissues and species of Arthropoda. *J. biol. Chem.* 246, 7159–7167.
Kurtti, T. J. and Brooks, M. A. (1976) Preparation of mycetocytes for culture *in vitro. J. Invert. Pathol.* 27, 209–214.
Laird, T. B. and Winston, P. W. (1975) Water and osmotic pressure regulation in the cockroach, *Leucophaea maderae. J. Insect Physiol.* 21, 1055–1060.
Lake, R. C. and Mills, R. R. (1975) *In vitro* biosynthesis of oothecal sclerotization agents from tyrosine by haemolymph of *Periplaneta americana. Insect Biochem.* 5, 659–669.
Lake, C. R., Mills, R. R. and Koeppe, J. K. (1975) *In vivo* conversion of noradrenalin to 3-hydroxy-4-*0*-*β*-D-glucosidobenzoic acid by the American cockroach. *Insect Biochem.* 5, 223–229.
Landureau, J. C. (1966) Cultures *in vitro* de cellules embryonnaires de Blattes (insectes dictyopteres). *Exp. Cell Res.* 41, 545–556.
Lane, N. J. (1978) Tight junctions not septate junctions are occluding in the insect rectum – a freeze fracture and tracer uptake study. *J. Cell. Biol.* 79, 218A.
Lane, N. J., Skaer, H. Le B. and Swales, L. S. (1977) Intercellular junctions in the central nervous system of insects. *J. Cell Sci.* 26, 175–199.
Lane, N. J. and Treherne, J. E. (1972) Studies on perineural junctional complexes and the sites of uptake of microperoxidase and lanthanum in the cockroach central nervous system. *Tissue Cell* 4, 427–436.
Lane, N. J. and Treherne, J. E. (1973) The ultrastructural organization of peripheral nerves in two insect species (*Periplaneta americana* and *Schistocerca gregaria*). *Tissue Cell* 5, 703–714.

Lanham, U. N. (1968) The Blochmann bodies: hereditary intracellular symbionts of insects. *Biol. Rev.* 43, 269–286.
Lanzrein, B. (1975) Programming, induction or prevention of the breakdown of the prothoracic gland in the cockroach *Nauphoeta cinerea. J. Insect Physiol.* 21, 367–389.
Lanzrein, B. (1979) The activity and stability of injected juvenile hormones (JH I, JH II and JH III) in last-instar larvae and adult females of the cockroach *Nauphoeta cinerea. Gen. comp. Endocrinol.* 39, 69–78.
Lanzrein, B., Gentinetta, V., Fehr, R. and Lüscher, M. (1978) Correlation between haemolymph juvenile hormone titre, corpus allatum volume, and corpus allatum *in vivo* and *in vitro* activity during oocyte maturation in a cockroach (*Nauphoeta cinerea*). *Gen. comp. Endocrinol.* 36, 339–345.
Lanzrein, B., Hashimoto, M., Parmakovich, V., Nakanishi, K., Wilhelm, R. and Lüscher, M. (1975) Identification and quantification of juvenile hormones from different developmental stages of the cockroach *Nauphoeta cinerea. Life Sci.* 16, 1271–1284.
Laughlin, S. B. (1975) The function of the lamina ganglionaris. In: *The Compound Eye and Vision of Insects*, Horridge, G. A., Ed., pp. 341–358. Clarendon Press, London.
Lawrence, P. A. (1973) The development of spatial patterns in the integument of insects. In: *Developmental Systems: Insects*, Vol. 2, Counce, S. J. and Waddington, C. H., Eds., pp. 157–209. Academic Press, New York.
Lawson, F. (1951) Structural features of the oothecae of certain cockroaches. *Ann. ent. Soc. Am.* 44, 269–285.
Lawson, F. A. and Johnson, M. E. (1970) Coxsackie A-12 in *Periplaneta americana* – preliminary report (Blattaria: Blattidae). *J. Kansas ent. Soc.* 43, 435–440.
Lawson, F. A. and Johnson, M. E. (1971a) Coxsackie A-12 in *Periplaneta americana*. Part II. Early replication in digestive system organs. *J. Kansas ent. Soc.* 44, 253–262.
Lawson, F. A. and Johnson, M. E. (1971b) Coxsackie A-12 in *Periplaneta americana*. Part III. Reappearance of the virus in salivary glands and other organs. *J. Kansas ent. Soc.* 44, 263–275.
Leake, L. D. and Walker, R. J. (1980) *Invertebrate Neuropharmacology*. John Wiley and Sons, New York.
Lederer, G. (1952) Ein Beitrag zur Ökologie der amerikanischen Schabe *Periplaneta americana* (Linné 1758). *Anz. Schaedlingsk.* 25, 102–104.
Ledoux, A. (1945) Étude experimentale du gregarisme et de l'interattraction sociale chez blattides. *Ann. Sci. Nat. Zool. Ser.* 7 11, 75–104.
Lee, M. O. (1924) Respiration in Orthoptera. *Am. J. Physiol.* 68, 135.
Lee, R. F. (1968) The histology and histochemistry of the anterior mid-gut of *Periplaneta americana* L. (Dictyoptera: Blattidae) with reference to the formation of the peritrophic membrane. *Proc. R. ent. Soc. Lond. Ser. A* 43, 122–134.
Lefroy, H. M. (1909) *Indian Insect Life*. London.
Leong, L. and Paran, T. P. (1966) A study of the nematode parasites of cockroaches in Singapore. *Med. J. Malaya* 20, 349.
Lettau, J., Foster, W. A., Harker, J. E. and Treherne, J. E. (1977) Diel changes in potassium activity in the hemolymph of the cockroach *Leucophaea maderae. J. exp. Biol.* 71, 171–186.
Leuthold, R. (1966) Die Bewegungsaktivität der Weibiichen Schab *Leucophaea maderae* (F.) im laufe des Fortpflazungszyklus und ihre experimentelle Beeinflussung. *J. Insect Physiol.* 12, 1303–1333.
Levi-Montalcini, R. and Chen, J. S. (1971) Selective outgrowth of nerve fibers *in vitro* from embryonic ganglia of *Periplaneta americana. Arch. ital. Biol.* 109, 307–337.
Levi-Montalcini, R., Chen, J. S., Seshan, K. R. and Aloe, L. (1973) An *in vitro* approach to the insect nervous system. In: *Developmental Neurobiology of Arthropods*, Young, D., Ed., pp. 5–36. Cambridge University Press, London.
Levy, R. I. and Schneiderman, H. A. (1966) Discontinuous respiration in insects. II, III, IV. *J. Insect Physiol.* 12, 83–104; 105–121; 465–492.

Lewis, G. W., Miller, P. L. and Mills, P. S. (1973) Neuromuscular mechanisms of abdominal pumping in the locust. *J. exp. Biol.* 59, 149–168.

Linzen, B. (1974) The tryptophan – ommochrome pathway in insects. *Adv. Insect Physiol.* 10, 117–246.

Lipke, H. and Geoghegan, T. (1971) Enzymolysis of sclerotized cuticle from *Periplaneta americana* and *Sarcophaga bullata*. *J. Insect Physiol.* 17, 415–425.

Lipke, H., Grainger, M. M. and Siakotos, A. N. (1965a) Polysaccharide and glycoprotein formation in the cockroach. I. Identity and titer of bound monosaccharides. *J. biol. Chem.* 240, 594–599.

Lipke, H., Graves, B. and Leto, S. (1965b) Polysaccharide and glycoprotein formation in the cockroach. II. Incorporation of D-glucose-^{14}C into bound carbohydrate. *J. biol. Chem.* 240, 601–608.

Lipton, G. R. (1969) *Feeding and Activity Rhythms in the American Cockroach, Periplaneta americana* (L.). PhD Thesis, Rutgers University.

Lipton, G. R. and Sutherland, D. J. (1970a) Activity rhythms in the American cockroach *Periplaneta americana*. *J. Insect Physiol.* 16, 1555–1566.

Lipton, G. R. and Sutherland, D. J. (1970b) Feeding rhythms in the American cockroach *Periplaneta americana*. *J. Insect Physiol.* 16, 1757–1767.

Lisa, J. D. and Ludwig, D. (1959) Uricase, guanase and xanthine oxidase from the fat body of the cockroach *Leucophaea maderae*. *Ann. ent. Soc. Am.* 52, 548–551.

Locke, M. (1958) The formation of tracheae and tracheoles in *Rhodnius prolixus*. *Q. J. Micros. Sci.* 99, 29–46.

Locke, M. (1959) The cuticular pattern in an insect *Rhodnius prolixus* Stal. *J. exp. Biol.* 36, 459–477.

Locke, M. (1974) The structure and formation of the integument in insects. In: *Physiology of Insecta*, Rockstein, M., Ed., Vol. 6, pp. 123–213. Academic Press, New York.

Lockey, K. H. (1976) Cuticular hydrocarbons of *Locusta*, *Schistocerca* and *Periplaneta*, and their role in waterproofing. Insect Biochem. 6, 457–472.

Lockshin, R. A. and Beauleton, J. (1974) Programmed cell death. *Life Sci.* 15, 1549–1565.

Loftus, R. (1966) Cold receptors on the antenna of *Periplaneta americana*. *Z. vergl. Physiol.* 52, 380–385.

Loftus, R. (1968) The response of the antennal cold receptor of *Periplaneta americana* to rapid temperature changes and to steady temperature. *Z. vergl. Physiol.* 59, 413–455.

Loftus, R. (1969) Differential thermal components in the response of the antennal cold receptor of *Periplaneta americana* to slowly changing temperature. *Z. vergl. Physiol.* 63, 415–433.

Loftus, R. (1976) Temperature-dependent dry receptor on the antenna of *Periplaneta americana*: Tonic responses. *J. comp. Physiol.* 111, 153–170.

Loher, W. (1974) Circadian control of spermatophore formation in the cricket *Teleogryllus commodus* Walker. *J. Insect Physiol.* 20, 1155–1172.

Lohmann, M. (1967a) Ranges of circadian period length. *Experientia* 23, 788–790.

Lohmann, M. (1967b) Richtungstendenz bei circadianen Frequenzänderungen. *Biol. Zentralbl.* 86, 623–628.

Lohmann, M. (1967c) Zur Bedeutung der lokomotorischen Aktivität in circadianen Systemen. *Z. vergl. Physiol.* 55, 307–332.

Longo, N. (1964) Probability learning and habit reversal in the cockroach. *Am. J. Psychol.* 77, 29–41.

Louloudes, S. J., Kaplanis, J. N., Robbins, W. E. and Monroe, R. E. (1961) Lipogenesis from C^{14}-acetate by the American cockroach. *Ann. Ent. Soc. Am.* 54, 99–103.

Lovell, K. L. and Eisenstein, E. M. (1973) Dark avoidance learning and memory disruption by carbon dioxide in cockroaches. *Physiol. Behav.* 10, 835–840.

Luco, J. and Aranda, L. (1964) An electrical correlate to the process of learning. *Acta physiol. latinoam.* 14, 274–288.

Ludwig, D. and Gallagher, M. R. (1966) Vitamin synthesis by the symbionts in the fat body of the cockroach, *Periplaneta americana* (L.). *J. N.Y. ent. Soc.* 74, 134–139.

Ludwig, D., Tracey, K. M. and Burns, M. L. (1957) Ratios of ions required to maintain the heart beat of the American cockroach, *Periplaneta americana Linnaeus*. *Ann. ent. Soc. Am.* 50, 244–246.

Lukat, R. (1978) Circadian growth layers in the cuticle of behaviorally arrhythmic cockroaches (*Blaberus fuscus* Ins., Blattoidea) *Experientia* 34, 477.

Lüscher, M. (1968a) Hormonal control of respiration and protein synthesis in the fat body of the cockroach *Nauphoeta cinerea* during oocyte growth. *J. Insect Physiol.* 14, 499–511.

Lüscher, M. (1968b) Oocyte protection – a function of a corpus cardiacum hormone in the cockroach, *Nauphoeta cinerea*. *J. Insect Physiol.* 14, 685–688.

Lüscher, M. and Leuthold, R. (1965) Über die hormonale Beinflussung des respiratorischen Stoffwechsels bei der Schabe *Leucophaea madera* (F). *Rev. Suisse Zool.* 72, 618–623.

Maa, W. and Bell, W. J. (1977) An endogenous component of the mechanism controlling the American cockroach vitellogenic cycle. *J. Insect Physiol.* 23, 895–897.

Maddrell, S. H. P. (1971) The mechanisms of insect excretory systems. In: *Advances in Insect Physiology*. Beament, J. W. L., Treherne, J. E. and Wigglesworth, V. B., Eds., Vol. 8, 199–331. Academic Press, New York.

Maddrell, S. H. P. (1974) Neurosecretion. In: *Insect Neurobiology*, Treherne, J. E., Ed., pp. 307–357. North-Holland Publishing Co., Amsterdam.

Maddrell, S. H. P. (1977) Insect Malpighian tubules. In: *Transport of Ions and Water in Animals*, Gupta, B. L., Moreton, R. B., Oschman, J. L. and Wall, B. J., Eds., pp. 541–569. Academic Press, London.

Maddrell, S. H. P. and Casida, J. E. (1971) Mechanism of insecticide induced diuresis in *Rhodnius*. *Nature* 231, 55–56.

Maddrell, S. H. P. and Gardiner, B. O. C. (1976) Excretion of alkaloids by Malpighian tubules of insects. *J. exp. Biol.* 64, 267–281.

Maddrell, S. H. P., Pilcher, D. E. M. and Gardiner, B. O. C. (1969) Stimulatory effect of 5-hydroxytryptamine (serotonin) on secretion by Malpighian tubules of insects. *Nature* 222, 784–785.

Maki, T. (1938) Studies on the thoracic musculature of insects. *Mem. Fac. Sci. Agric., Taihoku* 24, 1–343.

Malke, H. and Schwartz, W. (1966a) Untersuchungen uber die Symbiose von Tieren mit Pilsen und Bakterein. XII. Die Bedentung der Blattiden – Symbiose. *Z. Allg. Mikrobiol.* 6, 34–68.

Malke, H. and Schwartz, W. (1966b) Untersuchungen uber die Symbiose von Tieren mit Pilzen und Bakterein. XI. Die Rolle des Wirtslysozums in der Blattidensymbiose. *Arch. Mikrobiol.* 53, 17–32.

Malzacher, P. (1968) Embryogenese des Gehirns von *Carausius* und *Periplaneta*. *Z. morph. Okol. Tiere* 62, 103–161.

Mancini, G. and Frontali, N. (1967) Fine structure of the mushroom body neuropile of the brain of the roach, *Periplaneta americana*. *Z. Zellforsch.* 83, 334–343

Mancini, G. and Frontali, N. (1970) On the ultrastructural localization of catecholamines in the Beta lobes (corpora pedunculata) of *Periplaneta americana*. *Z. Zellforsch.* 103, 341–350.

Marks, E. P. (1972) Effects of ecdysterone on the deposition of cockroach cuticle *in vitro*. *Biol. Bull.* 142, 293–301.

Marks, E. P. (1973a) Deposition of insect cuticle *in vitro*: Differential response to α- and β-ecdysone. *Gen. comp. Endocrinol.* 21, 472–477.

Marks, E. P. (1973b) Effects of β-ecdysone on molt-linked differentiation *in vitro*. *Biol. Bull.* 145, 171–179.

Marks, E. P. (1980) Insect tissue culture: an overview, 1971–1978. *A. Rev. Ent.* 25, 73–101.

Marks, E. P. and Holman, G. M. (1974) Release from the brain and acquisition by the corpus cardiacum of a neurohormone *in vitro*. *J. Insect Physiol.* 20, 2087–2093.
Marks, E. P., Holman, G. M. and Borg, T. K. (1973) Synthesis and storage of an insect neurohormone in insect brains *in vitro*. *J. Insect Physiol.* 19, 471–477.
Marks, E. P., Ittycheriah, P. I. and Leloup, A. M. (1972) The effect of β-ecdysone on insect neurosecretion *in vitro*. *J. Insect Physiol.* 18, 847–850.
Marks, E. P. and Leopold, R. A. (1971) Deposition of cuticular substances *in vitro* by leg regeneration from the cockroach, *Leucophaea maderae* (F.). *Biol. Bull.* 140, 73–80.
Marks, E. P. and Reinecke, J. P. (1964) Regenerating tissues from the cockroach leg: a system for study *in vitro*. *Science* 143, 961–963.
Marks, E. P. and Reinecke, J. P. (1965) Regenerating tissues from the cockroach *Leucophaea maderea*: Effects of endocrine glands *in vitro*. *Gen. comp. Endocrinol.* 5, 241–247.
Marks, E. P., Reinecke, J. P. and Leopold, R. A. (1968) Regenerating tissues from the cockroach *Leucophaea maderae*: Nerve regeneration *in vitro*. *Biol. Bull.* 135, 520–529.
Mason, C. A. (1973) New features of the brain-retrocerebral neuroendocrine complex of the locust *Schistocerca vaga* (Scudder). *Z. Zellforsch.* 141, 19–32.
Matthews, J. R. and Downer, R. G. H. (1973) Hyperglycemia induced by anesthesia in the American cockroach, *Periplaneta americana* (L.). *Can. J. Zool.* 51, 395–397.
Matthews, J. R. and Downer, R. G. H. (1974) Origin of trehalose in stress-induced hyperglycaemia in the American cockroach, *Periplaneta americana*. *Can. J. Zool.* 52, 1005–1010.
Matthews, J. R., Downer, R. G. H. and Morrison, P. E. (1976) Estimation of glucose in haemolymph of the American cockroach, *Periplaneta americana*. *Comp. Biochem. Physiol.* 53A, 165–168.
Matthews, J. R., Downer, R. G. H. and Morrison, P. E. (1976) α-Glucosidase activity in haemolymph of the American cockroach, *Periplaneta americana*. *J. Insect Physiol.* 22, 157–163.
McAllan, J. W. and Chefurka, W. (1961a) Properties of transaminases and glutamic dehydrogenase in the cockroach *Periplaneta americana*. *Comp. Biochem. Physiol.* 3, 1–19.
McAllen, J. W. and Chefurka, W. (1961b) Some physiological aspects of glutamate–aspartate transamination in insects. *Comp. Biochem. Physiol.* 2, 290–299.
McArthur, J. M. (1929) An experimental study of the functions of the different spiracles in certain Orthoptera. *J. exp. Zool.* 53, 117–128.
McEnroe, W. D. (1966) *In vivo* preferential oxidation of ureide carbon No. 2 of uric acid by *Periplaneta americana*. *Ann. ent. Soc. Am.* 59, 1011.
McEnroe, W. D. and Forgash, A. J. (1957) The *in vivo* incorporation of C^{14} formate in the ureide groups of uric acid by *Periplaneta americana* (L.). *Ann. ent. Soc. Am.* 50, 429–431.
McEnroe, W. D. and Forgash, A. J. (1958) Formate metabolism in the American cockroach, *Periplaneta americana*. *Ann. ent. Soc. Am.* 51, 126–129.
McIndoo, N. E. (1939) Segmental blood vessels of the American cockroach (*Periplaneta americana* (L.)). *J. Morphol.* 65, 323–351.
McKittrick, F. A. (1964) Evolutionary studies on cockroaches. *Cornell Univ. Agric. exp. St. Mem.* 389, 1–197.
Mellanby, K. (1940) The daily rhythm of activity of the cockroach, *Blatta orientalis* (L.). II. Observations and experiments on a natural infestation. *J. exp. Biol.* 17, 278–285.
Menon, M. (1963) Endocrine influences on protein and fat in the haemolymph of the cockroach, *Periplaneta americana*. *Proc. 16th Int. Cong. Zool.* 1, 297.
Mercer, E. H. and Brunet, P. C. J. (1959) The electron microscopy of the left colleterial gland of the cockroach. *J. biophys. biochem. Cytol.* 5, 257–262.
Mercer, E. H. and Day, M. F. (1952) The fine structure of the peritrophic membranes of certain insects. *Biol. Bull.* 103, 384–394.

Mercer, E. H. and Nicholas, W. L. (1967) The ultrastructure of the capsule of the larval stages of *Moniliformis dubius* (Acanthocephala) in the cockroach *Periplaneta americana. Parasitology* 57, 169–174.
Metcalf, R. L. (1943) The storage and interaction of water soluble vitamins in the Malpighian system of *Periplaneta americana* (L.). *Arch. Biochem.* 2, 55–62.
Metcalf, R. L. (1971) Structure-activity relationships for insecticidal carbamates. *Bull. W.H.O.* 44, 43–78.
Metzger, R. and Trier, K. H. (1975) The significance of the aggregation pheromones of *Blattella germanica* and *Blatta orientalis. Angew Parasitol.* 16, 16–27.
Meyer, D. J. and Walcott, B. (1979) Differences in the response of identified motoneurons in the cockroach: role in the motor program for stepping. *Brain Res.* 178, 600–605.
Miall, L. C. and Denny, A. (1886) *The structure and life history of the cockroach Blatta orientalis* (L.). Lowell Reeve, London.
Milburn, N. S. (1966) Fine structure of the pleomorphic bacteroids in the mycetocytes and ovaries of several genera of cockroaches. *J. Insect Physiol.* 12, 1245–1254.
Milburn, N. S. and Bentley, D. R. (1971) On the dendritic topology and activation of cockroach giant interneurons. *J. Insect Physiol.* 17, 607–623.
Milburn, N. S. and Roeder, K. B. (1962) Control of efferent activity in the cockroach terminal abdominal ganglion by extracts of corpora cardiaca. *Gen. comp. Endocrinol.* 2, 70–76.
Milburn, N. S., Weiant, E. A. and Roeder, K. B. (1960) The release of efferent nerve activity in the roach *Periplaneta americana* by extracts of the corpus cardiacum. *Biol. Bull.* 118, 111–119.
Mill, P. J. and Lowe, D. A. (1971) Ultrastructure of the respiratory and non-respiratory dorso-ventral muscles of the larva of a dragonfly. *J. Insect Physiol.* 17, 1947–1960.
Miller, J. R. and Roelofs, W. L. (1978) Sustained-flight tunnel for measuring insect responses to wind-borne sex pheromones. *J. chem. Ecol.* 4, 187–198.
Miller, P. L. (1962) Spiracle control in adult dragonflies (Odonata). *J. exp. Biol.* 39, 513–535.
Miller, P. L. (1964) Factors altering spiracle control in adult dragonflies: hypoxia and temperature. *J. exp. Biol.* 41, 345–357.
Miller, P. L. (1966) The regulation of breathing in insects. *Adv. Insect Physiol.* 3, 279–344.
Miller, P. L. (1969) Inhibitory nerves to insect spiracles. *Nature* 221, 171–173.
Miller, P. L. (1973) Spatial and temporal changes in the coupling of cockroach spiracles to ventilation. *J. exp. Biol.* 59, 137–148.
Miller, P. L. (1974a) Respiration – Aerial Gas Transport. In: *Physiology of Insecta*, Rockstein, M., Ed., Vol. 6, pp. 345–402. Academic Press, New York and London.
Miller, P. L. (1974b) The neural basis of behaviour. In: *Insect Neurobiology*, Treherne, J. E., Ed., pp. 359–430. North-Holland Publishing Co., Amsterdam.
Miller, T. A. (1967) A Physiological and Cytochemical Investigation of the Cholinergic Response of the Heart of the American cockroach, *Periplaneta americana* Linn. PhD Thesis, University of California, Riverside, xii and 115 pp.
Miller, T. A. (1968a) Role of cardiac neurons in the cockroach heartbeat. *J. Insect Physiol.* 14, 1265–1275.
Miller, T. A. (1968b) Response of cockroach cardiac neurons to cholinergic compounds. *J. Insect Physiol.* 14, 1713–1717.
Miller, T. A. (1969) Initiation of activity in the cockroach heart. In: *Comparative Physiology of the Heart: Current Trends*, McCann, F. V., Ed., pp. 206–218. Birkhäuser Verlag, Basel.
Miller, T. A. (1973) Regulation of heartbeat of the American cockroach. In: *Neurobiology of Invertebrates*, Salanki, J., Ed., pp. 193–210. Akademiai Kiado, Budapest.
Miller, T. A. (1974) Electrophysiology of the insect heart. In: *The Physiology of Insecta*, Vol. 4, Rockstein, M., Ed., pp. 169–200. Academic Press, New York.
Miller, T. A. (1975a) Insect visceral muscle. In: *Insect Muscle*, Usherwood, P. N. R., Ed., pp. 545–606. Academic Press, London.

Miller, T. A. (1975b) Neurosecretion and the control of visceral organs in insects. *A. Rev. Ent.* 20, 133–149.
Miller, T. A. (1978) The insect neuromuscular system as a site of insecticide action. In: *Pesticide and Venom Neurotoxicity*, Shankland, D. L., Hollingworth, R. M. and Smyth, Jr, T., Eds., pp. 95–111. Plenum Press, New York.
Miller, T. A. (1979a) Nervous versus neurohormonal control of insect heartbeat. *Am. Zool.* 19, 77–86.
Miller, T. A. (1979b) *Insect Neurophysiological Techniques*, Springer-Verlag, New York.
Miller, T. A. and Metcalf, R. L. (1968a) Site of action of pharmacologically active compounds on the heart of *Periplaneta americana* (L.). *J. Insect Physiol.* 14, 383–394.
Miller, T. A. and Metcalf, R. L. (1968b). The cockroach heart as a bioassay organ. *Ent. Exp. Appl.* 11, 455–463.
Miller, T. A. and Usherwood, P. N. R. (1971) Studies of cardio-regulation in the cockroach, *Periplaneta americana. J. Exp. Biol.* 54, 329–348.
Mills, R. R. (1965) Hormonal control of tanning in the American cockroach. II. Assay for the hormone and the effect of wound healing. *J. Insect. Physiol.* 11, 1269–1275.
Mills, R. R. (1966) Hormonal control of tanning in the American cockroach. III. Hormone stability and postecydysial changes in hormone titre. *J. Insect Physiol.* 12, 275–280.
Mills, R. R. (1967) Hormonal control of excretion in the American cockroach. I. Release of a diuretic hormone from the terminal abdominal ganglion. *J. exp. Biol.* 46, 35–41.
Mills, R. R., Androuny, S. and Fox, F. R. (1968) Correlation of phenoloxidase activity with ecdysis and tanning hormone release in the American cockroach. *J. Insect Physiol.* 14, 603–611.
Mills, R. R. and Fox, F. R. (1972) The sclerotization process by the American cockroach: contribution of melanin. *Insect Biochem.* 2, 23–28.
Mills, R. R., Greenslade, F. C. and Couch, E. F. (1966) Studies on vitellogenesis in the American cockroach. *J. Insect Physiol.* 12, 767–779.
Mills, R. R. and Lake, C. R. (1966) Hormonal control of tanning in the American cockroach. IV. Preliminary purification of the hormone. *J. Insect Physiol.* 12, 1395–1401.
Mills, R. R. and Lake, C. R. (1971) Metabolism of tyrosine to *p*-hydroxyphenyl propionic acid and to *p*-hydroxyphenylacetic acid by the American cockroach. *Insect Biochem.* 1, 264–270.
Mills, R. R., Lake, C. R. and Alworth, W. L. (1967) Biosynthesis of *N*-acetyldopamine by the American cockroach. *J. Insect Physiol.* 13, 1539–1546.
Mills, R. R., Mathur, R. B. and Guerra, A. A. (1965) Studies on the hormonal control of tanning in the American cockroach. I. Release of an activation factor from the terminal abdominal ganglion. *J. Insect Physiol.* 11, 1047–1053.
Mills, R. R. and Nielsen, D. J. (1967) Hormonal control of tanning in the American cockroach. V. Some properties of the purified hormone. *J. Insect Physiol.* 13, 273–280.
Mills, R. R. and Whitehead, D. L. (1970) Hormonal control of tanning in the American cockroach: Changes in blood cell permeability during ecdysis. *J. Insect Physiol.* 16, 331–340.
Mills, R. R., Wright, R. D. and Sauer, J. R. (1970) Midgut epithelium of the American cockroach: probable mechanisms of water secretion. *J. Insect Physiol.* 16, 417–427.
Mittelstaedt, H. and Mittelstaedt, M.-L. (1973) Mechanismen der Orientierung ohne richtende Aussenreize. *Fortschraft Zool.* 21, 46–58.
Mittelstaedt-Burger, M.-L. (1972) Der Anteil der propriozeptiven Erregung an der Kurskontrolle bei Arthropoden (Diplopoden und Insekten). *Verh. Deut. Zool. Ges. Helgoland* 65, 220–225.
Mittelstaedt-Burger, M.-L. (1973) Idiothetic course control and visual orientation. In: *Information Processing in the Visual System of Arthropods*, Wehner, R., Ed., pp. 275–279. Springer-Verlag, New York.
Montgomery, T. H. (1898) Comparative cytological studies, with especial regard to the morphology of the nucleolus. *J. Morph.* 15, 266–582.

Moran, D. T. (1971) Fine structure of cockroach blood cells. *Tissue Cell* 3, 413–422.
Moran, D. T., Chapman, K. M. and Ellis, R. A. (1971) The fine structure of cockroach campaniform sensilla. *J. Cell Biol.* 48, 155–173.
Moran, D. T. and Rowley, J. C. III (1975a) High voltage and scanning electron microscopy of the site of stimulus reception of an insect mechanoreceptor. *J. ultrastruct. Res.* 50, 38–46.
Moran, D. T. and Rowley, J. C. III (1975b) The fine structure of the cockroach subgenual organ. *Tissue Cell* 7, 91–106.
Mordue, W. and Stone, J. V. (1977) Relative potencies of locust adipokinetic hormone and prawn red pigment-concentrating hormone in insect and crustacean systems. *Gen. comp. Endocrinol.* 33, 103–108.
Morgan, C. R. and Stokes, D. R. (1979) Ultrastructural heterogeneity of the mesocoxal muscles of *Periplaneta americana. Cell Tissue Res.* 201, 305–314.
Mosconi-Bernadini, P. (1966) Différents aspects de la dégénérescence de la glande prothoracique chez *Leucophaea maderae* et *Pycnoscelus surinamensis. Ann. Endocrinol., Paris* 27, 367–370.
Mote, M. J. and Goldsmith, T. H. (1970) Spectral sensitivities of color receptors in the compound eye of the cockroach *Periplaneta. J. exp. Zool.* 173, 137–146.
Moulins, M. (1971) La cavité préorale de *Blabera craniifer* Burm. (Insecte, Dictyoptère) et son innervation: étude anatomo-histologique de l'épipharynx et l'hypopharynx. *Zool. Jahrb. Abt. Anat. u. Ontog.* 88, 527–586.
Müller, H. P. and Engelmann, F. (1968) Studies on the endocrine control of metabolism in *Leucophaea maderae.* II. The effect of the corpora cardiaca on fat body respiration. *Gen. comp. Endocrinol.* 11, 43–50.
Müller, P. J., Masner, P., Trautmann, K. H., Suchy, M. and Wipf, H.-K. (1975) The isolation and identification of juvenile hormone from cockroach corpora allata *in vitro. Life Sci.* 15, 915–921.
Mullins, D. E. (1974) Nitrogen metabolism in the American cockroach: an examination of whole body ammonium and other cations excreted in relation to water requirements. *J. exp. Biol.* 61, 541–556.
Mullins, D. E. (1979a) Isolation and partial characterization of uric acid spherules obtained from cockroach tissues (Dictyoptera). *Comp. Biochem. Physiol.* 62A, 699–705.
Mullins, D. E. (1979b) The effects of dietary potassium on fat body and whole body urate storage in *Periplaneta americana. Insect Biochem.* 9, 61–67.
Mullins, D. E. and Cochran, D. G. (1972) Nitrogen excretion in cockroaches: uric acid is not a major product. *Science* 177, 699–701.
Mullins, D. E. and Cochran, D. G. (1973a) Nitrogenous excretory materials from the American cockroach. *J. Insect Physiol.* 19, 1007–1018.
Mullins, D. E. and Cochran, D. G. (1973b) Tryptophan metabolite excretion by the American cockroach. *Comp. Biochem. Physiol.* 44B, 549–555.
Mullins, D. E. and Cochran, D. G. (1974) Nitrogen metabolism in the American cockroach: an examination of whole body and fat body regulation of cations in response to nitrogen balance. *J. exp. Biol.* 61, 557–570.
Mullins, P. E. and Cochran, D. G. (1976) A comparative study of nitrogen excretion in twenty-three cockroach species. *Comp. Biochem. Physiol.* 53A, 393–399.
Mullins, D. E. and Cochran, D. G. (1975a) Nitrogen metabolism in the American cockroach. I. An examination of positive nitrogen balance with respect to uric acid stores. *Comp. Biochem. Physiol.* 50A, 489–500.
Mullins, D. E. and Cochran, D. G. (1975b) Nitrogen metabolism in the American cockroach. II. An examination of negative nitrogen balance with respect to mobilization of uric acid stores. *Comp. Biochem. Physiol.* 50A, 501–510.
Mullins, D. E. and Keil, C. B. (1980) Paternal investment of urates in cockroaches. *Nature* 283, 567–569.

Mundall, E. C., Tobe, S. S. and Stay, B. (1979) Induction of vitellogenin and growth of implanted oöcytes in male cockroaches. *Nature* 282, 97–98.

Munson, S. C. and Yeager, J. F. (1949) Blood volume and chloride normality in roaches (*Periplaneta americana* (L.)) injected with sodium chloride solutions. *Ann. ent. Soc. Am.* 42, 165–173.

Murdock, L. L., Hopkins, T. L. and Wirtz, R. A. (1970a) Phenylalanine metabolism in cockroaches, *Periplaneta americana*: intracellular symbionts and aromatic ring cleavage. *Comp. Biochem. Physiol.* 34, 143–146.

Murdock, L. L., Hopkins, T. L. and Wirtz, R. A. (1970b) Tyrosine metabolism *in vivo* in teneral and mature cockroaches, *Periplaneta americana. J. Insect Physiol.* 16, 555–560.

Murphy, T. A. and Wyatt, G. R. (1965) The enzymes of glycogen and trehalose synthesis in silk moth fat body. *J. biol. Chem.* 240, 1500–1508.

Murr, L. (1930) Uber den Geruchsinn der Mehlmottenschlupfwespe Habrobracon juglandis Ashmead. Zugleich ein Beitrag zun Orientierungsproblem. *Z. vergl. Physiol.* 11, 210–270.

Myers, T. B. and Retzlaff, E. (1963) Localization and action of the respiratory centre of the Cuban burrowing cockroach. *J. Insect Physiol.* 9, 607–614.

Nagai, T. (1970) Insect visceral muscle. Responses of the proctodeal muscles to mechanical stretch. *J. Insect Physiol.* 16, 437–448.

Nagai, T. (1972) Insect visceral muscle. Ionic dependence of electrical potentials in the proctodeal muscle fibres. *J. Insect Physiol.* 18, 2299–2318.

Nagai, T. (1973) Insect visceral muscle. Excitation and conduction in the proctodeal muscles. *J. Insect Physiol.* 19, 1753–1764.

Nagai, T. and Brown, B. E. (1969) Insect visceral muscle. Electrical potentials and contraction in fibres of the cockroach proctodeum. *J. Insect Physiol.* 15, 2151–2167.

Nagai, T. and Graham, W. G. (1974) Insect visceral muscle. Fine structure of the proctodeal muscle fibres. *J. Insect Physiol.* 20, 1999–2013.

Nakatsugawa, T. and Dahm, P. A. (1962) Activation of Guthion by tissue preparations from the American cockroach. *J. econ. Ent.* 55, 594–599.

Nakatsugawa, T. and Dahm, P. A. (1965) Parathion activation enzymes in the fat body microsomes of the American cockroach. *J. econ. Ent.* 58, 500–509.

Narahashi, T. (1962) Nature of the negative after-potential increased by the insecticide allethrin in cockroach giant axons. *J. cell. comp. Physiol.* 59, 67–76.

Narahashi, T. (1971) Effects of insecticides on excitable tissues. *Adv. Insect Physiol.* 8, 1–93.

Narahashi, T. (1974) Chemicals as tools in the study of excitable membranes. *Physiol. Rev.* 54, 813–889.

Narahashi, T. (1976) Effects of insecticides on nervous conduction and synaptic transmission. In: *Insecticide Biochemistry and Physiology*, Wilkinson, C. F., Ed., pp. 327–352. Plenum Press, New York.

Narahashi, T. (1979) Nerve membrane ionic channels as the target site of insecticides. In: *Neurotoxicology of Insecticides and Pheromones*, Narahashi, T., Ed., pp. 211–243. Plenum Press, New York.

Narahashi, T. and Haas, H. G. (1967) DDT: interaction with nerve membrane conductance changes. *Science* 157, 1438–1440.

Narahashi, T. and Yamasaki, T. (1960) Mechanism of increase in negative after-potential by dicophanum (DDT) in the giant axons of the cockroach. *J. Physiol.* 152, 122–140.

Natalizi, G. M. and Frontali, N. (1966) Purification of insect hyperglycaemic and heart accelerating hormones. *J. Insect Physiol.* 12, 1279–1287.

Natalizi, G. M., Pansa, M. C., d'Adjello, V., Casaglia, O., Bettini, S. and Frontali, N. (1970) Physiologically active factors from corpora cardiaca of *Periplaneta americana. J. Insect Physiol.* 16, 1827–1836.

Nath, V., Gupta, B. L. and Lal, B. (1958) Histochemical and morphological studies of the lipids in oogenesis. I. *Periplaneta americana. Q. J. Micros. Sci.* 99, 315–322.

Nath, V., Gupta, B. L. and Sehgal, P. (1957) Mitochondria and Golgi bodies in the spermatogenesis of *Periplaneta americana* as studied under the phase contrast microscope. *Res. Bull., Punjab Univ. Sci.* 112, 317–326.

Nath, V. and Mohan, P. (1929) Studies in the origin of yolk. IV. Oogenesis of *Periplaneta americana. J. Morph.* 48, 253–280.

Nayar, K. K. (1962) Effects of injecting juvenile hormone extracts on the neurosecretory system of adult male cockroaches (*Periplaneta americana*). *Mem. Soc. Endocrinol.* 12, 371–378.

Necheles, H. (1927) Observations on the causes of night activity in some insects. *Chin. J. Physiol.* 1, 143–156.

Neder, R. (1959) Allometrisches Wachstum von Hirnteilen bei drei verschieden grossen Schabenarten. *Zool. Jahrb., Abt. Allg. Zool. Physiol. Tiere* 77, 411–464.

Nelson, D. R. (1969) Hydrocarbon synthesis in the American cockroach. *Nature* 221, 854–855.

Nelson, D. R., Terranova, A. C. and Sukkestad, D. R. (1967) Fatty acid composition of diglyceride and free fatty acid fractions of the fat body and haemolymph of the cockroach, *Periplaneta americana. Comp. Biochem. Physiol.* 20, 907–917.

Nelson, M. C. (1979) Sound production in the cockroach *Gromphadorhina portentosa*. The sound production apparatus. *J. comp. Physiol.* 132, 27–38.

Nelson, W., Scheving, L. and Halberg, F. (1975) Circadian rhythms in mice fed a single daily meal at different stages of lighting regimen. *J. Nutr.* 105, 171–184.

Neville, A. C. (1965) Circadian organization of chitin in some insect skeletons *Q. J. micros. Sci.* 106, 315–325.

Neville, A. C. (1967) Chitin orientation in cuticle and its control. *Adv. Insect Physiol.* 4, 213–286.

Neville, A. C. (1970) Cuticle ultrastructure in relation to the whole insect. *Symp. R. ent. Soc., Lond.* 5, 17–39.

Neville, A. C. (1975) *Biology of the Arthropod Cuticle*. Springer-Verlag, Berlin.

Newcomer, W. S. (1954) The occurrence of β-glucosidase in digestive juice of the cockroach *Periplaneta americana. J. cell. comp. Physiol.* 43, 79–86.

Nicklaus, R. (1965) Die Erregung einzelner Fadenhaare von *Periplaneta americana* in Abhangigkeit von der Grosse und Richtung der Auslenkung. *Z. vergl. Physiol.* 50, 331–362.

Nicklaus, R., Lundquist, P. G. and Wersall, J. (1967) Elektronenmikroskopie am sensorischen Apparat der Fadenhaare auf den Cerci der Schabe *Periplaneta americana. Z. vergl. Physiol.* 56, 412–415.

Nielsen, D. J. and Mills, R. R. (1968) Changes in electrophoretic properties of haemolymph and terminal oöcyte proteins during vitellogenesis in the American cockroach. *J. Insect Physiol.* 14, 163–170.

Nigam, L. N. (1933) The life-history of a common cockroach (*Periplaneta americana* Linneus). *Ind. J. Agric. Sci.* 3, 530–543.

Nijenhuis, E. D. and Dresden, D. (1952) A micromorphological study on the sensory supply of the mesothoracic leg of the American cockroach, *Periplaneta americana. Proc. Kon. Ned. Akad. Wetensch. Ser. C.* 55, 300–310.

Nijenhuis, E. D. and Dresden, D. (1955) On the topographical anatomy of the nervous system of the mesothoracic leg of the American cockroach (*Periplaneta americana*). I. *Proc. Kon. Ned. Akad. Wetensch. Ser. C* 58, 121–136.

Nijhout, H. F. and Williams, C. M. (1974) Control of molting and metamorphosis in the tobacco hornworm, *Manduca sexta* (L.): cessation of juvenile hormone secretion as a trigger for pupation. *J. exp. Biol.* 61, 493–501.

Nijhout, M. M. and Koeppe, J. K. (1978) Ovarian-produced steroid in *Leucophaea maderae. Am. Zool.* 18, 626.

Nishiitsutsuji-Uwo, J., Petropulos, S. F. and Pittendrigh, C. S. (1967) Central nervous system control of circadian rhythmicity in the cockroach. I. Role of the pars intercerebralis. *Biol. Bull.* 133, 679–696.

Nishiitsutsuji-Uwo, J. and Pittendrigh, C. S. (1968) Central nervous system control of circadian rhythmicity in the cockroach. III. The optic lobes, locus of the driving oscillation? *Z. vergl. Physiol.* 58, 14–46.

Nishino, C., Tobin, T. R. and Bowers, W. S. (1977) Electro-antennogram responses of the American cockroach to germacrene D sex pheromone mimic. *J. Insect Physiol.* 23, 415–420.

Nishino, C. and Washio, H. (1976) Electro-antennograms of the American cockroach to odorous straight-chain compounds. *Appl. Ent. Zool.* 11, 222–228.

Noirot, C. and Noirot-Timothée, C. (1976) Fine structure of the rectum in cockroaches (Dictyoptera): general organization and intercellular junctions. *Tissue Cell* 8, 345–368.

Noirot, C., Smith, D. S., Cayer, M. L. and Noirot-Timothée, C. (1979) The organization and isolating function of insect rectal sheath cells: A freeze fracture study. *Tissue Cell* 11, 325–336.

Noirot-Timothée, C., Smith, D. S., Cayer, M. L. and Noirot, C. (1978) Septate junctions in insects: Comparisons between intercellular and intramembranous structures. *Tissue Cell* 10, 125–136.

Noland, J. L. and Baumann, C. A. (1949) Requirements of the German cockroach for choline and related compounds. *Proc. Soc. exp. Biol. Med.* 70, 198–201.

Norris, D. M. (1976) A molecular and submolecular mechanism of insect perception of certain chemical information in their environment. *Coll. Intern du C.N.R.S.* 265, 81–102.

Norris, D. M. and Chu, H. (1974) Morphology and ultrastructure of male *Periplaneta americana* as related to chemoreception. *Cell Tissue Res.* 150, 1–9.

Novak, F. J. (1981) Anatomy and fine structure of the retrocerebral nerves associated with the corpus allatum and foregut musculature of the American cockroach, *Periplaneta americana. J. Morph.* (In press.)

Nutting, W. L. (1951) A comparative anatomical study of the heart and accessory structures of the orthopteroid insects. *J. Morphol.* 89, 501–598.

Nutting, W. L. (1955) Extirpation of roach prothoracic glands. *Science* 122, 30–31.

O'Brien, R. D. (1978) The biochemistry of toxic action of insecticides. In: *Biochemistry of Insects*, Rockstein, M., Ed., pp. 515–539. Academic Press, New York.

O'Connor, A. K., O'Brien, R. D. and Salpeter, M. M. (1965) Pharmacology and fine structure of peripheral muscle innervation in the cockroach, *Periplaneta americana. J. Insect Physiol.* 11, 1351–1358.

O'Farrell, A. F. (1959) Trachael patterns in regenerated legs of cockroaches. In: *Physiology of Insect Development*. Campbell, F. L., Ed., pp. 75–76, 142–151. University Chicago Press, Chicago.

O'Farrell, A. F. and Stock, A. (1953) Regeneration and the molting cycle in *Blattella germanica* (L.). I. Single regeneration in the first instar. *Aust. J. biol. Sci.* 6, 485–500.

O'Farrell, A. F. and Stock, A. (1956a) Some aspects of regeneration in cockroaches. *Proc. 10th Int. Cong. Ent. Montreal* 2, 253–259.

O'Farrell, A. F. and Stock, A. (1956b) Regeneration and the molting cycle in *Blattella germanica* (L.). IV. Single and repeated regeneration and metamorphosis. *Aust. J. biol. Sci.* 9, 406–422.

O'Farrell, A. F. and Stock, A. (1958) Some aspects of regeneration in cockroaches. *Proc. 10th Int. Cong. Ent. Montreal.* 2, 253–259.

O'Farrell, A. F., Stock, A., Rae, C. A. and Morgan, J. A. (1960) Regeneration and development in the cockroach *Blattella germanica. Acta Soc. Ent. Csl.* 57, 317–324.

Oliver, G. W. O., Taberner, P. V., Rick, J. T. and Kerkut, G. A. (1971) Changes in GABA level, GAD and ChE activity in CNS of an insect during learning. *Comp. Biochem. Physiol.* 38B, 529–36.

Olomon, C. M., Breed, M. D. and Bell, W. J. (1976) Ontogenetic and temporal aspects of agonistic behavior in a cockroach, *Periplaneta americana. Behav. Biol.* 17, 243–248.

Olsen, W. P. (1973) Dieldrin transport in the insect: an examination of Gerolt's hypothesis. *Pest Biochem. Physiol.* 3, 384–392.

O'Riordan, A. M. (1968) *The Absorption of Electrolytes in the Midgut of the Cockroach.* PhD Thesis. University of Cambridge.

O'Riordan, A. M. (1969) Electrolyte movement in the isolated midgut of the cockroach (*Periplaneta americana* L.). *J. exp. Biol.* 51, 699–714.

Osborne, M. P. (1975) The ultrastructure of nerve–muscle synapses. In: *Insect Muscle*, Usherwood, P. N. R., Ed., pp. 151–205. Academic Press, London.

Oschman, J. L. and Wall, B. J. (1969) The structure of the rectal pads in *Periplaneta americana* with regard to fluid transport. *J. Morph.* 127, 475–510.

Oschman, J. L. and Wall, B. J. (1972) Calcium binding to intestinal membranes. *J. Cell Biol.* 55, 58–73.

O'Shea, M. and Evans, P. D. (1979) Potentiation of neuromuscular transmission by an octopaminergic neurone in the locust. *J. exp. Biol.* 79, 169–190.

Osinchak, J. (1966) Ultrastructural localization of some phosphatases in the prothoracic gland of the insect *Leucophaea maderae. Z. Zellforsch.* 72, 236–248.

Ozbas, S. and Hodgson, E. S. (1958) Action of insect neurosecretion upon central nervous system *in vitro* and upon behavior. *Proc. natn. Acad. Sci., U.S.A.* 44, 825–830.

Page, T. L. (1978) Interaction between bilaterally paired components of the cockroach circadian system. *J. comp. Physiol.* 124, 224–236.

Page, T. L. and Block, G. D. (1980) Circadian rhythmicity in cockroaches: effects of early post-embryonic development and ageing. *Physiol. Ent.* 5, 271–281.

Page, T. L., Caldarola, P. C. and Pittendrigh, C. S. (1977) Mutual entrainment of bilaterally distributed circadian pacemakers. *Proc. natn. Acad. Sci., U.S.A.* 74, 1277–1281.

Pak, K.-Y. and Harris, C. L. (1975) Evidence for a molecular code for learned shock-avoidance in cockroaches. *Comp. Biochem. Physiol.* 52A, 141–144.

Palka, J. (1979) Theories of pattern formation in insect neural development. *Adv. Insect Physiol.* 14, 251–349.

Pan, M. L., Bell, W. J. and Telfer, W. H. (1969) Vitellogenic blood protein synthesis by insect fat body. *Science* 165, 393–394.

Pan, M. L. and Wyatt, G. R. (1971) Juvenile hormone induces vitellogenin synthesis in the monarch butterfly. *Science* 174, 503–505.

Parnas, E. and Dagan, D. (1971) Functional organizations of giant axons in the central nervous systems of insects: new aspects. *Adv. Insect Physiol.* 8, 95–144.

Parnas, I. and Grossman, Y. (1973) Presynaptic inhibition in the phallic neuromuscular system of the cockroach *Periplaneta americana. J. comp. Physiol.* 82, 23–32.

Patel, N. G. and Cutkomp, K. L. (1967) Physiological responses of cockroaches to immobilization, DDT and dieldrin. *J. econ. Ent.* 60, 783–788.

Patel, N. G., Cutkomp, L. K. and Halberg, F. (1974) Rhythmicity in oxygen consumption and heart beat rate of the American cockroach, *Periplaneta americana* following immobilization and DDT treatments. *Int. J. Chronobiol.* 2, 197–208.

Patterson, J. A. and Goodman, L. J. (1974) Relationships between ocellar units in the ventral nerve cord and ocellar pathways in the brain of *Schistocerca gregaria. J. comp. Physiol.* 95, 251–262.

Pau, R. N. and Acheson, R. M. (1968) The identification of 3-Hydroxy-4-*O*-β-D-glucoside of benzyl alcohol in the left colleterial gland of *Blaberus discoidalis. Biochim. Biophys. Acta* 158, 206–211.

Pau, R. N., Brunet, P. C. J. and Williams, M. J. (1971) The isolation and characterization of proteins from the left colleterial gland of the cockroach, *Periplaneta americana* (L.). *Proc. R. Soc. Lond. Ser. B* 177, 565–579.

Paulpandian, A. (1959) Cyclic ventilation movement in the common cockroach, *Periplaneta americana. Curr. Sci.* 33, 404–405.

Pawlik, J. (1966) Control of the nematode *Leidynema appendiculata* (Leidy) (Nemata:

Rhabditida: Thelastomatidae) in laboratory cultures of the American cockroach. *J. econ. Ent.* 59, 468–469.

Pearson, K. G. (1972) Central programming and reflex control of walking in the cockroach. *J. exp. Biol.* 56, 173–193.

Pearson, K. G. (1973) Function of peripheral inhibitory axons in insects. *Am. Zool.* 13, 321–330.

Pearson, K. G. (1977) Interneurons in the ventral nerve cord of insects. In: *Identified Neurons and Behavior of Arthropods*, Hoyle, G., Ed., pp. 329–337. Plenum Press, New York.

Pearson, K. G. and Bergman, S. J. (1969) Common inhibitory motoneurones in insects. *J. exp. Biol.* 50, 445–473.

Pearson, K. G. and Bradley, A. B. (1972) Specific regeneration of excitatory motoneurons to leg muscles in the cockroach. *Brain Res.* 47, 492–496.

Pearson, K. G. and Fourtner, C. R. (1973) Identification of the somata of common inhibitory motoneurons in the metathoracic ganglion of the cockroach. *Can. J. Zool.* 51, 859–866.

Pearson, K. G. and Fourtner, C. R. (1975) Nonspiking interneurons in walking system of the cockroach. *J. Neurophysiol.* 38, 33–52.

Pearson, K. G. and Iles, J. F. (1970) Discharge patterns of coxal levator and depressor motoneurons of the cockroach, *Periplaneta americana. J. exp. Biol.* 52, 139–165.

Pearson, K. G. and Iles, J. F. (1971) Innervation of coxal depressor muscles in the cockroach *Periplaneta americana. J. exp. Biol.* 54, 215–232.

Pearson, K. G. and Iles, J. F. (1973) Nervous mechanisms underlying intersegmental co-ordination of leg movements during walking in the cockroach. *J. exp. Biol.* 58, 725–744.

Pearson, K. G. and Rowell, C. H. F. (1977) Functions of tonic sensory input in insects. *Ann. N.Y. Acad. Sci.* 290, 114–123.

Pearson, K. G., Stein, R. B. and Malhotra, S. K. (1970) Properties of action potentials from insect motor nerve fibres. *J. exp. Biol.* 53, 299–316.

Pemberton, C. E. (1945) Entomology. *Rep. Comm. exp. St. Hawaiian Sug. Planter's Ass.* 1943/1944, 22–26.

Penzlin, H. (1963) Uber die Regeneration bei Schaben (Blattaria). I. Das Regenerations-vermogen und die Genese des Regenerats. *Roux Arch. Entwmech. Org.* 154, 434–465.

Penzlin, H. (1964) The significance of the nervous system for regeneration in insects. *Roux. Arch. Entwmech. Org.* 155, 152–161.

Penzlin, H. (1965) Die bedeutung von hormonen fur die regeneration bei insekten. *Zool. Jahrb. Physiol.* 71, 584–594.

Penzlin, H. (1971) Zur Rolle des Frontalganglions bei Larven der Schabe *Periplaneta americana. J. Insect Physiol.* 17, 559–573.

Penzlin, H. and Stolzner, W. (1971) Frontal ganglion and water balance in *Periplaneta americana* (L.). *Experientia* 24, 390–391.

Persoons, C. J. (1977) *Structure elucidation of some insect pheromones*. PhD Thesis. Landbouwhogeschool te Wageningen. Pp. 83.

Persoons, C. J., Verwiel, P. E. J., Ritter, F. J., Talman, E., Nooijen, P. J. F. and Nooijen, W. J. (1976) Sex pheromones of the American cockroach, *Periplaneta americana*: A tentative structure of periplanone-B. *Tetrahedron Letters* 24, 2055–2058.

Persoons, C. J. Verwiel, P. E. J., Talman, E. and Ritter, F. J. (1979) Sex pheromone of the American cockroach, *Periplaneta americana.* Isolation and structure elucidation of Periplanone D. *J. chem. Ecol.* 5, 221–236.

Peterson, M. (1977) Reach for the New World. *Natn. Geograph.* 152, 724–767.

Phillips, J. E. (1964) Rectal absorption in the desert locust, *Schistocerca gregaria* Forskal. I. Water. *J. exp. Biol.* 41, 15–38.

Phillips, J. E. and Dockrill, A. A. (1968) Molecular sieving of hydrophilic molecules by the rectal intima of the desert locust (*Schistocerca gregaria*). *J. exp. Biol.* 48, 521–532.

Pichon, Y. (1963) La teneur en ions Na^+, K^+ et Ca^{2+} de l'hemolymphe de *Periplaneta americana* L., ses variations. *Bull. Soc. sci. Bretagne* 38, 147.
Pichon, Y. (1969a) Effets de la tétrodotoxine (T.T.X.) sur les caractéristiques du perméabilité membranaire de la fibre nerveuse isolée d'insecte. *C. R. Acad. Sci., Paris, Ser. D* 268, 1095–1097.
Pichon, Y. (1969b) Effets des ions tetraéthylammonium (T.E.$A.^+$) sur la membrane de l'axone géant d'insecte. *C. R. Soc. Biol.* 163, 952–958.
Pichon, Y. (1970) Ionic content of haemolymph in the cockroach, *Periplaneta americana*: A critical analysis. *J. exp. Biol.* 53, 195–209.
Pichon, Y. (1974) Axonal conduction in insects. In: *Insect Neurobiology*, Treherne, J. E., Ed., pp. 73–117. Elsevier, New York.
Pichon, Y. and Boistel, J. (1963) Modifications of the ionic content of the haemolymph and of the activity of *Periplaneta americana* in relation to diet. *J. Insect Physiol.* 9, 887–891.
Pichon, Y. and Boistel, J. (1967) Microelectrode study of the resting and action potentials of the cockroach giant axon with special reference to the role played by the nerve sheath. *J. exp. Biol.* 47, 357–373.
Pichon, Y. and Boistel, J. (1968) Ionic composition of hemolymph and nervous function in the cockroach, *Periplaneta americana* (L.). *J. exp. Biol.* 49, 31–38.
Pichon, Y. and Treherne, J. E. (1970) Extraneuronal potentials and potassium depolarization in cockroach giant axons. *J. exp. Biol.* 53, 485–493.
Piek, T. (1974) Ion barriers in muscle fibres. *Arch. Int. Physiol. Biochem.* 82, 337–339.
Piek, T. and Njio, K. D. (1979) Morphology and electrochemistry of insect muscle fibre membrane. *Adv. Insect Physiol.* 14, 185–250.
Piek, T., Visser, B. J. and Mantel, P. (1979) Effect of proctolin, BPP_{5a} and related peptides on rhythmic contractions in *Locusta migratoria. Comp. Biochem. Physiol.* 62C, 151–154.
Pierre, L. L. (1964) Uricase activity of isolated symbionts and the aposymbiotic fat body of a cockroach. *Nature* 201, 54–55.
Pihan, J. C. (1971) Mise en evidence d'un facteur tissulaire intervenant au cours de la morphogenèse du système trachéen chex les insectes Diptères. *J. Embryol. exp. Morph.* 26, 497–521.
Pihan, J. C. (1972) Facteurs intervenant au cours de la morphogenèse du système trachéen chez les insectes Diptères. *Bull. Soc. zool. France* 97, 351–361.
Pilgrim, R. L. C. (1954) Waste of carbon and of energy in nitrogen excretion. *Nature* 173, 491.
Pillai, M. K. K. and Saxena, K. N. (1959) Fate of fructose in the alimentary canal of the cockroach. *Physiol. Zool.* 32, 293–298.
Pipa, R. L. (1961) Studies on the hexapod nervous sytem. III. Histology and histochemistry of cockroach neuroglia. *J. comp. Neurol.* 116, 15–25.
Pipa, R. L. (1963) Studies on the hexapod nervous system. VI. Ventral nerve cord shortening: A metamorphic process in *Galleria mellonella* (L.) (Lepidoptera Pryallidae). *Biol. Bull.* 124, 293–302.
Pipa, R. L. (1973) Proliferation, movement, and regression of neurons during the post-embryonic development of insects. In: *Developmental Neurobiology of Arthropods*, Young, D., Ed., pp. 105–130. Cambridge University Press, Cambridge.
Pipa, R. L. (1978a) Locations and central projections of neurons associated with the retro-cerebral neuroendocrine complex of the cockroach *Periplaneta americana* (L.). *Cell Tissue Res.* 193, 443–455.
Pipa, R. L. (1978b) Patterns of neural reorganization during the postembryonic development of insects. *Int. Rev. Cytol., Suppl.* 7, 403–438.
Pipa, R. L. and Cook, E. F. (1959) Studies on the hexapod nervous system. I. The peripheral distribution of the thoracic nerves of the adult cockroach, *Periplaneta americana. Ann. ent. Soc. Am.* 52, 695–710.
Pipa, R. L., Cook, E. F. and Richards, A. G. (1959) Studies on the hexapod nervous system.

II. The histology of the thoracic ganglia of the adult cockroach, *Periplaneta americana* (L.). *J. comp. Neurol.* 113, 401–434.

Pipa, R. L. and Novak, F. J. (1979) Pathways and fine structure of neurons forming the nervi corporis allati II of the cockroach, *Periplaneta americana* (L.). *Cell Tissue Res.* 201, 227–237.

Piper, G. L. and Frankie, G. W. (1978) Integrated management of urban cockroach populations. In: *Perspectives in Urban Entomology*, Frankie, G. W. and Kochler, C. S. Eds., pp. 249–266. Academic Press, New York.

Piper, G. L., Frankie, G. W. and Loehr, J. (1978) Incidence of cockroach egg parasites in urban environments in Texas and Louisiana. *Environ. Ent.* 7, 289–293.

Piquette, P. G. and Fales, J. H. (1952) Rearing cockroaches for experimental purposes. *U.S. Dept. Agric. Bur. Ent. Plant Quar. ET*-301, 12 pp.

Pitman, R. M. (1971) Transmitter substances in insects: a review. *Comp. gen. Pharmacol.* 2, 347–371.

Pitman, R. M. (1975a) The ionic dependence of action potentials induced by colchicine in an insect motoneuron cell body. *J. Physiol.* 247, 511–520.

Pitman, R. M. (1975b) Calcium-dependent action potentials in the cell body of an insect motoneuron. *J. Physiol.* 251, 62P–63P.

Pitman, R. M. and Kerkut, G. A. (1970) Comparison of the actions of iontophoretically applied acetylcholine and γ-aminobutyric acid with the EPSP and IPSP in cockroach central neurons. *Comp. gen. Pharmacol.* 1, 221–230.

Pitman, R. M., Tweedle, C. D. and Cohen, M. J. (1972a) Branching of central neurons: intracellular cobalt injection for light and electron microscopy. *Science* 176, 412–414.

Pitman, R. M., Tweedle, C. D. and Cohen, M. J. (1972b) Electrical responses of insect central neurons: agumentation by nerve section or colchicine. *Science* 178, 507–509.

Pitman, R. M., Tweedle, C. D. and Cohen, M. J. (1973) The form of nerve cells: determination by cobalt impregnation. In: *Intracellular Staining in Neurobiology*. Kater, S. B. and Nicholson, C., Eds., pp. 83–97. Springer-Verlag, New York.

Pittendrigh, C. (1960a) Comment *after* Harker, J. E. (1960) Endocrine and nervous factors in insect circadian rhythms. *Cold Spring Harbor Symp. Quant. Biol.* 25, 279–287.

Pittendrigh, C. (1960b) Circadian rhythms and the circadian organization of living systems. *Cold Spring Harbor Symp. Quant. Biol.* 25, 159–184.

Pittendrigh, C. S. and Caldarola, P. C. (1973) General homeostasis of the frequency of circadian oscillations. *Proc. natn. Acad. Sci., U.S.A.* 70, 2697–2701.

Pittendrigh, C. S. and Dann, S. (1976a) A functional analysis of circadian pacemakers in nocturnal rodents. I. The stability and lability of spontaneous frequency. *J. comp. Physiol.* 106, 223–252.

Pittendrigh, C. S. and Dann, S. (1976b) A functional analysis of circadian pacemakers in nocturnal rodents. V. Pacemaker structure: a clock for all seasons. *J. comp. Physiol.* 106, 333–355.

Pohley, H.-J. (1959) Experimentelle beitrage zur lenkung der organentwicklung des hautungsrhythmus und der metamorphose bei der Schabe, *Periplaneta americana. Roux Archiv Dev. Biol.* 151, 323–380.

Pohley, H.-J. (1965) Regeneration and the molting cycle in *Ephestia kuhniella*. In: *Regeneration in Animals*, Kiortsis and Trampusch, Eds. North-Holland Publishing Co., Amsterdam.

Poláček, I. and Kubišta, V. (1960) Metabolism of the cockroach, *Periplaneta americana*, during flight. *Physiol. Bohem.* 9, 228–234.

Pope, P. (1953) Studies of the life histories of some Queensland Blattidae (Orthoptera). Part I. The domestic species. *Proc. R. Soc. Queensl.* 63, 23–46.

Post, L. C. (1972) Bursicon: its effect on tyrosine permeation into insect haemocytes. *Biochem. Biophys. Acta* 290, 424–428.

Powning, R. F., Day, M. F. and Irzykiewicz, H. (1951) Studies on the digestion of wool by insects. II. The properties of some insect proteinases. *Aust. J. sci. Res.* (B) 4, 49–63.

Powning, R. F. and Irzykiewicz, H. (1967) Lysozyme-like action of enzymes from the cockroach *Periplaneta americana* and from some other sources. *J. Insect Physiol.* 13, 1293–1299.
Prabhu, V. K. K. and Hema, P. (1969) Influence of recurrent nerve severence on haemolymph proteins in *Periplaneta americana* (L.). *Experientia* 25, 1115.
Prabhu, V. K. K. and Nayar, K. K. (1972a) Changes in blood proteins in the cockroach, *Periplaneta americana*, after chemosterilization with metepa. *Ent. exp. appl.* 15, 417–422.
Prabhu, V. K. K. and Nayar, K. K. (1972b) Haemolymph protein electrophoretic pattern in *Periplaneta americana* after administration of farnesyl methyl ether. *J. Insect Physiol.* 18, 1435–1440.
Pratt, G. E., Hamnett, A. F., Finney, J. R. and Weaver, R. J. (1978) The biosynthesis of radiolabeled cecropia (C_{18}) juvenile hormone and several analogs by isolated corpora allata of the American cockroach (*Periplaneta americana*) during incubation with methyl-labeled methionine and analogs of farnesenic acid. *Gen. comp. Endocrinol.* 34, 113.
Pratt, G. E. and Tobe, S. S. (1974) Juvenile hormone radiobiosynthesised by corpora allata of adult female locusts *in vitro*. *Life Sci.* 14, 575–586.
Pratt, G. E., Tobe, S. S. and Weaver, R. J. (1975a) Relative oxygenase activities in juvenile hormone biosynthesis of corpora allata of an African locust (*Schistocerca gregaria*) and American cockroach (*Periplaneta americana*). *Experientia* 31, 120–122.
Pratt, G. E., Tobe, S. S., Weaver, R. J. and Finney, J. R. (1975b) Spontaneous synthesis and release of C_{16} juvenile hormone by isolated corpora allata of female locust *Schistocerca gregaria* and female cockroach *Periplaneta americana*. *Gen. comp. Endocrinol.* 26, 478–484.
Pratt, G. E. and Weaver, R. J. (1975) Juvenile hormone biosynthesis by cultured cockroach corpora allata. *J. Endocrinol.* 64, 67P.
Pratt, G. E. and Weaver, R. J. (1978) The rate of juvenile hormone biosynthesis by insect corpora allata *in vitro*. In: *Comparative Endocrinology*, Gaillard, P. J. and Boer, H. H., Eds., pp. 503–506. Elsevier/North Holland, Amsterdam.
Pratt, G. E., Weaver, R. J. and Hamnett, A. F. (1976) Continuous monitoring of juvenile hormone release in superfused corpora allata of *Periplaneta americana*. In: *The Juvenile Hormones*, Gilbert, L. I., Ed., pp. 164–178. Plenum Press, New York.
Price, C. D. and Ratcliffe, N. A. (1974) A reappraisal of insect haemocyte classification by the examination of blood from fifteen insect orders. *Z. Zellforsch.* 147, 537–549.
Price, G. M. (1973) Protein and nucleic acid metabolism in insect fat body. *Biol. Rev.* 48, 333–375.
Prigent, J.-P. (1966) La structure du deutocerebron de l'adulte de *Periplaneta americana* (L.). *Bull. Soc. Zool. France* 91, 365–374.
Princis, K. (1966) *Orthopterorum Catalogus*, Beier, M., Ed., Pars 8, Fam.: Blattidae, Nocticolidae. pp. 404–456. Junk, 's-Gravenhagen.
Princis, K. (1971) *Orthopterorum Catalogus*, Beier, M., Ed., Pars 14, Fam.: Ectobiidae. pp. 1144–1145. Junk, 's-Gravenhagen.
Pringle, J. W. S. (1938a) Proprioception in insects. II. The action of the campaniform sensilla on the leg. *J. exp. Biol.* 15, 114–131.
Pringle, J. W. S. (1938b) Prorioception in insects. III. The function of the hair sensilla at the joints. *J. exp. Biol.* 15, 467–473.
Pringle, J. W. S. (1940) The reflex mechanism of the insect leg. *J. exp. Biol.* 17, 8–17.
Pringle, J. W. S. (1961) Proprioception in arthopods. In: *The Cell and Organism*. Ramsey, J. H. and Wigglesworth, V. B., Eds., pp. 256–282. Cambridge University Press, Cambridge.
Pritchatt, D. (1968) Avoidance of electric shock by the cockroach *Periplaneta americana*. *Animal Behav.* 16, 178–185.
Prosser, C. L. (1973) Water: Osmotic balance; Hormonal regulation. In: *Comparative*

Richter, K. and Stürzebecher, J. (1969) Elektronenphysiologie Untersuchungen zum Wirkungsmechanismus von Neurohormon D am lateralen Herznerven von *Periplaneta americana* (L.). *Zeitschr. Wiss. Zool., A* 180, 148–163.

Rick, J. T., Oliver, G. W. O. and Kerkut, G. A. (1972) Acquisition, extinction and re-acquisition of a conditioned response in the cockroach: the effects of orotic acid. *Q. J. exp. Psych.* 24, 282–286.

Riddiford, L. M. (1980) Insect endocrinology: action of hormones at the cellular level. *A. Rev. Physiol.* 42, 511–528.

Riddiford, L. M. and Truman, J. W. (1978) Biochemistry of insect hormones and insect growth regulators. In: *Biochemistry of Insects*, Rockstein, M., Ed., pp. 308–357. Academic Press, New York.

Ritter, H. (1961) Glutathione-controlled anaerobiosis in *Cryptocercus*, and its detection by polarography. *Biol. Bull.* 121, 330–346.

Ritter, Jr, H. (1964) Defense of mate and mating chamber in a wood roach. *Science* 143, 1459–1460.

Ritzmann, R. E. and Camhi, J. M. (1978) Excitation of leg motor neurons by giant interneurons in the cockroach *Periplaneta americana. J. comp. Physiol.* 125, 305–316.

Ritzmann, R. E., Tobias, M. L. and Fourtner, C. R. (1980) Flight activity initiated via giant interneurons of the cockroach: evidence for bifunctional trigger interneurons. *Science.* 210, 443–445.

Rivault, C. (1976) The role of the eyes and ocelli in the initiation of circadian activity rhythms in cockroaches. *Physiol. Ent.* 1, 227–286.

Robbins, W. E., Dutky, R. C., Monroe, R. E. and Kaplanis, J. N. (1962). The metabolism of H^3-β-sitosterol by the German cockroach. *Ann. ent. Soc. Am.* 55, 102–104.

Robbins, W. E., Kaplanis, J. N., Svoboda, J. A. and Thompson, M. J. (1971) Steroid metabolism in insects. *A. Rev. Ent.* 16, 53–72.

Roberts, B. (1973) The relative growth of normal and regenerated legs of the cockroach *Blattella germanica. Trans. R. ent. Soc, Lond.* 124.

Roberts, S. K. (1959) *Circadian Activity Rhythms in Cockroaches*. PhD Thesis, Princeton University.

Roberts, S. K. (1960) Circadian activity rhythms in cockroaches. I. The free-running rhythms in steady-state. *J. cell. comp. Physiol.* 55, 99–110.

Roberts, S. K. (1962) Circadian activity rhythms in cockroaches. II. Entrainment and phase shifting. *J. cell. comp. Physiol.* 59, 175–186.

Roberts, S. K. (1965a) Photoreception and entrainment of cockroach activity rhythms. *Science* 148, 958–959.

Roberts, S. K. (1965b) Significance of endocrines and central nervous system in circadian rhythms. In: *Circadian Clocks*, Aschoff, J., Ed., pp. 198–213. North-Holland Publishing Co., Amsterdam.

Roberts, S. K. (1966) Circadian activity rhythms in cockroaches. III. The role of endocrine and neural factors. *J. cell. Physiol.* 67, 473–486.

Roberts, S. K. (1974) Circadian rhythms in cockroaches. Effects of optic lobe lesions. *J. comp. Physiol.* 88, 21–30.

Roberts, S. K., Skopik, S. D. and Driskill, R. J. (1971) Circadian rhythms in cockroaches: does brain hormone mediate the locomotor cycle? In: *Biochronometry*, Menaker, M., Ed., pp. 505–516. National Academy of Science, U.S.A. Washington, D.C.

Robertson, J. and Pipa, R. L. (1973) Metamorphic shortening of interganglionic connectives of *Galleria mellonella* (Lepidoptera) *in vitro*: stimulation by ecdysone analogues. *J. Insect Physiol.* 19, 673–679.

Roeder, K. D. (1948) Organization of the ascending giant fiber system in the cockroach (*Periplaneta americana*). *J. exp. Zool.* 108, 243–262.

Roeder, K. D. (1959) A physiological approach to the relation between prey and predator. *Smithson. Misc. Coll.* 137, 287–306.

Roome, R. E. (1968) *The Function of the Stomatogastric Nervous System as a Link between Feeding, Endocrine Secretion and Growth in Insects*. PhD Thesis. University of Nottingham.

Roosen-Runge, E. C. (1977) *The Process of Spermatogenesis in Animals*. Cambridge University Press, Cambridge.

Rosenfeld, A. H. (1910) Blattid notes. *J. econ. Ent.* 3, 100–101.

Roth, L. M. (1969) The evolution of male tergal glands in the Blattaria. *Ann. ent. Soc. Am.* 62, 176–208.

Roth, L. M. (1970) Evolution and taxonomic significance of reproduction in *Blattaria. A. Rev. Ent.* 15, 75–96.

Roth, L. M. and Alsop, D. W. (1978) Toxins of Blattaria. In: *Arthropod Venoms*, Bettini, S., Ed. *Handbook exp. Pharmacol.* 48, 465–487.

Roth, L. M. and Cohen, S. (1973) Aggregation in Blattaria. *Ann. ent. Soc. Am.* 66, 1315–1323.

Roth, L. M. and Eisner, T. (1962) Chemical defenses of arthropods. *A. Rev. Ent.* 7, 107–136.

Roth, L. M. and Stay, B. (1958) The occurrence of paraquinones in some arthropods with emphasis on the quinone-secreting tracheal glands of *Diploptera punctata* (Blattaria). *J. Insect Physiol.* 1, 305–318.

Roth, L. M. and Willis, E. R. (1952) A study of cockroach behaviour. *Am. Midl. Nat.* 47, 66–129.

Roth, L. M. and Willis, E. R. (1955a) Water relations of cockroach oothecae. *J. econ. Ent.* 48, 33–36.

Roth, L. M. and Willis, E. R. (1955b) Water content of cockroach eggs during embryogenesis in relation to oviposition behavior. *J. exp. Zool.* 128, 489–510.

Roth, L. M. and Willis, E. R. (1956) Parthenogenesis in cockroaches. *Ann. ent. Soc. Am.* 49, 195–204.

Roth, L. M. and Willis, E. R. (1957) The medical and veterinary importance of cockroaches. *Smithson. Misc. Coll.* 134, 1–147.

Roth, L. M. and Willis, E. R. (1958) An analysis of oviparity and viviparity in the *Blattaria. Trans. Am. ent. Soc.* 83, 221–238.

Roth, L. M. and Willis, E. R. (1960) The biotic associations of cockroaches. *Smithson. Misc. Coll.* 141, 470 pp.

Roth, R. L. and Sokolove, P. G. (1975) Histological evidence for direct connections between the optic lobes of the cockroach *Leucophaea maderae*. *Brain Res.* 87, 23–39.

Rounds, H. D. (1968) Diurnal variation in the effectiveness of extracts of cockroach midgut in the release of intestinal proteinase activity. *Comp. Biochem. Physiol.* 25, 1125–1128.

Rounds, H. D. (1975) A lunar rhythm in the occurrence of bloodborne factors in cockroaches. *Comp. Biochem. Physiol.* 50, 193–197.

Rounds, H. D. and Gardner, Jr, F. E. (1968) A quantitative comparison of the activity of cardio-acceleratory extracts from various portions of the cockroach nerve cord. *J. Insect Physiol.* 14, 495–497.

Rounds, H. D. and Riffel, F. A. (1974) Life- and death-promoting factors from the subesophageal ganglion of the American cockroach. *J. Insect Physiol.* 20, 1403–1410.

Rowell, H. F. (1976) The cells of the insect neurosecretory system: constancy, variability, and the concept of the unique identifiable neuron. *Adv. Insect Physiol.* 12, 63–123.

Ruck, P. (1957) The electrical responses of dorsal ocelli in cockroaches and grasshoppers. *J. Insect Physiol.* 1, 109–123.

Ruck, P. (1958a) Dark adaptation of the ocellus in *Periplaneta americana*: A study of the electrical response to illumination. *J. Insect Physiol.* 2, 189–198.

Ruck, P. (1958b) A comparison of the electrical responses of compound eyes and dorsal ocelli in four insect species. *J. Insect Physiol.* 2, 261–274.

Ruck, P. (1961) Electrophysiology of the insect dorsal ocellus. I. Origin of the components in the electoretinogram. *J. gen. Physiol.* 44, 605–627.

Rudall, K. M. (1955) The distribution of collagen and chitin. *Symp. Soc. exp. Biol.* 9, 49–71.
Rudall, K. M. (1965) Skeletal structure in insects. In: *Aspects of Insect Biochemistry. Symp. Biochem. Soc.* 25, 83–92.
Rudall, K. M. (1967) Conformation in chitin-protein complexes. In: *Conformation of Biopolymers*, Vol. 2, Ramachandran, G. N., Ed., pp. 751–765. Academic Press, London.
Rueger, M. E. and Olson, T. A. (1969) Cockroaches (Blattaria) as vectors of food poisoning and food infection organisms. *J. med. Ent.* 6, 185–189.
Rüfenacht, H. and Lüscher, M. (1976) *In vitro*-Nachweiss Juvenilhormon (JH) abbauender Enzymaktivität während der Eireifung der Schabe *Nauphoeta cinerea. Rev. Suisse Zool.* 83, 934–939.
Runion, H. I. and Usherwood, P. N. R. (1966) A new approach to neuromuscular analysis in the intact free-walking insect preparation. *J. Insect Physiol.* 12, 1255–1263.
Rust, M. K. (1976) Quantitative analysis of male responses released by female sex pheromone in *Periplaneta americana. Animal Behav.* 24, 681–685.
Rust, M. K. and Bell, W. J. (1976) Chemo-anemotaxis: a behavioral response to sex pheromone in non-flying insects. *Proc. natn. Acad. Sci., U.S.A.* 73, 2524–2526.
Rust, M. K., Burk, T. and Bell, W. J. (1976) Pheromone-stimulated locomotory and orientation responses in the American cockroach. *Animal Behav.* 24, 52–67.
Ruth, E. (1976) Elektrophysiologie der Sensilla chaetica auf den Antennen von *Periplaneta americana. J. comp. Physiol.* 105, 55–64.
Ryan, M. and Nicholas, W. L. (1972) The reaction of the cockroach, *Periplaneta americana*, to the injection of foreign particulate material. *J. Invert. Pathol.* 19, 299–307.
Ryerse, J. S. and Locke, M. (1978) Ecdysterone-mediated cuticle deposition and the control of growth in insect tracheae. *J. Insect Physiol.* 24, 541–550.
Sacchi, V. F. and Giordana, B. (1979) Ions contribution to the transepithelial electrical potential difference across the isolated midgut of the cockroach, *Leucophaea maderae. Comp. Biochem. Physiol.* 64A, 419–425.
Sacktor, B. (1976) Biochemical adaptations for flight in the insect. *Biochem. Soc. Symp.* 41, 111–131.
Sacktor, B. and Bodenstein, D. (1952) Cytochrome *c* oxidase activity of various tissues of the American cockroach, *Periplaneta americana. J. cell. comp. Physiol.* 40, 157–161.
Sacktor, B. and Thomas, G. M. (1955) Succino-cytochrome *c* reductase activity of tissues of the American cockroach, *Periplaneta americana. J. cell. comp. Physiol.* 45, 241–245.
Sakai, M. (1969) Nereistoxin and cartap: their mode of action as insecticides. *Rev. Plant Protection Res.* 2, 17–29.
Salanki, J. (1973) Neural mechanisms in rhythm regulation of invertebrates. In: *Neurobiology of Invertebrates*. Salanki, J., Ed., pp. 17–31. Akadēmiai Kiadō, Budapest.
Salpeter, M. M. and Faeder, I. R. (1971) The role of sheath cells in glutamate uptake by insect nerve muscle preparations. *Prog. Brain Res.* 34, 103–114.
Samaranayaka, M. (1974) Insecticide-induced release of hyperglycaemic and adipokinetic hormones of *Schistocerca gregaria. Gen. comp. Endocrinol.* 24, 424–436.
Sams, G. R. (1977) *Juvenile Hormone Initiation of Yolk Deposition in vitro*. PhD Thesis, University of Kansas.
Sams, G. R. and Bell, W. J. (1977) Juvenile hormone control of yolk deposition *in vitro. Adv. Invert. Repro.* 1, 404–413.
Sams, G. R., Cocchiaro, G. R. and Bell, W. J. (1978) Metabolism of juvenile hormone in cultures of ovaries and fat body in the cockroach, *Periplaneta americana. In Vitro* 14, 956–960.
Sams, G. R., Weaver, R. and Bell, W. J. (1981) Characterization of vitellogenin in the American cockroach. *Biochem. biophys. Acta*, 609, 121–135.
Sanburg, L. L., Kramer, K. J., Kēzdy, F. J. and Law, J. H. (1975) Juvenile hormone-specific esterases in the haemolymph of the tobacco hornworm, *Manduca sexta. J. Insect Physiol.* 21, 873–887.

Sanchez, B. (1975) Respiration of ten species of insect at different temperatures. *Folia ent. mex.* 33, 93–94.
Sander, K. (1976) Specification of the basic body pattern in insect embryogenesis. *Adv. Insect Physiol.* 12, 125–238.
Sandler, M. B. and Solomon, J. (1976) The effect of environmental illumination on embryonic development in *Periplaneta americana*. *Ann. ent. Soc. Am.* 69, 889–890.
Sanford, E. W. (1918) Experiments in the physiology of digestion in the Blattidae. *J. exp. Zool.* 25, 355–411.
Sass, H. (1973) Das Zusammenspiel mehrerer Rezeptortypen bei der nervosen Codierung von Geruchsqualitaten. *Verh. dtsch. Zool. Ges., Mainz* 66, 198–201.
Sass, H. (1976) Zur nervosen Codierung von Geruchsreizen bei *Periplaneta americana*. *J. comp. Physiol.* 107, 49–65.
Sass, H. (1978) Olfactory receptors on the antenna of *Periplaneta*: Response constellations that encode food odors. *J. comp. Physiol.* 128, 227–233.
Sass, H. (1980) Physiological and morphological identification of antennal olfactory receptors of insects. *Joint Congress on chemoreception ECRO IV/ISOT VII, Noordwijkerhout, Holland.*
Sauer, J. R., Levy, J. J., Smith, D. W. and Mills, R. R. (1970) Effect of rectal lumen concentration on the reabsorption of ions and water by the American cockroach. *Comp. Biochem. Physiol.* 36, 601–614.
Sauer, J. R. and Mills, R. R. (1969a) Movement of calcium and magnesium across the midgut epithelium of the American cockroach. *J. Insect Physiol.* 15, 789–797.
Sauer, J. R. and Mills, R. R. (1969b) Movement of potassium and sodium across the midgut epithelium of the American cockroach. *J. Insect Physiol.* 15, 1489–1498.
Sauer, J. R. and Mills, R. R. (1971) Midgut epithelium of the American cockroach: control of lumen volume by substances from the terminal abdominal ganglion. *J. Insect Physiol.* 17, 1–8.
Sauer, J. R., Schlenz-True, R. and Mills, R. R. (1969) Salt and water transport by the *in vitro* cockroach midgut. *J. Insect Physiol.* 15, 483–493.
Sauerländer, S. and Ehrhardt, P. (1961) Reduzierung der Proteinfraktionen in der Hämolymphe bakterienkranker Blattiden. *Naturwissenschaft* 48, 674–675.
Sauerländer, S. and Köhler, F. (1961) Erhöhung der Körpertemperatur von *Periplaneta americana* L. im Verlauf zweier Bakteriosen. *Experientia* 17, 397–398.
Saunders, D. S. (1976) *Insect Clocks*. Pergamon Press, New York.
Saunders, D. S. (1977) Insect clocks. *Comp. Biochem. Physiol.* 56A, 1–5.
Saunders, D. S. and Thompson, E. J. (1977) 'Strong' phase response curve for the circadian rhythm of locomotor activity in a cockroach *Nauphoeta cinerea*. *Nature* 270, 241–243.
Schafer, R. (1973) Postembryonic development in the antenna of the cockroach, *Leucophaea maderae*: growth, regeneration and the development of the adult pattern of sense organs. *J. exp. Zool.* 183, 353–364.
Schafer, R. (1977a) The nature and development of sex attractant specificity in cockroaches of the genus *Periplaneta*. III. Normal intra- and interspecific behavioral responses and responses of insects with juvenile hormone-altered antennae. *J. exp. Zool.* 199, 73–84.
Schafer, R. (1977b) The nature and development of sex attractant specificity in cockroaches of the genus *Periplaneta*. IV. Electrophysiological study of attractant specificity and its determination by Juvenile Hormone. *J. exp. Zool.* 199, 189–208.
Schafer, R. and Sanchez, T. V. (1973) Antennal sensory system of the cockroach *Periplaneta americana*: Postembryonic development and morphology of the sense organs. *J. comp. Neurol.* 149, 335–354.
Schafer, R. and Sanchez, T. V. (1974) Juvenile hormone inhibits differentiation of olfactory sense organs during postembryonic development of cockroaches. *J. Insect Physiol.* 20, 965–974.
Schafer, R. and Sanchez, T. V. (1976) The nature and development of sex attractant

specificity in cockroaches of the genus *Periplaneta*. II. Juvenile hormone regulates sexual dimorphism in the distribution of antennal olfactory receptors. *J. exp. Zool.* 198, 323–336.

Schaller, D. (1978) Antennal sensory system of *Periplaneta americana*. Distribution and frequency of morphologic types of sensilla and their sex-specific changes during postembryonic development. *Cell Tissue Res.* 191, 121–139.

Scharrer, B. (1939) The differentiation between neuroglia and connective tissue sheath in the cockroach (*Periplaneta americana*). *J. comp. Neurol.* 70, 77–88.

Scharrer, B. (1946) The relationship between corpora allata and reproductive organs in adult *Leucophaea maderae* (Orthoptera). *Endocrinology* 38, 46–55.

Scharrer, B. (1948) The prothoracic glands of *Leucophaea maderae* (Orthoptera). *Biol. Bull.* 95, 186–198.

Scharrer, B. (1952) Neurosecretion. XI. The effects of nerve section on the intercerebralis–cardiacum allatum system of the insect *Leucophaea maderae*. *Biol. Bull.* 102, 261–272.

Scharrer, B. (1963) Neurosecretion. XIII. The ultrastructure of the corpus cardiacum of the insect *Leucophaea maderae*. *Z. Zellforsch.* 60, 761–796.

Scharrer, B. (1964a) Histophysiological studies on the corpus allatum of *Leucophaea maderae*. IV. Ultrastructure during normal activity cycle. *Z. Zellforsch.* 62, 125–148.

Scharrer, B. (1964b) The fine structure of blattarian prothoracic glands. *Z. Zellforsch.* 64, 301–326.

Scharrer, B. (1966) Ultrastructural study of the regressing prothoracic glands of blattarian insects. *Z. Zellforsch.* 69, 1–21.

Scharrer, B. (1968) Neurosecretion. XIV. Ultrastructural study of sites of release of neurosecretory material in Blattarian insects. *Z. Zellforsch.* 89, 1–16.

Scharrer, B. (1969) Neurohumors and neurohormones: Definitions and terminology. *J. Neuro-Visceral Relations*, Suppl. IX, 1–20.

Scharrer, B. (1971) Histophysiological studies on the corpus allatum of *Leucophaea maderae*. V. Ultrastructure of sites of origin and release of a distinctive cellular product. *Z. Zellforsch.* 120, 1–16.

Scharrer, B. (1972) Cytophysiological features of hemocytes in cockroaches. *Z. Zellforsch.* 129, 301–319.

Scharrer, B. (1978) Histophysiological studies on the corpus allatum of *Leucophaea maderae*. VI. Ultrastructural characteristics in gonadectomized females. *Cell Tissue Res.* 194, 533–545.

Scharrer, B. and Kater, S. B. (1969) Neurosecretion. XV. An electron microscopic study of the corpora cardiaca of *Periplaneta americana* after experimentally induced hormone release. *Z. Zellforsch.* 95, 177–186.

Scharrer, B. and Wurzelmann, S. (1974) Observations on synaptoid vesicles in insect neurons. *Zool. Jahrb. Physiol.* 78, 387–396.

Scharrer, B. and Wurzelmann, S. (1978) Neurosecretion. XVII. Experimentally induced release of neurosecretory material by exocytosis in the insect *Leucophaea maderae*. *Cell Tissue Res.* 190, 173–180.

Scheie, P. O., Smyth, T. and Greer, R. T. (1968) Perforations in cockroach cuticle as viewed by a scanning electron microscope. *Ann. ent. Soc. Am.* 61, 1346–1348.

Scheving, L. E. and Pauly, J. E. (1974) Circadian rhythms: some examples and comments on clinical application. *Chronobiologia* 1, 3–21.

Schlapfer, W. T., Haywood, P. and Barondes, S. H. (1972) Cholinesterase and choline acetyltransferase activities develop in whole explant but not in dissociated cell cultures of cockroach brain. *Brain Res.* 39, 540–544.

Schlue, W. R. (1974) Zur Zentralnervösen Organisation des Cercal-Flucht-Reflexes der Schabe (*Periplaneta americana* L.). I. Die Unsetzung cercaler Erregungen in Bein-Motoneuron-Efferenzen. *Z. Tierpsychol.* 34, 172–207.

Schlumberger, H. G. (1952) A comparative study of the reaction to injury. I. The cellular

response to methylcholanthrene and to talc in the body cavity of the cockroach, *Periplaneta americana. Arch. Path.* 24, 98–113.
Schmidt, B. A. (1979a) Ultrastructure of differentiated Malpighian tubules from cockroach nymphs during the molting cycle. *J. Morph.* 162, 361–388.
Schmidt, B. A. (1979b) Growth and differentiation of secondary Malpighian tubules in the cockroach *Periplaneta americana. J. Morph.* 162, 389–412.
Schmidt, K. (1969) Die companiformen Sensillen im Pedicellus der Florfliege (Chrysopa, Planipennia). *Z. Zellforsch.* 96, 478–489.
Schmidt, K. (1973) Vergleichende morphologische Untersuchungen an Mechanorezeptoren der Insekten. *Verh. Dtsch. Zool. Ges.*, Stuttgart 1972, 15–25.
Schmitt, J. B. (1954) The nervous system of the pregenital segments of some Orthoptera. *Ann. ent. Soc. Am.* 47, 677–682.
Schneiderman, H. A. (1960) Discontinuous respiration in insects: role of the spiracles. *Biol. Bull.* 119, 494–528.
Schneirla, T. C. (1953) Modifiability in insect behavior. In: *Insect Physiology*, Roeder, K. D., Ed., pp. 723–747. John Wiley & Sons, Inc., New York.
Schnorbus, H. (1971) Die Subgenualen Sinnesorgane von *Periplaneta americana*: Histologie und Vibrationsschwellen. *Z. vergl. Physiol.* 71, 14–48.
Schofield, P. K. (1979) Ionic permeability of the blood-brain barrier system of an insect *Carausius morosus. J. exp. Biol.* 82, 385–388.
Schofield, P. K. and Treherne, J. E. (1978) Kinetics of sodium and lithium movements across the blood-brain barrier of an insect. *J. exp. Biol.* 74, 239–251.
Schooley, D. A., Judy, K. J., Bergot, B. J., Hall, M. S. and Siddall, J. B. (1973) Biosynthesis of the juvenile hormone of *Manduca sexta*: Labeling pattern from mevalonate, propionate, and acetate. *Proc. natn. Acad. Sci., U.S.A.* 70, 2921–2925.
Schreuder, J. E. and de Wilde, J. (1952) Analysis of the dyspnoeic action of CO_2 in the cockroach, with some remarks on the honeybee. *Physiol. comp. Oecol.* 2, 355–361.
Schulz, H. and Schwarzberg, H. (1971) Possible control of endogenous nerve activity in the cockroach *Periplaneta americana*. In: *Neurobiology of Invertebrates*, Salanki, J., Ed., pp. 315, 318. Akadémiai Kaidó, Budapest.
Schwink, I. (1954) Experimentelle Untersuchungen neber Geruchssin und Stroemungswahrnehmung in der Orientierung bei Nachtschmetterlingen. *Z. vergl. Physiol.* 37, 19–56.
Seaman, G. R. and Clement, J. J. (1970) Tricarboxylic acid cycle enzymes in *Tetrahymena pyriformis* after infection of the cockroach, *Periplaneta americana. J. Protozool.* 17, 287–290.
Seaman, G. R. and Robert, N. L. (1968) Immunological response of male cockroaches to injection of *Tetrahymena pyriformis. Science* 161, 1359–1361.
Seaman, G. R. and Tosney, T. J. (1967) Alterations in morphology of *Tetrahymena pyriformis* S after facultative parasitism of the cockroach *Periplaneta americana. J. Protozool.* (*Suppl.*) 14, 23–24.
Seaman, G. R., Tosney, T., Berglund, R. and Goldberg, G. (1972) Infectivity and recovery of *Tetrahymena pyriformis* strain S from adult female cockroaches (*Periplaneta americana*). *J. Protozool.* 19, 644–647.
Seelinger, G. (1981) Behavioral responses of *Periplaneta americana* males to female sex pheromone extracts. (In preparation.)
Seelinger, U. (1979) *Die Mechanorezeptoren auf Scapus und Pedicellus von Periplaneta americana* (L.). Diplomarbeit beim Fachbereich Biologie der Universitat Regensburg.
Seidel, F. (1932) Die Potenzen der Furchungskerne in Libellenei und ihre Rolle bei der Aktivierung des Bildungszentrums. *Roux Arch.* 123, 213–276.
Seidel, F. (1936) Entwicklungsphysiologie des Insekten-Keims. *Verh. Deutsch. Zool. Ges.* 38, 291–336.

Seidel, F., Bock, E. and Krause, G. (1940) Die Organisation des Insekteneies. *Naturwissenschaft* 28, 433–446.
Seligman, I. M. and Doy, F. A. (1972) Studies on cyclic AMP mediation of hormonally induced cytolysis of the alary hypodermal cells and of hormonally controlled DOPA synthesis in *Lucilia cuprina Israel J. Ent.* 7, 129–142.
Selzer, R. (1979) Morphological and physiological identification of food-specific neurons in the deutocerebrum of *Periplaneta americana. J. comp. Physiol.* 134, 159–163.
Senff, R. E. (1966) The Electrophysiology of the Adult Heart of *Periplaneta americana*: Evidence for a Neurogenic Heart. PhD Thesis, Purdue University, Lafayette, Indiana, viii and 70 pp.
Seshan, K. R. and Levi-Montalcini, R. (1971) *In vitro* analysis of corpora cardiaca and corpora allata from nymphal and adult specimens of *Periplaneta americana. Arch. ital. Biol.* 109, 81–109.
Seshan, K. R. and Levi-Montalcini, R. (1973) Neuronal properties of nymphal and adult insect neurosecretory cells *in vitro. Science* 182, 291–293.
Seshan, K. R., Provine, R. R. and Levi-Montalcini, R. (1974) Structural and electrophysiological properties of nymphal and adult insect medial neurosecretory cells: An *in vitro* analysis. *Brain Res.* 78, 359–376.
Shaaya, E. (1978) Ecdysone and juvenile hormone activity in the larvae of the cockroach *Periplaneta americana. Insect Biochem.* 8, 193–195.
Shaaya, E. and Bodenstein, D. (1969). The function of the accessory sex glands in *Periplaneta americana* (L.). II. The role of the juvenile hormone in the synthesis of protein and protocatechuic acid glucoside. *J. exp. Zool.* 170, 281–292.
Shafer, S. C. and Mills, R. R. (1981) Cyclic AMP-dependant protein kinase during the ecdysial cycle of the American cockroach. (In preparation.)
Shah, J. and Bailey, E. (1976) Enzymes of ketogenesis in the fat body and the thoracic muscle of the adult cockroach. *Insect Biochem.* 6, 251–254.
Shambaugh, G. (1969) Toxicity and effects on motor co-ordination of some neurotropic drugs on the cockroach *Nauphoeta cinerea. Ann. ent. Soc. Am.* 62, 370–375.
Shankland, D. L. (1965) Nerves and muscles of the pregenital abdominal segments of the American cockroach, *Periplaneta americana* (L.). *J. Morph.* 117, 353–386.
Shankland, D. L. (1979), Action of dieldrin and related compounds on synaptic transmission. In: *Neurotoxicology of Insecticides and Pheromones*, Narahashi, T., Ed., pp. 139–153. Plenum Press, New York.
Shankland, D. L., Hollingworth, R. M. and Smyth, Jr, T. (Eds.) (1978) *Pesticide and Venom Neurotoxicity*. Plenum Press, New York.
Shankland, D. L., Rose, J. A. and Donninger, C. (1971) The cholinergic nature of the cercal nerve-giant fiber synapse in the sixth abdominal ganglion of the cockroach, *Periplaneta americana* (L.). *J. Neurobiol.* 2, 247–262.
Shankland, D. L. and Schroeder, M. E. (1973) Pharmacological evidence for a discrete neurotoxic action of dieldrin (HEOD) in the American cockroach, *Periplaneta americana* (L.). *Pest. Biochem. Physiol.* 3, 77–86.
Shapiro, E. (1976) Nerve–muscle interactions during limb regeneration in an insect. Diss. Abst. 37, 3225.
Shay, D. E. (1946) Observations on the cellular enclosures of the mid-gut epithelium of *Periplaneta americana. Ann. ent. Soc. Am.* 39, 165–169.
Shelton, P. M. J. (1979) Postembryonic determination of the ecdysial line in the cockroach: evidence for pattern regulation in the mediolateral axis. *J. Embryol exp. Morph.* 49, 27–46.
Shelton, P. M. J., Anderson, H. J. and Eley, S. (1977) Cell lineage and cell determination in the developing compound eye of the cockroach *Periplaneta americana. J. Embryol. exp. Morph.* 39, 235–252.
Shepard, M. and Keeley, L. L. (1972) Circadian rhythmicity and capacity for enforced activity

in the cockroach, *Blaberus discoidalis*, after cardiacectomy-allatectomy. *J. Insect Physiol.* 18, 595–601.
Shepheard, P. (1974) Control of head movement in the locust, *Schistocerca gregaria. J. exp. Biol.* 60, 735–767.
Shipp, E. and Otton, J. (1976) Irradiation-induced changes in circadian DDT-susceptibility and metabolic rhythms in the housefly. *Int. J. Chronob.* 4, 71–81.
Siakotos, A. N. (1960a) The conjugated plasma proteins of the American cockroach. I. The normal state. *J. gen. Physiol.* 43, 999–1013.
Siakotos, A. N. (1960b) The conjugated plasma proteins of the American cockroach. II. Changes during the molting and clotting processes. *J. gen. Physiol.* 43, 1015–1030.
Siegrist, S., Shafer, S. C. and Mills, R. R. (1981) *N*-acetyltransferase activity during the ecdysial cycle of the American cockroach. (In preparation.)
Silverman, J. M. (1977) Patterns of response to sex pheromone by young and mature adult male cockroaches, *Periplaneta americana. J. Insect Physiol.* 23, 1015–1019.
Silverman, J. M. (1978) Orientation behavior of the American cockroach. PhD Thesis, University of Kansas.
Silverman, J. M. and Bell, W. J. (1979) Role of strato and horizontal object orientation on mate finding and predator avoidance by the American cockroach. *Animal Behav.* 27, 652–657.
Simon, D. and Barth, R. H. (1977a) Sexual behavior in the cockroach genera *Periplaneta* and *Blatta*. I. Descriptive aspects. *Z. Tierpsychol.* 44, 80–107.
Simon, D. and Barth, R. H. (1977b) Sexual behavior in the cockroach genera *Periplaneta* and *Blatta*. II. Sex pheromones and behavioral responses. *Z. Tierpsychol.* 44, 162–177.
Simon, D. and Barth, R. H. (1977c) Sexual behavior in the cockroach genera *Periplaneta* and *Blatta*. III. Aggression and sexual behavior. *Z. Tierpsychol.* 44, 305–322.
Singer, G. and Norris, D. M. (1973) Comparative disc electrophoretic study of different proteinaceous extracts from *Periplaneta americana* males. *Comp. Biochem. Physiol.* 46B, 43–56.
Singh, S. P. and Das, A. B. (1978) Cytological and cytochemical alterations in fat body due to thermal acclimation of *Periplaneta americana* (Linn). *Comp. Physiol. Ecol.* 3, 187–194.
Skalsky, H. L. and Guthrie, F. E. (1975) Binding of insecticides to macromolecules in the blood of the rat and the American cockroach. *Pest. Biochem. Physiol.* 5, 27–34.
Smalley, K. N. (1963) The neural regulation of respiration in the cockroach, *Blaberus craniifer. Diss. Abst.* 24, 2629–2630.
Smalley, K. N. (1970) Median nerve neurosecretory cells in the abdominal ganglia of the cockroach, *Periplaneta americana. J. Insect Physiol.* 16, 241–250.
Smit, W. A., Becht, G. and Beenakkers, A. M. T. (1967) Structure, fatigue, and enzyme activities in 'fast' insect muscles. *J. Insect Physiol.* 13, 1857–1868.
Smith, D. S. (1966) The structure of intersegmental muscle fibers in an insect, *Periplaneta americana* (L.). *J. Cell Biol.* 29, 449–459.
Smith, D. S. (1968) *Insect cells: Their structure and function.* Oliver and Boyd, Edinburgh.
Smith, D. S. and Treherne, J. E. (1963) Functional aspects of the organization of the insect nervous system. *Adv. Insect Physiol.* 1, 401–484.
Smith, D. S. and Treherne, J. E. (1965) The electron microscope localization of cholinesterase activity in the central nervous system of an insect, *Periplaneta americana* (L.). *J. Cell Biol.* 26, 445–465.
Smith, N. A. (1969) Observations on the neural rhythmicity in the cockroach. In: *Comparative Physiology of the Heart: Current Trends*, McCann, F. V., Ed., pp. 200–205. Birkhäuser Verlag, Basel.
Smith, R. K. and House, C. R. (1977) Fluid secretion by isolated cockroach salivary glands. *Experientia* 33, 1182–1184.
Smyth, Jr, T., Greer, M. H. and Griffiths, D. J. G. (1973) Insect neuromuscular systems. *Am. Zool.* 13, 315–319.

Smyth, Jr, T., Ornberg, R. L. and Meyer, R. M. (1978) Actions of some neurotoxic proteins of black widow spider venom. In: *Pesticide and Venom Toxicity*, Shankland, D. L., Hollingworth, R. M. and Smyth, Jr, T., Eds., pp. 265–270. Plenum Press, New York.

Snipes, B. T. and Tauber, O. E. (1937) Time required for food passage through the alimentary tract of the cockroach *Periplaneta americana. Ann. ent. Soc. Am.* 30, 277–284.

Snodgrass, R. E. (1944) The feeding apparatus of biting and sucking insects affecting man and animals. *Smithsonian Misc. Coll.* 104 No. 7, 1–113.

Snyder, A. W. and Horridge, G. A. (1973) The optical function of changes in the medium surrounding the cockroach rhabdom. *J. comp. Physiol.* 81, 1–8.

Sokolove, P. G. (1975) Localization of the cockroach optic lobe circadian pacemaker with micro-lesions. *Brain Res.* 87, 13–22.

Solomon, J., Sandler, M. B., Cocchia, M. A. and Lawrence, A. (1977) Effect of environmental illumination on nymphal development, maturation rate and longevity of *Periplaneta americana. Ann. ent. Soc. Am.* 70, 409–413.

Sommer, E. W. and Wehner, R. (1975) The retina-lamina projection in the visual system of the bee, *Apis mellifera. Cell Tissue Res.* 163, 45–61.

Sommer, S. H. (1974) Aggregationsverhalten bei Schaben. *Angew. Parasitol.* 15, 10–30.

Sommer, S. H. (1975) Aggregation behavior of *Blattella germanica. Angew. Parasitol.* 16, 135–141.

Sommer, S. H. (1975) Experimentelle Untersuchungen zur circadianen lokomotorischen Aktivität von *Blattella germanica* L. (Dictyopt., Blattellidae). *Biol. Zentralb.* 94, 455–467.

Sonan, H. (1924) Observations upon *Periplaneta americana* Linnaeus, and *Periplaneta australasiae* Fabricius. *Trans. Nat. Hist. Soc. Formosa* 14, 4–21. (In Japanese.)

Sowa, B. A. and Borg, T. K. (1975) Density gradient centrifugation isolation of hormone-containing neurosecretory granules from the cockroach *Leucophaea maderae. J. Insect Physiol.* 21, 511–516.

Sowa, B. A. and Marks, E. P. (1975) An *in vitro* system for the quantitative measurement of chitin synthesis in the cockroach: Inhibition by TH-6040 and Polyoxin D. *Insect Biochem.* 5, 855–859.

Spencer, H. J. (1974) Analysis of the electrophysiological responses of the trochanteral hair receptors in the cockroach. *J. exp. Biol.* 60, 233–240.

Spinola, S. M. and Chapman, K. M. (1975) Proprioceptive indentation of the campaniform sensilla of cockroach legs. *J. comp. Physiol.* 96, 257–272.

Spira, M. E., Parnas, I. and Bergmann, F. (1969) Histological and electrophysiological studies on the giant axons of the cockroach, *Periplaneta americana. J. exp. Biol.* 50, 629–634.

Spring, J. H., Matthews, J. R. and Downer, R. G. H. (1977) Fate of glucose in haemolymph of the American cockroach, *Periplaneta americana. J. Insect Physiol.* 23, 525–529.

Srivastava, B. B. L. and Hopkins, T. L. (1975) Bursicon release and activity in haemolymph during metamorphosis of the cockroach, *Leucophaea maderae. J. Insect Physiol.* 21, 1985–1993.

Srivastava, J. P. and Saxena, S. C. (1967) On the alkaline and acid phosphatase in the alimentary tract of *Periplaneta americana* L. (Blattaria: Blattidae) *Appl. ent. Zool.* 2, 85–92.

Srivastava, R. P. (1962) On the nature of the cell extrusions from the epithelium in midgut and hepatic caeca of certain insects. *Proc. natn. Acad. Sci., Ind.* 32, 65–71.

Starratt, A. N. (1979) Proctolin, an insect neuropeptide. *Trends Neurosci.* 2, 15–17.

Starratt, A. N. and Brown, B. E. (1975) Structure of the pentapeptide proctolin, a proposed neurotransmitter in insects. *Life Sci.* 17, 1253–1256.

Starratt, A. N. and Brown, B. E. (1977) Synthesis of proctolin, a pharmacologically active pentapeptide in insects. *Can. J. Chem.* 55, 4238–4242.

Stay, B. and Clark, J. K. (1971) Fluctuation of protein granules in the fat body of the viviparous cockroach, *Diploptera punctata*, during the reproductive cycle. *J. Insect Physiol.* 17, 1747–1762.

Stay, B., Friedel, T., Feyereisen, R. and Tobe, S. S. (1980a) Ecdysteroids in ovary and hemolymph of the cockroach, *Diploptera punctata. Proc. XVI Int. Cong. Ent.*, Kyoto, 1980.

Stay, B., Friedel, T. Tobe, S. S. and Mundall, E. C. (1980b) Feedback control of juvenile hormone synthesis in cockroaches: Possible role for ecdysterone. *Science* 207, 898–900.

Stay, B., King, A. and Roth, L. M. (1960) Calcium oxalate in the oöthecae of cockroaches. *Ann. ent. Soc. Am.* 53, 79–86.

Stay, B. and Roth, L. M. (1962) The colleterial glands of cockroaches. *Ann. ent. Soc. Am.* 55, 124–130.

Stay, B. and Tobe, S. S. (1977) Control of juvenile hormone biosynthesis during the reproductive cycle of a viviparous cockroach. I. Activation and inhibition. *Gen. comp. Endocrinol.* 33, 531–540.

Stay, B. and Tobe, S. S. (1978) Control of juvenile hormone biosynthesis during the reproductive cycle of a viviparous cockroach. II. Effects of unilateral allatectomy, implantation of supernumerary corpora allata, and ovariectomy. *Gen. comp. Endocrinol.* 34, 276–286.

Steel, C. G. H. (1973) Humoral regulation of the cerebral neurosecretory system of *Rhodnius prolixus* (Stal.) during growth and moulting. *J. exp. Biol.* 58, 177–187.

Steel, C. G. H. and Morris, G. P. (1977) A simple technique for selective staining of neurosecretory products in epoxy sections with paraldehyde-fuchsin. *Can. J. Zool.* 55, 1571–1575.

Steele, J. E. (1961) Occurrence of a hyperglycaemic factor in the corpus cardiacum of an insect. *Nature* 192, 680–681.

Steele, J. E. (1963) The site of action of insect hyperglycaemic hormone. *Gen. comp. Endocrinol.* 3, 46–52.

Steele, J. E. (1964) The activation of phosphorylase in an insect by adenosine 3:5′-phosphate and other agents. *Am. Zool.* 4, 328.

Steele, J. E. (1969) A relationship between ionic environment and hormonal activation of phosphorylase in an insect. *Comp. Biochem. Physiol.* 29, 755–763.

Steele, J. E. (1976) Hormonal control of metabolism in insects. In: *Advances in Insect Physiology*. Treherne, J. E., Berridge, M. J. and Wigglesworth, V. B., Eds., Vol. 12, pp. 239–323. Academic Press, London.

Steinhaus, E. A. and Marsh, G. A. (1962) Report of diagnoses of diseased insects 1951–1961. Hilgardia 33, 349–490.

Sternburg, J. G. (1963) Autointoxication and some stress phenomena. *A. Rev. Ent.* 8, 19–38.

Sternburg, J. G. and Kearns, C. W. (1952) The presence of toxins other than DDT in the blood of DDT-poisoned roaches. *Science* 116, 144–147.

Still, W. C. (1979) (±)-Periplanone-B. Total synthesis and structure of the sex-excitant pheromone of the American cockroach. *J. Am. chem. Soc.* 101, 2493–2495.

Stobbart, R. H. and Shaw, J. (1974) Salt and water balance. In: *The Physiology of Insecta*, 2nd edition, Rockstein, M., Ed., Vol. V, pp. 361–446. Academic Press, New York.

Stock, A. and O'Farrell, A. F. (1954) Regeneration and the molting cycle in *Blattella germanica* (L.). II. Simultaneous regeneration of both metathoracic legs. *Aust. J. biol. Sci.* 7, 302–307.

Stokes, D. R., Vitale, A. J. and Morgan, C. R. (1979) Enzyme histochemistry of the mesocoxal muscles of *Periplaneta americana. Cell Tissue Res.* 198, 175–189.

Stone, J. V., Mordue, W., Batley, K. E. and Morris, H. R. (1976) Structure of locust adipokinetic hormone, a neurohormone that regulates lipid utilisation during flight. *Nature* 263, 207–211.

Storch, R. H. and Chadwick, L. E. (1967) The embryonal and nymphal cervicothoracic musculature of the American cockroach, *Periplaneta americana* (Blattaria: Blattidae). *Can. Ent.* 99, 113–145.

Storey, K. B. and Bailey, E. (1978a) The intracellular distribution of enzymes of carbohydrate degradation in the fat body of the adult male cockroach. *Insect Biochem.* 8, 75–81.

Storey, K. B. and Bailey, E. (1978b) Intracellular distribution of enzymes associated with lipogenesis and gluconeogenesis in fat body of the adult cockroach, *Periplaneta. Insect Biochem.* 8, 125–131.

Subramoniam, T. (1973) Incorporation of ^{14}C-leucine into the fat body of the cockroach *Periplaneta americana* during oocyte development. *J. Insect Physiol.* 19, 2209–2213.

Sudarsanam, B. P. (1980) *Vitellogenin Synthesis and Uptake by the Ovary in Periplaneta americana.* PhD Thesis, University of Kansas.

Sullivan, W. N., Cawley, B., Hayes, D. K., Rosenthal, J. and Halberg, F. (1970) Circadian rhythm in susceptibility of house flies and Madeira cockroaches to pyrethrum. *J. econ. Ent.* 63, 159–163.

Sutherland, D. J. (1969) Nerve severance and tumor induction in *Periplaneta americana* (L.). *Natn. Cancer Inst. Mono.* 31, 399–418.

Sutherland, D. J. and Chillseyzn, J. M. (1968) Function and operation of the cockroach salivary reservoir. *J. Insect Physiol.* 14, 21–31.

Suzuki, H. (1975) Antennal movements induced by odour and central projection of the antennal neurones in the honey-bee. *J. Insect Physiol.* 21, 831–847.

Svoboda, J. A., Kaplanis, J. N., Robbins, W. E. and Thompson, M. J. (1975) Recent developments in insect steroid metabolism. *A. Rev. Ent.* 20, 205–220.

Svoboda, J. A. and Robbins, W. E. (1971) The inhibitive effects of azasterols on sterol metabolism and growth and development in insects with special reference to the tobacco hornworm. *Lipids* 6, 113–119.

Szymanski, J. S. (1912) Modification of the innate behaviour of cockroaches. *Animal Behav.* 2, 81–90.

Takagi, M. (1977) The aggregation of cockroaches. *Iden* 31, 36–41. (In Japanese.)

Takahashi, R. (1924) Life-history of Blattidae. *Dobutsugaku Zasshi, Zool. Mag. Tokyo* 36, 215–230. (In Japanese.)

Takahashi, S. and Kitamura, C. (1972) Bioassay procedure of the sex-stimulant of the American cockroach, *Periplaneta americana* (Orthoptera: Blattidae). *Appl. Ent. Zool.* 7, 133–141.

Takahashi, S. and Kitamura, C. (1976) Role of sex pheromone in mating behavior of the cockroaches. *Proc. Symp. Insect Pheromones and their Applications*, pp. 77–88. Nagaoka & Tokyo, Japan.

Takashashi, S., Kitamura, C. and Horibe, I. (1978) Sex stimulant activity of sesquiterpenes to the males of the American cockroach. *Agric. biol. Chem.* 42, 79–82.

Talman, E., Verwiel, P. E. J., Ritter, F. J. and Persoons, C. J. (1978) The female sex pheromones of the American cockroach, *Periplaneta americana. Israel J. Chem.* 17, 227–235.

Tartivita, K. and Jackson, L. L. (1970) Cuticular lipids of insects. 1. Hydrocarbons of *Leucophaea maderae* and *Blatta orientalis. Lipids* 5, 35–37.

Tarver, R. V. and Pierre, L. L. (1967) Activity of malic dehydrogenase of the symbionts and fat bodies of a cockroach. *Nature* 213, 208–209.

Taylor, N. and Mills, R. R. (1976) Evidence for the presence of mandelic acid dehydrogenase and benzoyl formate decarboxylase in the American cockroach. *Insect Biochem.* 6, 85–87.

Teutsch-Felber, D. (1970) Experimentalle und histologische Untersuchungen an der thoraxmuskulatur von *Periplaneta americana* (L.). *Rev. Suisse Zool.* 77, 481–523.

Thomas, A. and Mesnier, M. (1973) Le rôle du systéme nerveux central sur les mécanismes de l'oviposition chez *Carausius morosus* et *Clitumnus extradentatus. J. Insect Physiol.* 19, 383–396.

Thomas, K. K. and Nation, J. L. (1966) Control of a sex-linked haemolymph protein by corpora allata during ovarian development in *Periplaneta americana. Biol. Bull.* 130, 254–265.

Thompson, M. J., Robbins, W. E., Dutky, S. R., Marks, E. P., Filipi, P. A. and Finegold, H. (1978) Isolation and identification of the metabolites of 22,25-dideoxyecdysone from cockroach fat body cultures. *Lipids* 13, 783–790.

Thurm, U. (1974) Mechanisms of electrical membrane responses in sensory receptors, illustrated by mechanoreceptors. In: *Biochemistry of Sensory Functions*, Jaenicke, L., Ed., pp. 367–390. Springer-Verlag, Heidelburg.

Tietz, A. and Weintraub, H. (1978) Hydrolysis of glycerides by lipases of the fat body of the locust, *Locustra migratoria. Insect Biochem.* 8, 11–16.

Tobe, S. S. (1980) Regulation of the corpora allata in adult female insects. In: *Insect Biology in the Future*, Locke, M. and Smith, D. S., Eds. Academic Press, New York, pp. 345–367.

Tobe, S. S., Musters, A. and Stay, B. (1979) Corpus allatum function during sexual maturation of male *Diploptera punctata. Physiol. Ent.* 4, 79–86.

Tobe, S. S. and Pratt, G. E. (1974a) The influence of substrate concentrations on the rate of insect juvenile hormone biosynthesis by corpora allata of the desert locust *in vitro. Biochem. J.* 144, 107–113.

Tobe, S. S. and Pratt, G. E. (1974b) Dependence of juvenile hormone release from corpus allatum on intraglandular content. *Nature* 252, 474–476.

Tobe, S. S. and Pratt, G. E. (1975) Corpus allatum activity *in vitro* during ovarian maturation in the desert locust, *Schistocerca gregaria. J. exp. Biol.* 62, 611–627.

Tobe, S. S. and Pratt, G. E. (1976) Farnesenic acid stimulation of juvenile hormone biosynthesis as an experimental probe in corpus allatum physiology. In: *The Juvenile Hormones*, Gilbert, L. I., Ed., pp. 147–163. Plenum Press, New York.

Tobe, S. S. and Saleuddin, A. S. M. (1977) Ultrastructural localization of juvenile hormone biosynthesis by insect corpora allata. *Cell Tissue Res.* 183, 25–32.

Tobe, S. S. and Stay, B. (1977) Corpus allatum activity *in vitro* during the reproductive cycle of the viviparous cockroach, *Diploptera punctata* (Eschscholtz). *Gen. comp. Endocrinol.* 31, 138–147.

Tobe, S. S. and Stay, B. (1979) Modulation of juvenile hormone synthesis by an analogue in the cockroach. *Nature* 281, 481–482.

Tobe, S. S. and Stay, B. (1980) Control of juvenile hormone biosynthesis during the reproductive cycle of a viviparous cockroach. III. Effects of denervation and age on compensation with unilateral allatectomy and supernumerary corpora allata. *Gen. comp. Endocrinol.* 40, 89–98.

Tobias, J. M. (1948) Potassium, sodium and water interchange in irritable tissues and haemolymph of an omnivorous insect, *Periplaneta americana. J. cell. comp. Physiol.* 31, 125–142.

Tobin, T. (1981) Responses of the American cockroach to Periplanone-B in a wind tunnel. *J. Insect Physiol.* (In press.)

Tobin, T., Seelinger, G. and Bell, W. J. (1981) Comparison of responses elicited by Periplanone-B and extract of the American cockroach sex pheromone. *J. Chem. Ecol.* (In press.)

Toh, Y. (1977) Fine structure of antennal sense organs of the male cockroach, *Periplaneta americana. J. ultrastruct. Res.* 60, 373–394.

Tolman, J. H. and Steele, J. E. (1976) An ouabain-sensitive (Na^+-K^+)-activated ATPase in the rectal epithelium of the American cockroach, *Periplaneta americana. Insect Biochem.* 6, 513–517.

Tourtellot, M. K. and Franklin, R. F. (1978) Movements of appendages studied with an electromagnetic induction micro-activity meter. *Animal Behav. Soc. Mtg.*, Seattle (Abst.).

Traina, M. E., Bellino, M., Serpietri, L., Massa, A. and Frontali, N. (1976) Heart-accelerating peptides from cockroach corpora cardiaca. *J. Insect Physiol.* 22, 323–329.

Trautmann, K. H. (1972) *In vitro* Studium der Trägerproteine von ³H-markierten juvenilhormonwirksamen Verbindungen in der Hämolymphe von *Tenebrio molitor* L. larven. *Z. Naturforsch.* 27B, 263–273.

Treherne, J. E. (1957) Glucose absorption in the cockroach. *J. exp. Biol.* 34, 478–485.

Treherne, J. E. (1958a) The absorption of glucose from the alimentary canal of the locust *Schistocerca gregaria* (Forsk.) *J. exp. Biol.* 35, 297–306.

Treherne, J. E. (1958b) The absorption and metabolism of some sugars in the locust *Schistocerca gregaria* (Forsk.) *J. exp. Biol.* 35, 611–625.

Treherne, J. E. (1958c) The digestion and absorption of tripalmitin in the cockroach *Periplaneta americana*. *J. exp. Biol.* 35, 862–870.

Treherne, J. E. (1959) Amino acid absorption in the locust (*Schistocerca gregaria* Forsk.) *J. exp. Biol.* 36, 533–545.

Treherne, J. E. (1960) The nutrition of the central nervous system in the cockroach, *Periplaneta americana* L. The exchange and metabolism of sugars. *J. exp. Biol.* 37, 513–533.

Treherne, J. E. (1961a) Sodium and potassium fluxes in the abdominal nerve cord of the cockroach, *Periplaneta americana* L. *J. exp. Biol.* 38, 315–322.

Treherne, J. E. (1961b) The movements of sodium ions in the isolated abdominal nerve cord of the cockroach, *Periplaneta americana*. *J. exp. Biol.* 38, 629–636.

Treherne, J. E. (1965) The chemical environment of the insect central nervous system. In: *The Physiology of the Insect Central Nervous System*, Treherne, J. E. and Beament, J. W. L., Eds., pp. 21–29. Academic Press, New York.

Treherne, J. E. (1974) The environment and function of insect nerve cells. In: *Insect Neurobiology*, Treherne, J. E., Ed., pp. 187–244. Elsevier, New York.

Treherne, J. E., Buchan, P. B. and Bennett, R. R. (1975) Sodium activity of insect blood; physiological significance and relevance to the design of physiological saline. *J. exp. Biol.* 62, 721–732.

Treherne, J. E., Lane, N. J., Moreton, R. B. and Pichon, Y. (1970) A quantitative study of potassium movements in the central nervous system of *Periplaneta americana*. *J. exp. Biol.* 53, 109–136.

Treherne, J. E. and Moreton, R. B. (1970) The environment and function of invertebrate nerve cells. *Int. Rev. Cytol.* 28, 45–88.

Treherne, J. E., and Schofield, J. E. (1979) Ionic homeostasis of the brain microenvironment in insects. *Trends Neurosci.* 2, 227–230.

Treherne, J. E. and Willmer, P. G. (1975) Hormonal control of integumentary water loss: evidence for a novel neuroendocrine control system in an insect, *Periplaneta americana*. *J. exp. Biol.* 63, 143–159.

Truman, J. W. (1976) Extraretinal photoreception in insects. *Photochem. Photobiol.* 23, 215–225.

Truman, J. W. (1979) Interaction between abdominal ganglia during the performance of hormonally triggered behavioral programmes in moths. *J. exp. Biol.* 83, 239–253.

Truman, J. W. and Sokolove, P. G. (1972) Silkworm eclosion: Hormonal triggering of a centrally programmed pattern of behavior. *Science* 175, 1491–1493.

Tucker, F. and Pichon, Y. (1972) Sodium efflux from the central nervous connectives of the cockroach. *J. exp. Biol.* 56, 441–458.

Tucker, L. E. (1977a) Effect of dehydration and rehydration on the water content and Na^+ and K^+ balance in the adult male *Periplaneta americana*. *J. exp. Biol.* 71, 49–66.

Tucker, L. E. (1977b) The influence of diet, age and state of hydration on Na^+, K^+ and urate balance in the fat body of the cockroach, *Periplaneta americana*. *J. exp. Biol.* 71, 67–79.

Tucker, L. E. (1977c) The influence of age, diet and lipid content on survival, water balance and Na^+ and K^+ regulation in dehydrating cockroaches. *J. exp. Biol.* 71, 81–93.

Tucker, L. E. (1977d) Regulation of ions in the hemolymph of the cockroach, *Periplaneta americana*, during dehydration and rehydration. *J. exp. Biol.* 71, 95–110.

Turner, C. H. (1912) An experimental investigation of an apparent reversal of the responses to light of the roach (*Periplaneta orientalis* L.). *Biol. Bull.* 23, 371–386.
Turner, R. B. and Acree, Jr, F. (1967) The effect of photoperiod on the daily fluctuation of haemolymph hydrocarbons in the American cockroach. *J. Insect Physiol.* 13, 519–522.
Turnquist, R. L. and Brindley, W. A. (1975) Microsomal oxidase activities in relation to age and chlorocyclizine induction in American cockroach, *Periplaneta americana*, fat body, midgut and hindgut. *Pest Biochem. Physiol.* 5, 211–220.
Twarog, B. M. and Roeder, K. D. (1957) Pharmacological observations on the desheathed last abdominal ganglion of the cockroach. *Ann. Ent. Soc. Am.* 50, 231–237.
Tweedle, C. D., Pitman, R. M. and Cohen, M. J. (1973) Dendritic stability of insect central neurons subjected to axotomy and de-afferentation. *Brain Res.* 60, 471–476.
Tyrer, N. M. (1971) Innervation of the abdominal intersegmental muscles in the grasshopper. II. Physiological analysis. *J. exp. Biol.* 55, 315–324.
Tyrer, N. M., Bacon, J. P. and Davies, C. A. (1979) Sensory projections from the wind-sensitive head hairs of the locust *Schistocerca gregaria*. Distribution in the central nervous system. *Cell Tissue Res.* 203, 79–92.
Tyrer, N. M. and Bell, E. M. (1974) The intensification of cobalt-filled neurone profiles using a modification of Timm's sulphide-silver method. *Brain Res.* 73, 151–155.
Tyrer, N. M. and Johnson, K. A. (1977) Does electrophoresis of cockroach muscle proteins detect recognition molecules? *Nature* 268, 759–761.
Ude, J. and Agricola, H. (1979) Synaptic connections of the nervus connectivus in the frontal ganglion of *Periplaneta americana* L. (Insecta): An electron microscopic and iontophoretic study. *Cell Tissue Res.* 204, 155–159.
Unger, R. H. (1957) Untersuchungen zur neurohormonalen Steuerung der Herztätigkeit bei Schaben (*Periplaneta orientalis*, *P. americana* und *Phyllodromia germanica*). *Biol. Centralbl.* 76, 204–225.
Urvoy, J. (1970) Study of leg grafts with intact connection to the corresponding nerve center in the place of antennae in *Blabera craniifer* Burm. *Ann. embryol. Morph.* 3, 193–196.
Usherwood, P. N. R. (1962a) The nature of 'slow' and 'fast' contractions in the coxal muscles of the cockroach. *J. Insect Physiol.* 8, 31–52.
Usherwood, P. N. R. (1962b) The action of the alkaloid ryanodine on insect skeletal muscles. *Comp. Biochem. Physiol.* 6, 181–199.
Usherwood, P. N. R. (1963) Spontaneous miniature potentials from insect skeletal muscle fibres. *J. Physiol.* 169, 149–160.
Usherwood, P. N. R. (1967) Insect neuromuscular mechanisms. *Am. Zool.* 7, 553–582.
Usherwood, P. N. R. (1968) A critical study of the evidence for peripheral inhibitory axons in insects. *J. exp. Biol.* 49, 201–222.
Usherwood, P. N. R. (1969a) Electrochemistry of insect muscle. *Adv. Insect Physiol.* 6, 205–278.
Usherwood, P. N. R. (1969b) Glutamate sensitivity of denervated insect muscle fibres. *Nature* 223, 411–413.
Usherwood, P. N. R. (1972) Transmitter release from insect excitatory motor nerve terminals. *J. Physiol.* 227, 527–552.
Usherwood, P. N. R. (1974) Nerve–muscle transmission. In: *Insect Neurobiology*, Treherne, J. E., Ed., pp. 245–305. Elsevier, New York.
Usherwood, P. N. R., Ed. (1975) *Insect Muscle*. Academic Press, London.
Usherwood, P. N. R. (1978) Amino acids as neurotransmitters. *Adv. comp. Physiol. Biochem.* 7, 227–309.
Usherwood, P. N. R. and Cull-Candy, S. G. (1975) Pharmacology of somatic nerve–muscle synapses. In: *Insect Muscle*, Usherwood, P. N. R., Ed., pp. 207–280. Academic Press, London.
Usherwood, P. N. R. and Grundfest, H. (1965) Peripheral inhibition in skeletal muscle of insects. *J. Neurophysiol.* 58, 497–518.

Usherwood, P. N. R. and Machili, P. (1968) Pharmacological properties of excitatory neuromuscular synapses in the locust. *J. exp. Biol.* 49, 341–361.
Usherwood, P. N. R., Machili, P. and Leaf, G. (1968) L-Glutamate at insect excitatory nerve-muscle synapses. *Nature* 219, 1169–1172.
Van Asperen, K. and Van Esch, I. (1956) The chemical composition of the haemolymph in *Periplaneta americana* with special reference to the mineral constituents. *Arch. Neerl. Zool.* 11, 342–360.
Van Cassel, M. (1968) Contribution à l'étude du rôle des yeaux et des ocelles dans la prise et le maintien du rhythme d'activité locomotrice circadien chez *Periplaneta americana. Ann. Epiphyt.* 19, 159–164.
Vandenberg, R. D. and Mills, R. R. (1974) Hormonal control of tanning by the American cockroach: Cyclic AMP as a probable intermediate. *J. Insect Physiol.* 20, 623–627.
Vandenberg, R. D. and Mills, R. R. (1975) Adenyl cyclase in the haemocytes of the American cockroach. *J. Insect Physiol.* 21, 221–229.
Vandenheuvel, W. J. A., Robbins, W. E., Kaplanis, J. N., Louloudes, S. J. and Horning, E. C. (1962) The major sterol from cholesterol-fed American cockroaches. *Ann. ent. Soc. Am.* 55, 723.
Van der Driessche, T. (1975) Circadian rhythms and molecular biology. *Biosystems* 6, 188–201.
Van Handel, E. (1978) Trehalose turnover and the co-existence of trehalose and active trahalase in cockroach haemolymph. *J. Insect Physiol.* 24, 151–154.
Van Herrewege, C. (1971) Consommation alimentaire chez les mâles adultes de *Blattella germanica* (L.): influence de l'âge, de la nourriture larvaire et du jeûne. *Arch. Sci. Physiol.* 25, 401–413.
Vardanis, A. (1963) Glycogen synthesis in the insect fat body. *Biochem. biophys. Acta* 73, 565–573.
Vea, E. V., Cutkomp, L. K. and Halberg, F. (1977) Interrelationships of oxygen consumption and insecticide sensitivity. *Chronobiologia* 4, 313–323.
Verrett, J. M. and Mills, R. R. (1973) Water balance during vitellogenesis by the American cockroach: translocation of water during the cycle. *J. Insect Physiol.* 19, 1889–1901.
Verrett, J. M. and Mills, R. R. (1975a) Water balance during vitellogenesis by the American cockroach: hydration of the oocytes. *J. Insect Physiol.* 21, 1061–1064.
Verrett, J. M. and Mills, R. R. (1975b) Water balance during vitellogenesis by the American cockroach: distribution of water during the six-day cycle. *J. Insect Physiol.* 21, 1841–1845.
Verrett, J. M. and Mills, R. R. (1976) Water balance during vitellogenesis by the American cockroach: effect of frontal ganglionectomy. *J. Insect Physiol.* 22, 251–257.
Vijayalekshmi, V. R. (1976) *Some Aspects of the Reproductive Physiology of Male Cockroaches.* PhD Thesis, Calicut University.
Vijayalekshmi, V. R. and Adiyodi, K. G. (1973a) Accessory sex glands of male *Periplaneta americana* (L.). I. Qualitative analysis of some non-enzymic components. *Ind. J. exp. Biol.* 11, 512–514.
Vijayalekshmi, V. R. and Adiyodi, K. G. (1973b) Accessory sex glands and their secretions in male *Periplaneta americana* (L.). II. Secretory behaviour and maturation of the mushroom gland complex. *Ind. J. exp. Biol.* 11, 515–520.
Vijayalekshmi, V. R. and Adiyodi, K. G. (1973c) Accessory sex glands of male *Periplaneta americana* (L.). III. Histochemistry of the mushroom-shaped and conglobate glands. *Ind. J. exp. Biol.* 11, 521–524.
Vijayalakshmi, S., Kumar, T. P. and Babu, K. S. (1978) Rhythmicity of amino transferase in the cockroach, *Periplaneta americana. Experientia* 34, 192–193.
Vijayalakshmi, S., Mohan, P. M. and Babu, K. S. (1977) Circadian rhythmicity in the nervous system of the cockroach *Periplaneta americana. J. Insect Physiol.* 23, 195–202.
Vilchez, C. A., de Vilchez, I. S. and Lobarbo, S. (1975) Relative value of light and food as synchronizers of liver phosphorylase circadian rhythm. *Chronobiologia* 2, 145–152.

Vroman, H. E., Kaplanis, J. N. and Robbins, W. E. (1964) Cholesterol turnover in the American cockroach, *Periplaneta americana* L. *J. Lipid Res.* 5, 418–421.

Vroman, H. E., Kaplanis, J. N. and Robbins, W. E. (1965) Effect of allatectomy on lipid biosynthesis and turnover in the female American cockroach, *Periplaneta americana* (L.). *J. Insect Physiol.* 11, 897–904.

Wade, N. (1978) Send not to know for whom the Nobel tolls: it's not for thee. *Science* 202, 295–296.

Walcott, B. (1975) Anatomical changes during light adaptation in insect compound eyes. In: *The Compound Eye and Vision of Insects*, Horridge, G. A., Ed., pp. 20–36. Clarendon Press, London.

Waldow, U. (1975) Multimodale Neurone im Deutocerebrum von *Periplaneta americana. J. comp. Physiol.* A. 101, 329–341.

Waldow, U. (1977) CNS units in the cockroach (*Periplaneta americana*): Specificity of response to pheromones and other stimuli. *J. comp. Physiol.* 116, 1–17.

Walker, P. A. (1965) The structure of the fat body in normal and starved cockroaches as seen with the electron microscope. *J. Insect Physiol.* 11, 1625–1631.

Wall, B. J. (1965) Regulation of water metabolism by the Malpighian tubules and rectum in the cockroach, *Periplaneta americana* (L.). *Zool Jahrb. Physiol.* 71, 702–709.

Wall, B. J. (1967) Evidence for antidiuretic control of rectal water absorption in the cockroach, *Periplaneta americana* (L.). *J. Insect Physiol.* 13, 565–578.

Wall, B. J. (1970) Effects of dehydration and rehydration on *Periplaneta americana. J. Insect Physiol.* 16, 1027–1042.

Wall, B. J. (1971) Local osmotic gradients in the rectal pads of an insect. *Fedn Proc. fedn Am. Socs exp. Biol.* 30, 42–48.

Wall, B. J. (1977) Fluid transport in the cockroach rectum. In: *Transport of Ions and Water in Animals*, Gupta, B. L., Moreton, R. B., Oschman, J. L. and Wall, B. J., Eds. pp. 599–612. Academic Press, London.

Wall, B. J. and Oschman, J. L. (1970) Water and solute uptake by rectal pads of *Periplaneta americana. Am. J. Physiol.* 218, 1208–1215.

Wall, B. J., Oschman, J. L. and Schmidt, B. A. (1975) Morphology and function of Malpighian tubules and associated structures in the cockroach, *Periplaneta americana. J. Morph.* 146, 265–306.

Wall, B. J. and Ralph, C. L. (1962) Responses of specific neurosecretory cells of the cockroach, *Blaberus giganteus* to dehydration. *Biol. Bull.* 122, 431–438.

Wall, B. J. and Ralph, C. L. (1964) Evidence for humoral regulation of Malpighian tubule excretion in the insect, *Periplaneta americana* L. *Gen. comp. Endocrinol.* 4, 452–456.

Wall, B. J. and Ralph, C. L. (1965) Possible regulation of rectal water resorption in the insect, *Periplaneta americana* (L.). *Am. Zool.* 5, 211.

Walther, J. B. (1958a) Untersuchungen am Belichtungspotential des Komplexauges von *Periplaneta*, mit farbigen Reizen und selektiver Adaptation. *Biol. Zentralblat* 77, 63–104.

Walther, J. B. (1958b) Changes in spectral sensitivity and form of retinal action potential of the cockroach eye by selective adaptation. *J. Insect Physiol.* 2, 142–157.

Walther, J. B. and Dodt, E. (1959) Die spektrale Sensitivitat von Insekten-Komplexaugen im Ultraviolett bis 290 nm. *Z. Naturforsch* 14b, 273–278.

Wang, C. M., Narahashi, T. and Yamada, M. (1971) The neurotoxic action of dieldrin and its derivatives in the cockroach. *Pest. Biochem. Physiol.* 1, 84–91.

Wareham, A. C., Duncan, C. J. and Bowler, K. (1974a) The resting potential of cockroach muscle membrane. *Comp. Biochem. Physiol.* 48A, 765–797.

Wareham, A. C., Duncan, C. J. and Bowler, K. (1974b) Electrogenesis in cockroach muscle. *Comp. Biochem. Physiol.* 48A, 799–813.

Warhurst, D. C. (1963) The biology of *Endolimax blattae* Lucas, an endocommensal of cockroaches. *J. Protozool.* 10 (Suppl.), 28.

Warhurst, D. (1964) *Growth and Survival, in vitro and in vivo, of Endolimax blattae, an Entozoic Amoeba of Cockroaches.* PhD Thesis, University of Leicester.
Warhurst, D. C. (1966) A note on *Polymastix* Bütschli, 1884, from the colon of *Periplaneta americana* (Insecta: Dictyoptera). *Parasitology* 56, 21–23.
Warhurst, D. C. (1967) Cultivation *in vitro* of *Endolimax blattae* Lucas, 1927 from the cockroach hind gut. *Parasitology* 57, 181–187.
Washio, H. M. and Inouye, S. T. (1975) The mode of spontaneous transmitter release at the insect neuromuscular junction. *Can. J. physiol. Pharmacol.* 53, 679–682.
Washio, H. M. and Inouye, S. T. (1978) The effect of calcium and magnesium on the spontaneous release of transmitter at insect motor nerve terminals. *J. exp. Biol.* 75, 101–112.
Washio, H. and Nishino, C. (1976) Electro-antennogram responses to the sex pheromone and other odours in the American cockroach. *J. Insect Physiol.* 22, 735–741.
Waterhouse, D. F. (1957) Digestion in insects. *A. Rev. Ent.* 2, 1–18.
Waterhouse, D. F. and McKellar, J. W. (1961) The distribution of chitinase activity in the body of the American cockroach. *J. Insect Physiol.* 6, 185–195.
Weaver, R. J. (1979) Comparable activities of left and right corpora allata consistent with humoral control of juvenile hormone biosynthesis in the cockroach. *Can. J. Zool.* 57, 343–345.
Weaver, R. J. and Pratt, G. E. (1976) Biosynthetic activity of the corpora allata related to reproductive development in adult female *Periplaneta americana. Gen. comp. Endocrinol.* 29, 250–251.
Weaver, R. J. and Pratt, G. E. (1977) The effect of enforced virginity and subsequent mating on the activity of the corpus allatum of *Periplaneta americana* measured *in vitro*, as related to changes in the rate of ovarian maturation. *Physiol. Ent.* 2, 59–76.
Weaver, R. J., Pratt, G. E. and Finney, J. R. (1975) Cyclic activity of the corpus allatum related to gonotrophic cycles in adult female *Periplaneta americana. Experientia* 31, 597–598.
Weber, C. and Renner, M. (1976) The ocellus of the cockroach *Periplaneta americana* (Blattaria): receptory area. *Cell Tissue Res.* 168, 209–222.
Weber, K., Pringle, J. R. and Osborn, M. (1972) Measurements of molecular weights by electrophoresis on SDS-acrylamide gel. *Meth. Enzymol.* 26, 3–27.
Weidler, D. J. and Sieck, G. C. (1977) A study of ion binding in the hemolymph of *Periplaneta americana. Comp. Biochem. Physiol.* 56A, 11–14.
Weirich, G. and Wren, J. (1976) Juvenile-hormone esterase in insect development: a comparative study. *Physiol. Zool.* 49, 341–350.
Weis-Fogh, T. (1967) Respiration and tracheal ventilation in locusts and other flying insects. *J. exp. Biol.* 47, 561–587.
Weis-Fogh, T. (1970) Structure and formation of insect cuticle. In: *Insect Ultrastructure*, Neville, A. C., Ed., pp. 165–85. Blackwell, Oxford.
Weiss, M. J. (1974) Neuronal connections and the function of the corpora pedunculata in the brain of the American cockroach, *Periplaneta americana* (L.). *J. Morph.* 142, 21–69.
Weiss, P. A. (1955) Nervous system (Neurogenesis). In: *Analysis of Development*, Willier, B. H., Weiss, P. A. and Hamburger, V. Eds., pp. 346–401. Saunders, Philadelphia.
Weiss, P. A. (1959) Cellular dynamics. *Rev. Mod. Physics* 31, 11–20.
Welbers, P. (1976) Analysis of the oxygen consumption of *Periplaneta americana* at constant and diurnally alternating temperatures. *Oecologia* 24, 175–192.
Wendt, R. B. and Weidler, D. J. (1973) Effect of external chloride replacement on action potentials of giant axons in *Periplaneta americana. Comp. Biochem. Physiol.* 44A, 1303–1311.
Werman, R., McCann, F. V. and Grundfest, H. (1961) Graded and all-or-none electrogenesis in arthropod muscle. I. The effects of alkali-earth cations on the neuromuscular system of *Romalea microptera. J. gen. Physiol.* 44, 979–995.
Westin, J. (1979) Responses to wind recorded from the cercal nerve of the cockroach

Periplaneta americana. I. Response properties of single sensory neurons. *J. comp. Physiol.* 133, 97–102.
Westin, J., Langberg, J. J. and Camhi, J. M. (1977) Responses of giant interneurons of the cockroach *Periplaneta americana* to wind puffs of different directions and velocities. *J. comp. Physiol.* 121, 307–324.
Wharton, D. R. A. and Lola, J. E. (1970) Blood conditions and lysozyme action in the aposymbiotic cockroach. *J. Insect Physiol.* 16, 199–209.
Wharton, D. R. A., Lola, J. E. and Wharton, M. L. (1967) Population density, survival, growth, and development of the American cockroach. *J. Insect Physiol.* 13, 699–716.
Wharton, D. R. A., Lola, J. E. and Wharton, M. L. (1968) Growth factors and population density in the American cockroach, *Periplaneta americana*. *J. Insect Physiol.* 14, 637–653.
Wharton, D. R. A., Miller, G. L. and Wharton, M. L. (1954a) The odorous attractant of the American cockroach, *Periplaneta americana* (L.). I. Quantitative aspects of the response to the attractant. *J. gen. Physiol.* 37, 461–469.
Wharton, D. R. A., Miller, G. L. and Wharton, M. L. (1954b) The odorous attractant of the American cockroach, *Periplaneta americana* (L.). II. A bioassay method for the attractant. *J. gen. Physiol.* 37, 471–481.
Wharton, D. R. A., Wharton, M. L. and Lola, J. E. (1965a) Blood volume and water content of the male American cockroach, *Periplaneta americana* L. Methods and the influence of age and starvation. *J. Insect Physiol.* 11, 391–404.
Wharton, D. R. A., Wharton, M. L. and Lola, J. E. (1965b) Cellulase in the cockroach, with special reference to *Periplaneta americana*. *J. Insect Physiol.* 11, 947–959.
Wharton, M. L. and Wharton, D. R. A. (1957) Production of sex attractant substance and of oothecae by the normal and irradiated American cockroach, *Periplaneta americana*. *J. Insect Physiol.* 1, 229–239.
Wheeler, R. E. (1963) Studies on the total haemocyte count and haemolymph volume in *Periplaneta americana* (L.) with special reference to the last molting cycle. *J. Insect Physiol.* 9, 223–235.
Wheeler, W. M. (1889) The embryology of *Blattella germanica* and *Doryphora decemlineata*. *J. Morph.* 3, 291–386.
Wheeler, W. M. (1893) A contribution to insect embryology. *J. Morph.* 8, 1–160.
Whitehead, A. T. (1971) The innervation of the salivary gland in the American cockroach: light and electron microscopic observation. *J. Morph.* 135, 483–506.
Whitehead, D. L. (1969) New evidence for the control mechanism of sclerotization in insects. *Nature* 224, 721–723.
Whitehead, D. L., Brunet, P. C. J. and Kent, P. W. (1960) Specificity *in vitro* of a phenoloxidase system from *Periplaneta americana* (L.). *Nature* 185, 610.
Whitehead, D. L., Brunet, P. C. J. and Kent, P. W. (1965a) Observations on the nature of the phenoloxidase system in the secretion of the left colleterial gland of *Periplaneta americana* (L.). I. The specificity. *Proc. cent. Afr. Sci. Med. Cong.* 351–364. Pergamon, London.
Whitehead, D. L., Brunet, P. C. J. and Kent, P. W. (1965b) Observations on the nature of the phenoloxidase system in the secretion of the left colleterial gland of *Periplaneta americana* (L.). II. Inhibition activation and particulate nature of the enzyme. *Proc. cent. Afr. Sci. Med. Cong.* 365–383. Pergamon, London.
Whitington, P. M. (1977) Incorrect connections made by regenerating cockroach motoneuron. *J. exp. Zool.* 201, 339–344.
Whitmore, E. and Gilbert, L. I. (1972) Haemolymph lipoprotein transport of juvenile hormone. *J. Insect Physiol.* 18, 1153–1167.
Whitten, J. M. (1972) Comparative anatomy of the tracheal system. *A. Rev. Ent.* 17, 373–402.
Wieczorek, H. (1978) Biochemical and behavioral studies of sugar reception in the cockroach. *J. comp. Physiol.* 124, 353–356.

Wiedenmann, G. (1977a) Activity peaks in the circadian rhythm of the cockroach *Leucophaea maderae*. *J. interdiscip. Cycle Res.* 8, 378–383.

Wiedenmann, G. (1977b) Weak and strong phase shifting in the activity rhythm of *Leucophaea maderae* (Blaberidae) after light pulses of high intensity. *Z. Naturforsch.* C 32, 464–465.

Wiedenmann, G. (1977c) No 'point of singularity' in the circadian activity rhythm of the cockroach, *Leucophaea maderae*. *Chronobiologia* 4, 165.

Wielgus, J. J., Bollenbacher, W. E. and Gilbert, L. I. (1979) Correlations between epidermal DNA synthesis and haemolymph ecdysteroid titre during the last larval instar of the tobacco hornworm, *Manduca sexta*. *J. Insect Physiol.* 25, 9–16.

Wiens, A. W. and Gilbert, L. I. (1967) Regulation of carbohydrate mobilisation and utilisation in *Leucophaea maderae*. *J. Insect Physiol.* 13, 779–794.

Wigglesworth, V. B. (1927) Digestion in the cockroach. II. The digestion of carbohydrates. *Biochem. J.* 21, 797–811.

Wigglesworth, V. B. (1945) Transpiration through the cuticle of insects. *J. exp. Biol.* 21, 97–114.

Wigglesworth, V. B. (1947) The epicuticle in an insect, *Rhodnius prolixus*. *Proc. R. Soc. Lond. Ser. B.* 134, 163–181.

Wigglesworth, V. B. (1959) The role of the epidermal cells in the 'migration' of tracheoles in *Rhodnius prolixus* (Hemiptera). *J. exp. Biol.* 36, 632–640.

Wigglesworth, V. B. (1960) The nutrition of the central nervous system in the cockroach *Periplaneta americana* L. The role of perineurium and glial cells in the mobilization of reserves. *J. exp. Biol.* 37, 500–513.

Wigglesworth, V. B. (1972) *The Principles of Insect Physiology*, 7th Edn. Chapman and Hall, London.

Wigglesworth, V. B. (1977) Structural changes in the epidermal cells of *Rhodnius* during tracheole capture. *J. Cell Sci.* 26, 161–174.

Wigglesworth, V. B. and Beament, J. W. (1950) The respiratory structures of some insect eggs. *Q. J. micros. Sci.* 9, 429–450.

Wilhelm, R. and Lüscher, M. (1974) On the relative importance of juvenile hormone and vitellogenin for oöcyte growth in the cockroach *Nauphoeta cinerea*. *J. Insect Physiol.* 20, 1887–1894.

Wilkins, M. B. (1960) A temperature-dependent endogenous rhythm in the rate of carbon dioxide output of *Periplaneta americana*. *Nature* 185, 481–482.

Wille, J. (1920) Biologie und Bekampfung der deutschen Schabe (*Phyllodromia germanica* L.). *Mono. zur angew. Ent. Beihefte* I. 7, 1–140.

Willey, R. B. (1961) The morphology of the stomodaeal nervous system in *Periplaneta americana* (L.) and other Blattaria. *J. Morph.* 108, 219–262.

Williams, F. X. (1942) The New Caledonian cockroach wasp (*Ampulex compressa*) in Hawaii. *Hawaiian Planter's Rec.* 46, 43–48.

Williams, J. L. D. (1975) Anatomical studies of the insect central nervous system: a ground plan of the midbrain and an introduction to the central complex in the locust, *Schistocerca gregaria* (Orthoptera). *J. Zool.* 176, 67–86.

Willis, E. R. and Lewis, N. (1957) The longevity of starved cockroaches. *J. econ. Ent.* 50, 438–440.

Willis, E. R., Riser, G. R. and Roth, L. M. (1958) Observations on reproduction and development in cockroaches. *Ann. ent. Soc. Am.* 51, 53–69.

Willis, J. H. and Brunet, P. C. J. (1966) The hormonal control of colleterial gland secretion. *J. exp. Biol.* 44, 363–378.

Willner, P. (1978) What does the headless cockroach remember? *Animal Learn. Behav.* 6, 249–257.

Willner, P. and Mellanby, J. (1974) Cholinesterase activity in the cockroach CNS does not change with training. *Brain Res.* 66, 481–490.

Wilson, D. M. (1965) Proprioceptive leg reflexes in cockroaches. *J. exp. Biol.* 43, 397–409.
Wilson, M. H. and Rounds, H. D. (1972) Stress-induced changes of glucose levels in cockroach hemolymph. *Comp. Biochem. Physiol.* 43A, 941–947.
Winston, P. W. and Beament, J. W. L. (1969) An active reduction of water level in insect cuticle. *J. exp. Biol.* 50, 541–546.
Winter, C. E., Gianotti, O. and Holzhacker, E. L. (1975) DDT-lipoprotein complex in the American cockroach hemolymph: A possible way of insecticide transport. *Pest. Biochem. Physiol.* 5, 155–162.
Wirtz, R. A., Hopkins, T. L. (1972) DOPA and tyrosine decarboxylation in the cockroach *Leucophaea maderae* in relation to cuticle formation and ecdysis. *Insect Biochem.* 7, 45–49.
Wirtz, R. A. and Hopkins, T. L. (1974) Tyrosine and phenylalanine concentrations in haemolymph and tissues of the American cockroach, *Periplaneta americana* during metamorphosis. *J. Insect Physiol.* 20, 1143–1154.
Wirtz, R. A. and Hopkins, T. L. (1977) Tyrosine and phenylalanine concentrations in the cockroaches *Leucophaea maderae* and *Periplaneta americana* in relation to cuticle formation and ecdysis. *Comp. Biochem. Physiol.* 56A, 263–266.
Wlodawer, P. and Baranska, J. (1965) Lipolytic activity of the fat body of the silkworm larvae. II. Characteristics of the two different lipases in the waxmoth fat body. *Acta Biochim. Polon.* 12, 39–47.
Wobus, U. (1966a) Der Einflusz der Lichtintensität auf die circadiene laufaktivitat der Schabe *Blaberus craniifer* Burm. (Insecta: Blattariae). *Biol. Zentralbl.* 85, 305–323.
Wobus, U. (1966b) Der Einfluss der Lichtintensitat auf die Resynchronization der circadianen laufaktivität der Schabe *Blaberus craniifer* Burm. (Insecta: Blattariae). *Z. vergl. Physiol.* 52, 276–289.
Wong, R. K. S. and Pearson, K. G. (1976) Properties of the trochanteral hair plate and its function in the control of walking in the cockroach. *J. exp. Biol.* 64, 233–249.
Wood, D. W. (1963) The sodium and potassium composition of some insect skeletal muscle fibres in relation to their membrane potentials. *Comp. Biochem. Physiol.* 9, 151–159.
Wood, D. W. (1965) The relationship between chloride ions and resting potential in skeletal muscle fibres of the locust and cockroach. *Comp. Biochem. Physiol.* 15, 303–312.
Woodland, D. J., Hall, B. K. and Calder, J. (1968) Gross bioenergetics of *Blattella germanica. Physiol. Zool.* 41, 424–431.
Woodson, P. B. J., Schlapfer, W. T. and Barondes, S. H. (1972) Postural avoidance learning in the headless cockroach without detectable changes in ganglionic cholinesterase. *Brain Res.* 37, 348–352.
Wright, B. R. (1976) Limb and wing receptors in insects, chelicerates and myriapods. In: *Structure and Function of Priopriceptors in the Invertebrates*, Mill, P. J., Ed., pp. 323–386. Chapman and Hall, London.
Wright, R. D., Sauer, J. R. and Mills, R. R. (1970) Midgut epithelium of the American cockroach: possible sites of neurosecretory release. *J. Insect Physiol.* 16, 1485–1491.
Wuest, J. (1978) Histological and cytological studies on the fat body of the cockroach *Nauphoeta cinerea* during the first reproductive cycle. *Cell Tissue Res.* 188, 481–490.
Wyatt, G. R. (1967) The biochemistry of sugars and polysaccharides in insects. *Adv. Insect Physiol.* 4, 287–360.
Wyatt, G. R. (1975) Regulation of protein and carbohydrate metabolism in insect fat body. *Verh. Dtsch. Zool. Ges.* 1974, 209–226.
Yamada, M., Tshui, S. and Kuwahara, Y. (1970) Odour discrimination: ''Sex pheromone specialists'' in the olfactory lobe of the cockroach. *Nature* 227, 855.
Yamasaki, T. and Narahashi, T. (1959a) Electrical properties of the cockroach giant axon. *J. Insect Physiol.* 3, 230–242.
Yamasaki, T. and Narahashi, T. (1959b) The effects of potassium and sodium ions on the resting and action potentials of the cockroach giant axon. *J. Insect Physiol.* 3, 146–158.

Yamasaki, T. and Narahashi, T. (1960) Synaptic transmission in the last abdominal ganglion of the cockroach. *J. Insect Physiol.* 4, 1–13.

Yeager, J. G. (1931) Observations on crop and gizzard movements in the cockroach *Periplaneta fulginosa* (Serv.) *Ann. ent. Soc. Am.* 24, 739–745.

Yeager, J. F. and Hendrickson, G. O. (1934) Circulation of blood cells in wings and wing pads of the cockroach, *Periplaneta americana* Linn. *Ann. ent. Soc. Am.* 27, 257–272.

Yeager, J. F. (1939) Electrical stimulation of isolated heart preparations from *Periplaneta americana. J. Agric. Res.* 59, 121–137.

Yeager, J. F. and Hager, A. (1934) On the rates of contraction of the isolated heart and Malpighian tube of the insect *Periplaneta orientalis*: method. *Iowa State Coll. J. Sci.* 8, 391–393.

Yokohari, F. (1978) Hygroreceptor mechanism in the antenna of the cockroach *Periplaneta. J. comp. Physiol.* 124, 53–60.

Yokohari, F. and Tateda, H. (1976) Moist and dry hygroreceptors for relative humidity of the cockroach, *Periplaneta americana* L. *J. comp. Physiol.* 106, 137–152.

Yokohari, F., Tominaga, Y., Ando, M. and Tateda, H. (1975) An antennal hygroreceptive sensillum of the cockroach. *J. electron Micros.* 24, 291–293.

Young, D. (1969) The motor neurons of the mesothoracic ganglion of *Periplaneta americana. J. Insect Physiol.* 15, 1175–1179.

Young, D. (1970) The structure and function of a connective chordotonal organ in the cockroach leg. *Phil. Trans. R. Soc. Ser. B.* 256, 401–428.

Young, D. (1972) Specific re-innervation of limbs transplanted between segments in the cockroach, *Periplaneta americana. J. exp. Biol.* 57, 305–316.

Young, D. (1973) Specificity and regeneration in insect motor neurons. In: *Developmental Neurobiology of Arthropods*, Young, D., Ed., pp. 179–202. Cambridge University Press, London.

Young, D., Ashhurst, D. E. and Cohen, M. J. (1970) Injury response of the neurons of *Periplaneta americana. Tissue Cell* 2, 387–398.

Zabinski, J. (1936) Inconstancy of the number of moults during postembryonic development of certain Blattidae. *Ann. Mus. Zool. Pol.* 11, 237–240.

Zacharuk, R. Y. (1980) Ultrastructure and function of insect chemosensilla. *A. Rev. ent.* 25, 27–47.

Zajonc, R. B. (1965) Social facilitation. *Science* 149, 269–274.

Zajonc, R. B., Heingartner, A. and Herman, E. M. (1969) Social enhancement and impairment of performance in the cockroach. *J. Pers. Soc. Psych.* 13, 83–92.

Zalokar, M. (1968) Effect of corpora allata on protein and RNA synthesis in colleterial glands of *Blattella germanica. J. Insect Physiol.* 14, 1177–1184.

Zanforlin, M., Cervato, O. and Cescon, L. (1973) Home effect on aggressive behaviour in *Periplaneta americana. Bull. Zool.* 40, 81–85.

Zebe, H., Sanwald, R. and Ritz, E. (1972) Insect vectors in serum hepatitis. *Lancet* 20, 1117–1118.

Zerhan, K. (1977) Potassium transport in insect midgut. In: *Transport of Ions and Water in Animals*, Gupta, B. L., Moreton, R. B., Oschman, J. L. and Wall, B. J., Eds., pp. 381–401. Academic Press, London.

Zill, S. N., Varela, F. J. and Moran, D. T. (1977) Modulation of locomotor activity by adjacent campaniform sensilla varies with sensillum orientation. *Neurosci Abst.* 3, 191.

Zuberi, R. I., Hafiz, S. and Ashrafi, S. H. (1969) Bacterial and fungal isolates from laboratory-reared *Aedes aegypti* (Linnaeus), *Musca domestica* (Linnaeus) and *Periplaneta americana* (Linnaeus). *Pak. J. Sci. Ind. Res.* 12, 77–82.

Species Index

Subject Index